THE EXPERIMENTAL FOUNDATIONS OF PARTICLE PHYSICS
Second Edition

Our current understanding of elementary particles and their interactions emerged from break-through experiments. This book presents these experiments, beginning with the discoveries of the neutron and positron, and following them through mesons, strange particles, antiparticles, and quarks and gluons. This second edition contains new chapters on the W and Z, the top quark, B-meson mixing and CP violation, and neutrino oscillations.

This book provides an insight into particle physics for researchers, advanced undergraduate and graduate students. Throughout the book, the fundamental equations required to understand the experiments are derived clearly and simply. Each chapter is accompanied by reprinted articles and a collection of problems with a broad range of difficulty.

ROBERT CAHN is a Senior Physicist at the Lawrence Berkeley National Laboratory. His theoretical work has focused on the Standard Model, and, together with his collaborators, he developed one of the most promising methods for discovering the Higgs boson. As a member of the BaBar Collaboration, he participated in the measurement of CP violation in B mesons.

GERSON GOLDHABER is a Professor in the Graduate School at the University of California at Berkeley, and Faculty Senior Physicist at the Lawrence Berkeley National Laboratory. He is co-discoverer of the antiproton annihilation process, the Bose–Einstein nature of pions, the J/Psi particle and psion spectroscopy, charmed mesons, and dark energy.

THE EXPERIMENTAL FOUNDATIONS OF PARTICLE PHYSICS

Second Edition

ROBERT N. CAHN
Lawrence Berkeley National Laboratory

GERSON GOLDHABER
*Lawrence Berkeley National Laboratory and
University of California at Berkeley*

CAMBRIDGE UNIVERSITY PRESS
Cambridge, New York, Melbourne, Madrid, Cape Town, Singapore, São Paulo, Delhi

Cambridge University Press
The Edinburgh Building, Cambridge CB2 8RU, UK

Published in the United States of America by Cambridge University Press, New York

www.cambridge.org
Information on this title: www.cambridge.org/9780521521475
First edition © Cambridge University Press 1989
Second edition © R. Cahn and G. Goldhaber 2009

This publication is in copyright. Subject to statutory exception
and to the provisions of relevant collective licensing agreements,
no reproduction of any part may take place without
the written permission of Cambridge University Press.

First published 1989
Paperback edition 1991
Reprinted 1993, 1995
Second edition 2009

Printed in the United Kingdom at the University Press, Cambridge

A catalog record for this publication is available from the British Library

Articles and figures reprinted by permission of the authors and the publishers:
American Institute of Physics: all articles from *Physical Review* and *Physical Review Letters*
Elsevier Science Publishers B. V.: all articles from *Nuclear Physics* and *Physics Letters*
Comptes Rendus de l'Académie des Sciences de Paris: C. R. Acad. Sciences, séance du 13 déc. 1944, p. 618
Nature, copyright Macmillan Journals Limited: Vol. 129, No. 3252, p. 312; Vol. 159, No. 4030, pp. 126-7;
Vol. 160, No. 4066, pp. 453-6; Vol. 160, No. 4077, pp. 855-7; Vol. 163, No. 4133, p. 82
Il Nuovo Cimento
Philosophical Magazine
Springer Verlag: *Zeitschrift für Physik*

ISBN 978-0-521-52147-5 hardback

Cambridge University Press has no responsibility for the persistence or
accuracy of URLs for external or third-party internet websites referred to
in this publication, and does not guarantee that any content on such
websites is, or will remain, accurate or appropriate.

For
our grandchildren
Zachary, Jakob, Mina, and Eve

and
Benjamin, Charles, and Samuel

Contents

Preface to the Second Edition		page ix
Preface to the First Edition		xi
1	The Atom Completed and a New Particle	1
2	The Muon and the Pion	13
3	Strangeness	49
4	Antibaryons	80
5	The Resonances	99
6	Weak Interactions	147
7	The Neutral Kaon System	185
8	The Structure of the Nucleon	209
9	The J/ψ, the τ, and Charm	247
10	Quarks, Gluons, and Jets	293
11	The Fifth Quark	323
12	From Neutral Currents to Weak Vector Bosons	357
13	Testing the Standard Model	395
14	The Top Quark	416
15	Mixing and CP Violation in Heavy Quark Mesons	434
16	Neutrino Masses and Oscillations	489
17	Epilogue	544
Index		546

Preface to the Second Edition

In the twenty years since the first edition, the promise of the Standard Model of Particle Physics has been fulfilled. The detailed behavior of the W and Z bosons did conform to expectations. The sixth quark finally arrived. The pattern of CP violation in B mesons fit convincingly the predictions based on the Kobayashi–Maskawa model. These three developments require three new chapters. The big surprise was the observation of neutrino oscillations. Neutrino masses and oscillations were not required by the Standard Model but are easily accommodated within it. An extensive fourth new chapter covers this history.

Though the neutrino story is not yet fully known, the basics of the Standard Model are all in place and so this is an appropriate time to update the Experimental Foundations of Particle Physics. We fully anticipate that the most exciting times in particle physics lie just ahead with the opening of the Large Hadron Collider at CERN. This Second Edition provides a recapitulation of some 75 years of discovery in anticipation of even more profound revelations.

Not only physics has changed, but technology, too. The bound journals we dragged to the xerox machine are now available from the internet with a few keystrokes on a laptop. Nonetheless, we have chosen to stick with our original format of text alternating with reprinted articles, believing Gutenberg will survive Gates and that there is still great value in having the physical text in your hands.

Choosing articles to reprint has become more difficult with the proliferation of experiments aimed at the most promising measurements. In some cases we have been forced to make an arbitrary selection from competing experiments with comparable results.

We would like to acknowledge again the physicists whose papers we reprint here. We have benefited from the advice of many colleagues for this Second Edition and would like to mention, in particular, Stuart Freedman, Fred Gilman, Dave Jackson, Zoltan Ligeti, Kerstin Tackmann, Frank Tackmann, George Trilling, and Stan Wojcicki.

R. N. C.
G. G.
Berkeley, California, 2008

Preface to the First Edition

Fifty years of particle physics research has produced an elegant and concise theory of particle interactions at the subnuclear level. This book presents the experimental foundations of that theory. A collection of reprints alone would, perhaps, have been adequate were the audience simply practicing particle physicists, but we wished to make this material accessible to advanced undergraduates, graduate students, and physicists with other fields of specialization. The text that accompanies each selection of reprints is designed to introduce the fundamental concepts pertinent to the articles and to provide the necessary background information. A good undergraduate training in physics is adequate for understanding the material, except perhaps some of the more theoretical material presented in smaller print and some portions of Chapters 6, 7, 8, and 12, which can be skipped by the less advanced reader.

Each of the chapters treats a particular aspect of particle physics, with the topics given basically in historical order. The first chapter summarizes the development of atomic and nuclear physics during the first third of the twentieth century and concludes with the discoveries of the neutron and the positron. The two succeeding chapters present weakly decaying non-strange and strange particles, and the next two the antibaryons and the resonances. Chapters 6 and 7 deal with weak interactions, parity and CP violation. The contemporary picture of elementary particles emerges from deep inelastic lepton scattering in Chapter 8, the discovery of charm and the tau lepton in Chapter 9, quark and gluon jets in Chapter 10, and the discovery of the b-quark in Chapter 11. The synthesis of all this is given in Chapter 12, beginning with neutral current interactions and culminating in the discovery of the W and Z.

A more efficient presentation can be achieved by working in reverse, starting from the standard model of QCD and electroweak interactions and concluding with the hadrons. This, however, leaves the reader with the fundamentally false impression that particle physics is somehow derived from an *a priori* theory. It fails, too, to convey the standard model's real achievement, which is to encompass the enormous wealth of data accumulated over the last fifty years.

Our approach, too, has its limitations. Devoting pages to reprinting articles has forced sacrifices in the written text. The result cannot be considered a complete textbook. The reader should consult some of the additional references listed at the end of each chapter.

The text by D. H. Perkins provides an excellent supplement. A more fundamental problem is that, quite naturally, we have reprinted (we believe) correct experiments and provided (we hope!) the correct interpretations. However, at any time there are many contending theories and sometimes contradictory experiments. By selecting those experiments that have stood the test of time and ignoring contemporaneous results that were later disproved, this book inevitably presents a smoother view of the subject than would a more historically complete treatment. Despite this distortion, the basic historical outline is clear. In the reprinted papers the reader will see the growth of the field, from modest experiments performed by a few individuals at cosmic-ray laboratories high atop mountains, to monumental undertakings of hundreds of physicists using apparatus weighing thousands of tons to measure millions of particle collisions. The reader will see as well the development of a description of nature at the most fundamental level so far, a description of elegance and economy based on great achievements in experimental physics.

Selecting articles to be reprinted was difficult. The sixty or so experimental papers ultimately selected all played important roles in the history of the field. Many other important articles have not been reprinted, especially when there were two nearly simultaneous discoveries of the same particle or effect. In two instances, for the sake of brevity, we chose to reprint just the first page of an article. By choosing to present usually the first paper on a subject often a later paper that may have been more complete has been neglected. In some cases, through oversight or ignorance we may simply have failed to include a paper that ought to be present. Some papers were not selected simply because they were too long. We extend our apologies to our colleagues whose papers have not been included for any of these reasons. The reprinted papers are referred to in boldface, while other papers are listed in ordinary type. The reprinted papers are supplemented by numerous figures taken from articles that have not been reprinted and which sometimes represent more recent results. Additional references, reviews or textbooks, are listed at the end of each chapter.

Exercises have been provided for the student or assiduous reader. They are of varying difficulty; the most difficult and those requiring more background are marked with an asterisk. In addition to a good standard textbook, the reader will find it helpful to have a copy of the most recent *Review of Particle Properties*, which may be obtained as described at the end of Chapter 2.

G. G. would like to acknowledge 15 years of collaboration in particle physics with Sulamith Goldhaber (1923–1965).

We would like to thank the many particle physicists who allowed us to reproduce their papers, completely or in part, that provide the basis for this book. We are indebted, as well, to our many colleagues who have provided extensive criticism of the written text. These include F. J. Gilman, J. D. Jackson, P. V. Landshoff, V. Lüth, M. Suzuki, and G. H. Trilling. The help of Richard Robinson and Christina F. Dieterle is also acknowledged. Of course, the omissions and inaccuracies are ours alone.

R. N. C.
G. G.
Berkeley, California, 1988

1

The Atom Completed and a New Particle

The origins of particle physics: The atom, radioactivity,
and the discovery of the neutron and the positron, 1895–1933.

The fundamental achievement of physical science is the atomic model of matter. That model is simplicity itself. All matter is composed of atoms, which themselves form aggregates called molecules. An atom contains a positive nucleus very much smaller than the full atom. A nucleus with atomic mass A contains Z protons and $A - Z$ neutrons. The neutral atom has, as well, Z electrons, each with a mass only 1/1836 that of a proton. The chemical properties of the atom are determined by Z; atoms with equal Z but differing A have the same chemistry and are known as isotopes.

This school-level description did not exist at all in 1895. Atoms were the creation of chemists and were still distrusted by many physicists. Electrons, protons, and neutrons were yet to be discovered. Atomic spectra were well studied, but presented a bewildering catalog of lines connected, at best, by empirical rules like the Balmer formula for the hydrogen atom. Cathode rays had been studied, but many regarded them as uncharged, electromagnetic waves. Chemists had determined the atomic weights of the known elements and Mendeleev had produced the periodic table, but the concept of atomic number had not yet been developed.

The discovery of X-rays by W. C. Röntgen in 1895 began the revolution that was to produce atomic physics. Röntgen found that cathode-ray tubes generate penetrating, invisible rays that can be observed with fluorescent screens or photographic film. This discovery caused a sensation. Royalty vied for the opportunity to have their hands X-rayed, and soon X-rays were put to less frivolous uses in medical diagnosis.

The next year, Henri Becquerel discovered that uranium emitted radiation that could darken photographic film. While not creating such a public stir as did X-rays, within two years radioactivity had led to remarkable new results. In 1898, Marie Curie, in collaboration with her husband, Pierre, began her monumental work, which resulted in the discovery of two new elements, polonium and radium, whose level of activity far exceeded that of uranium. This made them invaluable sources for further experiments.

A contemporaneous achievement was the demonstration by J. J. Thomson that cathode rays were composed of particles whose ratio of charge to mass was very much greater than that previously measured for ions. From his identification of electrons as a universal constituent of matter, Thomson developed his model of the atom consisting of many, perhaps thousands of electrons in a swarm with balancing positive charge. In time, however, it became clear that the number of electrons could not be so great without conflicting with data on the scattering of light by atoms.

The beginning of the new century was marked by Planck's discovery of the blackbody radiation law, which governs emission from an idealized object of a specified temperature. Having found empirically a functional form for the energy spectrum that satisfied both theoretical principles and the high-quality data that had become available, Planck persisted until he had a physical interpretation of his result: An oscillator with frequency ν has energy quantized in units of $h\nu$. In one of his three great papers of 1905, Einstein used Planck's constant, h, to explain the photoelectric effect: Electrons are emitted by illuminated metals, but the energy of the electrons depends on the frequency of the light, not its intensity. Einstein showed that this could be explained if light of frequency ν were composed of individual quanta of energy $h\nu$.

Investigations of radioactivity were pursued by others besides Becquerel and the Curies. A young New Zealander, Ernest Rutherford came to England after initiating his own research on electromagnetic waves. He was soon at the forefront of the investigations of radioactivity, identifying and naming alpha and beta radiation. At McGill University in Montreal, he and Frederick Soddy showed that radioactive decay resulted in the transmutation of elements. In 1907, Rutherford returned to England to work at Manchester, where his research team determined the structure of the atom.

Rutherford's favorite technique was bombardment with alpha particles. At McGill, Rutherford had found strong evidence that the alpha particles were doubly ionized helium atoms. At Manchester, together with Thomas Royds, he demonstrated this convincingly in 1909 by observing the helium spectrum produced in a region surrounding a radioactive source. Hans Geiger and Ernest Marsden, respectively aged 27 and 20, carried out an experiment in 1909 under Rutherford's direction in which alpha particles were observed to scatter from a thin metal foil. Much to their surprise, many of the alpha particles were scattered through substantial angles. This was impossible to reconcile with Thomson's model of the atom. In 1911, Rutherford published his analysis of the experiment showing that the atom had a small, charged nucleus.

This set the stage for the efforts of Niels Bohr. The atom of J. J. Thomson did not a priori have any particular size. The quantities of classical nonrelativistic physics did not provide dimensionful quantities from which a size could be constructed. In addition to the electron mass, m_e, there was the electron's charge squared, e^2, with dimensions mass $\times$ length^3/time^2. Bohr noted that Planck's constant had dimensions mass $\times$ length^2/time. In a somewhat ad hoc way, Bohr managed to combine m_e, e^2, and h to obtain as a radius for the hydrogen atom $a_0 = \hbar^2/(m_e e^2)$, where $\hbar = h/2\pi$, and derived the Balmer formula for the hydrogen spectrum, and the Rydberg constant which appears in it.

Despite this great achievement, the structure of atoms with higher values of Z remained obscure. In 1911, Max von Laue predicted that X-rays would show diffraction characteristics when scattered from crystals. This was demonstrated in short order by Friedrich and Knipping and in 1914 Moseley was able to apply the technique to analyze X-rays emitted by the full list of known elements. He found that certain discrete X-ray lines, the K lines, showed a simple behavior. Their frequencies were given by $\nu = \nu_0(n-a)^2$, where ν_0 was a fixed frequency and a was a constant near 1. Here n took on integral values, a different value for each element. Moseley immediately understood that n gave the positive charge of the nucleus. In a stroke, he had brought complete order to the table of elements. The known elements were placed in sequence and gaps identified for the missing elements.

While the atomic number was an integer, the atomic weights measured relative to hydrogen were sometimes close to integers and sometimes not, depending on the particular element. Soddy first coined the term isotopes to refer to chemically inseparable versions of an element with differing atomic weights. By 1913, J. J. Thomson had demonstrated the existence of neon isotopes with weights 20 and 22. The high-precision work of F. W. Aston using mass spectrometry established that each isotope had nearly integral atomic weight. The chemically observed nonintegral weights were simply due to the isotopic mixtures. It was generally assumed that the nucleus contained both protons and electrons, with their difference determining the chemical element.

The story of the years 1924–7 is well-known and needs no repeating here. Quantum mechanics developed rapidly, from de Broglie's waves through Heisenberg's matrix mechanics to its mature expression in the Schrödinger equation and Dirac's formulation of transition amplitudes. The problem of the electronic structure of the atom was reduced to a set of differential equations, approximations to which explained not just hydrogen, but all the atoms. Only the nucleus remained a mystery.

While the existence of the neutron was proposed by Rutherford as early as 1920, until its actual discovery both theorists and experimenters continued to speak of the nucleus as having A protons and $A - Z$ electrons. The development of quantum mechanics compounded the problems of this model. It was nearly impossible to confine the electron inside a space as small as a nucleus, since by the uncertainty principle this would require the electron to have very large momentum.

By 1926 it was understood that all particles were divided into two classes according to their angular momentum. The total angular momentum (spin) of a particle is always an integral or half-integral multiple of $\hbar$. Those with half-integral angular momentum (in units of $\hbar$) are called *fermions*, while those with integral angular momentum are called *bosons*. The quantum mechanical wave function of a system (e.g. an atom) must be antisymmetric under the interchange of identical fermions and symmetric under the interchange of identical bosons. Electrons, protons, and neutrons all have spin 1/2 (angular momentum $\hbar/2$) and are thus fermions. The alpha particle with spin 0 and the deuteron with spin 1 are bosons.

These fundamental facts about spin could not be reconciled with the prevailing picture of the nucleus N_7^{14}. If it contains 14 protons and 7 electrons, it should be a fermion and have half-integral spin. In fact, it was shown to have spin 1 by Ornstein and van Wyk, who studied the intensities of rotational bands in the spectrum of N_2^+, and shown to be a boson by measurements of its Raman spectrum by Rasetti. These results were consistent with each other, but not with the view that N_7^{14} contained 14 protons and 7 electrons.

Walter Bothe and Herbert Becker unknowingly observed neutrons when they used polonium as an alpha source to bombard beryllium. They produced the reaction:

$$He_2^4 + Be_4^9 \rightarrow C_6^{12} + n_0^1.$$

Bothe and Becker observed neutral "penetrating radiation" that they thought was X-rays. In 1931, Irène Curie and her husband, Frédéric Joliot, studied the same process and showed that the radiation was able to knock protons out of paraffin. Unfortunately, Joliot and Curie misinterpreted the phenomenon as scattering of gamma rays on protons. James Chadwick knew at once that Joliot and Curie had observed the neutral version of the proton and set out to prove it. His results were published in 1932 (**Ref. 1.1**, Ref. 1.2).

Chadwick noted that the proton ejected by the radiation had a velocity about one-tenth the speed of light. A photon capable of causing this would have an energy of about 50 MeV, an astonishingly large value since gamma rays emitted by nuclei usually have energies of just a few MeV. Furthermore, Chadwick showed that the same neutral radiation ejected nitrogen atoms with much more energy than could be explained by the hypothesis that the incident radiation consisted of photons, even if it were as energetic as 50 MeV. All these difficulties vanished if it was assumed that the incident radiation was due to a neutral partner of the proton. The problem with the statistics of the N_7^{14} nucleus was also solved. It consisted simply of seven neutrons and seven protons. It had integral spin and was thus a boson. With the discovery of the neutron, the last piece was in place: The modern atom was complete.

The neutron provided the key to understanding nuclear beta decay. In 1930, Wolfgang Pauli had postulated the existence of a light, neutral, feebly interacting particle, the neutrino (ν). Pauli did this to explain measurements demonstrating the apparent failure of energy conservation when a radioactive nucleus emitted an electron (beta ray). The unobserved energy was ascribed to the undetected neutrino. As described in Chapter 6, Enrico Fermi provided a quantitative theory based on the fundamental process $n \rightarrow pe\nu$.

In the same year as Chadwick found the final ingredient of tangible matter, C. D. Anderson began his exploration of fundamental particles that are not found ordinarily in nature. The explorations using X-rays and radioactive sources were limited to energies of a few MeV. To obtain higher energy particles it was necessary to use cosmic rays. The first observations of cosmic rays were made by the Austrian, Victor Hess, who ascended by balloon with an electrometer to an altitude of 5000 m. Pioneering measurements were made by the Soviet physicist Dimitry Skobeltzyn who used a cloud chamber to observe tracks made by cosmic rays. As described in greater detail in the next chapter, charged particles passing through matter lose energy by ionizing atoms in the medium. A cloud chamber contains a supersaturated vapor that forms droplets along the trail of ionization. When properly illuminated these tracks are visible and can be photographed. The momenta of the charged particles can be measured if the cloud chamber is placed in a magnetic field, where the curvature of the track is inversely proportional to the momentum.

Anderson was studying cosmic-ray particles in his cloud chamber built together with R. A. Millikan at the California Institute of Technology (**Ref. 1.3**) when he discovered the positron, a particle with the same mass as the electron but with the opposite charge. The cloud chamber had a 15-kG field. A 6-mm plate of lead separated the upper and lower portions of the chamber. Surprisingly, the first identified positron track observed entered from below. It was possible to prove this was a positive track entering from below rather than a negative track entering from above by noting the greater curvature above the plate. The greater curvature indicated lower momentum, the result of the particle losing energy when it passed through the lead plate. Having disposed of the possibility that there were two independent tracks, Anderson concluded that he was dealing with a new positive particle with a charge less than twice that of the electron and a mass much less than that of a proton. Indeed, if the charge was assumed equal in magnitude to that of the electron, the mass had to be less than 20 times the mass of the electron.

Just a few years before, P. A. M. Dirac had presented his relativistic wave equation for electrons, which predicted the existence of particles with a charge opposite that of the electron. Originally, Dirac identified these as protons, but J. Robert Oppenheimer and others showed that the predicted particles must have the same mass as the electron and hence must be distinct from the proton. Anderson had discovered precisely the particle required by the Dirac theory, the antiparticle of the electron, the positron.

While the discovery was fortuitous, Anderson had, of course, been aware of the predictions of the Dirac theory. Oppenheimer was then splitting his time between Berkeley and Caltech, and he had discussed the possibility of there being a particle of electronic mass but opposite charge. What was missing was an understanding of the mechanism that would produce these particles. Dirac had proposed the collision of two gamma rays giving an electron and a positron. This was correct in principle, but unrealizable in the laboratory. The correct mechanism of pair production was proposed after Anderson's discovery by Blackett and Occhialini. An incident gamma ray interacts with the electromagnetic field surrounding a nucleus and an electron–positron pair is formed. This is simply the mechanism proposed by Dirac with one of the gamma rays replaced by a virtual photon from the electromagnetic field near the nucleus. In fact, Blackett and Occhialini had evidence for positrons before Anderson, but were too cautious to publish the result (Ref. 1.4).

Anderson's positron (e^+), Thomson's electron (e^-), and Einstein's photon (γ) filled all the roles called for in Dirac's relativistic theory. To calculate their interactions in processes like $e^-e^- \to e^-e^-$ (Møller scattering), $e^+e^- \to e^+e^-$ (Bhabha scattering), or $\gamma e^- \to \gamma e^-$ (Compton scattering) was a straightforward task, when considered to lowest order in the electromagnetic interaction. It was clear, however, that in the Dirac theory there must be corrections in which the electromagnetic interaction acted more than the minimal number of times. Some of these corrections could be calculated. Uehling and Serber calculated the deviation from Coulomb's law that must occur for charged particles separated by distances comparable to the Compton wavelength of the electron, $\hbar/m_e c \approx 386$ fm (1 fm = 1 fermi = 10^{-15} m). Other processes, however, proved intractable because the corrections turned out to be infinite!

In the simple version of the Dirac theory, the $n = 2$ s-wave and p-wave states (orbital angular momentum 0 and 1, respectively) of hydrogen with total angular momentum (always measured in units of $\hbar$) $J = 1/2$ are degenerate. In 1947, Lamb and Retherford demonstrated that the $2S_{1/2}$ level lay higher than the $2P_{1/2}$ level by an amount equivalent to a frequency of about 1000 MHz. An approximate calculation of the shift, which was due to the emission and reabsorption of virtual photons by the bound electron, was given by Hans Bethe.

A complete formulation of quantum electrodynamics (QED) was given by Richard Feynman and independently by Julian Schwinger, whose work paralleled that done earlier in Japan by Sin-itiro Tomonaga. The achievement of Tomonaga, Feynman, and Schwinger was to show that the infinities found in the Dirac theory did not occur in the physical quantities of the theory. When the results were written in terms of the physical couplings and masses, all the other physical quantities were finite and calculable.

A test of the new theory was the magnetic moment of the electron. In the simple Dirac theory, the magnetic moment was $\mu = e\hbar/2m_e c = 2\mu_0 J_e$, where $J_e = 1/2$ is the electron spin and $\mu_0 = e\hbar/2m_e c$ is the Bohr magneton. More generally, we can write $\mu = g_e \mu_0 J_e$. Because of quantum corrections to the Dirac theory, g_e is not precisely 2. In 1948, by studying the Zeeman splittings in indium, gallium, and sodium, Kusch found that $g_e = 2(1 + 1.19 \times 10^{-3})$, while Schwinger calculated $g_e = 2(1 + \alpha/2\pi) = 2(1 + 1.16 \times 10^{-3})$. The currently accepted experimental value is $2(1 + 1.15965218111(74) \times 10^{-3})$ while the theoretical prediction is $2(1 + 1.15965218279(771) \times 10^{-3})$. The brilliant successes of QED made it the standard for what a physical theory should achieve, a standard emulated three decades later in theories formulated to describe the nonelectromagnetic interactions of fundamental particles.

Exercises

1.1 Confirm Chadwick's statement that if the protons ejected from the hydrogen were due to a Compton-like effect, the incident gamma energy would have to be near 50 MeV and that such a gamma ray would produce recoil nitrogen nuclei with energies up to about 400 keV. What nitrogen recoil energies would be expected for the neutron hypothesis?

1.2 The neutron and proton bind to produce a deuteron of intrinsic angular momentum 1. Given that the spins of the neutron and proton are 1/2, what are the possible values of the spin, $S = S_n + S_p$ and orbital angular momentum, L, in the deuteron? There is only one bound state of a neutron and a proton. For which L is this most likely? The deuteron has an electric quadrupole moment. What does this say about the possible values of L?

1.3 A positron and an electron bind to form positronium. What is the relationship between the energy levels of positronium and those of hydrogen?

1.4 The photodisintegration of the deuteron, $\gamma d \to pn$, was observed in 1934 by Chadwick and M. Goldhaber (Ref. 1.5). They knew the mass of ordinary hydrogen to be 1.0078 amu and that of deuterium to be 2.0136 amu. They found that the 2.62 MeV gamma ray from thorium C″ (Th_{81}^{208}) was powerful enough to cause the disintegration, while the 1.8 MeV γ from thorium C (Bi_{83}^{212}) was not. Show that this requires the neutron mass to be between 1.0077 and 1.0086 amu.

1.5 * In quantum electrodynamics there is a symmetry called charge conjugation that turns electrons into positrons and vice versa. The "wave function" of a photon changes sign under this symmetry. Positronium with spin S (0 or 1) and angular momentum L has charge conjugation $C = (-1)^{L+S}$. Thus the state $^3S_1$ ($S = 1$, $L = 0$) has $C = -1$ and the state $^1S_0$ ($S = 0$, $L = 0$) has $C = +1$. The $^1S_0$ state decays into two photons, the $^3S_1$ into three photons. Using dimensional arguments, estimate crudely the lifetimes of the $^1S_0$ and $^3S_1$ states and compare with the accepted values. [For a review of both theory and experiment, see M. A. Stroscio, *Phys. Rep.*, **22**, 215 (1975).]

Further Reading

The history of this period in particle physics is treated superbly by Abraham Pais in *Inward Bound*, Oxford University Press, New York, 1986.

A fine discussion of the early days of atomic and nuclear physics is given in E. Segrè, *From X-rays to Quarks: Modern Physicists and Their Discoveries*, W. H. Freeman, New York, 1980.

Personal recollections of the period 1930–1950 appear in *The Birth of Particle Physics*, L. M. Brown and L. Hoddeson eds., Cambridge University Press, New York, 1983. See especially the article by C. D. Anderson, p. 131.

Sir James Chadwick recounts the story of the discovery of the neutron in *Adventures in Experimental Physics*, β, B. Maglich, ed., World Science Education, Princeton, NJ, 1972.

References

1.1 J. Chadwick, "Possible Existence of a Neutron." *Nature*, **129**, 312 (1932).
1.2 J. Chadwick, "Bakerian Lecture." *Proc. Roy. Soc.*, **A142**, 1 (1933).
1.3 C. D. Anderson, "The Positive Electron." *Phys. Rev.*, **43**, 491 (1933).
1.4 P. M. S. Blackett and G. P. S. Occhialini, "Some Photographs of the Tracks of Penetrating Radiation." *Proc. Roy. Soc.*, **A 139**, 699 (1933).
1.5 J. Chadwick and M. Goldhaber, "A 'Nuclear Photo-effect': Disintegration of the Diplon by γ-rays." *Nature*, **134**, 237 (1934).

Letters to the Editor

[*The Editor does not hold himself responsible for opinions expressed by his correspondents. Neither can he undertake to return, nor to correspond with the writers of, rejected manuscripts intended for this or any other part of* NATURE. *No notice is taken of anonymous communications.*]

Possible Existence of a Neutron

It has been shown by Bothe and others that beryllium when bombarded by α-particles of polonium emits a radiation of great penetrating power, which has an absorption coefficient in lead of about 0·3 (cm.)$^{-1}$. Recently Mme. Curie-Joliot and M. Joliot found, when measuring the ionisation produced by this beryllium radiation in a vessel with a thin window, that the ionisation increased when matter containing hydrogen was placed in front of the window. The effect appeared to be due to the ejection of protons with velocities up to a maximum of nearly 3×10^9 cm. per sec. They suggested that the transference of energy to the proton was by a process similar to the Compton effect, and estimated that the beryllium radiation had a quantum energy of 50×10^6 electron volts.

I have made some experiments using the valve counter to examine the properties of this radiation excited in beryllium. The valve counter consists of a small ionisation chamber connected to an amplifier, and the sudden production of ions by the entry of a particle, such as a proton or α-particle, is recorded by the deflexion of an oscillograph. These experiments have shown that the radiation ejects particles from hydrogen, helium, lithium, beryllium, carbon, air, and argon. The particles ejected from hydrogen behave, as regards range and ionising power, like protons with speeds up to about $3\cdot2 \times 10^9$ cm. per sec. The particles from the other elements have a large ionising power, and appear to be in each case recoil atoms of the elements.

If we ascribe the ejection of the proton to a Compton recoil from a quantum of 52×10^6 electron volts, then the nitrogen recoil atom arising by a similar process should have an energy not greater than about 400,000 volts, should produce not more than about 10,000 ions, and have a range in air at N.T.P. of about 1·3 mm. Actually, some of the recoil atoms in nitrogen produce at least 30,000 ions. In collaboration with Dr. Feather, I have observed the recoil atoms in an expansion chamber, and their range, estimated visually, was sometimes as much as 3 mm. at N.T.P.

These results, and others I have obtained in the course of the work, are very difficult to explain on the assumption that the radiation from beryllium is a quantum radiation, if energy and momentum are to be conserved in the collisions. The difficulties disappear, however, if it be assumed that the radiation consists of particles of mass 1 and charge 0, or neutrons. The capture of the α-particle by the Be^9 nucleus may be supposed to result in the formation of a C^{12} nucleus and the emission of the neutron. From the energy relations of this process the velocity of the neutron emitted in the forward direction may well be about 3×10^9 cm. per sec. The collisions of this neutron with the atoms through which it passes give rise to the recoil atoms, and the observed energies of the recoil atoms are in fair agreement with this view. Moreover, I have observed that the protons ejected from hydrogen by the radiation emitted in the opposite direction to that of the exciting α-particle appear to have a much smaller range than those ejected by the forward radiation.

This again receives a simple explanation on the neutron hypothesis.

If it be supposed that the radiation consists of quanta, then the capture of the α-particle by the Be^9 nucleus will form a C^{13} nucleus. The mass defect of C^{13} is known with sufficient accuracy to show that the energy of the quantum emitted in this process cannot be greater than about 14×10^6 volts. It is difficult to make such a quantum responsible for the effects observed.

It is to be expected that many of the effects of a neutron in passing through matter should resemble those of a quantum of high energy, and it is not easy to reach the final decision between the two hypotheses. Up to the present, all the evidence is in favour of the neutron, while the quantum hypothesis can only be upheld if the conservation of energy and momentum be relinquished at some point.

J. CHADWICK.

Cavendish Laboratory,
Cambridge, Feb. 17.

The Positive Electron

CARL D. ANDERSON, *California Institute of Technology, Pasadena, California*
(Received February 28, 1933)

Out of a group of 1300 photographs of cosmic-ray tracks in a vertical Wilson chamber 15 tracks were of positive particles which could not have a mass as great as that of the proton. From an examination of the energy-loss and ionization produced it is concluded that the charge is less than twice, and is probably exactly equal to, that of the proton. If these particles carry unit positive charge the curvatures and ionizations produced require the mass to be less than twenty times the electron mass. These particles will be called positrons. Because they occur in groups associated with other tracks it is concluded that they must be secondary particles ejected from atomic nuclei.

Editor

ON August 2, 1932, during the course of photographing cosmic-ray tracks produced in a vertical Wilson chamber (magnetic field of 15,000 gauss) designed in the summer of 1930 by Professor R. A. Millikan and the writer, the tracks shown in Fig. 1 were obtained, which seemed to be interpretable only on the basis of the existence in this case of a particle carrying a positive charge but having a mass of the same order of magnitude as that normally possessed by a free negative electron. Later study of the photograph by a whole group of men of the Norman Bridge Laboratory only tended to strengthen this view. The reason that this interpretation seemed so inevitable is that the track appearing on the upper half of the figure cannot possibly have a mass as large as that of a proton for as soon as the mass is fixed the energy is at once fixed by the curvature. The energy of a proton of that curvature comes out 300,000 volts, but a proton of that energy according to well established and universally accepted determinations[1] has a total range of about 5 mm in air while that portion of the range actually visible in this case exceeds 5 cm without a noticeable change in curvature. The only escape from this conclusion would be to assume that at exactly the same instant (and the sharpness of the tracks determines that instant to within about a fiftieth of a second) two independent electrons happened to produce two tracks so placed as to give the impression of a single particle shooting through the lead plate. This assumption was dismissed on a probability basis, since a sharp track of this order of curvature under the experimental conditions prevailing occurred in the chamber only once in some 500 exposures, and since there was practically no chance at all that two such tracks should line up in this way. We also discarded as completely untenable the assumption of an electron of 20 million volts entering the lead on one side and coming out with an energy of 60 million volts on the other side. A fourth possibility is that a photon, entering the lead from above, knocked out of the nucleus of a lead atom two particles, one of which shot upward and the other downward. But in this case the upward moving one would be a positive of small mass so that either of the two possibilities leads to the existence of the positive electron.

In the course of the next few weeks other photographs were obtained which could be interpreted logically only on the positive-electron basis, and a brief report was then published[2] with due reserve in interpretation in view of the importance and striking nature of the announcement.

MAGNITUDE OF CHARGE AND MASS

It is possible with the present experimental data only to assign rather wide limits to the

---

[1] Rutherford, Chadwick and Ellis, *Radiations from Radioactive Substances*, p. 294. Assuming $R \propto v^3$ and using data there given the range of a 300,000 volt proton in air S.T.P. is about 5 mm.

[2] C. D. Anderson, Science **76**, 238 (1932).

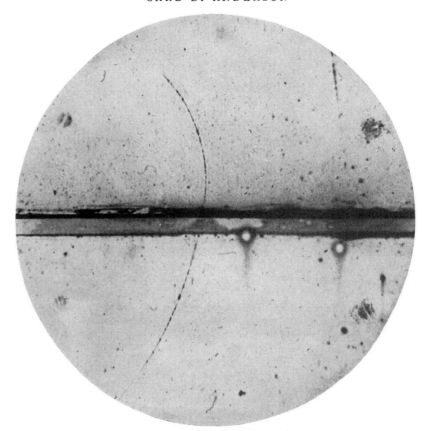

FIG. 1. A 63 million volt positron ($H\rho = 2.1 \times 10^5$ gauss-cm) passing through a 6 mm lead plate and emerging as a 23 million volt positron ($H\rho = 7.5 \times 10^4$ gauss-cm). The length of this latter path is at least ten times greater than the possible length of a proton path of this curvature.

magnitude of the charge and mass of the particle. The specific ionization was not in these cases measured, but it appears very probable, from a knowledge of the experimental conditions and by comparison with many other photographs of high- and low-speed electrons taken under the same conditions, that the charge cannot differ in magnitude from that of an electron by an amount as great as a factor of two. Furthermore, if the photograph is taken to represent a positive particle penetrating the 6 mm lead plate, then the energy lost, calculated for unit charge, is approximately 38 million electron-volts, this value being practically independent of the proper mass of the particle as long as it is not too many times larger than that of a free negative electron. This value of 63 million volts per cm energy-loss for the positive particle it was considered legitimate to compare with the measured mean of approximately 35 million volts[3] for negative electrons of 200–300 million volts energy since the rate of energy-loss for particles of small mass is expected to change only very slowly over an energy range extending from several million to several hundred million volts. Allowance being made for experimental uncertainties, an upper limit to the rate of loss of energy for the positive particle can then be set at less than four times that for an electron, thus fixing, by the usual relation between rate of ionization and

---

[3] C. D. Anderson, Phys. Rev. **43**, 381A (1933).

charge, an upper limit to the charge less than twice that of the negative electron. It is concluded, therefore, that the magnitude of the charge of the positive electron which we shall henceforth contract to positron is very probably equal to that of a free negative electron which from symmetry considerations would naturally then be called a negatron.

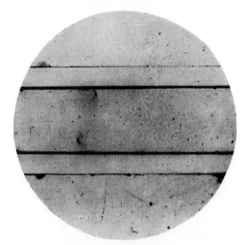

FIG. 2. A positron of 20 million volts energy ($H\rho = 7.1 \times 10^4$ gauss-cm) and a negatron of 30 million volts energy ($H\rho = 10.2 \times 10^4$ gauss-cm) projected from a plate of lead. The range of the positive particle precludes the possibility of ascribing it to a proton of the observed curvature.

It is pointed out that the effective depth of the chamber in the line of sight which is the same as the direction of the magnetic lines of force was 1 cm and its effective diameter at right angles to that line 14 cm, thus insuring that the particle crossed the chamber practically normal to the lines of force. The change in direction due to scattering in the lead,[3] in this case about 8° measured in the plane of the chamber, is a probable value for a particle of this energy though less than the most probable value.

The magnitude of the proper mass cannot as yet be given further than to fix an upper limit to it about twenty times that of the electron mass. If Fig. 1 represents a particle of unit charge passing through the lead plate then the curvatures, on the basis of the information at hand on ionization, give too low a value for the energy-loss unless the mass is taken less than twenty times that of the negative electron mass. Further determinations of $H\rho$ for relatively low energy particles before and after they cross a known amount of matter, together with a study of ballistic effects such as close encounters with electrons, involving large energy transfers, will enable closer limits to be assigned to the mass.

To date, out of a group of 1300 photographs of cosmic-ray tracks 15 of these show positive particles penetrating the lead, none of which can be ascribed to particles with a mass as large as that of a proton, thus establishing the existence of positive particles of unit charge and of mass small compared to that of a proton. In many other cases due either to the short section of track available for measurement or to the high energy of the particle it is not possible to differentiate with certainty between protons and positrons. A comparison of the six or seven hundred positive-ray tracks which we have taken is, however, still consistent with the view that the positive particle which is knocked out of the nucleus by the incoming primary cosmic ray is in many cases a proton.

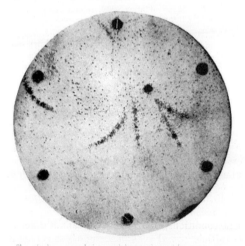

FIG. 3. A group of six particles projected from a region in the wall of the chamber. The track at the left of the central group of four tracks is a negatron of about 18 million volts energy ($H\rho = 6.2 \times 10^4$ gauss-cm) and that at the right a positron of about 20 million volts energy ($H\rho = 7.0 \times 10^4$ gauss-cm). Identification of the two tracks in the center is not possible. A negatron of about 15 million volts is shown at the left. This group represents early tracks which were broadened by the diffusion of the ions. The uniformity of this broadening for all the tracks shows that the particles entered the chamber at the same time.

From the fact that positrons occur in groups associated with other tracks it is concluded that they must be secondary particles ejected from an atomic nucleus. If we retain the view that a nucleus consists of protons and neutrons (and α-

FIG. 4. A positron of about 200 million volts energy ($H\rho = 6.6 \times 10^5$ gauss-cm) penetrates the 11 mm lead plate and emerges with about 125 million volts energy ($H\rho = 4.2 \times 10^5$ gauss-cm). The assumption that the tracks represent a proton traversing the lead plate is inconsistent with the observed curvatures. The energies would then be, respectively, about 20 million and 8 million volts above and below the lead, energies too low to permit the proton to have a range sufficient to penetrate a plate of lead of 11 mm thickness.

particles) and that a neutron represents a close combination of a proton and electron, then from the electromagnetic theory as to the origin of mass the simplest assumption would seem to be that an encounter between the incoming primary ray and a proton may take place in such a way as to expand the diameter of the proton to the same value as that possessed by the negatron. This process would release an energy of a billion electron-volts appearing as a secondary photon. As a second possibility the primary ray may disintegrate a neutron (or more than one) in the nucleus by the ejection either of a negatron or a positron with the result that a positive or a negative proton, as the case may be, remains in the nucleus in place of the neutron, the event occurring in this instance without the emission of a photon. This alternative, however, postulates the existence in the nucleus of a proton of negative charge, no evidence for which exists. The greater symmetry, however, between the positive and negative charges revealed by the discovery of the positron should prove a stimulus to search for evidence of the existence of negative protons. If the neutron should prove to be a fundamental particle of a new kind rather than a proton and negatron in close combination, the above hypotheses will have to be abandoned for the proton will then in all probability be represented as a complex particle consisting of a neutron and positron.

While this paper was in preparation press reports have announced that P. M. S. Blackett and G. Occhialini in an extensive study of cosmic-ray tracks have also obtained evidence for the existence of light positive particles confirming our earlier report.

I wish to express my great indebtedness to Professor R. A. Millikan for suggesting this research and for many helpful discussions during its progress. The able assistance of Mr. Seth H. Neddermeyer is also appreciated.

2
The Muon and the Pion

The discoveries of the muon and charged pions in cosmic-ray experiments and the discovery of the neutral pion using accelerators, 1936–51.

The detection of elementary particles is based on their interactions with matter. Swiftly moving charged particles produce ionization and it is this ionization that is the basis for most techniques of particle detection. During the 1930s cosmic rays were studied primarily with cloud chambers, in which droplets form along the trails of ions left by the cosmic rays. If the cloud chamber is in a region of magnetic field, the tracks show curvature. According to the Lorentz force law, the component of the momentum in the plane perpendicular to the magnetic field is given by $p(\text{MeV}/c) = 0.300 \times 10^{-3} B(\text{gauss}) r(\text{cm})$ or $p(\text{GeV}/c) = 0.300 \times B(\text{T}) r(\text{m})$, where r is the radius of curvature. By measuring the track of a particle in a cloud chamber it is possible to deduce the momentum of the particle.

The energy of a charged particle can be deduced by measuring the distance it travels before stopping in some medium. The charged particles other than electrons slow primarily because they lose energy through the ionization of atoms in the medium, unless they collide with a nucleus. The range a particle of a given energy will have in a medium is a function of the mass density of the material and of the density of electrons.

The collisional energy loss per unit path length of a charged particle of velocity v depends essentially linearly on the density of electrons in the material, $\rho_e = \rho N_A Z/A$, where ρ is the mass density of the material, N_A is Avogadro's number, and Z and A represent the atomic number and mass of the material. The force between the incident particle of charge ze and each electron is proportional to $z\alpha$, where $\alpha \approx 1/137$ is the fine structure constant. The energy transferred to the electron in a collision is proportional to $(z\alpha)^2$. A good representation of the final result for the energy loss is

$$\frac{dE}{dx} = \frac{N_A Z}{A} \frac{4\pi z^2 \alpha^2 (\hbar c)^2}{m_e v^2} \left[\ln \frac{2 m_e v^2 \gamma^2}{I} - \frac{v^2}{c^2} \right] \quad (2.1)$$

where $x = \rho l$ measures the path length in g cm^{-2}. Here $\gamma^2 = (1 - v^2/c^2)^{-1}$ and $I \approx 16 Z^{0.9}$ eV is a measure of the ionization potential. A practical feeling for the result is obtained by using $N_A = 6.02 \times 10^{23}$ g^{-1} and $\hbar c = 197$ MeV fm $= 197$ MeV 10^{-13} cm to obtain the relation $4\pi N_A \alpha^2 \hbar^2 / m_e = 0.307$ MeV/(g cm^{-2}). The expression for dE/dx has a minimum when γ is about 3 or 4. Typical values of minimum ionization are 1 to 2 MeV/(g cm^{-2}).

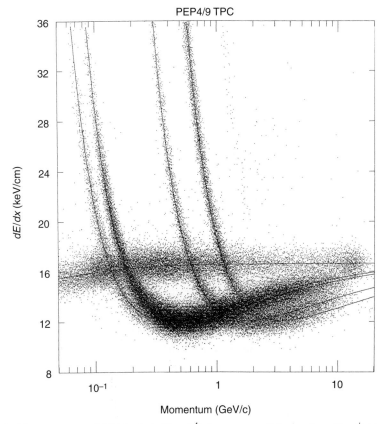

Figure 2.1. Measurements of dE/dx (in keV cm^{-1}) for many particles produced in e^+e^- collisions at a center of mass energy of 29 GeV. Each dot represents a single particle. Bands are visible for several distinct particle types. The flat band consists of electrons. The vertical bands, from left to right, show muons, charged pions, charged kaons, and protons. There is also a faint band of deuterons. The curves show the predicted values of dE/dx. The data were obtained with the Time Projection Chamber (TPC) developed by D. Nygren and co-workers at the Lawrence Berkeley Laboratory. The ionization measurements are made in a mixture of argon and methane gases at 8.5 atmospheres pressure. The data were taken at the Stanford Linear Accelerator Center. [TPC/Two-Gamma Collaboration, *Phys. Rev. Lett.*, **61**, 1263, (1988)]

Since the value of dE/dx depends on the velocity of the charged particle, it is possible to distinguish different particles with the same momentum but different masses by a careful measurement of dE/dx. In Figure 2.1 we show an application of this principle.

Energy loss by electrons is not dominated by the ionization process. In addition to losing energy by colliding with electrons in the material through which they pass, electrons lose energy by radiating photons whenever they are accelerated, a process called *bremsstrahlung* (braking radiation). Near the nuclei of heavy atoms there are intense electric fields. Electrons passing by nuclei undergo large accelerations. Although this in itself results in little energy loss directly (because the nuclei are heavy and recoil very little), the acceleration produces a good deal of bremsstrahlung and thus energy loss by the electrons. This mechanism is peculiar to electrons: Other incident charged particles do not lose much energy by bremsstrahlung because their greater mass reduces the acceleration they receive from the electric field around the nucleus. The modern theory of energy loss by electrons and positrons was developed by Bethe and Heitler in 1934.

The energy loss by an electron passing through a material is proportional to the density of nuclei, $\rho N_A/A$. The strength of the electrostatic force between the electron and a nucleus is proportional to $Z\alpha$ where Z is the atomic number of the material. The energy loss is proportional to $(Z\alpha)^2\alpha$, where the electromagnetic radiation by the electron accounts for the final factor of α. A good representation of the energy loss through bremsstrahlung is

$$\frac{dE}{dx} = \frac{N_A}{A} \frac{4Z(Z+1)\alpha^3(\hbar c)^2}{m_e^2 c^4} E \ln \frac{183}{Z^{1/3}} \equiv E/X_0. \tag{2.2}$$

Of course this represents the energy loss, so the energy varies as $\exp(-x/X_0)$ where x is the path length (in g/cm^2) and X_0 is called the radiation length. A radiation length in lead is 6.37 g cm^{-2} which, using the density of lead, is 0.56 cm. For iron the corresponding figures are 13.86 g cm^{-2} and 1.76 cm.

If a photon produced by bremsstrahlung is sufficiently energetic, it may contribute to an electromagnetic shower. The photon can "convert," that is, turn into an electron–positron pair as discussed in the previous chapter. The newly created particles will themselves lose energy and create more photons, building up a shower. Eventually the energy of the photons created will be less than that necessary to create additional pairs and the shower will cease to grow. The positrons eventually slow down and annihilate with atomic electrons to produce photons. Thus all the energy in the initial electron is ultimately deposited in the material through ionization and excitation of atoms.

In 1937, Anderson, together with S. H. Neddermeyer, made energy loss measurements by placing a 1-cm platinum plate inside a cloud chamber. By measuring the curvature of the tracks on both sides of the plate, they were able to determine the loss in momentum. Since they observed particles in the 100–500 MeV/c momentum range, if the particles were electrons or positrons, they were highly relativistic and their energy was given simply by $E = pc$. According to the Bethe–Heitler theory, the particles should have lost in the plate an amount of energy proportional to their incident energy. Moreover, the particles with this energy should have been associated with an electromagnetic shower. What Neddermeyer and Anderson observed was quite different. The particles could be separated into two classes. The first class behaved just as the Bethe–Heitler theory predicted. The particles of the second class, however, lost nearly no energy in the platinum plate: They were "penetrating." Moreover, they were not associated with electromagnetic showers.

Since the Bethe–Heitler theory predicted large energy losses for electrons because they were light and could easily emit radiation, Neddermeyer and Anderson (**Ref. 2.1**) were led to consider the possibility that the component of cosmic rays that did not lose much energy consisted of particles heavier than the electron. On the other hand, the particles in question could not be protons because protons of the momentum observed would be rather slow and would ionize much more heavily in the cloud chamber than the observed particles, whose ionization was essentially the same as that of the electrons. Neddermeyer and Anderson gave as their explanation

there exist particles of unit charge with a mass larger than that of a normal free electron and much smaller than that of a proton .... [That they] occur with both positive and negative charges suggests that they might be created in pairs by photons.

While the penetrating component of cosmic rays had been observed by others before Neddermeyer and Anderson, the latter were able to exclude the possibility that this component was due to protons. Moreover, Neddermeyer and Anderson observed particles of energy low enough to make the application of the Bethe–Heitler theory convincing. At the time, many doubted that the infant theory of quantum electrodynamics, still plagued with perplexing infinities, could be trusted at very high energies. The penetrating component of cosmic rays could be ascribed to a failure of the Bethe–Heitler theory when the penetrating particles were extremely energetic. Neddermeyer and Anderson provided evidence for penetrating particles at energies for which the theory was believed to hold.

At nearly the same time, Street and Stevenson reported similar results and soon improved upon them (**Ref. 2.2**). To determine the mass of the newly discovered particle, they sought to measure its momentum and ionization at the same time. Since the ionization is a function of the velocity, the two measurements would in principle suffice to determine the mass. However, the ionization is weakly dependent on the velocity except when the velocity is relatively low, that is, when the particle is near the end of its path and the ionization increases dramatically. To obtain a sample of interesting events, Street and Stevenson used counters in both coincidence and anticoincidence: The counters fired only if a charged particle passed through them and the apparatus was arranged so that the chamber was expanded to create supersaturation and a picture taken only if a particle entered the chamber (coincidence) but was not detected exiting (anticoincidence). This method of triggering the chamber was invented by Blackett and Occhialini. In addition, a block of lead was placed in front of the apparatus to screen out the showering particles. In late 1937, Street and Stevenson reported a track that ionized too much to be an electron with the measured momentum, but traveled too far to be a proton. They measured the mass crudely as 130 times the rest mass of the electron, an answer smaller by a factor 1.6 than later, improved results, but good enough to place it clearly between the electron and the proton.

In 1935, before the discovery of the penetrating particles, Hideki Yukawa predicted the existence of a particle of mass intermediate between the electron and the proton. This particle was to carry the nuclear force in the same way as the photon carries the electromagnetic force. In addition, it was to be responsible for beta decay. Since the range of nuclear forces is about 1 fm, the mass of the particle predicted by Yukawa was about $(\hbar/c)/10^{-13}$ cm $\approx$ 200 MeV/c^2. When improved measurements were made, the mass of the new particle was determined to be about 100 MeV/c^2, close enough to the theoretical estimate to make natural the identification of the penetrating particle with the Yukawa particle.

How could this identification be confirmed? In 1940, Tomonaga and Araki showed that positive and negative Yukawa particles should produce very different effects when they came to rest in matter. The negative particles would be captured into atomic-like orbits, but with very small radii. As a result, they would overlap the nucleus substantially. Given that the Yukawa particle was designed to explain nuclear forces, it would certainly interact extremely rapidly with the nucleus, being absorbed long before it could decay directly. On the other hand, the positive Yukawa particles would come to rest between the atoms and would decay.

The lifetime of the penetrating particle was first measured by Franco Rassetti who found a value of about 1.5×10^{-6} s. Improved results, near 2.2×10^{-6} s were obtained by Rossi and Nereson, and by Chaminade, Freon, and Maze. Working under very difficult circumstances in Italy during World War II, Conversi, Pancini, and Piccioni (**Ref. 2.3**) investigated further the decays of positive and negative penetrating particles that came to rest in various materials. Using a magnetic-focusing arrangement that Rossi had developed, Conversi, Pancini, and Piccioni were able to select either positive or negative penetrating particles from the cosmic rays and then determine whether they decayed or not when stopped in matter. The positive particles did indeed decay, as predicted by Tomonaga and Araki. When the absorber was iron, the negative particles did not decay, but were absorbed by the nucleus, again in accordance with the theoretical prediction. However, when the absorber was carbon, the negative particles decayed. This meant that the Tomonaga–Araki prediction as applied to the penetrating particles was wrong by many orders of magnitude: These could not be the Yukawa particles.

Shortly thereafter, D. H. Perkins (**Ref. 2.4**) used photographic emulsions to record an event of precisely the type forecast by Tomonaga and Araki. Photographic emulsions provide a direct record of cosmic-ray events with extremely fine resolution. Perkins was able to profit from advances in the technology of emulsion produced by Ilford Ltd. The event in question had a slow negative particle that came to rest in an atom, most likely a light atom like carbon, nitrogen, or oxygen. After the particle was absorbed by the nucleus, the nucleus was blasted apart and three fragments were observed in the emulsion. This single event apparently showed the behavior predicted by Tomonaga and Araki, contrary to the results of the Italian group.

The connection between the results of Conversi, Pancini, Piccioni and the observation of Perkins was made by the Bristol group of Lattes, Occhialini, and Powell (**Ref. 2.5**) in one of several papers by the group, again using emulsions. Their work established that there were indeed two different particles, one of which decayed into the other. The observed decay product appeared to have fixed range in the emulsion. That is, it appeared always to be produced with the same energy. This indicated that the decay was into two bodies and not more. Because of inaccurate mass determinations, at first it was believed that the unseen particle in the decay could not be massless. Quickly, the picture was corrected and completed: The pion, π, decayed into a muon, μ (the names given by Lattes *et al.*), and a very light particle, presumably Pauli's neutrino. The π (which Perkins had likely seen) was much like Yukawa's particle except that it was not the origin of beta decay, since beta decays produce electrons rather than muons. The μ (which Anderson and Neddermeyer had found) was just like an electron, only heavier. The pion has two charge states, π^+ and

π^- that are charge conjugates of each other and which yield μ^+ and μ^-, respectively, in their decays.

In modern parlance, bosons (particles with integral spin) like the pion that feel nuclear forces are called *mesons*. More generally, all particles that feel nuclear forces, including fermions like the proton and neutron are called *hadrons*. Fermions (particles with half-integral spin) like the muon and electron that are not affected by these strong forces are called *leptons*. While a negative pion would always be absorbed by a nucleus upon coming to rest, the absorption of the negative muon was much like the well-known radioactive phenomenon of K-capture in which an inner electron is captured by a nucleus while a proton is transformed into a neutron and a neutrino is emitted. In heavy atoms, the negative muon could be absorbed (because it largely overlapped with the nucleus) with small nuclear excitation and the emission of a neutrino, while in the light atoms it would usually decay, because there was insufficient overlap between the muon and the nucleus.

Cosmic rays were the primary source of high energy particles until a few years after World War II. Although proton accelerators had existed since the early 1930s, their low energies had restricted their applications to nuclear physics. The early machines included Robert J. Van de Graaff's electrostatic generators, developed at Princeton, the voltage multiplier proton accelerator built by J. D. Cockroft and E. T. S. Walton at the Cavendish Laboratory, and the cyclotron built by Ernest O. Lawrence and Stanley Livingston in Berkeley.

The cyclotron incorporated Lawrence's revolutionary idea, resonant acceleration of particles moving in a circular path, giving them additional energy on each circuit of the machine. The particles moved in a plane perpendicular to a uniform magnetic field. Cyclotrons typically contain two semi-circular "dees" and the particles are given a kick by an electric field each time they pass from one dee to the other, though the original cyclotron of Lawrence and Livingston contained just one dee. The frequency of the machine was determined by the Lorentz force law, $F = evB$, and the formula for the centripetal acceleration, $v^2/r = F/m = evB/m$ so that angular frequency is given by

$$\omega = \frac{eB}{m}. \tag{2.3}$$

The cyclotron frequency is independent of the radius of the trajectory: As the energy of the particle increases, so does the radius in just such a way that the rotational frequency is constant. It was thus possible to produce a steady stream of high energy particles spiraling outward from a source at the center.

Cyclotrons of ever-increasing size were constructed by Lawrence and his team in an effort to achieve higher and higher energies. Ultimately the technique was limited by relativistic effects. The full equation for the frequency is actually

$$\omega = \frac{eB}{\gamma m} \tag{2.4}$$

where γ is the factor describing the relativistic mass increase, $\gamma = E/mc^2$. When protons were accelerated to relativistic velocities, the required frequency decreased.

The synchrocyclotron solved this problem by using bursts of particles, each of which was accelerated with an RF system whose frequency decreased in just the right way to compensate for the relativistic effect. The success of the synchrocyclotron was due to the development of the theory of "phase stability" developed by E. McMillan and independently by V. I. Veksler. In 1948, the 350-MeV, 184-inch proton synchrocyclotron at Berkeley became operational and soon thereafter Lattes and Gardner observed charged pions in photographic emulsions.

It was already known that cosmic-ray showers had a "soft" component, consisting primarily of electromagnetic radiation. Indeed, Lewis, Oppenheimer, and Wouthuysen had suggested that this component could be due to neutral mesons that decayed into pairs of photons. Such neutral mesons, partners of the charged pions, had been proposed by Nicholas Kemmer in 1938 in a seminal paper on isospin invariance, the symmetry relating the proton to the neutron.

Strong circumstantial evidence for the existence of a neutral meson with a mass similar to that of the charged pion was obtained by Bjorklund, Crandall, Moyer, and York using the 184-inch synchrocyclotron (Ref. 2.6). See Figure 2.2. Bjorklund *et al.* used a pair spectrometer to measure the photons produced by the collisions of protons on targets of carbon and beryllium. The pair spectrometer consisted of a thin tantalum radiator in which photons produced electron–positron pairs whose momenta were measured in a magnetic field. When the incident proton beam had an energy less than 175 MeV, the observed yield of photons was consistent with the expectations from bremsstrahlung from the proton. However, when the incident energy was raised to 230 MeV, many more photons were observed and with an energy spectrum unlike that for bremsstrahlung. The most likely explanation of the data was the production of a neutral meson decaying into two photons.

Evidence for these photons was also obtained in a cosmic-ray experiment by Carlson, Hooper, and King working at Bristol (Ref. 2.7). The photons were observed by their conversions into e^+e^- pairs in photographic emulsion. See Figure 2.3. This experiment placed an upper limit on the lifetime of the neutral pion of 5×10^{-14} s. The technique used was a new one. The direction of the converted photon was projected back towards the primary vertex of the event. The impact parameter, the distance of closest approach of that line to the primary vertex, was measured. Because the neutral pion decayed into two photons, the direction of a single one, in principle, did not point exactly to the primary vertex. In fact, the lifetime could not be measured in this experiment since it turned out to be about 10^{-16} s, far less than the limit obtainable at the time.

Direct confirmation of the two-photon decay was provided by Steinberger, Panofsky, and Steller **(Ref. 2.8)** using the electron synchrotron at Berkeley. The synchrotron relied on the principle of phase stability underlying the synchrocyclotron, but differed in that the beam was confined to a small beam tube, rather than spiraling outward between the poles of large magnets. In the electron synchroton, the strength of the magnetic field varied as the particles were accelerated.

The electron beam was used to generate a beam of gamma rays with energies up to 330 MeV. Two photon detectors were placed near a beryllium target. Events were accepted only if photons were seen in both detectors. The rate for these coincidences was studied as

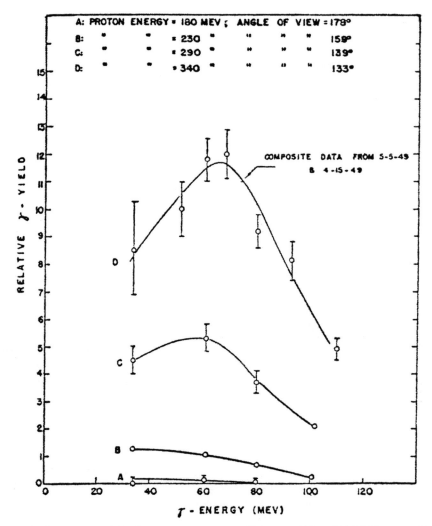

Figure 2.2. Gamma-ray yields from proton–carbon collisions at 180 to 340 MeV proton kinetic energy. The marked increase with increasing proton energy is the result of passing the π^0 production threshold. The π^0 decays into two photons. (Ref. 2.6)

a function of the angle between the photons and the angle between the plane of the final state photons and the incident beam direction. The data were consistent with the decay of a neutral meson into two photons with a production cross section for the neutral meson similar to that known for the charged mesons. The two-photon decay proved that the neutral meson could not have spin one since Yang's theorem forbids the decay of a spin-1 particle into two photons.

The proof of Yang's theorem follows from the fundamental principle of linear superposition in quantum mechanics, which requires that the transition amplitude, a scalar quantity, depend linearly on the

2. The Muon and the Pion

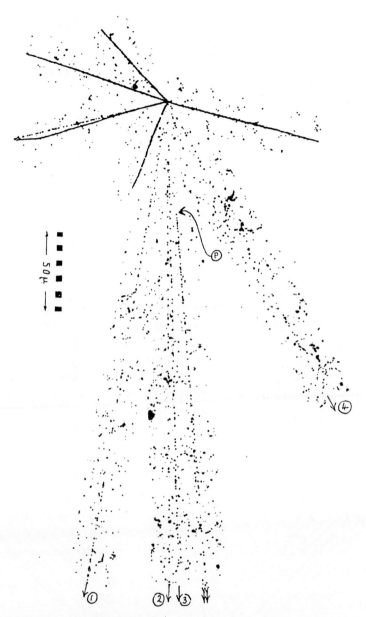

Figure 2.3. An emulsion event showing an e^+e^- pair created by conversion of a photon from π^0 decay. The conversion occurs at the point marked P. (Ref. 2.7)

spin orientation of each particle in the process. The decay amplitude for a spin-1 particle into two photons would have to be linear in the polarization vector of the initial particle and each of the two final-state photons. The polarization vector for a photon points in the direction of the electric field, which is perpendicular to the momentum. For a massive spin-1 state it is similar, except that it can point in any spatial direction, not just perpendicular to the direction of the momentum. In addition, the amplitude would have to be even under interchange of the two photons since they are identical bosons. Since real photons are transversely polarized, if the momentum and polarization vectors of a photon are $\mathbf{k}$ and ϵ, then $\mathbf{k} \cdot \epsilon = 0$. Let the polarization vector of the initial particle in its rest frame be η and those of the photons be ϵ_1 and ϵ_2. Let the momentum of photon 1 be $\mathbf{k}$ so that of photon 2 is $-\mathbf{k}$. We must construct a scalar from these vectors.

If we begin with $\epsilon_1 \cdot \epsilon_2$ the only non-zero factor including η is $\eta \cdot \mathbf{k}$, but $\epsilon_1 \cdot \epsilon_2 \, \eta \cdot \mathbf{k}$ is odd under the interchange of 1 and 2 since this takes $\mathbf{k}$ into $-\mathbf{k}$. If we start with $\epsilon_1 \times \epsilon_2$ we have as possible scalars $\epsilon_1 \times \epsilon_2 \cdot \eta$, $(\epsilon_1 \times \epsilon_2) \cdot (\eta \times \mathbf{k})$, and $\epsilon_1 \times \epsilon_2 \cdot \mathbf{k} \, \eta \cdot \mathbf{k}$. The first and third are odd under the interchange of 1 and 2 and the second vanishes identically since $(\epsilon_1 \times \epsilon_2) \cdot (\eta \times \mathbf{k}) = \epsilon_1 \cdot \eta \, \epsilon_2 \cdot \mathbf{k} - \epsilon_2 \cdot \eta \, \epsilon_1 \cdot \mathbf{k}$.

A year later, in 1951, Panofsky, Aamodt, and Hadley (Ref. 2.9) published a study of negative pions stopping in hydrogen and deuterium targets. Their results greatly expanded knowledge of the pions. The experiment employed a more sophisticated pair spectrometer, as shown in Figure 2.4. The reactions studied with the hydrogen target were

$$\pi^- p \to \pi^0 n$$
$$\pi^- p \to \gamma n$$

The latter process gave a monochromatic photon whose energy yielded $275.2 \pm 2.5 \, m_e$ as the mass of the π^-, an extremely good measurement. See Figure 2.5. The photons produced by the decay of the π^0 were Doppler-shifted by the motion of the decaying π^0. From the spread of the observed photon energies, it was possible to deduce the mass difference between the neutral and charged pion. Again, an excellent result, $m_{\pi^-} - m_{\pi^0} = 10.6 \pm 2.0 \, m_e$, was obtained. The capture of the π^- is assumed to occur from an s-wave state since the cross section for the lth partial waves is suppressed by k^{2l}, where k is the momentum of the incident pion. If the final π^0 is produced in the s-wave, then the parity

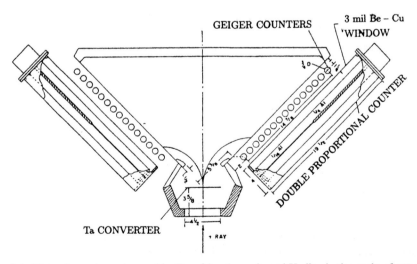

Figure 2.4. The pair spectrometer used by Panofsky, Aamodt, and Hadley in the study of $\pi^- p$ and $\pi^- d$ reactions. A magnetic field of 14 kG perpendicular to the plane shown bent the positrons and electrons into the Geiger counters on opposite sides of the spectrometer. (Ref. 2.9)

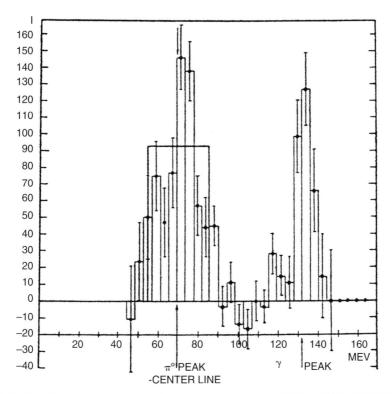

Figure 2.5. The photon energy spectrum for $\pi^- p$ reactions at rest. The band near 70 MeV is due to photons from π^0 decay. The line near 130 MeV is due to $\pi^- p \to n\gamma$. (Ref. 2.9)

of the neutral and charged pions must be the same. The momentum of the produced π^0, however, is not terribly small so this argument is not unassailable.

Parity is the name given to the reflection operation $\mathbf{r} \to -\mathbf{r}$. Its importance was first emphasized by Wigner in connection with Laporte's rule, which says that atomic states are divided into two classes and electric dipole transitions always take a state from one class into a state in the other. In the hydrogen atom, a state with orbital angular momentum l has the property

$$P\psi(\mathbf{r}) = \psi(-\mathbf{r}) = (-1)^l \psi(\mathbf{r}). \tag{2.5}$$

The state is unchanged except for the multiplicative factor of modulus unity. We therefore say that the parity is $(-1)^l$. This result is not general. Consider a two-electron atom with electrons in states with angular momentum l and l'. The parity is $(-1)^{l+l'}$, but the total angular momentum, L, is constrained only by $|l-l'| \leq L \leq l+l'$. Thus, in general the parity need not be $(-1)^L$. Electric dipole transitions take an atom in a state of even parity ($P = +1$) to a state with odd parity ($P = -1$), and vice versa.

Elementary particles are said to have an "intrinsic" parity, $\eta = \pm 1$. The parity operation changes the wave function by a factor η, in addition to changes resulting from the explicit position dependence. By convention, the proton and neutron each have parity $+1$. Having established this convention, the parity of the pion becomes an experimental question. The deuteron is a state of total angular momentum one. The total angular momentum comes from the combined spin angular momentum, which takes the value 1, and the orbital angular momentum, which is mostly 0 (s-wave), but partly 2 (d-wave). The deuteron thus has parity +1. The standard notation gives the total angular momentum, J, and parity, P, in the form $J^P = 1^+$. The spin, orbital, and total angular momentum are displayed in spectroscopic notation as $^{2S+1}L_J$, that is $^3S_1$ and $^3D_1$ for the components of the deuteron.

With the deuterium target, the reactions that could be observed in the same experiment were

$$\pi^- d \to nn$$
$$\pi^- d \to nn\gamma$$
$$\pi^- d \to nn\pi^0$$

In fact, the third was not seen, and the presence of the first had to be inferred by comparison to the data for $\pi^- p$. (See Figure 2.6). This inference was important because it established that the π^- could not be a scalar particle. If the π is a scalar and is absorbed from the s-wave orbital (as is reasonable to assume), the initial state also has $J^P = 1^+$. However, because of the exclusion principle, the only $J = 1$ state of two neutrons is $^3P_1$, which has odd parity. Thus if $\pi^- d \to nn$ occurs, the π^- cannot be a scalar. The absence of the third reaction was to be expected if the π^- and π^0 had the same parity. The two lowest nn states are ^1S and ^3P. The former cannot be produced if the charged and neutral pions have the same parity. If the nn state is ^3P, then parity conservation requires that the π^0 be in a p-wave. The presence of two p-waves in a process with such little phase space would greatly inhibit its production.

Subsequent experiments determined additional properties of the pions. The spin of the charged pion was obtained by comparing the reactions $pp \to \pi^+ d$ and $\pi^+ d \to pp$. The

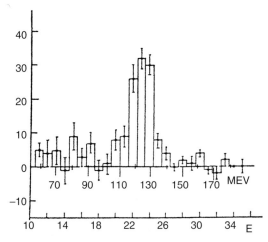

Figure 2.6. The photon energy spectrum from $\pi^- d$ reactions at rest. The line near 130 MeV is due to $\pi^- d \to nn\gamma$. (Ref. 2.9)

cross section for a scattering process with two final state particles is related to the Lorentz invariant matrix element, $\mathcal{M}$, by

$$\frac{d\sigma}{d\Omega} = \frac{1}{64\pi^2 s} \frac{p'}{p} |\mathcal{M}|^2. \qquad (2.6)$$

In this relation s is the square of the total energy in the center of mass, p and p' are the center-of-mass momenta in the initial state and final states, and $d\Omega$ is the solid angle element in the center of mass. The matrix element squared is to be averaged over the spin configurations of the initial state and summed over those of the final state.

The reactions $pp \to \pi^+ d$ and $\pi^+ d \to pp$ have the same scattering matrix elements (provided time reversal invariance is assumed), so their rates (at the same center-of-mass energy) differ only by phase space factors (p/p') and by the statistical factors resulting from the spins:

$$\frac{d\sigma(\pi^+ d \to pp)/d\Omega}{d\sigma(pp \to \pi^+ d)/d\Omega} = \frac{(2s_p + 1)^2}{(2s_d + 1)(2s_\pi + 1)} \frac{p_{pp}^2}{p_{\pi d}^2} \qquad (2.7)$$

where s_π is the spin of the π^+ and $p_{\pi d}$ and p_{pp} are the center-of-mass momenta for the πd and pp at the same center-of-mass energy. The proton and deuteron spins, s_p and s_d, were known. The pp reaction was measured by Cartwright, Richman, Whitehead, and Wilcox. The reverse reaction was measured independently by Clark, Roberts, and Wilson (Ref. 2.10) and then by Durbin, Loar, and Steinberger (**Ref. 2.11**).

The comparison showed the π^+ to have spin 0. Since the Panofsky, Aamodt, and Hadley paper had excluded $J^P = 0^+$ for the π^- and thus for its charge conjugate, the π^+ necessarily had $J^P = 0^-$. Since the π^0 decays into two photons it has integral spin and is thus a boson. Since it cannot have spin 1, it is reasonable to expect it has spin 0. Then, since its parity has been shown to be the same as that of the π^-, it follows that it, too, is 0^-. It is, however, possible to measure the parity directly. A small fraction of the time, about 1/80, the neutral pion will decay into $\gamma e^+ e^-$, the latter two particles being called a Dalitz pair. About $(1/160)^2$ of the time it decays into two Dalitz pairs. By studying the correlations between the planes of the Dalitz pairs, it is possible to show directly that the π^0 has $J^P = 0^-$, as was demonstrated in 1959 by Plano, Prodell, Samios, Schwartz, and Steinberger (Ref. 2.12).

The π^0 completed the triplet of pions: π^-, π^0, π^+. The approximate equality of the charged and neutral pion masses was reminiscent of the near equality of the masses of the neutron and proton. Nuclear physicists had observed an approximate symmetry, isotopic spin or isospin. This symmetry explains the similarity between the spacing of the energy levels in ^{13}C ($6p, 7n$) and ^{13}N ($6n, 7p$). Just as the nucleons represent an isospin doublet, the pions represent an isospin triplet.

Isospin is so named because its mathematical description is entirely analogous to ordinary spin or angular momentum in quantum mechanics. The isospin generators satisfy

$$[I_x, I_y] = iI_z \quad \text{etc.} \qquad (2.8)$$

and states can be classified by $\mathbf{I}^2 = I(I+1)$ and I_z. Thus $I_z(p) = 1/2$, $I_z(n) = -1/2$, $I_z(\pi^+) = 1$, $I_z(\pi^0) = 0$, etc. The rules for addition of angular momentum apply, so a state of a pion ($I = 1$) and a nucleon ($I = 1/2$) can be either $I = 3/2$ or $I = 1/2$. The state $\pi^+ p$ has $I_z = 3/2$ and is thus purely $I = 3/2$, whereas $\pi^+ n$ has $I_z = 1/2$ and is partly $I = 1/2$ and partly $I = 3/2$.

The isospin and parity symmetries contrast in several respects. Parity is related to space-time, while isospin is not. For this reason, isospin is termed an "internal" symmetry. Parity is a discrete symmetry, while isospin is a continuous symmetry since it is possible to consider rotations in isospin space by any angle. Isospin is an approximate symmetry since, for example, the neutron and proton do not have exactly the same mass. Parity was believed, until 1956, to be an exact symmetry.

Exercises

2.1 Determine the expected slope of the line in Fig. 1 of Neddermeyer and Anderson, Ref. 2.1 assuming the particles are electrons and positrons.

2.2 Verify the estimate of the mass of the particle seen by Street and Stevenson, Ref. 2.2, using the measurement of $H\rho$ and the ionization.

2.3 Assume for simplicity that $dE/dx = (dE/dx)_{min}/\beta^2 \equiv C/\beta^2$. Prove that the range of a particle of initial energy $E_0 = m\gamma_0$ is $R = mc^2(\gamma_0 - 1)^2/(C\gamma_0)$. Find the range of a muon in iron ($C = 1.48$ MeV cm^2 g^{-1}) for initial momentum between 0.1 GeV/c and 1 TeV/c. Do the same for a proton. Compare with the curves in the *Review of Particle Physics*.

2.4 What is the range in air of a typical α particle produced in the radioactive decay of a heavy element?

2.5 How is the mass of the π^- most accurately determined? The mass of the π^0? The *Review of Particle Physics* is an invaluable source of references for measurements of this sort.

2.6 How is the lifetime of the π^0 measured?

2.7 * Use dimensional arguments to estimate very crudely the rate for π^- absorption by a nucleus from a bound orbital. Assume any dimensionless coupling is of order 1.

2.8 * Use classical arguments to estimate the time required for a μ^- to fall from the radius of the lowest electron orbit to the lowest μ orbit in iron. Assume the power is radiated continuously in accordance with the results of classical electrodynamics.

2.9 * The π^0 decays at rest isotropically into two photons. Find the energy and angular distributions of the photons if the π^0 has a velocity β along the z axis.

Further Reading

For the early history of particle physics, especially cosmic-ray work, see *Colloque international sur l'histoire de la physique des particules, Journal de Physique* **48**, supplement au no. 12. Dec. 1982. Les Editions de Physique, Paris 1982 (in English).

Reminiscences of early work on the muon and the pion are contained in many of the articles in *The Birth of Particle Physics*, edited by L. M. Brown and L. Hoddeson, Cambridge University Press, Cambridge, 1983. See especially the article by S. Hayakawa for information on the independent developments in Japan that paralleled those discussed in this chapter.

For a flavor of particle physics around 1950 and for the opportunity to learn physics from one of the great masters, see *Nuclear Physics*, from a course taught by Enrico Fermi, notes taken by J. Orear, A. H. Rosenfeld, and R. A. Schluter, University of Chicago Press, Chicago, 1949.

For fundamentals of the interaction of elementary particles with matter and an early perspective on experimental particle physics, see *High Energy Particles*, by Bruno Rossi, Prentice-Hall, New York, 1952.

For a complete classical treatment of the interaction of charged particles with matter, see *Classical Electrodynamics*, Third Edition, by J. D. Jackson, Wiley, New York, 1999.

For information on particle masses, quantum numbers, and so on, and concise treatments of the behavior of high energy particles in matter, see *Review of Particle Physics*, written by the Particle Data Group and published biennially. A shortened version, the *Particle Properties Data Booklet*, is available for free by writing to the Particle Data Group, Lawrence Berkeley National Laboratory, 1 Cyclotron Road, Berkeley, CA 94720, USA or to CERN Scientific Information Service, CH-1211, Geneva 23, Switzerland.

For the development of accelerators, see M. Stanley Livingston *Particle Accelerators: A Brief History*, Harvard University Press, Cambridge, Mass., 1969. Fundamental papers are reprinted in *The Development of High Energy Accelerators*, Classics of Science, v. III, edited by M. S. Livingston, Dover, 1966.

References

2.1 S. H. Neddermeyer and C. D. Anderson, "Note on the Nature of Cosmic Ray Particles." *Phys. Rev.*, **51**, 884 (1937).
2.2 J. C. Street and E. C. Stevenson, "New Evidence for the Existence of a Particle of Mass Intermediate between the Proton and Electron." *Phys. Rev.*, **52**, 1003 (1937).
2.3 M. Conversi, E. Pancini, and O. Piccioni, " On the Disintegration of Negative Mesons." *Phys. Rev.*, **71**, 209 (1947).
2.4 D. H. Perkins, "Nuclear Disintegration by Meson Capture." *Nature*, **159**, 126 (1947).
2.5 C. M. G. Lattes, G. P. S. Occhialini, and C. F. Powell, "Observations on the Tracks of Slow Mesons in Photographic Emulsions." *Nature*, **160**, 453 (1947). Also Part II, *ibid.* p. 486. See also C. M. G. Lattes, H. Muirhead, G. P. S. Occhialini, and C. F. Powell, *Nature*, **159**, 694 (1947).
2.6 R. Bjorklund, W. E. Crandall, B. J. Moyer, and H. F. York, "High Energy Photons from Proton-Nucleus Collisions." *Phys. Rev.*, **77**, 213 (1950).
2.7 A. G. Carlson, J. E. Hooper, and D. T. King, "Nuclear Transmutations Produced by Cosmic-Ray Particles of Great Energy – Part V. The Neutral Meson." *Phil. Mag.*, **41**, 701 (1950).

2.8 J. Steinberger, W. K. H. Panofsky, and J. Steller, "Evidence for the Production of Neutral Mesons by Photons." *Phys. Rev.*, **78**, 802 (1950).

2.9 W. K. H. Panofsky, R. L. Aamodt, and J. Hadley, "The Gamma-Ray Spectrum Resulting from Capture of Negative π-Mesons in Hydrogen and Deuterium." *Phys. Rev.*, **81**, 565 (1951).

2.10 D. L. Clark, A. Roberts, and R. Wilson, "Cross section for the reaction $\pi^+ d \rightarrow pp$ and the spin of the π^+ meson." *Phys. Rev.*, **83**, 649 (1951).

2.11 R. Durbin, H. Loar, and J. Steinberger, "The Spin of the Pion via the Reaction $\pi^+ + d \leftrightarrows p + p$." *Phys. Rev.*, **83**, 646 (1951).

2.12 R. Plano *et al.*, "Parity of the Neutral Pion." *Phys. Rev. Lett.*, **3**, 525 (1959).

Note on the Nature of Cosmic-Ray Particles

SETH H. NEDDERMEYER AND CARL D. ANDERSON
California Institute of Technology, Pasadena, California
(Received March 30, 1937)

MEASUREMENTS[1] of the energy loss of particles occurring in the cosmic-ray showers have shown that this loss is proportional to the incident energy and within the range of the measurements, up to about 400 Mev, is in approximate agreement with values calculated theoretically for electrons by Bethe and Heitler. These measurements were taken using a thin plate of lead (0.35 cm), and the observed individual losses were found to vary from an amount below experimental detection up to the whole initial energy of the particle, with a mean fractional loss of about 0.5. If these measurements are correct it is evident that in a much thicker layer of heavy material multiple losses should become much more important, and the probability of observing a particle loss less than a large fraction of its initial energy should be very small. For the purpose of testing this inference and also for checking our previous measurements[2] which had shown the presence of some particles less massive than protons but more penetrating than electrons obeying the Bethe-Heitler theory, we have taken about 6000 counter-tripped photographs with a 1 cm plate of platinum placed across the center of the cloud chamber. This plate is equivalent in electron thickness to 1.96 cm of lead, and to 1.86 cm of lead for a Z^2 absorption. The results of 55 measurements on particles in the range below 500 Mev are given in Fig. 1, and in Fig. 2 the distribution of particles is shown as a function of the fraction of energy lost. The shaded part of the diagram represents particles which either enter the chamber accompanied by other particles or else themselves produce showers in the bar of platinum. It is clear that the particles separate themselves into two rather well-defined groups, the one consisting largely of shower particles and exhibiting a high absorbability, the other consisting of particles entering singly which in general lose a relatively small fraction of their initial energy, although there are four cases in which the loss is more than 60 percent. A considerable part of the spread on the negative abscissa can be accounted for by errors; it seems likely, however, that the case plotted at the extreme left represents a particle moving upward. Particles of both signs are distributed over the whole diagram, and moreover, the initial energies of the particles of each group are distributed over the whole measured range.

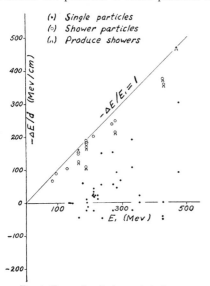

FIG. 1. Energy loss in 1 cm of platinum.

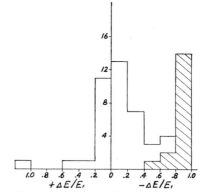

FIG. 2. Distribution of fractional losses in 1 cm of platinum.

---
[1] Anderson and Neddermeyer, Phys. Rev. **50**, 263 (1936).
[2] Anderson and Neddermeyer, Report of London Conference, Vol. 1 (1934), p. 179.

The chief source of error in these experiments lies not in the curvature measurements themselves, but in the track distortions produced by irregular motions of the gas in the chamber. The distortions are much larger when a thick plate is inside the chamber than when it is left unobstructed. These distortions are not sufficient to alter essentially the distribution of observed losses for the nonpenetrating group, but they could have a very serious effect in the part of the distribution representing small losses. This is especially true inasmuch as this group represents a small percentage of the total number of tracks, selected solely on the basis that they should exhibit a measurable curvature and at the same time be free from obvious distortion. The problem of measuring small energy losses is then evidently an extremely difficult one compared to that of measuring energy distributions in an unobstructed chamber. While it is possible in many cases to distinguish a distortion as such when a magnetic field is present, it is necessary to obtain independent criteria as to the reliability of the measurements; it is not a satisfactory procedure to try to do this simply by measuring curvatures of tracks taken with no field and comparing the curvature distribution thus found with the one obtained when the field is present. Observations made with no magnetic field indicate that serious distortions occur on about 5 percent of the photographs, and show that they are by no means a uniform function of the orientation and position of the track in the chamber. It is therefore not possible to correct for distortion in individual cases. When large distortions do occur, however, they are likely to obey one or both of the following correlations: (a) a curvature concave upward when the track makes a considerable angle with the vertical; (b) a curvature concave toward the center of the chamber. The observed percentages of measured single tracks obeying the correlations

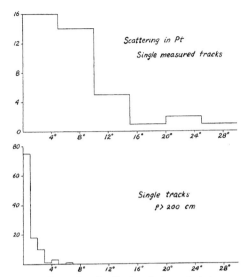

FIG. 3. Scattering distributions in 1 cm of platinum.

TABLE I. *Correlations between track curvatures and positions and orientations of tracks.*

Designation (See text)	Percentage correlation		
	Observed	Expected	Observed (excluding apparent gains)
(a)	52	50	55
(b)	67	50	55
(a) and (b)	33	25	27
neither	15	25	18

(a) and (b) are compared in Table I with the percentages expected if the observed curvatures have no relation to the positions and orientations of the tracks. If the 11 cases of apparent gains in energy are left out of consideration the observed percentages are brought somewhat closer to the expected values as shown in the last column.

A second independent check on the validity of the measurements can be obtained by measuring the scattering of the particles which show apparent curvatures, and comparing this with the scattering exhibited by those single tracks whose curvatures are just outside the range of measurability. In Fig. 3 are shown the distributions of scattering angles (the angles projected on the plane of the chamber) for the measured single tracks and for single tracks with a radius of curvature, $\rho \gtrsim 200$ cm (475 Mev). As it is scarcely conceivable that distortions could influence the scattering measurements by as much as 5°, these distributions constitute strong independent evidence that the measured tracks actually lie in the energy range indicated by the curvature determinations.

It has been known for a long time that there exist particles of both penetrating and nonpenetrating types. Crussard and Leprince-Ringuet[3] have recently made measurements of

[3] Crussard and Leprince-Ringuet, C. R. **204**, 240 (1937).

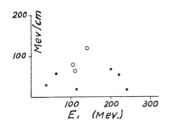

FIG. 4. Early measurements of energy loss in 0.7–1.5 cm of Pb. Dots indicate single particles; circles, shower particles.

energy loss in a range mainly above that covered in our experiments. They have concluded from their data that either the absorption law changes with energy or else that there is a difference in character among the particles. This same conclusion has already been twice stated by the writers.[4, 5] The present data appear to constitute the first experimental evidence for the existence of particles of both penetrating and nonpenetrating character in the energy range extending below 500 Mev. Moreover, the penetrating particles in this range do not ionize perceptibly more than the nonpenetrating ones, and cannot therefore be assumed to be of protonic mass. The lowest $H\rho$ among the penetrating group is 4.5×10^5 gauss cm. A proton of this curvature would ionize at least 25 times as strongly as a fast electron. It is interesting that our early measurements[2] of the energy loss in thicknesses of lead from 0.7 to 1.5 cm show a similar tendency to separate into two groups. They are reproduced in Fig. 4. If reinterpreted in the light of our present data they provide no evidence against high absorbability for electrons.

The nonpenetrating particles are readily interpreted as free positive and negative electrons. Interpretations of the penetrating ones encounter very great difficulties, but at present appear to be limited to the following hypotheses: (a) that an electron (+ or −) can possess some property other than its charge and mass which is capable of accounting for the absence of numerous large radiative losses in a heavy element; or (b) that there exist particles of unit charge, but with a mass (which may not have a unique value) larger than that of a normal free electron[6]

---
[4] Reference 2, p. 182.
[5] Reference 1, p. 268.
[6] The energies referred to throughout are, of course, calculated on the assumption of electronic mass. For a mass $m \lesssim 50 m_e$ the actual energy is very roughly $E = E_{el} - mc^2$ in the range of curvature here considered.

and much smaller than that of a proton; this assumption would also account for the absence of numerous large radiative losses, as well as for the observed ionization. Inasmuch as charge and mass are the only parameters which characterize the electron in the quantum theory, assumption (b) seems to be the better working hypothesis. If the penetrating particles are to be distinguished from free electrons by a greater mass, and since no evidence for their existence in ordinary matter obtains, it seems likely that there must exist some very effective process for removing them.

The experimental fact that penetrating particles occur both with positive and negative charges suggests that they might be created in pairs by photons, and that they might be represented as higher mass states of ordinary electrons.

Independent evidence indicating the existence of particles of a new type has already been found, based on range, curvature and ionization relations; for example, Figs. 12 and 13 of our previous publication.[1] In particular the strongly ionizing particle of Fig. 13 cannot readily be explained except in terms of a particle of e/m greater than that of a proton. The large value of e/m apparently is not due to an e greater than the electronic charge since above the plate the particle ionizes imperceptibly differently from a fast electron, whereas below the plate its ionization definitely exceeds that of an electron of the same curvature in the magnetic field; the effects, however, are understandable on the assumption that the particle's mass is greater than that of a free electron. We should like to suggest, merely as a possibility, that the strongly ionizing particles of the type of Fig. 13, although they occur predominantly with positive charge, may be related with the penetrating group above.

We wish to express our gratitude to Professor Millikan for his helpful discussions and encouragement. These experiments have been made possible by the Baker Company, who very generously loaned us the bar of platinum; and by funds supplied by the Carnegie Institution of Washington.

Note added in proof: Excellent experimental evidence showing the existence of particles less massive than protons, but more penetrating than electrons obeying the Bethe-Heitler theory has just been reported by Street and Stevenson, Abstract no. 40, Meeting of American Physical Society, Apr. 29, 1937.

NOVEMBER 1, 1937　　　　PHYSICAL REVIEW　　　　VOLUME 52

LETTERS TO THE EDITOR

Prompt publication of brief reports of important discoveries in physics may be secured by addressing them to this department. Closing dates for this department are, for the first issue of the month, the eighteenth of the preceding month, для the second issue, the third of the month. Because of the late closing dates for the section no proof can be shown to authors. The Board of Editors does not hold itself responsible for the opinions expressed by the correspondents.

Communications should not in general exceed 600 words in length.

New Evidence for the Existence of a Particle of Mass Intermediate Between the Proton and Electron

Anderson and Neddermyer[1] have shown that, for energies up to 300 and 400 Mev, the cosmic-ray shower particles have energy losses in lead plates corresponding to those predicted by theory for electrons. Recent studies of range[2] and energy loss[3] indicate that the singly occurring cosmic-ray corpuscles, even in the energy range below 400 Mev, are more penetrating than shower particles of corresponding magnetic deflection. Thus the natural assumptions have been expressed: the shower particles are electrons, the theory describing their energy losses is satisfactory, and the singly occurring particles are not electrons. The experiments cited above have shown from consideration of the specific ionization that the penetrating rays are not protons. The suggestion has been made that they are particles of electronic charge, and of mass intermediate between those of the proton and electron. If this is true, it should be possible to distinguish clearly such a particle from an electron or proton by observing its track density and magnetic deflection near the end of its range, although it is to be expected that the fraction of the total range in which the distinction can be made is very small. To examine this possibility experimentally we have used the arrangement of apparatus of Fig. 1. The three-counter telescope consisting of tubes 1, 2, and 3 and a lead filter L for removing shower particles, selects penetrating rays directed toward the cloud chamber C which is in a magnetic field of 3500 gauss. The type of track desired is one so near the end of its range as it enters the chamber that there is no chance of emergence below. In order to reduce the number of photographs of high energy particles, the tube group 4 was used as a cut-off counter with a circuit so arranged that the chamber would be set off only in those cases when a coincident discharge of counters 1, 2, and 3 was unaccompanied by a discharge of 4. The tripping of the cloud chamber valve was delayed about one sec. to facilitate determination of the drop count along a track. Because of geometrical imperfections of the arrangement and of counter inefficiency the cut-off circuit prevented

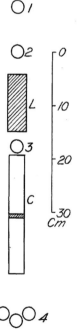

Fig. 1. Geometrical arrangement of apparatus.

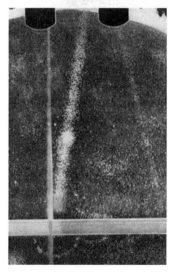

Fig. 2. Track A.

1003

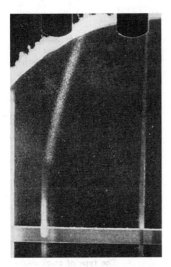

FIG. 3. Track B.

FIG. 4. Photograph of the track of a penetrating particle of high energy for comparison with A and B.

expansion for only $\frac{3}{4}$ of the discharges of the telescope. At the present time 1000 photos have been taken (equivalent to 4000 if the cut-off counter had not been used). Two tracks of interest, in that they have ionization densities definitely greater than usual, have been obtained: one A (see Fig. 2) is believed due to a proton and the other B (see Fig. 3) to a particle of mass approximately 130 times the rest mass of an electron. Track A which terminated in the lead strip at the center of the chamber exhibited an ionization density 2.4 times as great as the usual thin tracks and an $H\rho$ value approximately 2×10^5 gauss cm in a direction to indicate a positive particle. Track B which passed out of the lighted region above the lead plate had an ionization density about six times as great as normal thin tracks (the ion density was too great to permit an accurate ion count) and an $H\rho$ value of 9.6×10^4 gauss cm. If it is assumed, as seems reasonable, that the particle entered from above, the sign is negative. If it is taken that the ionization density varies inversely as the velocity squared, the rest mass of the particle in question is found to be approximately 130 times the rest mass of the electron. Because of uncertainty in the ion count this determination has a probable error of some 25 percent. In any case it does not seem possible to explain this track as due to a proton traveling up, for the observed $H\rho$ value would indicate a proton of 4.4×10^5 electron volts energy and therefore with a range of approximately one cm in the chamber. The track is clearly visible for 7 cm in the chamber.

The only possible objection to the conclusions reached above is that the bending of track A is largely due to distortion, but this is very unlikely, for the deflection is quite uniform and has a maximum value greater than ten times any distortions usually encountered in the thin tracks of high energy particles.

J. C. STREET
E. C. STEVENSON

Research Laboratory of Physics,
Harvard University,
Cambridge, Massachusetts,
October 6, 1937.

[1] Anderson and Neddermeyer, Phys. Rev. 50, 263 (1936).
[2] Street and Stevenson, Phys. Rev. 51, 1005 (1937).
[3] Neddermeyer and Anderson, Phys. Rev. 51, 885 (1937).

Letters to the Editor

PUBLICATION of brief reports of important discoveries in physics may be secured by addressing them to this department. The closing date for this department is, for the issue of the 1st of the month, the 8th of the preceding month and for the issue of the 15th, the 23rd of the preceding month. No proof will be sent to the authors. The Board of Editors does not hold itself responsible for the opinions expressed by the correspondents. Communications should not exceed 600 words in length.

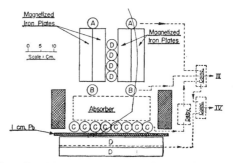

FIG. 1. Disposition of counters, absorber, and magnetized iron plates. All counters "D" are connected in parallel.

On the Disintegration of Negative Mesons

M. CONVERSI, E. PANCINI, AND O. PICCIONI*

Centro di Fisica Nucleare del C. N. R. Istituto di Fisica dell'Università di Roma, Italia

December 21, 1946

IN a previous Letter to the Editor,[1] we gave a first account of an investigation of the difference in behavior between positive and negative mesons stopped in dense materials. Tomonaga and Araki[2] showed that, because of the Coulomb field of the nucleus, the capture probability for negative mesons at rest would be much greater than their decay probability, while for positive mesons the opposite should be the case. If this is true, then practically all the decay processes which one observes should be owing to positive mesons.

Several workers[3] have measured the ratio η between the number of the disintegration electrons and the number of mesons stopped in dense materials. Using aluminum, brass, and iron, these workers found values of η close to 0.5 which, if one assumes that the primary radiation consists of approximately equal numbers of positive and negative mesons, support the above theoretical prediction. Auger, Maze, and Chaminade,[4] on the contrary, found η to be close to 1.0, using aluminum as absorber.

Last year we succeeded in obtaining evidence of different behavior of positive and negative mesons stopped in 3 cm of iron as an absorber by using magnetized iron plates to concentrate mesons of the same sign while keeping away mesons of the opposite sign (at least for mesons of such energy that would be stopped in 3 cm of iron). We obtained results in agreement with the prediction of Tomonaga and Araki. After some improvements intended to increase the counting rate and improve our discrimination against the "mesons of the opposite sign," we continued the measurements using, successively, iron and carbon as absorbers. The recording equipment was one which two of us had previously used in a measurement of the meson's mean life.[5] It gave threefold (III) and fourfold (IV) delayed coincidences. The difference (III)−(IV) (after applying a slight correction for the lack of efficiency of the fourfold coincidences) was owing to mesons stopped in the absorber and ejecting a disintegration electron which produced a delayed coincidence. The minimum detected delay was about 1 μsec. and the maximum about 4.5 μsec. Our calculations of the focusing properties of the magnetized plates (20 cm high; $\beta = 15,000$ gauss) and including roughly the effects of scattering, showed that we should expect almost complete cut-off for the "mesons of the opposite sign." This is confirmed by our results, since otherwise it would be very hard to explain the almost complete dependence on the sign of the meson observed in the case of iron.

The results of our last measurements with two different absorbers are given in Table I. In this table "Sign" refers to the sign of the meson concentrated by the magnetic field. $M = (\text{III}) - (\text{IV}) - P(\text{IV})$, the number of decay electrons, is corrected for the lack of efficiency (p) in our fourfold coincidences (~ 0.046).

The value M_{-} (5 cm Fe) is but slightly greater than the correction for the lack of efficiency in our counting, so that we can say that perhaps no negative mesons and, at most, only a few (~ 5) percent undergo β-decay with the accepted half-life.

The results with carbon as absorber turn out to be quite inconsistent with Tomonaga and Araki's prediction. We used cylindrical graphite rods having a mean effective thickness of 4 cm because we were unable to procure a graphite plate. In addition, when concentrating negative mesons, we placed above the graphite a 5-cm thick plate of iron to guard against the scattering of very low energy mesons which might destroy the concentrating effect of our magnets. We alternated the following three measurements:

A. Negative mesons with 4 cm C and 5 cm Fe,
B. Negative mesons with 6.2 cm Fe (6.2 cm Fe is approximately equivalent to 4 cm C+5 cm Fe as far as energy loss is concerned.
C. Positive mesons with 4 cm C.

TABLE I. Results of measurements on β-decay rates for positive and negative mesons.

Sign	Absorber	III	IV	Hours	M/100 hours
(a) +	5 cm Fe	213	106	155.00'	67 ±6.5
(b) −	5 cm Fe	172	158	206.00'	3
(c) −	none	71	69	107.45'	−1
(d) +	4 cm C	170	101	179.20'	36 ±4.5
(e) −	4 cm C+5 cm Fe	218	146	243.00'	27 ±3.5
(f) −	6.2 cm Fe	128	120	240.00'	0

The comparison between A and B gave the difference in behavior between Fe and C, once we had established the fact that practically no disintegration electrons came from negative mesons in the 5-cm iron plate. The comparison between A and C gives the difference in behavior between negative and positive mesons in carbon. This must be considered as a qualitative comparison because of the slightly different action of the magnetic field in concentrating mesons of different ranges (4 cm C+5 cm Fe in one case and 4 cm of C in the other). We could not, of course, add 5 cm of Fe for the positive mesons too, since positive mesons do decay in Fe.

The great yield of negative decay electrons from carbon shows a marked difference between it and iron as absorbers. Tomonaga and Araki's calculation also give for carbon a much higher ratio of capture to decay probability for negative mesons, so we are forced to doubt their estimation. It is possible that a suitable dependence of the capture cross section, σ_c, on the nuclear charge, Z, might explain these results; however, if the ratio of the capture to decay probability also depends on the density as Tomonaga and Araki pointed out, then it would require a very irregular dependence on Z to also explain the cloud-chamber pictures of some authors[6] showing negative mesons stopped in the chamber without any decay electrons coming out.

Concerning the difference between M_+ and M_- in carbon, we should like to point out that it is not necessary to assume that σ_c for carbon has an appreciable value for negative mesons. A positive excess, $(H_+ - H_-)/(H_+ + H_-)$ of 20 percent in the hard component, as it seems to be[7] is sufficient to explain our results since this gives $H_+/H_- = 1.5$ which is greater than M_+/M_- for carbon. Impurities in the graphite could also explain some preference for M_+, with a suitable dependence of σ_c on Z.

Further experiments on this subject are now in progress, in an attempt to calculate the capture cross section, and to know how it depends on Z.

* Now Visiting Research Associate at Massachusetts Institute of Technology, Cambridge, Massachusetts.
[1] M. Conversi, E. Pancini, and O. Piccioni, Phys. Rev. 68, 232 (1945).
[2] S. Tomonaga and G. Araki, Phys. Rev. 58, 90 (1940).
[3] F. Rasetti, Phys. Rev. 60, 198 (1941); B. Rossi and N. Nereson, Phys. Rev. 62, 417 (1942); M. Conversi and O. Piccioni, Nuovo Cimento 2, 71 (1944); Phys. Rev. 70, 874 (1946).
[4] P. Auger, R. Maze, and Chaminade, Comptes rendus 213, 381 (1941).
[5] M. Conversi and O. Piccioni, Nuovo Cimento 2, 40 (1944); Phys. Rev. 70, 859 (1946).
[6] Y. Nishina, M. Takeuchi, and T. Ichimiya, Phys. Rev. 55, 585 (1939); H. Maier-Leibnitz, Zeits f. Physik 112, 569 (1939); T. H. Johnson and R. P. Shutt, Phys. Rev. 61, 380 (1942).
[7] H. Jones, Rev. Mod. Phys. 11, 235 (1939); D. J. Hughes, Phys. Rev. 57, 592 (1940); G. Bernardini, M. Conversi, E. Pancini, E. Scrocco, and G. C. Wick, Phys. Rev. 68, 109 (1945).

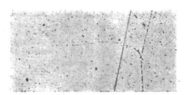

Fig. 1 a. PHOTOMICROGRAPH OF CENTRE OF STAR, SHOWING TRACK OF MESON PRODUCING DISINTEGRATION. (LEITZ 2 MM. OIL-IMMERSION OBJECTIVE. × 500)

Nuclear Disintegration by Meson Capture

RECENTLY, multiple nuclear disintegration 'stars', produced by cosmic radiation, have been investigated by the photographic emulsion technique. Plates coated with 50 μ Ilford B.1 emulsions[1] were exposed in aircraft for several hours at 30,000 ft. One of these disintegrations was of particular interest, for whereas all stars previously observed had been initiated by radiation not producing ionizing tracks in the emulsion, the one in question appears to be due to nuclear capture of a charged particle, presumably a slow meson.

The star consists of four tracks A, B, C and D (Fig. 1). A, B and D lie almost in the plane of the emulsion, whereas C dips steeply (at about 40°) and ends in the glass. D is due to a proton of energy 3·7 MeV., and C also corresponds to a proton, of more than 3 MeV., and most likely about 5 MeV. Track B was most probably produced by a triton of 5·6 MeV. A short track, about 1 μ long, between A and B is apparently due to the residual recoil nucleus.

Track A appears to enter the emulsion surface about 150 μ from the star centre. On account of the relatively large distances between consecutive grains at this range, the track cannot be distinguished at all easily against the spontaneous background of grains, and only the last 100 μ of track (below arrow) can be traced with certainty. Assuming it to be singly charged, the mass of the particle producing track A has been roughly evaluated by the following methods.

(1) Both ionization and scattering increase towards the origin of the star, hence the particle was definitely travelling *towards* the disintegration point.

An electron is discounted because the observed ionization is far too high (an electron track of this range would, in fact, not be detected at all), and the scattering too small. On the other hand, a proton is discounted since the observed scattering is too great (Fig. 2). We must, therefore, conclude that the particle had a mass intermediate between that of electron and proton.

Fig. 1 b. TRACE OF COMPLETE STAR ON SCREEN OF PROJECTION MICROSCOPE, SHOWING PROJECTION OF THE TRACKS IN THE PLANE OF THE EMULSION. TRACK A CANNOT BE TRACED WITH CERTAINTY BEYOND THE ARROW

The grain density along track A does, in fact, agree well with that to be expected of a meson of the observed range of about one tenth of the proton mass. The range-energy curve for mesons in the emulsion has been obtained from that for protons (kindly lent by Dr. C. F. Powell), using the ratio of the masses of the two particles.

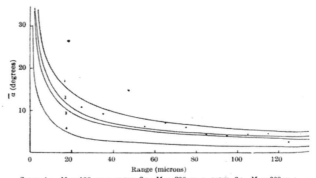

Curve 1: $M = 100\ m_e$; curve 2: $M = 200\ m_e$; curve 3: $M = 300\ m_e$; curve 4: $M = 1,800\ m_e$

Fig. 2. MULTIPLE SCATTERING OF MESONS AND PROTON IN THE EMULSION. $\bar{a}$ DENOTES THE MEAN ANGLE OF SCATTERING FOR A 10 μ LAYER OF THE EMULSION. CURVES CALCULATED FOR PARTICLES OF MASS 100 m_e, 200 m_e, 300 m_e AND 1,800 m_e (PROTON). m_e BEING THE ELECTRON MASS. EXPERIMENTAL POINTS, FROM MEASUREMENTS ON TRACK A, INDICATED BY DOTS

(2) The elastic scattering of mesons and protons can be calculated from the known emulsion constitution, using Williams's formula[2] for the mean square angle of multiple scattering, and is shown plotted against range in Fig. 2. Bearing in mind the fact that the single scattering will introduce large fluctuations, the experimental points appear to be consistent with a meson mass between 100 m_e and 300 m_e, m_e being the electron mass. Only the horizontal projection of the scattering along the track can actually be measured; on the average, multiplying this by a factor $\sqrt{2}$ will give the true value. The scattering of fast electrons cannot be found from any simple formula, but it is certainly much greater than the observed values.

(3) A value of the meson mass can also be arrived at from the energy required in the nuclear disintegration. In this process, the emitted particles must have sufficient energy to surmount the potential barrier, so that the disintegrating nucleus in this case cannot be that of silver or bromine, and must, therefore, be either carbon, oxygen or nitrogen (in the gelatin).

An approximate calculation of the momentum of the recoil nucleus indicates that at least one neutron must be postulated to conserve momentum; the kinetic energy of this neutron must be around 4 MeV. If we assume that a negative meson is captured by a nucleus nearly at the end of its range and is annihilated, so that the rest energy of the meson becomes nuclear excitation energy, we could have disintegration schemes such as

$$Y^- + O_8^{16} \rightarrow N_7^{16*} \rightarrow H_1^3 + 2H_1^1 + n_o^1 + Be_4^{10*}$$
$$(\text{or } 2n_o^1 + Be_4^9)$$
$$Y^- + N_7^{14} \rightarrow C_6^{14*} \rightarrow H_1^3 + 2H_1^1 + n_o^1 + Li_3^{6*}$$
$$Y^- + C_6^{12} \rightarrow B_5^{12*} \rightarrow H_1^3 + 2H_1^1 + n_o^1 + He_2^{6*}$$

As the recoil nucleus would be expected to have a fairly high excitation energy (5–10 MeV.) above the ground-state, it must be relatively stable against further disintegration into charged particles. With this limitation, there are still a large number of possible reactions (considering all isotopes of carbon, oxygen and nitrogen), but it appears that in general the mass excess of the recoil nucleus $\sim$ 15 MeV., whereas that of the initial one $\sim$ 5 MeV. or less. The negative Q value of the reaction, allowing for excitation energy, is then found to be $\sim$ 60 MeV., and the total kinetic energy of the ejected particles $\sim$ 20 MeV. The total excitation energy of the original nucleus would then be $\sim$ 80 MeV., probably with an error of $\pm$ 20 MeV. (to allow for the various numbers of neutrons emitted, etc.).

On the above hypothesis, the meson should, therefore, have a rest energy of 60–100 MeV., that is, a mass of between 120 m_e and 200 m_e.

Near the end of the meson track, a small number of grains are observed slightly off the main track. If these are due to fast secondary electrons, their ranges appear to be considerably greater than would be expected from the energy of the primary.

I am indebted to the A.O.C., Royal Air Force, Benson, Oxon., for kindly exposing the plates.

D. H. PERKINS

Imperial College of Science and Technology,
London, S.W.7.
Jan. 8.

Powell, Occhialini, Livesey and Chilton, *J. Sci. Instr.*, **23**, 102 (1946).
Williams, *Proc. Roy. Soc.*, A, **169**, 531 (1938).

OBSERVATIONS ON THE TRACKS OF SLOW MESONS IN PHOTOGRAPHIC EMULSIONS*

By C. M. G. LATTES, Dr. G. P. S. OCCHIALINI and Dr. C. F. POWELL

H. H. Wills Physical Laboratory, University of Bristol

INTRODUCTION. In recent experiments, it has been shown that charged mesons, brought to rest in photographic emulsions, sometimes lead to the production of secondary mesons. We have now extended these observations by examining plates exposed in the Bolivian Andes at a height of 5,500 m., and have found, in all, forty examples of the process leading to the production of secondary mesons. In eleven of these, the secondary particle is brought to rest in the emulsion so that its range can be determined. In Part 1 of this article, the measurements made on these tracks are described, and it is shown that they provide evidence for the existence of mesons of different mass. In Part 2, we present further evidence on the production of mesons, which allows us to show that many of the observed mesons are locally generated in the 'explosive' disintegration of nuclei, and to discuss the relationship of the different types of mesons observed in photographic plates to the penetrating component of the cosmic radiation investigated in experiments with Wilson chambers and counters.

Part I. Existence of Mesons of Different Mass

As in the previous communications[1], we refer to any particle with a mass intermediate between that of a proton and an electron as a meson. It may be emphasized that, in using this term, we do not imply that the corresponding particle necessarily has a strong interaction with nucleons, or that it is closely associated with the forces responsible for the cohesion of nuclei.

We have now observed a total of 644 meson tracks which end in the emulsion of our plates. 451 of these were found, in plates of various types, exposed at an altitude of 2,800 m. at the Observatory of the Pic du Midi, in the Pyrenees; and 193 in similar plates exposed at 5,500 m. at Chacaltaya in the Bolivian Andes. The 451 tracks in the plates exposed at an altitude of 2,800 m. were observed in the examination of 5 c.c. emulsion. This corresponds to the arrival of about 1·5 mesons per c.c. per day, a figure which represents a lower limit, for the tracks of some mesons may be lost through fading, and through failure to observe tracks of very short range. The true number will thus be somewhat higher. In any event, the value is of the same order of magnitude as that we should expect to observe in delayed coincidence experiments at a height of 2,800 m., basing our estimates on the observations obtained in similar experiments at sea-level, and making reasonable assumptions about the increase in the number of slow mesons with altitude. It is therefore certain that the mesons we observe are a common constituent of the cosmic radiation.

Photomicrographs of two of the new examples of secondary mesons, Nos. III and IV, are shown in Figs. 1 and 2. Table 1 gives details of the characteristics of all events of this type observed up to the time of writing, in which the secondary particle comes to the end of its range in the emulsion.

TABLE 1

Event No.	Range in emulsion in microns of	
	Primary meson	Secondary meson
I	133	613
II	84	565
III	1040	621
IV	133	591
V	117	638
VI	49	595
VII	460	616
VIII	900	610
IX	239	666
X	256	637
XI	81	590

Mean range $614 \pm 8\,\mu$. Straggling coefficient $\sqrt{\Sigma \Delta_i^2/n} = 4\cdot3$ per cent, where $\Delta_i = R_i - \bar{R}$, R_i being the range of a secondary meson, and $\bar{R}$ the mean value for n particles of this type.

---

* This article contains a summary of the main features of a number of lectures given, one at Manchester on June 18 and four at the Conference on Cosmic Rays and Nuclear Physics, organised by Prof. W. Heitler, at the Dublin Institute of Advanced Studies, July 5–12. A complete account of the observations, and of the conclusions which follow from them, will be published elsewhere.

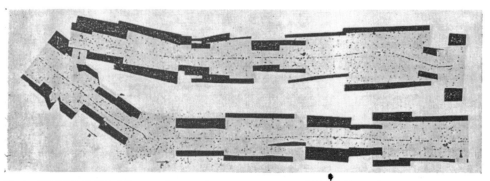

Fig. 1. OBSERVATION BY MRS. I. POWELL. COOKE × 95 ACHROMATIC OBJECTIVE; C2 ILFORD NUCLEAR RESEARCH EMULSION LOADED WITH BORON. THE TRACK OF THE μ-MESON IS GIVEN IN TWO PARTS, THE POINT OF JUNCTION BEING INDICATED BY a AND AN ARROW

meson, produced in a process of the type which we observe, will remain within the emulsion, of thickness 50 μ, for a distance greater than 500 μ. If we assume, as a first approximation, that the trajectories are rectilinear, we obtain a value for the probability of 1 in 20. The marked Coulomb scattering of mesons in the Nuclear Research emulsions will, in fact, increase the probability of 'escape'. The six events which we observe in plates exposed at 2,800 m., in which the secondary particle remains in the emulsion for a distance greater than 500 μ, therefore correspond to the occurrence in the emulsion of 120 $\pm$ 50 events of this particular type. Our observations, therefore, prove that the production of a secondary meson is a common mode of decay of a considerable fraction of those mesons which come to the end of their range in the emulsion.

Second, there is remarkable consistency between the values of the range of the secondary mesons, the variation among the individual values being similar to that to be expected from 'straggling', if the particles are always ejected with the same velocity. We can therefore conclude that the secondary mesons are all of the same mass and that they are emitted with constant kinetic energy.

If mesons of lower range are sometimes emitted in an alternative type of process, they must occur much less frequently than those which we have observed; for the geometrical conditions, and the greater average grain-density in the tracks, would provide much more favourable conditions for their detection. In fact, we have found no such mesons of shorter range. We cannot, however, be certain that mesons of greater range are not sometimes produced. Both the lower ionization in the beginning of the trajectory, and the even more unfavourable conditions of detection associated with the greater lengths of the tracks, would make such a group, or groups, difficult to observe. Because of the large fraction of the mesons which, as we have seen, can be attributed to the observed process, it is reasonable to assume that alternative modes of decay, if they exist, are much less frequent than that which we have observed. There is, therefore, good evidence for the production of a single homogeneous group of secondary mesons, constant in mass and kinetic energy. This strongly suggests a fundamental process, and not one involving an interaction of a primary meson with a particular type of nucleus in the emulsion. It is convenient to refer to this process

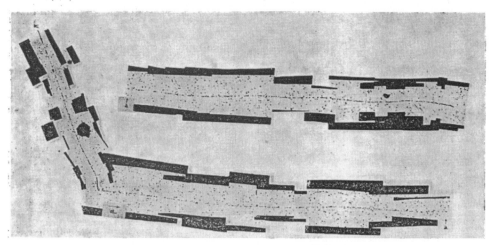

Fig. 2. COOKE × 95 ACHROMATIC OBJECTIVE. C2 ILFORD NUCLEAR RESEARCH EMULSION LOADED WITH BORON

No. 4066 October 4, 1947 NATURE

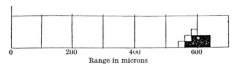

Fig. 3. DISTRIBUTION IN RANGE OF TEN SECONDARY MESONS. THOSE MARKED ■ STOP IN THE EMULSION; THE THREE MARKED ☐ LEAVE THE EMULSION WHEN NEAR THE END OF THEIR RANGE. MEAN RANGE OF SECONDARY MESONS, 606 MICRONS. THE RESULTS FOR EVENTS NOS. VIII TO XI ARE NOT INCLUDED IN THE FIGURE

in what follows as the μ-decay. We represent the primary mesons by the symbol π, and the secondary by μ. Up to the present, we have no evidence from which to deduce the sign of the electric charge of these particles. In every case in which they have been observed to come to the end of their range in the emulsion, the particles appear to stop without entering nuclei to produce disintegrations with the emission of heavy particles.

Knowing the range-energy relation for protons in the emulsion, the energy of ejection of the secondary mesons can be deduced from their observed range, if a value of the mass of the particles is assumed. The values thus calculated for various masses are shown in Table 2.

TABLE 2

Mass in m_e	100	150	200	250	300
Energy in MeV.	3·0	3·6	4·1	4·5	4·85

No established range-energy relation is available for protons of energies above 13 MeV., and it has therefore been necessary to rely on an extrapolation of the relation established for low energies. We estimate that the energies given in Table 2 are correct to within 10 per cent.

Evidence of a Difference in Mass of π- and μ-Mesons

It has been pointed out[1] that it is difficult to account for the μ-decay in terms of an interaction of the primary meson with the nucleus of an atom in the emulsion leading to the production of an energetic meson of the same mass as the first. It was therefore suggested that the observations indicate the existence of mesons of different mass. Since the argument in support of this view relied entirely on the principle of the conservation of energy, a search was made for processes which were capable of yielding the necessary release of energy, irrespective of their plausibility on other grounds. Dr. F. C. Frank has re-examined such possibilities in much more detail, and his conclusions are given in an article to follow. His analysis shows that it is very difficult to account for our observations, either in terms of a nuclear disintegration, or of a 'building-up' process in which, through an assumed combination of a negative meson with a hydrogen nucleus, protons are enabled to enter stable nuclei of the light elements with the release of binding energy. We have now found it possible to reinforce this general argument for the existence of mesons of different mass with evidence based on grain-counts.

We have emphasized repeatedly[1] that it is necessary to observe great caution in drawing conclusions about the mass of particles from grain-counts. The main source of error in such determinations arises from the fugitive nature of the latent image produced in the silver halide granules by the passage of fast particles. In the case of the μ-decay process, however, an important simplification occurs. It is reasonable to assume that the two meson tracks are formed in quick succession, and are subject to the same degree of fading. Secondly, the complete double track in such an event is contained in a very small volume of the emulsion, and the processing conditions are therefore identical for both tracks, apart from the variation of the degree of development with depth. These features ensure that we are provided with very favourable conditions in which to determine the ratio of the masses of the π- and μ-mesons, in some of these events.

In determining the grain density in a track, we count the number of individual grains in successive intervals of length 50 μ along the trajectory, the observation being made with optical equipment giving large magnification (× 2,000), and the highest available resolving power. Typical results for protons and mesons are shown in Fig. 4. These results were obtained from observations on the tracks in a single plate, and it will be seen that there is satisfactory resolution between the curves for particles of different types. The 'spread' in the results for

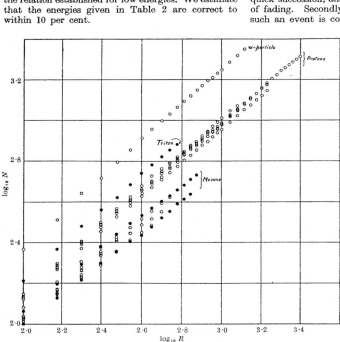

Fig. 4. N IS TOTAL NUMBER OF GRAINS IN TRACK OF RESIDUAL RANGE R (SCALE-DIVISIONS). 1 SCALE-DIVISION = 0·85 MICRONS

different particles of the same type can be attributed to the different degrees of fading associated with the different times of passage of the particles through the emulsion during an exposure of six weeks.

Applying these methods to the examples of the μ-decay process, in which the secondary mesons come to the end of their range in the emulsion, it is found that in every case the line representing the observations on the primary meson lies above that for the secondary particle. We can therefore conclude that there is a significant difference in the grain-density in the tracks of the primary and secondary mesons, and therefore a difference in the mass of the particles. This conclusion depends, of course, on the assumption that the π- and μ-particles carry equal charges. The grain-density at the ends of the tracks, of particles of both types, are consistent with the view that the charges are of magnitude $|e|$.

A more precise comparison of the masses of the π- and μ-mesons can only be made in those cases in which the length of the track of the primary meson in the emulsion is of the order of 600 μ. The probability of such a favourable event is rather small, and the only examples we have hitherto observed are those listed as Nos. III and VIII in Table 1. A mosaic of micrographs of a part only of the first of these events is reproduced in Fig. 1, for the length of the track of the μ-meson in the emulsion exceeds 1,000 μ. The logarithms of the numbers of grains in the tracks of the primary and secondary mesons in this event are plotted against the logarithm of the residual range in Fig. 5. By comparing the residual ranges at which the grain-densities in the two tracks have the same value, we can deduce the ratio of the masses. We thus obtain the result $m_\pi/m_\mu = 2.0$. Similar measurements on event No. VIII give the value 1·8. In considering the significance which can be attached to this result, it must be noticed that in addition to the standard deviations in the number of grains counted, there are other possible sources of error. Difficulties arise, for example, from the fact that the emulsions do not consist of a completely uniform distribution of silver halide grains. 'Islands' exist, in which the concentration of grains is significantly higher, or significantly lower, than the average values, the variations being much greater than those associated with random fluctuations. The measurements on the other examples of μ-decay are much less reliable on account of the restricted range of the π-mesons in the emulsion; but they give results lower than the above values. We think it unlikely, however, that the true ratio is as low as 1·5.

The above result has an important bearing on the interpretation of the μ-decay process. Let us assume that it corresponds to the spontaneous decay of the heavier π-meson, in which the momentum of the μ-meson is equal and opposite to that of an emitted photon. For any assumed value of the mass of the μ-meson, we can calculate the energy of ejection of the particle from its observed range, and thus

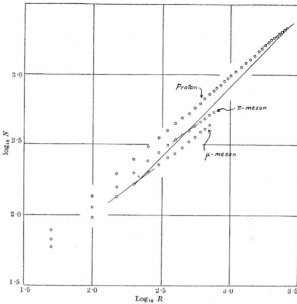

Fig. 5. N IS TOTAL NUMBER OF GRAINS IN TRACK OF RESIDUAL RANGE R (SCALE-DIVISIONS). 1 SCALE-DIVISION = 0·85 MICRONS
THE 45°-LINE CUTS THE CURVES OF THE MESONS AND PROTON IN THE REGION OF THE SAME GRAIN DENSITY

determine its momentum. The momentum, and hence the energy of the emitted photon, is thus defined; the mass of the π-meson follows from the relation

$$c^2 m_\pi = c^2 m_\mu + E_\mu + h\nu.$$

It can thus be shown that the ratio m_π/m_μ is less than 1·45 for any assumed value of m_μ in the range from 100 to 300 m_e, m_e being the mass of the electron (see Table 3). A similar result is obtained if it is assumed that a particle of low mass, such as an electron or a neutrino, is ejected in the opposite direction to the μ-meson.

TABLE 3

Assumed mass m_μ	E (MeV.)	$h\nu$ (MeV.)	m_π	$m_\pi/m_\mu \pm 3$ per cent
100 m	3·0	17	140 m_e	1·40
150	3·6	23	203	1·35
200	4·1	29	264	1·32
250	4·5	34	325	1·30
300	4·85	39	387	1·29

On the other hand, if it is assumed that the momentum balance in the μ-decay is obtained by the emission of a neutral particle of mass equal to the μ-meson mass, the calculated ratio is about 2·1 : 1.

Our preliminary measurements appear to indicate, therefore, that the emission of the secondary meson cannot be regarded as due to a spontaneous decay of the primary particle, in which the momentum balance is provided by a photon, or by a particle of small rest-mass. On the other hand, the results are consistent with the view that a neutral particle of approximately the same rest-mass as the μ-meson is emitted. A final conclusion may become possible when further examples of the μ-decay, giving favourable conditions for grain-counts, have been discovered.

[1] *Nature*, **159**, 93, 186, 694 (1947).

Evidence for the Production of Neutral Mesons by Photons*

J. STEINBERGER, W. K. H. PANOFSKY, AND J. STELLER
Radiation Laboratory, Department of Physics, University of California, Berkeley, California
(Received April 28, 1950)

In the bombardment of nuclei by 330-Mev x-rays, multiple gamma-rays are emitted. From their angular correlation it is deduced that they are emitted in pairs in the disintegration of neutral particles moving with relativistic velocities and therefore of intermediate mass. The neutral mesons are produced with cross sections similar to those for the charged mesons and with an angular distribution peaked more in the forward direction. The production cross section in hydrogen and the production cross section per nucleon in C and Be are comparable.

I. INTRODUCTION

NEUTRAL mesons which are coupled strongly to nuclei must be expected to be unstable against decay into two or more gamma-rays. The modes of decay, and expected lifetimes, have been discussed extensively.[1] These gamma-rays are then supposed to be responsible for the soft showers which often accompany energetic cosmic-ray nuclear events.[2] The evidence in favor of the existence of the neutral meson has recently been greatly strengthened by the discovery at Berkeley[3] of gamma-rays which behave in all ways as if they were due to the disintegration of a neutral meson. They are produced by proton bombardment of various nuclei and have a production cross section which depends on proton energy much like that of charged mesons. Their energy is approximately 70 Mev on the average, half that of the charged π-meson, and the energy spread is in agreement with the Doppler shift due to the velocity of the parent mesons. The lifetime of the mesons is less than 10^{-13} sec., which is in agreement with the theoretical expectations.

The evidence is therefore already much in favor of the existence of a gamma-unstable neutral meson. However, until now, coincidences between the two gamma-rays have never been observed. We report here the detection of such coincidences, produced by the bombardment of various nuclei in the x-ray beam of the Berkeley synchrotron. This must be regarded as strong additional evidence supporting the existence of the neutral meson.

II. EXPERIMENTAL ARRANGEMENT

The apparatus is sketched in Fig. 1. The synchrotron x-ray beam of 330-Mev maximum energy is collimated in two successive collimators. The second collimator serves only to intercept some of the electrons produced at the edge of the first collimator. The beam then strikes a target, which, for most of the experiment, is a cylinder of beryllium, $1\frac{1}{2}$ inches long and 2 inches in diameter. The particles produced in the target are detected in two telescopes, each consisting of three scintillation counters. A converter, usually $\frac{1}{4}$ inch of lead, is inserted between the two crystals nearest the target in each telescope. An event is recorded if simultaneous (resolving time 10^{-7} sec.) pulses are recorded in the outer four crystals, but none in the two near the target. That is, we require that there be two particles, one in each telescope, neutral at first which are converted into charged particles by the lead, and which penetrate one crystal and enter the next. With a beam intensity of about 10^{11} Mev/min. the counting rate for such coincidences at favorable orientations of the telescopes is about 10 counts/min.

III. NATURE OF THE COINCIDENCES

Let us first describe the experiments which identify the particles as gamma-rays, indicate their energy and show that their origin is the nuclear rather than the Coulomb field. In Table I we list the relative detection

---

* This work was performed under the auspices of the AEC.
[1] Y. Tanikawa, Proc. Phys. Math. Soc. Japan 24, 610 (1940). R. J. Finkelstein, Phys. Rev. 72, 414 (1947). H. Fukuda and Y. Miamoto, Prog. Theor. Phys. 4, 347 (1949). Ozaki, Oneda, and Sasaki, Prog. Theor. Phys. 4, 524 (1949). J. Steinberger, Phys. Rev. 76, 1180 (1949). C. N. Yang, Phys. Rev. 77, 243 (1950).
[2] The implications of the gamma-decay of neutral mesons for the soft component in the cosmic radiation were pointed out by J. R. Oppenheimer (Phys. Rev. 71, 462 (T) (1947). It was assumed that in high energy nuclear events neutral mesons are emitted with multiplicities similar to those for charged mesons. The neutral mesons decay into photons and account for the early development of extensive showers, as well as the large total amount of soft radiation. These bursts of soft radiation accompanying energetic nuclear events were actually observed in the cloud chamber by W. Fretter, Phys. Rev. 73, 41 (1948), 76, 511 (1949); C. Y. Chao, Phys. Rev. 75, 581 (1949); Gregory, Rossi, and Tinlot, Phys. Rev. 77, 299 (1949); and J. Green, Thesis, University of California, 1950. They were found in photographic plates by Kaplan, Peters, and Bradt, Phys. Rev. 76, 1735 (1949). Both the cloud-chamber pictures and the photographic star show that the showers begin with gamma-rays rather than electrons.
[3] Bjorklund, Crandall, Moyer, and York, Phys. Rev. 77, 213 (1950).

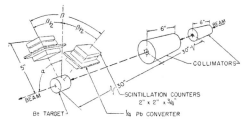

FIG. 1. Experimental arrangement.

efficiency for various converter materials and thicknesses. Without converters the counting rate is almost zero, then increases as the converter thickness in each arm is increased to $\frac{1}{4}$ inch of lead, and only slightly from $\frac{1}{4}$ inch to $\frac{1}{2}$ inch. This is as expected from shower theory for about 100-Mev photons. Copper of $\frac{1}{4}$ inch thickness has approximately the same conversion efficiency as has $\frac{1}{16}$ inch of lead, again in agreement with shower theory, since the number of shower units is the same for these thicknesses.

The coincidences are attenuated by a factor of four when $\frac{1}{4}$ inch of lead is inserted between the target and the anticoincidence crystals. This again is as expected for photons. Furthermore, it can be seen from Table I that both telescopes require converters, so that both particles must be photons.

To measure the energy of the conversion electrons, aluminum absorbers were inserted between the last two crystals of one of the telescopes. Unfortunately, at these energies the radiation losses are important, and therefore the straggling large. We have plotted in Fig. 2 the coincidence counting rate as a function of the average energy required to traverse the telescope. Because the photons originate in moving mesons, the average gamma-ray energy is expected to be approximately 100 Mev, and the average electron energy 50 Mev, quite in agreement with the observed attenuation.

The nuclear origin of the photons is demonstrated by the fact that the cross section for these coincidences is only six times as big for a lead nucleus as for beryllium, which is less than the ratio of the nuclear areas. On the other hand, ordinary shower cross sections increase by a factor of 400.

Finally, we have looked for coincidences with the beam energy reduced to about 175 Mev with angles α and β of the telescope both 90°. The cross section per Q (the number Q for a bremsstrahlung beam is equal to the total energy divided by the maximum energy of the spectrum) is at least 50 times smaller here than at 330 Mev. This steep excitation function is also observed for charged meson production.

We believe, therefore, that it is demonstrated that the observed coincidences are caused by gamma-rays of about 100-Mev average energy, of non-Coulombic origin, and with a threshold similar to that for charged mesons.

IV. ANGULAR CORRELATION AND DISTRIBUTION OF THE GAMMA-RAYS

To study further the properties of these coincidences, we have measured their rate as a function of the angle, α, between the beam direction and the plane of the telescopes and of the correlation angle β (see Fig. 1). Consider first the variation with β at a fixed α, say 90°. 180° coincidences are rare. The counting rate increases with decreasing β to a maximum at 90°, and then drops sharply. This behavior must actually be expected of gamma-rays which are the decay products of neutral

TABLE I. Relative detection efficiency as a function of absorber material and thickness.

Converter in telescope 1	Converter in telescope 2	Relative counting rate $\alpha=\beta=90°$
none	none	0.01±0.005
$\frac{3}{32}$-in. Pb	$\frac{3}{32}$-in. Pb	0.17±0.013
$\frac{1}{16}$-in. Pb	$\frac{1}{16}$-in. Pb	0.3 ±0.02
$\frac{1}{8}$-in. Pb	$\frac{1}{8}$-in. Pb	0.67±0.08
$\frac{1}{4}$-in. Pb	$\frac{1}{4}$-in. Pb	1.00±0.06
$\frac{1}{4}$-in. Cu	$\frac{1}{4}$-in. Cu	0.39±0.03
none	$\frac{1}{4}$-in. Pb	0.15±0.05
$\frac{1}{16}$-in. Pb	$\frac{1}{4}$-in. Pb	0.62±0.07
$\frac{1}{2}$-in. Pb	$\frac{1}{4}$-in. Pb	1.07±0.1
$\frac{1}{4}$-in. Pb	$\frac{1}{4}$-in. Pb	0.28±0.05
$\frac{1}{4}$-in. Pb absorbers placed in front of both telescopes.		

mesons, because of the motion of the decaying mesons. A meson at rest decaying into two gamma-rays, emits them in opposite direction. But when this is seen from a system in which the meson has a total energy E, then the included angle β varies between π and $2\sin^{-1}(1/E)$ with a probability which favors the small angles tremendously. The median angle is $2\sin^{-1}[2/(3E^2+1)^{\frac{1}{2}}]$. E is the total meson energy in units of its rest energy.

For 70-Mev mesons the minimum angle of β is 84° and the median angle 92°. Since the distribution is so heavily peaked, not much error is introduced if one assumes, as is done in the following, that to an angle β corresponds a unique energy, that of the median angle. Therefore a measurement of the distribution in β is a measure of the distribution in energy of the neutral mesons, although the angular resolution of our telescopes is insufficient to give more than a glimpse of the energy distribution. We have included in Fig. 3 curves in which the observed[4] energy distributions of the π^+-meson made by the same x-rays on hydrogen are transformed into distributions in β and arbitrarily normalized. All corrections due to scattering and angular resolution are omitted. The general shape of the curves is certainly well reproduced by the experiment. It is

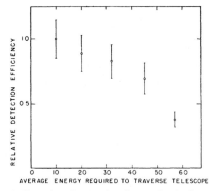

FIG. 2. Absorption of conversion electrons in aluminum. The energy includes the average radiation loss.

---
[4] Bishop, Cook, and Steinberger (to be published).

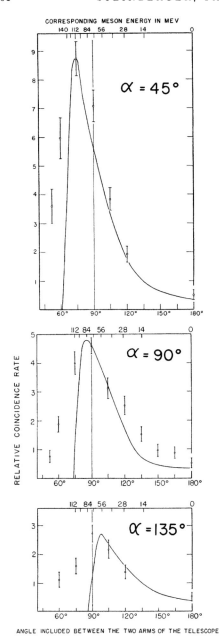

FIG. 3. Variation of coincidence rate with the included angle β between the two arms of the telescope. The curves are those expected on the assumption that the gamma-rays are the decay products of a neutral meson, emitted with the same energy distribution as are π^+-mesons from hydrogen. The curves are arbitrarily normalized for each angle α.

therefore clear that if the gamma-rays are the decay product of intermediate particles, these particles must move with velocities of the order of $v/c \simeq 0.8$. Excited nucleons of this velocity cannot be produced by x-rays of 330 Mev; the particles must therefore have an intermediate mass. Furthermore, it is possible to see that the decay must be into only 2 photons, since the expected angular distributions for a decay into more than two photons would not show a valley for small angles β.

The distribution in the angle α of the beam with the plane of the telescope shown in Fig. 4, is interesting chiefly because of the difference between this distribution and the angular distribution of π^+-photo-mesons[5] from either carbon or hydrogen targets. This is not particularly surprising, since various theories also give quite different results for charged and neutral mesons.

V. HYDROGEN CROSS SECTION AND TOTAL CROSS SECTION

At one setting of the telescopic angles, $\alpha = \beta = 90°$, we have compared the cross sections of hydrogen and carbon. This was done by comparing the count from a polyethylene (CH_2) block and a perforated carbon block of the same size and carbon content as the CH_2. The result is: $\sigma_{H\pi^0}/\sigma_{C\pi^0} = 0.12 \pm 0.03$. This again differs from the results for positive mesons, where $\sigma_{H\pi^+}/\sigma_{C\pi^+} \simeq 0.55$. The difference is probably in part caused by the fact that both neutrons and protons can contribute to neutral meson production, but only protons to π^+-production. In part, it may also be possible to ascribe this to the same phenomenon which, according to Chew,[6] is responsible for the large hydrogen-carbon ratio for the positive mesons. In the case of π^+-production, the reaction is inhibited by the fact that, when the proton is changed into a neutron, there is an oversupply of neutrons in the immediate neighborhood and the number of states available to it is small because of the Pauli principle. This is not significant in the neutral case because the nucleon's charge does not change.

The curves in Fig. 3 can be integrated to yield a total cross section for beryllium. $\sigma_{Be} = 7.5 \times 10^{-28}$ cm² per Q, while for carbon and hydrogen, assuming the same angular distribution, $\sigma_C = 10 \times 10^{-28}$ and $\sigma_H = 1.3 \times 10^{-28}$ cm² per Q. The absolute x-ray intensity is known[7] to about 10 percent, but the efficiency of the detecting system only to within a factor of two, so that there is a corresponding error in the above cross sections. The hydrogen cross section is approximately the same as that for π^+-production;[5] those for carbon and beryllium are somewhat higher.[8]

One might assume that the charge of the meson would play an important role in the production of mesons in the electromagnetic field of the photon. This

---

[5] J. Steinberger and A. S. Bishop, Phys. Rev. **78**, 493 (1950).
[6] G. Chew (private communication).
[7] Blocker, Kenney, and Panofsky (to be published).
[8] McMillan, Peterson, and White, Science **110**, 579 (1949).

is contradicted by the observed angular distribution of π^+-mesons produced by photons in H_2. The angular distribution indicates that the principal process responsible for charged meson production is the interaction of the photon with the spin of the nucleon. If the neutral meson has the same transformation properties as the charged, it then appears plausible that the production cross sections in hydrogen should be comparable, as seems to be the case. However, actual calculations on the basis of pseudoscalar theory, both in the classical and in the perturbation theory approximation, which give a reasonable angular distribution for the π^+-production, give smaller values for neutral meson production. Whether or not this is a new difficulty in a theory which already has several, is not clear. From a less restricted point of view it is not a surprising result.

VI. SUMMARY

In the bombardment of various nuclei by 330-Mev x-rays, photons with the following properties are emitted:

(1) At least two are emitted in coincidence.
(2) They each have an average energy of about 100 Mev.
(3) The Z dependence of the production indicates that they have their origin in a nuclear interaction, and not in the Coulomb field.
(4) The threshold for their production is at least 150 Mev.
(5) The angular correlation of the photons shows that they are emitted in pairs as the only decay products of particles moving with velocities of the order of $v/c=0.8$, and therefore of intermediate mass.
(6) The total cross section for production from hydrogen is about the same as that for production of π^+-mesons; other light nuclei cross sections are somewhat higher than those for the positive mesons.

It is clear from these properties that the gamma-rays are the decay products of neutral mesons. Since spin $\frac{1}{2}$,

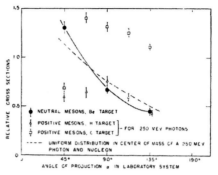

FIG. 4. Variation with the angle α between the plane of the telescope and the beam. Each point represents an integral over the angle β.

and spin 1 mesons are forbidden to decay into two photons,[1] the spin must be zero, excluding the possibility of very high intrinsic angular momenta. It seems reasonable, and it is in good agreement with all observations, to assume that both charged and neutral mesons are of the same type. It then follows from the angular distribution of the x-ray produced π^+-mesons, and the high cross sections for making neutral mesons by x-rays, that the π-meson is a pseudoscalar. This remark applies, of course, only to the character of the meson, and not to any particular field theory for the interaction of mesons with nucleons.

All phases of this experiment have been discussed with Professor Edwin McMillan and his advice has been of great help. The bombardments were carried out by the synchrotron crew under the direction of W. Gibbons.

The Spin of the Pion via the Reaction $\pi^+ + d \rightleftharpoons p + p$

R. Durbin, H. Loar, and J. Steinberger
*Columbia University, New York, New York**
(Received June 21, 1951)

It is possible to determine the spin of the pion by comparing the forward and backward rates of the reaction $\pi^+ + d \rightleftharpoons p + p$. The backward rate has been measured in Berkeley. We have measured the forward rate. Comparison of the two results shows the spin to be zero. In the light of other recent experimental results the meson is then pseudoscalar.

DURING the past few years there has accumulated an increasing amount of evidence that the pion is pseudoscalar. This consists chiefly of the following:

(a) The pion has integral spin. This follows from star formation in the capture of π^- mesons in photographic emulsions, as well as from angular momentum conservation in such reactions as $p + p \rightarrow \pi^+ + d$; $p + h\nu \rightarrow \pi^+ + n$.

(b) The neutral pion does not have spin one, since it decays into two γ-rays and such a transition is forbidden for systems with angular momentum $\hbar$.[1]

(c) Because of their similar masses and their similar nuclear production cross sections, as well as from the evidence on charge independence of nuclear forces, it is likely that neutral and charged mesons have the same transformation properties, so that charged pions also cannot have spin one.

(d) The pion is not scalar. This follows from the experiment on the capture of stopped negative pions in deuterium,[2] as well as the results on the production of pions, both charged and uncharged, by γ-rays.[3] Both experiments give best theoretical agreement in the pseudoscalar meson theory. It has therefore appeared quite probable that the pion is pseudoscalar; but the evidence, especially against a spin of two or greater, is poor.

It has been pointed out by Cheston and Marshak[4] that the reaction $\pi^+ + d \rightleftharpoons p + p$ lends itself to a determination of the spin of the pion. The forward and backward reactions are related by a detailed balancing argument, if one assumes initially unpolarized particles, so that

$$\frac{d\sigma(\rightarrow)}{d\Omega} \bigg/ \frac{d\sigma(\leftarrow)}{d\Omega} = \frac{4}{3} \frac{p^2}{q^2(2s+1)},$$

where s is the spin of the meson, and p, q, are the momenta of the proton and meson in the center-of-mass system. The cross sections, of course, are also in the center-of-mass system. The argument is rigorous, independent of meson theory, and in this rests its chief contribution. The reaction $p + p \rightarrow \pi^+ + d$ has been measured by Cartwright, Richman, Whitehead, and Wilcox[5] for 340-Mev protons, corresponding to a meson energy of 21 Mev in the center-of-mass system. We present here results on the inverse reaction.

The experimental arrangement is shown in Fig. 1. The Nevis cyclotron delivers a beam of approximately 20 positive mesons per square centimeter per second outside the concrete shielding. These are produced when the 380-Mev protons strike an internal Be target. The mesons are magnetically analyzed in the fringing field of the cyclotron, and by a small magnet outside the shielding. The energy resolution is ± 4 Mev at 75 Mev and the composition of the beam is 90 percent π^+, and 10 percent μ^+ mesons.[6] The beam is defined by the

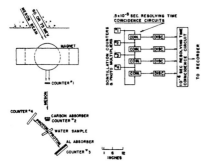

Fig. 1. Arrangement of the beam collimation, water sample, and detectors. Block diagram of circuits.

---

* This research has been supported by a joint program of the ONR and AEC.
[1] Steinberger, Panofsky, and Steller, Phys. Rev. **78**, 802 (1950); C. N. Yang, Phys. Rev. **77**, 242 (1950).
[2] Panofsky, Aamodt, and Hadley, Phys. Rev. **82**, 97 (1951); Brueckner, Serber, and Watson, Phys. Rev. **81**, 575 (1951).
[3] Bishop, Steinberger, and Cook, Phys. Rev. **80**, 291 (1950).

[4] R. E. Marshak, Phys. Rev. **82**, 313 (1951); W. B. Cheston, to be published.
[5] Cartwright, Richman, Whitehead, and Wilcox, Phys. Rev. **81**, 652 (1951); V. Peterson, Phys. Rev. **79**, 407 (1950); C. Richman and M. H. Whitehead, to be published. We are most indebted to Professors Richman and Wilcox, Mr. Cartwright, and Miss Whitehead for the privilege of quoting as yet unpublished results, and in particular, the fine meson spectrum shown in Fig. 5.
[6] The beam is analyzed in the following way: Heavy particles and electrons are detected in a measurement of the velocity distribution of the beam particles by means of a time of flight measurement. The mesons have velocities $\sim 0.7c$, the electrons which penetrate the counters have the velocity of light, and heavy particles have smaller velocities. The μ-mesons are measured at the end of their range by means of the delayed coincidences of their electron decay product. This is only possible in the π^- beam, since π^+ mesons at the end of their range are not captured but produce μ^+ mesons and interfere. We have assumed that, since the μ-mesons in the beam are the result of the decay in flight of the π-meson and since π^- and π^+ mesons have at least approximately the same lifetime [Lederman, Booth, Byfield, and Kessler,

crystal scintillation counters 1 and 2, and the energy reduced in the carbon absorber. The liquid scintillation counters 3 and 4 are set with respect to the water sample so that protons emitted 180° apart in the center-of-mass system of the meson and deuteron will be detected.[7] The water sample is 2.5 g/cm² thick. The aluminum absorber in front of counter No. 3 is thin enough to transmit the protons emitted in the meson absorption under study, but stops scattered protons or deuterons. The difference in counting rate using heavy water and light water targets is entirely due to the reaction $\pi^+ + d \to p + p$. It is in principle quite easy to check that this is so. In actuality, the counting rate is small and only a limited number of checks have been made, to wit: (1) Plateau. In the beginning of each experiment the counters are placed so that the meson beam penetrates all four, and the voltages (amplification) of the phototubes adjusted so that mesons are detected with full efficiency. But the proton pulses in counters 3 and 4 should be larger by a factor two or three than the

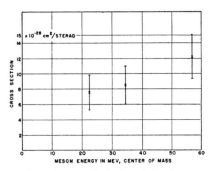

Fig. 3. Differential cross section for the emission of a proton at 45° (and one at 135°) to the meson beam in the c.m. system.

The results on the differential cross section at three angles are shown in Fig. 4. They are the combined results of three determinations, under conditions which varied somewhat. For instance, counters No. 3 and No. 4 were sometimes 4½ in. and at other times 8 in. in diameter. The average meson energy in the target for the three runs was 28 Mev. The data are corrected for the geometrical efficiency of the detecting system which varied from 0.48 to 0.84, for the beam composition (90 percent π^+, 10 percent μ^+), for the nuclear absorption of the mesons and the protons in the target (7 percent), and for an inefficiency of 8 percent in the circuits due to blocking.

The cross sections expected under the assumption of spins zero and one, on the basis of the results of Cartwright, Richman, Whitehead, and Wilcox, and of Peterson[5] are also shown. The Berkeley group has measured the energy distribution of mesons produced in the bombardment of hydrogen by 340 Mev protons.

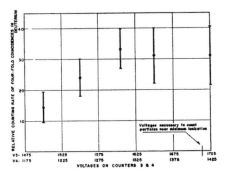

Fig. 2. Counting rate of fourfold events after subtraction, as a function of phototube voltage (amplification) in the proton detection counters 3 and 4.

meson pulses. Figure 2 shows that the pulses responsible for the subtracted fourfold coincidences are large. The $D_2O - H_2O$ difference is counted with full efficiency at voltages 100 volts (i.e., a factor of two in gain) below those necessary to count mesons. (2) When counters 3 and 4 were moved out of line to angles improper for the detection of protons with 180° c.m. angular correlation, no events were observed within statistical accuracy.

Figure 3 shows the energy dependence of the reaction at 45° in the c.m. system. It is sufficiently flat that errors due to energy spread of the meson beam and finite target thickness are small.

Phys. Rev. 83, 686 (1951)], the μ-contamination is independent of charge.
[7] The center-of-mass transformation is small, approximately 0.05c. The correlation angles in the laboratory system do not differ from 180° by more than 15°. Angular distribution measurements are easier in this reaction than in the inverse, where the large center-of-mass transformation results in a large variation of experimental conditions with angle.

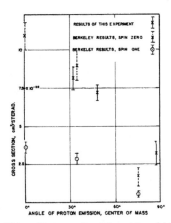

Fig. 4. Differential cross section of the reaction $\pi^+ + d \rightleftharpoons p + p$ at three angles of emission of the proton in the c.m. system. The average meson energy is 28 Mev in the c.m. system. The dotted points show the cross sections expected for spin one and spin zero pions on the basis of the Berkeley results (see reference 5).

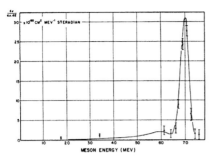

FIG. 5. The spectrum of mesons produced in the collision of 340 Mev protons in the forward direction. This experiment has been performed by Cartwright, Richman, Whitehead, and Wilcox (see reference 5).

(See Fig. 5.) This spectrum consists of a continuum due to the reaction $p+p \rightarrow \pi^+ + n + p$, and a sharp peak at an energy which exceeds the theoretical limit of the continuum and is due to the reaction $p+p \rightarrow \pi^+ + d$.[8] The cross section is obtained by integrating the energy spectrum under the peak.

Comparison of the two results shows that the π^+ meson spin is zero, quite outside the possible limits of error in the two experiments. Combining this result with those of Panofsky[2] and those on the photomeson production,[3] the meson is very likely pseudoscalar.

It is necessary to point out that the same ratio of cross sections could be obtained also for non-zero spin mesons provided that pions were completely polarized both in the $p+\text{Be}$ and $p+p$ meson production reactions. Such polarization is theoretically possible, but only in the longitudinal mode, that is, in the mode with zero component of angular momentum along the propagation axis. This seems a rather remote possibility.

A similar experiment has been performed by D. L. Clark, A. Roberts, and R. Wilson, who have reached the same conclusions. We are indebted to them for a pre-publication copy of their results.

The experiment is being continued. The reaction is interesting also in other connections. As Bethe[9] has pointed out, the angular distribution is quite perplexing. Furthermore, the energy dependence will shed some light on the momentum dependence of the meson nucleon interaction. We are therefore in the process of measuring the angular distribution at several energies.

We wish to acknowledge our indebtedness to the engineering staff of the Nevis Cyclotron Laboratory, especially Mr. Harrison Edwards and Mr. Julius Spiro.

---

[8] The actual production of deuterons in coincidence with mesons has been observed by Crawford, Crowe, and Stevenson, Phys. Rev. **82**, 97 (1951).

[9] H. A. Bethe, letter to R. E. Marshak with copies to C. Richman and J. Steinberger. We wish to express our thanks to Professor Bethe for this communication.

3
Strangeness

The discoveries of the strange particles, 1943–1959.

The elucidation of the $\pi \to \mu\nu$ decay sequence left particle physics in a relatively simple state. Yukawa's particle had been found and the only unanticipated particle was the muon, of which I. I. Rabi is said to have remarked "Who ordered that?" The question remains unanswered. The cosmic-ray experiments of the next few years quickly and thoroughly destroyed the simplicity that had previously prevailed. The proliferation of new particles, many with several patterns of decay, produced great confusion. The primary source of confusion was whether each new decay mode represented a new particle or was simply an alternative decay for a previously observed particle. Continued experimentation with improved accuracy and statistics eventually resolved these ambiguities, but basic uncertainties remained. What was the nature of these particles? How were they related to the more familiar particles? The examination of these questions led to the development of the concepts of associated production, strangeness, and ultimately, parity violation and $SU(3)$.

Remarkably, another meson seems to have been discovered before the pion. Working in the French Alps in 1943, Leprince-Ringuet and L'héritier took 10,000 triggered pictures in a 75 cm x 15 cm x 10 cm cloud chamber placed inside a magnetic field of 2500 gauss (**Ref. 3.1**). This permitted careful measurements of the momenta of the charged tracks. One of the pictures showed an incident positive particle of about 500 MeV/c momentum produce a secondary of about 1 MeV/c. By assuming the incident particle had scattered elastically on an electron and using the measured angles, Leprince-Ringuet and L'héritier determined the mass of the incident particle to be $990\, m_e \pm 12\%$ (506 ± 61 MeV), astonishingly close to the mass of the K^+. It was impossible that this could have been a π (whose mass was known shortly after the French result was finally published in 1946). Hans Bethe showed that the data were consistent with the incident particle being a proton only if extreme errors were assigned to the measurements.

Cosmic-ray research just after World War II centered in a few laboratories, including Bristol, whose group was led by Powell; Manchester, led by Blackett; Ecole Polytechnique headed by Leprince-Ringuet; Caltech, headed by Anderson; and Berkeley, led by Brode and Fretter. In 1947, the year of the $\pi \to \mu\nu$ paper of Lattes, Occhialini, and Powell, G. D. Rochester and C. C. Butler published two cloud chamber pictures showing forked

Table 3.1. *Comparison of old and current nomenclature for selected decays.*

Old	Current	
τ	$K^+ \to \pi^+\pi^+\pi^-$	$(K_{\pi 3})$
V_1^0	$\Lambda^0 \to p\pi^-$	
$V_2^0 (\theta^0)$	$K_S^0 \to \pi^+\pi^-$	
κ	$K^+ \to \mu^+\nu_\mu$	$(K_{\mu 2})$
	$K^+ \to \mu^+\pi^0\nu_\mu$	$(K_{\mu 3})$
$\chi(\theta^+)$	$K^+ \to \pi^+\pi^0$	$(K_{\pi 2})$
V^+, Λ^+	$\Sigma^+ \to p\pi^0, n\pi^+$	

tracks (**Ref. 3.2**). One proved to be the decay of a neutral particle into two charged particles and the other, the decay of a charged particle into another charged particle and at least one neutral. Whereas the event of Leprince-Ringuet and L'héritier may have established the existence of a particle with mass between the pion and the proton, the discovery of Rochester and Butler was much more revealing. It showed there were unstable particles decaying into other particles, perhaps pions. These unstable particles could be either charged or neutral, and had lifetimes on the scale of 10^{-9} to 10^{-10} s.

Surprisingly, the discovery of Rochester and Butler was not confirmed for over two years. Before that occurred, the Bristol group, using emulsions of increased sensitivity, observed the decay of a charged particle into three charged particles (**Ref. 3.3**). This particular decay came to be known as the tau meson. A guide to some of the old notation for the unstable particle decays is given in Table 3.1.

Confirmation of the events of Rochester and Butler was produced by the group at Caltech, which included C. D. Anderson, R. B. Leighton, and E. W. Cowan. Both neutral- and charged-particle decays were observed in their cloud chamber exposures, but no accurate estimate of the masses of the decaying particles was possible. A year later, in 1951, the Manchester group published results they obtained by taking their cloud chamber to the Pic-du-Midi in the Pyrenees. Studying the neutral decays, they were able to infer the existence of two distinct neutral particles, V_1^0 and V_2^0.

The progress on the charged-particle decays was slower. There was confirmation of the tau meson decay. In addition, O'Ceallaigh, working at Bristol, produced emulsion evidence for the decay of a charged particle into a μ^+ and one or more neutrals, the κ decay (Ref. 3.4). In one exposure, the μ^+ was convincingly identified through its decay into e^+. (See Figure 3.1).

While the tau meson mass had been measured quite well, the mass of the V_2^0 or θ^0 was not determined until the work of R. W. Thompson and co-workers at Indiana University (**Ref. 3.5**). They were able to establish a Q value for the decay of 214 MeV, in good agreement with the present value ($M_K - 2M_\pi = 219$ MeV). This indicated that the tau and

3. Strangeness

Figure 3.1. A κ (K) meson stops at P, decaying into a muon and neutrals. The muon decays at Q to an electron and neutrals. The muon track is shown in two long sections. Note the lighter ionization produced by the electron, contrasted with the heavy ionization produced by the muon near the end of its range. The mass of the κ was measured by scattering and grain density to be 562 ± 70 MeV (Ref. 3.4).

theta mesons had just about the same mass and set the stage for the famous puzzle about the parities of these particles.

The year 1953 marked a turning point in the investigation of the new V-particles. The great achievements of cosmic-ray physics in exploring the new particles was summarized

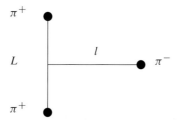

Figure 3.2. Diagram for the angular momentum in τ^+ meson decay. The $\pi^+\pi^+$ angular momentum, L, must be even. The orbital angular momentum, l, of the π^- must be added to L to obtain the total angular momentum (that is, the spin) of the tau.

in a meeting at Bagnères-de-Bigorre in France. The V_1^0 was well established, as was the tau. There were indications of both positive and negative hyperons (particles heavier than a proton). The negative hyperon was observed in a cascade that produced a neutral hyperon that itself decayed (Refs. 3.6, 3.7) There was a κ, which decayed into a muon plus neutrals, and a χ, which decayed into a charged pion plus neutrals. The $\theta \to \pi^+\pi^-$ was established, too.

At the Bagnères Conference, Richard Dalitz presented his analysis of the tau that was designed to determine its spin and parity through its decay into three pions. Some immediate observations about the spin and parity of the tau are possible. If there is no orbital angular momentum in the decay, the spin is zero and the parity is $(-1)^3$ because the parity of each pion is -1, and thus $J^P = 0^-$. The system of $\pi^+\pi^+$ can have only even angular momentum because of Bose statistics. Dalitz indicated this angular momentum by L and the orbital angular momentum of the system consisting of the π^- and the $(\pi^+\pi^+)$ by l. See Figure 3.2. Then the total angular momentum, J, was the vector sum of L and l. If $L = 0$, then $J = l$, and $P = (-1)^{J+1}$. For $L = 2$, other combinations were possible. Dalitz noted that since the sum of the pion energies was a constant, $E_1 + E_2 + E_3 = Q$, each event could be specified by two energies and indicated on a two-dimensional plot. (Here we are using kinetic energies, that is relativistic energies less rest masses.) If E_1 corresponds to the more energetic π^+ and E_2 to the less energetic π^+, all the points fall on one half of the plot. See Figure 3.3. If the decay involves no angular momentum and there are no effects from interactions between the produced pions, the points will be evenly distributed on the plot. Deviations from such a distribution give indications of the spin and parity. For example, as $E_3 \to 0$, the π^- is at rest and thus has no angular momentum. Thus $l = 0$, $J = L$ and $P = (-1)^{J+1}$. Hence if the tau is *not* in the sequence $0^-, 2^-, 4^-, \ldots$ there should be a depletion of events near $E_3 = 0$. As data accumulated in 1953 and 1954, it became apparent that there was no such depletion and thus it was established that τ^+ had J^P in the series $0^-, 2^-, \ldots$

The decay distribution for a two-body decay is given by Fermi's Golden Rule (which is actually due to Dirac) in relativistic form:

$$d\Gamma = \frac{1}{32\pi^2}|\mathcal{M}|^2 \frac{p_{cm} d\Omega}{M^2}. \tag{3.1}$$

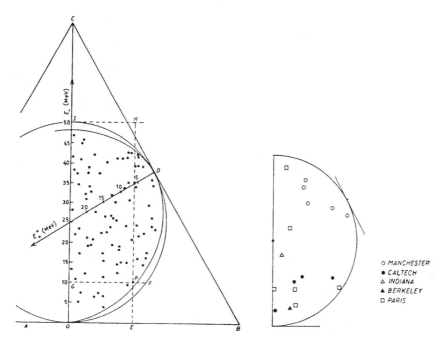

Figure 3.3. Dalitz plots showing worldwide compilations of tau meson decays ($\tau^+ \to \pi^+\pi^+\pi^-$) as reported by E. Amaldi at the Pisa Conference in June 1955 [*Nuovo Cimento Sup.* IV, 206 (1956)]. On the left, data taken in emulsions. On the right, data from cloud chambers. There is no noticeable depletion of events near $E_3 = 0$, i.e. near the bottom center of the plot. Parity conservation would thus require the tau to have $J^P = 0^-, 2^-$.

Here $d\Gamma$ is the decay rate, p_{cm} is the center-of-mass momentum of either final state particle, M is the mass of the decaying particle and $d\Omega$ is the solid angle element into which one final state particle passes. $\mathcal{M}$ is the Lorentz-invariant amplitude for the process. The amplitude $\mathcal{M}$ will involve the momenta of the various particles and factors to represent the spins of the particles.

For three-body decays there are more final state variables. If the particles are spinless or if polarization is ignored, however, there are only two variables necessary to specify the final state. They may be chosen to be the energies of the final state particles. The Golden Rule then takes the form

$$d\Gamma = \frac{1}{64\pi^3 M}|\mathcal{M}|^2 dE_1 dE_2. \tag{3.2}$$

Thus if $\mathcal{M}$ is constant, $d\Gamma \sim dE_1 dE_2$ and the events fall evenly on the Dalitz plot.

By examining the Dalitz plot, inferences can be drawn about spin and parity. Consider the $\tau \to 3\pi$. If the tau is spinless and the values of L and l are zero, $\mathcal{M}$ should be nearly constant. (Actually, it need not be absolutely constant. It may still depend on the Lorentz-invariant products of the momenta in the problem.) Suppose, on the contrary, tau has spin 1. Then it will be represented by a polarization vector, ϵ. The amplitude must be linear in ϵ. If we treat the pions as nonrelativistic, it suffices to consider just three-momenta rather than four-momenta. The amplitude, in order to be rotationally invariant, must be the dot product of ϵ with a vector made from the various pion momenta. In addition, because of Bose statistics, the amplitude must be invariant under interchange of the two

π^+'s, particles 1 and 2. Two examples are

$$\epsilon \cdot \mathbf{p}_3$$
$$\epsilon \cdot (\mathbf{p}_1 - \mathbf{p}_2) \times \mathbf{p}_3 \ (\mathbf{p}_1 - \mathbf{p}_2) \cdot \mathbf{p}_3$$

Both represent spin-1 decays. The parity of the decaying object, assuming parity is conserved in the decay, is determined by examining the behavior of the quantity dotted into ϵ. In the first case, the single momentum contributes (-1) to the parity since the momenta are reversed by the operation. In addition, the intrinsic parities of the three pions contribute $(-1)^3$. Altogether, the parity is even, so the state is $J^P = 1^+$. In the second instance, there are four factors of momentum and the parity is finally odd. In both cases, the amplitude vanishes as p_3 goes to zero in accordance with the earlier argument.

Dalitz's analysis led ultimately to the τ–θ puzzle: were the θ^+ (which decayed into $\pi^+\pi^0$) and τ^+, whose masses and lifetimes were known to be similar, the same particle? Of course this would require them to have the same spin and parity. But the parity of the $\theta^+ \to \pi^+\pi^0$ was necessarily $(-1)^J$ if its spin was J. These values were incompatible with the results for the tau showing that it had J^P in the sequence $0^-, 2^-, \ldots$. How this contradiction was resolved will be seen in Chapter 6.

Cosmic-ray studies had found evidence for hyperons besides the $\Lambda = V_1^0$. Positive particles of a similar mass were observed and initially termed V_1^+ or Λ^+. Evidence for this particle, now called the Σ^+, was observed by Bonetti et al. (Ref. 3.8) in photographic emulsion, and by York et al. (Ref. 3.9) in a cloud chamber. See Figure 3.4. Furthermore, a hyperfragment, which is a Λ or Σ^+ bound in a nucleus, was observed by Danysz and Pniewski in photographic emulsion (Ref. 3.10). See Figure 3.5. Working at Caltech, E. W. Cowan confirmed the existence of a negative hyperon (now called the Ξ^-) that itself decayed into $\Lambda^0\pi^-$ (Ref. 3.11).

By the end of the year 1953, the Cosmotron at Brookhaven National Laboratory was providing pion beams that quickly confirmed the cosmic-ray results and extended them. The existence of the V_1^+ (Σ^+) was verified and the V_1^- (Σ^-) was discovered. An especially important result was the observation of four events in which a pair of unstable particles was observed (**Ref. 3.12**). Such events were expected on the basis of theories that Abraham Pais and Murray Gell-Mann developed to explain a fundamental problem posed by the unstable particles. These unstable particles were clearly produced with a large cross section, some percent of the cross section for producing ordinary particles, pions and nucleons. The puzzle was this: The new particles were produced in strong interactions and decayed into strongly interacting particles, but if the decays involved strong interactions, the particle lifetimes should have been ten orders of magnitude less than those observed.

The first step in the resolution was made by Pais, who suggested that the new particles could only be made in pairs. One could assign a multiplicative quantum number, a sort of parity, to each particle, with the pion and nucleon carrying a value $+1$ and the new particles, K, Λ, etc. carrying -1. The product of these numbers was required to be the same in the initial and final state. Thus $\pi^- p \to K^0 \Lambda$ would be allowed, but $\pi^- p \to K^0 n$ would be forbidden. The Cosmotron result on the production of pairs of unstable particles was

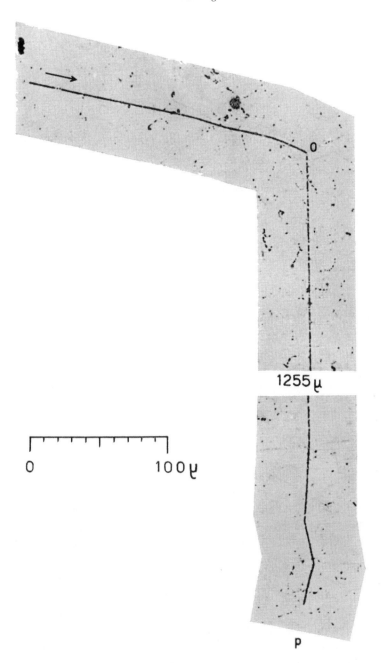

Observed by R. LEVI SETTI

Figure 3.4. An emulsion event with a Σ^+ entering from the left. The decay is $\Sigma^+ \to p\pi^0$. The p is observed to stop after 1255 μm. (Ref. 3.8)

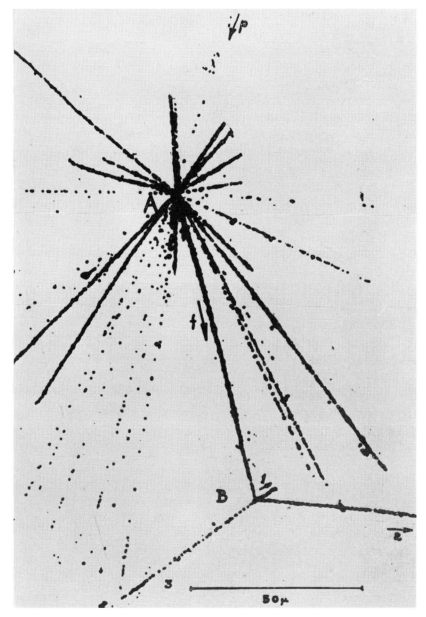

Figure 3.5. The star at A is caused by a cosmic-ray (marked p) incident from above colliding with a silver or bromine atom in the emulsion. The track f is due to a nuclear fragment with charge about 5. Its decay at point B shows that it contained a hyperon. The scale at the bottom indicates 50 μm. (Ref. 3.10)

consistent with Pais's explanation. Pais's parity was to be conserved only in the strong (nuclear) interactions. The weak interactions were not to obey this rule, so weak decays like $\Lambda \to \pi^- p$ were allowed. However, because the weak interactions are quite feeble, the lifetimes of the unstable particles could be much longer than would have been the case if the decays went through the strong interaction.

The associated-production proposal of Pais was only a partial explanation. The full solution was given by Gell-Mann. In Gell-Mann's proposal, the new quantum number that was introduced was not multiplicative, but additive. Each strongly interacting particle has an additive quantum number called *strangeness*. For the old particles (pion and nucleon) the strangeness, S, is 0. For the K^+ the strangeness is +1, while for the Λ and Σ's it is -1. Pairs of mesons with identical masses but opposite electric charges are antiparticles of each other, just as the positron is the antiparticle of the electron. Each antiparticle is assigned the opposite strangeness from the particle. Thus the K^- has strangeness -1. While Gell-Mann's proposal allowed $\pi^- p \to K^+ \Sigma^-$ but not $\pi^- p \to K^0 n$, just as the scheme of Pais, some of its predictions were different. For example, Gell-Mann's rules forbid $nn \to \Lambda\Lambda$ while Pais's allow it. An especially important distinction was $\pi^- p \to K^- \Sigma^+$. This is forbidden by Gell-Mann's proposal (the final state has strangeness -2) but allowed by that of Pais. Gell-Mann proposed that the strong interactions conserved isospin and strangeness, and that electromagnetism conserved strangeness, but allowed a unit change of isospin. The weak interactions violated isospin and allowed a unit change of strangeness.

The proposal of Gell-Mann initially met severe opposition. His classification of the particles placed the K meson into two isospin doublets: (K^+, K^0) and $(\overline{K}^0, K^-)$. Two objections were raised: First he was requiring that a neutral meson not be its own antiparticle. Though Kemmer had shown years before that there was nothing wrong with this, it still seemed odd. Moreover, many thought it was impossible to have isodoublet bosons (the K's) and isovector fermions (the Σ's), rather than the better known isodoublet fermions (nucleons) and isovector bosons (the pions). The objections, of course, eventually gave way, as did the resistance to the name strangeness.

The proposal of Gell-Mann was arrived at independently by Nakano and Nishijima. The strangeness S, baryon number B ($B = 1$ for nucleons and the hyperons Λ, Σ, Ξ), the third component of isospin, I_z, and charge, Q, were linked by the Gell-Mann–Nishijima relation,

$$Q = I_z + (B + S)/2. \tag{3.3}$$

Since the masses of the Σ^+ and Σ^- were not close enough to the mass of the Λ for them to form an isotriplet, a new hyperon, Σ^0, was predicted that would decay into Λ and a γ. Since the Ξ^- decayed weakly into $\Lambda \pi^-$ it was assigned $S = -2$. Using the Gell-Mann–Nishijima equation, we see that the Ξ^- must have $I_z = -1/2$. Thus a Ξ^0 with $I_z = 1/2$ is required. These predictions of Gell-Mann were subsequently verified.

In 1954 the Bevatron started operating with proton energies up to 6 GeV at the Radiation Laboratory in Berkeley. The early emulsion work at the Bevatron concentrated on K^+ (that is, θ^+, χ^+, or κ^+) and τ^+ studies. This work considerably augmented the cosmic-ray data on mass equality (**Ref. 3.13**, Ref. 3.14) and lifetime equality (Ref. 3.15, 3.16) between the K^+ and the τ^+. If these were different particles, they had to be a very close doublet in mass with very similar lifetimes as well! Subsequent counter experiments at the Bevatron and Cosmotron (Refs. 3.17, 3.18) gave even closer agreement for the lifetimes of the various K decay modes and the tau.

Just as data from accelerators began to supplant those from cosmic rays, a major effort, the G-Stack (for "giant") experiment, was mounted by the groups from Bristol, Milan, and Padua. A volume of 15 liters of emulsion was flown at a height of 27,000 meters for six hours. The emulsion stack was thick enough to stop many of the particles produced by decays at rest. Tracing microscopic tracks through 250 sheets of emulsion was an enormous task, but the reward was also great: the clear identification of the decays $K_{\mu 2}$, $K_{\pi 2}$ and K_{e3}.

In 1955, W. D. Walker measured two cloud chamber events apparently of the form $\pi^- p \to K^0 \Lambda$ (Ref. 3.19). One event was consistent with the interpretation that there were no additional unobserved particles. The other, however, was inconsistent with this hypothesis and instead fitted better the supposition that a γ or ν had been produced as well. Walker argued that the best interpretation was that the Λ was a decay product. The deduced mass of the decaying object agreed very well with the known masses of the Σ^+ and Σ^-. It was natural to conclude that the actual process was $\pi^- p \to K^0 \Sigma^0$, followed by $\Sigma^0 \to \Lambda \gamma$. Indeed, Walker showed that that hypothesis explained some discrepancies in the events reported earlier by Fowler *et al.*

The discovery of the Ξ^0 did not take place until 1959. Since the Ξ has strangeness -2, its production by pions is quite infrequent: the minimal process would be $\pi^- p \to K^0 K^0 \Xi^0$. A more effective means is to start with a particle with strangeness -1. This was accomplished by L. Alvarez and co-workers using a hydrogen bubble chamber and a mass-separated beam of K^- mesons of momentum about 1 GeV/c produced by the Bevatron. Using the great analytical power of the bubble chamber technique, they were able to identify an event $K^- p \to K^0 \Xi^0$ (Ref. 3.20). The K^0 decayed into $\pi^+ \pi^-$. The Ξ^0 decayed into $\Lambda^0 \pi^0$. Both the decay of the K^0 and the decay of the Ξ^0 gave noticeable gaps in the bubble chamber pictures. The Λ^0 was identified by its charged decay mode, $\Lambda \to p\pi^-$. The last hyperon, Ω^-, was not discovered until 1964, as discussed in Chapter 5.

The bubble chamber was invented by Donald Glaser in 1953. The first chambers used propane and other liquid hydrocarbons. The idea was rapidly adapted by Luis Alvarez and his group who used liquid hydrogen (and later also deuterium) as the working liquid. They also developed methods for building increasingly large chambers. The bubble chamber works by producing a superheated liquid by rapid expansion just before (about 10 ms) the arrival of the particles to be studied. Bubbles are formed when boiling starts around the ions produced by the passage of the charge particles through the liquid. These bubbles are allowed to grow for about 2 ms at which time lights are flashed and the bubbles are photographed. The properties of bubble chambers are ideally suited for use with accelerators. At an accelerator, the arrival time of a particle beam is known. This allows one to expand the chamber before the arrival of the charged particles, which is not possible in cosmic-ray experiments.

Exercises

3.1 Suppose that in an experiment like that of Leprince-Ringuet and L'héritier a singly charged particle of mass $M \gg m_e$ scatters elastically from an electron. Let the incident particle's momentum be p and the scattered electron's (relativistic) energy be E. Further, let χ be the angle the electron makes with the incident particle (which is nearly undeflected). Show that

$$M = p \left[\frac{E + m_e}{E - m_e} \cos^2 \chi - 1 \right]^{1/2}. \tag{3.4}$$

For the event of Leprince-Ringuet and L'héritier, the cloud chamber was in a magnetic field of about 2450 gauss. The incident particle had a radius of curvature of 700 cm while that of the electron was 1.5 cm. Take $\chi = 20°$ and assume the scattering plane was perpendicular to the magnetic field. Estimate M.

3.2 Using the data from Table 1 of Rochester and Butler, Ref. 3.2, and the current values for the π, K, p, and Λ masses, determine whether their photograph 1 is $K^0 \to \pi^- \pi^+$ or $\Lambda \to p\pi^-$. Are the errors in the measurements small enough to permit a confident choice?

3.3 Suppose a neutral particle decays into a positive of mass m^+ and a negative of mass m^-. Assume the angular distribution in the initial particle's rest frame is isotropic. Let p_z^+ be the component of the positive particle's momentum along the direction of the incident particle measured in the lab and similarly for p_z^-. Define

$$\alpha = \frac{p_z^+ - p_z^-}{p_z^+ + p_z^-}. \tag{3.5}$$

Show that the points (α, p_t), where p_t is the momentum of a decay product perpendicular to the line of flight of the initial particle, lie on an ellipse. Discuss how this could be used to separate $\Lambda \to p\pi^-$ from $K^0 \to \pi^+\pi^-$. See R. W. Thompson in the *Proceedings of the 3rd Rochester Conference*.

3.4 Using the data of Thompson *et al.*, and the mass of the charged pion, determine the mass of the K^0 and the associated uncertainty. Compare with the Q value quoted by these authors.

3.5 Carry out the construction of the Dalitz plot for $\tau \to 3\pi$. Assume the pions are nonrelativistic with energies E_1, E_2, E_3. Let $M_K - 3M_\pi = Q$ and define $\epsilon_i = E_i/Q$. Construct an equilateral triangle with center $x = 0$, $y = 0$ and base along $x = -1/3$. Then each side is of length $2/\sqrt{3}$. Now for each point inside the triangle, let ϵ_3 be the distance to the base, ϵ_1 the distance to the right leg of the triangle and ϵ_2 the distance to the left leg. Using the nonrelativistic approximation, show that the physical points lie inside the circle

$$x^2 + y^2 = 1/9.$$

Make plots showing the contours of equal probability density for the decay of the τ for the two possibilities, $J^P = 1^-$ and $J^P = 1^+$, using the matrix elements given in the text.

3.6 Consider the decay $K^+ \to \mu^+\pi^0\nu_\mu$. What is the relation between the energy of the muon in the K^+ rest frame and the invariant mass squared of the $\pi^0 - \nu$ system? What is the maximum energy the muon can have, again in the K rest frame? If the energy of the muon is E, what is the range of energies possible for the π^0? Use this and the relation

$$d\Gamma \sim |\mathcal{M}|^2 dE_1 dE_2$$

to determine the muon energy spectrum assuming the matrix element $\mathcal{M}$ is constant. Use the result to evaluate the likelihood that the two events discussed by C. O'Ceallaigh, Ref. 3.4, are $K \to \mu^+\pi^0\nu_\mu$. Assume the neutrino, ν_μ, is massless.

Further Reading

A. Pais gives a first hand account in *Inward Bound*, Oxford University Press, New York, 1986.

For a fine historical review, see the articles by C. Peyrou, R. H. Dalitz, M. Gell-Mann and others in *Colloque International sur l'Histoire de la Physique des Particules, Journal de Physique*, **48**, supplement au no. 12. Dec. 1982. Les Editions de Physique, Paris, 1982. (In English)

References

3.1 L. Leprince-Ringuet and M. L'héritier, *Comptes Rendus Acad. Sciences de Paris, séance du 13 Dec. 1944, p. 618* "Existence probable d'une particule de masse 990 m_e dans le rayonnement cosmique."

3.2 G. D. Rochester and C. C. Butler, "Evidence for the Existence of New Unstable Elementary Particles." *Nature*, **160**, 855 (1947).

3.3 R. Brown, U. Camerini, P. H. Fowler, H. Muirhead, C. F. Powell, and D. M. Ritson, "Observations with Electron-Sensitive Plates Exposed to Cosmic Radiation." *Nature*, **163**, 82 (1949). Only the first page is reproduced here.

3.4 C. O'Ceallaigh, "Masses and Modes of Decay of Heavy Mesons - Part I." *Phil. Mag.*, **XLII**, 1032 (1951).

3.5 R. W. Thompson, A. V. Buskirk, L. R. Etter, C. J. Karzmark and R. H. Rediker, "An Unusual Example of V^0 Decay." *Phys. Rev.*, **90**, 1122 (1953).

3.6 R. Armenteros *et al.*, "The Properties of Charged V Particles." *Phil. Mag.*, **43**, 597 (1952).

3.7 C. D. Anderson *et al.*, "Cascade Decay of V Particles." *Phys. Rev.*, **92**, 1089 (1953).

3.8 A. Bonetti, R. Levi Setti, M. Panetti, and G. Tomasini, "On the Existence of Unstable Particles of Hyperprotonic Mass." *Nuovo Cimento*, **10**, 1 (1953).

3.9 C. M. York, R. B. Leighton, and E. K. Bjornerud, "Direct Experimental Evidence for the Existence of a Heavy Positive V Particle." *Phys. Rev.*, **90**, 167 (1953).

3.10 M. Danysz and J. Pniewski, "Delayed Disintegration of a Heavy Nuclear Fragment." *Phil. Mag.*, **44**, 348 (1953).

3.11 E. W. Cowan, "A V-Decay Event with a Heavy Negative Secondary, and Identification of the Secondary V-Decay Event in a Cascade." *Phys. Rev.*, **94**, 161 (1954).

3.12 W. B. Fowler, R. P. Shutt, A. M. Thorndike, and W. L. Whittemore, "Production of Heavy Unstable Particles by Negative Pions." *Phys. Rev.*, **93**, 861 (1954).

3.13 R. W. Birge et al., "Bevatron K Mesons." *Phys. Rev.*, **99**, 329 (1955).

3.14 W. W. Chupp et al., "K meson mass from K-hydrogen scattering event." *Phys. Rev.*, **99**, 1042 (1955).

3.15 E. L. Iloff et al., "Mean Lifetime of Positive K Mesons." *Phys. Rev.*, **99**, 1617 (1955).

3.16 L. W. Alvarez and S. Goldhaber, "Lifetime of the τ Meson." *Nuovo Cimento IV*, **X,2**, 33 (1956).

3.17 L. W. Alvarez et al., "Lifetime of K mesons." *Phys. Rev.*, **101**, 503 (1956).

3.18 V. Fitch and R. Motley, "Mean Life of K^+ Mesons." *Phys. Rev.*, **101**, 496 (1956).

3.19 W. D. Walker, "$\Lambda^0 - \theta^0$ Production in $\pi^- p$ Collisions at 1 BeV." *Phys. Rev.*, **98**, 1407 (1955).

3.20 L. Alvarez et al. , "Neutral Cascade Hyperon Event." *Phys. Rev. Lett.*, **2**, 215 (1959).

PHYSIQUE NUCLÉAIRE. — *Existence probable d'une particule de masse* $990\ m_0$ *dans le rayonnement cosmique.* Note ([1]) de MM. **Louis Leprince-Ringuet** et **Michel Lhéritier**.

Nous avons pris, au cours de l'année 1943, dans le laboratoire de Largentière (Hautes-Alpes) situé à 1000^m d'altitude, une série de 10000 clichés de trajectoires cosmiques commandées par compteurs. Les rayons, filtrés par 10^{cm} de plomb, traversaient une chambre de Wilson de 75^{cm} de hauteur, placée dans un champ magnétique H de 2500 gauss environ. Nous nous sommes placés dans les conditions expérimentales les plus favorables [discutées précédemment ([2]), ([3]), ([4])] pour profiter au mieux des clichés de collision entre particules pénétrantes et électrons du gaz de la chambre, dans le but de déterminer la masse au repos de la particule incidente.

Nous avons obtenu une dizaine de clichés intéressants. Le plus remarquable représente une collision dans le gaz pour laquelle d'excellentes conditions sont réalisées : le secondaire fait avec le plan médian de la chambre un angle ζ tel que $\tang\zeta = 0,32$ et son rayon de courbure projeté (1^{cm},6), ainsi que la flèche dont il s'écarte du primaire sont mesurables avec précision. Le (Hρ) du primaire $= 1,7 \times 10^6$ gauss $\times$ cm. La formule de collision élastique donne pour le primaire, qui est positif, la masse au repos

$$\mu_0 = 990 \pm 12\ \%\ (\text{limites extrêmes de l'erreur})\quad ([5]).$$

La masse ainsi obtenue peut surprendre. Les indications suivantes, qui donnent des garanties de la validité de la mesure, nous ont poussés à publier ce résultat.

([1]) Les enroulements sont doublés et montés de façon que la somme des ampères-tours produits par le courant des tubes électroniques soit pratiquement annulée à l'état de repos.

([1]) Séance du 17 juin 1944.

([2]) L. Leprince-Ringuet, S. Gorodetzky, E. Nageotte, R. Richard-Foy, *Comptes rendus*, **211**, 1940, p. 382; *Phys. Rev.*, **59**, 1941, p. 460; *Journal de Physique*, **2**, 1941, p. 63.

([3]) R. Richard-Foy, *Comptes rendus*, **213**, 1941; *Cahiers de Physique*, 2e série, 1942, p. 65.

([4]) S. Gorodetzky, *Thèse*, Paris, 1942; *Ann. de Physique*, **19**, 1944, pp. 5-70.

([5]) Nous n'avons pas tenu compte de l'erreur que peut introduire une courbure naturelle des trajectoires, due à la diffusion coulombienne; cette erreur peut être de l'ordre de 5%.

1° Il ne peut y avoir, dans l'ensemble de notre expérience, de grossière erreur : en effet nous sommes conduits, pour déduire la masse des données du cliché, à effectuer deux opérations successives : tout d'abord lire sur l'abaque de Richard-Foy (³) la valeur d'une quantité Z fonction seulement des données du *secondaire* et de la valeur du champ; ensuite calculer la masse μ_0 par la relation $\mu_0 = Z(\rho_1/\rho_0)$, ρ_1 étant la projection du rayon de courbure du *primaire*, sur le plan perpendiculaire au champ, et ρ_0 la quantité ρ_0 cm $= 1700/\text{H}$ gauss : le primaire n'intervient que dans cette seconde opération.

Or certains clichés, notamment six clichés remarquables, donnent des collisions avec un secondaire se présentant dans de bonnes conditions, mais avec un primaire trop peu courbé pour pouvoir être mesuré. Ces clichés, inutilisables pour fournir une valeur de la masse, sont du plus haut intérêt pour donner une confirmation de la méthode et de la validité de l'expérience. Ils doivent en effet donner un Z voisin de zéro, puisque le rapport ρ_1/ρ_0 est alors parfois supérieur à 10000 et que μ_0 ne peut dépasser plusieurs milliers. On trouve ce résultat sur l'abaque à partir des données mesurées du secondaire, et de la valeur du champ : les limites extrêmes trouvées pour Z encadrent bien la valeur zéro, de façon parfois extrêmement précise. Or ceci n'est possible que si d'une part la collision est élastique, et si d'autre part il n'y a pas d'erreur grave sur les mesures.

2° De grandes précautions ont été prises pour s'assurer du caractère secon-

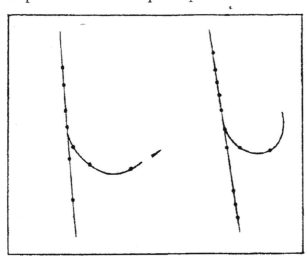

Dessin stéréoscopique de la collision.

daire de l'électron éjecté (notamment stéréoscopie triple) et pour obtenir une restitution dans l'espace donnant les meilleures garanties d'exactitude et de séparation.

3° Le cliché donnant la masse 990 se présente particulièrement bien : des erreurs même notables sur la mesure de la flèche, sur celle de l'angle d'éjection dans l'espace et sur la valeur du champ affecteraient à peu près linéairement la valeur de la masse, ce qui est le cas le plus favorable. On pourrait supposer, pour expliquer ce résultat, que le secondaire a subi, peu après le choc, une diffusion notable invisible à l'œil; mais ce ne peut être le cas : même si l'on supposait une diffusion brusque de 6 degrés au départ (ce qui serait fort improbable), la valeur de la masse ne serait à modifier que de 80 unités.

4° Signalons enfin que trois autres collisions, observées dans la série de nos clichés, permettent d'encadrer la valeur de la masse du mésoton normal, sans toutefois autoriser, à cause de la valeur élevée du moment réduit ou de l'imprécision sur la flèche, une mesure précise de sa masse.

En résumé, nous possédons une mesure se rapportant à une particule positive de masse $(990 \pm 12 \%) m_0$. On peut remarquer que cette valeur est à peu près quadruple de celle du mésoton normal; une particule ayant une masse moitié de celle du proton (925) entrerait d'ailleurs aussi dans l'intervalle de nos mesures; cela sous réserve que la collision soit élastique, fait que nous ne pouvons naturellement pas affirmer pour le cliché envisagé, mais qui est vérifié pour tous les autres clichés remarquables présentant une collision à primaire peu courbé dans la même série d'expériences, sous réserve également que les charges électriques des particules incidente et heurtée soient celles de l'électron, ce qui est probable; la charge double de l'une des particules est exclue par l'observation de l'ionisation.

EVIDENCE FOR THE EXISTENCE OF NEW UNSTABLE ELEMENTARY PARTICLES

By Dr. G. D. ROCHESTER
AND
Dr. C. C. BUTLER
Physical Laboratories, University, Manchester

AMONG some fifty counter-controlled cloud-chamber photographs of penetrating showers which we have obtained during the past year as part of an investigation of the nature of penetrating particles occurring in cosmic ray showers under lead, there are two photographs containing forked tracks of a very striking character. These photographs have been selected from five thousand photographs taken in an effective time of operation of 1,500 hours. On the basis of the analysis given below we believe that one of the forked tracks, shown in Fig. 1 (tracks a and b), represents the spontaneous transformation in the gas of the chamber of a new type of uncharged elementary particle into lighter charged particles, and that the other, shown in Fig. 2 (tracks a and b), represents similarly the transformation of a new type of charged particle into two light particles, one of which is charged and the other uncharged.

The experimental data for the two forks are given in Table 1; H is the value of the magnetic field, α the angle between the tracks, p and Δp the measured momentum and the estimated error. The signs of the particles are given in the last column of the table, a plus sign indicating that the particle is positive if moving down in the chamber. Careful re-projection of the stereoscopic photographs has shown that each pair of tracks is copunctal. Moreover, both tracks occur in the middle of the chamber in a region of uniform illumination, the presence of background fog surrounding the tracks indicating good condensation conditions.

Though the two forks differ in many important respects, they have at least two essential features in common: first, each consists of a two-pronged fork with the apex in the gas; and secondly, in neither case is there any sign of a track due to a third ionizing particle. Further, very few events at all similar to these forks have been observed in the 3-cm. lead plate, whereas if the forks were due to any type of collision process one would have expected several hundred times as many as in the gas. This argument indicates, therefore, that the tracks cannot be due to a collision process but must be due to some type of spontaneous process for which the probability depends on the distance travelled and not on the amount of matter traversed.

This conclusion can be supported by detailed arguments. For example, if either forked track were due to the deflexion of a charged particle by collision with a nucleus, the transfer of momentum would be so large as to produce an easily visible recoil track. Then, again, the attempt to account for Fig. 2 by a collision process meets with the difficulty that the incident particle is deflected through 19° in a single collision in the gas and only 2·4° in traversing 3 cm. of lead—a most unlikely event. One specific collision process, that of electron pair production by a high-energy photon in the field of the nucleus, can be excluded on two grounds: the observed angle between the tracks would only be a fraction of a degree, for example, 0·1° for Fig. 1, and a large amount of electronic component should have accompanied the photon, as in each case a lead plate is close above the fork.

We conclude, therefore, that the two forked tracks do not represent collision processes, but do represent spontaneous transformations. They represent a type of process with which we are already familiar in the decay of the meson into an electron and an assumed neutrino, and the presumed decay of the heavy meson recently discovered by Lattes, Occhialini and Powell[1].

TABLE 1. EXPERIMENTAL DATA

Photograph	H (gauss)	α (deg.)	Track	p (eV./c.)	Δp (eV./c.)	Sign
1	3500	66·6	a	$3·4 \times 10^8$	$1·0 \times 10^8$	+
			b	$3·5 \times 10^8$	$1·5 \times 10^8$	−
2	7200	161·1	a	$6·0 \times 10^8$	$3·0 \times 10^8$	+
			b	$7·7 \times 10^8$	$1·0 \times 10^8$	+

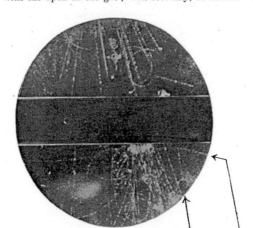 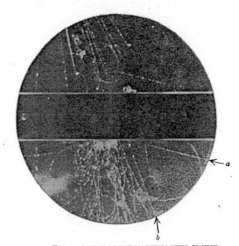

Fig. 1. STEREOSCOPIC PHOTOGRAPHS SHOWING AN UNUSUAL FORK ($a\,b$) IN THE GAS. THE DIRECTION OF THE MAGNETIC FIELD IS SUCH THAT A POSITIVE PARTICLE COMING DOWNWARDS IS DEVIATED IN AN ANTICLOCKWISE DIRECTION

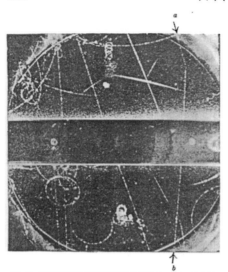 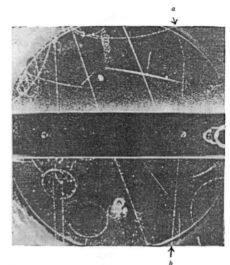

Fig. 2. STEREOSCOPIC PHOTOGRAPHS SHOWING AN UNUSUAL FORK (a b). THE DIRECTION OF THE MAGNETIC FIELD IS SUCH THAT A POSITIVE PARTICLE COMING DOWNWARDS IS DEVIATED IN A CLOCKWISE DIRECTION

The Masses of the Incident Particles

Let us assume that a particle of mass M and initial momentum P is transformed spontaneously into two particles of masses m_1 and m_2, momenta p_1 and p_2 at angles of θ and φ with the direction of the incident particle. Then the following relations must hold:

$$\sqrt{M^2c^4+P^2c^2} = \sqrt{m_1^2c^4+p_1^2c^2} + \sqrt{m_2^2c^4+p_2^2c^2} \quad (1)$$

$$P = p_1 \cos \theta + p_2 \cos \varphi \quad (2)$$

$$p_1 \sin \theta = p_2 \sin \varphi. \quad (3)$$

These general relations may be used to obtain the mass of the incident particle as a function of the assumed masses of the secondary particles.

The value of M must be greater than that obtained by taking the rest masses of the secondary particles as small compared with their momenta; thus the minimum value M_{min} is given by the following equation:

$$M_{min}c^2 = c\sqrt{(p_1+p_2)^2 - P^2}. \quad (4)$$

Applying this equation to the forked track of Fig. 1, after calculating P from the observed values of p_1 and p_2, it is found that M_{min} is $(770 \pm 200)m$, where m is the mass of the electron. The application of equation (4) to the forked track of Fig. 2, however, after calculating p_2 from the observed values of P and p_1, shows that $M_{min} = (1,700 \pm 150)m$. This value of the mass would require an ionization for the incident particle of twice minimum, which is inconsistent with the observed ionization. We are therefore justified in assuming that the real value of P is greater than the observed value which, as indicated in Table 1, has a large error. If larger values of P are assumed, then M_{min} is reduced in value. The lowest value of M_{min} is $(980 \pm 150)m$ if P is 14.5×10^8 eV./c. Beyond this value of P the mass increases slowly with increasing momentum. No choice of incident momentum will bring the mass of the incident particle below $980\ m$.

In the special case where the incident particle disintegrates transversely into two particles of equal mass m_0, giving a symmetrical fork, equation (1) reduces to the following expression,

$$\frac{M}{m} = \frac{2m_0}{m}\left(1 + \frac{p^2c^2}{m_0^2c^4}\cdot \sin^2\theta\right)^{1/2}, \quad (5)$$

where p is the momentum of each of the secondary particles. Some typical results for different assumed secondary particles, calculated from equation (5), are given in Table 2. On the reasonable assumption that the secondary particles are light or heavy mesons, that is, with masses of $200m$ or $400m$, we find that the incident particle in each photograph has a mass of the order of $1,000m$.

TABLE 2. MASS OF INCIDENT PARTICLE AS A FUNCTION OF MASS OF SECONDARY PARTICLE

Photograph	Assumed secondary particle m_0/m	Momentum of observed secondary particle (eV./c.)	Incident particle M/m
1	0	$3.5 \times 10^8 \pm 1.0 \times 10^8$	770 ± 200
	200	"	870 ± 200
	400	"	1110 ± 150
	1837	"	3750 ± 50
2	0	$7.7 \times 10^8 \pm 1.0 \times 10^8$	980 ± 150
	200	"	1080 ± 100
	400	"	1280 ± 100
	1837	"	3820 ± 50

Upper values of the masses of the incident particles may also be obtained from the values of the ionization and the momenta. Thus for each of the observed particles in Fig. 1, the ionization is indistinguishable from that of a very fast particle. We conclude, therefore, that $\beta = v/c \gg 0.7$. Since the momentum of the incident particle may be found from the observed momenta of the secondary particles, we can apply equation (1) to calculate M. In this way we find $M/m < 1,600$. Again, since the ionization of the incident particle in Fig. 2 is light, $\beta \gg 0.7$, from which it can be shown that $M/m < 1200$. This last result, however, must be taken with caution because of the uncertainty in the measured value of the momentum of the incident particle.

One further general comment may be made. This is that the observation of two spontaneous disintegrations in such a small number of penetrating showers suggests that the life-time of the unstable particles is much less than the life-time of the ordinary meson. An approximate value of this life-time may be derived as follows. The probability of an unstable particle of life-time τ_0 decaying in a short distance D is given by

$$p = \frac{D(1-\beta^2)^{1/2}}{\tau_0 c \beta}. \quad (6)$$

Since the total number of penetrating particles in the penetrating showers so far observed is certainly less than 50, we must assume that the number of our new unstable particles is unlikely to have been greater than 50. Since one particle of each type has been observed to decay, we can therefore put $p \approx 0.02$. Setting $D \approx 30$ cm., and $\beta = 0.7$, we find from equation (6) that $\tau_0 = 5.0 \times 10^{-8}$ sec.

We shall now discuss possible alternative explanations of the two forks.

Photograph 1. We must examine the alternative possibility of Photograph 1 representing the spontaneous disintegration of a charged particle, coming up from below the chamber, into a charged and an uncharged particle. If we apply the argument which led to equation (4) to this process, it is readily seen that the incident particle would have a minimum mass of $1,280m$. Thus the photograph cannot be explained by the decay of a back-scattered ordinary meson. Bearing in mind the general direction of the other particles in the shower, it is thought that assumption of the disintegration of a neutral particle moving downwards into a pair of particles of about equal mass is more probable. Further, it can be stated with some confidence that the observed ionizing particles are unlikely to be protons because the ionization of a proton of momentum 3.5×10^8 eV./c. would be more than four times the observed ionization.

Photograph 2. In this case we must examine the possibility of the photograph representing the spontaneous decay of a neutral particle coming from the right-hand side of the chamber into two charged particles. The result of applying equation (4) to this process is to show that the minimum mass of the neutral particle would be about $3,000m$. In view of the fact that the direction of the neutral particle would have to be very different from the direction of the main part of the shower, it is thought that the original assumption of the decay of a charged penetrating particle and an assumed neutral particle is the more probable.

We conclude from all the evidence that Photograph 1 represents the decay of a neutral particle, the mass of which is unlikely to be less than $770m$ or greater than $1,600m$, into the two observed charged particles. Similarly, Photograph 2 represents the disintegration of a charged particle of mass greater than $980m$ and less than that of a proton into an observed penetrating particle and a neutral particle. It may be noted that no neutral particle of mass $1,000m$ has yet been observed; a charged particle of mass $990m \pm 12$ per cent has, however, been observed by Leprince-Ringuet and L'héritier[2].

Peculiar cloud-chamber photographs taken by Jánossy, Rochester and Broadbent[3] and by Daudin[4] may be other examples of Photograph 2.

It is a pleasure to record our thanks to Prof. P. M. S. Blackett for the keen interest he has taken in this investigation and for the benefit of numerous stimulating discussions. We also wish to acknowledge the help given us by Prof. L. Rosenfeld, Mr. J. Hamilton and Mr. H. Y. Tzu of the Department of Theoretical Physics, University of Manchester. We are indebted to Mr. S. K. Runcorn for his assistance in running the cloud chamber in the early stages of the work.

[1] Lattes, C. M. G., Occhialini, G. P. S., and Powell, C. F., *Nature*, 160, 453, 486 (1947).
[2] Leprince-Ringuet, L., and L'héritier, M., *J. Phys. Radium.* (Sér. 8), 7, 66, 69 (1946). Bethe, H. A., *Phys. Rev.*, 70, 821 (1946).
[3] Jánossy, L., Rochester, G. D., and Broadbent, D., *Nature*, 155, 142 (1945). (Fig. 2. Track at lower left-hand side of the photograph.)
[4] Daudin, J., *Annales de Physique*, 11ᵉ Série, 19 (Avril–Juin), 1944 (Planche IV, Cliché 16).

OBSERVATIONS WITH ELECTRON-SENSITIVE PLATES EXPOSED TO COSMIC RADIATION*

By Miss R. BROWN, U. CAMERINI, P. H. FOWLER, H. MUIRHEAD
and Prof. C. F. POWELL
H. H. Wills Physical Laboratory, University of Bristol

and D. M. RITSON
Clarendon Laboratory, Oxford

PART 2. FURTHER EVIDENCE FOR THE EXISTENCE OF UNSTABLE CHARGED PARTICLES, OF MASS $\sim 1,000\ m_e$, AND OBSERVATIONS ON THEIR MODE OF DECAY

ONE of the first events found in the examination of electron-sensitive plates exposed at the Jungfraujoch is represented in the mosaic of photomicrographs shown in Fig. 8. There are two centres, A and B, from which the tracks of charged particles diverge, and these are joined by a common track, t. Because of the short duration of the exposure, and the small number of disintegrations occurring in the plate, the chance that the observation corresponds to a fortuitous juxtaposition of the tracks of unrelated events is very small—of the order 1 in 10^7. It is therefore reasonable to exclude it as a serious possibility. Further observations in support of this assumption are presented in a later paragraph.

that it carried the elementary electronic charge; and that it had reached, or was near, the end of its range at the point A. We therefore assume that the particle k initiated the train of events represented by the tracks radiating from A and B. It follows that the particle producing track t originated in star A, and produced the disintegration B. In order to analyse the event, we first attempted to determine the mass of the particle k.

Mass Determinations by Grain-Counts

About a year ago, experiments were made in this Laboratory to determine the ratio, m_π/m_μ, of the masses of π- and μ-mesons, by the method of grain-counting[5], and by studying the small-angle scattering of the particles in their passage through the emulsion[4]. The values obtained by the two methods were $m_\pi/m_\mu = 1\cdot 65 \pm 0\cdot 11$, and $m_\pi/m_\mu = 1\cdot 35 \pm 0\cdot 10$*, respectively. Recent experiments at

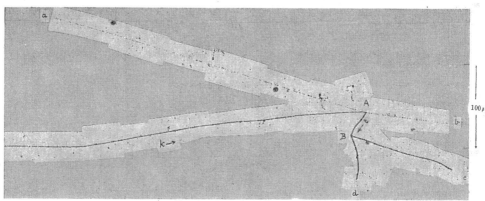

Observer: Mrs. W. J. van der Merwe
Fig. 8

An inspection of the track k shows that the particle producing it approached the centre of disintegration A. The range of the particle in the emulsion exceeds $3,000\ \mu$, and there is continuous increase in the grain-density along the track in approaching A. Near A, the grain-density is indistinguishable from that of particles of charge e, recorded in the same plate, near the end of their range.

The evidence for the direction of motion of the particle based on grain-counts is supported by observations on the small-angle deviations in the track due to Coulomb scattering. These deviations are most frequent near A, and the scattering is less marked at points remote from it.

From these observations, it is reasonable to conclude that the particle k approached the point A;

Berkeley[6] suggest that the true value is $1\cdot 33 \pm 0\cdot 02$, a result which throws serious doubt on the reliability of the method based on grain-counts. Because of the advantage of this method, and of the important conclusions which have been based on it, experiments were made to determine the conditions in which reliable results can be obtained.

In the first experiments[5], the two most serious experimental difficulties arose from the fading of the latent image and from the variation of the degree of development with depth. This made it necessary to

---

* Continued from page 51.

* For the following reasons, the limits of error quoted above, in the determination of m_π/m_μ by observations on scattering, are less than those given in ref. 4. Previously, values for the mass of the different types of mesons, classified phenomenologically, were given separately. It is now known, however, that at least the majority of the σ-mesons are π^--particles; and the ρ-mesons, μ^+- and μ^--particles. The different results can therefore be combined to give a value for m_π/m_μ with a greater statistical weight.

LETTERS TO THE EDITOR

An Unusual Example of V^0 Decay*

R. W. THOMPSON, A. V. BUSKIRK, L. R. ETTER,
C. J. KARZMARK, AND R. H. REDIKER
Department of Physics, Indiana University, Bloomington, Indiana
(Received April 6, 1953)

THE object of this note is to report in detail a rather unusual V^0 decay recently observed in the new magnetic chamber.[1] Event R-118 is shown in Fig. 1. The V^0 particle, occurring with a penetrating shower, decays after traversing about one-eighth of the illuminated height of the chamber. The decay takes place

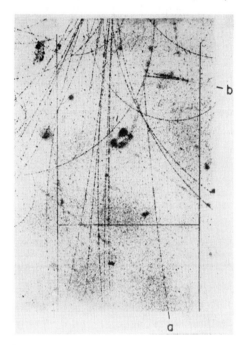

FIG. 1. Event R-118. The upper ⅔ of the chamber is shown. The obscuration of track b, by an unrelated old track, is less in the other two stereoscopic views.

very near the center of the illuminated depth; the positive fragment (track a) is projected almost vertically down, and the negative fragment (track b) almost horizontally to the right.

The track of the positive fragment is 43 cm long and is well illuminated throughout its entire length. The comparator plot for the left eye is shown in Fig. 2. The corrected momentum, derived from the three eyes, is 0.67 ± 0.02 Bev/c, and the ionization is indistinguishable from that of electronic tracks nearby. A proton of this momentum would be heavily ionizing by a factor 2.3, which it is believed would not easily escape detection under these conditions. Thus it is probable that the positive fragment is lighter than a proton, which means that the disintegration is not of the type $V_1{}^0 \rightarrow p + \pi$.

The track of the negative fragment, although relatively short, is highly curved and distinctly heavily ionizing. The track makes an angle of 26° with the plane of the chamber and crosses the front boundary of the illuminated depth near the right side of the chamber, as evidenced by the fading of track b in Fig. 1. The comparator plot for the left eye is shown in Fig. 2. The momentum is 0.094 ± 0.008 Bev/c, and the track is heavily ionizing by an estimated factor of 2 or 3. A pion of this momentum would be heavily ionizing by a factor 2.5; and a muon by a factor 1.8; thus it is likely that the negative fragment is a pion or nuon. A τ-meson of this momentum would be heavily ionizing by a factor in the neighborhood of 15 and would not easily escape detection. We thus consider it unlikely that this event represents a decay of the type suggested by the California Institute of Technology group,[2] namely $V_3^0 \rightarrow (\tau^-$ or $\kappa^-) + \pi^+ + 60$ Mev.

The corrected angle between the tracks is $79.2° \pm 0.4°$, the relatively large error arising primarily from the shortness of the negative track. Although the orientation of the event with respect to the associated shower suggests that it represents the decay of a neutral V particle (the angle of decay being so large), the possibility that the event is grossly misinterpreted must not be overlooked. The possibility that the event represents the decay of a charged particle (track b) which enters the chamber from the right can be excluded on energetic grounds since the total energy indicated by track b is considerably less than the kinetic energy of track a. The possibility that the event represents the decay of a charged particle (track a) which enters the chamber from below is a possible but relatively improbable interpretation.

The $Q(\pi, \pi)$ value is 215 ± 7. The next most accurate cases[1] R-32 and R-39 give 212 ± 10 and 216 ± 7, and a recent unpublished event (R-151) gives 219 ± 15. Confirmation of individual $Q(\pi, \pi)$ values in the neighborhood of 214 Mev has not been forthcoming

In view of the limited statistics (7 cases of V_4^0), we cannot, of course, exclude a splitting of the V_4^0 structure in the Q-curve plot,[1] or a three-body decay. The latter might account for part of the differences with the results of other investigators, and for events[1] 328 and R65B.

* Assisted by the U. S. Office of Ordnance Research and by grant of the Frederick Gardner Cottrell Fund of the Research Corporation.
[1] Proceedings of the Third Rochester Conference. See Thompson, Buskirk, Etter, Karzmark, and Rediker, Phys. Rev. 90, 329 (1953).
[2] Leighton, Wanlass, and Anderson, Phys. Rev. 89, 148 (1953).
[3] We are indebted to Dr. Butler for recently informing us that the Manchester data has been remeasured to give $Q(\pi, \pi)$ values in the neighborhood of 170 Mev instead of 122 Mev as previously reported.
[4] We are indebted to Dr. Bridge for sending us a preprint of a forthcoming paper of the Massachusetts Institute of Technology group.
[5] With the new magnetic chamber, we find roughly equal numbers of V_1^0 and V_4^0 (see below).
[6] The Manchester group and the Massachusetts Institute of Technology group use the symbol V_2^0 to indicate the class of all neutral V particles other than $V_1^0 \rightarrow p + \pi$. In that usage, the V_2^0 symbol seems inappropriate here, in view of the large variety of decay schemes in that class which have been reported (see reference 2), but which we have not as yet observed.
[7] In case one or both fragments are muons, slight modification of the Q value would be necessary.
[8] The kinetic energy per pion in the c.m. system would be 107 Mev, very near $E_\pi = 115 \pm 10$ Mev for the χ-decay. See the excellent review paper of D. H. Perkins in the *Proceedings of the Third Annual Rochester Conference on High Energy Physics* (Interscience Publishing Company, New York, 1953).

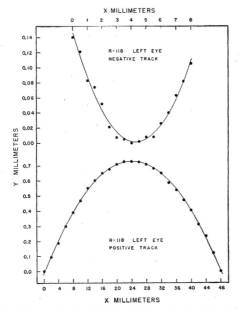

FIG. 2. Comparator plots of R-118 for the left eye. The plots for the other two stereoscopic views are of the same quality.

from other groups, and there is certainly no indication therefrom for the existence of either a line or a pronounced maximum in that neighborhood. We are inclined to consider that the values in the neighborhood of 214 determined as described above are incompatible with a value of 170 Mev,[3] limits of 115–185 Mev,[4] or a range of 100–130 Mev,[2] so that the present observations appear to be distinct from those of other investigators.

It may be that the experimental arrangement has biased[5] our observations in favor of a different type[6] of neutral V particle, $V_4^0 \rightarrow (\pi^+$ or $\mu^+) + (\pi^-$ or $\mu^-) + Q$, where $Q(\pi, \pi) = 214 \pm 5$ Mev.[7,8]

PHYSICAL REVIEW VOLUME 93, NUMBER 4 FEBRUARY 15, 1954

Production of Heavy Unstable Particles by Negative Pions*

W. B. FOWLER, R. P. SHUTT, A. M. THORNDIKE, AND W. L. WHITTEMORE
Brookhaven National Laboratory, Upton, New York
(Received November 10, 1953)

> In addition to two previously discussed cloud-chamber examples of V-particle production by 1.5-Bev π^- mesons from the Cosmotron, four further examples are discussed here. In two of the new examples a $\Lambda^0(V_1^0)$ and a $\vartheta^0(V_4^0)$ are seen to decay in a geometry indicating that they were produced together in a $\pi^- - p$ collision. A third example is best interpreted as production of a $\Lambda^-(V_1^-)$ together with a $K^+(V_2^+)$ by a π^- colliding with a proton. A fourth example shows a probable Λ^- decaying into a π^- and a neutron with a Q value of about 130 Mev. A cross section of ~ 1 millibarn for V-particle production is inferred from the number of $\pi^- - p$ collisions observed.

IN a previous letter[1] we have reported two examples of Λ^0 particles[2] produced in hydrogen by negative pions (π^-) of 1.5-Bev kinetic energy. For both examples it was shown that if there were only two resulting particles the second was a K^0 particle of a mass of about 650 Mev. Since, however, the K^0 was not seen to decay it was also possible to balance energies and momenta by assuming two lighter neutral particles (including π^0) instead of one K^0 in addition to the Λ^0.

Observations on heavy unstable particles in cosmic radiation indicate lifetimes as long as 10^{-10} to 10^{-9} second with production cross sections at least 10^{-2} times those for π mesons. These facts can be reconciled theoretically[3] if it is assumed that the particles must be produced doubly (two at a time). Different particles may be produced simultaneously, in addition to pairs of identical ones. It appears that no evidence for double production of such particles, exceeding purely statistical coincidence, has been reported in cosmic radiation although a few examples might perhaps be interpreted in this manner.[4] The two examples of production in hydrogen reported by this group indicate multiple production. This interpretation is uncertain, however, since the K^0 particles are not observed to decay and the computed masses are higher than the value of about 500 Mev most commonly found for K particles.

We now wish to describe three additional cases where double production of heavy unstable particles by 1.5-Bev π^- is indicated with more certainty then in the previous two examples. We shall refer to the previous examples as cases A and B. Case C shows a Λ^0 together with a ϑ^0 probably produced in a heavy nucleus. Case D shows the same combination produced in hydrogen. Case E shows what may well be a Λ^- together with a K^+ with mass of about 500 Mev produced in hydrogen. A few cases of Λ^+ have been observed in cosmic radiation[5,6] but evidence for Λ^- is less certain. In a fourth case, called F, we shall describe an additional example best interpreted as a Λ^-.

Details of the experimental method will be described in a later article. A diffusion cloud chamber filled with 20 atmospheres of hydrogen was exposed to a 1.5-Bev π^- beam produced in a carbon target by the 2.2-Bev circulating proton beam in the Cosmotron. The pions were selected and collimated by the field of the Cosmotron magnet and a channel in the concrete shielding around the machine. The beam thus obtained is quite monoenergetic with a spread probably less than ± 0.1 Bev. The beam was finally deflected by a magnet into a concrete house containing the cloud chamber mounted between the pole faces of a magnet providing an average field of 10 500 gauss. This magnetic field has been calibrated to 1 percent accuracy and its vertical and horizontal components have been mapped throughout the chamber.

The examples to be described were obtained out of a total of 26 000 photographs obtained at a rate of one every 7 to 8 seconds. In addition many examples of V events were found for which no associated V event or nuclear interaction in the gas was observed. Finally, we have observed about 170 pion interactions in hydrogen, most of which lead to single and multiple production of pions. These will be described in a later article.

CASE C: Λ^0 WITH ϑ^0

Figure 1 (case C) shows a photograph of two V events, of which (a) is considered to be a Λ^0, and (b) a ϑ^0. Data which lead to this identification are given in Table I. Row 2 gives the momenta as measured on the 35-mm film by means of a Cooke microscope. The measure-

---

* Work performed under the auspices of the U. S. Atomic Energy Commission.
[1] Fowler, Shutt, Thorndike, and Whittemore, Phys. Rev. **91**, 1287 (1953).
[2] We are using here the nomenclature suggested for V events at the International Congress on Cosmic Radiation, Bagnères-de-Bigorre, France. Accordingly $\Lambda^{0,+,-} \rightarrow \text{nucleon} + \text{pion} + Q_\Lambda$; $K^{0,+,-}$ is any particle whose mass falls between those of pion and proton; for example, $\vartheta^0(\rightarrow \pi^+ + \pi^- + Q_\vartheta)$ is a K^0.
[3] A. Pais, Phys. Rev. **86**, 663 (1952). References to previous work are cited in this article. A. Pais, Proceedings of the Lorentz-Kamerlingh Onnes Conference, Physica (to be published); M. Goldhaber, Phys. Rev. **92**, 1279 (1953); M. Gell-Mann, Phys. Rev. **92**, 833 (1953).
[4] Leighton, Wanlass, and Anderson, Phys. Rev. **89**, 148 (1953); Fretter, May, and Nakada, Phys. Rev. **89**, 168 (1953); G. D. Rochester and C. C. Butler, Repts. Progr. in Phys. **16**, 364 (1953), give a comprehensive review of cosmic-ray results.

[5] York, Leighton, and Bjornerud, Phys. Rev. **90**, 167 (1953).
[6] International Congress on Cosmic Radiation, Bagnères-de-Bigorre, France (unpublished).

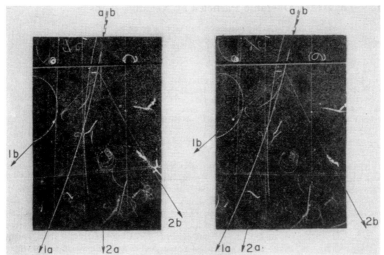

FIG. 1. Case C. Diffusion cloud-chamber photograph of two neutral V particles (a) and (b), whose lines of flight are almost colinear. (a) is believed to be a Λ^0 decaying into a proton (1a) and a negative π meson (2a). Tracks 1a and 2a practically coincide in the right view. (b) is probably a ϑ^0 decaying into π^+ (1b) and π^- (2b).

ments have been corrected for dip angle and space variations of magnetic field and magnification. The mass limits given in row 4 were deduced from measured momenta and estimated ionization densities. Row 5 shows what particle was assumed for further calculation. The angles between tracks in row 6 were obtained from a 3-dimensional reprojecting system as well as by geometrical reconstruction and calculation. The Q value for (a) is, within the given error, consistent with the usually accepted value of 37 Mev for a Λ^0. The Q value for (b) is rather large compared to the best value of 214 Mev given in the literature[7] for a ϑ^0.

From the measured momenta and angles in space for each track one can infer that the lines of flight of the Λ^0 and ϑ^0 practically coincide. From the total number of Λ^0 and ϑ^0 (originating in the walls) found in the 26 000 pictures one concludes that the probability for random association of the two particles in such a geometry is $\sim 10^{-9}$. We therefore should be justified in assuming that the Λ^0 and ϑ^0 were produced in one act. Since no incident particle is visible in the chamber and since the Λ^0 and ϑ^0 travel at an angle of about 24° upwards from the bottom of the chamber, we assume that they were produced in or below the bottom glass plate, though probably not much further down since the pion beam was fairly well collimated. The origin must, of course, lie on the line of intersection of the decay planes of the Λ^0 and ϑ^0. Both particles travel in directions between 0.5° and 1.5° to this line for a range of reasonable choices for the position of the origin.

The following procedure can now be used to determine the Q values more accurately than before. Since

[7] Thompson, Buskirk, Etter, Karzmark, and Rediker, Phys. Rev. **90**, 1122 (1953).

we know the directions of the lines of flight of Λ^0 and ϑ^0 with fair certainty we can deduce the momentum of one decay product in each V event from the momentum of the other. As row 2, Table I, shows, the momenta of 1a and 2b are known much better than those of 1b and 2a. We therefore have deduced the momenta of 1b and 2a. For the momenta of 1a and 2b we have chosen their lowest values within the given experimental errors. This choice will result in somewhat lower Q values than calculated above. The new momenta are given in row 8 and the recalculated Q values in row 9. The indicated remaining errors are now mainly due to the mentioned uncertainty in the directions of flight of Λ^0 and ϑ^0. The Q value for the ϑ^0 is still not in good agreement with the usual value of 214 Mev.

From the momenta in row 8 and the given angles one finds the momenta of Λ^0 and ϑ^0 given in row 10. Assuming that the particles were produced in a collision between a π^- and a nucleon and that no additional particle was involved $(\pi^- + p \rightarrow \Lambda^0 + \vartheta^0)$ one calculates from the resultant of the momenta of Λ^0 and ϑ^0 that the total energy of incident pion and nucleon must have been 2050 ± 25 Mev before the collision. On the other hand, with the masses of 1119 ± 10 and 546 ± 20 Mev for Λ^0 and ϑ^0, respectively, one finds for the sum of their energies a value of 2068 ± 30 Mev. This value must be equal to the initial energy just calculated, if no other particles are involved in the collision, and one indeed finds agreement. Therefore the Λ^0 and ϑ^0 could have been produced in a $\pi^- - p$ collision, the π^- previously having been scattered upward by 24°, with the required energy. Such repeated interactions could take place in a heavy nucleus. For this interpretation the almost colinear flight of the Λ^0 and ϑ^0 would mean that in the

TABLE I. Measurements and results for case C.

	Track	Event a 1a	Event a 2a	Event b 1b	Event b 2b
1	Sign of charge	+	−	+	−
2	Measured momenta (Mev/c)	272±8	205±40	451±70	391±15
3	Estimated ionization density	>5×min	<1.5×min	<1.5×min	<1.5×min
4	Mass limit (Mev)	>780	<170	<380	<330
5	Assumed particle	proton	pion	pion	pion
6	Angle between tracks (degrees)		31.1±1		70±1
7	Q values calculated from above data (Mev)		54±20		271±30
8	Momenta for (2a) and (1b) calculated from direction of line of flight, assuming lowest momenta for (1a) and (2b) (Mev/c)	264	189±20	441±15	376
9	Q values recalculated (Mev)		49±10		266±20
10	Momentum of unstable particle (Mev/c)		437		673

center-of-mass system (c.m.s.) of π^- and nucleon the Λ^0 went almost straight backward. The same will be found for the next example to be discussed. From the observed momenta and angles a π^- of the beam colliding with a nucleon could *not* have produced the Λ^0 and ϑ^0 traveling in the observed direction, with simultaneous production of some other particle going downward to balance transverse momenta.

CASE D: Λ^0 AND ϑ^0 PRODUCED IN π^-−p COLLISION

Figure 2 (case D) shows a photograph of a π^- track (marked π^-) disappearing abruptly in the hydrogen, and two nearby V events of which (a) is considered to be a Λ^0 and (b) a ϑ^0. Data which lead to this identification are given in Table II. The measured momentum of the incident particle (row 2) agrees well with the momentum of 1630 Mev/c for a 1.5-Bev π^-. The other momenta are not well determined, mostly because the tracks are of short length. The longest track (2b), furthermore, seems to show a slight deflection, perhaps due to scattering or to $\pi-\mu$ decay, which makes its momentum also unreliable. The given mass limits (row 4), however, seem to justify the assumed choices for the individual particles (row 5). In particular, since in a collision in hydrogen only one proton was present and since this proton is probably found in 1a, 2b could hardly be another proton in spite of the rather high upper mass limit.

The decay planes of both particles contain the end point of the incident track. The directions of flight of Λ^0 and ϑ^0 and incident π^- appear coplanar, though this is not accurately determined because of the small angle between π^- and ϑ^0 (b). Therefore no additional neutral particle has to be assumed, although calculation shows that kinematically this would be possible. One is therefore justified in assuming the same reaction as in case C ($\pi^-+p \rightarrow \Lambda^0+\vartheta^0$). All pertinent angles are given in rows 6 and 7. From these angles and the momentum of the incident π^- one can calculate the momenta of the decay products directly. The results (row 9) agree with the measured values (row 2) within the errors. With the momenta from row 9, Q values have been calculated (row 10). Two experimental errors are given separately with each Q value. The first was determined from the probable variation of Q with the possible variations of all of the angles (rows 6 and 7) used for the computation. The second error was found from the variation of Q with a possible uncertainty of ±100 Mev/c for the momentum of the incident π^-. One sees that within the errors both Q's may agree with the usual values of 37 and 214 Mev, respectively, although the Q for the ϑ^0 is again rather high and agrees with the value found for case C. It should be pointed out that the error for the Q for the Λ^0 depends most strongly on the measurement

FIG. 2. Case D. Photograph of a 1.5-Bev π^- producing two neutral V particles in a collision with a proton. Tracks 1a and 2a, believed to be proton and π^-, respectively, are the decay products of a Λ^0. A ϑ^0 is probably seen to decay into π^+ (1b) and π^- (2b). Because of the rather "foggy" quality of this picture tracks 1b, 2a, and 2b have been retouched for better reproduction.

TABLE II. Data for case D.

		Incident track	Event a 1a	2a	Event b 1b	2b
1	Sign of charge	−	+	−	+	−
2	Measured momenta (Mev/c)	1620±160	210^{+210}_{-70}	140^{+300}_{-60}	210^{+210}_{-70}	840±300
3	Estimated ionization density	<1.5×min	>5×min	<1.5×min	<1.5×min	<1.5×min
4	Mass limit (Mev)		>400	<370	<350	<870
5	Assumed particle	pion	proton	pion	pion	pion
6	Angle between incident π^- and direction of flight (degrees)		11.1±1		2.3±0.5	
7	Angle between direction of flight and decay products (degrees)		8.8±1	11.3±1.5	29.3±2	11.0±2
8	Momentum of (a) and (b) calculated from incident momentum and angles (Mev/c)	1630	282		1357	
9	Momenta of decay products from momenta of (a) and (b) and angles (Mev/c)		160	125	400	1027
10	Q values calculated from rows 7 and 9		27±11±3		258±35±21	

of the small angle (2.3°) between incident π^- and ϑ^0 (b). Modifying this angle to make Q_{Λ^0} equal to 37 Mev would reduce Q_{ϑ^0}, which depends on this angle less strongly, to a value of 244 Mev.

From the momenta (row 8) of the Λ^0 and ϑ^0 and their masses of 1097±12 and 538±40 Mev, respectively, one calculates that the kinetic energy of the incident π^- must have been 1.52±0.04 Bev which is quite consistent with the 1.5-Bev beam energy.

CASE E: POSSIBLE Λ^- AND K^+ PRODUCED IN $\pi^- - p$ COLLISION

Figure 3 (case E) shows a photograph of what on first sight appears to be a π^- scattered forward by a proton with a subsequent decay of the π^-. Closer inspection shows that the momentum of the decay product

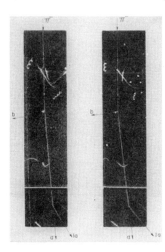

FIG. 3. Case E. Photograph of a $\pi^- - p$ collision event possibly resulting in a Λ^- (a) with a π^- (1a) as a decay product and a K^+ whose decay is not seen.

(1a) and its angle with respect to the track of the decaying particle (a) are much too large for a $\pi - \mu$ or $\mu - \beta$ decay. Therefore (a) must have been a heavy unstable particle.

As far as can be ascertained from the "scattering" event, particle (a) might have been produced at the Cosmotron target or at the cloud chamber wall, and scattered by a proton in the cloud chamber. Certainly the rate of production of heavy mesons would have to be large (~10 percent of that for pions) and their decay lifetime long (~10^{-8} sec) for the beam to contain an appreciable contamination of heavy mesons. No beam particles have shown a decay resembling that of (a). If (a) were produced in the target, it would be quite remarkable that it lives until it reaches the cloud chamber, is scattered, and then decays within the chamber. Such an origin is possible but seems unlikely. If (a) were a particle produced in the wall and scattered in the chamber, it would be remarkable that the incident track has both direction and momentum characteristic of beam tracks. For these reasons we assume that the incident particle is a beam π^- producing a charged unstable particle in a collision with a proton.

Data are given in Table III. Since the tracks are short and the momenta are high, the latter are not well determined. We therefore assume that the incident particle had the beam energy of 1.5 Bev, and compute the momenta of (a) and (b) by assuming that no additional neutral particle was produced. This assumption is very probably justified because (a), (b), and incident track are coplanar. One finds the momentum values given in row 8. If the mass of (a) is given, that of (b) is determined so that their total energies equal that of the incident π^-. In Table IV a number of such consistent values are given. The errors on the mass values of (b) include the uncertainties given for the angle measurements as well as an improbably large uncertainty of ±300 Mev for the energy of the incident π^-. To obtain a mass of 930 Mev (proton) for (b) does not seem to be possible unless (a) is a pion, without making

TABLE III. Data for case E.

		Incident track	Track a	Track 1a	Track b
1	Sign of charge	−	−	−	+
2	Measured momenta (Mev/c)	1800±600	1400±400	>120[a]	−
3	Estimated ionization density	<1.5×min	<1.5×min	<1.5×min	3 to 6×min
4	Mass limit (Mev)		<1500		
5	Assumed particle	pion		pion	
6	Angle between incident track and collision products (degrees)		8±1		79±3
7	Angle between (a) and its decay product (degrees)			36±1	
8	Momenta from incident momentum and angles (Mev/c)	1630	1604		227

[a] Track 1a is slightly distorted. Therefore only a lower limit can be given.

quite unreasonable assumptions for the uncertainties of the measured angles and of the incident momentum.[8] In the cosmic radiation Λ^+ particles have been found[6] with Q values of about 130 Mev, leading to a mass of 1200 Mev. Row 5 of Table IV shows that such a mass for (a) would lead to a mass for (b). which is quite consistent with that usually found for K particles. Case E can therefore be interpreted as the charged counterpart of case C and case D in which Λ^0 and K^0 were produced. For this interpretation the present photograph is an example of the reaction $\pi^- + p \rightarrow \Lambda^- + K^+$.

A calculation has been performed to investigate whether a neutrino (ν) or π^0 could have been produced in addition, which might then change the above conclusions. One finds that only the combinations $[\Lambda^-, K^+, \pi^0]$ or $[\Lambda^-, K^+, \nu]$ are possible (though unlikely because of the observed coplanarity) while the combinations $[\Lambda^+, K^-, \pi^0]$, $[\Lambda^+, K^-, \nu]$, $[p, K^-, \pi^0]$, or $[p, K^-, \nu]$ are kinematically not possible.

From the mass of (b) of 520 Mev and its momentum of 227 Mev/c one finds that (b) should show an ionization density of ~4×min which agrees with the estimated ionization density given in row 3.

Assuming a decay of $\Lambda^- \rightarrow n + \pi^-$, the calculated momentum for 1a of 1604 Mev/c and measured momentum for (b) of ≥120 Mev/c lead to a Q value ≥50 Mev. Track 1a may be somewhat distorted so that these figures probably represent lower limits only. If much of the apparent curvature of 1a is due to distortion, the Q value may be as high as 130 Mev, corresponding to a momentum of 440 Mev/c for 1a, as has been assumed in the previous discussion.

[8] For the present discussion it is of particular importance to exclude the possibility that the mass of (b), M_b, could be that of a proton (930 Mev) for the assumption that (a) is a K^- of mass 500 Mev. One finds that $\partial M_b/\partial p_0 = 8.1$ Mev/(100 Mev/c), where p_0 is the momentum of the incident π^- (so far assumed to be the beam momentum of 1630 Mev/c). Furthermore $\partial M_b/\partial \alpha_a = -11.3$ Mev/degree and $\partial M_b/\partial \alpha_b = -3.7$ Mev/degree, where α_a and α_b are the angles recorded in Table III, row 6, for (a) and (b), respectively. It is very hard to conceive how π^- of momenta >1700 Bev/c could be contained in the beam. With this upper limit and the uncertainties for α_a and α_b given in Table IV one sees that $M_b \leq 890$ Mev if all uncertainties are assumed to act together in a direction to increase M_b. Only by decreasing α_a by 3° and α_b by 9°, for example, could one obtain a value for M_b of 930 Mev. The values for M_a and M_b noted in rows 2 and 3 of Table IV are then not possible without destroying a nucleon.

CASE F: EXAMPLE OF A Λ^-

The photograph shown in Fig. 4 (case F) shows another V event which may be interpreted as a Λ^-. The picture shows a negative particle (a) of momentum 1190±170 Mev/c and of estimated ionization density ≤1.5×min, apparently produced in the wall of the chamber. The angle between (a) and the beam direction is 8°. Particle (a) decays into a negative particle (1a) of momentum 83±3 Mev/c and estimated ionization density 2 to 3×minimum.[9] The mass of the decay product thus lies between 110 and 150 Mev, identifying it as a π^-. The angle between (a) and 1a is 76°. If one additional neutral decay product is assumed one calculates Q values and mass values for (a) as given in Table V. One sees that only the assumption of a neutron (row 1) leads to Q and mass values compatible with those found in cosmic radiation. Particle (a) can also not be identified as a $\tau^- \rightarrow 2\pi^0 + \pi^- + 70$ Mev because 1a alone would have an energy of 230 Mev in the rest system of the τ^-. The assumption in row 4 ($K^- \rightarrow \mu^- + 2\nu$) is unlikely because the decay product is most probably

TABLE IV. Consistent masses for particle (a) and particle (b) of case E.

	Particle (a)	Mass of (a) (Mev)	Particle (b)	Mass of (b) (Mev)
1	π^-	140	proton	934±6
2	K^-	500	?	860±20
3	?	760	?	760±30
4	Λ^- ($Q=37$ Mev)	1107	K^+	570±50
5	Λ^- ($Q=130$ Mev)	1200	K^+	520±50

TABLE V. Q values and masses for case F, assuming different masses for the neutral decay product.

	Charged decay product	Neutral decay product	Mass of neutral decay prod. (Mev)	Q for (a) (Mev)	Mass of (a) (Mev)
1	π^-	n	930	130^{+25}_{-15}	1200
2	π^-	π^0	140	430 ±70	710
3	π^-	K^0	500	230±35	870
4	μ^-	2ν	0	>520±80	>620

[9] For the assumption of a Λ^0 traveling backwards the momentum of (a) is much too high.

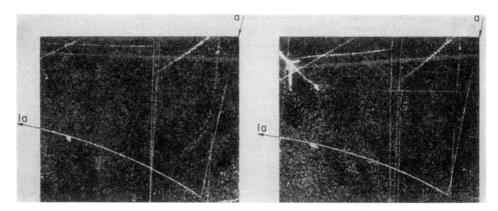

FIG. 4. Case F. Photograph of a negative unstable particle (a) best interpreted as a Λ^-. The decay product (1a) is identified as a π^- from momentum and ionization density.

identified as a π^-. We therefore are left with the conclusion that the present photograph shows the decay of a $\Lambda^- \rightarrow n + \pi^- + Q$, where $Q = 130_{-15}^{+25}$ Mev. The errors on the latter are due to the uncertainty of the measured momentum of (a).

It has been pointed out[10] that if the hypothesis of charge independence applies to $\Lambda^{+,-,0}$, then the Λ^- must have isotopic spin z component $(-3/2)$ with isotopic spin $3/2$, in which case a doubly-charged Λ^{++} should also exist. No Λ^{++} decays have been reported to date, but the number of observed Λ^- and Λ^+ is so small that no significant discrepancy exists. A track which might be interpreted as a Λ^{++} has been reported by Ascoli.[11]

LIFETIMES, CROSS SECTION, AND ANGULAR RELATIONSHIPS

The role played by chance in apparent lifetimes, cross sections, and angular relationships deduced from a small number of cases is very large. Nevertheless the few available observations will be compiled in the following paragraphs.

Table VI shows the lifetimes. All values are consistent with mean lifetimes of 10^{-10} to 3×10^{-10} sec cited in the literature.[4-6]

TABLE VI. Observed lifetimes of all particles in units of 10^{-10} sec.

Particle	A	B	C	D	E	F
Λ^0	0.4	0.3	6	9		
Λ^-					2	3
ϑ^0			2	0.1		
K^0	>4[a]	>3[a]				
K^+					>0.7	

[a] Not taking into account that decay may result in two neutral particles and thus be invisible.

[10] D. C. Peaslee, Phys. Rev. **86**, 127 (1952).
[11] G. Ascoli, Phys. Rev. **90**, 1079 (1953).

For a mean lifetime of 3×10^{-10} sec for the Λ^0 and 1.5×10^{-10} for the ϑ^0, the given cloud-chamber geometry, a π^--beam energy of 1.5 Bev, and isotropic angular distributions (in the c.m.s.) of Λ and K particles one can estimate that 60 percent of the Λ and 50 percent of of the K should be seen to decay inside the chamber. One can conclude that for 80 percent of all occurring cases one should see at least one of the two particles decay and for 20 percent of all cases one should see both decaying. (For shorter lifetimes or angular distributions peaked forwards or backwards the probabilities are even larger.) Therefore the 4 cases of unstable particle production in hydrogen (A, B, D, E) observed here may correspond to 5 ± 3 cases that actually happened.[12] This number can be compared with the other 170 $\pi^- - p$ interactions observed, including events with two and four outgoing prongs. Making use of Fermi's statistical theory of meson production[13] and of the isotopic spin formalism, Fermi[14] has calculated the probabilities for the different combinations. One finds that combinations resulting in no outgoing prongs are expected to occur in only 12 percent of all interactions. Therefore our 170 observed cases may correspond to 190 actual interactions. The total interaction cross section has been found to be 34 ± 3 millibarns.[15] Therefore the cross section for heavy unstable particle production by 1.5-Bev π^- in hydrogen is ~ 1 millibarn.

Table VII gives the angles between the decay planes of the particles and their production planes (the plane formed by the incident track and the line of flight of the unstable particle.) The absence of large angles for the Λ particles is surprising and might be taken as an indication for a large spin for these particles. In fact the

[12] Decays of Λ^0 and ϑ^0 into neutral particles cannot be observed in this experiment and are not taken into account here.
[13] E. Fermi, Progr. Theoret. Phys. (Japan) **5**, 570 (1950).
[14] E. Fermi (private communication).
[15] Cool, Madansky, and Piccioni, Phys. Rev. **93**, 637 (1954).

separation between the decay and production planes seems smaller than would be expected unless the spin is taken to be unreasonably large.

Table VIII shows the angles between incident π^- and direction of emission of Λ^0 in the c.m.s. Again the result is surprising because of the absence of angles near 90° for which the solid angle is largest. The solid angle between 170° and 180° amounts to only 1.5 percent of the hemisphere, yet both C and D fall into this region. This may indicate that large angular momentum states are involved in the production of these particles, which might be consistent with the possibility

TABLE VII. Angles between decay plane and production plane.

Particle	A	B	C	D	E
Λ^0	5°±5°	30°±20°	18°±7°	27°±10°	
Λ^-					7°±5°
ϑ^0			60°±6°	70°±5°	

of a large spin of the Λ^0. Since $\lambda_0 \approx 2 \times 10^{-14}$ cm in the c.m.s. for a π^- of 1.5-Bev laboratory energy, angular momentum states up to $L=6$ may be possible.

SUMMARY

These examples of the production of heavy unstable particles in $\pi^- - p$ collisions have been shown to be consistent with a double production process,

$$\pi^- + p \rightarrow \Lambda + K,$$

occurring with a cross section of about 1 millibarn for

TABLE VIII. Angle between incident π^- and direction of emission of $\Lambda^{0,-}$ in c.m.s.

Particle	A	B	C	D	E
Λ^0	141°	125°	177°	174°	
Λ^-					30°

1.5-Bev π^-. Further work is required to determine whether production is *always* double in these and nucleon-nucleon collisions.

In four cases the Λ is a Λ^0, and in two of these the K^0 is observed to decay and can be considered to be a ϑ^0, although the Q values of about 260 Mev are not quite consistent with the cosmic-ray value of 214 Mev. One case is interpreted as a Λ^-, and an additional charged decay originating outside the chamber is thought to be a Λ^- with Q value of 130 Mev.

The data suggest an angular correlation between Λ-decay planes and production planes, which may mean that Λ particles have large spin, and a preferred backward (or forward) emission in the c.m.s., which may mean that Λ particles are produced in states of high angular momentum. Many more data are needed to determine such angular relationships.

We wish to express our thanks to the many members of the Cosmotron Department whose efforts have created the opportunity for this work and whose cooperation is enabling us to pursue it. Our thanks are also due to the other members of the cloud-chamber group without whose support this work could not be continued.

Bevatron K-Mesons*

ROBERT W. BIRGE, ROY P. HADDOCK, LEROY T. KERTH,
JAMES R. PETERSON, JACK SANDWEISS,† DONALD H. STORK,
AND MARIAN N. WHITEHEAD

Radiation Laboratory, University of California, Berkeley, California
(Received May 9, 1955)

To facilitate the search for K-mesons from the Bevatron, two of us (L.T.K. and D.H.S.) have suggested the use of a strong-focusing spectrometer (Fig. 1),[1] consisting of a magnetic quadrupole focusing lens[2] followed by an analyzing magnet. Particles of any desired momentum can be brought to a focus, forming an image of the target at a point behind the analyzing magnet. Emulsion stacks are placed at this point. With this arrangement we have found examples of four types of heavy mesons first established in cosmic-ray work.[3] Particles of different mass can be separated according to their ranges in emulsion. For particles of momentum 360 Mev/c, the range of K's is 4.6 times the range of the protons, and pions pass through the emulsion stack at minimum ionization.

A stack of 107 Ilford G.5600-μ pellicles,[4] 3.5 in. by 3.5 in., has been exposed so that 114-Mev K-particles stopped in the center of the stack. The proper time of flight for such particles from the target to the emulsion is about 10^{-8} sec.

This stack has been scanned in a swath across the direction of the meson flux for tracks lying in the plane of the emulsion whose ionization is visibly greater than minimum. Particles stopping in the stack (beyond the position of the swath) have masses less than 1200 m_e. Particles that go all the way through the stack have masses less than 800 m_e.

The tracks are followed until they stop and the endings are examined for decays. To date 300 decays have been observed. Twenty of these are π^+ mesons whose unique decay into three charged pions is readily identifiable. Among the others, all of which decay into one lightly ionizing secondary, only those with a secondary that is flat or with an ionization obviously higher than minimum have been categorized. Three examples have been found of what is assumed to be the alternate decay of the τ into one charged and two neutral pions, with the pion stopping in the emulsion stack. Two events decay into low-energy muons (less than 55 Mev) and are presumably examples of the κ or $K_{\mu 3}$.

To establish the existence of the $K_{\pi 2}(\theta^+, \chi^+)$ or of the $K_{\mu 2}$ mesons, either very large emulsions are needed to stop the long-range secondary or else very accurate measurements of the multiple scattering and the ionization must be made. Measurements on four fortuitously flat secondaries at a distance of 5 cm from the decay point revealed that three of the primaries were $K_{\pi 2}$'s and one presumably a $K_{\mu 2}$, as determined by the tentative identification of the secondary as a high-energy muon. Excellent calibration on grain count is available from the π mesons of known energy traversing the same region of the emulsions. From the number of K-mesons found here compared to the number found at about 25 cm from the target,[5] it is unlikely that the mean life of any of the K's seen is less than 3×10^{-9} sec.

In the initial exposure, the momentum resolution as determined from the proton ranges allows a mass determination to $\pm 40\ m_e$ on each K-meson. With a few exceptions, all particles with lightly ionizing secondaries fall within a distribution of this width centered about 20 m_e below the average for τ mesons plotted separately (Fig. 2). In a subsequent exposure the momentum resolution has been improved. The scattered points on the high-mass side of the main distribution may be due to particles that suffered inelastic collisions, or scattered off the channel. A

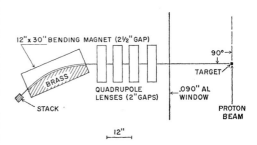

FIG. 1. Strong-focusing spectrometer.

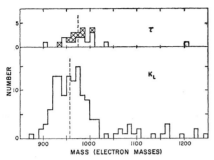

Fig. 2. Mass distributions: Upper histogram represents τ mesons. Crossed squares refer to τ's found nonsystematically. Lower histogram is made up of particles decaying into a single lightly ionizing secondary.

comparison of the measured τ meson mass of (974 ± 6) m_e with the accepted value[3] of 966 m_e indicates a possible systematic error.

This work was done with the encouragement and guidance of Professor Chaim Richman. Most of the scanning was performed by Mrs. Beverly Baldridge, Miss Irene d'Arche, Mrs. Marilynn Harbert, Mrs. Edith Goodwin, and Miss Kathryn Palmer.

It is with pleasure that we acknowledge the help and advice in nuclear emulsion techniques given us by Professor Powell's group at Bristol. The assistance of the Bevatron crew, under the direction of Dr. Edward J. Lofgren, and their skillful operation of the machine are greatly appreciated.

* This work was performed under the auspices of the U. S. Atomic Energy Commission.
† National Science Foundation Predoctoral Fellow.
[1] Kerth, Stork, Birge, Haddock, and Whitehead, Bull. Am. Phys. Soc., Vol. 30, No. 3, 41 (1955).
[2] Courant, Livingston, and Snyder, Phys. Rev. 88, 1190 (1952).
[3] Padua Conference, Nuovo cimento 11, Suppl. No. 2 (1954).
[4] Birge, Kerth, Richman, Stork, and Whetstone, University of California Radiation Laboratory Report No. UCRL-2690, September, 1954 (unpublished).
[5] Proceedings of the 5th Rochester Conference (to be published).

4

Antibaryons

The discovery of the antiproton and other antimatter, 1955–1959.

While the existence of antiparticles was established with Anderson's discovery of the positron in 1932, it was not clear in 1955 whether the pattern of each fermion having an antiparticle, suggested by the Dirac equation, would hold for baryons, the heavy particles $p, n, \Lambda, \Sigma,$ and Ξ. There were two arguments raising doubts about such particles. One was that nucleons had an anomalous magnetic moment that differed markedly from the Dirac moment. Measurements by Otto Stern in 1933, later improved by I. I. Rabi, had shown that the proton had a magnetic moment of 2.79 nuclear magnetons. [One nuclear magneton is $e\hbar/(2M_p c)$, where M_p is the nucleon mass.] The neutron's magnetic moment, which would be zero if the neutron were an ordinary Dirac particle, was measured by L. Alvarez and F. Bloch in 1940 to have a value of -1.91 nuclear magnetons. The second reason was based on a cosmological argument. Where were the antigalaxies one expected if the Universe had baryon–antibaryon symmetry?

One of the motivations for the choice of the energy for the Bevatron was the hope that the antiproton could be found. The momentum chosen, 6.5 GeV/c, was above threshold for antiproton production on free protons, $p + p \to p + p + p + \bar{p}$, to occur. In 1955, one year after the Bevatron became operational, there were a number of different plans to look for the antiproton, including two within the Segrè group at Berkeley, an experiment using electronic counters and a photographic emulsion experiment.

The detection of the antiproton was first achieved in 1955 by O. Chamberlain, E. Segrè, C. Wiegand, and T. Ypsilantis (**Ref. 4.1**). The primary obstacle to overcome was the background from the much more copiously produced π^- whose charge was the same as that of the antiproton. To separate the antiprotons, Chamberlain *et al.* measured both the momentum and velocity of the negative particles.

The beam from the Bevatron impinged on a copper target. Negative particles produced with a momentum near 1.19 GeV/c were focused by a quadrupole magnet on a first set of scintillators, which emitted light when charged particles passed through them. A second quadrupole focused the beam on a second set of scintillators 40 feet farther down the line. An antiproton with momentum 1.19 GeV/c and a velocity $v = 0.78\,c$ required 51 ns for the flight, while a π^- of this momentum needed only 40 ns.

4. Antibaryons

Additional verification was provided by using Cherenkov counters. Cherenkov counters detect the light emitted by charged particles passing through a medium when the velocity of the particle is greater than the velocity of light in the medium. Since that velocity is the usual velocity of light divided by the index of refraction, it is possible to fill the detector with a gas, possibly under pressure, so that the detector will respond only to particles with velocities exceeding some minimum value. To demonstrate the presence of antiprotons, one Cherenkov counter was set to count pions and was used in anticoincidence for the protonic-mass particles, that is, if the particle was determined to be a pion, it was rejected. A second one was a specially designed differential counter that only responded to particles in a narrow velocity band corresponding to the protonic mass. This counter was used in coincidence for the acceptance of the $\bar{p}$ candidates.

Some 60 antiproton candidates had been observed by October 1955. Calibrating the apparatus with ordinary protons allowed a determination of the mass of the negative particle and it was found to be the same as the proton's to within 5%. This was strong circumstantial evidence that this was the antiproton and not some other long-lived, negative particle. Still, the fundamental property of the antiproton, its ability to annihilate with a proton or neutron to produce a final state with no baryons in it, had not been confirmed.

The Bevatron's high energy proton beam provided the opportunity to look for antiprotons in other ways. With emulsions it is possible, in principle, to measure the large energy released when an antiproton annihilates with a proton or neutron, providing direct evidence for the antiparticle character of the annihilating particle.

While the experiment of Chamberlain *et al.* was being set up, an emulsion stack was exposed at the location of the first scintillator in a collaborative experiment between a Berkeley group under G. Goldhaber and E. Segrè and a Rome group under E. Amaldi. This exposure required a 132 g cm^{-2} copper absorber to slow the antiprotons so they would stop in the emulsion. After laborious scanning, both in Berkeley and in Rome, one stopping negative particle of protonic mass was observed by the Rome group (Ref. 4.2). The energy release observed was 850 MeV. (See Figure 4.1.)

In another effort to confirm the antiparticle nature of the new negative particles, Brabant *et al.* placed a lead glass Cherenkov counter at the end of the antiproton beam of the Chamberlain–Segrè team in order to look for evidence of annihilation (Ref. 4.3). While sizeable energy releases were observed, none was greater than the rest mass of the proton.

In December 1955, a second emulsion exposure was carried out at the Bevatron, this time with the momentum selected to be 700 MeV/c. This value was chosen so that the antiprotons entered the emulsion with tracks giving twice minimum ionization, making them readily distinguishable from the more numerous minimum ionizing pion tracks. This procedure turned out to be most effective. The first track of protonic mass that was followed through the emulsion stack until it came to rest released 1350 ± 50 MeV **(Ref. 4.4)**. This was unequivocal evidence for an antiproton–nucleon annihilation. The complete analysis turned up 35 antiproton annihilations, more than half of which had energy releases greater than the mass of the proton (Ref. 4.5).

The team consisting of Cork, Lambertson, Piccioni, and Wenzel in another experiment at the Bevatron established the existence of the antineutron **(Ref. 4.6)** by observing the

Figure 4.1. The first antiproton star observed in an emulsion. The incident antiproton is track L. The light tracks a and b are pions. Track c is a proton. The remaining tracks are protons or alpha particles. The exposure was made at the Bevatron. (Ref. 4.2)

Figure 4.2. An antiproton enters the bubble chamber from the top. Its track disappears at the arrow as it charge exchanges, $p\bar{p} \to n\bar{n}$. The antineutron produces the star seen in the lower portion of the picture. The energy released in the star was greater than 1500 MeV. (Ref. 4.7)

charge-exchange process, $\bar{p}p \to \bar{n}n$. This experiment used a highly efficient antiproton beam constructed with the aid of magnets using the principle of strong-focusing, which will be described in Chapter 6.

The antiproton beam was directed on a cube of liquid scintillator in which the charge-exchange process occurred. The produced antineutron continued forward into a lead glass Cherenkov counter that detected the annihilation of the antineutron. To demonstrate that antineutrons, not antiprotons, were responsible for the annihilation, counters were placed in front of the Cherenkov counter and events with charged particles were rejected. The liquid scintillator was also monitored to make sure that the reaction that took place there was indeed charge exchange rather than annihilation into mesons of the incident antiproton.

The final annihilations occurring in the Cherenkov counter were compared with those produced directly by antiprotons. Their similarity established that antineutrons had been observed.

The bubble chamber contributed as well to the discovery of antibaryons. An experiment by W. Powell and E. Segrè *et al.* using the Berkeley 30-inch propane bubble chamber at the Bevatron found a clear antiproton charge-exchange event showing an antineutron annihilation star. This event is reproduced in Figure 4.2. The anti-lambda ($\overline{\Lambda}$) was first seen in emulsions by D. Prowse and M. Baldo-Ceolin (Ref. 4.8). A classic picture of $\Lambda\overline{\Lambda}$ production observed in an antiproton exposure of the 72-inch hydrogen bubble chamber at the Bevatron is shown in Figure 4.3. The next few years witnessed the discoveries of the $\overline{\Sigma}$ (Refs. 4.10, 4.11), the $\overline{\Xi}$ (Ref. 4.12), and even the $\overline{\Omega}$ (Ref. 4.13). (The discovery of the Ω^- itself is discussed in Chapter 5.) Ultimately, all the stable baryons were shown to have antiparticles.

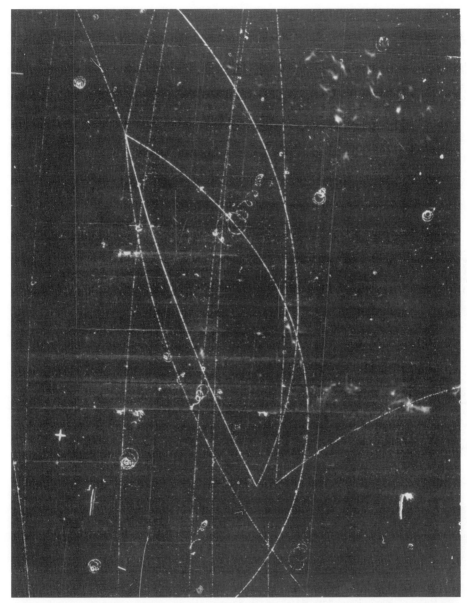

Figure 4.3. Production of a $\Lambda\bar{\Lambda}$ pair by an incident antiproton. The antiproton enters the chamber at the bottom and annihilates with a proton. The Λ and $\bar{\Lambda}$ decay nearby. The antiproton from the antilambda annihilates on the left-hand side of the picture and gives rise to a 4 prong star. The picture is from the 72-inch bubble chamber at the Bevatron. (Ref. 4.9)

Exercises

4.1 Show that a Fermi energy of 25 MeV lowers the threshold incident kinetic energy for antiproton production by a proton incident on a nucleus to 4.3 GeV.

4.2 Derive the half-angle of the cone into which Cherenkov radiation is projected in terms of the velocity of the radiating particle and the index of refraction of the medium.

4.3 Design a differential Cherenkov counter that can separate π^- and $\bar{p}$ as in Ref. 4.1. See the reference quoted therein.

4.4 Suppose positive and negative kaon beams are available for an exposure of a hydrogen bubble chamber. For which beam is the threshold lowest for the production of $\overline{\Sigma}^-$, $\overline{\Sigma}^0$, $\overline{\Sigma}^+$, $\overline{\Xi}^+$, $\overline{\Xi}^0$, and $\overline{\Omega}^+$? Give the reaction that has the lowest threshold and the incident momentum at threshold.

4.5 How was the magnetic moment of the neutron measured by L. Alvarez and F. Bloch [*Phys. Rev.* **57**, 111 (1940)]?

References

4.1 O. Chamberlain, E. Segrè, C. Wiegand, and T. Ypsilantis, "Observation of Antiprotons." *Phys. Rev.*, **100**, 947 (1955).

4.2 O. Chamberlain *et al.*, "Antiproton star observed in emulsion." *Phys. Rev.*, **101**, 909 (1956). Also *Nuovo Cimento* **X 3**, 447 (1956).

4.3 J. M. Brabant *et al.*, "Terminal Observations on 'Antiprotons'." *Phys. Rev.*, **101**, 498 (1956).

4.4 O. Chamberlain *et al.*, "Example of an Antiproton-Nucleon Annihilation." *Phys. Rev.*, **102**, 921 (1956).

4.5 W. H. Barkas *et al.*, "Antiproton-nucleon annihilation process." *Phys. Rev.*, **105**, 1037 (1957).

4.6 B. Cork, G. R. Lambertson, O. Piccioni, and W. A. Wenzel, "Antineutrons Produced from Antiprotons in Charge Exchange Collisions." *Phys. Rev.*, **104**, 1193 (1957).

4.7 L. Agnew *et al.*, "$\bar{p}p$ Elastic and Charge-Exchange Scattering at about 120 MeV." *Phys. Rev.*, **110**, 994 (1958).

4.8 D. Prowse and M. Baldo-Ceolin, "Anti-lambda Hyperon." *Phys. Rev. Lett.*, **1**, 179 (1958).

4.9 J. Button *et al.*, "The Reaction $\bar{p}p \to \overline{Y}Y$." *Phys. Rev.*, **121**, 1788 (1961).

4.10 J. Button *et al.*, "Evidence for the reaction $\bar{p}p \to \overline{\Sigma}^0 \Lambda$." *Phys. Rev. Lett.*, **4**, 530 (1960).

4.11 C. Baltay *et al.*, "Antibaryon Production in Antiproton–Proton Reactions at 3.7 BeV/c." *Phys. Rev.*, **140**, B1027 (1965).

4.12 H. N. Brown *et al.*, "Observation of Production of a $\Xi^-\overline{\Xi}^+$ Pair." *Phys. Rev. Lett.*, **8**, 255 (1962).; CERN-Ecole Polytechnique - Saclay Collaboration, "Example of anticascade ($\overline{\Xi}^+$) particle production in $\bar{p}p$ interactions at 3 GeV/c." *Phys. Rev. Lett.*, **8**, 257 (1962).

4.13 A. Firestone *et al.*, "Observation of the $\overline{\Omega}^+$." *Phys. Rev. Lett.*, **26**, 410 (1971).

Observation of Antiprotons*

OWEN CHAMBERLAIN, EMILIO SEGRÈ, CLYDE WIEGAND,
AND THOMAS YPSILANTIS

*Radiation Laboratory, Department of Physics, University of
California, Berkeley, California*
(Received October 24, 1955)

ONE of the striking features of Dirac's theory of the electron was the appearance of solutions to his equations which required the existence of an antiparticle, later identified as the positron.

The extension of the Dirac theory to the proton requires the existence of an antiproton, a particle which bears to the proton the same relationship as the positron to the electron. However, until experimental proof of the existence of the antiproton was obtained, it might be questioned whether a proton is a Dirac particle in the same sense as is the electron. For instance, the anomalous magnetic moment of the proton indicates that the simple Dirac equation does not give a complete description of the proton.

The experimental demonstration of the existence of antiprotons was thus one of the objects considered in the planning of the Bevatron. The minimum laboratory kinetic energy for the formation of an antiproton in a nucleon-nucleon collision is 5.6 Bev. If the target nucleon is in a nucleus and has some momentum, the

TABLE I. Characteristics of components of the apparatus.

$S1, S2$	Plastic scintillator counters 2.25 in. diameter by 0.62 in. thick.
$C1$	Čerenkov counter of fluorochemical 0-75, ($C_8F_{16}O$); $\mu_D = 1.276$; $\rho = 1.76$ g cm^{-3}. Diameter 3 in.; thickness 2 in.
$C2$	Čerenkov counter of fused quartz: $\mu_D = 1.458$; $\rho = 2.2$ g cm^{-3}. Diameter 2.38 in.; length 2.5 in.
$Q1, Q2$	Quadrupole focusing magnets: Focal length 119 in.; aperture 4 in.
$M1, M2$	Deflecting magnets 60 in. long. Aperture 12 in. by 4 in. $B \cong 13\,700$ gauss.

threshold is lowered. Assuming a Fermi energy of 25 Mev, one may calculate that the threshold for formation of a proton-antiproton pair is approximately 4.3 Bev. Another, two-step process that has been considered by Feldman[1] has an even lower threshold.

There have been several experimental events[2-4] recorded in cosmic-ray investigations which might be due to antiprotons, although no sure conclusion can be drawn from them at present.

With this background of information we have performed an experiment directed to the production and detection of the antiproton. It is based upon the determination of the mass of negative particles originating at the Bevatron target. This determination depends on the simultaneous measurement of their momentum and velocity. Since the antiprotons must be selected from a heavy background of pions it has been necessary to measure the velocity by more than one method. To date, sixty antiprotons have been detected.

Figure 1 shows a schematic diagram of the apparatus. The Bevatron proton beam impinges on a copper target and negative particles scattered in the forward direction with momentum 1.19 Bev/c describe an orbit as shown in the figure. These particles are deflected 21° by the field of the Bevatron, and an additional 32° by magnet $M1$. With the aid of the quadrupole focusing magnet $Q1$ (consisting of 3 consecutive quadrupole magnets) these particles are brought to a focus at counter $S1$, the first scintillation counter. After passing through counter $S1$, the particles are again focused (by $Q2$), and deflected (by $M2$) through an additional angle of 34°, so that they are again brought to a focus at counter $S2$.

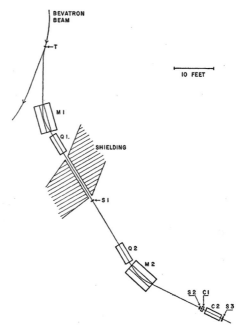

FIG. 1. Diagram of experimental arrangement.
For details see Table I.

The particles focused at $S2$ all have the same momentum within 2 percent.

Counters $S1$, $S2$, and $S3$ are ordinary scintillation counters. Counters $C1$ and $C2$ are Čerenkov counters. Proton-mass particles of momentum 1.19 Bev/c incident on counter $S2$ have $v/c = \beta = 0.78$. Ionization energy loss in traversing counters $S2$, $C1$, and $C2$ reduces the average velocity of such particles to $\beta = 0.765$. Counter $C1$ detects all charged particles for which $\beta > 0.79$. $C2$ is a Čerenkov counter of special design that counts only particles in a narrow velocity interval, $0.75 < \beta < 0.78$. This counter will be described in a separate publication. In principle, it is similar to some of the counters described by Marshall.[5] The requirement that a particle be counted in this counter represents one of the determinations of velocity of the particle.

The velocity of the particles counted has also been determined by another method, namely by observing the time of flight between counters $S1$ and $S2$, separated by 40 ft. On the basis of time-of-flight measurement the separation of π mesons from proton-mass particles is quite feasible. Mesons of momentum 1.19 Bev/c have $\beta=0.99$, while for proton-mass particles of the same momentum $\beta=0.78$. Their respective flight times over the 40-ft distance between $S1$ and $S2$ are 40 and 51 millimicroseconds.

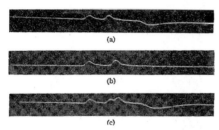

FIG. 2. Oscilloscope traces showing from left to right pulses from $S1$, $S2$, and $C1$. (a) meson, (b) antiproton, (c) accidental event.

The beam that traverses the apparatus consists overwhelmingly of π^- mesons. One of the main difficulties of the experiment has been the selection of a very few antiprotons from the huge pion background. This has been accomplished by requiring counters $S1$, $S2$, $C2$, and $S3$ to count in coincidence. Coincidence counts in $S1$ and $S2$ indicate that a particle of momentum 1.19 Bev/c has traversed the system with a flight time of approximately 51 millimicroseconds. The further requirement of a coincidence in $C2$ establishes that the particle had a velocity in the interval $0.75<\beta<0.78$. The latter requirement of a count in $C2$ represents a measure of the velocity of the particle which is essentially independent of the cruder electronic time-of-flight measurement. Finally, a coincident count in counter $S3$ was required in order to insure that the particle traversed the quartz radiator in $C2$ along the axis and suffered no large-angle scattering.

As outlined thus far, the apparatus has some shortcomings in the determination of velocity. In the first place, accidental coincidences of $S1$ and $S2$ cause some mesons to count, even though a single meson would be completely excluded because its flight time would be too short. Secondly, the Čerenkov counter $C2$ could be actuated by a meson (for which $\beta=0.99$) if the meson suffered a nuclear scattering in the radiator of the counter. About 3 percent of the mesons, which ideally should not be detected in $C2$, are counted in this manner. Both of these deficiencies have been eliminated by the insertion of the guard counter $C1$, which records all particles of $\beta>0.79$. A pulse from $C1$ indicates a particle (meson) moving too fast to be an antiproton of the selected momentum and indicates that this event should be rejected. In Table I, the characteristics of the components of the apparatus are summarized.

The pulses from counters $S1$, $S2$, and $C1$ were displayed on an oscilloscope trace and photographically recorded. From the separation of pulses from $S1$ and $S2$ the flight time of the particle could be measured with an accuracy of 1 millimicrosecond, and the pulse in the guard counter $C1$ could be measured. Figure 2 shows three oscilloscope traces, with the pulses from $S1$, $S2$, and $C1$ appearing in that order. The first trace (a) shows the pulses due to a meson passing through the system. It was recorded while the electronic circuits were adjusted for meson time of flight for calibration purposes. The second trace, Fig. 2(b), shows the pulses resulting from an antiproton. The separation of pulses from $S1$ and $S2$ indicates the correct antiproton time of

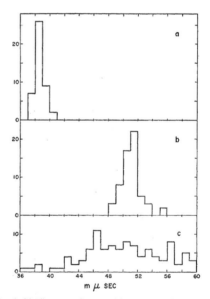

FIG. 3. (a) Histogram of meson flight times used for calibration. (b) Histogram of antiproton flight times. (c) Apparent flight times of a representative group of accidental coincidences. Times of flight are in units of 10^{-9} sec. The ordinates show the number of events in each 10^{-10}-sec intervals.

flight, and the absence of the $C1$ pulse shows that no meson passed through $C1$. The third trace, Fig. 2(c), shows the accidental coincidence of two mesons with a difference of time such as to register in the electronic circuits. Either the presence of a pulse from $C2$ or the presence of multiple pulses from $S1$ or $S2$ would be sufficient to identify the trace as due to one or more mesons.

An over-all test of the apparatus was obtained by changing the position of the target in the Bevatron, inverting the magnetic fields in $M1$, $M2$, $Q1$, and $Q2$, and detecting positive protons.

Each oscilloscope sweep of the type shown in Fig. 2 can be used to make an approximate mass measurement for *each* particle, since the magnetic fields determine the momentum of the particle and the separation of pulses $S1$ and $S2$ determine the time of flight. For protons of our selected momentum the mass is measured to about 10 percent, using this method only.

The observed times of flight for antiprotons are made more meaningful by the fact that the electronic gate time is considerably longer than the spread of observed antiproton flight times. The electronic equipment accepts events that are within ±6 millimicroseconds of the right flight time for antiprotons, while the actual antiproton traces recorded show a grouping of flight times to ±1 or 2 millimicroseconds. Figure 3(a) shows a histogram of meson flight times; Fig. 3(b) shows a similar histogram of antiproton flight times. Accidental coincidences account for many of the sweeps (about $\frac{2}{3}$ of the sweeps) during the runs designed to detect antiprotons. A histogram of the apparent flight times of accidental coincidences is shown in Fig. 3(c). It will be noticed that the accidental coincidences do not show the close grouping of flight times characteristic of the antiproton or meson flight times.

Mass measurement.—A further test of the equipment has been made by adjusting the system for particles of

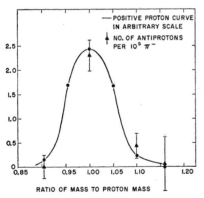

FIG. 4. The solid curve represents the mass resolution of the apparatus as obtained with protons. Also shown are the experimental points obtained with antiprotons.

different mass, in the region of the proton mass. A test for the reality of the newly detected negative particles is that there should be a peak of intensity at the proton mass, with small background at adjacent mass settings. By changing only the magnetic field values of $M1$, $M2$, $Q1$, and $Q2$, particles of different momentum may be chosen. Providing the velocity selection is left completely unchanged, the apparatus is then set for particles of a different mass. These tests have been made for both positive and negative particles in the vicinity of the proton mass. Figure 4 shows the curve obtained using positive protons, which is the mass resolution

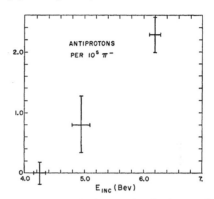

FIG. 5. Excitation curve for the production of antiprotons relative to meson production as a function of Bevatron beam energy.

curve of the instrument. Also shown in Fig. 4 are the experimental points obtained with antiprotons. The observations show the existence of a peak of intensity at the proton mass, with no evidence of background when the instrument is set for masses appreciably greater or smaller than the proton mass. This test is considered one of the most important for the establishment of the reality of these observations, since background, if present, could be expected to appear at any mass setting of the instrument. The peak at proton mass may further be used to say that the new particle has a mass within 5 percent of that of the proton mass. It is mainly on this basis that the new particles have been identified as antiprotons.

Excitation function.—A very rough determination has been made of the dependence of antiproton production cross section on the energy of the Bevatron proton beam. A more exact determination will be attempted in the future, but up to the present it has not been possible to monitor reliably the amount of beam actually striking the target. Furthermore, the solid angle of acceptance of the detection apparatus may not be independent of Bevatron energy since the shape of the orbit on which the antiprotons emerge depends somewhat on the magnetic field strength within the Bevatron magnet. It has, however, been possible to measure the ratio of antiprotons to mesons (both at momentum 1.19 Bev/c) emitted in the forward direction from the target as a function of Bevatron energy. The resulting approximate excitation function is shown in the form of three experimental points in Fig. 5.

Even at 6.2 Bev, the antiprotons appear only to the extent of one in 44 000 pions. Because of the decay of pions along the trajectory through the detecting apparatus, this number corresponds to one antiproton in 62 000 mesons generated at the target. It will be seen from Fig. 5 that there is no observed antiproton production at the lowest energy. Although the production of antiprotons does not seem to rise as sharply with increasing energy as might at first be expected, the data indicate a reasonable threshold for production of antiprotons. It must again be emphasized that Fig. 5 shows only the excitation function relative to the meson excitation function, hence the true excitation function is not known at this time. If and when detailed meson production excitation functions become known, data of the type shown in Fig. 5 may allow a true antiproton production excitation function to be determined. It should also be mentioned that the angle of emission from the target actually varies slightly with Bevatron energy. At 6.2 Bev, it is 3°, at 5.1 Bev it is 6°, and at 4.2 Bev it is 8° from the forward direction at the Bevatron target.

Possible spurious effects.—The possibility of a negative hydrogen ion being mistaken for an antiproton is ruled out by the following argument: It is extremely improbable that such an ion should pass through all the counters without the stripping of its electrons. It may be added that except for a few feet near the target the whole trajectory through the apparatus is though gas at atmospheric pressure, either in air or, near the magnetic lenses, in helium gas introduced to reduce multiple scattering.

None of the known heavy mesons or hyperons have the proper mass to explain the present observations. Moreover, no such particles are known that have a mean life sufficiently long to pass through the apparatus without a prohibitive amount of decay since the flight time through the apparatus of a particle of proton mass is 10.2×10^{-8} sec. However, this possibility cannot be strictly ruled out. In the description of the new particles as antiprotons, a reservation must be made for the possible existence of previously unknown negative particles of mass very close to 1840 electron masses.

The observation of pulse heights in counters $S1$ and $S2$ indicates that the new particles must be singly charged. No multiply charged particle could explain the experimental results.

Photographic experiments directed toward the detection of the terminal event of an antiproton are in progress in this laboratory and in Rome, Italy, using emulsions irradiated at the Bevatron, but to this date no positive results can be given. An experiment in conjunction with several other physicists to observe the energy release upon the stopping of an antiproton in a large lead-glass Čerenkov counter is in progress and its results will be reported shortly. It is also planned to try to observe the annihilation process of the antiproton in a cloud chamber, using the present apparatus for counter control.

The whole-hearted cooperation of Dr. E. J. Lofgren, under whose direction the Bevatron has been operated, has been of vital importance to this experiment. Mr. Herbert Steiner and Mr. Donald Keller have been very helpful throughout the work. Dr. O. Piccioni has made very useful suggestions in connection with the design of the experiment. Finally, we are indebted to the operating crew of the Bevatron and to our colleagues, who have cheerfully accepted many weeks' postponement of their own work.

* This work was done under the auspices of the U. S. Atomic Energy Commission.
[1] G. Feldman, Phys. Rev. **95**, 1967 (1954).
[2] Evans Hayward, Phys. Rev. **72**, 937 (1947).
[3] Amaldi, Castagnoli, Cortini, Franzinetti, and Manfredini, Nuovo cimento **1**, 492 (1955).
[4] Bridge, Courant, DeStaebler, and Rossi, Phys. Rev. **95**, 1101 (1954).
[5] J. Marshall, Ann. Rev. Nuc. Sci. **4**, 141 (1954).

Corrections supplied by the authors: In Fig. 3a the peak should be located at 40×10^{-9}s. In the caption for Fig. 3 the last sentence should read: "The ordinates show the number of events in each 10^{-9} sec interval."

Example of an Antiproton-Nucleon Annihilation

O. Chamberlain, W. W. Chupp, A. G. Ekspong, G. Goldhaber,
S. Goldhaber, E. J. Lofgren, E. Segrè, and C. Wiegand,
*Radiation Laboratory and Department of Physics,
University of California, Berkeley, California*

AND

E. Amaldi, G. Baroni, C. Castagnoli, C. Franzinetti,
and A. Manfredini, *Istituto Fisico dell'Università
Roma, Italy and Istituto Nazionale di Fisica
Nucleare Sez. di Roma, Italy*

(Received March 8, 1956)

THE existence of antiprotons has recently been demonstrated at the Berkeley Bevatron by a counter experiment.[1] The antiprotons were found among the momentum-analyzed (1190 Mev/c) negative particles emitted by a copper target bombarded by 6.2-Bev protons. Concurrently with the counter experiment, stacks of nuclear emulsions were exposed in the beam adjusted to accept 1090-Mev/c negative particles in an experiment designed to observe the properties of antiprotons when coming to rest. This required a 132-g/cm² copper absorber to slow down the antiprotons sufficiently to stop them in the emulsion stack. Only one antiproton was found[2] in stacks in which seven were expected, assuming a geometric interaction cross section for antiprotons in copper. It has now been found[3] that the cross section in copper is about twice geometric, which explains this low yield.

In view of this result a new irradiation was planned in which (1) no absorbing material preceded the stack,

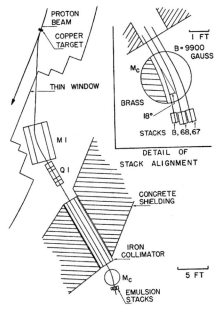

Fig. 1. Plan of the irradiation.

(2) the range of the antiprotons ended in the stack, and (3) antiprotons and mesons were easily distinguishable by grain density at the entrance of the stack. In order to achieve these three results it was necessary to select antiprotons of lower momentum, even if these should be admixed with a larger number of π^- than at higher momenta.

In the present experiment we exposed a stack in the same beam used previously, adjusted for a momentum of 700 Mev/c instead of 1090 Mev/c. Since the previous work[2] had indicated that the most troublesome background was due to ordinary protons, the particles were also passed through a clearing magnetic field just prior to their entrance into the emulsion stacks. The clearing magnet (M_c) had $B = 9900$ gauss, circular pole faces of diameter 76 cm, and a gap of 18 cm, so particles scattered from the pole faces of the clearing magnet could be ignored on the basis of their large dip angle in the emulsions. With this arrangement we have achieved conditions in which the negative particles enter the emulsions at a well-defined angle, and extremely few positive particles enter the emulsions within the same range of angles. For the first time we have obtained an exposure in which more antiprotons than protons enter the stacks with the proper entrance angles. Under these conditions it is relatively easy to find antiprotons in these stacks even though approximately 5×10^5 negative π mesons at minimum ionization accompany one antiproton. The exposure arrangement is shown in Fig. 1. The beam collimation was such that at any given position at the leading edge of the stacks the angular half-width of the pion entrance angles is less than 1° both in dip and in the plane of the emulsions. This very small angular spread allowed us to apply strict angular criteria for picking up antiproton tracks, and thus helped to reduce confusing background tracks to a negligible level. The antiproton tracks were picked up at the leading edge of the emulsions on the basis of a grain count ($\sim$twice minimum) and angular criteria (angle between track and average direction of pions less than 5°) and were then followed along the track.

A number of antiproton stars have been observed in these nuclear emulsions.[4] The one we will describe here was found in Berkeley and is of particular interest since it is the first example of a particle of protonic mass $(m/M_p=1.013\pm0.034)$ which on coming to rest gave rise to a star with a visible energy release greater than M_pc^2. This example thus constitutes a proof that the particles here observed undergo an annihilation process with a nucleon, a necessary requirement for Dirac's antiproton.

Description of the Event.—The particle marked P^- in Fig. 2 entered the emulsion stack at an angle of less than 1° from the direction defined by the π^- mesons in the beam. It came to rest in the stack and produced an 8-prong star. Its total range was $R=12.13\pm0.14$ cm. Table I gives the results of three independent mass measurements on the incoming particle. The first two methods listed in Table I use measurements made entirely in the emulsion stack. The third combines the range, as measured in the stack, with the momentum as determined by magnetic field measurements. For the position and entrance angle of this particle into the stack the momentum is $p=696$ Mev/c with an estimated 2% error. All three methods are in good agreement and give a mass of $m=1.013\pm0.034$ in proton mass units.

TABLE I. Mass measurements.

Method	Residual range cm of emulsion	Mass M/M_p
Ionization–range	2, 5.5, and 12	0.97 ±0.10
Scattering–range (constant sagitta)	0–1	0.93 ±0.14
Momentum–range	12.13+0.12 (air and helium equivalent)	1.025±0.037
Weighted mean		1.013±0.034

Of the particles forming the star, five came to rest in the emulsion stack, two left the stack (tracks numbered 4 and 8), and one disappeared in flight (track number 3). The tracks numbered 1, 4, and 6 in Fig. 2 were caused by heavy particles. Particle 4 was near the end of its range ($R_{res}=2$ mm) when it left the stack. Tracks 1 and 4 are probably due to protons and track 6 is a triton. However, owing to the large dip angles the assignments for tracks 1, 4, and 6 are not certain. Track 2 has the characteristics of a π meson and on coming to rest gives a 2-prong σ star. It is thus a negative π meson. Particle 5 came to rest and gave the typical π-μ-e decay, and was thus a positive π meson. From the measured range its energy would have been 18 Mev; however, after

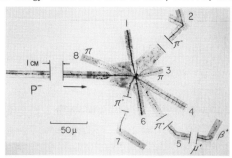

FIG. 2. Reproduction of the P^- star. The description of the prongs is given in Table II. The star was observed by A. G. Ekspong and the photomicrograph was made by D. H. Kouns.

0.22 mm it underwent a 22° scattering that appears to be inelastic. The initial energy as estimated from the grain density change was 30 ± 6 Mev. Track 7 is very steep (dip angle=83.5°). The particle came to rest as a typical light ρ meson after traversing 30 emulsions. At the end of the track there is a blob and possibly an associated slow electron. The most probable assignment is a negative π meson, although a negative μ meson cannot be ruled out.

In addition to the three stopping π mesons there are two other tracks which we know were caused by light particles, presumably π mesons. Track 8 had $p\beta=190\pm30$ Mev/c and $g/g_0=1.10\pm0.04$, which is consistent with a π meson of 125 ± 25 Mev energy, but is not consistent with a much heavier particle. After 16 mm it shows a 17° scattering with no detectable change in energy. Track 3 is very steep (dip angle=73.5°) and its ionization is about minimum. The particle traversed 81 plates and disappeared in flight after an observed range of 50 mm. The $p\beta$ has been determined by a new modification of the multiple scattering technique to be 250 ± 45 Mev/c. The new method, which is applicable to steep tracks, is based on measurements of the coordinates of the exit point of the track in each emulsion with reference to a well-aligned millimeter grid[5] printed on each pellicle in the stack. A detailed description of this method will be given in a subsequent paper.

TABLE II. Measurements and data on the eight prongs of the P^- star shown in Fig. 2.

Track number	Range mm	Number of plates traversed	Dip angle	Projected angle	$p\beta$ Mev/c	Ionization g/g_0	Identity	E_{kin} Mev	Total energy Mev
1	0.59	2	−56.5°	103°			$p(?)$	10	18
2	27.9	11	+6.5°	61.5°			π^-	43	183
3	>50	81	−73.5°	14.5°	250±45	1.10±0.04	$\pi(?)$	174±40	314±40
4	>14.2	16	+53°	318.5°			$p(?)$	70±5	78±5
5	6.2	3	+4°	305.5°			π^+	30±6	170±6
6	9.5	15	−63.5°	281°			$T(?)$	82	98
7	18.6	30	−83.5°	255°			π^-	34	174
8	>22.3	16	+33°	163°	190±30	~1	$\pi(?)$	125±25	265±25
							Total visible energy[a]:		1300±50 Mev
							For momentum balance:	≥100	Mev
							Total energy release:		≥1400±50 Mev

[a] To obtain the minimum possible value of the visible energy release, still consistent with our observations, one has to make the very unlikely assumptions about the identity of tracks 3, 6, 7, and 8: that tracks 3 and 8 are due to electrons, track 6 to a proton, and track 7 to a μ^- meson. The total visible energy release in this case becomes 1084±55 Mev. To this must be added at least 50 Mev to balance momentum, bringing the total energy release to ≥1134±55 Mev.

The observations do not allow us to rule out the possibility that tracks 3 and 8 are due to electrons. It is, however, very unlikely that a fast electron could travel 50 mm (1.7 radiation lengths) in the emulsion (as does track 3) without a great loss of energy due to bremsstrahlung. The energy (particle 3) deduced from the measured $p\beta(E=250 \text{ Mev})$ must be considered a lower limit.

In Table II, the pertinent data on the eight prongs are summarized. The last column gives the total visible energy per particle (E_{kin}+8-Mev binding energy per nucleon, or E_{kin}+140-Mev rest energy per π meson) for the most probable assignments as discussed above. The total visible energy is 1300±50 Mev, and the momentum unbalance is 750 Mev/c. To balance momentum, an energy of at least 100 Mev is required in neutral particles (i.e., about 5 neutrons with parallel and equal momenta), which brings the lower limit for the observed energy release to 1400±50 Mev.

However, as some of the identity assignments to the star prongs are not certain, we have also computed the energy release for the extreme and very unlikely assignments, given at the foot of Table II, which are chosen to give the minimum energy release. In this case the total visible energy is 1084±55 Mev and the resultant momentum is 380 Mev/c, which to be balanced requires at least 50 Mev in neutral particles (three or four neutrons). In this unrealistic case the lower limit for the observed energy release is 1134±55 Mev, which still exceeds the rest energy of the incoming particle by about three standard deviations.

We conclude that the observations made on this reaction constitute a conclusive proof that we are dealing with the antiparticle of the proton.

A second important observation is the high multiplicity of charged π mesons (one π^+, two π^-, and two π mesons with unknown charge). The fact that so many π mesons escaped from the nucleus where the annihilation took place, together with the low number of heavy particles emitted (three), may indicate that the struck nucleus was one of the light nuclei of the emulsion (C, N, O). Two of the outgoing heavy prongs carried rather high energies (70 Mev for the proton, 82 Mev for the triton), and they may have resulted from the reabsorption of another two π mesons.

We are greatly indebted to the Bevatron crew for their assistence in carrying out the exposure. We also wish to thank Mr. J. E. Lannutti for help with measurements and the analysis of the event.

* This work was performed under the auspices of the U. S. Atomic Energy Commission.
[1] Chamberlain, Segrè, Wiegand, and Ypsilantis, Phys. Rev. 100, 947 (1955).
[2] Chamberlain, Chupp, Goldhaber, Segrè, Wiegand, Amaldi, Baroni, Castagnoli, Franzinetti, and Manfredini, Phys. Rev. 101, 909 (1956), and Nuovo cimento (to be published).
[3] Chamberlain, Keller, Segrè, Steiner, Wiegand, and Ypsilantis, Phys. Rev. (to be published).
[4] Several stacks exposed in the 700 Mev/c beam are being studied in Berkeley by A. G. Ekspong and G. Goldhaber; W. W. Chupp and S. Goldhaber; R. Birge, D. H. Perkins, D. Stork, and L. van Rossum; W. Barkas, H. Heckman, and F. Smith; and in Rome by E. Amaldi, G. Baroni, C. Castagnoli, C. Franzinetti, and A. Manfredini.
[5] Goldhaber, Goldsack, and Lannutti, University of California Radiation Laboratory Report UCRL-2928 (unpublished).

Antineutrons Produced from Antiprotons in Charge-Exchange Collisions*

BRUCE CORK, GLEN R. LAMBERTSON, ORESTE PICCIONI,†
AND WILLIAM A. WENZEL

*Radiation Laboratory, University of California,
Berkeley, California*

(Received October 3, 1956)

THE principle of invariance under charge conjugation gained strong support when it was found that the Bevatron produces antiprotons.[1-3] Another prediction of the same theory which could be tested experimentally was the existence of the antineutron. Additional interest arises from the fact that charge conjugation has somewhat less obvious consequences when applied to neutral particles than it has when applied to particles with electric charge.

The purpose of this experiment was to detect the annihilation of antineutrons produced by charge exchange from antiprotons. Because the yield of antineutrons was expected to be low, a relatively large flux of antiprotons was required. Protons of 6.2-Bev energy bombarded an internal beryllium target of the Bevatron (Fig. 1). With a system of two deflecting magnets and five magnetic lenses a beam of 1.4-Bev/c negative particles was obtained. Six scintillation counters connected in coincidence distinguished antiprotons from negative mesons by time of flight. In Figs. 1 and 2, F is the last counter of this system, which counted 300 to 600 antiprotons per hour. Antiprotons interacting in the thick converter, X (Fig. 2), sometimes produce antineutrons which pass through the scintillators S_1 and S_2 without detection and finally interact in the lead glass Čerenkov Counter C, producing there a pulse of light so large as to indicate the annihilation of a nucleon and an antinucleon.

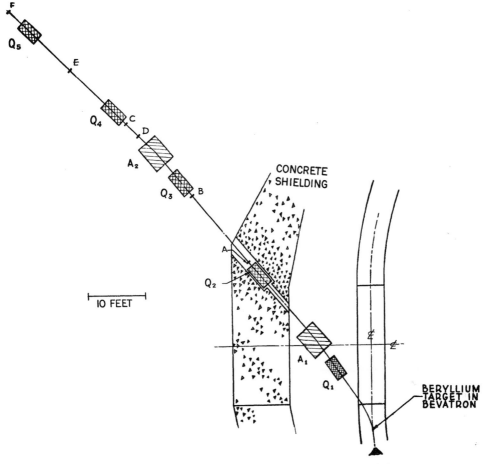

FIG. 1. Antiproton-selecting system. Q_1 through Q_5 are focusing quadrupoles. A_1 and A_2 are analyzing magnets. A through F are 4-by-4-by-$\frac{1}{4}$-inch scintillators.

The Čerenkov Counter C is a piece of lead glass, 13 by 13 by 14 in., density=4.8, index of refraction =1.8, viewed by 16 RCA 6655 photomultipliers. This instrument is similar to the one used in a previous experiment on antiprotons.[2] A 1-in. lead plate is placed between S_1 and S_2 to convert high-energy gamma rays which could otherwise be confused with the antineutrons. Ordinary neutrons and neutral mesons (heavier than pions) can be detected by the Čerenkov counter but their average light pulse is much smaller than that from the annihilation of an antineutron. However, a relatively small background of these neutral secondaries would distort the apparent spectrum of antineutrons.

To discriminate against these neutral secondaries, the charge-exchange converter, X, was made of a scintillating toluene-terphenyl solution, viewed by four photomultipliers connected in parallel. In this way neutral particles producing pulses in the Čerenkov counter ("neutral events") could be separated according to whether they originated in an annihilation, indicated by a large pulse in X, or in the less violent process expected to accompany the charge-exchange production of an antineutron. A quantitative criterion for this separation is derived from a comparison between the pulse-height spectra in X, shown in Fig. 3. The dashed curve, obtained in a separate experiment, is the spectrum produced by antiprotons passing through but not interacting in X. The sharp peak in the spectrum provides the calibration of X; the ionization loss for

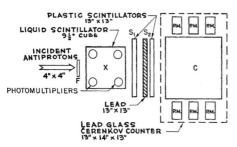

FIG. 2. Antineutron-detecting system. X is the charge-exchange scintillator; S_1 and S_2 are scintillation counters; C is a lead-glass Čerenkov counter (later a large scintillator).

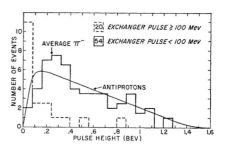

FIG. 4. Pulse-height spectrum in lead glass counter for neutral events. The solid histogram is for 54 antineutron events (energy loss in charge-exchange scintillator less than 100 Mev). Dashed histogram is for 20 other neutral events. Smooth solid curve is for antiprotons and is normalized to the solid histogram.

transmitted antiprotons is readily computed to be 50 Mev. The smooth solid curve of Fig. 3, obtained with the geometry of Fig. 2, represents all antiproton interactions in X from which no pulse was observed in S_1 and S_2, whether or not a pulse in C occurred. For those events in which a neutral particle produced a pulse in C, the histogram of Fig. 3 gives the pulse-height distribution.

The difference between the solid curve and the histogram is remarkable in that it shows that the rare interactions that produce neutral particles detected by the Čerenkov counter release much less energy in X than the other unselected interactions. In fact, the peak of the histogram is at a smaller pulse height than that which corresponds to the ionization loss of a noninteracting antiproton (50 Mev). This is what we should expect if the neutral particles were antineutrons, for in this case no nucleonic annihilation could take place in X. Conversely, production of other energetic neutrals should exhibit the characteristic large pulse of an annihilation event in X. The histogram suggests, therefore, that the apparatus detects a small background of events of this latter type. The pulse height of 100 Mev in Fig. 3 has been selected to separate this background from antineutron events. Figure 4 shows the separate pulse-height distributions in the Čerenkov counter for the events which produced in X a pulse less than 100 Mev (solid histogram), and for the events which produced a pulse larger than 100 Mev (dashed histogram). The great difference between the two histograms with respect to both average pulse height and shape confirms the interpretation by which the neutral events are divided into antineutrons and background.

The energy scale in Fig. 4 is obtained by relating the pulse height produced by π mesons going through the glass to the computed ionization energy loss of 240 Mev. This calibration was repeated every day.

The standard for annihilation pulses is provided by the smooth curve of Fig. 4, which is the pulse-height distribution for antiprotons entering the lead glass when S_1, S_2, and the lead plate are removed. Comparison of the solid histogram with this antiproton curve justifies our interpretation that the solid histogram is produced by antineutrons.

For comparison with the annihilation spectra of Fig. 4, Fig. 5 shows the spectra obtained with 750-Mev positive protons (solid curve) and with 600-Mev negative pions incident on the glass Counter C. These spectra indicate that large pulses are rarely produced by particles of such energies. From this it is apparent that even high-energy neutrons could not produce a spectrum like the solid histogram of Fig. 4.

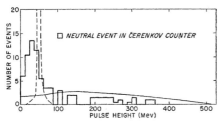

FIG. 3. Pulse-height spectrum in charge-exchange scintillator for 74 neutral events in lead glass. Histogram is for all neutral events. The smooth solid curve is for calibrating antiprotons for which no pulse occurred in S_1 or S_2. Smooth dashed curve is for noninteracting antiprotons. Smooth curves are each normalized to histogram.

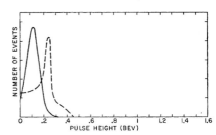

FIG. 5. Pulse-height spectrum in lead glass counter for π mesons (dashed curve) and for positive protons (solid curve). The curves are normalized.

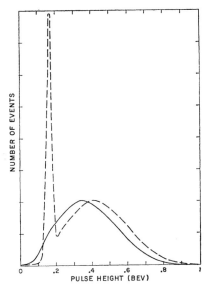

FIG. 6. Pulse-height spectrum of antiprotons in large scintillation counter. The dashed curve is for all incident antiprotons. The solid curve has had noninteracting antiprotons removed and includes a correction to permit comparison with antineutrons.

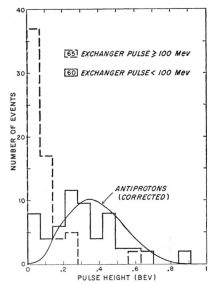

FIG. 7. Pulse-height spectrum in large scintillation counter for neutral events. Solid histogram is for 60 antineutrons (energy loss in charge-exchange scintillator less than 100 Mev). Dashed histogram is for 65 other neutral events. The smooth solid curve is the corrected antiproton curve from Fig. 6.

To determine the number of γ rays incident on S_1, the lead between S_1 and S_2 was removed. The number of neutral events per incident antiproton increased by a factor of 7. From the known probability that a single high-energy γ ray would be transmitted through 1 in. of lead without converting (3% for a γ-ray energy of 300 Mev), this observed increase shows that our neutral events contain at most 20% of γ-ray background before selection on the basis of pulse height in X.

The lead glass Counter C is very sensitive to γ rays and insensitive to ionization losses by slow particles. The desirability of comparing the spectra of antineutrons and antiprotons obtained with an entirely different type of detector led us to repeat the experiment with Counter C replaced by a liquid scintillator. This scintillator, 28 in. thick and 5 ft³ in volume, was large enough to detect a substantial part of the energy of an annihilation event. For this experiment the thickness of the lead converter between S_1 and S_2 was increased to 1.5 in. As before, the antineutron detector was calibrated with antiprotons. The pulse-height distribution of antiprotons in the large scintillator is given by Fig. 6. The noninteracting antiprotons produce the sharp peak.

The solid smooth curve in Fig. 7 is the solid curve of Fig. 6, obtained by correcting the pulse height by 70 Mev toward lower energy because antiprotons ionize before interacting in the scintillator. Sixty neutral events were obtained (Fig. 7) after selection with the same criterion as before on the pulse height in X. Again the selected neutral spectrum and the antiproton spectrum are in agreement, although not so strikingly as with the lead glass. The sixty selected events apparently include some contamination. This interpretation is confirmed by the shape of the spectrum in X for all neutral events (Fig. 8). There are now many more neutral secondaries from inelastic collisions of antiprotons than there were in the experiment with the lead glass, and the separation between antineutrons and background is therefore not so good. The larger number of neutral secondaries is probably to be attributed to the greater sensitivity of the scintillator to neutrons.

The lead glass and the scintillator are of nearly equal efficiency in detecting the antineutrons. The observed

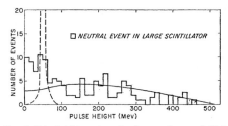

FIG. 8. Pulse-height spectrum in charge-exchange scintillator for 125 neutral events in large scintillator. The smooth curves are the same as in Fig. 3, each normalized to histogram.

yield from about 20 g/cm² of toluene is 0.0030±0.0005 antineutrons per antiproton with the lead glass, and 0.0028±0.0005 with the liquid scintillator. With the assumption that the interaction cross section for antineutrons is the same as for antiprotons, the inefficiency of the detector due to attenuation in S_1, S_2, and the lead converter, and to transmission of the detector can be calculated, and is found to be about 50%. From the observed antineutron yield the mean free path for charge exchange of the type detected is about 2300 g/cm² of toluene (C_7H_8); or, in other words, the exchange cross section is about 2% of the annihilation cross section for this material. This corresponds to a cross section of approximately 8 millibarns in carbon for this process.

The generous support of many groups, including the Bevatron operating group under Dr. Edward J. Lofgren, is greatly appreciated.

We thank Professor David Frisch of Massachusetts Institute of Technology for the loan of the lead glass used in the Čerenkov counter.

* This work was done under the auspices of the U. S. Atomic Energy Commission.
† On leave of absence from Brookhaven National Laboratory, Upton, New York.
[1] Chamberlain, Segrè, Wiegand, and Ypsilantis, Phys. Rev. **100**, 947 (1955).
[2] Brabant, Cork, Horwitz, Moyer, Murray, Wallace, and Wenzel, Phys. Rev. **101**, 498 (1956).
[3] Chamberlain, Chupp, Ekspong, Goldhaber, Goldhaber, Lofgren, Segrè, Wiegand, Amaldi, Baroni, Castagnoli, Franzinetti, and Manfredini, Phys. Rev. **102**, 921 (1956).

5

The Resonances

A pattern evolves, 1952–1964.

Most of the particles whose discoveries are described in the preceding chapters have lifetimes of 10^{-10} s or more. They travel a perceptible distance in a bubble chamber or emulsion before decaying. The development of particle accelerators and the measurement of scattering cross sections revealed new particles in the form of resonances. The resonances corresponded to particles with extremely small lifetimes as measured through the uncertainty relation $\Delta t \Delta E = \hbar$. The energy uncertainty, ΔE, was reflected in the width of the resonance, usually 10 to 200 MeV, so the implied lifetimes were roughly $\hbar/100$ MeV $\approx 10^{-25}$ s. As more and more particles and resonances were found, patterns appeared. Ultimately these patterns revealed a deeper level of particles, the quarks.

The first resonance in particle physics was discovered by H. Anderson, E. Fermi, E. A. Long, and D. E. Nagle, working at the Chicago Cyclotron in 1952 (**Ref. 5.1**). They observed a striking difference between the $\pi^+ p$ and $\pi^- p$ total cross sections. The $\pi^- p$ cross section rose sharply from a few millibarns and came up to a peak of about 60 mb for an incident pion kinetic energy of 180 MeV. The $\pi^+ p$ cross section behaved similarly except that for any given energy, its cross section was about three times as large as that for $\pi^- p$.

In two companion papers they investigated the three scattering processes:

(1) $\pi^+ p \to \pi^+ p$ elastic π^+ scattering
(2) $\pi^- p \to \pi^0 n$ charge-exchange scattering
(3) $\pi^- p \to \pi^- p$ elastic π^- scattering

They found that of the three cross sections, (1) was largest and (3) was the smallest. The data were very suggestive of the first half of a resonance shape. The π^+ cross section rose sharply but the data stopped at too low an energy to show conclusively a resonance shape. K. A. Brueckner, who had heard of these results, suggested that a resonance in the πp system was being observed and noted that a spin-3/2, isospin-3/2 πp resonance would give the three processes in the ratio 9:2:1, compatible with the experimental result. Furthermore, the spin-3/2 state would produce an angular distribution of the form $1 + 3\cos^2\theta$ for each of the processes, while a spin-1/2 state would give an isotropic distribution. The $\pi^+ p$ state must have total isospin $I = 3/2$ since it has $I_z = 3/2$. If the resonance were not in the $I = 3/2$ channel, the $\pi^+ p$ state would not participate. Fermi proceeded to show that a phase shift analysis gave the $J = 3/2$, $I = 3/2$ resonance. C. N. Yang, then a student of Fermi's, showed, however, that the phase shift analysis had ambiguities and that the resonant hypothesis was not unique. It took another two years to settle fully the matter with many measurements and phase shift analyses. Especially important was the careful work of J. Ashkin et al. at the Rochester cyclotron, which showed that there is indeed a resonance, what is now called the $\Delta(1232)$ (Ref. 5.2). A mature analysis of the $J = 3/2, I = 3/2$ pion–nucleon channel is shown in Figure 5.1.

The canonical form for a resonance is associated with the names of G. Breit and E. Wigner. A heuristic derivation of a resonance amplitude is obtained by recalling that for s-wave potential scattering, the scattering amplitude is given by

$$f = \frac{\exp(2i\delta) - 1}{2ik} \tag{5.1}$$

where δ is the phase shift and k is the center-of-mass momentum. For elastic scattering the phase shift is real. If there is inelastic scattering δ has a positive imaginary part. For the purely elastic case it follows that

$$Im(1/f) = -k \tag{5.2}$$

which is satisfied by

$$1/f = (r - i)k \tag{5.3}$$

where r is any real function of the energy. Clearly, the amplitude is biggest when r vanishes. Suppose this occurs at an energy E_0 and that r has only a linear dependence on E, the total center-of-mass energy. Then we can introduce a constant Γ that determines how rapidly r passes through zero:

$$f = \frac{1}{2k(E_0 - E)/\Gamma - ik} = \frac{1}{k} \cdot \frac{\Gamma/2}{(E_0 - E) - i\Gamma/2}. \tag{5.4}$$

The differential cross section is

$$d\sigma/d\Omega = |f|^2 \tag{5.5}$$

and the total cross section is

$$\sigma = 4\pi |f|^2 = \frac{4\pi}{k^2} \frac{\Gamma^2/4}{(E - E_0)^2 + \Gamma^2/4}. \tag{5.6}$$

The quantity Γ is called the full width at half maximum or, more simply, the width. This formula can be generalized to include spin for the resonance (J), the spin of two incident particles (S_1, S_2), and multichannel effects. The total width receives contributions from various channels,

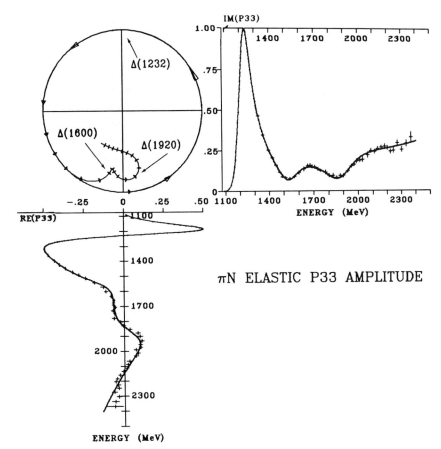

Figure 5.1. An analysis of the $J = 3/2, I = 3/2$ channel of pion–nucleon scattering. Scattering data have been analyzed and fits made to the various angular momentum and isospin channels. For each channel there is an amplitude, $a_{IJ} = (e^{i\delta_{IJ}} - 1)/2i$, where δ_{IJ} is real for elastic scattering and $\text{Im}\delta_{IJ} > 0$ if there is inelasticity. Elastic scattering gives an amplitude on the boundary of the Argand circle, with a resonance occurring when the amplitude reaches the top of the circle. In the figure, the elastic resonance at 1232 MeV is visible, as well as two inelastic resonances. Tick marks indicate 50 MeV intervals. The projections of the imaginary and real parts of the $J = 3/2, I = 3/2$ partial wave amplitude are shown to the right and below the Argand circle [Results of R. E. Cutkosky as presented in *Review of Particle Properties*, *Phys. Lett.* **170B**, 1 (1986)].

$\Gamma = \sum_n \Gamma_n$, where Γ_n is the partial decay rate into the final state n. If the partial width for the incident channel is Γ_{in} and the partial width for the final channel is Γ_{out}, the Breit–Wigner formula is

$$\sigma = \frac{4\pi}{k^2} \frac{2J+1}{(2S_1+1)(2S_2+1)} \frac{\Gamma_{in}\Gamma_{out}/4}{(E-E_0)^2 + \Gamma^2/4}. \tag{5.7}$$

In this formula, k is the center-of-mass momentum for the collision.

As higher pion energies became available at the Brookhaven Cosmotron, more πp resonances (this time in the $I = 1/2$ channel and hence seen only in $\pi^- p$) were observed,

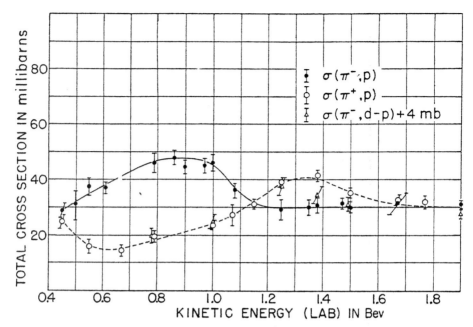

Figure 5.2. Data from the Brookhaven Cosmotron for $\pi^+ p$ and $\pi^- p$ scattering. The cross section peak present for $\pi^- p$ and absent for $\pi^+ p$ demonstrates the existence of an $I = 1/2$ resonance (N^*) near 900 MeV kinetic energy (center-of-mass-energy 1685 MeV). A peak near 1350 MeV kinetic energy (center-of-mass-energy 1925 MeV) is apparent in the $\pi^+ p$ channel, indicating an $I = 3/2$ resonance, as shown in Figure 5.1. Ultimately, several resonances were found in this region. (Ref. 5.3)

as shown in Figure 5.2. Improved measurements of these resonances came from photoproduction experiments, $\gamma N \to \pi N$, carried out at Caltech and at Cornell (Ref. 5.4).

The full importance and wide-spread nature of resonances became clear only in 1960 when Luis Alvarez and a team that was to include A. Rosenfeld, F. Solmitz, and L. Stevenson began their work with separated K^- beams in hydrogen bubble chambers exposed at the Bevatron. The first resonance observed **(Ref. 5.5)** was the $I = 1$ $\Lambda\pi$ resonance originally called the Y_1^*, but now known as the $\Sigma(1385)$. The reaction studied in the Lawrence Radiation Laboratory's 15-inch hydrogen bubble chamber was $K^- p \to \Lambda \pi^+ \pi^-$ at 1.15 GeV/c. The tracks in the bubble chamber pictures were measured on semiautomatic measuring machines and the momenta were determined from the curvature and the known magnetic field. The measurements were refined by requiring that the fitted values conserve momentum and energy. The invariant masses of the pairs of particles,

$$M_{12}^2 = (p_1 + p_2)^2 = (E_1 + E_2)^2 - (\mathbf{p_1} + \mathbf{p_2})^2 \tag{5.8}$$

were calculated. For three-particle final states a Dalitz plot was used, with either the center-of-mass frame kinetic energies, or equivalently, two invariant masses squared, as variables. As for the τ-meson decay originally studied by Dalitz, in the absence of dynamical correlations, purely s-wave decays would lead to a uniform distribution over the Dalitz plot.

The most surprising result found by the Alvarez group was a band of high event density at fixed invariant mass, indicating the presence of a resonance.

The data showed resonance bands for both the $Y^{*+} \to \Lambda\pi^+$ and the $Y^{*-} \to \Lambda\pi^-$ processes. Since the isospins for Λ and π are 0 and 1 respectively, the Y^* had to be an isospin-1 resonance. The Alvarez group also tried to determine the spin and parity of the Y^*, but with only 141 events this was not possible.

This first result was followed rapidly by the observation of the first meson resonance, the $K^*(890)$, observed in the reaction $K^-p \to \overline{K}^0\pi^-p$, measured in the same bubble chamber exposure **(Ref. 5.6)**. This result was based on 48 identified events, of which 21 lay in the K^* resonance peak. The data were adequate to demonstrate the existence of the resonance, but provided only the limit $J < 2$ for the spin. The isospin was determined to be 1/2 on the basis of the decays $K^{*-} \to K^-\pi^0$ and $K^{*0} \to K^-\pi^+$.

A very important $J = 1$ resonance had been predicted first by Y. Nambu and later by W. Frazer and J. Fulco. This $\pi\pi$ resonance, the ρ, was observed by A. R. Erwin et al. using the 14-inch hydrogen bubble chamber of Adair and Leipuner at the Cosmotron **(Ref. 5.7)**. The reactions studied were $\pi^-p \to \pi^-\pi^0 p$, $\pi^-p \to \pi^-\pi^+n$, and $\pi^-p \to \pi^0\pi^0 n$. Events were selected so that the momentum transfer between the initial and final nucleons was small. For these events, there was a clear peak in the $\pi\pi$ mass distribution. From the ratio of the rates for the three processes, the $I = 1$ assignment was indicated, as required for a $J = 1$ $\pi\pi$ resonance ($J = 1$ makes the spatial wave function odd, so bose statistics require that the isospin wave function be odd, as well).

By requiring that the momentum transfer be small, events were selected that corresponded to the "peripheral" interactions, that is, those where the closest approach (classically) of the incident particles was largest. In these circumstances, the uncertainty principle dictates the reaction be described by the virtual exchange of the lightest particle available, in this instance, a pion. Thus the interaction could be viewed as a collision of an incident pion with a virtual pion emitted by the nucleon. The subsequent interaction was simply $\pi\pi$ scattering. This fruitful method of analysis was developed by G. Chew and F. Low. For the Erwin et al. experiment, the analysis showed that the $\pi\pi$ scattering near 770 MeV center-of-mass energy was dominated by a spin-1 resonance.

Shortly after the discovery of the ρ, a second vector (spin-1) resonance was found, this time in the $I = 0$ channel. B. Maglich, together with other members of the Alvarez group, studied the reaction $\overline{p}p \to \pi^+\pi^-\pi^+\pi^-\pi^0$ using a 1.61 GeV/c separated antiproton beam **(Ref. 5.8)**. After scanning, measurement, and kinematic fitting, distributions of the $\pi\pi\pi$ masses were examined. A clean, very narrow resonance was observed with a width $\Gamma < 30$ MeV. The peak occurred in the $\pi^+\pi^-\pi^0$ combination, but not in the combinations with total charges other than 0. This established that the resonance had $I = 0$. A Dalitz plot analysis showed that $J^P = 1^-$ was preferred, but was not a unique solution. The remaining uncertainty was eliminated in a subsequent paper (Ref. 5.9). The Dalitz plot proved an especially powerful tool in the analysis of resonance decays, especially of those into three pions. This was studied systematically by Zemach, who determined where zeros should occur for various spins and isospins, as shown in Figure 5.3.

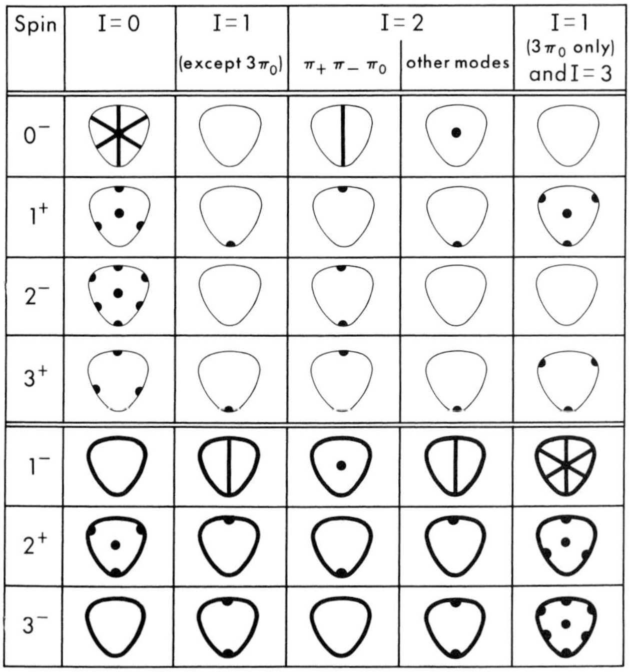

Figure 5.3. Zemach's result for the location of zeros in decays into three pions. The dark spots and lines mark the location of zeros. C. Zemach, *Phys. Rev.* **133**, B1201 (1964).

The discovery of the meson resonances took place in "production" reactions. The resonance was produced along with other final-state particles. The term "formation" is used to describe processes in which the resonance is formed from the two incident particles with nothing left over, as in the Δ resonance formed in πN collisions ($N = p$ or n).

The term "resonance" is applied when the produced state decays strongly, as in the ρ or K^*. States such as the Λ, which decay weakly, are termed particles. The distinction is, however, somewhat artificial. Which states decay weakly and which decay strongly is determined by the masses of the particles involved. The ordering of particles by mass may not be fundamental. Geoffrey Chew proposed the concept of "nuclear democracy," that all particles and resonances were on an equal footing. This view has survived and a resonance like K^* is regarded as no less fundamental than the K itself, even though its lifetime is shorter by a factor of 10^{14}.

The proliferation of particles and resonances called for an organizing principle more powerful than the Gell-Mann–Nishijima relation and one was found as a generalization of isospin. One way to picture isospin is to regard the proton and neutron as fundamental objects. The pion can then be thought of as a combination of a nucleon and an antinucleon, for example, $\bar{n}p \rightarrow \pi^+$. This is called the Fermi–Yang model. S. Sakata proposed to extend this by taking the n, p, and Λ as fundamental. In this way the strange mesons could be accommodated: $\overline{\Lambda}p \rightarrow K^+$. The hyperons like Σ could also be represented: $\bar{n}\Lambda p \rightarrow \Sigma^+$. Isospin, which can be represented by the n and p, has the mathematical structure of

$SU(2)$. Sakata's symmetry, based on n, p, and Λ, is $SU(3)$. Ultimately, Murray Gell-Mann and independently, Yuval Ne'eman proposed a similar but much more successful model.

Each isospin or $SU(3)$ multiplet must be made of particles sharing a common value of spin and parity. Without knowing the spins and parities of the particles it is impossible to group them into multiplets. Because the decays $\Lambda \to \pi^- p$ and $\Lambda \to \pi^0 n$ are weak and, as we shall learn in the next chapter, do not conserve parity, it is necessary to fix the parity of the Λ by convention. This is done by taking it to have $P = +1$ just like the nucleon. With this chosen, the parity of the K is an experimental issue.

The work of M. Block et al. (Ref. 5.10) studying hyperfragments produced by K^- interactions in a helium bubble chamber showed the parity of the K^- to be negative. The process observed was $K^- \text{He}^4 \to \pi^- \text{He}^4_\Lambda$. The He^4_Λ hyperfragment consists of $ppn\Lambda$ bound together. It was assumed that the hyperfragment had spin-zero and positive parity, as was subsequently confirmed. The reaction then had only spin-zero particles and the parity of the K^- had to be the same as that of the π^- since any parity due to orbital motion would have to be identical in the initial and final states.

The parity of the Σ was determined by Tripp, Watson, and Ferro-Luzzi (Ref. 5.11) by studying $K^- p \to \Sigma \pi$ at a center-of-mass energy of 1520 MeV. At this energy there is an isosinglet resonance with $J^P = 3/2^+$. The angular distribution of the produced particles showed that the parity of the Σ was positive. Thus it could fit together with the nucleon and Λ in a single multiplet. The Ξ was presumed to have the same J^P.

In the Sakata model the baryons p, n, and Λ formed a 3 of $SU(3)$, while the pseudoscalars formed an octet. In the version of Gell-Mann and Ne'eman the baryons were in an octet, not a triplet. The baryon octet included the isotriplet Σ and the isodoublet Ξ in addition to the nucleons and the Λ. The basic entity of the model of Gell-Mann and Ne'eman was the octet. All particles and resonances were to belong either to octets, or to multiplets that could be made by combining octets. The rule for combining isospin multiplets is the familiar law of addition of angular momentum. For $SU(3)$, the rule for combining two octets gives $1 + 8 + 8 + 10 + 10^* + 27$. (Here the 10 and 10* are two distinct ten-dimensional representations.) The "eightfold way" postulated that only these multiplets would occur. The baryon octet is displayed in Figure 5.4.

The pseudoscalar mesons known in 1962 were the π^+, π^0, π^-, the $K^+, K^0, \overline{K}^0$, and the K^-. Thus, there was one more to be found according to $SU(3)$. A. Pevsner of Johns Hopkins University and M. Block of Northwestern University, together with their coworkers found this particle, now called the η, by studying bubble chamber film from Alvarez's 72-inch bubble chamber filled with deuterium. The exposure was made with a π^+ beam of 1.23 GeV/c at the Bevatron (**Ref. 5.12**). The particle was found in the $\pi^+ \pi^- \pi^0$ channel at a mass of 546 MeV. No charged partner was found, in accordance with the $SU(3)$ prediction that the new particle would be an isosinglet. The full pseudoscalar octet is displayed in Figure 5.5 in the conventional fashion.

The η was established irrefutably as a pseudoscalar by M. Chrétien et al. (Ref. 5.13) who studied $\pi^- p \to \eta n$ at 1.72 GeV using a heavy liquid bubble chamber. The heavy liquid improved the detection of photons by increasing the probability of conversion. This enabled the group to identify the two-photon decay of the η. See Figure 5.6. By Yang's

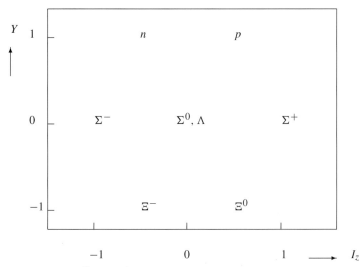

Figure 5.4. The baryon $J^P = 1/2^+$ octet containing the proton and the neutron. The horizontal direction measures I_z, the third component of isospin. The vertical axis measures the hypercharge, $Y = B + S$, the sum of baryon number and strangeness.

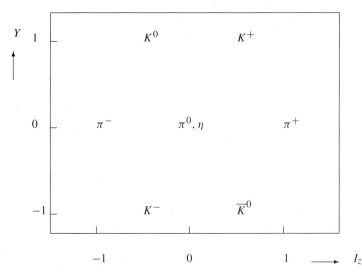

Figure 5.5. The pseudoscalar octet. The horizontal direction measures I_z while the vertical measures the hypercharge, $Y = B + S$.

theorem, this excluded spin-one as a possibility. The absence of the two-pion decay mode excluded the natural spin–parity sequence $0^+, 1^-, 2^+, \ldots$ If the possibility of spin two or higher is discounted, only 0^- remains.

Surprisingly the decay of the η into three pions is an electromagnetic decay. The η has three prominent decay modes : $\pi^+\pi^-\pi^0$, $\pi^0\pi^0\pi^0$, and $\gamma\gamma$. The last is surely electro-

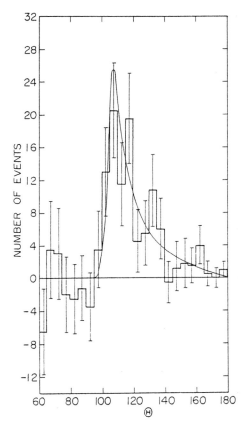

Figure 5.6. A histogram of the opening angle between the two photons in the decay $\eta \to \gamma\gamma$. The solid curve is the theoretical expectation corresponding to the mass of the η (Ref. 5.13).

magnetic, and since it is comparable in rate to the others, they cannot be strong decays. The absence of a strong decay is most easily understood in terms of G-parity, a concept introduced by R. Jost and A. Pais, and independently, by L. Michel.

G-parity is defined to be the product of charge conjugation, C, with the rotation in isospin space $e^{-i\pi I_y}$. Since the strong interactions respect both charge conjugation and isospin invariance, G-parity is conserved in strong interactions. The non-strange mesons are eigenstates of G-parity and for the neutral members like ρ^0 ($I = 1$, $C = +1$), ω^0 ($I = 0$, $C = -1$), η^0 ($I = 0$, $C = +1$), and π^0 ($I = 1$, $C = +1$), the G-parity is simply $C(-1)^I$. All members of the multiplet have the same G-parity even though the charged particles are not eigenstates of C. Thus the pions all have $G = -1$. The ρ has even G-parity and decays into an even number of pions. The ω has odd G-parity and decays into an odd number of pions.

The η has $G = +1$ and cannot decay strongly into an odd number of pions. On the other hand, it cannot decay strongly into two pions since the $J = 0$ state of two pions must have even parity, while the η is pseudoscalar. Thus the strong decay of the η must be into four pions. Now this is at the edge of kinematic possibility (if two of the pions are neutral), but to obtain $J^P = 0^-$, the pions must have some orbital angular momentum. This is scarcely possible given the very small momenta the pions would have in such a decay. As a result, the 3π decay, which violates G-parity and thus must be electromagnetic, is a dominant mode.

The $SU(3)$ symmetry is not exact. Just as the small violations of isospin symmetry lead to the proton–neutron mass difference, the larger deviations from $SU(3)$ symmetry break the mass degeneracy among the particles in the meson and baryon octets. By postulating a simple form for the symmetry breaking, Gell-Mann and subsequently, S. Okubo were able to predict the mass relations

$$\frac{1}{2}(m_p + m_\Xi) = \frac{1}{4}(m_\Sigma + 3m_\Lambda), \tag{5.9}$$

$$m_K^2 = \frac{1}{4}(m_\pi^2 + 3m_\eta^2). \tag{5.10}$$

The use of m for the baryons and m^2 for the mesons relies on dynamical considerations and does not follow from $SU(3)$ alone. The relations are quite well satisfied.

The baryon and pseudoscalar octets are composed of particles that are stable, that is, decay weakly or electromagnetically, if at all. In addition, the resonances were also found to fall into $SU(3)$ multiplets in which each particle had the same spin and parity. The vector meson multiplet consists of the ρ^+, ρ^0, ρ^-, K^{*+}, K^{*0}, $\overline{K}^{*0}$, K^{*-}, and ω. The spin of the $K^*(890)$ was determined in an experiment by W. Chinowsky et al. (Ref. 5.14) who observed the production of a pair of resonances, $K^+ p \to K^* \Delta$. They found that $J > 0$ for the K^*, while Alston et al. found $J < 2$. The result was $J^P = 1^-$. An independent method, due to M. Schwartz, was applied by R. Armenteros et al. (Ref. 5.15) who reached the same conclusion.

An additional vector meson, ϕ, decaying predominantly into $K\overline{K}$ was discovered by two groups, a UCLA team under H. Ticho (Ref. 5.16) and a Brookhaven–Syracuse group, P. L. Connolly et al. (**Ref. 5.17**), the former using an exposure of the 72-inch hydrogen bubble chamber to K^- mesons at the Bevatron, the latter using the 20-inch hydrogen bubble chamber at the Cosmotron.

The reactions studied were

(1) $\quad K^- p \to \Lambda K^0 \overline{K}^0$

(2) $\quad K^- p \to \Lambda K^+ K^-$

A sharp peak very near the $K\overline{K}$ threshold was observed and it was demonstrated that the spin of the resonance was odd, and most likely $J = 1$.

The analysis relies on the combination of charge conjugation and parity, CP. From the decay $\phi \to K^+ K^-$ we know that if the spin of the ϕ is J, then $C = (-1)^J$, $P = (-1)^J$, and so it has $CP = +1$. As discussed in Chapter 7, the neutral kaon system has very special properties. The K^0 and $\overline{K}^0$ mix to produce a short-lived state, K_S^0 and a longer-lived K_L^0. These are very nearly eigenstates of CP with $CP(K_S^0) = +1$, $CP(K_L^0) = -1$. M. Goldhaber, T. D. Lee, and C. N. Yang noted that a state of angular momentum J composed of a K_S^0 and a K_L^0 thus has $CP = -(-1)^J$. Thus the observation of the $K_S^0 K_L^0$ in the decay of the CP even ϕ would show the spin to be odd. Conversely, the observation of $K_L^0 K_L^0$ or $K_S^0 K_S^0$ would, because of Bose statistics, show the state to have even angular momentum. The long-lived K is hard to observe because it exits from the bubble chamber before decaying. Thus when the experiment of Connolly et al. observed 23 ΛK_S^0, but no events $\Lambda K_S^0 K_S^0$, it was concluded that the spin was odd, and probably $J = 1$.

With the addition of the ϕ there were nine vector mesons. This filled an octet multiplet and a singlet (a one-member multiplet). The isosinglet members of these two multiplets have the same quantum numbers, except for their $SU(3)$ designation. Since $SU(3)$ is an approximate rather than an exact symmetry, these states can mix, that is, neither the ω nor the ϕ is completely singlet or completely octet. The same situation arises for the pseudoscalars, where there is in addition an η' meson, which mixes with the η.

The octet of spin-1/2 baryons including the nucleons consisted of the $p, n, \Lambda, \Sigma^+, \Sigma^0, \Sigma^-, \Xi^0, \Xi^-$. This multiplet was complete. The Δ had spin 3/2 and could not be part of this multiplet. An additional spin-3/2 baryon resonance was known, the $Y^*(1385)$ or $\Sigma(1385)$. Furthermore, another baryon resonance was found by the UCLA group (**Ref. 5.18**) and the Brookhaven–Syracuse collaboration (Ref. 5.19) that discovered the ϕ. They observed the reactions

$$K^- p \to \Xi^- \pi^0 K^+$$
$$K^- p \to \Xi^- \pi^+ K^0$$

and found a resonance in the $\Xi\pi$ system with a mass of about 1530 MeV. Its isospin must be 3/2 or 1/2. If it is the former, the first reaction should be twice as common as the second, while experiment found the second dominated. The spin and parity were subsequently determined to be $J^P = (3/2)^+$.

The $J^P = (3/2)^+$ baryon multiplet thus contained 4Δs, 3Σ*s, and 2Ξ*s. The situation came to a head at the 1962 Rochester Conference. According to the rules of the eightfold way, this multiplet could only be a 10 or a 27. The 27 would involve baryons of positive strangeness. None had been found. Gell-Mann, in a comment from the floor, declared the multiplet was a 10 and that the tenth member had to be an $S = -3, I = 0, J^P = (3/2)^+$ state with a mass of about 1680 MeV. It was possible to predict the mass from the pattern of the masses of the known members of the multiplet. For the 10, it turns out that there should be equal spacing between the multiplets. From the known differences $1385 - 1232 = 153$, $1530 - 1385 = 145$, the mass was predicted to be near 1680. The startling aspect of the prediction was that the particle would decay weakly, not strongly since the lightest $S = -3$ state otherwise available is $\Lambda \overline{K}^0 K^-$ with a mass of more than 2100 MeV. Thus the new state would be a particle, not a resonance. The same conclusion had been reached independently by Y. Ne'eman, who was also in the audience.

Bubble chamber physicists came home from the conference and started looking for the Ω^-, as it was called. Two years later, a group including Nick Samios and Ralph Shutt working with the 80-inch hydrogen bubble chamber at Brookhaven found one particle with precisely the predicted properties **(Ref. 5.20)**. The decay sequence they observed was

$$K^- p \rightarrow \Omega^- K^+ K^0$$
$$\Omega^- \rightarrow \Xi^0 \pi^-$$
$$\Xi^0 \rightarrow \Lambda \pi^0$$
$$\Lambda \rightarrow p \pi^-$$

The π^0 was observed through the conversion of its photons. The complete $J^P = 3/2^+$ decuplet is shown in Figure 5.7.

This was a tremendous triumph for both theory and experiment. With the establishment of $SU(3)$ pseudoscalar and vector octets, a spin-1/2 baryon octet, and finally a spin-3/2 baryon decuplet, the evidence for the eightfold way was overwhelming. Other multiplets were discovered, the tensor meson $J^{PC} = 2^{++}$, octet [$f_2(1270)$, $K_2(1420)$, $a_2(1320)$, $f_2'(1525)$], $J^{PC} = 1^{++}$ and $J^{PC} = 1^{+-}$ meson octets, and numerous baryon octets and decuplets. The discoveries filled the ever-growing editions of the *Review of Particle Physics*.

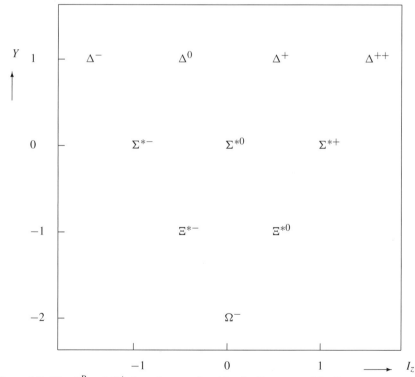

Figure 5.7. The $J^P = 3/2^+$ decuplet completed by the discovery of the Ω^-.

A clearer understanding of $SU(3)$ emerged when Gell-Mann and independently, G. Zweig proposed that hadrons were built from three basic constituents, "quarks" in Gell-Mann's nomenclature. Now called u ("up"), d ("down"), and s ("strange"), these could explain the eightfold way. The mesons were composed of a quark (generically, q) and an antiquark ($\bar{q}$). The Sakata model was resurrected in a new and elegant form. The $SU(3)$ rules dictate that the nine combinations formed from $q\bar{q}$ produce an octet and a singlet. This can be displayed graphically in "weight diagrams," where the horizontal distance is I_z, while the vertical distance is $\sqrt{3}Y/2 = \sqrt{3}(B+S)/2$. The combinations $q\bar{q}$, which make an octet and a singlet of mesons, are represented as sums of vectors, one from q and one from $\bar{q}$.

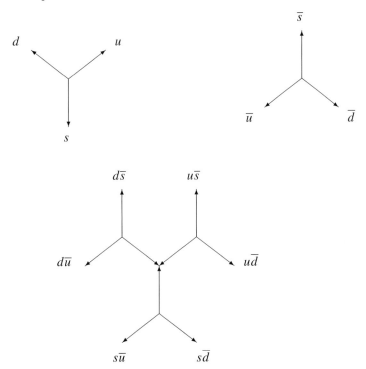

In the $q\bar{q}$ diagram there are three states at the origin ($u\bar{u}$, $d\bar{d}$, $s\bar{s}$) and one state at each of the other points. The state $(u\bar{u} + d\bar{d} + s\bar{s})/\sqrt{3}$ is completely symmetric and forms the singlet representation. The eight other states form an octet. For the pseudoscalar mesons the octet is π^+, π^0, π^-, K^+, K^0, $\bar{K}^0$, K^-, η and the singlet is η'. Actually, since $SU(3)$ is not an exact symmetry, it turns out that there is some mixing of the η and η', as mentioned earlier.

Baryons are produced from three quarks. The $SU(3)$ multiplication rules give $3 \times 3 \times 3 = 10 + 8 + 8 + 1$, so only decuplets, octets, and singlets are expected. The $J^P = (3/2)^+$ decuplet shown in Figure 5.7 contains states like $\Delta^{++} = uuu$ and $\Omega^- = sss$. The $J^P = (1/2)^+$ octet contains the proton (uud), the neutron (udd), etc. There are baryons that are primarily $SU(3)$ singlets, like the $\Lambda(1405)$, which has $J^P = (1/2)^-$, and the $\Lambda(1520)$, with $J^P = (3/2)^-$.

The simplicity and elegance of the quark description of the fundamental particles was most impressive. Still, the quarks seemed even to their enthusiasts more shorthand notation than dynamical objects. After all, no one had observed a quark. Indeed, no convincing evidence was found for the existence of free quarks during the years following their introduction by Gell-Mann and Zweig. Their later acceptance as the physical building blocks of hadrons came as the result of a great variety of experiments described in Chapters 8–11.

Exercises

5.1 Predict the value of the $\pi^+ p$ cross section at the peak of the $\Delta(1232)$ resonance and compare with the data.

5.2 Show that for an $I = 3/2$ resonance the differential cross sections for $\pi^+ p \to \pi^+ p$, $\pi^- p \to \pi^0 n$, and $\pi^- p \to \pi^- p$ are in the ratio 9:2:1. Show that the $\Delta(1232)$ produced in πp scattering yields a $1 + 3\cos^2\theta$ angular distribution in the center-of-mass frame.

5.3 For the $\Delta^{++}(1232)$ and the $Y^{*+}(1385)$, make Argand plots of the elastic amplitudes for $\pi^+ p \to \pi^+ p$ and $\pi^+ \Lambda \to \pi^+ \Lambda$ using the resonance energies and widths given in Table II of Alston *et al.* (Ref. 5.5).

5.4 Verify the ratios expected for $I(\pi\pi) = 0, 1, 2$ in Table I of Erwin *et al.* (**Ref. 5.7**).

5.5 Verify that isospin invariance precludes the decay $\omega \to 3\pi^0$.

5.6 What is the width of the η? How is it measured? Check the *Review of Particle Physics*.

5.7 Verify the estimate of Connolly *et al.* (**Ref. 5.17**) that if $J(\phi) = 1$, then

$$\frac{BR(\phi \to K_S^0 K_L^0)}{BR(\phi \to K_S^0 K_L^0) + BR(\phi \to K^+ K^-)} = 0.39.$$

5.8 How was the parity of the Σ determined? See (Ref. 5.11).

Further Reading

The authoritative compilation of resonances is compiled by the Particle Data Group. A new *Review of Particle Physics* is published in even numbered years and updated annually on the web.

References

5.1 H. L. Anderson, E. Fermi, E. A. Long, and D. E. Nagle, "Total Cross Sections of Positive Pions in Hydrogen." *Phys. Rev.*, **85**, 936 (1952). and *ibid.* p. 934.

5.2 J. Ashkin *et al.*, " Pion Proton Scattering at 150 and 170 MeV." *Phys. Rev.*, **101**, 1149 (1956).

5.3 R. Cool, O. Piccioni, and D. Clark, "Pion-Proton Total Cross Sections from 0.45 to 1.9 BeV." *Phys. Rev.*, **103**, 1082 (1956).
5.4 H. Heinberg et al., "Photoproduction of π^+ Mesons from Hydrogen in the Region 350 - 900 MeV." *Phys. Rev.*, **110**, 1211 (1958). Also F. P. Dixon and R. L. Walker, "Photoproduction of Single Positive Pions from Hydrogen in the 500 – 1000 MeV Region." *Phys. Rev. Lett.*, **1**, 142 (1958).
5.5 M. Alston et al., "Resonance in the $\Lambda\pi$ System." *Phys. Rev. Lett.*, **5**, 520 (1960).
5.6 M. Alston et al., "Resonance in the $K\pi$ System." *Phys. Rev. Lett.*, **6**, 300 (1961).
5.7 A. R. Erwin, R. March, W. D. Walker, and E. West, "Evidence for a $\pi-\pi$ Resonance in the $I=1$, $J=1$ State." *Phys. Rev. Lett.*, **6**, 628 (1961).
5.8 B. C. Maglić, L. W. Alvarez, A. H. Rosenfeld, and M. L. Stevenson, "Evidence for a $T=0$ Three Pion Resonance." *Phys. Rev. Lett.*, **7**, 178 (1961).
5.9 M. L. Stevenson, L. W. Alvarez, B. C. Maglić, and A. H. Rosenfeld, "Spin and Parity of the ω Meson." *Phys. Rev.*, **125**, 687 (1962).
5.10 M. M. Block et al., "Observation of He^4 Hyperfragments from K^- $-$ He Interactions; the K^- $-$ Λ Relative Parity." *Phys. Rev. Lett.*, **3**, 291 (1959).
5.11 R. D. Tripp, M. B. Watson, and M. Ferro-Luzzi, "Determination of the Σ Parity." *Phys. Rev. Lett.*, **8**, 175 (1962).
5.12 A. Pevsner et al., "Evidence for a Three Pion Resonance Near 550 MeV." *Phys. Rev. Lett.*, **7**, 421 (1961).
5.13 M. Chrétien et al., "Evidence for Spin Zero of the η from the Two Gamma Ray Decay Mode." *Phys. Rev. Lett.*, **9**, 127 (1962).
5.14 W. Chinowsky, G. Goldhaber, S. Goldhaber, W. Lee, and T. O'Halloran, "On the Spin of the K^* Resonance." *Phys. Rev. Lett.*, **9**, 330 (1962).
5.15 R. Armenteros et al., "Study of the K^* Resonance in $p\bar{p}$ Annihilations at Rest." Proc. Int. Conf. on High Energy Nuclear Physics, Geneva, 1962, p. 295 (CERN Scientific Information Service)
5.16 P. Schlein et al., "Quantum Numbers of a 1020-MeV $K\bar{K}$ Resonance." *Phys. Rev. Lett.*, **10**, 368 (1963).
5.17 P. L. Connolly et al., "Existence and Properties of the ϕ Meson." *Phys. Rev. Lett.*, **10**, 371 (1963).
5.18 G. M. Pjerrou et al., "Resonance in the $\Xi\pi$ System at 1.53 GeV." *Phys. Rev. Lett.*, **9**, 114 (1962).
5.19 L. Bertanza et al., "Possible Resonances in the $\Xi\pi$ and $K\bar{K}$ Systems." *Phys. Rev. Lett.*, **9**, 180 (1962).
5.20 V. E. Barnes et al., "Observation of a Hyperon with Strangeness Minus Three." *Phys. Rev. Lett.*, **12**, 204 (1964).

Total Cross Sections of Positive Pions in Hydrogen*

H. L. ANDERSON, E. FERMI, E. A. LONG,† AND D. E. NAGLE
*Institute for Nuclear Studies, University of Chicago,
Chicago, Illinois*
(Received January 21, 1952)

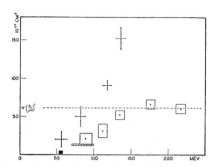

FIG. 1. Total cross sections of negative pions in hydrogen (sides of the rectangle represent the error) and positive pions in hydrogen (arms of the cross represent the error). The cross-hatched rectangle is the Columbia result. The black square is the Brookhaven result and does not include the charge exchange contribution.

IN a previous letter,[1] measurements of the total cross sections of negative pions in hydrogen were reported. In the present letter, we report on similar experiments with positive pions.

The experimental method and the equipment used in this measurement was essentially the same as that used in the case of negative pions. The main difference was in the intensity, which for the positives was much less than for the negatives, the more so the higher the energy. This is due to the fact that the positive pions which escape out of the fringing field of the cyclotron magnet are those which are emitted in the backward direction with respect to the proton beam, whereas the negative pions are those emitted in the forward direction. The difficulty of the low intensity was in part compensated by the fact that the cross section for positive pions turned out to be appreciably larger than for negative pions. The results obtained thus far are summarized in Table I.

In Fig. 1 the total cross sections of positive and negative pions are collected. It is quite apparent that the cross section of the positive particles is much larger than that of the negative particles, at least in the energy range from 80 to 150 Mev.

In this letter and in the two preceding ones,[1,2] the three processes: (1) scattering of positive pions, (2) scattering of negative pions with exchange of charge, and (3) scattering of negative pions without exchange of charge have been investigated. It appears that over a rather wide range of energies, from about 80 to 150 Mev, the cross section for process (1) is the largest, for process (2) is intermediate, and for process (3) is the smallest. Furthermore, the cross sections of both positive and negative pions increase rather rapidly with the energy. Whether the cross sections level off at a high value or go through a maximum, as might be expected if there should be a resonance, is impossible to determine from our present experimental evidence.

Brueckner[3] has recently pointed out that the existence of a broad resonance level with spin 3/2 and isotopic spin 3/2 would give an approximate understanding of the ratios of the cross sections for the three processes (1), (2), and (3). We might point out in this connection that the experimental results obtained to date are also compatible with the more general assumption that in the energy interval in question the dominant interaction responsible for the scattering is through one or more intermediate states of isotopic spin 3/2, regardless of the spin. On this assumption, one finds that the ratio of the cross sections for the three processes should be (9:2:1), a set of values which is compatible with the experimental observations. It is more difficult, at present, to say anything specific as to the nature of the intermediate state or states. If there were one state of spin 3/2, the angular distribution for all three processes should be of the type $1+3\cos^2\theta$. If the dominant effect were due to a state of spin 1/2, the angular distribution should be isotropic. If states of higher spin or a mixture of several states were involved, more complicated angular distributions would be expected. We intend to explore further the angular distribution in an attempt to decide among the various possibilities.

Besides the angular distribution, another important factor is the energy dependence. Here the theoretical expectation is that, if there is only one dominant intermediate state of spin 3/2 and isotopic spin 3/2, the total cross section of negative pions should at all points be less than $(8/3)\pi\lambda^2$. Apparently, the experimental cross section above 150 Mev is larger than this limit, which indicates that other states contribute appreciably at these energies. Naturally, if a single state were dominant, one could expect that the cross sections would go through a maximum at an energy not far from the energy of the state involved. Unfortunately, we have not been able to push our measurements to sufficiently high energies to check on this point.

Also very interesting is the behavior of the cross sections at low energies. Here the energy dependence should be approximately proportional to the 4th power of the velocity if only states of spin 1/2 and 3/2 and even parity are involved and if the pion is pseudoscalar. The experimental observations in this and other laboratories seem to be compatible with this assumption, but the cross section at low energy is so small that a precise measurement becomes difficult.

* Research sponsored by the ONR and AEC.
† Institute for the Study of Metals, University of Chicago.
[1] Anderson, Fermi, Long, Martin, and Nagle, Phys. Rev., this issue.
[2] Fermi, Anderson, Lundby, Nagle, and Yodh, preceding Letter, this issue, Phys. Rev.
[3] K. A. Brueckner (private communication).

TABLE I. Total cross sections of positive pions in hydrogen.

Energy (Mev)	Cross section (10^{-27} cm²)
56 ±8	20 ±10
82 ±7	50 ±13
118 ±6	91 ± 6
136 ±6	152 ±14

RESONANCE IN THE $\Lambda\pi$ SYSTEM*

Margaret Alston, Luis W. Alvarez, Philippe Eberhard,[†] Myron L. Good,[‡]
William Graziano, Harold K. Ticho,[||] and Stanley G. Wojcicki
Lawrence Radiation Laboratory and Department of Physics, University of California, Berkeley, California
(Received October 31, 1960)

We report a study of the reaction

$$K^- + p \to \Lambda^0 + \pi^+ + \pi^- \quad (1)$$

produced by 1.15-Bev/c K^- mesons and observed in the Lawrence Radiation Laboratory's 15-in. hydrogen bubble chamber. A preliminary report of these results was presented at the 1960 Rochester Conference.[1] The beam was purified by two velocity spectrometers.[2] A Ξ^0 hyperon observed during the run[3] and the preliminary cross sections[4] for various K^- reactions at 1.15 Bev/c have been reported previously. Reaction (1) was the first one selected for detailed study, because it appeared to take place with relatively large probability and because the event, a 2-prong interaction accompanied by a V, was easily identified. In a volume of the chamber sufficiently restricted so that the scanning efficiency was near 100%, 255 such events were found. These events were measured, and the track data supplied to a computer which tested each event for goodness of fit to various kinematic hypotheses. The possible reactions, the distribution of events, and the corresponding cross sections are given in Table I. An event was placed in a given category of Table I if the χ^2 probability for the other hypotheses was < 1%. It appears likely that the majority of the events in group (e) are also reactions of type (1). This belief is based on the following arguments:

1. Since the kinematics of a $\Lambda\pi\pi$ fit (four constraints) are more overdetermined than those of a $\Sigma^0\pi\pi$ fit (two constraints), it is relatively easy for a $\Lambda\pi\pi$ reaction to fit a $\Sigma^0\pi\pi$ reaction, but only

Table I. Distribution of events among different reactions.

Reaction	No. of events	Cross section (mb)
(a) $K^- + p \to \bar{K}^0 + p + \pi^-$	48	2.0 ±0.3
(b) $K^- + p \to (\Lambda$ or $\Sigma^0) + \pi^+ + \pi^- + \pi^0$	39	1.1 ±0.2
(c) $K^- + p \to \Sigma^0 + \pi^+ + \pi^-$	27	4.1 ±0.4
(d) $K^- + p \to \Lambda + \pi^+ + \pi^-$	49	
(e) $K^- + p \to (\Lambda$ or $\Sigma^0) + \pi^+ + \pi^-$	92	
Total	255	7.2 ±0.5

very few Σ^0 configurations can fit the $\Lambda\pi\pi$ reactions.

2. The events of group (e) when treated as $\Sigma^0\pi\pi$ reactions give a χ^2 distribution which is much worse than that obtained when they are treated as $\Lambda\pi\pi$ reactions.

In what follows, the 141 events of groups (d) and (e) are treated as examples of reaction (1). We estimate that 10 to 15% are actually Σ^0 events.

The energy distribution of the two pions in the K^--p barycentric system is shown in Fig. 1. If the cross section were dominated by phase space alone, the distribution of the points on the two-dimensional plot of Fig. 1 should be uniform. This is clearly not the case. On the contrary, both the π^+ and the π^- distributions have peaks near 285 Mev, such as would be expected from a quasi-two-body reaction of the type

$$K^- + p \to Y^{*\pm} + \pi^\mp, \qquad (2)$$

the Y^* having a mass spectrum peaking at ~1380 Mev. If the Y^* of mass 1380 Mev breaks up according to

$$Y^{*\pm} \to \Lambda^0 + \pi^\pm, \qquad (3)$$

the pions from this breakup are expected to have energies ranging from 58 to 175 Mev in the K^-p rest system. Those pions from (3) are well separated from the pions arising from reaction (2) in the energy histograms.

The isotopic spin of this excited hyperon must be one, since it breaks up into a Λ and a π. Since the Y^* is produced with a pion, also of isotopic spin one, the reaction could proceed either in the $I=0$ or the $I=1$ state. Therefore the ratio of Y^{*+} to Y^{*-} will depend on the relative magnitude and phase of the two isotopic-spin amplitudes and thus could differ from unity. We observed 59 Y^{*+} events and 82 Y^{*-} events, using the criterion for separation that the high-momentum π meson is the pion from reaction (2).

Figure 2 shows the distribution in mass of the Y^* state (both Y^{*+} and Y^{*-}) including all 141 events, again using the higher energy pion in each event to calculate the Y^* mass. The experimental uncertainty in the mass for each event is small compared to the observed width of the curve. The curves of Fig. 2 are discussed later.

Figure 3 shows production angular distributions for Y^{*+} and Y^{*-} in the K^-p rest system. Partial waves with $l > 0$ appear to be present, as would be expected since $\hbar k/m_\pi c$ approximately equals 3. The difference between the Y^{*+} and Y^{*-} angular distributions may reflect the different superposi-

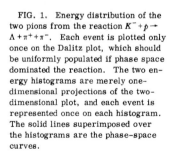

FIG. 1. Energy distribution of the two pions from the reaction $K^- + p \to \Lambda + \pi^+ + \pi^-$. Each event is plotted only once on the Dalitz plot, which should be uniformly populated if phase space dominated the reaction. The two energy histograms are merely one-dimensional projections of the two-dimensional plot, and each event is represented once on each histogram. The solid lines superimposed over the histograms are the phase-space curves.

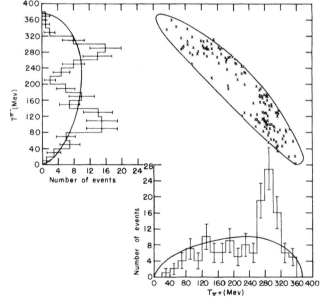

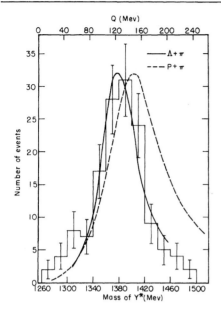

FIG. 2. Mass distribution for Y^* and fitted curves for $\pi\Lambda$ and πp resonances. The lower scale refers only to the $\pi\Lambda$ resonance. Q is the kinetic energy released when either isobar dissociates. The curve for the $\pi\Lambda$ resonances is fitted to the center eight histogram intervals of our data. The πp curve is the fit obtained by Gell-Mann and Watson,[7] to πp scattering data. Both fits are to the formula $\sigma \propto \lambdabar^2 \Gamma^2/[(E-E_0)^2+\frac{1}{4}\Gamma^2]$, where $\Gamma = 2b(a/\lambdabar)^3/[1+(a/\lambdabar)^2]$.

tions of the isotopic-spin zero and one amplitudes for the two cases.

The following two methods were used in an effort to determine the spin of Y^*.

(a) The angular momentum of Y^* was investigated by means of an Adair analysis.[5] We first restricted ourselves to production angles with $|\cos\theta| \geq 0.8$. For this angular range the Adair analysis should be valid if only S and P waves are present in the production process. We then computed η for each event, where

$$\eta = \vec{P}_{K^-} \cdot \vec{P}_\Lambda / (|\vec{P}_{K^-}| \| \vec{P}_\Lambda |).$$

Of the 29 events with $|\cos\theta| \geq 0.8$, the fraction 0.62 ± 0.09 has $|\eta| \geq 0.5$. If the above-mentioned restriction on the angular interval is sufficient to insure the validity of the Adair analysis, this ratio is expected to be 0.50 for $j = 1/2$ and 0.73 for $j = 3/2$. The experimental result is thus ~1.3 standard deviations from both possibilities, and no conclusion may be drawn from the data. Similar results were obtained for several larger values of the cutoff angle. Presence of D waves, however, cannot be excluded by the production angular distributions (Fig. 3). If they are present, indeed, then none of these choices of angle would be sufficiently restrictive to guarantee the success of the Adair analysis.

(b) Since Y^* may be polarized perpendicular to its plane of production, correlations can exist between the decay angle of the Y^* and the polarization of the resulting Λ. Also, a net Λ polarization

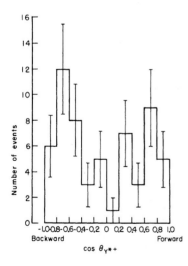
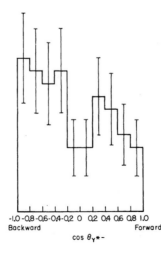

FIG. 3. Angular distribution of $Y^{*\pm}$ in the K^-p barycentric system for the reactions $K^- + p \to Y^{*\pm} + \pi^\mp$.

can result. With our limited data, we see no statistically significant Λ polarization or angular correlations.

However, one can also look for anisotropy, i.e., a polar-to-equatorial ratio, in the decay angle of Y^* with respect to the normal to the plane of production. For spin 3/2, the distribution must be of the form $A + B\xi^2$ by the Sachs-Eisler theorem,[6] independent of the Y^* parity, where we have

$$\xi = (\vec{P}_{K^-} \times \vec{P}_{Y^*}) \cdot \vec{P}_\Lambda / (|\vec{P}_{K^-} \times \vec{P}_{Y^*}||\vec{P}_\Lambda|),$$

and $\vec{P}_a$ is the momentum of the particle a in the K^-p barycentric system.

Since the coefficient B is a function of the production angle, we want to restrict ourselves to that range of the solid angle where the polar-to-equatorial anisotropy is probably greatest along the normal to the production plane. For production angles near 0 deg and 180 deg (Adair-analysis region), one expects the polar-to-equatorial ratio to be most different from unity in another direction (namely along the direction of the beam). Thus the equatorial region of production angles is more likely to show a large anisotropy along the direction in question. Therefore the production-angle range $\sin\theta \geq 0.866$ was selected for study. We find the ratio of events with $|\xi| > 0.5$ to all events is 0.355. If the distribution is isotropic, as is required for spin 1/2, we expect 0.500 ± 0.063 for our 62 events. The result is thus 2.3 standard deviations from isotropy. The 45-to-1 odds against isotropy overstate the case for higher spin because this is the fourth anisotropy looked for.

Since Y^* may be regarded as a hyperon isobar, which decays into a π and a Λ, it evidently corresponds to a resonance in pion-hyperon scattering. The mass distribution of Fig. 2 then invites a comparison to the cross section for pion-nucleon scattering in the 3/2-3/2 state. For this purpose a p-wave resonance formula employed by Gell-Mann and Watson[7] for pion-nucleon scattering was fitted to our $\pi\Lambda$ data by using the eight central histogram intervals of Fig. 2. In fitting the curve, it was found that the interaction radius (a) could be varied over a wide range without changing the goodness of fit appreciably, provided that the reduced width (b) was also changed appropriately. The radius parameter was therefore fixed arbitrarily at $\hbar/m_\pi c$. Table II summarizes our results for Y^*, along with those of Gell-Mann and Watson for the 3-3 resonance.

Even if Y^* does turn out to be a p-wave reso-

Table II. Parameters for π-Λ and π-p resonance fitted to $\sigma \propto \lambdabar^2 \Gamma^2/[(E-E_0)^2 + \frac{1}{4}\Gamma^2]$, where $\Gamma = 2b(a/\lambdabar)^3/[1+(a/\lambdabar)^2]$.

Parameter	π-p	π-Λ
Interaction radius a in units of $\hbar/m_\pi c$	0.88	1
Reduced width b (Mev)	58	33.4
Resonance energy E_0 (Mev)	159	129.3
Full width at half maximum (Mev)	100	64
Lifetime (sec)		$\sim 10^{-23}$

nance, there are still many reasons why the π-Λ resonance parameters must not be taken too literally: (a) There is a small contamination of $\Sigma^0\pi\pi$ events in our data. (b) A nonresonant background may be present. (c) The production matrix element for reaction (2) might well depend on the outgoing momentum, and hence distort the mass distribution of Y^*. (d) Two thresholds for other possible decay modes of Y^* appear within the mass interval covered by the resonance curve; i.e., the $\Sigma\pi$ mode threshold around 1330 Mev and the $\bar{K}N$ threshold around 1435 Mev. This must have some effect on the shape of the mass spectrum as observed via the $\Lambda\pi$ decay mode. (e) Final-state pion-pion interaction could disturb the spectrum. (f) Even when the two resonances, Y^{*+} and Y^{*-}, are well resolved in terms of intensity—as in our experiment—there can still be an appreciable interference between the amplitude in which the π^+ arises from reaction (2) and the π^- from reaction (3) and the amplitude in which the roles of the two pions are reversed.

If we bear all these uncertainties in mind, the resemblance to the 3-3 resonance is certainly remarkable (Fig. 2). The resonance energies when expressed in terms of barycentric kinetic energies differ by only 30 Mev, which is much less than the width of either resonance. Furthermore, the widths are at least comparable.

These results are strongly reminiscent of the concept of global symmetry which predicts two spin 3/2 pion-hyperon resonances, one with $T=1$, the other with $T=2$.[8] These are the hyperon counterparts of the $J = T = 3/2$ resonance of the pion-nucleon system. On the other hand, the possibility that Y^* is a $J=1/2$ resonance cannot be excluded on the basis of our data. The concept of pion-hyperon resonance in either $J = 1/2$ or 3/2 state has been discussed recently by several authors.[9]

A study of $\Sigma^{\mp}\pi^{\pm}\pi^0$ events in our experiment is under way at present. The results, however, are too incomplete for us to be able to draw any definite conclusions.

The authors are greatly indebted to the bubble chamber crew under the direction of James D. Gow for their fine job in operating the chamber, especially Robert D. Watt and Glen J. Eckman for their invaluable help with the velocity spectrometers. We also gratefully acknowledge the cooperation of Dr. Edward J. Lofgren and the Bevatron crew, as well as the skilled work and cooperation of our scanning and measuring staff. Special thanks are due the many colleagues in our group who developed the PANG and KICK computer programs—especially Dr. Arthur H. Rosenfeld, and to Dr. Frank Solmitz for many helpful discussions.

One of us (P.E.) is grateful to the Philippe's Foundation Inc. and to the Commisariat à l'Energie Atomique for a fellowship.

---

*This work was done under the auspices of the U. S. Atomic Energy Commission.

†Presently at Laboratoire de Physique Atomique, College de France, Paris, France.

‡Presently at University of Wisconsin, Madison, Wisconsin.

∥Presently at University of California at Los Angeles, Los Angeles, California.

[1]Margaret Alston, L. W. Alvarez, P. Eberhard, M. L. Good, W. Graziano, H. K. Ticho, and S. Wojcicki, paper presented at the Tenth Annual Rochester Conference on High-Energy Nuclear Physics, 1960 (to be published).

[2]P. Eberhard, M. L. Good, and H. K. Ticho, Lawrence Radiation Laboratory Report UCRL-8878 Rev, December, 1959 (unpublished); also Rev. Sci. Instr. (to be published).

[3]L. W. Alvarez, P. Eberhard, M. L. Good, W. Graziano, H. K. Ticho, and S. Wojcicki, Phys. Rev. Letters 2, 215 (1959).

[4]L. W. Alvarez, in Proceedings of the 1959 International Conference on High-Energy Physics at Kiev (unpublished); also Lawrence Radiation Laboratory Report UCRL-9354, August, 1960 (unpublished).

[5]R. K. Adair, Phys. Rev. 100, 1540 (1955).

[6]E. Eisler and R. G. Sachs, Phys. Rev. 72, 680 (1947).

[7]M. Gell-Mann and K. Watson, Annual Review of Nuclear Science (Annual Reviews, Inc., Palo Alto, California, 1954), Vol. 4.

[8]M. Gell-Mann, Phys. Rev. 106, 1297 (1957).

[9]R. H. Capps, Phys. Rev. 119, 1753 (1960); R. H. Capps and M. Nauenberg, Phys. Rev. 118, 593 (1960); R. H. Dalitz and S. F. Tuan, Ann. Phys. 10, 307 (1960); M. Nauenberg, Phys. Rev. Letters 2, 351 (1959); A. Komatsuzawa, R. Sugano, and Y. Nogami, Progr. Theoret. Phys. (Kyoto) 21, 151 (1959); Y. Nogami, Progr. Theoret. Phys. (Kyoto) 22, 25 (1959); D. Amati, A. Stanghellini, and B. Vitale, Nuovo cimento 13, 1143 (1959); L. F. Landovitz and B. Margolis, Phys. Rev. Letters 2, 318 (1959); M. H. Ross and C. L. Shaw, Ann. Phys. (to be published).

RESONANCE IN THE K-π SYSTEM*

Margaret Alston, Luis W. Alvarez, Philippe Eberhard,[†] Myron L. Good,[‡]
William Graziano, Harold K. Ticho,[∥] and Stanley G. Wojcicki
Lawrence Radiation Laboratory and Department of Physics, University of California, Berkeley, California
(Received February 16, 1961)

In a continuation of the study of the interaction of 1.15-Bev/c K^- mesons in hydrogen by means of the Lawrence Radiation Laboratory 15-inch hydrogen bubble chamber, we now report a study of the reaction[1]

$$K^- + p \to \overline{K}^0 + \pi^- + p. \quad \text{(A)}$$

Examples of this reaction were easily identified in those cases in which the $\overline{K}^0$ decayed into charged pions and appeared in the chamber as a two-prong interaction associated with a V. A kinematic analysis isolated 48 events of reaction (A) from other events with similar topology.[2] In only one case was the identification not unique. Correcting for neutral decays of the $\overline{K}^0$ and for escape from the chamber, we find a total cross section of 2.0 ± 0.3 mb for Reaction (A).

The events are shown on a Dalitz plot in Fig. 1. If the reaction were entirely dominated by phase space, the Dalitz plot would be uniformly populated. Instead, a strong clumping around proton kinetic energy of 20 Mev is observed. This effect cannot be explained by an interaction matrix element that increases monotonically with decreasing proton energy. Whereas an extrapolation from the region 15 Mev $\leq T_p \leq$ 25 Mev would lead one to expect a minimum of 16 events in the region $T_p \leq 15$ Mev, only three are found there. No experimental bias against very low energy protons in the K-p center-of-mass system can exist, since such protons have laboratory-system momenta of approximately 600 Mev/c, and are easily identified. The observed distribution can best be explained by a quasi-two-body reaction of the type

$$K^- + p \to K^{*-} + p, \quad \text{(B)}$$

followed by a decay,

$$K^{*-} \to \overline{K}^0 + \pi^-. \quad \text{(C)}$$

The 3-3 resonance of the pion-nucleon system would show itself on the Dalitz plot as a concentration of points along the diagonal line drawn through Fig. 1. The absence of any evidence for this resonance in our data can be explained if Reaction (A) proceeds primarily through the $I = 0$ channel, which cannot produce a p-π^- system in the $I = 3/2$ state. Further, even in the $I = 1$ channel, the 3-3 resonance favors $(n + \pi^0) + \overline{K}^0$ over $(p + \pi^-) + \overline{K}^0$, and hence provides additional suppression of this resonance.

The mass distribution of the K^{*-} is shown in

300

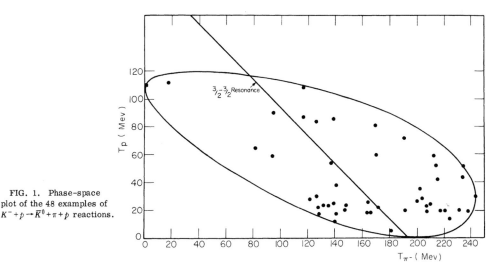

FIG. 1. Phase-space plot of the 48 examples of $K^- + p \to \bar{K}^0 + \pi + p$ reactions.

Fig. 2. The mean value is 885 ± 3 Mev. After removing the number of background events estimated from the phase-space distribution and unfolding the experimental error on each of the remaining 22 events (typically 3 to 4 Mev), we obtained a full width at half maximum of 16 Mev, corresponding to a lifetime of 4×10^{-23} second.

The angular distribution for Reaction (B) is consistent with isotropy. Assuming that the K^{*-} system is produced predominantly in the s state (which appears likely both because of the closeness to the threshold and because of the isotropic

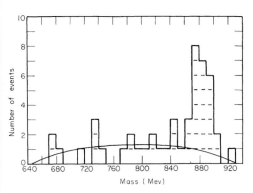

FIG. 2. Mass spectrum of the $\bar{K}^0 - \pi^-$ system. The solid line represents the phase-space curve normalized to background events.

distribution), we can obtain an upper limit of the $\bar{K}^*$ spin S. If $S = 0$, then the reaction can be produced only through the $p_{1/2}$ ingoing channel; if $S = 1$, through $s_{1/2}$ and $d_{3/2}$, etc. In any case the decay angular distribution[3] is given by

$$I(\theta) = |aY_{S,0}(\theta, \phi)|^2 + |bY_{S,1}(\theta, \phi)|^2, \quad (1)$$

with $|a|^2 + |b|^2 = 1$, where $a = 1$ for $S = 0$. Here θ is the angle of the $\bar{K}^0$ in the K^{*-} rest system with respect to the incoming K^- direction. The mean value of $\cos^2\theta$, based on the distribution function (1), is

$$\langle \cos^2\theta \rangle = \frac{(2S^2 + 2S - 3) + 2|a|^2}{4S^2 + 4S - 3}. \quad (2)$$

Experimentally, $\langle \cos^2\theta \rangle$ for the 21 events lying in the K^{*-} mass range between 870 and 900 Mev is 0.275. Using $S = 2$ in Eq. (2), we find the expected value of $\langle \cos^2\theta \rangle \geq 0.429$, with a standard deviation of 0.051. The experimental result thus deviates from the range of values expected for $S = 2$ by three standard deviations. For $S > 2$ the discrepancy is even greater. On the other hand, the experimental result is consistent (within errors) with $S = 0$ or $S = 1$. It is worth noting that an isotropic decay distribution is obtained both for $S = 0$ and for $S = 1$ if the $d_{3/2}$ input channel does not contribute. For this reason, if $S = 0$, experiments at several momenta will be required to settle this problem.

Assuming that the K^- and $\bar{K}^0$ are an isotopic-spin doublet, and $I = 1/2$ for the K^{*-} isotopic spin, the branching ratio $R = (K^{*-} \to K^- + \pi^0)/(K^{*-} \to \overline{K}^0 + \pi^-)$ equals $1/2$; for $I = 3/2$, $R = 2$. In either case another charge state of $\overline{K}^*$, namely $\overline{K}^{*0}$, should exist and decay into $K^- + \pi^+$ or $\overline{K}^0 + \pi^0$. The value of this branching ratio, also, depends on the isotopic spin of the $\overline{K}^*$.

To investigate the isotopic spin properties of the $\overline{K}^*$, we searched for examples of the following two reactions:

$$K^- + p \to K^- + \pi^0 + p, \quad \text{(D)}$$

$$K^- + p \to K^- + \pi^+ + n. \quad \text{(E)}$$

These events appear as two-prong interactions and are much more difficult to identify than the events already discussed, since there are usually several possible interpretations for each inelastic two-prong event. In particular, there was a pion contamination of about 10% in our incident K^- beam, and the inelastic pion interactions are kinematically very similar to Reactions (D) and (E). Both the kinematic fits and the ionization of the tracks were used to identify the events, but these criteria were not always sufficient to distinguish between various hypotheses. At present we have processed only about 2/3 of our two-prong interactions, but we feel that the data obtained are reasonably unbiased.

In both Reactions (D) and (E) there are peaks in the nucleon kinetic energy distribution in the $\overline{K}^*$ resonance region. On the basis of the number of events in the proton peak of Reaction (D), our present data allow us to make a crude estimate of the branching ratio: $R = 0.75 \pm 0.35$. The data thus strongly favor the $I = 1/2$ state.

The experimental production ratio of $\overline{K}^*$ via Reactions (A) and (E) is about 1, and is thus consistent with the production of the $\overline{K}^*$ through a pure isotopic spin state.

An $I = 1/2$ particle, called the K', with negative parity with respect to the K meson, has been invoked by Tiomno to explain the backward Λ peaking in associated production.[4] Gell-Mann postulated the existence of such a particle to permit the construction of a strangeness-violating weak-interaction axial current.[5] The K^* that we have observed has properties consistent with those postulated for the K', but, as discussed above, the K^* spin and parity remain to be established.

As in our previous communication, we acknowledge gratefully the assistance of the many people who helped us to obtain and analyze these data. One of us (P.E.) is grateful to the Philippe's Foundation, Incorporated, and to the Commisariat a l' Énergie Atomique for a fellowship.

---

*Work done under the auspices of the U. S. Atomic Energy Commission.

†Presently at Laboratoire de Physique Atomique, Collège de France, Paris, France.

‡Presently at University of Wisconsin, Madison, Wisconsin.

‖Presently at University of California at Los Angeles, Los Angeles, California.

[1]Preliminary versions of this report were given by M. L. Good at the meeting of the American Physical Society, Chicago, November, 1960 [Bull. Am. Phys. Soc. 5, 414 (1960)]; and by Margaret H. Alston at the Topical Conference on Strong Interactions, Berkeley, December, 1960 (unpublished).

[2]M. Alston, L. W. Alvarez, P. Eberhard, M. L. Good, W. Graziano, H. K. Ticho, and S. Wojcicki, Phys. Rev. Letters 5, 520 (1960).

[3]Richard Spitzer and Henry P. Stapp, Phys. Rev. 109, 540 (1958); a more complete version of this paper is contained in University of California Radiation Laboratory Report UCRL-3796 (Rev.), 1957 (unpublished); also Henry P. Stapp, Lawrence Radiation Laboratory (private communication). We are grateful to Dr. Stapp for several illuminating discussions on this subject.

[4]Jayme Tiomno, in Proceedings of the 1960 Annual International Conference on High-Energy Physics at Rochester (Interscience Publishers, New York, 1960), pp. 466, 467, and 513; also J. Tiomno, A. L. L. Videira, and N. Zagury, Phys. Rev. Letters 6, 120 (1961).

[5]M. Gell-Mann, in Proceedings of the 1960 Annual International Conference on High-Energy Physics at Rochester (Interscience Publishers, New York, 1960), p. 508.

EVIDENCE FOR A π-π RESONANCE IN THE $I=1$, $J=1$ STATE*

A. R. Erwin, R. March, W. D. Walker, and E. West
Brookhaven National Laboratory, Upton, New York and University of Wisconsin, Madison, Wisconsin
(Received May 11, 1961)

Since the earliest data became available on pion production by pions, certain features have been quite clear. The main feature which is strongly exhibited above energies of 1 Bev is that collisions are preferred in which there is a small momentum transfer to the nucleon.[1] This is shown by the nucleon angular distributions which are sharply peaked in the backward direction. These results suggest that large-impact-parameter collisions are important in such processes. The simplest process that could give rise to such collisions is a pion-pion collision with the target pion furnished in a virtual state by the nucleon. The quantitative aspects of such collisions have been discussed by a number of authors. Goebel, Chew and Low, and Salzman and Salzman[2] discussed means of extracting from the data the π-π cross section.

Holladay and Frazer and Fulco[3] deduced from electromagnetic data that indeed there must be a strong pion-pion interaction. In particular, Frazer and Fulco deduced that there probably was a resonance in the $I=1$, $J=1$ state. A qualitative set of π-p phase shifts in the 400-600 Mev[4] region were used by Bowcock et al.[5] to deduce an energy of about 660 Mev in the π-π system for the resonance. The work of Pickup et al.[6] showed an indication of a peak in the π-π spectrum at an energy of about 600 Mev.

The present experiment was designed to explore the π-π system up to an energy of about 1 Bev. The π^- beam was produced by the external proton beam No. 1 at the Cosmotron. A suitable set of quadrupole and bending magnets focussed the pion beam on a Hevimet slit about 10 ft from the Adair-

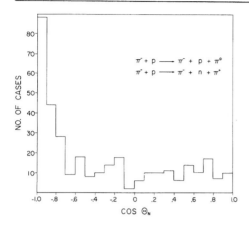

FIG. 1. The angular distribution of the nucleons from the processes $\pi^-+p\to \pi^-+\pi^0+p$ and $\pi^-+p\to \pi^-+\pi^++n$.

Table I. Ratios of final states.

	$I=0$	$I=1$	$I=2$	Experiment ($\Delta \leq 400$ Mev/c)
$\pi^-\pi^+n$	2	2	2/9	1.7 ± 0.3
$\pi^-\pi^0 p$	0	1	1	1
$\pi^0\pi^0 n$	1	0	4/9	$<0.25 \pm 0.25$

Leipuner 14-in. H_2 bubble chamber. The pions were guided into the chamber by another bending magnet. The measured momentum was 1.89 ± 0.07 Bev/c.

Events selected for measurement were taken in a fiducial volume of the chamber. The forward-going track was required to be at least 10 cm long. Measurements were made on a digitized system and the output was analyzed by use of an IBM-704. The events were analyzed by means of a program based on the "Guts" routine written by members of the Alvarez bubble chamber group.

Figure 1 shows the combined angular distribution for the nucleons from the two processes, $\pi^-+p\to \pi^-+\pi^0+p$ and $\pi^-+p\to \pi^-+\pi^++n$, which appear to be identical within statistics. The results indicate a large number of events with small momentum transfer to the nucleons.

We concentrate our interest on those events with small momentum transfer since these events satisfy the qualitative criterion of being examples of π-π collisions. Somewhat arbitrarily, we center our attention on cases in which the momentum transfer to the nucleon is less than 400 Mev/c. Table I gives the ratios of the three possible final states $\pi^-\pi^+n$, $\pi^-\pi^0 p$, and $\pi^0\pi^0 n$, assuming the π-π scattering to be dominated, respectively, by the $I=0$, 1, 2 scattering states of the π-π system.

The experimental results in the last column indicate a strong domination by $I=1$ state. For the $I=1$ state the basic π-π scattering cross sections $\sigma(\pi^-\pi^0 \to \pi^-\pi^0)$ and $\sigma(\pi^-\pi^+ \to \pi^-\pi^+)$ are equal.

The nucleon four-momentum transfer spectrum seems to be in qualitative agreement with the theory for the process in which a π is knocked out of the cloud. Figure 2 shows ideograms for the mass spectrum of the di-pions for cases with $\Delta \leq 400$ Mev/c and $\Delta > 400$ Mev/c, where Δ is the four-momentum transfer to the nucleon. The curve for $\Delta \leq 400$ Mev/c clearly shows a peak at 765 Mev/c. In the ideogram for $\Delta > 400$ Mev/c the peak is still present but seems to be smeared to higher values of the di-pion mass, m^*. One worries that diagrams other than the one involving

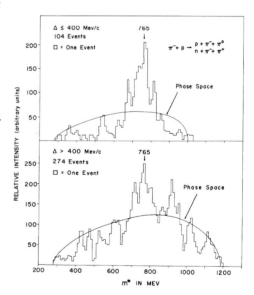

FIG. 2. The combined mass spectrum for the $\pi^-\pi^0$ and $\pi^-\pi^+$ system. The smooth curve is phase space as modified for the included momentum transfer and normalized to the number of events plotted. Events used in the upper distribution are not contained in the lower distribution.

one-pion exchange might be contributing to the observed peak in this m^* spectrum. In particular, an important contribution at lower energies comes from a diagram in which one of the π's rescatters off the nucleon and ends up in the 3-3 state with respect to the nucleon. If one restricts the data to cases with $\Delta \leq 400$ Mev/c this diagram does not seem to be very important, but if one takes cases with $\Delta > 400$ Mev/c many cases consistent with rescattering are found.

In order to deduce values of the π-π cross section, we use the formula[2]

$$\frac{d^2\sigma}{dm^* d\Delta^2} = \frac{3f^2}{\pi} \frac{\Delta^2}{(\Delta^2+1)^2}\left(\frac{m^*}{q_{iL}}\right)^2 K\bar{\sigma}_{\pi-\pi},$$

where $\bar{\sigma}_{\pi-\pi}$ is the mean of $\sigma(\pi^-\pi^0 \rightarrow \pi^-\pi^0)$ and $\sigma(\pi^-\pi^+ \rightarrow \pi^-\pi^+)$. In the above formula all momenta and energies are measured in units of pion masses. q_{iL} = momentum of the incoming pion measured in units of the pion mass. K = momentum of the pions in the di-pion center-of-mass system. Then

$$\delta\sigma = \frac{3f^2}{\pi}\left(\frac{m^*}{q_{iL}}\right)^2 K\bar{\sigma}_{\pi-\pi}\, \delta m^* \int_{\Delta_{\min}(m^*)}^{\Delta_{\max}} \frac{\Delta^2 d\Delta^2}{(\Delta^2+1)^2}.$$

The results of this calculation using the experimentally determined $\delta\sigma$'s are shown in Fig. 3. The results indicate a peak in the neighborhood of 750 Mev with a width of 150-200 Mev, which is about 3/4 of what it would be $(12\pi\lambda^2)$ for a resonance in the $I=1$, $J=1$ state. Since this cross section was determined off the energy shell, it is difficult to estimate the effect of the interference of other diagrams and also the effect of line broadening.[7] Whether or not the other peak and the S-wave scattering indicated in Fig. 3 are real will have to await better statistics for verification.

We wish to acknowledge with gratitude the help and cooperation of R. K. Adair and L. Leipuner, in the use of their bubble chamber, and to the latter also for his assistance in adapting the "Guts" routine to our use. We also acknowledge the help of J. Boyd, J. Bishop, P. Satterblom, R. P. Chen, C. Seaver, and K. Eggman in measuring, scanning, and tabulating. We were greatly aided by Dr. J. Ballam and Dr. H. Fechter in setting up the beam. We have had helpful conversations with Dr. R. K. Adair, Dr. C. J. Goebel, Dr. M. L. Good, and in particular Dr. G. Takeda.

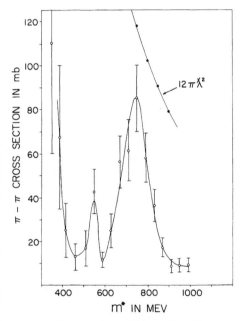

FIG. 3. The π-π cross section as deduced from cases with the four-momentum transfer less than 400 Mev/c.

EVIDENCE FOR A $T = 0$ THREE-PION RESONANCE*

B. C. Maglić, L. W. Alvarez, A. H. Rosenfeld, and M. L. Stevenson

Lawrence Radiation Laboratory and Department of Physics, University of California, Berkeley, California

(Received August 14, 1961)

The existence of a heavy neutral meson with $T = 0$ and $J = 1^-$ was predicted by Nambu[1] in an attempt to explain the electromagnetic form factors of the proton and neutron. Chew[2] has pointed out that such a vector meson should exist on dynamical grounds as a three-pion resonance or a bound state. Such a particle is also expected in the vector meson theory of Sakurai[3] and, as a member of an octet of mesons, according to the unitary symmetry theory[4]; and for other reasons.[5] We will refer to it as ω.

Previous searches[6] for ω have primarily been confined to the mass region $m_\omega < 3\mu$, with μ = the pion mass, where only the following radiative decay modes are allowed: $\omega \to \pi^0 + \gamma$, $\omega \to 2\pi^0 + \gamma$, and $\omega \to \pi^+ + \pi^- + \gamma$. The ω cannot decay into two pions.

The present search was made assuming $m_\omega > 3m_\pi$, where the decay

$$\omega \to \pi^+ + \pi^- + \pi^0 \qquad (1)$$

is possible.[7] We have searched for such a 3-pion decay mode by studying the effective mass distribution of triplets of pions in the reaction

$$\bar{p} + p \to \pi^+ + \pi^+ + \pi^- + \pi^- + \pi^0. \qquad (2)$$

We have measured 2500 four-prong events produced by antiprotons of 1.61 Bev/c in the 72-inch hydrogen bubble chamber.[8] The c.m. energy is 2.29 Bev. Upon fitting these 2500 four-prong events by using our kinematics program KICK, 800 four-prong events had a $\chi^2 \leq 6.5$ for hypothesis (2) and would not fit the hypothesis that no π^0 was produced (610 of these 800 had a $\chi^2 < 2.5$).

The 800 four-prong events must have some small contamination of events in which two π^0's were produced, but inspection of the "missing mass" distribution convinces us that it is <7%. Other tests confirm this low contamination. For example, the angular distribution of the π^0 is symmetric within statistics, and the momentum of the π^0 resembles the momentum distribution of the charged pions.

We have evaluated the 3-body effective mass,

$$M_3 = [(E_1 + E_2 + E_3)^2 - (\vec{P}_1 + \vec{P}_2 + \vec{P}_3)^2]^{1/2}, \qquad (3)$$

for each pion triplet in Reaction (2). Each of the 800 four-prong events yields ten such quantities corresponding to the following charge states:

$|Q| = 0$: $\pi^+\pi^-\pi^0$ (800×4 combinations), (4)

$|Q| = 1$: $\pi^\pm\pi^\pm\pi^\mp$ (800×4 combinations), (4')

and

$|Q| = 2$: $\pi^\pm\pi^\pm\pi^0$ (800×2 combinations). (4'')

For each value of M_3 as given by Eq. (4) we can calculate an uncertainty δM_3, by using the variance-covariance matrix of the fitted track variables, which is evaluated by KICK. By using these δM_3 we have formed the resolution function of M_3, and find that it has a half-width at half-maximum, $\Gamma_{\text{resol}}/2$, equal to 8.7 Mev. However, our input errors to KICK allow only for Coulomb scattering and estimated measurement accuracy, and do not account for optical distortion and unknown systematic errors. For example, our distributions have the correct shape but are too wide by a scale factor of about 2. This suggests that our average input error is too small by about $\sqrt{2}$. Hence, our estimate of δM_3 must be increased by about $\sqrt{2}$, and of $\Gamma_{\text{resol}}/2$ to 12 Mev. We chose 20-Mev histogram intervals for plotting our M_3 distribution.

In Fig. 1 we have plotted the M_3 distributions for the 800 Reactions (2). Distributions 1(A) and 1(B) are for charge combinations $|Q| = 1$ and 2, respectively. The solid curves are an approximation to phase space.

The neutral M_3 distribution, 1(C), shows a peak centered at 787 Mev that contains 93 pion triplets above the phase-space estimate of 98. To contrast the difference between the neutral M_3 distribution and that for $|Q| > 1$, we have replotted at the bottom of Fig. 1 both the neutral distribution and $\frac{2}{3}$ the sum of the $|Q| = 1$ and $|Q| = 2$ distributions.

Figure 2 shows the M_3 spectra with phase space subtracted. The absence of the peak in the $|Q| > 0$ distributions determines the isotopic spin of the resonance,

$$T_\omega = 0.$$

The χ^2 distribution of the events in the "peak region" was compared with the χ^2 distribution of the events in the adjacent "control region," ranging from $M_3 \geq 820$ to $M_3 < 900$ Mev. These distributions agree with each other, which indicates that the events in the peak are genuine, rather than being caused by some unknown background reaction which was misinterpreted as Reaction

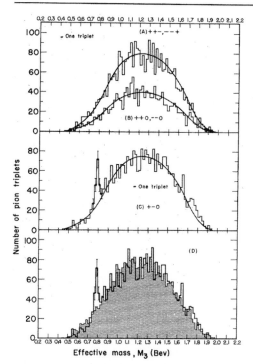

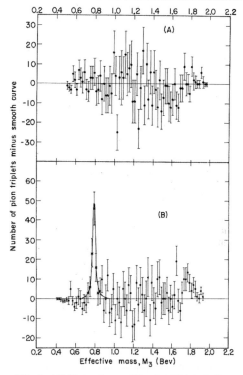

FIG. 1. Number of pion triplets versus effective mass (M_3) of the triplets for reaction $\bar{p}+p \rightarrow 2\pi^+ + 2\pi^- + \pi^0$. (A) is the distribution for the combination (4'), $|Q|=1$; (B) is for the combination (4"), $|Q|=2$; and (C) for (4), $Q=0$, with 3200, 1600, and 3200 triplets, respectively. Full width of one interval is 20 Mev. In (D), the combined distributions (A) and (B) (shaded area) are contrasted with distribution (C) (heavy line).

FIG. 2. (A) M_3 spectrum of the pion triplets in the combined distributions 1(A) and 1(B), with the smooth curve subtracted. (B) M_3 spectrum of the neutral pion triplets in distribution 1(C), again with the smooth background subtracted; a resonance curve is drawn through the peak at 787 Mev with $\Gamma/2 = 15$ Mev. The error flags are $\sqrt{N}$, where N is the total number of triplets per 20-Mev interval before subtraction of the smooth background curve.

(2). The missing-mass distributions in the two regions also agree with each other, thus supporting the above conclusion.

The peak in Fig. 2(B) appears to have a half-width $\Gamma/2 < 15$ Mev. This is so close to our resolution, $\Gamma_{\text{resol}}/2$, of 12 Mev that we cannot unfold it without further study and at present can only conclude that

$$M_\omega = 787 \text{ Mev},$$

and

$$\Gamma/2 < 15 \text{ Mev}. \quad (5)$$

By using the uncertainty principle, we see that this half-width implies a mean life $\tau > 4 \times 10^{-23}$ sec. Our ω's are produced with a typical c.m. momentum of 800 Mev/c, so that in a mean life they travel farther than 13 f.

We now assume that the ω peak is real, and want to estimate how many ω mesons it contains. As shown in Fig. 1(C), 191 triplets have M_3 values between 740 and 820 Mev. (We call this the "peak region.") However, these 191 triplets come from only 170 different four-prong events [i.e., 21 Reactions (2) have two values of M_3 in the peak region]. We use the charged M_3 distribution to estimate the background in the interval as 98 triplets, and then calculate a production of 83 ± 16 ω mesons out of 800 Reactions (2); i.e.,

$(10 \pm 2)\%$ of Reactions (2) proceed via

$$\bar{p} + p \to \pi^+ + \pi^- + \omega. \tag{6}$$

Among the same 800 five-pion events, we have searched for—and found—the $T = J = 1$ pion-pion resonance (ρ meson).[9] We found that approximately 30% of them proceed via

$$\bar{p} + p \to 2\pi + \rho. \tag{7}$$

We have checked whether there is any correlation between the observed ρ mesons and the ω mesons. For each triplet inside the peak region, $740 \le M_3 < 820$ Mev, we have evaluated the effective mass of the remaining $\pi^+\pi^-$ doublet, M_2. The M_2 distribution is consistent with a continuum, starting from about 300 Mev, that has $(10 \pm 2)\%$ of the doublets with values of M_2 in the region of ρ, which we took to be 750 ± 50 Mev. There is no evidence that the ω and the ρ are produced in association.

Although the masses of ω and ρ differ by only 35 Mev, we believe that they cannot be the same particle, because of their different widths (the $\Gamma/2$ for ρ being 40 Mev), isotopic spin, and G-conjugation parity—which forbids $2\pi \to 3\pi$.

In referring to the $T = 0$ 3π resonance as ω, we have tacitly supposed that it is in fact a vector state with $J = 1^-$. However, the spin and parity must be decided by experiment. Even if we assume the spin is < 2, there are left three possibilities which are listed in Table I. A $T = 0$ state of three pions must be antisymmetric in all pairs; hence all three pions must have different charges, i.e., $\pi^0\pi^0\pi^0$ is forbidden. The matrix element of the $\pi^+\pi^-\pi^0$ state is conveniently analyzed in terms of a single pion plus a di-pion. The pions of the di-pion are assigned momentum $\vec{P}$ and angular momentum $\vec{L}$ (in the di-pion rest frame). Then another pair of variables, $\vec{p}$ and $\vec{l}$, describe the remaining pion in the 3π rest frame. Because the state is antisymmetric in any pair, $\vec{L}$ must be odd; henceforth, we assume $L = 1$. Then if $l = 0$ we have a $J = 1^-$ (i.e., vector) matrix element, as listed on the bottom line of Table I. Since three pions are involved, there is an intrinsic parity of $(-1)^3$, so that the corresponding "meson" is not V, but A.

If $l = 1$, the matrix element can be 1+ (axial) or 0+ (scalar) corresponding, respectively, to a vector meson (ω) or a pseudoscalar (PS) meson.

Do we have enough data to distinguish between totally antisymmetric A vs S vs V matrix elements? It is convenient to make a Dalitz plot[10] [Fig. 3(D) for the peak region events, 3(A) for the control region events] that displays the threefold symmetry of three pions in an antisymmetric state. Unit area on a Dalitz plot is proportional to the corresponding Lorentz-invariant phase space, so that the density of plotted points is proportional to the square of the matrix element. It is easily shown that the size of the figure is proportional to $T_1 + T_2 + T_3 = Q = m_\omega - (2m_{\pi^\pm} + m_{\pi^0})$.[11] Because of the finite width of the peak and the control regions, Q varies from event to event, so we use normalized variables, T_i/Q. The antisymmetry allows the plot to be folded about any median, so that in Figs. 3(C) and 3(B) all the data have been concentrated into $\frac{1}{6}$ of the plot area; the statistical distribution of the events is then more evident.

All three competing matrix elements, being antisymmetric, must vanish where any two pions "touch" in momentum space. If two pions touch, the third must have its maximum kinetic energy [regions (d), (f), and (b) on the plot]. The resonance region points [Figs. 3(C) and 3(D)] seem to show the required depopulation at points (d), (f), and (b).

More evident, however, on the plot is the fact that near $p = 0$ [points (a), (c), and (e)] the density of peak-region points is only one half of that on the control plot. This is all the more suggestive when it is remembered that even the peak-region data contain only $(43 \pm 7)\%$ resonance events. This depopulation at $p = 0$ suggests an angular momen-

Table I. Possible three-pion resonances with $T = 0$, $J \le 1$.

"Meson" Type, J	$\vec{l}, \vec{L}$	Matrix element Type, J	Simple example	Vanishes at:
$V, 1-$	1, 1	$A, 1+$	$E_-(\vec{p}_0 \times \vec{p}_+) + E_0(\vec{p}_+ \times \vec{p}_-) + E_+(\vec{p}_- \times \vec{p}_0)$	whole boundary
$PS, 0-$	1, 1	$S, 0+$	$(E_- - E_0)(E_0 - E_+)(E_+ - E_-)$	$a, c, e + b, d, f$
$A, 1+$	0, 1	$V, 1-$	$E_-(\vec{p}_0 - \vec{p}_+) + E_0(\vec{p}_+ - \vec{p}_-) + E_+(\vec{p}_- - \vec{p}_0)$	b, d, f only

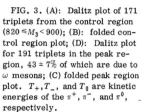
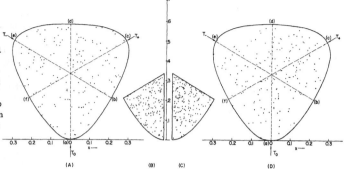

FIG. 3. (A): Dalitz plot of 171 triplets from the control region ($820 \leq M_3 < 900$); (B): folded control region plot; (D): Dalitz plot for 191 triplets in the peak region, $43 \pm 7\%$ of which are due to ω mesons; (C) folded peak region plot. T_+, T_-, and T_0 are kinetic energies of the π^+, π^-, and π^0, respectively.

tum barrier ($l > 0$) and constitutes mild evidence against a V matrix element (A meson).

The two stronger remaining candidates have A vs S matrix elements. The dashed lines in Fig. 3(D), as well as the two straight lines of the folded distribution in Fig. 3(C), correspond to equal energies of two pions. The scalar matrix element (S) of Table I vanishes when any two pions have the same energy, and therefore would require depopulation along these lines. This is not observed. An A matrix element has terms in $\vec{p}_i \times \vec{p}_j$, which vanish for collinear pions. The boundary of the plot represents collinearity, and seems indeed to be depopulated; although clearly more statistics and more detailed analysis such as investigation of polarization and alignment are needed.

We conclude that the data fit the qualitative criteria for an axial vector matrix element (ω meson); there is reasonable evidence against both an A meson and a PS meson.

The film used in this measurement was obtained in collaboration with J. Button, P. Eberhard, G. Kalbfleisch, J. Lannutti, G. Lynch, and N. H. Xuong; and this experiment would not have been possible without their help. It is a pleasure to thank Professor Murray Gell-Mann for his theoretical discussions. We wish to acknowledge the active participation of C. Tate, L. Champomier, A. Hussain, C. Rinfleisch, and F. Richards in the final stages of this experiment.

---

*Work done under the auspices of the U. S. Atomic Energy Commission.

[1]Y. Nambu, Phys. Rev. 106, 1366 (1957). See also S. Bergia, A. Stanghellini, S. Fubini, and C. Villi, Phys. Rev. Letters 6, 367 (1961); R. Blankenbecler and J. Tarski (to be published); G. Chew, R. Karplus S. Gasiorowicz, and F. Zachariasen, Phys. Rev. 110, 265 (1958); P. Federbush, M. Goldberger, and S. Treiman, Phys. Rev. 112, 642 (1958).

[2]G. F. Chew, Phys. Rev. Letters 4, 142 (1960). See also V. de Alfaro and B. Vitale, Phys. Rev. Letters 7, 72 (1961).

[3]J. Sakurai, Ann. Phys. 11, 1 (1960); Nuovo cimento 16, 388 (1960). See also V. I. Ogievetski and I. V. Polubarinov, Dubna Report D-676, 1961, submitted to J. Exptl. Theoret. Phys. (U.S.S.R.) for publication.

[4]M. Ikeda, S. Ogawa, and Y. Ohnuki, Progr. Theoret. Phys. (Kyoto) 22, 715 (1959); see also Y. Ohnuki, in Proceedings of the 1960 Annual International Conference on High-Energy Physics at Rochester (Interscience Publishers, Inc., New York, 1960); Y. Yamaguchi, Progr. Theoret. Phys. (Kyoto), Suppl. No. 11 (1959); J. Wess, Nuovo cimento 15, 52 (1960); Y. Neeman, Nuclear Phys. (to be published); A. Salam and J. C. Ward, Nuovo cimento (to be published); M. Gell-Mann (to be published); M. Gell-Mann and F. Zachariasen, Phys. Rev. (to be published).

[5]S. Sawada and M. Yonezawa, Progr. Theoret. Phys. (Kyoto) 22, 610 (1959). A somewhat similar particle was predicted by M. Johnson and E. Teller, Phys. Rev. 98, 783 (1955); H. Duerr and E. Teller, Phys. Rev. 101, 494 (1956); E. Teller, Proceedings of the Sixth Annual Rochester Conference on High-Energy Physics, 1956 (Interscience Publishers, New York, 1956); H. Duerr, Phys. Rev. 103, 469 (1956).

[6]A. Alberigi, C. Bernardini, R. Querzoli, G. Salvini, A. Silverman, and G. Stoppini (reported by G. Bernardini), in Ninth Annual International Conference on High-Energy Physics at Kiev, July, 1959 (Academy of Science U.S.S.R., Moscow, 1960), Vol. 1, p. 42; R. Gomez, H. Burkhardt, M. Daybell, H. Ruderman, M. Sands, and R. Talman, Phys. Rev. Letters 5, 170 (1960). A. Abashian, N. Booth, and K. Crowe, Phys. Rev. Letters 5, 258 (1960); 7, 35 (1961). J. Button, P. Eberhard, G. R. Kalbfleisch, J. Lannutti, S. Limentani, G. Lynch, B. Maglić, M. L. Stevenson, and N. H. Xuong (reported by F. Solmitz), in Proceedings of the 1960 Annual International Conference on High-Energy Physics at Rochester (Interscience Publishers, Inc., New York, 1960), p. 166; K. Berkelman, G. Cortellessa, and A. Reale, Phys. Rev. Letters 6, 234 (1961).

E. Pickup, F. Ayer, and E. O. Salant, Phys. Rev. Letters 5, 161 (1960); J. G. Rushbrooke and D. Radojčić, Phys. Rev. Letters 5, 567 (1960); J. Anderson, V. Bang, P. Burke, D. Carmony, and N. Schmitz, Phys. Rev. Letters 6, 365 (1961); A. R. Erwin, R. March, W. D. Walker, and E. West, Phys. Rev. Letters 6, 628 (1961); D. Stonehill, C. Baltay, H. Courant, and N. H. Xuong, Phys. Rev. 121, 1788 (1961).

[7]According to G. Sudarshan (University of Rochester, private communication) m_ω is expected to be within the limits $m_\rho < m_\omega < m_\rho + m_\pi$, where $m_\rho = 750$ Mev is the mass of the $T = J = 1$ $\pi\pi$ resonance.

[8]J. Button, P. Eberhard, G. R. Kalbfleisch, J. E. Lannutti, G. R. Lynch, B. C. Maglić, M. L. Stevenson, and N. H. Xuong, Phys. Rev. 121, 1788 (1961).

[9]M. L. Stevenson, G. R. Kalbfleisch, B. C. Maglić, and A. H. Rosenfeld, Lawrence Radiation Laboratory Report UCRL-9814 (to be published). For previous evidence see: I. Derado, Nuovo cimento 15, 853 (1961); W. Fickinger, E. C. Fowler, H. Kraybill, J. Sandweiss, J. Sanford, and H. Taft, Phys. Rev. Letters 6, 624 (1961).

[10]R. Dalitz, Phil. Mag. 44, 1068 (1953). See also E. Fabri, Nuovo cimento 11, 479 (1954).

[11]Specifically, Q equals the height of the exscribed triangle.

EVIDENCE FOR A THREE-PION RESONANCE NEAR 550 Mev*

A. Pevsner, R. Kraemer, M. Nussbaum, C. Richardson, P. Schlein, R. Strand, and T. Toohig
The Johns Hopkins University, Baltimore, Maryland

and

M. Block, A. Engler, R. Gessaroli, and C. Meltzer
Northwestern University, Evanston, Illinois
(Received November 10, 1961)

A study has been under way of multipion resonances in $\pi^+ + d$ reactions observed in the Lawrence Radiation Laboratory 72-in. bubble chamber exposed to a 1.23-Bev/c pion beam from the Bevatron. A preliminary report on this research was given at the Aix-en-Provence Conference on Elementary Particles[1] where the existence of the ω^0 meson reported by the Berkeley group[2] was confirmed. Since then these data have been substantially increased, although the experiment is still in progress. The existence of a second neutral 3-pion resonance with a mass of approximately 550 Mev is indicated by this larger sample of events.

Many authors[3] have speculated on the existence of neutral, strongly interacting bosons of mass of the order of $3\text{-}4 m_\pi$, in order to fit the data for nucleon form factors obtained from electron scattering experiments. These bosons could be readily identified experimentally in the reaction

$$\pi^+ + d \rightarrow p + p + X^0. \quad (1)$$

In order to observe the possible decay mode,

$$X^0 \rightarrow \pi^+ + \pi^- + \pi^0, \quad (2)$$

we consider the reaction

$$\pi^+ + d \rightarrow p + p + \pi^+ + \pi^- + \pi^0. \quad (3)$$

Only events where both protons are visible and at least one proton stops in the chamber with a range less than 15 cm were accepted for analysis.[4]

The events were measured with a digitized microscope and reconstructed by the Berkeley PANG program. A kinematic fit[5] was obtained for the assumed π^0 using the KICK program, and the effective mass of the fitted 3-pion system was then calculated. In order to check the identification of the π^0, we have calculated the missing neutral mass for events which fit our criteria. An ideogram[6] for this missing neutral mass is given in Fig. 1 for the first 199 of our events.

Figure 2 is a histogram of the effective mass of the 3-pion system for our 233 events. An average mass uncertainty on a given event is $\sim \pm 20$ Mev. The large peak near 770 Mev is clearly identifiable as the ω^0. Another large peak in the 3-pion mass plot of Fig. 2 is seen near 550 Mev, which strongly suggests the existence of a second 3-pion resonance (or particle). We shall hereafter refer to this particle as η.

In order to estimate the number of events in this peak which are reasonably due to the η particle, we make the following interpretation of our data. We believe the impulse approximation is reasonably valid because of the loose structure of the deuteron. Thus the basic reaction we are looking at is

$$\pi^+ + n \rightarrow p + X^0, \quad (4a)$$

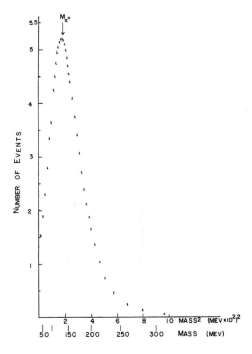

FIG. 1. Ideogram of the missing mass in the reaction $(\pi^+ + d \rightarrow p + p + \pi^+ + \pi^- + \text{missing mass})$ for 199 events which meet the selection criteria.

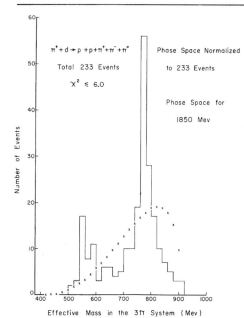

FIG. 2. Histogram of the effective mass of the three-pion system for 233 events.

where
$$X^0 \to \pi^+ + \pi^- + \pi^0. \tag{4b}$$

We have calculated the Lorentz-invariant phase space[7] for the 3-pion mass from the background reaction to (4a), i.e.,

$$\pi^+ + n \to p + \pi^+ + \pi^- + \pi^0, \tag{5}$$

using the experimental average of the total energy in the p-3π center-of-mass system (1850 Mev). This curve, normalized to the total number of events, is plotted in Fig. 2.

Clearly, because of the presence of the ω^0 particle at 770 Mev, such a normalization of phase space yields a gross overestimate of events expected near 550 Mev. Between 540 and 600 Mev there are 36 events in the experimental distribution, whereas the overestimated phase space would account for 12.

An analysis of the data, which takes into account the spread in errors on the individual events on the histogram, gives a mass of approximately 764 Mev with a half-width at half maximum of ≤20 Mev for the ω^0 and a mass of ~546 Mev with a half-width at half maximum of ≤25 Mev for the η.

An attempt is being made to determine the isotopic spin for both peaks by studing the reaction

$$\pi^+ + d \to p + \pi^+ + \pi^+ + \pi^- + n. \tag{6}$$

Only 61 events were found in an analysis of one-half the film represented by Fig. 2. The low yield is probably indicative of the lack of any resonance in the isotopic spin states 1 and 2. This is in accord with the Berkeley assignment of $T = 0$ to the ω^0.

A search for the $\pi^0 + \gamma$ decay mode of the ω^0 and η is being carried out by a study of events of the type

$$\pi^+ + d \to p + p + \text{(neutrals)}. \tag{7}$$

The results will be available shortly.

The proton form factor F_{1p} obtained from electron scattering experiments[8] cannot be fitted using only the ω^0 and ρ particles.[9] However, a three-pion resonance of mass ≤$4m\pi$ having $T = 0$ and spin 1^- would make a fit to the data possible.[10] With the film on hand we expect to more than double our statistics, so that a determination of the isotopic spin and spin of the η may be possible to see whether it fits these theories.

The authors wish to thank Dr. Luis Alvarez, Dr. Edwin McMillan, Dr. Frank Crawford, and the staff of the Lawrence Radiation Laboratory for their cooperation and for the facilities which made this experiment possible. We also wish to thank Dr. Walter Selove of the University of Pennsylvania and Dr. Leon Madansky and Dr. Gordon Feldman of The Johns Hopkins University for helpful discussions, and Dr. A. Rosenfeld, Dr. P. Berge, and Mr. R. Harvey of the Lawrence Radiation Laboratory, Mrs. Doris Ellis of The Johns Hopkins University, and Dr. David Onley and Mr. Arthur Kovacs of Duke University, for their invaluable help with the computer programs.

---

*Work supported by the U. S. Air Force Office of Scientific Research, the National Science Foundation, and Office of Naval Research.

[1]A. Pevsner, R. Kraemer, M. Nussbaum, P. Schlein, T. Toohig, M. Block, A. Kovacs, and C. Meltzer, Proceedings of the 1961 Conference on Elementary Particles, Aix-en-Provence (to be published).

[2]B. Maglić, L. Alvarez, A. Rosenfeld, and M. L. Stevenson, Phys. Rev. Letters 7, 178 (1961).

[3]Y. Nambu, Phys. Rev. 106, 1366 (1957); G. F. Chew, Phys. Rev. Letters 4, 142 (1960); J. J. Sakurai, Ann. Phys. 11, 1 (1960); S. Bergia, A. Stanghellini, S. Fubini, and C. Villi, Phys. Rev. Letters 6, 367 (1961).

[4] In a sample representing one-fourth of the film reported here, the requirement that one of the protons had to stop with a range ≤15 cm was removed. The results agree within statistics with those reported here.

[5] Events were accepted for analysis which fit the following criteria: (a) $\chi^2 \leq 6$ for the hypothesis $\pi^+ + d \to p + p + \pi^+ + \pi^- + \pi^0$. (b) $\chi^2 \geq 25$ for the hypothesis $\pi^+ + d \to p + p + \pi^+ + \pi^-$. (c) If the nonstopping proton had a momentum ≥700 Mev/c, where it becomes difficult to differentiate a proton from a π^+ by ionization in this chamber, then the χ^2 had to be greater than 15 for the hypothesis $\pi^+ + d \to p + n + \pi^+ + \pi^+ + \pi^-$, which is another background reaction under these circumstances.

[6] The ideogram was calculated in units of mass squared since our experimental errors are Gaussian in this representation. Each event was given a constant-area Gaussian distribution.

[7] M. M. Block, Phys. Rev. 101, 796 (1956); P. Srivastava and G. Sudarshan, Phys. Rev. 110, 765 (1958).

[8] R. Hofstadter and R. Herman, Phys. Rev. Letters 6, 293 (1961); R. M. Littauer, H. F. Schopper, and R. R. Wilson, Phys. Rev. Letters 7, 141 (1961).

[9] S. Fubini, Proceedings of the 1961 Conference on Elementary Particles, Aix-en-Provence (to be published); P. T. Matthews, ibid.; G. Breit, Proc. Natl. Acad. Sci. U. S. 46, 746 (1960); Y. Fujii, Progr. Theoret. Phys. (Kyoto) 21, 232 (1959).

[10] G. Feldman, T. Fulton, and K. C. Wali (private communication); see also J. Sakurai, Phys. Rev. Letters 7, 355 (1961).

EXISTENCE AND PROPERTIES OF THE φ MESON*

P. L. Connolly, E. L. Hart, K. W. Lai, G. London,[†] G. C. Moneti,[‡] R. R. Rau,
N. P. Samios, I. O. Skillicorn, and S. S. Yamamoto
Brookhaven National Laboratory, Upton, New York

and

M. Goldberg, M. Gundzik, J. Leitner, and S. Lichtman
Syracuse University, Syracuse, New York
(Received 27 March 1963)

In a previous publication[1] we reported evidence for the existence of a resonance in the $K\bar{K}$ system (which we shall call the φ meson[2]) with a mass of ~1020 MeV and a width ≤20 MeV. The purpose of this Letter is to report additional data, the analysis of which confirms the existence of the resonance[3] and provides a conclusive determination of the mass, width, parity, spin, isospin, and branching ratios of this resonance. In particular we find $M = 1019 \pm 1$ MeV, $\Gamma = 1^{+2}_{-1}$ (with $\Gamma > 0$), parity $(P) = -1$, spin $(J) = 1$, isotopic spin $(I) = 0$, and charge conjugation $(C) = -1$. The vector nature of the φ clearly establishes that it is not related to the mass enhancement in the $K_1 K_1$ system observed by other groups.[4]

This experiment is part of a continuing study of the $K^- \text{-} p$ interaction[5] at 2.23 BeV/c. The data were collected in two exposures of the 20-

inch BNL hydrogen chamber at Brookhaven's AGS. Our study of the $K\bar{K}$ system is based upon the reactions

$$K^- + p \to \Lambda + K^0 + \bar{K}^0, \quad (1)$$

$$K^- + p \to \Lambda + K^+ + K^-. \quad (2)$$

We shall refer to (1) as the "neutral channel" and to (2) as the "charged channel." At present, the total analyzed data consist of channels (1) and (2) from the first exposure[6] and channel (1) from the second exposure; the numbers of events involved are summarized in Table I. A total of 36 events in the neutral channel and 22 events in the charged channel have been analyzed.

The general analysis procedure used to identify events on the basis of χ^2 fitting and ionization information is described in detail in a previous publication.[5] Background contamination and detection bias are negligible. Owing to the difference in neutral missing mass, channel (1) events are easily distinguished from the competing reactions, $K^- + p \to \Xi^0 + K^0$ and $\pi^- + p \to (\Lambda \text{ or } \Sigma^0) + K^0 (+ \pi^0)$. Competition from $\pi^- + p \to \Lambda + K^0 + 2\pi^0$ is negligible due to the small pion contamination in the beam and the small phase space available for this reaction. This is further substantiated by the paucity of events of the type $\pi^- + p \to \Lambda + K^0 + \pi^+ + \pi^-$, only two such events occurring in the first run. The only significant background is due to the reaction $K^- + p \to \Sigma^0 + K^0 + \bar{K}^0$, which we estimate on the basis of missing mass studies and other information to be ~10%. Because of severe competition from other modes, candidates for channel (2) are used only if the Λ decays visibly. Due to the more frequent occurrence of the topologically similar reactions $K^- + p \to \Lambda + \pi^+ + \pi^-$ and $K^- + p \to \Lambda + \pi^+ + \pi^- + \pi^0$, it was necessary to measure all events consisting of a V and two charged prongs. For those cases which were kinematically ambiguous it was always possible to determine the $\Lambda K^+ K^-$ events due to the difference in predicted ionization for the K's and π's. We therefore believe the charged channel to be free from bias.

The Dalitz plot of the (squared) $K\bar{K}$ and ΛK[7] effective masses of 58 events from both the neutral and charged $K\bar{K}$ channels is shown in Fig. 1. The enhancement over the phase-space distribution in the region $M^2(K\bar{K}) = 1.04$ BeV^2 is ap-

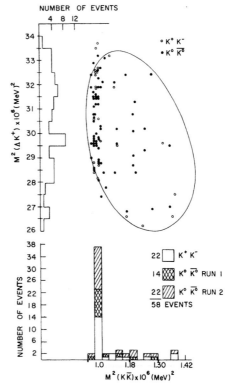

FIG. 1. Dalitz plot for the reaction $K^- + p \to \Lambda + K + \bar{K}$. The effective-mass distribution for $K\bar{K}$ and for ΛK^+ are projected on the abscissa and ordinate (see reference 7).

Table I. Summary of events from both exposures.

	No. of events Channel (1)	No. of events Channel (2)	No. of events in peak Channel (1)		No. of events in peak Channel (2)	
			Obs.	Corrected	Obs.	Corrected
First exposure	14	22	9	19	14	23
Second exposure	22	Not analyzed	13	Not Useful	Not analyzed	

parent,[8] as is the lack of any significant irregularity in the $M^2(\Lambda K)$ distribution, indicating the absence of any appreciable ΛK-$K\bar{K}$ interference. A comparison of the observed φ peak with the $K_1 K_1$ enhancement found by Alexander et al.[4] clearly indicates on the basis of widths alone that the two effects are different.

We have examined the enhancement region of Fig. 1 by plotting an ideogram of $M(K\bar{K})$ for events between 1000 and 1040 MeV. This distribution peaks at 1019 ± 1 MeV and has an observed width of ~10 MeV. We defer a discussion of the true width Γ since its determination is to some extent influenced by the spin quantum number assigned to the φ.

First, we consider the parity of the φ, or that which is equivalent for a $K\bar{K}$ system, the charge conjugation number, C. Determination of C rests upon the observation by Goldhaber et al.[9] that if the $K\bar{K}$ system has $C = +1$, it may decay into $K_1 K_1$ or $K_2 K_2$, while if it has $C = -1$, it may decay only into $K_1 K_2$. Taking account of the space correlations we may compute, for both $C = +1$ and $C = -1$ hypotheses, the expected rates of φ decay into the experimentally observable topological variations of final states of channel (1), which we label by their visible V's as follows: $\Lambda K_1 K_1$, ΛK_1, and $K_1 K_1$. The predicted[10] relative rates of $\Lambda K_1 K_1$, ΛK_1, and $K_1 K_1$ are given in Table II. If these are compared with the observed rates, also given in Table II, one sees that the $C = -1$ hypothesis is in excellent agreement with the data, while the $C = +1$ hypothesis is in disagreement with the data by 12 standard deviations; this is just a reflection of the fact that there are 23 events in the peak which are ΛK_1, and not one $\Lambda K_1 K_1$ or $K_1 K_1$. We conclude, therefore, that the φ has $C = P = -1$, or equivalently that its spin is odd, the most likely values being $J = 1$ or 3.

Information concerning the spin can be obtained from a consideration of the (relative) decay rate α_J, where

$$\alpha_J = \frac{\varphi \to K_1 K_2}{\varphi \to K_1 K_2 + K^+ K^-}.$$

In the absence of a $K^+ K^0$ mass difference and charge effects, the ratio α is clearly independent of J (in fact $\alpha = 0.5$). The spin dependence of α_J arises from the different angular momentum and Coulomb barriers appropriate to the $(K_1 K_2)$ and $(K^+ K^-)$ systems. Using an interaction radius[11] of $(2M_\pi)^{-1}$, a Coulomb correction of 4% for $J = 1$ and 3% for $J = 3$, and center-of-mass momenta of $P_\pm = 125$ MeV/c and $P_0 = 107$ MeV/c, we estimate $\alpha_{J=1} = 0.39$ and $\alpha_{J=3} = 0.26$. From the (first exposure) data of Table I, after correcting for neutral modes and fiducial region differences, we find $\alpha_{\text{expt}} = 0.45 \pm 0.10$. Comparing this with our theoretical estimates we see that the observed ratio of relative kaon decay modes is in good agreement with the $J = 1$ hypothesis and disagrees with that of $J = 3$ by ~2 standard deviations. In principle further information on the spin can be obtained from the shape of the φ peak. If the φ meson has a finite width, and if our experimental resolution is symmetric, then the observed shape of $M(K\bar{K})$ should exhibit an asymmetry depending on the spin J. The $M(K^+ K^-)$ and $M(K_1^0 K_2^0)$ distributions in the form of Gaussian ideograms as well as their respective resolution functions[12] are shown in Fig. 2. In both instances there appear similar asymmetries. Since the phase-space background is extremely small ($1\frac{1}{2}$ events) it is probable that the $M(K\bar{K})$ asymmetry[13] reflects the spin dependence of the φ decay. The experimental data was then fitted with both p- and d-wave modified Breit-Wigner formula including the experimental mass resolution. Both

Table II. Predicted and observed relative rates for different topological types of channel (1).

Topological Type of Channel (1)	Predicted Relative Rates $C = -1$ $(K_1 K_2)$	$C = +1$ $(K_1 K_1)$ $(K_2 K_2)$	Observed Relative Rates For Events in Peak
$\Lambda K_1 K_1$	0	0.4	0 ± 0.04
ΛK_1	1	0.4	1 ± 0.2
$K_1 K_1$	0	0.2	0 ± 0.04

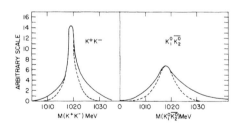

FIG. 2. The solid curve is the Gaussian ideogram of the $(K\bar{K})$ mass spectrum after subtraction of phase space. The dashed curve is the Gaussian ideogram of the $(K\bar{K})$ mass resolution function.

spin-1 and spin-3 cases fit the data for slightly different values of the true width Γ_T. Since the charged (K^+K^-) masses are considerably better determined $(\Gamma_{+-}{}^{\text{res}} = 3 \pm 1$ MeV) than their neutral $(K_1{}^0 K_2{}^0)$ counterpart $(\Gamma_{00}{}^{\text{res}} = 8 \pm 2$ MeV), they constitute a more suitable sample from which to obtain Γ_T. Taking into account background and statistical uncertainties we find $\Gamma_T = 1^{+2}_{-1}$ MeV, and as argued above we believe $\Gamma_T > 0$.

Next, we consider the isotopic spin of the φ. The strongest evidence concerning isospin comes from a determination of the G-parity (G). Since we have established that the φ spin is odd, then $G = -(-1)^I$. If the G-parity of the φ were $+1$, the G-allowed 2π decay mode would predominate over the $K\bar{K}$ mode. From the ratio of phase space and barrier penetration factors (for $J = 1$) we find that $(Q \to 2\pi)/(Q \to K\bar{K}) \cong 10$ for an interaction radius $R = (1/2m_\pi)$ and increases to $\cong 20$ for $R = (1/2m_k)$. We have searched for a 2-pion decay mode by investigating the $M(\pi^+\pi^-)$ distribution from the reaction

$$K^- + p \to \Lambda + \pi^+ + \pi^-. \qquad (3)$$

The $\Lambda\pi^+\pi^-$ final state frequently results from the decay of a $Y_1{}^*$ intermediate state,[14] a circumstance which would complicate the search for $\varphi \to 2\pi$, therefore the $Y_1{}^*$ production events were removed on the basis of their $M(\Lambda\pi)$ values (taken as 1385 ± 40 MeV for present purposes).[15] The $M(\pi^+\pi^-)$ distribution of the remaining events is shown as the solid curve of Fig. 3. There is no evidence of enhancement at 1020 MeV which would indicate $\Lambda + \varphi$ production. We estimate that the 300-event sample of Fig. 3 (from the first exposure only) does not contain more than five $\varphi \to 2\pi$ events. From this, and the $K\bar{K}$ data of the first exposure (see Table I), taking into account corrections for unobservable modes,

we find an upper limit to the relative 2π rate of

$$\left(\frac{\varphi \to 2\pi}{\varphi \to K\bar{K}}\right) = \frac{5}{19 + 23} < 0.2.$$

Since the discrepancy between this upper limit and the predicted lower limit of this $G = +1$ decay is about 2 orders of magnitude, we conclude[16] that the φ has negative G-parity, which implies isotopic spin 0.

For a φ with negative G-parity and spin 1, the main competition to the $K\bar{K}$ decay mode is expected to come from[17] 3π decay, and in particular from the mode $\varphi \to \rho + \pi$. We have searched for the latter decay mode in the final state

$$\Lambda + \pi^- + \pi^+ + \pi^0. \qquad (4)$$

As discussed in an earlier publication,[5] only ~60% of the final states (4) arise from the nonresonant reaction $K^- + p \to \Lambda + \pi^- + \pi^+ + \pi^0$. The remainder of the $\Lambda + \pi^+ + \pi^- + \pi^0$ final states result from the decay of the following resonant intermediate states: (5) $Y_1{}^* + \pi + \pi$ or (6) $\Lambda + \omega$, (7) $\Lambda + \varphi$, (8) $\Lambda + \rho + \pi$. In order to avoid correlations in the 3π system due to $Y_1{}^*$ decay, the intermediate states (5), which may be recognized by means of a $\Lambda\pi$ effective mass[15] of 1385 ± 30, are removed from the sample of $\Lambda + \pi^+ + \pi^- + \pi^0$ final states. The 2π mass spectrum is searched for events with a ρ mass, taken[15] to be 750 ± 75 MeV. For events satisfying these criteria, their 3π effective mass spectrum, shown as the solid curve of Fig. 4, is examined for evidence of a

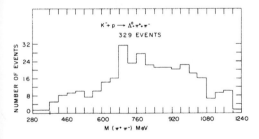

FIG. 3. The $M(\pi^+\pi^-)$ distribution from the reaction $K^- + p \to \Lambda + \pi^+ + \pi^-$ after removing $Y_1{}^*$ production events (see text).

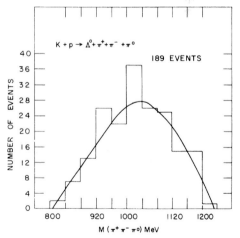

FIG. 4. The $M(\pi^+\pi^-\pi^0)$ distribution from the reaction $K^- + p \to \Lambda + \pi^+ + \pi^- + \pi^0$ after removing $Y_1{}^*$ production events (see text).

peak at the φ mass. There is a deviation at $M(3\pi) = 1020$ MeV of about 1.5 standard deviations above background and of width consistent with the experimental resolution at this mass. The sample of Fig. 4 representing all available data from the first exposure contains ~10 $\varphi \rightarrow \rho + \pi$ events. From this and the relevant $K\bar{K}$ data (see Table I) we find, correcting for neutral modes,

$$\beta = \frac{\varphi \rightarrow \rho + \pi}{\varphi \rightarrow K + \bar{K}} \approx 0.35 \pm 0.2.$$

One can estimate β_J either from the ratio of phase space, barrier penetration, spin, and isospin factors which give $\beta_{J=1} \approx 4$ for an interaction radius of $(2m_\pi)^{-1}$ or from a dynamical approach as done by Sakurai[2] giving $\beta_{J=1} \approx 3$. The observed rate is lower than these predicted values by one order of magnitude; however the above estimates are uncertain[18] by at least this amount so that this discrepancy need not be disconcerting.

It should be noted that the φ, being a 1^{--} meson, may be accommodated within a number of theories of elementary particles. In the unitary symmetry model of Gell-Mann,[19] the φ could be the heretofore absent (singlet) partner of the vector meson octet. It has been noted that the pseudoscalar meson octet and baryon supermultiplets satisfy the generalized mass formula[19,20] to a high accuracy. However, the same mass formula applied to the vector meson octet predicts an isosinglet mass of 930 MeV, which is different from the observed ω mass of 780 MeV. This discrepancy might be explained by the mixing of the ω and φ,[2] a possibility which arises since they have the same quantum numbers. Within the framework of Sakurai's "vector theory of strong interactions"[21] the φ may play the role of the vector meson B_Y coupled to the hypercharge current. Finally, with respect to the Chew-Frautschi conjecture,[22] the φ presumably starts a new trajectory, as do the other vector mesons.

We would like to acknowledge the invaluable assistance of the AGS staff, the BNL 20-inch crew, and many technical assistants. Two of us (J. L. and N. P. S.) are indebted to Professor J. Sakurai for stimulating conversations and correspondence.

---

*Research carried out under the auspices of the U. S. Atomic Energy Commission.

†Graduate student on leave from University of Rochester, Rochester, New York.

‡Presently at the Istituto Nazionale di Fisica Nucleare, Rome, Italy and The University of Roma, Rome, Italy.

[1]L. Bertanza et al., Phys. Rev. Letters 9, 180 (1962).

[2]This was suggested by J. J. Sakurai, Phys. Rev. Letters 9, 472 (1962).

[3]The existence of this resonance has also been confirmed by P. Schlein, W. E. Slater, L. T. Smith, D. H. Stork, and H. K. Ticho, Phys. Rev. Letters 10, 368 (1963).

[4]A. R. Erwin et al., Phys. Rev. Letters 9, 34 (1962); G. Alexander et al., Phys. Rev. Letters 9, 460 (1962).

[5]L. Bertanza et al., Proceedings of the International Conference on High-Energy Nuclear Physics, Geneva, 1962 (CERN Scientific Information Service, Geneva, Switzerland, 1962), pp. 279-284.

[6]The data published in reference 1 consisted of a partial sample at 2.23 BeV/c from the first exposure along with 8 events at 2.50 BeV/c.

[7]Since the K^0 cannot be distinguished from the $\bar{K}^0$, both the ΛK^0 and $\Lambda \bar{K}^0$ masses are plotted in the Dalitz plot. However, each one of these points is counted only as 1/2 event in the $M^2(\Lambda K)$ projection, so that it is properly normalized.

[8]As determined by a χ^2 test, the probability that the observed $M^2(K\bar{K})$ distribution is due to phase space production is less than 10^{-6}.

[9]M. Goldhaber, T. D. Lee, and C. N. Yang, Phys. Rev. 112, 1796 (1958).

[10]In determining these rates and all other K_1^0 decay rates we have taken $\Gamma(K_1^0 \rightarrow \pi^0 + \pi^0)/\Gamma(K_1^0 \rightarrow \text{total}) = 1/3$. See Proceedings of the International Conference on High-Energy Nuclear Physics, Geneva, 1962 (CERN Scientific Information Service, Geneva, Switzerland, 1962), p. 836.

[11]The choice $M = 2M_\pi$ is clearly open to question. It should be emphasized however that the predicted rates are not sensitive to the choice of M so long as $M \geq 2M_\pi$. The values of α_J for $M = 2M_K$ are $\alpha_{J=1} = 0.38$, $\alpha_{J=3} = 0.26$.

[12]A study of the shape of our mass resolution was carried out by measuring the $M(\pi^- p)$ distribution from Λ^0 decay, which is known to have 0 width and a Q value similar to φ decay. The observed $M(\pi^- p)$ distribution was indeed symmetric, being characterized by $M_\Lambda = 1116 \pm 0.5$ MeV.

[13]The $\Lambda^0 K^+ K^-$ sample is completely free from contamination, in particular from $\Sigma^0 K^+ K^-$. Although $\Sigma^0 K^0 \bar{K}^0$ contamination is present in the neutral sample, the similarity in the asymmetry of both distributions allows us to conclude that the contamination does not appreciably contribute to the observed asymmetry.

[14]L. Bertanza et al., Phys. Rev. Letters 10, 176 (1963).

[15]The data were studied using several values of the resonance width and no significant difference resulted.

[16]Additional direct evidence for $I = 0$ which is of an entirely preliminary nature comes from the absence of a φ-type enhancement in the $M^2(K\bar{K})$ distribution of

[16] $I=1$ combinations, $K^0 K^-$ and $\bar{K}^0 K^+$ produced in the reactions $K^- + p \to \Sigma^- + K^+ + \bar{K}^0$ and $K^- + p \to \Sigma^+ + K^- + K^0$. Further, if one assumes $I = 1$, the triangular inequality relating these reactions to their neutral counterparts $K^- + p \to \Sigma^0 + K^0 + \bar{K}^0$ and $\Sigma^0 + K^+ + K^-$, $[\sigma(\Sigma^+)]^{1/2} + [\sigma(\Sigma^-)]^{1/2} \geq 2[\sigma(\Sigma^0)]^{1/2}$, is violated to the extent $(3 \pm 3)^{1/2} + (0 \pm 1)^{1/2} \geq 2(22 \pm 5)^{1/2}$.

[17] We estimate, for example, that $\varphi \to 2\pi$ via an electromagnetic transition is $\sim 2 \times 10^{-3}$ less frequent than $\varphi \to K\bar{K}$. The 3π rate is dominated by $\varphi \to \rho + \pi$ because only two-body phase space is involved.

[18] Due to these uncertainties the ratio β provides essentially no new information on the φ spin.

[19] M. Gell-Mann, Phys. Rev. 125, 1067 (1962); California Institute of Technology Report CTSL-20, 1961 (unpublished). For appropriate remarks, also see reference 2.

[20] S. Okubo, Progr. Theoret. Phys. 27, 949 (1962).

[21] J. J. Sakurai, Ann. Phys. 11, 1 (1960); Phys. Rev. Letters 7, 335 (1961).

[22] G. F. Chew and S. C. Frautschi, Phys. Rev. Letters 8, 41 (1962).

RESONANCE IN THE ($\Xi\pi$) SYSTEM AT 1.53 GeV*

G. M. Pjerrou, D. J. Prowse, P. Schlein, W. E. Slater, D. H. Stork, and H. K. Ticho
Department of Physics, University of California at Los Angeles, Los Angeles, California
(Received June 27, 1962)

We wish to report the existence of a narrow resonance in the ($\Xi\pi$) system which we observed in the study of the interactions of negative K mesons in the LRL 72-in. hydrogen bubble chamber. The separated incident K^- beam, originating in the Bevatron, had a momentum of 1.80 ± 0.08 GeV/c; the uncertainty includes both the 6% momentum spread of the beam and the momentum loss in the chamber. The film was scanned for events with the following topology: "one positive secondary, one negative secondary with a decay, and one or two associated V's." This topology includes the reactions

$$K^- + p \rightarrow \Xi^- + K^+, \quad (1)$$
$$\rightarrow \Xi^- + K^+ + \pi^0, \quad (2)$$
$$\rightarrow \Xi^- + K^0 + \pi^+. \quad (3)$$

The events located by the scanners were analyzed using the kinematic fitting programs PANG and KICK. Based on a second scan of 80% of the film, the scanning efficiency was 99%. Only one possible example of multiple pion production was found. Table I gives the observed numbers of events of types (1), (2), and (3). Four events with charged K_1^0 decay were also consistent with the reaction $\pi^- + p \rightarrow \Sigma^- + K^0 + \pi^+$ due to a (presumed but as yet not studied) pion background. However, the χ^2 for this hypothesis is quite abnormal. Furthermore, the expected ratios of events of type (3) with charged Λ decay only,

Table I. Summary of Ξ^- production events at 1.80 GeV/c.

Reaction		Events	σ (μb)
$\Xi^- K^+$		94	135 ± 22
$\Xi^- K^+ \pi^0$		20	30 ± 8
$\Xi^- K^0 \pi^+$:	Λ decay	38	
	K_1^0 decay	10	
	Λ and K_1^0 decay	18	
	Total $\Xi^- K^0 \pi^+$	66	75 ± 13

charged K_1^0 only, and with both charged Λ and charged K_1^0 decays, 4:1:2, agree well with the experimental numbers. These observations, coupled with the fact that the effective-mass distribution for ($\Xi\pi$) systems for events with charged K_1^0 agrees with those for the other cases, suggest strongly that all of these events are, in fact, examples of Reaction (3). Preliminary cross sections, based on a τ count and including corrections for neutral decays and for scanning bias of events with Ξ's shorter than 0.5 cm, are given in the third column of Table I. The detailed results of a study of our examples of Reaction (1) will be communicated in the near future.

Figure 1 shows Dalitz plots of the invariant mass squared of the ($\Xi\pi$) systems, $M_{\Xi\pi}^2$, vs $M_{K\pi}^2$ for the observed examples of Reactions (2) and (3). In both neutral ($\Xi^-\pi^+$) and charged

114

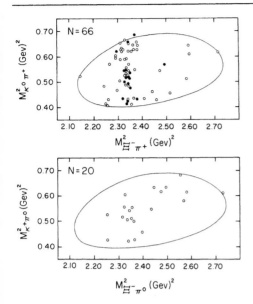

FIG. 1. Invariant-mass-squared Dalitz plots for the examples of Reactions (2) and (3). Solid dots represent events with both charged Λ and K_1^0 decays.

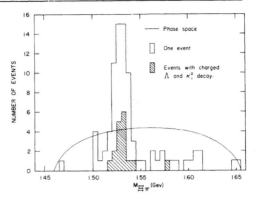

FIG. 2. Effective-mass plot for all $\Xi\pi$ systems observed in this experiment. The line is the relativistic three-body phase space.

($\Xi^-\pi^0$) systems, but especially the former, there is evidence for a pronounced resonance. There is no evidence for a resonance in other particle pairs. Figure 2 shows an effective-mass plot for all ($\Xi\pi$) pairs, regardless of charge. Defining all events between 1.515 and 1.545 GeV as within the resonance and neglecting the small nonresonant background, the resonance peaks at $M_{\Xi\pi} = 1.529 \pm 0.005$ GeV. The uncertainty is due primarily to as yet incompletely studied systematic errors. The resonance is quite sharp. The observed $\Gamma/2$ is comparable with our experimental errors which are typically ~5 MeV. In the case of events of type (3) where both Λ and K_1^0 decay via charged modes (shown shaded in Fig. 2), the momenta of all final reaction products may be individually measured. The experimental errors are accordingly somewhat smaller, ~2.5 MeV. The calculated Γ for these events is ~7 MeV. Within our precision, the width of the resonance can therefore not be determined, but it is not likely to be larger than $\Gamma = 7$ MeV.

The production amplitudes for the ($\Xi\pi K$) reaction, starting with K^- on protons, may be written as follows:

$$\Xi^-\pi^0 K^+ \to [-\sqrt{2}a_{3/2,1} + a_{1/2,1} - a_{1/2,0}], \quad (a)$$

$$\Xi^-\pi^+ K^0 \to [a_{3/2,1} + \sqrt{2}a_{1/2,1} + \sqrt{2}a_{1/2,0}], \quad (b)$$

$$\Xi^0\pi^0 K^0 \to [\sqrt{2}a_{3/2,1} - a_{1/2,1} - a_{1/2,0}], \quad (c)$$

$$\Xi^0\pi^- K^+ \to [-a_{3/2,1} - \sqrt{2}a_{1/2,1} + \sqrt{2}a_{1/2,0}], \quad (d)$$

where $a_{t,T}$ is the amplitude for the production of the ($\Xi\pi$) system in isospin state t from an overall isospin state T. Assuming that the resonance occurs in a unique isospin state and neglecting interference effects with nonresonant backgrounds, (a) and (b) predict a production ratio $(\Xi^-\pi^0 K^+)/(\Xi^-\pi^+ K^0) = 2$ for $t = \frac{3}{2}$. The observed ratio in the $M_{\Xi\pi}$ region between 1.515 and 1.545 GeV is, in fact, 0.21 ± 0.07 which suggests strongly that the ($\Xi\pi$) resonance occurs in the $t = \frac{1}{2}$ state where the $a_{1/2,1}$ and $a_{1/2,0}$ amplitudes may interfere such as to give any $(\Xi^-\pi^0 K^+)/(\Xi^-\pi^+ K^0)$ production ratio. If the $a_{3/2,1}$ amplitude is indeed negligible, then the reaction

$$K^- + p \to \Xi^0 + \pi^- + K^+ \quad (4)$$

should occur twice as often as Reaction (2). Accordingly, we expect to find ~40 events in our film. We have searched 25% of our total exposure for events with the topology "two prong plus V" and found six events which fit only Hypothesis (4) while three additional ones fit (4) and other hypotheses, although with very high χ^2's. The observed number of examples of (4) is thus not in disagreement with the expected

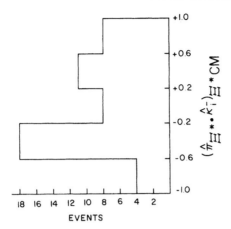
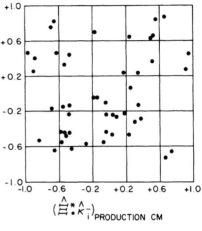
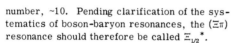
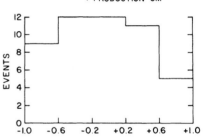

FIG. 3. Scatter plot and angular distributions for the Ξ^* production angle and the decay angle both relative to the incident K^- direction and each in the appropriate c. m. system.

number, ~10. Pending clarification of the systematics of boson-baryon resonances, the $(\Xi\pi)$ resonance should therefore be called $\Xi_{1/2}^*$.

Figure 3 shows a scatter diagram of the cosine of the Ξ^* production angle in the production c.m. system versus the cosine of the emission angle of the pion from Ξ^* decay with respect to the direction of the incident K^-. Only $\Xi^-\pi^+K^0$ events in the $M_{\Xi\pi}$ range from 1.515 to 1.545 GeV have been used. As only 14% of the events are expected to show charged K_1^0 decay alone, the detection of the events depends primarily on the existence of a Ξ track of observable length. At our cutoff of 0.50 cm, the probability of missing an event varies from 9% for extreme forward Ξ's to 24% for extreme backward Ξ's. Hence, large systematic biases in Fig. 3 are not likely. The angular distribution at production appears isotropic. At a $\Gamma \leq 7$ MeV, the mean separation of the Ξ^* and K^0 at the time of Ξ^* decay is ≥ 20F; hence final-state interactions between the K^0 and the other particles do not appear likely. This seems to be borne out by the fact that there appears to be no significant asymmetry about 90° in the Ξ^* decay relative to the Ξ^* direction of motion. The outgoing momentum in the two-body process

$$K^- + p \rightarrow \Xi^{*0} + K^0 \qquad (5)$$

is 0.32 GeV/c and large P-wave contributions are not expected but cannot be ruled out experimentally. At corresponding outgoing momenta the Ξ^-K^+ reaction shows large contributions from higher partial waves.[1] The distribution of the Ξ^* decay angle relative to the incident K^- also appears isotropic within statistics. According to the well-known argument of Adair,[2] S-wave production coupled with a flat decay distribution along the production direction would establish the Ξ^* spin to be $\frac{1}{2}$.

However, in view of the poor statistics, it is impossible to rule out some suitably chosen combination of S and P waves, coupled with a $\Xi_{1/2}^*$ with spin $\frac{3}{2}$. Hence, the Ξ^* spin is as yet not determined. Further work on this question is in progress.

We acknowledge with gratitude the assistance of many members of the hydrogen bubble chamber group at Lawrence Radiation Laboratory and, in particular, the support of Professor Luis Alvarez. Without the cooperation of the Bevatron personnel under the direction of Dr. E. J. Lofgren this work could not have been carried out. We are indebted to Professor N. Byers for many stimulating discussions and for clarification of questions regarding the Ξ^* spin. Last, but not least, we wish to acknowledge the painstaking and patient work of our scanners.

─────────

*Work supported in part by the U. S. Atomic Energy Commission.

[1]Luis W. Alvarez et al. (to be published).

[2]R. K. Adair, Phys. Rev. **100**, 1540 (1955).

OBSERVATION OF A HYPERON WITH STRANGENESS MINUS THREE*

V. E. Barnes, P. L. Connolly, D. J. Crennell, B. B. Culwick, W. C. Delaney,
W. B. Fowler, P. E. Hagerty,† E. L. Hart, N. Horwitz,† P. V. C. Hough, J. E. Jensen,
J. K. Kopp, K. W. Lai, J. Leitner,† J. L. Lloyd, G. W. London,‡ T. W. Morris, Y. Oren,
R. B. Palmer, A. G. Prodell, D. Radojičić, D. C. Rahm, C. R. Richardson, N. P. Samios,
J. R. Sanford, R. P. Shutt, J. R. Smith, D. L. Stonehill, R. C. Strand, A. M. Thorndike,
M. S. Webster, W. J. Willis, and S. S. Yamamoto
Brookhaven National Laboratory, Upton, New York
(Received 11 February 1964)

It has been pointed out[1] that among the multitude of resonances which have been discovered recently, the $N_{3/2}*(1238)$, $Y_1*(1385)$, and $\Xi_{1/2}*(1532)$ can be arranged as a decuplet with one member still missing. Figure 1 illustrates the position of the nine known resonant states and the postulated tenth particle plotted as a function of mass and the third component of isotopic spin. As can be seen from Fig. 1, this particle (which we call Ω^-, following Gell-Mann[1]) is predicted to be a negatively charged isotopic singlet with strangeness minus three.[2] The spin and parity should be the same as those of the $N_{3/2}*$, namely, $3/2^+$. The 10-dimensional representation of the group SU_3 can be identified with just such a decuplet. Consequently, the existence of the Ω^- has been cited as a crucial test of the theory of unitary symmetry of strong interactions.[3,4] The mass is predicted[5] by the Gell-Mann–Okubo mass formula to be about 1680 MeV/c^2. We wish to report the observation of an event which we believe to be an example of the production and decay of such a particle.

The BNL 80-in. hydrogen bubble chamber was exposed to a mass-separated beam of 5.0-BeV/c K^- mesons at the Brookhaven AGS. About 100 000 pictures were taken containing a total K^- track length of ~10^6 feet. These pictures have been partially analyzed to search for the more characteristic decay modes of the Ω^-.

The event in question is shown in Fig. 2, and the pertinent measured quantities are given in Table I. Our interpretation of this event is

$$K^- + p \rightarrow \Omega^- + K^+ + K^0$$
$$\quad \hookrightarrow \Xi^0 + \pi^-$$
$$\qquad \hookrightarrow \Lambda^0 + \pi^0$$
$$\qquad\quad \hookrightarrow \gamma_1 + \gamma_2$$
$$\qquad\qquad \hookrightarrow e^+ + e^-$$
$$\qquad\qquad \hookrightarrow e^+ + e^-$$
$$\qquad \hookrightarrow \pi^- + p. \qquad (1)$$

From the momentum and gap length measurements, track 2 is identified as a K^+. (A bubble density of 1.9 times minimum was expected for this track while the measured value was 1.7 ± 0.2.) Tracks 5 and 6 are in good agreement with the decay of a Λ^0, but the Λ^0 cannot come from the primary interaction. The Λ^0 mass as calculated from the measured proton and π^- kinematic quantities is 1116 ± 2 MeV/c^2. Since the bubble density from gap length measurement of track 6 is 1.52 ± 0.17, compared to 1.0 expected for a π^+ and 1.4 for a proton, the interpretation of the V as a K^0 is unlikely. In any case, from kinematical considerations such a K^0 could not come from the production vertex. The Λ^0 appears six decay lengths from the wall of the bubble chamber, and there is no other visible origin in the chamber.

The event is unusual in that two gamma rays, apparently associated with it, convert to electron-positron pairs in the liquid hydrogen. From measurements of the electron momenta and angles, we determine that the effective mass of the two gamma rays is 135.1 ± 1.5 MeV/c^2, consistent with a π^0 decay. In a similar manner, we have used the calculated π^0 momentum and angles, and the values from the fitted Λ^0 to deter-

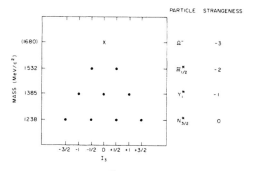

FIG. 1. Decuplet of $\frac{3}{2}^+$ particles plotted as a function of mass versus third component of isotopic spin.

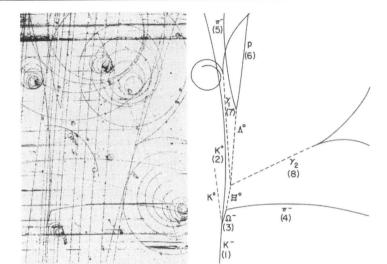

FIG. 2. Photograph and line diagram of event showing decay of Ω^-.

mine the mass of the neutral decaying hyperon to be 1316 ± 4 MeV/c^2 in excellent agreement with that of the Ξ^0. The projections of the lines of flight of the two gammas and the Λ^0 onto the XY plane (parallel to the film) intersect within 1 mm and in the XZ plane within 3 mm. The calculated momentum vector of the Ξ^0 points back to the decay point of track 3 within 1 mm and misses the production vertex by 5 mm in the XY plane. The length of the Ξ^0 flight path is 3 cm with a calculated momentum of 1906 ± 20 MeV/c. The transverse momenta of the Ξ^0 and of track 4 balance within the errors, indicating that no other particle is emitted in the decay of particle 3.

We will now discuss the decay of particle 3. From the momentum and gap length measurements on track 4, we conclude that its mass is less than that of a K. Using the Ξ^0 momentum and assuming particle 4 to be a π^-, the mass of particle 3 is computed to be 1686 ± 12 MeV/c^2 and its momentum to be 2015 ± 20 MeV/c. Note that the measured transverse momentum of track 4, 248 ± 5 MeV/c, is greater than the maximum momentum for the possible decay modes of the known particles (given in Table II), except for $\Xi^- \to e^- + n + \nu$. We reject this hypothesis not only because it involves $\Delta S = 2$, but also because it disregards the previously established associations of the Λ and two gammas with the event.

Table I. Measured quantities.

Track	Azimuth (deg)	Dip (deg)	Momentum (MeV/c)
1	4.2 ± 0.1	1.1 ± 0.1	4890 ± 100
2	6.9 ± 0.1	3.3 ± 0.1	501 ± 5.5
3	14.5 ± 0.5	-1.5 ± 0.6	...
4	79.5 ± 0.1	-2.7 ± 0.1	281 ± 6
5	344.5 ± 0.1	-12.0 ± 0.2	256 ± 3
6	9.6 ± 0.1	-2.5 ± 0.1	1500 ± 15
7	357.0 ± 0.3	3.9 ± 0.4	82 ± 2
8	63.3 ± 0.3	-2.4 ± 0.2	177 ± 2

Table II. Maximum transverse momentum of the negative decay product for various particle decays.

Decay modes	Maximum transverse momentum (MeV/c)
$\pi^- \to \mu^- + \nu$	30
$K^- \to \mu^- + \nu$	236
$K^- \to \pi^- + \pi$	205
$K^- \to e^- + \pi^0 + \nu$	229
$\Sigma^- \to \pi^- + n$	192
$\Sigma^- \to e^- + \Lambda^0 + \nu$	78
$\Sigma^- \to e^- + n + \nu$	229
$\Xi^- \to \pi^- + \Lambda^0$	139
$\Xi^- \to e^- + \Lambda^0 + \nu$	190
$\Xi^- \to e^- + n + \nu$	327

The proper lifetime of particle 3 was calculated to be 0.7×10^{-10} sec; consequently we may assume that it decayed by a weak interaction with $\Delta S = 1$ into a system with strangeness minus two. Since a particle with $S = -1$ would decay very rapidly into $Y + \pi$, we may conclude that particle 3 has strangeness minus three. The missing mass at the production vertex is calculated to be 500 ± 25 MeV/c^2, in good agreement with the K^0 assumed in Reaction (1). Production of the event by an incoming π^- is excluded by the missing mass calculated at the production vertex, and would not alter the interpretation of the decay chain starting with track 3.

In view of the properties of charge ($Q = -1$), strangeness ($S = -3$), and mass ($M = 1686 \pm 12$ MeV/c^2) established for particle 3, we feel justified in identifying it with the sought-for Ω^-. Of course, it is expected that the Ω^- will have other observable decay modes, and we are continuing to search for them. We defer a detailed discussion of the mass of the Ω^- until we have analyzed further examples and have a better understanding of the systematic errors.

The observation of a particle with this mass and strangeness eliminates the possibility which has been put forward[6] that interactions with $\Delta S = 4$ proceed with the rates typical of the strong interactions, since in that case the Ω^- would decay very rapidly into $n + K^0 + \pi^-$.

We wish to acknowledge the excellent cooperation of the staff of the AGS and the untiring efforts of the 80-in. bubble chamber and scanning and programming staffs.

---

*Work performed under the auspices of the U. S. Atomic Energy Commission and partially supported by the U. S. Office of National Research and the National Science Foundation.
†Syracuse University, Syracuse, New York.
‡University of Rochester, Rochester, New York.
[1]M. Gell-Mann, Proceedings of the International Conference on High-Energy Nuclear Physics, Geneva, 1962 (CERN Scientific Information Service, Geneva, Switzerland, 1962), p. 805; R. Behrends, J. Dreitlein, C. Fronsdal, and W. Lee, Rev. Mod. Phys. 34, 1 (1962); S. L. Glashow and J. J. Sakurai, Nuovo Cimento 25, 337 (1962).
[2]A possible example of the decay of this particle was observed by Y. Eisenberg, Phys. Rev. 96, 541 (1954).
[3]M. Gell-Mann, Phys. Rev. 125, 1067 (1962); Y. Ne'eman, Nucl. Phys. 26, 222 (1961).
[4]See, however, R. J. Oakes and C. N. Yang, Phys. Rev. Letters 11, 174 (1963).
[5]M. Gell-Mann, Synchrotron Laboratory, California Institute of Technology, Internal Report No. CTSL-20, 1961 (unpublished); S. Okubo, Progr. Theoret. Phys. (Kyoto) 27, 949 (1962).
[6]G. Racah, Nucl. Phys. 1, 302 (1956); H. J. Lipkin, Phys. Letters 1, 68 (1962).

6
Weak Interactions

From parity violation to two neutrinos, 1956–1962.

In 1930, Wolfgang Pauli postulated the existence of the neutrino, a light, feebly interacting particle. Pauli did this to account for the electron spectrum seen in beta decay. If the electron were the only particle emitted in beta decay, it would always have an energy equal to the difference between the initial and final nuclear state energies. Measurements showed, however, that the electron's energy was variable and calorimetric measurements confirmed that some of the energy was being lost. So disturbing was this problem that Bohr even suggested that energy might only be conserved on average!

Beta decay could not be understood without a successful model of the nucleus and that came after the discovery of the neutron by Chadwick in 1932. The neutrino and the neutron provided the essential ingredients for Fermi's theory of weak interactions. He saw that the fundamental process was $n \to pe\nu$. Using the language of quantized fields Fermi could write this as an interaction:

$$p^\dagger(x)n(x)e^\dagger(x)\nu(x)$$

where each letter stands for the operator that destroys the particle represented or creates its antiparticle. Thus the $n(x)$ destroys a neutron or creates an antineutron. The dagger makes the field into its adjoint, for which destruction and creation are interchanged. Thus $p^\dagger(x)$ creates protons and destroys antiprotons. The position at which the creation and destruction take place is x.

Fermi wrote the theory in terms of a Hamiltonian. It had to be invariant under translations in space. This is achieved by writing something like

$$H \propto \int d^3x \; p^\dagger(x)n(x)e^\dagger(x)\nu(x) \tag{6.1}$$

suitably modified to be Lorentz invariant.

The relativistic theory of fermions was developed by Dirac. Each fermion is represented by a column vector of four entries (essentially for spin up and down, for both particle and antiparticle). For a nonrelativistic particle, the first two entries are much larger than the last two. These "large components" are equivalent to Pauli's two component spinor representation of a nonrelativistic spin one-half particle. Explicitly, a particle of mass m, three-momentum $\mathbf{p}$, and energy $E = \sqrt{m^2 + p^2}$ and with spin orientation indicated by a two-component spinor χ is represented by a Dirac spinor

$$u(p) = \sqrt{E+m} \begin{pmatrix} \chi \\ \frac{\sigma \cdot \mathbf{p}}{E+m} \chi \end{pmatrix}. \tag{6.2}$$

Indeed, in the nonrelativistic limit where $p \ll m, E$, the lower two components are much smaller than the upper two. Thus if $\chi = \begin{pmatrix} 1 \\ 0 \end{pmatrix}$ and $\mathbf{p}$ has components p_x, p_y, p_z, then

$$u(p) = \begin{pmatrix} \sqrt{E+m} \\ 0 \\ p_z/\sqrt{E+m} \\ (p_x + ip_y)/\sqrt{E+m} \end{pmatrix}. \tag{6.3}$$

Despite their appearance, these spinors are not four-vectors because they transform in a completely different way. It is possible to make Lorentz scalars and four-vectors from pairs of spinors. Ordinary four-vectors, $a = (a_0, \mathbf{a})$ and $b = (b_0, \mathbf{b})$, can be combined to make a Lorentz-invariant product $a \cdot b = a_0 b_0 - \mathbf{a} \cdot \mathbf{b} = a_\mu b^\mu$, where $a_0 = a^0$, $a_i = -a^i$, for $i = 1, 2, 3$. Pairs of spinors are combined with the Dirac matrices which can be expressed as

$$\gamma^0 = \begin{pmatrix} I & 0 \\ 0 & -I \end{pmatrix}, \quad \gamma^1 = \begin{pmatrix} 0 & \sigma_1 \\ -\sigma_1 & 0 \end{pmatrix}, \quad \gamma^2 = \begin{pmatrix} 0 & \sigma_2 \\ -\sigma_2 & 0 \end{pmatrix}, \quad \gamma^3 = \begin{pmatrix} 0 & \sigma_3 \\ -\sigma_3 & 0 \end{pmatrix}. \tag{6.4}$$

where σ_i are the usual 2×2 Pauli spin matrices, and I is the 2×2 unit matrix. In this convention one writes $\gamma_0 = \gamma^0$, $\gamma_i = -\gamma^i$, $i = 1, 2, 3$. A Lorentz invariant is obtained by placing a γ^0 between a spinor (a column vector) and an adjoint spinor, $\psi^\dagger$, which is the row vector obtained by taking the complex conjugate of each component:

$$\psi^\dagger \gamma^0 \psi \equiv \overline{\psi} \psi \tag{6.5}$$

where ψ is a four component spinor and $\overline{\psi} = \psi^\dagger \gamma^0$, or equivalently, $\psi^\dagger = \overline{\psi} \gamma^0$. The combination $\overline{\psi} \gamma_\mu \psi$, $\mu = 0, 1, 2, 3$, transforms as a four-vector. Thus $\psi^\dagger \psi = \overline{\psi} \gamma_0 \psi$ is not a scalar, but the zeroth component of a vector quantity.

Rather than using ψ or u for each spinor, it is often clearer to indicate the particle type, so a spinor for a proton is indicated simply by p, one for a neutrino by ν, and so on. Thus $\nu(x)$ is the neutrino field at x, a field that destroys neutrinos or creates antineutrinos. Similarly $p^\dagger$ and $\overline{p} = p^\dagger \gamma^0$ create protons or destroy antiprotons. An operator like $e(x)$ can be expressed in terms of momentum through a Fourier transform. For example, if $e(x)$ acts on a state with an electron of momentum p, a factor of the spinor $u(p)$ is produced.

A possible interaction that is Lorentz-invariant is of the form

$$H \propto \int d^3x \, \overline{p}(x) n(x) \overline{e}(x) \nu(x). \tag{6.6}$$

6. Weak Interactions

This is not what Fermi chose. He noted that the usual electromagnetic current for an electron, which receives contributions from the motion of the charge and the magnetic moment, can be written in Dirac notation as

$$J_\mu(x) = \bar{e}(x)\gamma_\mu e(x). \tag{6.7}$$

This object transforms as a relativistic four-vector. Electrodynamics can be viewed as the interaction of such currents. By analogy, Fermi wrote

$$H = g \int d^3x \; \bar{p}(x)\gamma^\mu n(x) \; \bar{e}(x)\gamma_\mu \nu(x) \tag{6.8}$$

where g was a constant. There also had to be an interaction that was the Hermitian conjugate of this and would describe e^+ emission, a process discovered by Irène Curie and Frédéric Joliot in 1933:

$$H = g \int d^3x \; \bar{n}(x)\gamma^\mu p(x) \; \bar{\nu}(x)\gamma_\mu e(x). \tag{6.9}$$

Many consequences of Fermi's theory can be obtained without detailed computation, which is often prevented by lack of detailed information on the nuclear wave functions. By the Golden Rule, the decay rate is governed by

$$\Gamma \propto \int d^3 p_e \, d^3 p_\nu \, \delta(Q - E_e - E_\nu) |H_{fi}|^2 \tag{6.10}$$

where p_e is the electron's momentum and E_e is its energy and similarly for the neutrino. The total energy available in the decay is Q, the mass difference between the initial and final nuclei, minus the electron mass. The Dirac delta function guarantees energy conservation. The recoiling nucleus balances the momentum, but contributes negligibly to the energy. If we ignore the dependence of the matrix element, H_{fi}, on the momentum, we find

$$\frac{d\Gamma}{dp_e} \propto p_e^2 (Q - E_e)^2 |H_{fi}|^2. \tag{6.11}$$

Thus $(1/p_e)(d\Gamma/dp_e)^{1/2}$ should be a linear function of E_e. A plot of these quantities is called a Kurie plot and the expectation of linearity is borne out in many decays. The high energy portion of a Kurie plot for tritium decay is shown in Figure 6.1.

Looking at the Fermi theory in greater detail, we consider the term

$$\bar{p}(x)\gamma^\mu n(x) = p^\dagger(x)\gamma^0 \gamma^\mu n(x) \tag{6.12}$$

involving the nucleons only. This operator changes the initial nuclear state to the final one, transforming a neutron into a proton. The nucleons can be considered nonrelativistic. Of their four components, only the first two are important and these represent spin-1/2 in the usual way. Since γ^1, γ^2 and γ^3 connect large components to small components, only $\bar{p}\gamma^0 n$

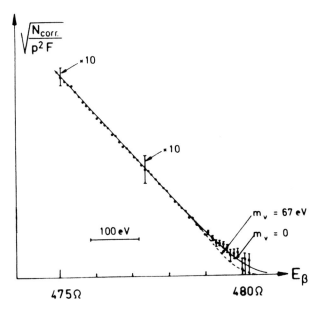

Figure 6.1. The Kurie plot for the beta decay of tritium showing the portion of the electron spectrum near the end point at 18.6 keV. As pointed out by Fermi in his 1934 paper setting out the principles of beta decay, if the neutrino mass is non-zero there will be a deviation of the plot from linearity near the end point. By studying this region with extreme care, Bergkvist was able to set an upper limit of 60 eV on the mass of the neutrino (more precisely, the electron-antineutrino) [K. E. Bergkvist, *Nucl. Phys.* **B39**, 317 (1972)]. The x-axis of the Figure shows the magnet setting of the spectrometer. The interval corresponding to 100 eV is indicated, as well as two sample error bars with a magnification of 10. The curves expected, including the effects of the apparatus resolution, for neutrino masses of 67 eV and 0 eV are shown. Without the resolution effects, the curve for 0 eV would be a straight line, while the 67 eV curve would fall more abruptly to zero.

will be important. Thus $\bar{p}\gamma^\mu n$ reduces to $p^\dagger n$ where in the final expression we consider the spinors to have just two components. This operator changes a neutron into a proton without changing its location or affecting its spin. It cannot change the angular momentum: It is a $\Delta J = 0$ operator. Moreover, it cannot change the parity. These are the selection rules analogous to, but different from, those familiar in radiative transitions between atomic states.

In fact, it is found that not all beta decays occur between nuclear states with identical angular momenta, so the Fermi interaction cannot be a complete description. To generalize it, we consider the possible forms made from two (four-component) fermion fields and combinations of Dirac matrices:

$$\bar{p}n \qquad S \quad (scalar)$$
$$\bar{p}\gamma_5 n \qquad P \quad (pseudoscalar)$$
$$\bar{p}\gamma^\mu n \qquad V \quad (vector)$$
$$\bar{p}\gamma^\mu \gamma_5 n \qquad A \quad (axial\ vector)$$
$$\bar{p}\sigma^{\mu\nu} n \qquad T \quad (tensor)$$

Here we have introduced

$$\sigma^{\mu\nu} = \frac{i}{2}[\gamma^\mu, \gamma^\nu] = \frac{i}{2}(\gamma^\mu\gamma^\nu - \gamma^\nu\gamma^\mu) \qquad (6.13)$$

and

$$\gamma_5 = \begin{pmatrix} 0 & I \\ I & 0 \end{pmatrix}. \qquad (6.14)$$

The names "scalar," "vector," etc., describe the behavior of the bilinears under the Lorentz group and parity. Lorentz invariant quantities can be obtained by combining with the corresponding forms like $\bar{e}v$, $\bar{e}\gamma_5 v$, etc. Before 1956, it was presumed that parity was conserved in weak interactions. This allowed combinations like $\bar{p}n\,\bar{e}v$ but forbade $\bar{p}\gamma_5 n\,\bar{e}v$, $\bar{p}\gamma_\mu\gamma_5 n\,\bar{e}\gamma^\mu v$, etc.

Using the forms of the Dirac matrices and the rule that only the two upper two components of a spinor are important for a nonrelativistic particle, it is easy to see what kinds of terms are available for the nuclear part of the beta-decay amplitude:

$$
\begin{array}{lll}
S: & \bar{p}n & \to \quad p^\dagger n \\
P: & \bar{p}\gamma_5 n & \to \quad 0 \\
V: & \bar{p}\gamma^\mu n & \to \quad p^\dagger n \quad \text{for } \mu = 0, \text{ zero otherwise} \\
A: & \bar{p}\gamma^\mu\gamma_5 n & \to \quad p^\dagger \sigma^i n \quad \text{for } \mu = i = 1, 2, 3, \text{ zero if } \mu = 0 \\
T: & \bar{p}\sigma^{\mu\nu} n & \to \quad p^\dagger \sigma^i n \quad \text{if } \mu = j, \nu = k\ (j,k = 1,2,3) \\
& & \qquad\qquad \text{and } i, j, k \text{ cyclic, zero otherwise}
\end{array}
$$

In the right-hand column, the p and n represent two-component spinors and σ^i is a Pauli matrix.

Thus we see that two kinds of nuclear transitions are possible, ones like those in the original Fermi theory, due to $p^\dagger n$, and those due to $p^\dagger \sigma n$. The former are called Fermi transitions and the latter Gamow–Teller transitions. Because of the σ, the Gamow–Teller transitions can change the angular momentum of the nucleus by one unit. However, the operator still does not change parity. In summary, S and V give Fermi transitions, while T and A give Gamow–Teller transitions. Fermi transitions in which the angular momentum of the nucleus changes are not allowed. Thus from the existence of transitions like $O^{14} \to N^{14*} + e^+ + v$ ($0^+ \to 0^+$) and $He^6 \to Li^6 + e^- + v$ ($0^+ \to 1^+$) we know that there must be at least one of S and V as well as at least one of T and A. It was also possible to show that if we have both S and V, or both T and A in a parity conserving theory, the Kurie plot would not be straight, in contradiction with the data. Thus the nuclear part of the transition was thought to be either S or V, together with T or A, and parity conserving.

Distinguishing between these choices required observing more than the electron energy spectrum. The angle between the electron and neutrino directions could be inferred by measuring the recoil of the nucleus. The dependence on this angle measured the relative amounts of V versus S and A versus T. The results before 1957 indicated a preference for T over A, especially in the $He^6 \to Li^6 + e^- + v$ decay.

In addition to nuclear beta decay, information on weak interactions was available from decays of strongly interacting particles, especially kaons, and from the decay of the muon. A thorough analysis of the decay $\mu \to e\nu\nu$ was given by L. Michel in 1950, assuming parity conservation. He found that the shape of the energy spectrum was determined up to a single parameter, ρ, that was a function of the relative amounts of S, P, V, A, and T. With $x = 2p_e/m_\mu$, the intensity of the spectrum is

$$dN/dx \propto x^2 [1 - x + (2/3)\rho(4x/3 - 1)]. \tag{6.15}$$

A measurement in 1955 gave $\rho = 0.64 \pm 0.10$. The currently accepted value is 0.7509 ± 0.0010, consistent with the maximal value allowed, 3/4. Two examples of the electron spectrum from muon decay are displayed in Figure 6.2.

About the same time, the universality of the weak interaction was becoming evident. By universality one means that the interaction is of the same form and strength in all situations. Tiomno and Wheeler suggested that the pairs (e, ν), (μ, ν), and (n, p) entered into the weak interaction in an equivalent way. Nuclear beta decay involves (n, p) and (e, ν). The charged pion can be viewed as a bound state of a nucleon and an antinucleon. In this way, the weak interaction responsible for charged pion decay involves (n, p) and (μ, ν). The decay of the muon depends on (μ, ν) and (e, ν).

The giant step in understanding weak interactions came in 1956 when T. D. Lee and C. N. Yang pointed out that there was no evidence in favor of parity conservation in weak interactions. The precipitating issue was the $\tau - \theta$ puzzle. As described in Chapter 3, the τ was the 3π decay of the K^+. The analysis begun by Dalitz had shown that the 3π system had J^P in the series $0^-, 2^-, \ldots$ On the other hand, the θ^+ (or χ^+) decayed into $\pi^0 \pi^+$ and had "natural" spin-parity: $J^P = 0^+, 1^-, \ldots$ Measurements showed that the masses and lifetimes of the θ and τ were very similar, perhaps equal. The θ and the τ seemed to be the same particle, except that they had different values of J^P. The Proceedings of the Sixth Annual Rochester Conference in April 1956 record that after Yang's talk,

Feynman brought up a question of [Martin] Block's: Could it be that the θ and τ are different parity states of the same particle which has no definite parity, *i.e.* that parity is not conserved. That is, does nature have a way of defining right or left-handedness uniquely. Yang stated that he and Lee looked into this matter without arriving at any definite conclusions.

A few months later, there were conclusions. Despite the overwhelming prejudice that parity must be a good symmetry because it was a symmetry of space itself just as rotational invariance is, Lee and Yang demonstrated that there was no evidence for or against parity conservation in weak interactions. To test for possible violation of parity it was necessary to observe a dependence of a decay rate (or cross section) on a term that changed sign under the parity operation. Parity reverses momenta and positions, but not angular momentum (or spins). In a nuclear decay, the momenta available are $\mathbf{p_e}$, $\mathbf{p_\nu}$, and $\mathbf{p_N}$, the momenta of the electron, neutrino and recoil nucleus. Terms like $\mathbf{p_e} \cdot \mathbf{p_\nu}$ cannot show parity violation. The invariant formed from the three momenta, $\mathbf{p_e} \cdot \mathbf{p_\nu} \times \mathbf{p_N}$, would change sign under parity, but vanishes because the momenta are coplanar. To test for parity violation in nuclear beta decay required consideration of spin. If the decaying nucleus were oriented, it would be

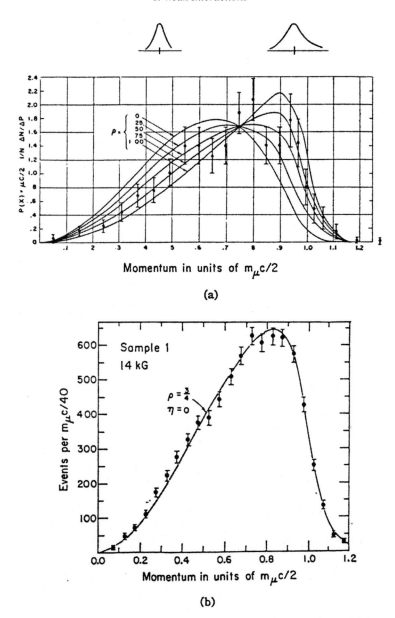

Figure 6.2. Two examples of the electron momentum spectrum in muon decay. (a) An early measurement made in a high pressure cloud chamber at the Columbia University Nevis Cyclotron which gave the value $\rho = 0.64 \pm 0.10$. The variation with the parameter ρ of the spectrum shape, including the experimental resolution, is shown in the curves. The bell-shaped curves show the resolution of the experiment at two values of the electron momentum [C. P. Sargent et al., *Phys. Rev.* **99**, 885 (1955)]. (b) A later spectrum obtained with a hydrogen bubble chamber which, when combined with earlier spark chamber measurements, gave $\rho = 0.752 \pm 0.003$ [S. E. Derenzo, *Phys. Rev.* **181**, 1854 (1969)].

possible to measure the angular dependence of the decay, looking for a term proportional to $<\mathbf{J}> \cdot \mathbf{p_e}$, where $<\mathbf{J}>$ was the average nuclear spin. This was achieved by C. S. Wu in collaboration with E. Ambler and co-workers at the National Bureau of Standards, who had the necessary low temperature facility **(Ref. 6.1)**.

Wu and her co-workers chose to work with Co^{60}, whose ground state has $J^P = 5^+$. It beta-decays through a Gamow–Teller transition with a half-life of 5.2 y, yielding Ni^{60} in the 4^+ state. The excited Ni state decays through two successive γ emissions to 2^+ and then 0^+, with γ energies 1.173 and 1.332 MeV, respectively. The NBS team included experts in producing nuclear polarization through adiabatic demagnetization. The degree of polarization of the Co sample was monitored by observing the anisotropy of the gamma radiation. The polarization of the Co^{60} was transmitted to the Ni^{60}, giving a difference between the rates for gamma emission in the polar and equatorial directions, relative to the axis of the applied polarizing magnetic field.

The beta-decay rate along the direction of the magnetic field, that is, along the nuclear polarization direction was monitored. Reversing the magnetic field reversed the direction of $<\mathbf{J}>$. The counting rate indeed showed a dependence on $<\mathbf{J}> \cdot \mathbf{p_e}$. Not only was the rate different for the two magnetic field orientations, but as the sample warmed, the dependence of the rate on the field orientation disappeared at the same speed as the polarization itself disappeared, showing the connection of the decay angular distribution was with the nuclear orientation, not simply with the applied magnetic field.

Word of this *tour de force* spread rapidly and new experiments were undertaken even before the results of Wu's team appeared in print. Indeed, two further experiments appeared in rapid succession showing parity violation in the sequence $\pi^+ \to \mu^+ \to e^+$ **(Refs. 6.2, 6.3)**. Rather than beginning with a polarized beam, these experiments exploited the prediction of Lee and Yang that parity violation would lead to polarization of the μ along its line of flight in the $\pi \to \mu\nu$ decay. The polarization of the μ is retained when it slows down in matter. A distribution of decay electrons relative to the incident beam direction of the form $1 + a\cos\theta$ is then expected, where a depends on the degree of polarization of the μ. Garwin, Lederman, and Weinrich, working with the Nevis Cyclotron at Columbia University, applied a magnetic field to the region where the muons stopped. This caused the spin of the muon to precess. In this elegant fashion, they demonstrated parity violation, measured its strength and simultaneously measured the magnetic moment of the μ^+ by measuring the rate of precession. At the same time, Friedman and Telegdi, at the University of Chicago, also found parity violation by observing the same decay sequence, but working in emulsions and without a magnetic field. The emulsion experiment was started before the others, but took longer to complete because of the laborious scanning procedure.

With the violation of parity, the number of terms to be considered in nuclear beta decay doubled. A general interaction could be written

$$H = \frac{G_F}{\sqrt{2}} \int d^3x \left(C_S \bar{p}n \, \bar{e}\nu + C'_S \bar{p}n \, \bar{e}\gamma_5\nu + C_V \bar{p}\gamma^\mu n \, \bar{e}\gamma_\mu\nu + C'_V \bar{p}\gamma^\mu n \, \bar{e}\gamma_\mu\gamma_5\nu + \cdots \right)$$
(6.16)

where $G_F = 1.166 \times 10^{-5}$ GeV^{-2} is known as the Fermi constant. The terms with coefficients C_i are parity conserving, while those with coefficients C'_i are parity violating. The years 1957 and 1958 brought a wealth of experiments aimed at determining the constants $C_S, C'_S, C_V, C'_V, \ldots$ Parity violation allowed rates to depend on $\sigma_e \cdot \mathbf{p}_e$, i.e. longitudinal polarization of the electron emitted in beta decay. Frauenfelder and co-workers (**Ref. 6.4**) found a large electron polarization, $<\sigma_e \cdot \mathbf{p}_e>/p_e \approx -1$. This result was consistent with the proposal that the neutrino has a single handedness:

$$H = \frac{G_F}{\sqrt{2}} \int d^3x \left[C_S \bar{p}n \, \bar{e}(1 \pm \gamma_5)\nu + C_V \bar{p}\gamma_\mu n \, \bar{e}\gamma^\mu(1 \pm \gamma_5)\nu + \cdots \right] \quad (6.17)$$

If the negative sign is used, the neutrinos participating in the interaction are left-handed, that is, their spins are antiparallel to their momenta (helicity $-1/2$). If the positive sign is taken, they are right-handed (helicity $+1/2$). The experiment of Frauenfelder et al. showed that the electrons were mostly left-handed. This would follow from, say, $\bar{e}\gamma^\mu(1-\gamma_5)\nu$ or from $\bar{e}(1+\gamma_5)\nu$. More completely, if the neutrino has a single handedness and the nuclear part is V or A, then the neutrino should be left-handed, while if the nuclear part is S or T, the neutrino should be right-handed. Remarkably, it was possible to do an experiment to measure the handedness of the neutrino!

This was accomplished by M. Goldhaber, L. Grodzins, and A. W. Sunyar (**Ref. 6.5**). The experiment is based on a subtle point, the strong energy dependence of resonant scattering of X-rays. When an excited nucleus emits an X-ray, the energy of the X-ray is not exactly equal to the difference of the nuclear levels because the recoiling nucleus carries some energy. However, if the emitting nucleus is moving in the direction of the X-ray emission, the Doppler shift makes up for some of the energy loss. The resonant scattering of such X-rays is then much stronger since the X-ray's energy is closer to the energy of excitation of the nucleus. This could be exploited in $Eu^{152\,m}$ which decays by electron capture, with a half-life of about 9 hours. In electron capture, an inner shell electron interacts with the nucleus according to $e^- p \rightarrow n\nu$. In this case, the overall reaction was $e^- + Eu^{152\,m} \rightarrow Sm^{152*} + \nu$. The initial nucleus has $J = 0$ and the final nucleus, $J = 1$. The latter decays very rapidly by γ emission to the $J = 0$ ground state. If we take the neutrino direction as the z axis and assume the captured electron is in an s-wave, the intermediate Sm^{152*} state has $J_z = 1$ or 0 if the ν has $J_z = -1/2$ and $J_z = -1$ or 0 if the ν has $J_z = 1/2$. Now if a gamma ray is emitted in the negative z direction (where resonant scattering is greatest because the motion of the nucleus compensates for the energy lost in recoil), it has $J_z = 1$ or -1, and in fact its helicity has the same sign as that of the neutrino. By measuring the circular polarization of the gamma ray with magnetized iron, the neutrino helicity is measured. The result found was that the neutrino is left-handed.

The outcome of this and many of the experiments at the time were in agreement with the V-A theory proposed by Marshak and Sudarshan and by Feynman and Gell-Mann. The V and A terms for the nuclear beta decay were coupled to the $\bar{e}\gamma_\mu(1-\gamma_5)\nu$ term:

$$H = \frac{G_F}{\sqrt{2}} \int d^3x \, \bar{p}(x)\gamma^\mu(g_v + g_a\gamma_5)n(x) \, \bar{e}(x)\gamma_\mu(1-\gamma_5)\nu(x) \quad (6.18)$$

where g_v and g_a are the vector and axial vector couplings of the weak current to the nucleons. The value of g_v is very nearly one. It can be measured in pure Fermi transitions like

O^{14} decay, in which the nuclear matrix element is calculable because the initial and final nuclei are members of the same isomultiplet. The axial coupling constant can be measured in neutron decay, either from the neutron lifetime or from more detailed measurements of the decay. By studying the decay of free polarized neutrons, Telegdi and co-workers were able to confirm the V-A form of the interaction and measure sign as well as the magnitude of g_a/g_v (Ref. 6.6). The currently accepted value of g_a/g_v is -1.2695 ± 0.0029.

More generally, for processes with an electron and a neutrino in the final state, like $K^- \to \pi^0 e^- \nu$, the V-A theory postulates an interaction

$$H = g \int d^3x \, J^\dagger_{\mu\,\text{had}}(x) J^\mu_{\text{lep}}(x) + \text{Hermitian conjugate} \tag{6.19}$$

where

$$J^\mu_{\text{lep}}(x) = \bar{e}(x)\gamma^\mu(1-\gamma_5)\nu(x). \tag{6.20}$$

The hadronic current, J^μ_{had} cannot be specified so precisely. For nuclear beta decay one can limit the possible forms since the nucleons are nonrelativistic. For decays like $\pi^- \to \pi^0 e^- \nu$ and $n \to p e^- \nu$, Feynman and Gell-Mann proposed that the vector part of the hadronic currents that raised or lowered the charge of the hadrons by one unit and did not change strangeness was part of an isotriplet of currents. The third, or charge-nonchanging, component of the triplet was the isovector part of the electromagnetic current, that is, the part responsible for the difference in the electromagnetic behavior of the neutron and proton. Since the electromagnetic current is conserved, so would be the vector part of the hadronic weak current. This proposal was known as the conserved vector current hypothesis (CVC) and was actually first given by the Soviet physicists S. S. Gershtein and Ya. B. Zeldovich.

CVC has been tested in pion beta decay, $\pi^+ \to \pi^0 e^+ \nu$ and in a comparison of the weak decays $B^{12} \to C^{12} e^- \bar{\nu}$, $N^{12} \to C^{12} e^+ \nu$ with the electromagnetic decay $C^{12*} \to C^{12} \gamma$. The three nuclei B^{12}, C^{12*}, and N^{12} form an isotriplet and C^{12} is the isosinglet ground state. In these processes, the weak decay rates can be calculated because the decay depends on the vector current and the weak vector current matrix elements can be obtained from the isovector electromagnetic current matrix elements measured in C^{12*} decay.

The V-A theory proved very successful and has survived as the low energy description of weak interactions. The weak hadronic current has two pieces, $\Delta S = 0$ (e.g. $n \to p e^- \nu$) and a $\Delta S = 1$ piece (e.g. $K \to \mu\nu$, $K \to \pi\mu\nu$). The strengths of the strangeness-changing and the strangeness-nonchanging interactions are not the same. N. Cabibbo described this by proposing that while in leptonic decays (like $\mu \to e\nu\nu$) the interaction could be written as

$$\frac{G_F}{\sqrt{2}} J^\mu_{\text{lep}}(x) J^\dagger_{\text{lep}\,\mu}(x), \tag{6.21}$$

in semileptonic decays, in which both hadrons and leptons participate, it should be

$$\frac{G_F}{\sqrt{2}} \left[\cos\theta_c J^\mu_{\Delta S=0} + \sin\theta_c J^\mu_{\Delta S=1}(x)\right] J^\dagger_{\text{lep}\,\mu}(x) + \text{Herm. conj.} \tag{6.22}$$

The Cabibbo angle, θ_c, expresses a rotation between the $\Delta S = 0$ and $\Delta S = 1$ currents. The cosine of the Cabibbo angle can be determined by measuring beta decays in $0^+ \to 0^+$ transitions in which the nuclei belong to the same isospin multiplet and comparing with G_F as measured in muon decay. In these circumstances, CVC determines the relevant nuclear matrix element. The results give $\cos\theta \approx 0.970 - 0.977$, so $\theta_c \approx 13°$. Values of $\sin\theta_c$ derived from $\Delta S = 1$ decays are consistent with this value. The significance of the Cabibbo angle became clearer in subsequent years, as we shall see in Chapters 9 and 11.

A regularity noted by Gell-Mann when he invented strangeness was that in semileptonic decays $\Delta S = \Delta Q$. Thus in $K^+ \to \pi^0 \mu^+ \nu$, the hadronic system loses one unit of strangeness and one unit of charge. The decay $\Sigma^- \to n e^- \nu$ ($\Delta S = 1$, $\Delta Q = 1$) is observed while $\Sigma^+ \to n e^+ \nu$ ($\Delta S = 1$, $\Delta Q = -1$) is not. Even more striking is the absence of processes in which the strangeness of the hadronic system changes, but its charge does not. Thus $K^+ \to \pi^+ \nu \bar{\nu}$ and $K^+ \to \pi^+ e^+ e^-$ are extremely rare. The absence of strangeness changing neutral weak currents was to play a profound role in later developments.

The success of the Fermi theory was convincing evidence for the existence of the neutrino. Still, although the helicity of the neutrino was indirectly measured, there had been no detection of interactions initiated by the neutrinos themselves. This was first achieved by Cowan and Reines using antineutrinos produced in beta decays inside a nuclear reactor. When Reines began to think about means for detecting them, he began by considering the neutrinos that would be emitted from a fission bomb. The nuclear reactor turned out to be much more practical.

The enormous number of beta decays from neutron-rich radionuclei produced by fission provide a prolific source of antineutrinos. However, the environment around a reactor is far from ideal. Reines' idea was to show that his signal for neutrino-induced processes was greater when the reactor was on than when it was off. Early results were obtained in 1956, but a greatly improved experiment was reported in 1958 (**Ref. 6.7**). In the 1958 version of the experiment, the process $\bar{\nu}_e p \to e^+ n$ was observed by detecting both the e^+ and the neutron. The positron annihilation produced two photons, which were detected as a prompt signal using liquid scintillator. The neutrons slowed down by collisions with hydrogen and then were captured by cadmium, whose subsequent gamma decay was observed. The positron and neutron signatures were required to be in coincidence, with allowance for the time required for the neutron to slow down. The experiment is displayed schematically in Figure 6.3.

Bruno Pontecorvo and Melvin Schwartz independently proposed studying neutrino interactions with accelerators, using the decays $\pi \to \mu\nu$ and $K \to \mu\nu$ as neutrino sources. The cross sections for neutrino reactions are fantastically small, on the order of $\sigma \propto G_F^2 s$, where s is the center-of-mass energy squared. Thus for $s = 1$ GeV^2, using the convenient approximations, $G_F \approx 10^{-5}$ GeV^{-2}, 0.389 mb GeV ≈ 1 with $\hbar=c=1$, $\sigma \approx 10^{-10} \times 0.4$ mb, some 12 orders of magnitude smaller than hadronic cross sections. Still, with a sufficiently large target and neutrino flux, such experiments are possible.

Neutrino beams could not be effectively produced at the accelerators available in the mid-1950s. These included the 3-GeV Cosmotron at Brookhaven and the 6-GeV Bevatron

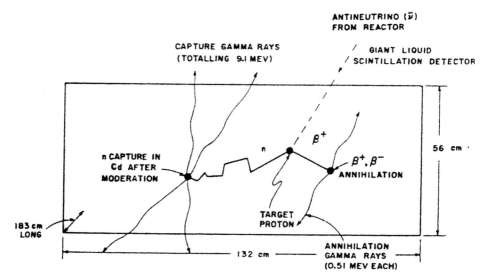

Figure 6.3. A schematic diagram of the experiment of Reines and Cowan in which antineutrinos from a nuclear reactor were detected. The dashed line entering from above indicates the antineutrino. The antineutrino transmutes a proton into a neutron and a positron. The annihilation of the positron produces two prompt gamma rays, which are detected by the scintillator. The neutron is slowed in the scintillator and eventually captured by cadmium, which then also emits delayed gamma rays. The combination of the prompt and delayed gamma rays is the signature of the antineutrino interaction (**Ref. 6.7**).

at Berkeley, and the 10-GeV machine at Dubna in the Soviet Union, all of which were proton synchrotrons. The next generation of machines were based on a new principle, strong focusing. In 1952, E. Courant, M. S. Livingston, and H. Snyder at Brookhaven discovered that by arranging bending magnets so that the gradients of successive magnets alternated between increasing radially and decreasing radially, the overall effect was to focus the beam in both the vertical and horizontal directions. Moreover, the beam excursions away from the central orbit were much decreased in amplitude. As a result, it was possible to make much smaller beam tubes and magnets with much smaller apertures.

Strong focusing can also be done with pairs of quadrupole magnets, one focusing in the horizontal plane and the next in the vertical plane. It is this arrangement that is most often employed in proton accelerators. This strong focusing principle was employed as early as 1955 (**Refs. 3.13, 4.1, 4.4, 4.6**) in the construction of beam lines. Subsequent to the work of Courant, Livingston, and Snyder, it was learned that the principle had been discovered earlier by N. Christofilos, working independently and alone in Athens. His idea had been communicated to the Lawrence Radiation Laboratory in Berkeley where it languished in the files unnoticed.

Strong focusing led to the construction of much higher energy proton machines. The first, the 28-GeV Proton Synchrotron (PS), was completed at CERN, the European Nuclear Research Center in Geneva, in 1959. A similar machine, the Alternating Gradient Synchrotron (AGS), was completed at Brookhaven in 1960.

In 1962, a team including Schwartz, Lederman, and Steinberger **(Ref. 6.8)** reported results from an accelerator experiment in which neutrino interactions were observed. The neutrino beam was generated by directing the 15-GeV proton beam from the AGS on a beryllium target. Secondary π's and K's produced the neutrinos by decay in flight.

Since the interaction rate of the neutrinos was expected to be minute, extreme care was taken to prevent extraneous backgrounds from reaching the detector. Shielding included a 13.5-m iron wall. Detection was provided by a 10-ton spark chamber with aluminum plates separated around the edges by lucite spacers. The detector was surrounded on top, back, and front by anticoincidence counters to exclude events initiated by charged particles. Background was reduced by accepting only those events that coincided with the 20-ns bursts of particles from the accelerator, separated by 220-ns intervals. Even with these precautions, many triggered events were due to muons or neutrons that made their way into the detector. Most of these could be rejected by scanning the photographic record of the spark chamber output.

Of the remaining events, those showing a single charged particle with momentum less than 300 MeV (assuming the track to be that of a muon) were rejected as possibly due to background including neutron-induced events. This left 34 events apparently with single muons of energy greater than 300 MeV, candidates for $\nu n \to p\mu^-$ and $\bar{\nu} p \to n\mu^+$. In addition, there were 22 events with more than one visible track. These were candidates for $\nu n \to n\pi^+ \mu^-$ and $\nu n \to p\mu^-$. Eight other events appeared "showerlike." Careful analysis showed that only a few of these were likely to be due to electrons.

The substantial difference between the number of muons produced and the number of electrons produced showed clearly that the neutrinos obtained from $\pi \to \mu\nu$ (which is vastly more frequent than the decays $\pi \to e\nu$ or $K \to \pi^0 e\nu$) generated muons rather than electrons. In this way, it was shown that there were two neutrinos, ν_μ and ν_e, and two conserved quantum numbers, muon number (+1 for μ^- and ν_μ) and electron number (+1 for e^- and ν_e). The ν_μ is created in $\pi^+ \to \mu^+ \nu_\mu$, the $\bar{\nu}_\mu$ in $\pi^- \to \mu^- \bar{\nu}_\mu$, the ν_e in $\pi^+ \to e^+ \nu_e$, and the $\bar{\nu}_e$ in $n \to p e^- \bar{\nu}_e$. The process $\nu_\mu n \to p e^-$ was forbidden by these rules. Separately conserved electron and muon numbers also forbid the unobserved decay $\mu \to e\gamma$. In addition to establishing the existence of two distinct neutrinos, the experiment demonstrated the feasibility of studying high energy neutrino interactions at accelerators. Subsequent neutrino experiments played a critical role in the development of particle physics.

The V-A theory provided a comprehensive phenomenological picture of weak interactions. The leptonic, semileptonic, and nonleptonic weak interactions were encompassed. The $\Delta S = 0$ and $\Delta S = 1$ processes were described by Cabibbo's proposal. Nevertheless, it was clear that the theory was incomplete. The Fermi interaction occurred at a point and was thus an s-wave interaction. The cross section for an s-wave interaction is limited by unitarity to be no greater than $4\pi/p_{cm}^2$. However, we have seen that in the V-A theory cross sections grow as $G_F^2 s \propto G_F^2 p_{cm}^2$. A contradiction occurs roughly when $p_{cm} = 300$ GeV. This circumstance can be improved, though not completely cured, by supposing that the Fermi interaction does not occur at a point, but is transmitted by a massive vector boson, the W. The idea goes back to Yukawa who had hoped his meson would explain both strong and weak interactions. If the W were heavy, it would produce a factor in the beta-decay amplitude of roughly f^2/m_W^2, where m_W is the W mass and f is its coupling to the nucleon and $e\nu$. Crudely then, $G_F \approx f^2/m_W^2$. The smallness of G_F could be due to f being small or m_W being large, or both. Experimental searches for the W in the mass range up to a few GeV were unsuccessful.

Exercises

6.1 Tritium, H^3, decays to $He^3 + e^- + \bar{\nu}_e$ with a half-life of 12.33 y. The maximum electron energy is close to 18.6 keV. Show what the high energy end of the Kurie plot would look like if the neutrino were (a) massless and if (b) it had a mass of 67 eV. Compare with Fig. 6.1.

6.2 What is the source of the dependence of Mott scattering, which was used by Frauenfelder *et al.*, on the polarization of the electron?

6.3 The decay amplitude for $\mu \to e\nu\bar{\nu}$ is proportional to G_F, so the decay rate is proportional to G_F^2. By dimensional analysis, the decay rate is proportional to $G_F^2 m_\mu^5$. The complete result is

$$\Gamma(\mu \to e\nu\bar{\nu}) = \frac{G_F^2 m_\mu^5}{192\pi^3}$$

and the lifetime is 2.2×10^{-6} s. In 1975, a new lepton analogous to the μ, called the τ was discovered. What are the expected partial decay rates of $\tau \to \mu\nu\bar{\nu}$ and $\tau \to e\nu\bar{\nu}$ if $m_\tau = 1.8$ GeV? Compare with the data.

6.4 Estimate on dimensional grounds the lifetime of the neutron. Compare with experiment.

6.5 The branching ratios for $\Lambda \to p\pi^-$ and $\Lambda \to n\pi^0$ are 64.2% and 35.8%, respectively. What would we expect if the nonleptonic Hamiltonian were a $\Delta I = 1/2$ operator? A $\Delta I = 3/2$ operator?

6.6 * The decays $\pi \to \mu\nu$ and $\pi \to e\nu$ are governed by the V-A interaction

$$\mathcal{H} = \int d^3x \, \frac{G_F}{\sqrt{2}} J_\lambda^{had}(x) \, \bar{\nu}_e(x) \gamma^\lambda (1 - \gamma_5) e(x).$$

The hadronic matrix element

$$< 0 | J_\lambda^{had} | \pi >$$

must be proportional to the pion four-momentum, q_λ. Show that this means the decay amplitudes for the two processes are proportional to m_μ and m_e, respectively, and thus

$$\frac{\Gamma(\pi \to \mu\nu)}{\Gamma(\pi \to e\nu)} \propto \left(\frac{m_\mu}{m_e}\right)^2 \times \text{phase space}.$$

6.7 * The matrix element squared for the decay $\mu^- \to e^- \nu_\mu \bar{\nu}_e$ is

$$\mathcal{M}^2 = 64 G_F^2 (P + ms) \cdot p_{\bar{\nu}_e} \, p_e \cdot p_{\nu_\mu}$$

where P is the muon four-momentum, m is its mass, and s is the four-vector spin of the muon. In the rest frame of the muon, s has only space components and is a unit vector in the direction of the spin. Use the formula

$$d\Gamma = \frac{(2\pi)^4}{2M} |\mathcal{M}|^2 \frac{d^3 p_1}{(2\pi)^3 2 E_1} \frac{d^3 p_2}{(2\pi)^3 2 E_2} \frac{d^3 p_3}{(2\pi)^3 2 E_3} \delta^4(P - p_1 - p_2 - p_3)$$

to establish

(a) $\Gamma = \dfrac{G_F^2 M^5}{192 \pi^3}$,

(b) $d\Gamma/dx \propto x^2 (1 - 2x/3)$ where $x = 2 E_e / m$,

(c) $\dfrac{d\Gamma}{dx \, d\cos\theta} \propto x^2 [(3 - 2x) + (2x - 1)\cos\theta]$,

where θ is the angle between the muon spin and the electron direction.

(d) $\dfrac{d\Gamma}{d\cos\theta} \propto 1 + \tfrac{1}{3}\cos\theta$.

Further Reading

Weak interactions are covered quite thoroughly in the text by E. D. Commins and P. H. Bucksbaum, *Weak Interactions of Leptons and Quarks*, Cambridge University Press, Cambridge, 1983.

A personal recollection of the two-neutrino experiment by Melvin Schwartz appears in *Adventures in Experimental Physics*, α, B. Maglich, ed., World Science Education, Princeton, NJ, 1972.

References

6.1 C. S. Wu et al., "Experimental Test of Parity Conservation in Beta Decay." *Phys. Rev.*, **105**, 1413 (1957).

6.2 R. L. Garwin, L. M. Lederman, and M. Weinrich, "Observations of the Failure of Conservation of Parity and Charge Conjugation in Meson Decays: the Magnetic Moment of the Free Muon." *Phys. Rev.*, **105**, 1415 (1957).

6.3 J. I. Friedman and V. L. Telegdi, "Nuclear Emulsion Evidence for Parity Nonconservation in the Decay Chain $\pi^+ - \mu^+ - e^+$." *Phys. Rev.*, **105**, 1681 (1957). See also *Phys. Rev.* **106**, 1290 (1957).

6.4 H. Frauenfelder et al., "Parity and the Polarization of Electrons from Co^{60}." *Phys. Rev.*, **106**, 386 (1957).

6.5 M. Goldhaber, L. Grodzins, and A. W. Sunyar, "Helicity of Neutrinos." *Phys. Rev.*, **109**, 1015 (1958).

6.6 M. T. Burgy *et al.*, "Measurements of Spatial Asymmetries in the Decay of Polarized Neutrons." *Phys. Rev.*, **120**, 1829 (1960).

6.7 F. Reines and C. L. Cowan, Jr., "Free Anti Neutrino Absorption Cross Section. I. Measurement of the Free Anti Neutrino Absorption Cross Section by Protons." *Phys. Rev.*, **113**, 273 (1959). and R. E. Carter *et al.*, "Free Antineutrino Absorption Cross Section II. Expected Cross Section from Measurements of Fission Fragment electron Spectrum," ibid. p. 280. Only the first page of I is reproduced.

6.8 G. Danby *et al.*, "Observation of High Energy Neutrino Reactions and the Existence of Two Kinds of Neutrinos." *Phys. Rev. Lett.*, **9**, 36 (1962).

Experimental Test of Parity Conservation in Beta Decay*

C. S. Wu, *Columbia University, New York, New York*

AND

E. Ambler, R. W. Hayward, D. D. Hoppes, AND R. P. Hudson,
National Bureau of Standards, Washington, D. C.

(Received January 15, 1957)

IN a recent paper[1] on the question of parity in weak interactions, Lee and Yang critically surveyed the experimental information concerning this question and reached the conclusion that there is no existing evidence either to support or to refute parity conservation in weak interactions. They proposed a number of experiments on beta decays and hyperon and meson decays which would provide the necessary evidence for parity conservation or nonconservation. In beta decay, one could measure the angular distribution of the electrons coming from beta decays of polarized nuclei. If an asymmetry in the distribution between θ and $180° - \theta$ (where θ is the angle between the orientation of the parent nuclei and the momentum of the electrons) is observed, it provides unequivocal proof that parity is not conserved in beta decay. This asymmetry effect has been observed in the case of oriented Co^{60}.

It has been known for some time that Co^{60} nuclei can be polarized by the Rose-Gorter method in cerium magnesium (cobalt) nitrate, and the degree of polarization detected by measuring the anisotropy of the succeeding gamma rays.[2] To apply this technique to the present problem, two major difficulties had to be overcome. The beta-particle counter should be placed *inside* the demagnetization cryostat, and the radioactive nuclei must be located in a *thin surface* layer and polarized. The schematic diagram of the cryostat is shown in Fig. 1.

To detect beta particles, a thin anthracene crystal $\frac{3}{8}$ in. in diameter$\times\frac{1}{16}$ in. thick is located inside the vacuum chamber about 2 cm above the Co^{60} source. The scintillations are transmitted through a glass window and a Lucite light pipe 4 feet long to a photomultiplier (6292) which is located at the top of the cryostat. The Lucite head is machined to a logarithmic spiral shape for maximum light collection. Under this condition, the Cs^{137} conversion line (624 kev) still retains a resolution of 17%. The stability of the beta counter was carefully checked for any magnetic or temperature effects and none were found. To measure the amount of polarization of Co^{60}, two additional NaI gamma scintillation counters were installed, one in the equatorial plane and one near the polar position. The observed gamma-ray anisotropy was used as a measure of polarization, and, effectively, temperature. The bulk susceptibility was also monitored but this is of secondary significance due to surface heating effects, and the gamma-ray anisotropy alone provides a reliable measure of nuclear polarization. Specimens were made by taking good single crystals of cerium magnesium nitrate and growing on the upper surface only an additional crystalline layer containing Co^{60}. One might point out here that since the allowed beta decay of Co^{60} involves a change of spin of

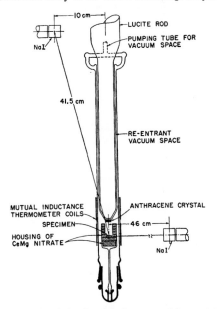

Fig. 1. Schematic drawing of the lower part of the cryostat.

one unit and no change of parity, it can be given only by the Gamow-Teller interaction. This is almost imperative for this experiment. The thickness of the radioactive layer used was about 0.002 inch and contained a few microcuries of activity. Upon demagnetization, the magnet is opened and a vertical solenoid is raised around the lower part of the cryostat. The whole process takes about 20 sec. The beta and gamma counting is then started. The beta pulses are analyzed on a 10-channel pulse-height analyzer with a counting interval of 1 minute, and a recording interval of about 40 seconds. The two gamma counters are biased to accept only the pulses from the photopeaks in order to discriminate against pulses from Compton scattering.

A large beta asymmetry was observed. In Fig. 2 we have plotted the gamma anisotropy and beta asymmetry vs time for polarizing field pointing up and pointing down. The time for disappearance of the beta asymmetry coincides well with that of gamma anisotropy. The warm-up time is generally about 6 minutes, and the warm counting rates are independent of the field direction. The observed beta asymmetry does not change sign with reversal of the direction of the demagnetization field, indicating that it is not caused by remanent magnetization in the sample.

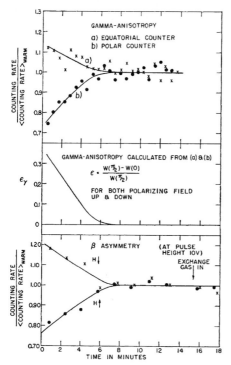

FIG. 2. Gamma anisotropy and beta asymmetry for polarizing field pointing up and pointing down.

The sign of the asymmetry coefficient, α, is negative, that is, the emission of beta particles is more favored in the direction opposite to that of the nuclear spin. This naturally implies that the sign for C_T and C_T' (parity conserved and parity not conserved) must be opposite. The exact evaluation of α is difficult because of the many effects involved. The lower limit of α can be estimated roughly, however, from the observed value of asymmetry corrected for backscattering. At velocity $v/c \approx 0.6$, the value of α is about 0.4. The value of $\langle I_z \rangle / I$ can be calculated from the observed anisotropy of the gamma radiation to be about 0.6. These two quantities give the lower limit of the asymmetry parameter β ($\alpha = \beta \langle I_z \rangle / I$) approximately equal to 0.7. In order to evaluate α accurately, many supplementary experiments must be carried out to determine the various correction factors. It is estimated here only to show the large asymmetry effect. According to Lee and Yang[3] the present experiment indicates not only that conservation of parity is violated but also that invariance under charge conjugation is violated.[4] Furthermore, the invariance under time reversal can also be decided from the momentum dependence of the asymmetry parameter β. This effect will be studied later.

The double nitrate cooling salt has a highly anisotropic g value. If the symmetry axis of a crystal is not set parallel to the polarizing field, a small magnetic field will be produced perpendicular to the latter. To check whether the beta asymmetry could be caused by such a magnetic field distortion, we allowed a drop of $CoCl_2$ solution to dry on a thin plastic disk and cemented the disk to the bottom of the same housing. In this way the cobalt nuclei should not be cooled sufficiently to produce an appreciable nuclear polarization, whereas the housing will behave as before. The large beta asymmetry was not observed. Furthermore, to investigate possible internal magnetic effects on the paths of the electrons as they find their way to the surface of the crystal, we prepared another source by rubbing $CoCl_2$ solution on the surface of the cooling salt until a reasonable amount of the crystal was dissolved. We then allowed the solution to dry. No beta asymmetry was observed with this specimen.

More rigorous experimental checks are being initiated, but in view of the important implications of these observations, we report them now in the hope that they may stimulate and encourage further experimental investigations on the parity question in either beta or hyperon and meson decays.

The inspiring discussions held with Professor T. D. Lee and Professor C. N. Yang by one of us (C. S. Wu) are gratefully acknowledged.

* Work partially supported by the U. S. Atomic Energy Commission.
[1] T. D. Lee and C. N. Yang, Phys. Rev. **104**, 254 (1956).
[2] Ambler, Grace, Halban, Kurti, Durand, and Johnson, Phil. Mag. **44**, 216 (1953).
[3] Lee, Oehme, and Yang, Phys. Rev. (to be published).

[4] Their arguments are as follows: From the He⁶ recoil experiment and from Eq. (A-4) of reference 1 one concludes that $(|C_A|^2+|C_A'|^2)/(|C_T|^2+|C_T'|^2) \lesssim \frac{1}{3}$. Hence, by comparing Eq. (16) of reference 3 [see also Eq. (A-6) of reference 1], one concludes that the present large asymmetry is possible only if both conservation of parity and invariance under charge conjugation are violated.

Observations of the Failure of Conservation of Parity and Charge Conjugation in Meson Decays: the Magnetic Moment of the Free Muon*

RICHARD L. GARWIN,† LEON M. LEDERMAN, AND MARCEL WEINRICH

Physics Department, Nevis Cyclotron Laboratories, Columbia University, Irvington-on-Hudson, New York, New York

(Received January 15, 1957)

LEE and Yang[1-3] have proposed that the long held space-time principles of invariance under charge conjugation, time reversal, and space reflection (parity) are violated by the "weak" interactions responsible for decay of nuclei, mesons, and strange particles. Their hypothesis, born out of the $\tau-\theta$ puzzle,[4] was accompanied by the suggestion that confirmation should be sought (among other places) in the study of the successive reactions

$$\pi^+ \to \mu^+ + \nu, \quad (1)$$

$$\mu^+ \to e^+ + 2\nu. \quad (2)$$

They have pointed out that parity nonconservation implies a polarization of the spin of the muon emitted from stopped pions in (1) along the direction of motion and that furthermore, the angular distribution of electrons in (2) should serve as an analyzer for the muon polarization. They also point out that the longitudinal polarization of the muons offers a natural way of determining the magnetic moment.[5] Confirmation of this proposal in the form of preliminary results of β decay of oriented nuclei by Wu et al. reached us before this experiment was begun.[6]

By stopping, in carbon, the μ^+ beam formed by forward decay in flight of π^+ mesons inside the cyclotron, we have performed the meson experiment, which establishes the following facts:

I. A large asymmetry is found for the electrons in (2), establishing that our μ^+ beam is strongly polarized.

II. The angular distribution of the electrons is given by $1+a\cos\theta$, where θ is measured from the velocity vector of the incident μ's. We find $a=-\frac{1}{3}$ with an estimated error of 10%.

III. In reactions (1) and (2), parity is not conserved.

IV. By a theorem of Lee, Oehne, and Yang,[2] the observed asymmetry proves that invariance under charge conjugation is violated.

V. The g value (ratio of magnetic moment to spin) for the (free) μ^+ particle is found to be $+2.00\pm0.10$.

VI. The measured g value and the angular distribution in (2) lead to the very strong probability that the spin of the μ^+ is $\frac{1}{2}$.[7]

VII. The energy dependence of the observed asymmetry is not strong.

VIII. Negative muons stopped in carbon show an asymmetry (also leaked backwards) of $a \sim -1/20$, i.e., about 15% of that for μ^+.

IX. The magnetic moment of the μ^-, bound in carbon, is found to be negative and agrees within limited accuracy with that of the μ^+.[8]

X. Large asymmetries are found for the e^+ from polarized μ^+ beams stopped in polyethylene and calcium. Nuclear emulsion (as a target in Fig. 1) yields an asymmetry of about half that observed in carbon.

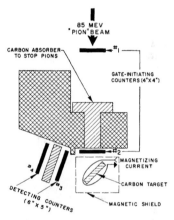

FIG. 1. Experimental arrangement. The magnetizing coil was close wound directly on the carbon to provide a uniform vertical field of 79 gauss per ampere.

The experimental arrangement is shown in Fig. 1. The meson beam is extracted from the Nevis cyclotron in the conventional manner, undergoing about 120° of magnetic deflection in the cyclotron fringing field and about $-30°$ of deflection and mild focusing upon emerging from the 8-ft shielding wall. The positive beam contains about 10% of muons which originate principally in the vicinity of the cyclotron target by pion decay-in-flight. Eight inches of carbon are used in the entrance telescope to separate the muons, the mean range of the "85-"Mev pions being ~ 5 in. of carbon. This arrangement brings a maximum number of muons to rest in the carbon target. The stopping of a muon is signalled by a fast 1–2 coincidence count. The subsequent beta decay of the muon is detected by the electron telescope 3–4 which normally requires a particle of range >8 g/cm²($\sim$25-Mev electrons) to register. This arrangement has been used to measure the lifetimes of μ^+ and μ^- mesons in a vast number of elements.[9] Counting rates are normally $\sim$20 electrons/

min in the μ^+ beam and ~150 electrons/min in the μ^- beam with background of the order of 1 count/min.

In the present investigation, the 1-2 pulse initiates a gate of duration $T = 1.25$ μsec. This gate is delayed by $t_1 = 0.75$ μsec and placed in coincidence with the electron detector. Thus the system counts electrons of energy >25 Mev which are born between 0.75 and 2.0 μsec after the muon has come to rest in carbon. Consider now the possibility that the muons are created in reaction (1) with large polarization in the direction of motion. If the gyromagnetic ratio is 2.0, these will maintain their polarization throughout the trajectory. Assume now that the processes of slowing down, stopping, and the microsecond of waiting do not depolarize the muons. In this case, the electrons emitted from the target may have an angular asymmetry about the polarization direction, e.g., for spin $\frac{1}{2}$ of the form $1 + a\cos\theta$. In the absence of any vertical magnetic field, the counter system will sample this distribution at $\theta = 100°$. We now apply a small vertical field in the magnetically shielded enclosure about the target, which causes the muons to precess at a rate of $(\mu/s\hbar)H$ radians per sec. The probability distribution in angle is carried around with the μ-spin. In this manner we can, with a fixed counter system, sample the entire distribution by plotting counts as a function of magnetizing current for a given time delay. A typical run is shown in Fig. 2. As an example of a systematic check, we have

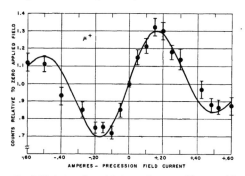

FIG. 2. Variation of gated 3–4 counting rate with magnetizing current. The solid curve is computed from an assumed electron angular distribution $1 - \frac{1}{3} \cos\theta$, with counter and gate-width resolution folded in.

reduced the absorber in the telescope to 5 in. so that the end-of-range of the main pion beam occurred at the carbon target. The electron rate rose accordingly by a factor of 10, indicating that now electrons were arising from muons isotropically emitted by pions at rest in the carbon. No variation in counting rate with magnetizing current was then observed, the ratio of the rate for $I = +0.170$ amp to that for $I = -0.150$ amp, for example, being 0.989 ± 0.028. The highest field produced at the target was ~50 gauss which generates a stray field outside of the magnetic shield of $< \frac{1}{10}$ the cyclotron fringing field of 20 gauss. The only conceivable effect of the magnetizing current is the precession of muon spins and we are, therefore, led to conclusions I–IV as necessary consequences of these observations.

The solid curve in Fig. 2 is a theoretical fit to a distribution $1 - \frac{1}{3}\cos\theta$, where

(1) the gyromagnetic ratio is taken to be $+2.00$;[10]

(2) the angular breadth of the electron telescope and the gate-width smearing are folded in, as well as (to first order) the exponential decay rate of muons within the gate;

(3) the small residual cyclotron stray field (*up* for Fig. 2, the positive magnetizing current producing a *down* field) is included. This has the accidental effect of converting the 100° initial angle ($H=0$) to 89° as in Fig. 2. We note that this experiment establishes only a lower limit to the magnitude of a, since the percent polarization at the time of decay is not known. If polarization is complete, $a = -0.33 \pm 0.03$.

Proof of the 2π symmetry of the distribution and the sign of the moment was obtained by shifting the electron counters to 65° with respect to the incident muon direction. The repetition of a magnetizing run yielded a curve as in Fig. 2 but shifted to the right by 0.075 ampere (5.9 gauss) corresponding to a precession angle of 37°, in agreement with the spatial rotation of the counter system. Thus we are led to conclusions V and VI.

A specific model, the two-component neutrino theory, has been proposed by Lee and Yang[3] in an attempt to introduce parity nonconservation naturally into elementary particle theory. This theory predicts, for our experimental arrangement and on the basis of 1.86 for the integrated spectrum (Fig. 2), a ratio of the order of 2.5 for energies greater than 35 Mev. We have increased the amount of absorber in the electron telescope to exclude electrons of less than ~35 Mev. The resulting peak-to-valley ratio was then observed to be 1.92 ± 0.19.[11]

We have also detected asymmetry in negative muon decay and have verified that the moment is negative and roughly equal to that of the positive muon.[7] The asymmetry in this case is also peaked backwards.

Various other materials were investigated for μ^+ mesons. Nuclear emulsion as a target was found to have a significantly weaker asymmetry (peak-to-valley ratio of 1.40 ± 0.07) and it is interesting to note that this did not increase with reduced delay and gate width. Neither was there any evidence for an altered moment. It seems possible that polarized positive and negative muons will become a powerful tool for exploring magnetic fields in nuclei (even in Pb, 2% of the μ^- decay into electrons[9]), atoms, and interatomic regions.

The authors wish to acknowledge the essential role of Professor Tsung-Dao Lee in clarifying for us the papers of Lee and Yang. We are also indebted to Professor C. S. Wu[6] for reports of her preliminary results in the Co^{60} experiment which played a crucial part in the

Columbia discussions immediately preceding this experiment.

* Research supported in part by the joint program of the Office of Naval Research and the U. S. Atomic Energy Commission.

† Also at International Business Machines, Watson Scientific Laboratories, New York, New York.

[1] T. D. Lee and C. N. Yang, Phys. Rev. **104**, 254 (1956).
[2] Lee, Oehme, and Yang, Phys. Rev. (to be published).
[3] T. D. Lee and C. N. Yang, Phys. Rev. (to be published).
[4] R. Dalitz, Phil. Mag. **44**, 1068 (1953).
[5] T. D. Lee and C. N. Yang (private communication).
[6] Wu, Ambler, Hudson, Hoppes, and Hayward, Phys. Rev. **105**, 1413 (1957), preceding Letter.
[7] The Fierz-Pauli theory for spin $\frac{3}{2}$ particles predicts a g value of $\frac{2}{3}$. See F. J. Belinfante, Phys. Rev. **92**, 997 (1953).
[8] V. Fitch and J. Rainwater, Phys. Rev. **92**, 789 (1953).
[9] M. Weinrich and L. M. Lederman, *Proceedings of the CERN Symposium, Geneva, 1956* (European Organization of Nuclear Research, Geneva, 1956).
[10] The field interval, ΔH, between peak and valley in Fig. 2 gives the magnetic moment directly by $(\mu \Delta H/s\hbar)(t_1 + \frac{1}{2}T)\delta = \pi$, where $\delta = 1.06$ is a first-order resolution correction which takes into account the finite gate width and muon lifetime. The 5% uncertainty comes principally from lack of knowledge of the magnetic field in carbon. Independent evidence that $g=2$ (to $\sim 10\%$) comes from the coincidence of the polarization axis with the velocity vector of the stopped μ's. This implies that the spin precession frequency is identical to the μ cyclotron frequency during the 90° net magnetic deflection of the muon beam in transit from the cyclotron to the 1-2 telescope. We have designed a magnetic resonance experiment to determine the magnetic moment to $\sim 0.03\%$.

[11] *Note added in proof.*—We have now observed an energy dependence of a in the $1+a \cos\theta$ distribution which is somewhat less steep but in rough qualitative agreement with that predicted by the two-component neutrino theory ($\mu \to e + \nu + \bar{\nu}$) without derivative coupling. The peak-to-valley ratios for electrons traversing 9.3 g/cm², 15.6 g/cm², and 19.8 g/cm² of graphite are observed to be 1.80 ± 0.07, 1.84 ± 0.11, and 2.20 ± 0.10, respectively.

Nuclear Emulsion Evidence for Parity Nonconservation in the Decay Chain $\pi^+ - \mu^+ - e^{+*}$†

Jerome I. Friedman and V. L. Telegdi

Enrico Fermi Institute for Nuclear Studies, University of Chicago, Chicago, Illinois
(Received January 17, 1957)

LEE and Yang[1] recently re-examined the problem as to whether parity is conserved in nature and emphasized the fact that one actually lacks experimental evidence in support of this most natural hypothesis in the case of weak interactions (such as β decay). Violation of parity conservation can be inferred essentially only by measuring the probability distribution of some *pseudoscalar* quantity, e.g., of the projection of a polar vector along an axial vector, and measurements of this kind had not been reported. Lee and Yang suggested several experiments in which a spin direction is available as a suitable axial vector; in particular, they pointed out that the initial direction of motion of the muon in the process $\pi \to \mu + \nu$ can serve for this purpose, as the muon will be produced with its spin axis along its initial line of motion if the Hamiltonian responsible for this process does not have the customary invariance properties. If parity is further not conserved in the process $\mu \to e + 2\nu$, then a forward-backward asymmetry in the distribution of angles $W(\theta)$ between this initial direction of motion and the momentum, $\mathbf{p}_e$, of the decay electron is predicted.

It is easy to observe the pertinent correlation by bringing π^+ mesons to rest in a nuclear emulsion in which the μ^+ meson also stops. One has only to bear in mind two facts: (1) even weak magnetic fields, such as the fringing field of a cyclotron, can obliterate a real effect, as the precession frequency of a Dirac μ meson is $(2.8/207) \times 10^6$ sec^{-1}/gauss; (2) μ^+ can form "muonium," i.e., $(\mu^+ e^-)$, and the formation of this atom can be an additional source of depolarization, both through its internal hyperfine splitting and the precession of its total magnetic moment around the external field. In the absence of specific experiments on muonium formation, one can perhaps be guided by analogous data on positronium in solids.[2,3]

With these facts in mind, we exposed (in early October, 1956) nuclear emulsion pellicles (1 mm thick) to a π^+ beam of the University of Chicago synchrocyclotron. The pellicles were contained inside three concentric tubular magnetic shields and subject to $\leq 4 \times 10^{-3}$ gauss. Over 1300 complete $\pi - \mu - e$ decays have been recorded to date, and the space angle θ defined above has been calculated for each. From these preliminary data we find[4]

$$\left\{ \int_{90°}^{180°} |W(\theta)| d\Omega - \int_{0}^{90°} |W(\theta)| d\Omega \right\} \Big/ \int_{0}^{180°} W(\theta) d\Omega$$

$$= 0.062 \pm 0.027,$$

or
$$W(\theta) = 1 - 0.12 \cos\theta,$$
i.e., the presence of an excess in the backward hemisphere to a 95% confidence level. This effect agrees in sign and magnitude with one observed in a recent analogous experiment[5] performed electronically at Columbia University.

In connecting our result with basic theoretical principles, one has to remember (a) that the asymmetry observed here is only a lower limit owing to the possibility of muonium formation[6] and other conceivable depolarization effects; (b) that existence of an asymmetry implies the joint violation of parity conservation and charge conjugation invariance rather than of parity conservation alone.[7]

In view of the intrinsic importance of the subject, we consider it worthwhile to present our data at this preliminary stage.

We would like to thank the Columbia workers, in particular R. L. Garwin, for communicating their unpublished results to us. We are grateful to R. Oehme for illuminating theoretical discussions and to R. Levi-Setti for criticism of the experimental techniques.

[*] This work was supported by a joint program of the Office of Naval Research and the U. S. Atomic Energy Commission.

[†] For technical reasons, this Letter could not be published in the same issue as that of Garwin, Lederman, and Weinrich, Phys. Rev. **105**, 1415 (1957).

[1] T. D. Lee and C. N. Yang, Phys. Rev. **104**, 254 (1956).

[2] S. Berko and F. Hereford, Revs. Modern Phys. **28**, 299 (1956).

[3] Telegdi, Sens, Yovanovitch, and Warshaw, Phys. Rev. **104**, 867 (1956).

[4] *Note added in proof.*—From 2000 events, we get for this ratio 0.091 ± 0.022.

[5] Garwin, Lederman, and Weinrich (private communication from R. L. Garwin, January 13, 1957).

[6] The Columbia workers find that the observed asymmetry appears to depend on the material stopping the μ^+ mesons. This is consistent with $(\mu^+ e^-)$ formation and suggests the use of μ^- mesons for absolute measurements, in low-Z materials with no nuclear magnetic moment.

[7] R. Oehme (private communication); Lee, Oehme, and Yang, Phys. Rev. (to be published).

Parity and the Polarization of Electrons from Co^{60}†

H. Frauenfelder, R. Bobone, E. von Goeler, N. Levine,
H. R. Lewis, R. N. Peacock, A. Rossi, and G. De Pasquali

University of Illinois, Urbana, Illinois
(Received March 1, 1957)

LEE and Yang[1] recently proposed that parity may not be conserved in weak interactions and suggested various experiments to verify their hypothesis. Two of the experiments have since been performed with positive result—the asymmetry of the electron emission from aligned nuclei[2] and the polarization of muons.[3,4] In a second paper,[5] Lee and Yang discuss a two-component theory of the neutrino and consider some more experimental tests. Among these, they list the measurement of momentum and polarization of electrons emitted in beta decay. If parity is not conserved, the electrons should be longitudinally polarized. For tensor and scalar interaction, the degree of polarization is simply equal to (v/c).[6,7] We have found this polarization in the case of Co^{60}.

The observation of the expected longitudinal polarization of the electrons is difficult. However, by means of an electrostatic deflector, the longitudinal polarization can be transformed into a transverse one.[8] The transverse polarization can be measured by scattering the electrons with a thin foil of a high-Z material (Mott scattering). Because of the spin-orbit interaction, the elastically scattered electrons show a strong left-right asymmetry, especially at scattering angles between 90° and 150°.[9] From this measurable asymmetry, the initial longitudinal polarization can be calculated.

The experimental arrangement is housed in a cylindrical vacuum chamber of 30-cm diameter. The electrons from a Co^{60} source are deflected in a cylindrical electrostatic field (radius of curvature 6 cm) by about 108° and then impinge on the scattering foil. The left-right asymmetry of electrons scattered into the angular interval 95° to 140° is measured with two end-window Geiger counters (3.5 mg/cm² mica windows). Two electroplated Co^{60} sources are used, one of about 1 mC strength on aluminum (1.7 mg/cm²), the other of 6 mC strength on a silver-covered rubber hydrochloride film (0.6 mg/cm²). The electrostatic deflector is designed so that electrons of about 100-kev energy completely change their polarization from longitudinal to transverse. The scattering foils (0.05 mg/cm² gold, 0.15 mg/cm² gold, 1.7 mg/cm² aluminum, all backed by 0.9 mg/cm² Mylar) can be interchanged from the outside.

For an ideal arrangement, the left-right asymmetry in the counters would be $L/R = [1 + Pa(\theta)]/[1 - Pa(\theta)]$.[10] P is the initial longitudinal polarization of the electrons and $a(\theta)$ the polarization asymmetry factor after scattering by an angle θ in the analyzer foil. In the actual experiment, however, the determination of P from L/R involves corrections for (1) the asymmetry of the two counters, (2) the finite extension of scatterer and counters, and (3) incomplete transformation from longitudinal to transverse polarization. The first correction was performed experimentally by using the nearly isotropic scattering from aluminum foils; the second and third corrections were calculated in a first approximation. A correction for depolarization in the source and the analyzer was neglected completely.

The results of some runs are given in Table I. Even though these data are only very preliminary, some conclusions can be drawn.

TABLE I. The polarization of electrons from Co^{60}.

Electron energy kev	$\beta = v/c$	Gold scattering foil mg/cm²	Left-right asymmetry L/R	Longitudinal polarization P
50	0.41	0.15	1.03±0.03	−0.04
68	0.47	0.15	1.13±0.02	−0.16
77	0.49	0.05	1.35±0.06	−0.40
77	0.49	0.15	1.30±0.09	−0.35

1. The violation of parity conservation is obvious. Every run (about five in addition to the ones shown in Table I) shows a definite left-right asymmetry.

2. The negative sign of the polarization P indicates that the beta particles are polarized in the direction opposite to their momentum. This conclusion agrees with the experiment of Wu et al.[2]

3. The values of P are not in disagreement with the two-component theory, which gives $P = -v/c$. The deviations, especially at lower energies, can easily be due to depolarization in the source and in the analyzer. More accurate measurements and further investigation of the corrections are required for a detailed comparison between theory and experiment.

We are very much indebted to Dr. J. Weneser for many illuminating discussions.

† Assisted by the joint program of the Office of Naval Research and the U. S. Atomic Energy Commission.
[1] T. D. Lee and C. N. Yang, Phys. Rev. **104**, 254 (1956).
[2] Wu, Ambler, Hayward, Hoppes, and Hudson, Phys. Rev. **105**, 1413 (1957).
[3] J. I. Friedman and V. L. Telegdi, Phys. Rev. **105**, 1681 (1957).
[4] Garwin, Lederman, and Weinrich, Phys. Rev. **105**, 1415 (1957).
[5] T. D. Lee and C. N. Yang, Phys. Rev. **105**, 1671 (1957).
[6] L. Landau (to be published).
[7] Jackson, Treiman, and Wyld (to be published).
[8] H. A. Tolhoek, Revs. Modern Phys. **28**, 277 (1956).
[9] N. Sherman, Phys. Rev. **103**, 1601 (1956).
[10] We use $P = (I_+ - I_-)/(I_+ + I_-)$, where I_+ is the intensity of electrons polarized along their initial momenta and I_- is the intensity of electrons polarized in the opposite direction. We define "left" by $\mathbf{p}_3 \cdot (\mathbf{p}_1 \times \mathbf{p}_2) > 0$, where $\mathbf{p}_1$, $\mathbf{p}_2$, and $\mathbf{p}_3$ are, respectively, the electron momenta immediately after emission from the source, before scattering from the analyzer, and after scattering from the analyzer.

Helicity of Neutrinos*

M. Goldhaber, L. Grodzins, and A. W. Sunyar

Brookhaven National Laboratory, Upton, New York
(Received December 11, 1957)

A COMBINED analysis of circular polarization and resonant scattering of γ rays following orbital electron capture measures the helicity of the neutrino. We have carried out such a measurement with Eu^{152m}, which decays by orbital electron capture. If we assume the most plausible spin-parity assignment for this isomer compatible with its decay scheme,[1] 0−, we find that the neutrino is "left-handed," i.e., $\sigma_\nu \cdot \hat{p}_\nu = -1$ (negative helicity).

Our method may be illustrated by the following simple example: take a nucleus A (spin $I=0$) which decays by allowed orbital electron capture, to an excited state of a nucleus $B(I=1)$, from which a γ ray is emitted to the ground state of $B(I=0)$. The conditions necessary for resonant scattering are best fulfilled for those γ rays which are emitted opposite to the neutrino, which have an energy comparable to that of the neutrino, and which are emitted before the recoil energy is lost. Since the orbital electrons captured by a nucleus are almost entirely s electrons ($K, L_\mathrm{I}, \cdots$ electrons of spin $S=\frac{1}{2}$), the substates of the daughter nucleus

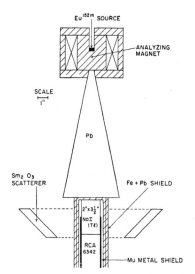

FIG. 1. Experimental arrangement for analyzing circular polarization of resonant scattered γ-rays. Weight of Sm_2O_3 scatterer: 1850 grams.

B, formed when a neutrino is emitted in the Z direction, are $m=-1$, 0 if the neutrino has positive helicity, and $m=+1$, 0 if the neutrino has negative helicity. In either case, the helicity of the γ ray emitted in the $(-Z)$ direction is the same as that of the neutrino. Thus, a measurement of the circular polarization of the γ rays which are resonant-scattered by the nucleus B, yields directly the helicity of the neutrino, if one assumes only the well-established conservation laws of momentum and angular momentum.

To carry out this measurement we have used a nucleus which appears to have the properties postulated in the example given: $_{63}Eu^{152m}$(9.3 hr). It probably has spin 0 and odd parity.[1] It decays to an excited state of $_{62}Sm^{152}$(1−) with emission of neutrinos which have an energy of 840 kev in the most prominent case of K-electron capture. This is followed by an $E1$ γ-ray transition of 960 kev to the ground state (0+). The excited state has a mean life of $(3\pm1)\times10^{-14}$ sec, as determined by Grodzins.[1] Thus, even in a solid source most of the γ-ray emission takes place before the momentum of the recoil nucleus has changed appreciably.

The experimental arrangement used is shown in Fig. 1. The Eu^{152m} source is inserted inside an electromagnet which is alternately (every three minutes) magnetized in the up or down direction. The γ rays which pass through the magnet are resonant-scattered from a Sm_2O_3 scatterer (26.8% Sm^{152}), and detected in a 2-in.×3½-in. cylindrical NaI(Tl) scintillation counter. The photomultiplier (RCA 6342) is magnetically shielded by an iron cylinder and a mu-metal shield.

The effectiveness of this magnetic shield was demonstrated by check experiments with a Cs^{137} γ-ray source in a manner similar to that described previously.[2] No significant effect of magnetic field reversal on the photomultiplier output was noticed when two narrow acceptance channels were set on the steeply sloping low- and high-energy wings of the 661-kev photopeak, respectively.

The source was produced by bombarding ∼10 mg of Eu_2O_3 in the Brookhaven reactor. In typical runs the intensity varied from 50–100 mC. Nine runs varying in length from 3 to 9 hours were carried out. The scattered radiation is shown in Fig. 2. It contains both γ rays emitted from the 960-kev state (960 and 840 kev). Counts were accumulated simultaneously in 3 channels A, B, and C as shown in Fig. 2. A cycle of field reversals was used such that the decay corrections were negligible. No effects of field reversal or decay were noticed in channel C. Channel A exhibited a possible small magnetic field effect which was less than one-tenth of that observed in channel B. In channel B, which bracketed the photopeaks, a total of ∼3×10⁶ counts were accumulated. In 6 runs carried out in the arrangement shown in Fig. 1, an effect $\delta=(N_--N_+)/\frac{1}{2}(N_-+N_+)=+0.017\pm0.003$ was found in channel B after the nonresonant background had been subtracted. Here N_+ is defined as the counting rate with the magnetic field pointing up, and N_- as the counting rate with the field pointing down.

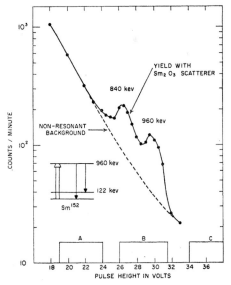

FIG. 2. Resonant-scattered γ rays of Eu^{152m}. Upper curve is taken with arrangement shown in Fig. 1 with unmagnetized iron. Lower curve shows nonresonant background (including natural background).

The magnet response was tested with the bremsstrahlung from a $Sr^{90}+Y^{90}$ source, for which the helicity is negative,[2] $\sigma_\gamma \cdot \hat{p}_\gamma = -1$. Eu^{152m} runs were made with both solid and dissolved sources (HCl solution), and both gave similar results. The effective path length of the 960-kev γ ray in the magnet was somewhat uncertain, partly because of the extent of the source and partly because of a lack of knowledge of the field distribution in the return path. We estimate that the average path is equivalent to 3 ± 0.3 mean free paths in fully magnetized iron. From this we expect an effect of $\delta = \pm 0.025$ with an accuracy of 10%, if the 960-kev γ rays are 100% circularly polarized, with the $-$ sign corresponding to positive helicity (spin parallel to their direction of propagation) and the $+$ sign to negative helicity. Thus we find that in our case the γ rays are $(68\pm 14\%)$ circularly polarized, and that their helicity is negative. As a further check, 3 runs were carried out with a shorter magnet of length $3\frac{1}{2}$ in. with the source on top of the magnet. In this case also a negative helicity was found, the circular polarization being $(66\pm 15\%)$.

From the energy of the neutrinos emitted, the width of the γ-ray line, and the fact that the circular polarization varies with the direction of emission of the γ ray relative to the neutrino as $\cos\theta$, we calculate that a circular polarization, which in the ideal case discussed above would be 100%, would be reduced to $\sim 75\%$. This should be further reduced to a slight extent because of the effect of thermal motion and because some γ rays are emitted after the recoil has changed momentum. Thus our result seems compatible with spin $0-$ for Eu^{152m}, and 100% negative helicity of the neutrinos emitted in orbital electron capture.[3]

In all formulations of β-decay theory no distinction is made between the neutrino emitted in orbital electron capture and that emitted in β^+ decay. Taken together with the fact that the helicity of the positrons in a Gamow-Teller transition is positive[4] or with the fact that positrons are emitted from oriented nuclei in the direction in which the nuclear spin is pointing,[5] our result indicates that the Gamow-Teller interaction is axial vector (A) for positron emitters, in agreement with the conclusions of Hermannsfeldt et al.[6] These authors show that all recoil experiments with β^+ emitters are compatible with AV, but not with TS interactions which have been reported for β^- emitters (largely based on He^6 recoil experiments).[7] The AV combination may be compatible with lepton conservation and a universal Fermi interaction as pointed out by Sudarshan and Marshak[8] and by Feynman and Gell-Mann.[9] This view is strengthened by the recent results showing positive helicity for the positrons from μ^+ decay.[10] It would therefore seem desirable to apply the method described here to a β^- emitter in order to determine the helicity of the antineutrino. Although the analysis of such an experiment is considerably more complicated, it may prove possible to reach a decision between A and T, which is independent of the "classical" recoil experiments.

We wish to thank J. Weneser for many valuable discussions.†

* Work performed under the auspices of the U. S. Atomic Energy Commission.

[1] L. Grodzins, Phys. Rev. **109**, 1015 (1958), preceding Letter.
[2] Goldhaber, Grodzins, and Sunyar, Phys. Rev. **106**, 826 (1957).
[3] It is worthwhile to inquire how our conclusions are affected if the less plausible spin-parity assignments of $1\mp$ are assumed for Eu^{152m}. For the case of a $1^- \xrightarrow{\epsilon} 1^- \xrightarrow{\gamma} 0^+$ transition, J. Weneser (unpublished) finds

$$\sigma_\gamma \cdot \hat{p}_\gamma = \frac{(\sigma_\nu \cdot \hat{p}_\nu)[\frac{1}{2}|G_{GT}|^2 \pm \sqrt{2}|G_{GT}| \cdot |G_F|]}{|G_{GT}|^2 + |G_F|^2},$$

where $G_{GT}=M_{GT}C_{GT}$ and $G_F=M_FC_F$. This has been calculated on the simplifying assumption that the two-component neutrino theory and time-reversal invariance hold [see T. D. Lee and C. N. Yang, Phys. Rev. **105**, 1671 (1957)]. For a neutrino helicity of -1 the photon helicity varies from $+0.5$ to -1.0, and for a neutrino helicity of $+1$ the photon helicity varies from -0.5 to $+1.0$. Considering the reduction factors discussed above, the experimentally found helicity of the γ rays is in agreement with the assumption of neutrinos of negative helicity, even if Eu^{152m} has spin-parity $1-$. In the other very unlikely case of a $1+$ assignment to Eu^{152m}, we could not at present draw a definite conclusion concerning the neutrino helicity. The theory for first forbidden transitions is being investigated by A. M. Bincer.
[4] L. A. Page and M. Heinberg, Phys. Rev. **106**, 1220 (1957).
[5] Ambler, Hayward, Hoppes, Hudson, and Wu, Phys. Rev. **106**, 1361 (1957); Postma, Huiskamp, Miedema, Steenland, Tolhoek, and Gorter, Physica **23**, 259 (1957).
[6] Hermannsfeldt, Maxson, Stähelin, and Allen, Phys. Rev. **107**, 641 (1957).
[7] B. M. Rustad and S. L. Ruby, Phys. Rev. **97**, 991 (1955).
[8] E. C. G. Sudarshan and R. Marshak, Phys. Rev. (to be published).
[9] R. P. Feynman and M. Gell-Mann, Phys. Rev. **109**, 193 (1958).
[10] Culligan, Frank, Holt, Kluyver, and Massam, Nature **180**, 751 (1957).

† *Note added in proof.*—According to a private communication from Professor V. L. Telegdi, a refinement of the experiment of Burgy, Epstein, Krohn, Novey, Raboy, Ringo, and Telegdi, [Phys. Rev. **107**, 1731 (1957)] favors $V-A$ for the β interaction.

Free Antineutrino Absorption Cross Section. I. Measurement of the Free Antineutrino Absorption Cross Section by Protons*

FREDERICK REINES AND CLYDE L. COWAN, JR.†

Los Alamos Scientific Laboratory, University of California, Los Alamos, New Mexico

(Received September 8, 1958)

The cross section for the reaction $p(\bar{\nu},\beta^+)n$ was measured using antineutrinos ($\bar{\nu}$) from a powerful fission reactor at the Savannah River Plant of the United States Atomic Energy Commission. Target protons were provided by a 1.4×10^3 liter liquid scintillation detector in which the scintillator solution (triethylbenzene, terphyenyl, and POPOP) was loaded with a cadmium compound (cadmium octoate) to allow the detection of the reaction by means of the delayed coincidence technique. The first pulse of the pair was caused by the slowing down and annihilation of the positron (β^+), the second by the capture of the neutron (n) in cadmium following its moderation by the scintillator protons. A second giant scintillation detector without cadmium loading was used above the first to provide an anticoincidence signal against events induced by cosmic rays. The antineutrino signal was related to the reactor by means of runs taken while the reactor was on and off. Reactor radiations other than antineutrinos were ruled out as the cause of the signal by a differential shielding experiment. The signal rate was 36 ± 4 events/hr and the signal-to-noise ratio was $\frac{1}{5}$, where half the noise was correlated and cosmic-ray associated and about half was due to non-reactor-associated accidental coincidences. The cross section per fission $\bar{\nu}$ (assuming 6.1 $\bar{\nu}$ per fission) for the inverse beta decay of the proton was measured to be $(11\pm 2.6) \times 10^{-44}$ cm$^2/\bar{\nu}$ or $(6.7\pm 1.5) \times 10^{-43}$ cm^2/fission. These values are consistent with prediction based on the two-component theory of the neutrino.

I. INTRODUCTION

A DETERMINATION of the cross section for the reaction: antineutrino ($\bar{\nu}$) on a proton (p^+) to yield a positron (β^+) and a neutron (n),

$$\bar{\nu}+p^+ \rightarrow \beta^+ + n, \quad (1)$$

permits a check to be made on the combination of fundamental parameters on which the cross section depends. Implicit in a theoretical prediction of the cross section are (1) the principle of microscopic reversibility, (2) the spin of the $\bar{\nu}$, (3) the particular neutrino theory employed: e.g., two- or four-component, (4) the neutron half-life and its decay electron spectrum, and (5) the spectrum of the incident $\bar{\nu}$'s.

An experiment which was performed to identify antineutrinos from a fission reactor[1] yielded an approximate value for this cross section. Following this work, however (and prior to the parity developments involved in point 3), the equipment was modified in order to obtain a better value of the cross section. The modification consisted in the addition of a cadmium salt of 2-ethylhexanoic acid to the scintillator solution[2] of one of the detectors of reference 1, utilizing the protons of the solution as targets for antineutrinos, and making the necessary changes in circuitry to observe both positrons and neutron captures in the detector resulting from antineutrino-induced beta decay in the detector. In addition, a second detector used in the experiment of reference 1 was now used as an anticoincidence shield against cosmic-ray-induced backgrounds, and static shielding was increased by provision of a water tank about 12-inches thick below the target detector. The delayed-coincidence count rate resulting from the positron pulse followed by the capture of the neutron was observed as a function of reactor power, and an analysis of the reactor-associated signal yielded, in addition to an independent identification of the free antineutrino, a measure of the cross section for the reaction and a spectrum of first-pulse (or $\bar{\nu}$) energies. Since the antineutrino spectrum is simply related to the β^+ spectrum, the measurement yields an antineutrino spectrum above the 1.8-Mev reaction threshold. The spectrum is, however, seriously degraded by edge effects in the detector.

This experiment was identical in principle with that performed at Hanford in 1953.[3] It was, however, definitive from the point of view of antineutrino identification (whereas the Hanford experiment was not) because of a series of technical improvements, coupled with the better shielding against cosmic rays achieved by going underground. The improvements consisted in the use of an isolated power supply to diminish electrical noise from nearby machinery, better shielding from the reactor gamma-ray and neutron background, a more complete anticoincidence shield against charged cosmic rays through the use of a liquid scintillation detector, and use of a large detector containing 6.5 times as many proton targets.[4] In addition, oscilloscopic presentation and photographic recording of the data assisted materially in analyzing the signals and rejecting electrical noise.

---

* Work performed under the auspices of the U. S. Atomic Energy Commission.
† Now at the Department of Physics, George Washington University, Washington, D. C.
[1] Cowan, Reines, Harrison, Kruse, and McGuire, Science **124**, 103 (1956).
[2] Ronzio, Cowan, and Reines, Rev. Sci. Instr. **29**, 146 (1958), describe the preparation and handling of liquid scintillators developed for the Los Alamos neutrino program.
[3] F. Reines and C. L. Cowan, Jr., Phys. Rev. **90**, 492 (1953).
[4] The gain, times 6.5, due to the increase in target protons was largely balanced by a decrease in the neutron detection efficiency, times $\frac{1}{8}$, made necessary by other experimental considerations.

OBSERVATION OF HIGH-ENERGY NEUTRINO REACTIONS AND THE EXISTENCE OF TWO KINDS OF NEUTRINOS*

G. Danby, J-M. Gaillard, K. Goulianos, L. M. Lederman, N. Mistry, M. Schwartz,[†] and J. Steinberger[†]

Columbia University, New York, New York and Brookhaven National Laboratory, Upton, New York
(Received June 15, 1962)

In the course of an experiment at the Brookhaven AGS, we have observed the interaction of high-energy neutrinos with matter. These neutrinos were produced primarily as the result of the decay of the pion:

$$\pi^\pm \to \mu^\pm + (\nu/\bar{\nu}). \qquad (1)$$

It is the purpose of this Letter to report some of the results of this experiment including (1) demonstration that the neutrinos we have used produce μ mesons but do not produce electrons, and hence are very likely different from the neutrinos involved in β decay and (2) approximate cross sections.

Behavior of cross section as a function of energy. The Fermi theory of weak interactions which works well at low energies implies a cross section for weak interactions which increases as phase space. Calculation indicates that weak interacting cross sections should be in the neighborhood of 10^{-38} cm^2 at about 1 BeV. Lee and Yang[1] first calculated the detailed cross sections for

$$\nu + n \rightarrow p + e^-,$$
$$\bar{\nu} + p \rightarrow n + e^+, \quad (2)$$
$$\nu + n \rightarrow p + \mu^-,$$
$$\bar{\nu} + p \rightarrow n + \mu^+, \quad (3)$$

using the vector form factor deduced from electron scattering results and assuming the axial vector form factor to be the same as the vector form factor. Subsequent work has been done by Yamaguchi[2] and Cabbibo and Gatto.[3] These calculations have been used as standards for comparison with experiments.

Unitarity and the absence of the decay $\mu \rightarrow e + \gamma$. A major difficulty of the Fermi theory at high energies is the necessity that it break down before the cross section reaches $\pi\lambda^2$, violating unitarity. This breakdown must occur below 300 BeV in the center of mass. This difficulty may be avoided if an intermediate boson mediates the weak interactions. Feinberg[4] pointed out, however, that such a boson implies a branching ratio $(\mu \rightarrow e + \gamma)/(\mu \rightarrow e + \nu + \bar{\nu})$ of the order of 10^{-4}, unless the neutrinos associated with muons are different from those associated with electrons.[5] Lee and Yang[6] have subsequently noted that any general mechanism which would preserve unitarity should lead to a $\mu \rightarrow e + \gamma$ branching ratio not too different from the above. Inasmuch as the branching ratio is measured to be $\lesssim 10^{-8}$,[7] the hypothesis that the two neutrinos may be different has found some favor. It is expected that if there is only one type of neutrino, then neutrino interactions should produce muons and electrons in equal abundance. In the event that there are two neutrinos, there is no reason to expect any electrons at all.

The feasibility of doing neutrino experiments at accelerators was proposed independently by Pontecorvo[8] and Schwartz.[9] It was shown that the fluxes of neutrinos available from accelerators should produce of the order of several events per day per 10 tons of detector.

The essential scheme of the experiment is as follows: A neutrino "beam" is generated by decay in flight of pions according to reaction (1). The pions are produced by 15-BeV protons striking a beryllium target at one end of a 10-ft long straight section. The resulting entire flux of particles moving in the general direction of the detector strikes a 13.5-m thick iron shield wall at a distance of 21 m from the target. Neutrino interactions are observed in a 10-ton aluminum spark chamber located behind this shield.

The line of flight of the beam from target to detector makes an angle of 7.5° with respect to the internal proton direction (see Fig. 1). The operating energy of 15 BeV is chosen to keep the muons penetrating the shield to a tolerable level.

The number and energy spectrum of neutrinos from reaction (1) can be rather well calculated, on the basis of measured pion-production rates[10] and the geometry. The expected neutrino flux from π decay is shown in Fig. 2. Also shown is

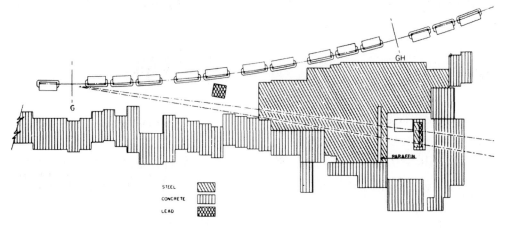

FIG. 1. Plan view of AGS neutrino experiment.

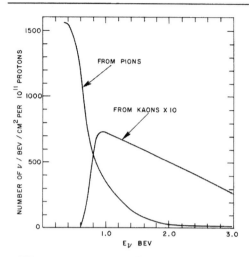

FIG. 2. Energy spectrum of neutrinos expected in the arrangement of Fig. 1 for 15-BeV protons on Be.

an estimate of neutrinos from the decay $K^{\pm} \to \mu^{\pm} + \nu(\bar{\nu})$. Various checks were performed to compare the targeting efficiency (fraction of circulating beam that interacts in the target) during the neutrino run with the efficiency during the beam survey run. (We believe this efficiency to be close to 70%.) The pion-neutrino flux is considered reliable to approximately 30% down to 300 MeV/c, but the flux below this momentum does not contribute to the results we wish to present.

The main shielding wall thickness, 13.5 m for most of the run, absorbs strongly interacting particles by nuclear interaction and muons up to 17 BeV by ionization loss. The absorption mean free path in iron for pions of 3, 6, and 9 BeV has been measured to be less than 0.24 m.[11] Thus the shield provides an attenuation of the order of 10^{-24} for strongly interacting particles. This attenuation is more than sufficient to reduce these particles to a level compatible with this experiment. The background of strongly interacting particles within the detector shield probably enters through the concrete floor and roof of the 5.5-m thick side wall. Indications of such leaks were, in fact, obtained during the early phases of the experiment and the shielding subsequently improved. The argument that our observations are not induced by strongly interacting particles will also be made on the basis of the detailed structure of the data.

The spark chamber detector consists of an array of 10 one-ton modules. Each unit has 9 aluminum plates 44 in. × 44 in. × 1 in. thick, separated by $\frac{3}{8}$-in. Lucite spacers. Each module is driven by a specially designed high-pressure spark gap and the entire assembly triggered as described below. The chamber will be more fully described elsewhere. Figure 3 illustrates the arrangement of coincidence and anticoincidence counters. Top, back, and front anticoincidence sheets (a total of 50 counters, each 48 in. ×11 in. ×$\frac{1}{2}$ in.) are provided to reduce the effect of cosmic rays and AGS-produced muons which penetrate the shield. The top slab is shielded against neutrino events by 6 in. of steel and the back slab by 3 ft of steel and lead.

Triggering counters were inserted between adjacent chambers and at the end (see Fig. 3). These consist of pairs of counters, 48 in. ×11 in. ×$\frac{1}{2}$ in., separated by $\frac{3}{4}$ in. of aluminum, and in fast coincidence. Four such pairs cover a chamber; 40 are employed in all.

The AGS at 15 BeV operates with a repetition period of 1.2 sec. A rapid beam deflector drives the protons onto the 3-in. thick Be target over a period of 20-30 μsec. The radiation during this interval has rf structure, the individual bursts being 20 nsec wide, the separation 220 nsec. This structure is employed to reduce the total "on" time and thus minimize cosmic-ray background. A Čerenkov counter exposed

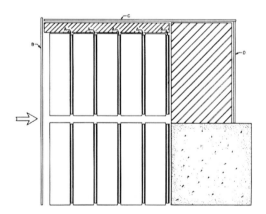

FIG. 3. Spark chamber and counter arrangement. A are the triggering slabs; B, C, and D are anticoincidence slabs. This is the front view seen by the four-camera stereo system.

to the pions in the neutrino "beam" provides a train of 30-nsec gates, which is placed in coincidence with the triggering events. The correct phasing is verified by raising the machine energy to 25 BeV and counting the high-energy muons which now penetrate the shield. The tight timing also serves the useful function of reducing sensitivity to low-energy neutrons which diffuse into the detector room. The trigger consists of a fast twofold coincidence in any of the 40 coincidence pairs in anticoincidence with the anticoincidence shield. Typical operation yields about 10 triggers per hour. Half the photographs are blank, the remainder consist of AGS muons entering unprotected faces of the chamber, cosmic rays, and "events." In order to verify the operation of circuits and the gap efficiency of the chamber, cosmic-ray test runs are conducted every four hours. These consist of triggering on almost horizontal cosmic-ray muons and recording the results both on film and on Land prints for rapid inspection (see Fig. 4).

A convenient monitor for this experiment is the number of circulating protons in the AGS machine. Typically, the AGS operates at a level of $2-4 \times 10^{11}$ protons per pulse, and 3000 pulses per hour. In an exposure of 3.48×10^{17} protons, we have counted 113 events satisfying the following geometric criteria: The event originates within a fiducial volume whose boundaries lie 4 in. from the front and back walls of the chamber and 2 in. from the top and bottom walls. The first two gaps must not fire, in order to exclude events whose origins lie outside the chambers. In addition, in the case of events consisting of a single track, an extrapolation of the track backwards (towards the neutrino source) for two gaps must also remain within the fiducial volume. The production angle of these single tracks relative to the neutrino line of flight must be less than 60°.

These 113 events may be classified further as follows:

(a) 49 short single tracks. These are single tracks whose visible momentum, if interpreted as muons, is less than 300 MeV/c. These presumably include some energetic muons which leave the chamber. They also include low-energy neutrino events and the bulk of the neutron produced background. Of these, 19 have 4 sparks or less. The second half of the run (1.7×10^{17} protons) with improved shielding yielded only three tracks in this category. We will not consider these as acceptable "events."

(b) 34 "single muons" of more than 300 MeV/c. These include tracks which, if interpreted as muons, have a visible range in the chambers such that their momentum is at least 300 MeV/c. The origin of these events must not be accompanied by more than two extraneous sparks. The latter requirement means that we include among "single tracks" events showing a small recoil. The 34 events are tabulated as a function of momentum in Table I. Figure 5 illustrates 3 "single muon" events.

(c) 22 "vertex" events. A vertex event is one whose origin is characterized by more than one track. All of these events show a substantial energy release. Figure 6 illustrates some of these.

(d) 8 "showers." These are all the remaining events. They are in general single tracks, too irregular in structure to be typical of μ mesons, and more typical of electron or photon showers. From these 8 "showers," for purposes of comparison with (b), we may select a group of 6 which are so located that their potential range within the chamber corresponds to μ mesons in excess of 300 MeV/c.

In the following, only the 56 energetic events of type (b) (long μ's) and type (c) (vertex events) will be referred to as "events."

Arguments on the neutrino origin of the ob-

FIG. 4. Land print of Cosmic-ray muons integrated over many incoming tracks.

Table I. Classification of "events."

Single tracks			
$p_\mu < 300$ MeV/c [a]	49	$p_\mu > 500$	8
$p_\mu > 300$	34	$p_\mu > 600$	3
$p_\mu > 400$	19	$p_\mu > 700$	2
Total "events" 34			
Vertex events			
Visible energy released < 1 BeV			15
Visible energy released > 1 BeV			7

[a] These are not included in the "event" count (see text).

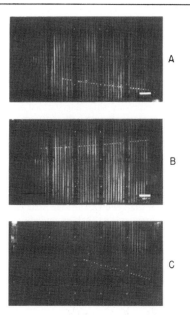
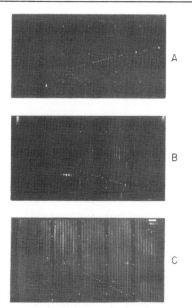

FIG. 5. Single muon events. (A) $p_\mu > 540$ MeV and δ ray indicating direction of motion (neutrino beam incident from left); (B) $p_\mu > 700$ MeV/c; (C) $p_\mu > 440$ with δ ray.

FIG. 6. Vertex events. (A) Single muon of $p_\mu > 500$ MeV and electron-type track; (B) possible example of two muons, both leave chamber; (C) four prong star with one long track of $p_\mu > 600$ MeV/c.

served "events."

1. The "events" are not produced by cosmic rays. Muons from cosmic rays which stop in the chamber can and do simulate neutrino events. This background is measured experimentally by running with the AGS machine off on the same triggering arrangement except for the Čerenkov gating requirement. The actual triggering rate then rises from 10 per hour to 80 per second (a dead-time circuit prevents jamming of the spark chamber). In 1800 cosmic-ray photographs thus obtained, 21 would be accepted as neutrino events. Thus 1 in 90 cosmic-ray events is neutrino-like. Čerenkov gating and the short AGS pulse effect a reduction by a factor of $\sim 10^{-6}$ since the circuits are "on" for only 3.5 μsec per pulse. In fact, for the body of data represented by Table I, a total of 1.6×10^6 pulses were counted. The equipment was therefore sensitive for a total time of 5.5 sec. This should lead to $5.5 \times 80 = 440$ cosmic-ray tracks which is consistent with observation. Among these, there should be 5 ± 1 cosmic-ray induced "events." These are almost evident in the small asym-

metry seen in the angular distributions of Fig. 7. The remaining 51 events cannot be the result of cosmic rays.

2. The "events" are not neutron produced. Several observations contribute to this conclusion.

(a) The origins of all the observed events are uniformly distributed over the fiduciary volume, with the obvious bias against the last chamber induced by the $p_\mu > 300$ MeV/c requirement. Thus there is no evidence for attenuation, although the mean free path for nuclear interaction in aluminum is 40 cm and for electromagnetic interaction 9 cm.

(b) The front iron shield is so thick that we can expect less than 10^{-4} neutron induced reactions in the entire run from neutrons which have penetrated this shield. This was checked by removing 4 ft of iron from the front of the thick shield. If our events were due to neutrons in line with the target, the event rate would have increased by a factor of one hundred. No such effect was observed (see Table II). If neutrons penetrate the shield, it must be from other di-

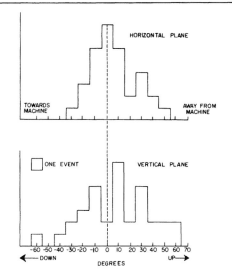

FIG. 7. Projected angular distributions of single track events. Zero degree is defined as the neutrino direction.

rections. The secondaries would reflect this directionality. The observed angular distribution of single track events is shown in Fig. 7. Except for the small cosmic-ray contribution to the vertical plane projection, both projections are peaked about the line of flight to the target.
(c) If our 29 single track events (excluding cosmic-ray background) were pions produced by neutrons, we would have expected, on the basis of known production cross sections, of the order of 15 single π^0's to have been produced. No cases of unaccompanied π^0's have been observed.

Table II. Event rates for normal and background conditions.

	Circulating protons × 10^{16}	No. of Events	Calculated cosmic-ray[c] contribution	Net rate per 10^{16}
Normal run	34.8	56	5	1.46
Background I[a]	3.0	2	0.5	0.5
Background II[b]	8.6	4	1.5	0.3

[a] 4 ft of Fe removed from main shielding wall.
[b] As above, but 4 ft of Pb placed within 6 ft of Be target and subtending a horizontal angular interval from 4° to 11° with respect to the internal proton beam.
[c] These should be subtracted from the "single muon" category.

3. *The single particles produced show little or no nuclear interaction and are therefore presumed to be muons.* For the purpose of this argument, it is convenient to first discuss the second half of our data, obtained after some shielding improvements were effected. A total traversal of 820 cm of aluminum by single tracks was observed, but no "clear" case of nuclear interaction such as large angle or charge exchange scattering was seen. In a spark chamber calibration experiment at the Cosmotron, it was found that for 400-MeV pions the mean free path for "clear" nuclear interactions in the chamber (as distinguished from stoppings) is no more than 100 cm of aluminum. We should, therefore, have observed of the order of 8 "clear" interactions; instead we observed none. The mean free path for the observed single tracks is then more than 8 times the nuclear mean free path.

Included in the count are 5 tracks which stop in the chamber. Certainly a fraction of the neutrino secondaries must be expected to be produced with such small momentum that they would stop in the chamber. Thus, none of these stoppings may, in fact, be nuclear interactions. But even if all stopping tracks are considered to represent nuclear interactions, the mean free path of the observed single tracks must be 4 nuclear mean free paths.

The situation in the case of the earlier data is more complicated. We suspect that a fair fraction of the short single tracks then observed are, in fact, protons produced in neutron collisions. However, similar arguments can be made also for these data which convince us that the energetic single track events observed then are also noninteracting.[12]

It is concluded that the observed single track events are muons, as expected from neutrino interactions.

4. *The observed reactions are due to the decay products of pions and K mesons.* In a second background run, 4 ft of iron were removed from the main shield and replaced by a similar quantity of lead placed as close to the target as feasible. Thus, the detector views the target through the same number of mean free paths of shielding material. However, the path available for pions to decay is reduced by a factor of 8. This is the closest we could come to "turning off" the neutrinos. The results of this run are given in terms of the number of events per 10^{16} circulating protons in Table II. The rate of "events" is reduced from 1.46 ± 0.2 to 0.3 ± 0.2 per 10^{16} in-

cident protons. This reduction is consistent with that which is expected for neutrinos which are the decay products of pions and K mesons.

Are there two kinds of neutrinos? The earlier discussion leads us to ask if the reactions (2) and (3) occur with the same rate. This would be expected if ν_μ, the neutrino coupled to the muon and produced in pion decay, is the same as ν_e, the neutrino coupled to the electron and produced in nuclear beta decay. We discuss only the single track events where the distinction between single muon tracks of $p_\mu > 300$ MeV/c and showers produced by high-energy single electrons is clear. See Figs. 8 and 4 which illustrate this difference.

We have observed 34 single muon events of which 5 are considered to be cosmic-ray background. If $\nu_\mu = \nu_e$, there should be of the order of 29 electron showers with a mean energy greater than 400 MeV/c. Instead, the only candidates which we have for such events are six "showers" of qualitatively different appearance from those of Fig. 8. To argue more precisely, we have exposed two of our one-ton spark chamber modules to electron beams at the Cosmotron. Runs were taken at various electron energies. From these we establish that the triggering efficiency for 400-MeV electrons is 67%. As a quantity characteristic of the calibration showers, we have taken the total number of observed sparks. The mean number is roughly linear with electron energy up to 400 MeV/c. Larger showers saturate the two chambers which were available. The spark distribution for 400 MeV/c showers is plotted in Fig. 9, normalized to the $\frac{2}{3} \times 29$ expected showers. The six "shower" events are also plotted. It is evident that these are not consistent with the prediction based on a universal theory with $\nu_\mu = \nu_e$. It can perhaps be argued that the absence of electron events could be understood in terms of the coupling of a single neutrino to the electron which is much weaker than that to the muon at higher momentum transfers, although at lower momentum transfers the results of β decay, μ capture, μ decay, and the ratio of $\pi \rightarrow \mu + \nu$ to $\pi \rightarrow e + \nu$ decay show that these couplings are equal.[13] However, the most plausible explanation for the absence of the electron showers, and the only one which preserves universality, is then that $\nu_\mu \neq \nu_e$; i.e., that there are at least two types of neutrinos. This also resolves the problem raised by the forbiddenness of the $\mu^+ \rightarrow e^+ + \gamma$ decay.

It remains to understand the nature of the 6 "shower" events. All of these events were obtained in the first part of the run during conditions in which there was certainly some neutron background. It is not unlikely that some of the events are small neutron produced stars. One or two could, in fact, be μ mesons. It should also be remarked that of the order of one or two electron events are expected from the neutrinos produced in the decays $K^+ \rightarrow e^+ + \nu_e + \pi^0$ and

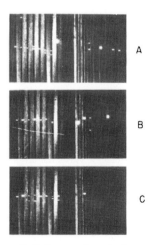

FIG. 8. 400-MeV electrons from the Cosmotron.

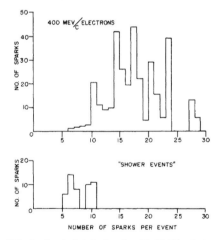

FIG. 9. Spark distribution for 400-MeV/c electrons normalized to expected number of showers. Also shown are the "shower" events.

$K_2^0 \rightarrow e^{\pm} + \nu_e + \pi^{\mp}$.

The intermediate boson. It has been pointed out[1] that high-energy neutrinos should serve as a reasonable method of investigating the existence of an intermediate boson in the weak interactions. In recent years many of the objections to such a particle have been removed by the advent of V-A theory[14] and the remeasurement of the ρ value in μ decay.[15] The remaining difficulty pointed out by Feinberg,[4] namely the absence of the decay $\mu \rightarrow e + \gamma$, is removed by the results of this experiment. Consequently it is of interest to explore the extent to which our experiment has been sensitive to the production of these bosons.

Our neutrino intensity, in particular that part contributed by the K-meson decays, is sufficient to have produced intermediate bosons if the boson had a mass m_W less than that of the mass of the proton (m_p). In particular, if the boson had a mass equal to $0.6 m_p$, we should have produced ~20 bosons by the process $\nu + p \rightarrow w^+ + \mu^- + p$. If $m_W = m_p$, then we should have observed 2 such events.[16]

Indeed, of our vertex events, 5 are consistent with the production of a boson. Two events, with two outgoing prongs, one of which is shown in Fig. 6(B), are consistent with both prongs being muons. This could correspond to the decay mode $w^+ \rightarrow \mu^+ + \nu$. One event shows four outgoing tracks, each of which leaves the chamber after traveling through 9 in. of aluminum. This might in principle be an example of $w^+ \rightarrow \pi^+ + \pi^- + \pi^+$. Another event, by far our most spectacular one, can be interpreted as having a muon, a charged pion, and two gamma rays presumably from a neutral pion. Over 2 BeV of energy release is seen in the chamber. This could in principle be an example of $w^+ \rightarrow \pi^+ + \pi^0$. Finally, we have one event, Fig. 6(A), in which both a muon and an electron appear to leave the same vertex. If this were a boson production, it would correspond to the boson decay mode $w^+ \rightarrow e^+ + \nu$. The alternative explanation for this event would require (i) that a neutral pion be produced with the muon; and (ii) that one of its gamma rays convert in the plate of the interaction while the other not convert visibly in the chamber.

The difficulty of demonstrating the existence of a boson is inherent in the poor resolution of the chamber. Future experiments should shed more light on this interesting question.

Neutrino cross sections. We have attempted to compare our observations with the predicted cross sections for reactions (2) using the theory.[1-3] To include the fact that the nucleons in (2) are, in fact, part of an aluminum nucleus, a Monte Carlo calculation was performed using a simple Fermi model for the nucleus in order to evaluate the effect of the Pauli principle and nucleon motion. This was then used to predict the number of "elastic" neutrino events to be expected under our conditions. The results agree with simpler calculations based on Fig. 2 to give, in terms of number of circulating protons,

from $\pi \rightarrow \mu + \nu$,   0.60 events/$10^{16}$ protons,

from $K \rightarrow \mu + \nu$,   0.15 events/10^{16} protons,

Total   0.75 events/10^{16} ± ~30%.

The observed rates, assuming all single muons are "elastic" and all vertex events "inelastic" (i.e., produced with pions) are

"Elastic": 0.84 ± 0.16 events/10^{16} (29 events),

"Inelastic": 0.63 ± 0.14 events/10^{16} (22 events).

The agreement of our elastic yield with theory indicates that no large modification to the Fermi interaction is required at our mean momentum transfer of 350 MeV/c. The inelastic cross section in this region is of the same order as the elastic cross section.

Neutrino flip hypothesis. Feinberg, Gursey, and Pais[17] have pointed out that if there were two different types of neutrinos, their assignment to muon and electron, respectively, could in principle be interchanged for strangeness-violating weak interactions. Thus it might be possible that

$\pi^+ \rightarrow \mu^+ + \nu_1$ while $K^+ \rightarrow \mu^+ + \nu_2$

$\pi^+ \rightarrow e^+ + \nu_2$ $K^+ \rightarrow e^+ + \nu_1$.

This hypothesis is subject to experimental check by observing whether neutrinos from $K_{\mu 2}$ decay produce muons or electrons in our chamber. Our calculation of the neutrino flux from $K_{\mu 2}$ decay indicates that we should have observed 5 events from these neutrinos. They would have an average energy of 1.5 BeV. An electron of this energy would have been clearly recognizable. None have been seen. It seems unlikely therefore that the neutrino flip hypothesis is correct.

The authors are indebted to Professor G. Feinberg, Professor T. D. Lee, and Professor C. N. Yang for many fruitful discussions. In particular, we note here that the emphasis by Lee and Yang on the importance of the high-energy behavior of weak interactions and the likelihood of the existence of two neutrinos played an important part in stimulating this research.

We would like to thank Mr. Warner Hayes for technical assistance throughout the experiment. In the construction of the spark chamber, R. Hodor and R. Lundgren of BNL, and Joseph Shill and Yin Au of Nevis did the engineering. The construction of the electronics was largely the work of the Instrumentation Division of BNL under W. Higinbotham. Other technical assistance was rendered by M. Katz and D. Balzarini. Robert Erlich was responsible for the machine calculations of neutrino rates, M. Tannenbaum assisted in the Cosmotron runs.

The experiment could not have succeeded without the tremendous efforts of the Brookhaven Accelerator Division. We owe much to the co-operation of Dr. K. Green, Dr. E. Courant, Dr. J. Blewett, Dr. M. H. Blewett, and the AGS staff including J. Spiro, W. Walker, D. Sisson, and L. Chimienti. The Cosmotron Department is acknowledged for its help in the initial assembly and later calibration runs.

The work was generously supported by the U. S. Atomic Energy Commission. The work at Nevis was considerably facilitated by Dr. W. F. Goodell, Jr., and the Nevis Cyclotron staff under Office of Naval Research support.

---

*This research was supported by the U. S. Atomic Energy Commission.

†Alfred P. Sloan Research Fellow.

[1]T. D. Lee and C. N. Yang, Phys. Rev. Letters 4, 307 (1960).

[2]Y. Yamaguchi, Progr. Theoret. Phys. (Kyoto) 6, 1117 (1960).

[3]N. Cabbibo and R. Gatto, Nuovo cimento 15, 304 (1960).

[4]G. Feinberg, Phys. Rev. 110, 1482 (1958).

[5]Several authors have discussed this possibility. Some of the earlier viewpoints are given by: E. Konopinski and H. Mahmoud, Phys. Rev. 92, 1045 (1953); J. Schwinger, Ann. Phys. (New York) 2, 407 (1957); I. Kawakami, Progr. Theoret. Phys. (Kyoto) 19, 459 (1957); M. Konuma, Nuclear Phys. 5, 504 (1958); S. A. Bludman, Bull. Am. Phys. Soc. 4, 80 (1959); S. Oneda and J. C. Pati, Phys. Rev. Letters 2, 125 (1959); K. Nishijima, Phys. Rev. 108, 907 (1957).

[6]T. D. Lee and C. N. Yang (private communications). See also Proceedings of the 1960 Annual International Conference on High-Energy Physics at Rochester (Interscience Publishers, Inc., New York, 1960), p. 567.

[7]D. Bartlett, S. Devons, and A. Sachs, Phys. Rev. Letters 8, 120 (1962); S. Frankel, J. Halpern, L. Holloway, W. Wales, M. Yearian, O. Chamberlain, A. Lemonick, and F. M. Pipkin, Phys. Rev. Letters 8, 123 (1962).

[8]B. Pontecorvo, J. Exptl. Theoret. Phys. (U.S.S.R.) 37, 1751 (1959) [translation: Soviet Phys.−JETP 10, 1236 (1960)].

[9]M. Schwartz, Phys. Rev. Letters 4, 306 (1960).

[10]W. F. Baker et al., Phys. Rev. Letters 7, 101 (1961).

[11]R. L. Cool, L. Lederman, L. Marshall, A. C. Melissinos, M. Tannenbaum, J. H. Tinlot, and T. Yamanouchi, Brookhaven National Laboratory Internal Report UP-18 (unpublished).

[12]These will be published in a more complete report.

[13]H. L. Anderson, T. Fujii, R. H. Miller, and L. Tau, Phys. Rev. 119, 2050 (1960); G. Culligan, J. F. Lathrop, V. L. Telegdi, R. Winston, and R. A. Lundy, Phys. Rev. Letters 7, 458 (1961); R. Hildebrand, Phys. Rev. Letters 8, 34 (1962); E. Bleser, L. Lederman, J. Rosen, J. Rothberg, and E. Zavattini, Phys. Rev. Letters 8, 288 (1962).

[14]R. Feynman and M. Gell-Mann, Phys. Rev. 109, 193 (1958); R. Marshak and E. Sudershan, Phys. Rev. 109, 1860 (1958).

[15]R. Plano, Phys. Rev. 119, 1400 (1960).

[16]T. D. Lee, P. Markstein, and C. N. Yang, Phys. Rev. Letters 7, 429 (1961).

[17]G. Feinberg, F. Gursey, and A. Pais, Phys. Rev. Letters 7, 208 (1961).

7

The Neutral Kaon System

From the discovery of the K_L^0 to CP violation, 1956–1964, and beyond

The development of the concept of strangeness created something of a puzzle: What is the nature of the K^0 and $\overline{K}^0$? They differ only in their strangeness, a quantity not conserved by the weak interactions, through which they decay. Thus, for example, they both can decay into $\pi^+\pi^-$ and $\pi^+\pi^-\pi^0$. The explanation was given by Gell-Mann and Pais before parity violation was discovered. We present their proposal modified to incorporate parity violation, but assuming at first that the combination, CP, of charge conjugation and parity inversion is a good symmetry of both the weak and strong interactions.

The K^0 is an eigenstate of the strong interactions, as is the $\overline{K}^0$. They are antiparticles of each other so they can be transformed into each other by charge conjugation and thus have opposite strangeness. If there were no weak interactions, the K^0 and $\overline{K}^0$ would be stable and equal in mass. The weak interactions break the degeneracy and make the neutral kaons unstable. The particles with well-defined masses and lifetimes are the physical states. These states are linear combinations of K^0 and $\overline{K}^0$, the strong interaction eigenstates.

Since the action of CP on a K^0 produces a $\overline{K}^0$ we can establish a phase convention by

$$CP|K^0\rangle = |\overline{K}^0\rangle. \tag{7.1}$$

If CP is conserved, the physical eigenstates are the eigenstates of CP. These are simply

$$|K_1^0\rangle = \frac{1}{\sqrt{2}}\left[|K^0\rangle + |\overline{K}^0\rangle\right], \tag{7.2}$$

$$|K_2^0\rangle = \frac{1}{\sqrt{2}}\left[|K^0\rangle - |\overline{K}^0\rangle\right], \tag{7.3}$$

where K_1^0 has $CP = +1$ and K_2^0 has $CP = -1$. The decays $K^0 \to \pi^+\pi^-$ and $\overline{K}^0 \to \pi^+\pi^-$ are both allowed by the weak-interaction selection rules. The $\pi^+\pi^-$ state with angular momentum zero necessarily has $P = (-1)^L = +1$, $C = (-1)^L = +1$ since both C and P interchange the two pions, which are in an s-wave, and thus $CP = +1$. It follows that the K_2^0 cannot decay into $\pi^+\pi^-$ if CP is conserved. On the other hand, a $\pi^+\pi^-\pi^0$ state that is entirely s-wave must have $CP = -1$ because the $\pi^+\pi^-$ part has $CP = +1$ by the above reasoning, while the remaining π^0 has $CP = -1$. Since the important decay channel $\pi\pi$ is closed to it, the K_2^0 has a longer lifetime than the K_1^0.

Because K^0 and $\overline{K}^0$ have well-defined strangeness and strangeness is conserved in hadronic collisions it is these states that are directly produced. Gell-Mann and Pais noted that basic quantum mechanics tells us to regard a produced K^0 as a superposition of the CP-even K_1^0 and the CP-odd K_2^0. The K_1^0 portion of the state dies much more rapidly than the K_2^0 portion, so that after a period of time only the latter is present if the particle has not yet decayed. While decays into $\pi\pi$ or $\pi\pi\pi$ are possible from either K^0 or $\overline{K}^0$, by the $\Delta S = \Delta Q$ rule, a decay to $e^+\nu\pi^-$ is possible only from K^0, while a decay to $e^-\bar{\nu}\pi^+$ must come from $\overline{K}^0$.

The K_2^0 was observed in 1956 by Lande et al. using a 3-GeV beam from the Brookhaven Cosmotron (**Ref. 7.1**). A cloud chamber filled 90% with helium and 10% with argon was placed six meters from the interaction point. All K_1^0s and Λs would have decayed by the time of their arrival at the cloud chamber. In the cloud chamber, forked tracks were observed that were kinematically unlike $\theta^0 \to \pi^+\pi^-$. It was concluded that they represented $\pi^\pm e^\mp \nu$, possibly $\pi^\pm \mu^\mp \nu$, and occasionally $\pi^+\pi^-\pi^0$. The lifetime was judged to be in the range 10^{-9} s $< \tau < 10^{-6}$ s, whereas the short-lived K^0 (θ) had a lifetime around 10^{-10} s. Additional evidence for a long-lived neutral K was obtained by W. F. Fry and co-workers using a K^- beam from the Bevatron with an emulsion target (**Ref. 7.2**).

These results were followed by a more complete report by Lande, Lederman, and Chinowsky showing clearly the $\mu\pi\nu$, $e\pi\nu$, and 3π modes (Ref. 7.3). They obtained further confirmation of the Gell-Mann–Pais prediction by noting a neutral K that interacted with a helium nucleus to produce $\Sigma^- pp n\pi^+$, a state with negative strangeness. The neutral K beam was overwhelmingly of positive strangeness initially since the threshold for $pn \to p\Lambda K^0$ is much lower than that for, say, $pn \to pn K^0 \overline{K}^0$. Thus there was strong evidence for the transformation $K^0 \to \overline{K}^0$.

In vacuum, the time development of the K_1^0 and K_2^0 is

$$|K_1^0(\tau)\rangle = e^{-im_1\tau - \Gamma_1\tau/2} \frac{1}{\sqrt{2}} \left[|K^0(0)\rangle + |\overline{K}^0(0)\rangle \right], \quad (7.4)$$

$$|K_2^0(\tau)\rangle = e^{-im_2\tau - \Gamma_2\tau/2} \frac{1}{\sqrt{2}} \left[|K^0(0)\rangle - |\overline{K}^0(0)\rangle \right], \quad (7.5)$$

where $m_{1,2}$ and $\Gamma_{1,2}$ are the masses and decay rates of the K_1^0 and K_2^0. Here τ is the proper time, $\tau = t/\gamma$, t is the time measured in the laboratory, and $\gamma = (1-\beta^2)^{-1/2}$, where $\beta = v/c$. Because of virtual weak transitions between the K^0 and $\overline{K}^0$, the masses m_1 and m_2 differ slightly. If a state, $|\Psi\rangle$, that is purely K^0 is produced at $\tau = 0$, it will oscillate

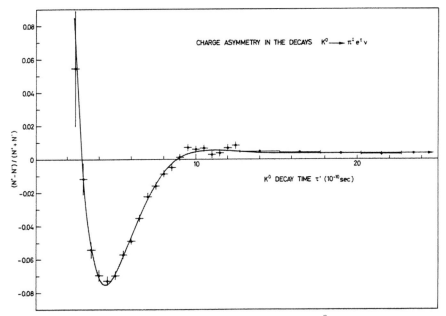

Figure 7.1. The charge asymmetry observed for $K^0 \to \pi^- e^+ \nu$ and $\overline{K}^0 \to \pi^+ e^- \bar{\nu}$ as a function of the proper time, starting from a predominantly K^0 beam. The number of observed positrons is N^+ and the number of observed electrons is N^-. The interference effect seen is sensitive to the $K_L - K_S$ mass difference. For large values of the proper time, the non-zero asymmetry is a CP violating effect and determines Re ϵ [S. Gjesdal et al., Phys. Lett. **52B**, 113 (1974)]. This CP violating effect was first observed in Refs. 7.6 and 7.7.

between K^0 and $\overline{K}^0$ with amplitudes

$$\langle K^0 | \Psi(\tau) \rangle = \frac{1}{2}(e^{-im_1\tau - \Gamma_1\tau/2} + e^{-im_2\tau - \Gamma_2\tau/2}), \tag{7.6}$$

$$\langle \overline{K}^0 | \Psi(\tau) \rangle = \frac{1}{2}(e^{-im_1\tau - \Gamma_1\tau/2} - e^{-im_2\tau - \Gamma_2\tau/2}). \tag{7.7}$$

These oscillations can be observed through semileptonic decays since the semileptonic decays are $K^0 \to \pi^- e^+ \nu$ and $\overline{K}^0 \to \pi^+ e^- \bar{\nu}$. An example is shown in Fig. 7.1. There the charge asymmetry in the decay of K^0 is shown as a function of the proper time. The ratio of "wrong sign" leptons (e^-) to "right sign" leptons (e^+) from a state that is initially a K^0, integrated over all time, is

$$\frac{\text{wrong sign}}{\text{right sign}} = \frac{(\Gamma_1 - \Gamma_2)^2 + 4(\Delta m)^2}{2(\Gamma_1 + \Gamma_2)^2 - (\Gamma_1 - \Gamma_2)^2 + 4(\Delta m)^2}, \tag{7.8}$$

where $\Delta m = m_1 - m_2$. Since the decay rate of the K_1^0, Γ_1, is much greater than the K_2^0 decay rate, Γ_2, this ratio is nearly unity.

Even more dramatic predictions had been made for the neutral-K system. Pais and Piccioni in 1955 predicted that K_2^0s passing through matter would regenerate a coherent

K_1^0 component. In matter, the time development is altered because the K^0 and $\overline{K}^0$ interact differently with nucleons. For example, $\overline{K}^0 p \to \pi^+ \Lambda$ is allowed while $K^0 p \to \pi^+ \Lambda$ is not. In fact, the elastic-scattering amplitudes, f and $\overline{f}$, for $K^0 p$ and $\overline{K}^0 p$ differ, just as those for $K^+ p \to K^+ p$ and $K^- p \to K^- p$ do. Forward-moving neutral kaons accumulate an extra phase from elastic scattering. As in ordinary electromagnetic interactions, this scattering can be translated into an index of refraction

$$n = 1 + \frac{2\pi N}{k^2} f(0), \qquad (7.9)$$

where N is the number density of scatterers, k is the wave number of the incident particles, and $f(0)$ is the (complex) elastic-scattering amplitude in the forward direction, which is related to the total cross section by the optical theorem

$$\sigma_{tot} = \frac{4\pi}{k} \operatorname{Im} f(0). \qquad (7.10)$$

Since K^0 and $\overline{K}^0$ have different total cross sections, they have different (complex) indices of refraction. In going a distance l, a particle picks up an extra phase $k(n-1)l$. The distance l is related to the proper time interval by $l = \tau \beta \gamma$.

To incorporate this effect, we write first the Schrödinger equation for propagation in free space. It is easy to guess what this is since we already have the solutions in the form of $|K_1^0(\tau)\rangle$ and $|K_2^0(\tau)\rangle$. If we let ψ be a column matrix whose upper entry gives the K^0 component and whose lower entry gives the $\overline{K}^0$ component, then in terms of the proper time

$$i \frac{\partial \psi}{\partial \tau} = \begin{pmatrix} M - i\frac{\Gamma}{2} & M_{12} - i\frac{\Gamma_{12}}{2} \\ M_{12} - i\frac{\Gamma_{12}}{2} & M - i\frac{\Gamma}{2} \end{pmatrix} \psi \equiv \mathcal{H} \psi, \qquad (7.11)$$

where $M = (m_1 + m_2)/2$, $\Gamma = (\Gamma_1 + \Gamma_2)/2$, $M_{12} = (m_1 - m_2)/2 = \Delta m/2$, and $\Gamma_{12} = (\Gamma_1 - \Gamma_2)/2 = \Delta \Gamma/2$. We indicate the elastic K^0-nucleus forward scattering amplitude by f_0 and that for $\overline{K}^0$ by $\overline{f}_0$. The effective Hamiltonian $\mathcal{H}$ includes the effects of weak decays and the interactions that are second order in weak interactions responsible for Δm and $\Delta \Gamma$. With the inclusion of the effects of the medium we have

$$i \frac{\partial \psi}{\partial \tau} = \begin{pmatrix} M - i\frac{\Gamma}{2} - \frac{2\pi N \beta \gamma f_0}{k} & M_{12} - i\frac{\Gamma_{12}}{2} \\ M_{12} - i\frac{\Gamma_{12}}{2} & M - i\frac{\Gamma}{2} - \frac{2\pi N \beta \gamma \overline{f}_0}{k} \end{pmatrix} \psi. \qquad (7.12)$$

This is a slight modification of the Hamiltonian $\mathcal{H}$, so the eigenstates – the states that propagate without turning into each other – are only slightly different from the eigenstates in vacuum, that is, K_1^0 and K_2^0. A bit of algebra reveals that these states may be written

$$|K_1^{0\prime}\rangle = |K_1^0\rangle + r|K_2^0\rangle, \tag{7.13}$$

$$|K_2^{0\prime}\rangle = |K_2^0\rangle - r|K_1^0\rangle, \tag{7.14}$$

where the regeneration parameter is a small number, typically of order 10^{-3},

$$r = -\frac{\pi N \beta \gamma}{k} \cdot \frac{f_0 - \bar{f}_0}{m_1 - m_2 - \frac{i}{2}\Gamma_1}. \tag{7.15}$$

The expression has been simplified by noting that since the K_1^0 decays much faster than the K_2^0, $\Gamma_1 \gg \Gamma_2$.

If a neutral kaon beam travels a long distance, only K_2^0s are left. If the K_2^0s traverse a medium, quantum mechanics tells us to analyze their propagation in terms of the eigenstates in that medium. The K_2^0 is mostly $K_2^{0\prime}$, but with a small component of $K_1^{0\prime}$. These two pieces will acquire slightly different phases passing through the medium. When they exit, the states must be reanalyzed in terms of K_1^0 and K_2^0. This will reintroduce a component of K_1^0 of order r. The result is that an amplitude for K_1^0 will be generated proportional to

$$r\left[1 - e^{(i\Delta m + \Delta\Gamma/2)L}\right], \tag{7.16}$$

where $L = l/\beta\gamma$. We see then, that the amount of K_1^0 regenerated depends on the difference of the masses. It is thus possible to measure this difference which is extremely small compared to mass splittings like those between isospin partners.

An early measurement of the mass difference was made by F. Muller *et al.* at the Bevatron using regeneration techniques (**Ref. 7.4**). In addition to K_1^0s produced coherently in exactly the forward direction, K_1^0s are produced through the ordinary scattering process $K_2^0 p \to K_1^0 p$. This "diffractive" process produces particles mostly in the forward direction also, but not with such pronounced forward peaking as the coherent regeneration. Through the reaction $\pi^- p \to K^0 \Lambda$, Muller *et al.* generated a 670-MeV/c neutral kaon beam. A 30-inch propane bubble chamber was placed downstream where the surviving beam was purely K_2^0. The K_1^0s produced by the K_2^0 beam were detected by looking for charged pion pairs that reconstructed to the proper mass. By measuring the angular distribution of these K_1^0s it was possible to demonstrate the existence of the coherently regenerated beam. A first measurement of the mass difference was obtained:

$$|m_2 - m_1|/\Gamma_1 = 0.85 \genfrac{}{}{0pt}{}{+0.3}{-0.25}. \tag{7.17}$$

The current values are $1/\Gamma_1 = 0.8953 \pm 0.0005 \times 10^{-10}$ s and $m_2 - m_1 = 0.5292 \pm 0.0009 \times 10^{10}$ s^{-1}, giving $(m_2 - m_1)/\Gamma_1 = 0.474$. The determination that the mass of the K_2^0 is greater than the mass of the K_1^0 required measuring interference with an additional phase known a priori. In practice this meant observing the scattering of neutral kaons with nucleons and looking at the interference between the K^0 and $\overline{K}^0$ contributions.

After the fall of parity invariance, it appeared that the combination of charge conjugation plus parity, CP, was still a good symmetry, as we assumed in the above analysis. There were (and are) solid theoretical reasons for believing that the combination of time reversal invariance, T, together with C and P gives a good symmetry, CPT. Thus if CP is a good symmetry, so is T.

If CP is a good symmetry, the longer-lived neutral kaon is strictly forbidden to decay into two pions. Nonetheless, in 1964 Christenson, Cronin, Fitch, and Turlay observed its decay to $\pi^+\pi^-$ (**Ref. 7.5**). Another supposed symmetry had fallen. The experiment, carried out at the Alternating Gradient Synchrotron (AGS) at Brookhaven found that the CP-violating decay had a branching ratio of about 2×10^{-3}. Since most of the prominent decays of the longer-lived neutral kaon (which we henceforth refer to as K_L^0) have two charged particles in the final state, just as in the decay being sought, careful momentum measurements and particle identification were essential to separating $K_L^0 \to \pi^+\pi^-$ from the background.

The apparatus was a two-armed spectrometer, each arm of which had a magnet for momentum determination, scintillator for triggering on charged particles, a Cherenkov counter for discriminating against $e^\pm$ simulating $\pi^\pm$, and spark chambers for tracking the charged particles. A small but convincing signal was obtained for the CP-violating decay. The experiment was soon repeated and confirmed at several laboratories.

If CP is conserved, with the convention $CP|K^0\rangle = |\overline{K}^0\rangle$ the off-diagonal matrix elements of $\mathcal{H}$ are $(M_{12} - i\Gamma_{12}/2) = \frac{1}{2}(\Delta m - i\Delta\Gamma/2)$. If we chose a new phase convention for the states, $|K^0\rangle \to e^{i\chi}|K^0\rangle$ and $|\overline{K}^0\rangle \to e^{-i\chi}|\overline{K}^0\rangle$, one off-diagonal matrix element would become $M'_{12} - i\Gamma'_{12}/2 = e^{-2i\chi}\frac{1}{2}(\Delta m - i\Delta\Gamma/2)$ and the other $M''_{12} - i\Gamma''_{12}/2 = e^{2i\chi}\frac{1}{2}(\Delta m - i\Delta\Gamma/2)$. The conservation of CP, which we have assumed to this point would be manifested by M'_{12} and Γ'_{12} being relatively real, and similarly for M''_{12} and Γ''_{12}. In addition, we would find $M'_{12} = M''^*_{12}$ and $\Gamma'_{12} = \Gamma''^*_{12}$.

When CP is violated, the equation for time development takes the more general form

$$i\frac{\partial \psi}{\partial \tau} = \begin{pmatrix} M - i\frac{\Gamma}{2} & M_{12} - i\frac{\Gamma_{12}}{2} \\ M^*_{12} - i\frac{\Gamma^*_{12}}{2} & M - i\frac{\Gamma}{2} \end{pmatrix} \psi, \qquad (7.18)$$

where M and Γ are still real, but M_{12} and Γ_{12} are complex. If M_{12} and Γ_{12} are not relatively real, CP is violated in the mass matrix. The off-diagonal M_{12} corresponds to virtual K^0-$\overline{K}^0$ transitions while Γ_{12} is due to real transitions, which are dominated by the $I = 0$ $\pi\pi$ state, as explained below.

Because CP violation in the neutral K system is a small effect, M_{12}/Γ_{12} is nearly real and we write

$$M_{12} = R\Gamma_{12}(1+i\kappa), \tag{7.19}$$

where R and κ are independent of the phase convention for the states and $|\kappa| \ll 1$. The small imaginary part has an insignificant effect on the masses and lifetimes and we can write

$$\Gamma_{12} = e^{-2i\phi}(\Gamma_S - \Gamma_L)/2 \equiv e^{-2i\phi}\Delta\Gamma/2, \tag{7.20}$$

$$2R|\Gamma_{12}| = m_S - m_L = \Delta m. \tag{7.21}$$

where the (convention-dependent) phase of Γ_{12} is displayed explicitly. In fact we already know that $R = \Delta m/\Delta\Gamma \approx -0.47$.

The physical states are

$$|K_S^0\rangle = \frac{1}{\sqrt{2}}\left[e^{-i\phi}(1+\epsilon)|K^0\rangle + e^{+i\phi}(1-\epsilon)|\overline{K}^0\rangle\right],$$

$$|K_L^0\rangle = \frac{1}{\sqrt{2}}\left[e^{-i\phi}(1+\epsilon)|K^0\rangle - e^{+i\phi}(1-\epsilon)|\overline{K}^0\rangle\right], \tag{7.22}$$

where

$$\epsilon = \frac{i\kappa R}{2R - i}. \tag{7.23}$$

This determines the phase of ϵ to be

$$\arg\epsilon = \tan^{-1}\frac{2(m_L - m_S)}{\Gamma_S - \Gamma_L} = 43.5°, \tag{7.24}$$

once we use $\kappa > 0$, which follows from the determination that $\Re\epsilon > 0$, described below.

Let us look at some of the details of the $K_L^0 \to \pi\pi$ decay. The 2π states can be decomposed into $I = 0$ and $I = 2$ components, since an $I = 1$ $\pi\pi$ state cannot have $J = 0$. Each decay amplitude has a strong phase and a weak phase. The strong phase is due to the interaction of the pions in the final state and is thus the same for the decays of K_S^0 and K_L^0. Roughly speaking, since the interaction is only in the final state, its effect is to introduce a factor $e^{i\delta}$, half that of the full scattering, $e^{2i\delta}$, where δ is the $\pi\pi$ phase shift at a center-of-mass energy equal to m_K.

The weak phase is intrinsic to the decay itself and arises from the weak interaction Hamiltonian. The Hamiltonian must be Hermitian so the weak amplitude for a $\overline{K}^0$ decay must be the complex conjugate of that for a K^0 decay. In terms of the real quantities $\mathcal{A}_I$ and the weak phases λ_I we have

$$\langle(2\pi)_I|H_{\text{wk}}|K^0\rangle = \mathcal{A}_I e^{i\lambda_I} e^{i\delta_I} \tag{7.25}$$

$$\langle(2\pi)_I|H_{\text{wk}}|\overline{K}^0\rangle = \mathcal{A}_I e^{-i\lambda_I} e^{i\delta_I}. \tag{7.26}$$

These results can be simplified by observing the following. First, the much faster decay of the K_S^0 compared to $K^+ \to \pi^+\pi^0$ shows that $|\mathcal{A}_0| \gg |\mathcal{A}_2|$. This is known as the $\Delta I = 1/2$ rule since the $\Delta I = 3/2$ interaction responsible for $K^+ \to \pi^+\pi^0$ (which has $I = 2$ in the final state) is weaker than the $\Delta I = 1/2$ operator responsible for $K_S^0 \to \pi^+\pi^-$. Now just as the hypothetical decay rate of a K^0 to a $\pi\pi$ state with isospin I would be proportional to $(\mathcal{A}_I e^{i\lambda_I} e^{i\delta_I})^*(\mathcal{A}_I e^{i\lambda_I} e^{i\delta_I})$, the transition matrix element Γ_{12} is proportional to $(\mathcal{A}_I e^{i\lambda_I} e^{i\delta_I})^*(\mathcal{A}_I e^{-i\lambda_I} e^{i\delta_I})$ and thus has the phase $e^{-2i\lambda_I}$. Since the $I = 0$ amplitude dominates, we have from Eq. (7.20), to a very good approximation, $\phi = \lambda_0$.

Violation of CP in the neutral K system arises from the small numbers ϵ and $\lambda_2 - \lambda_0$. Keeping terms of first order in these parameters

$$\langle(2\pi)_{I=0}|H_{\text{wk}}|K_S\rangle = \sqrt{2}\mathcal{A}_0 e^{i\delta_0}, \tag{7.27}$$

$$\langle(2\pi)_{I=2}|H_{\text{wk}}|K_S\rangle = \sqrt{2}\mathcal{A}_2 e^{i\delta_2}, \tag{7.28}$$

$$\langle(2\pi)_{I=0}|H_{\text{wk}}|K_L^0\rangle = \sqrt{2}\epsilon\mathcal{A}_0 e^{i\delta_0}, \tag{7.29}$$

$$\langle(2\pi)_{I=2}|H_{\text{wk}}|K_L^0\rangle = \sqrt{2}(\epsilon + i\lambda_2 - i\lambda_0)\mathcal{A}_2 e^{i\delta_2}. \tag{7.30}$$

Note that both ϵ and $\lambda_2 - \lambda_0$ are independent of the phase convention chosen for the neutral kaon states. The traditional CP-violation parameters for the neutral K system are defined by

$$\eta_{+-} = \frac{\langle\pi^+\pi^-|H_{\text{wk}}|K_L^0\rangle}{\langle\pi^+\pi^-|H_{\text{wk}}|K_S^0\rangle} = \epsilon + \epsilon', \tag{7.31}$$

$$\eta_{00} = \frac{\langle\pi^0\pi^0|H_{\text{wk}}|K_L^0\rangle}{\langle\pi^0\pi^0|H_{\text{wk}}|K_S^0\rangle} = \epsilon - 2\epsilon'. \tag{7.32}$$

Expanding in the small quantity $\mathcal{A}_2/\mathcal{A}_0$, we find

$$\epsilon' = \frac{i}{\sqrt{2}}e^{i(\delta_2-\delta_0)}(\lambda_2 - \lambda_0)\frac{\mathcal{A}_2}{\mathcal{A}_0} = \frac{i}{\sqrt{2}}e^{i(\delta_2-\delta_0)}\text{Im}\frac{\mathcal{A}_2 e^{i\lambda_2}}{\mathcal{A}_0 e^{i\lambda_0}}. \tag{7.33}$$

The mass-mixing matrix determines ϵ while ϵ' is due to CP violation in the decays. From its definition, the magnitude of $\eta_\pm$ can be rewritten in terms of the branching ratio for the CP-violating decay $K_L \to \pi^+\pi^-$ and three other well measured quantities:

$$|\eta_{+-}|^2 = \frac{\Gamma(K_L^0 \to \pi^+\pi^-)}{\Gamma(K_S^0 \to \pi^+\pi^-)} = \frac{B(K_L^0 \to \pi^+\pi^-)\Gamma(K_L^0 \to \text{all})}{B(K_S^0 \to \pi^+\pi^-)\Gamma(K_S^0 \to \text{all})} \approx (2.3 \times 10^{-3})^2. \tag{7.34}$$

The analogous measurement for the decay into neutral pions is of course more difficult.

To measure the phases of η_{+-} and η_{00} requires observing the interference between $K_L^0 \to \pi\pi$ and $K_S^0 \to \pi\pi$. This can be accomplished using a K_L^0 beam and regenerating a

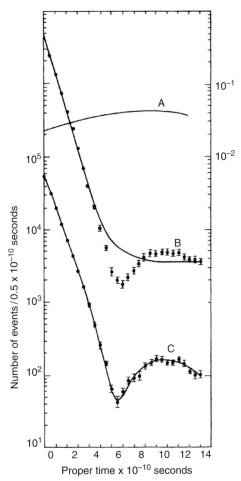

Figure 7.2. Data for $K_{L,S} \to \pi^+\pi^-$ as a function of the proper time after a K_L^0 beam has passed through a carbon regenerator. Curve A shows the detection efficiency as indicated on the right-hand scale. Curve B shows data for all values of the K momentum. The solid curve shows the shape expected in the absence of K_L^0-K_S^0 interference. The interference is apparent and can be used to determine ϕ_{+-}. Curve C shows the data for a restricted interval of K momenta. The solid curve shows a fit including interference. [W. C. Carithers et al., Phys. Rev. Lett. **34**, 1244 (1975)]

small amount of K_S^0, or by using a K^0 beam. In the latter case, one first sees the quickly decaying K_S^0 component. At the end, one sees only the CP-violating K_L^0 decay (if care is taken to observe only the $\pi\pi$ final state!). In between, there is an interval when the contributions from K_S^0 and K_L^0 are comparable, and the interference can be measured. In Fig. 7.2 data obtained using the regenerator method are shown.

CP violation has been observed in $K_L^0 \to \mu\pi\nu$ and $K_L^0 \to e\pi\nu$. Aside from phase space considerations, these decays should be similar. From the $\Delta S = \Delta Q$ rule, one anticipates that the allowed decays to $\pi\mu\nu$ should be $K^0 \to \pi^-\mu^+\nu$ and $\overline{K}^0 \to \pi^+\mu^-\nu$. It follows directly that

$$A_L = \frac{\Gamma(K_L^0 \to \pi^-\mu^+\nu) - \Gamma(K_L^0 \to \pi^+\mu^-\nu)}{\Gamma(K_L^0 \to \pi^+\mu^-\nu) + \Gamma(K_L^0 \to \pi^-\mu^+\nu)} = 2\Re\,\epsilon. \tag{7.35}$$

Unlike the $K_L^0 \to \pi\pi$ decay, the decay process here is allowed even without CP violation. It is the small difference between two allowed rates that is due to CP violation. Thus very high statistics are required. The result can be compared to the measurement of the real part of ϵ obtained in the $K_L^0 \to \pi\pi$ decays. An early measurement of $K \to \pi e \nu$ was obtained by a group headed by Steinberger (Ref. 7.6). The analogous process, $K \to \pi\mu\nu$ was measured by a team led by M. Schwartz (Ref. 7.7). Data from a later experiment are shown in Fig. 7.1.

In 1999, CP violation was observed in an additional decay, $K_L^0 \to \pi^+\pi^-e^+e^-$ (Ref. 7.8). The relevant observable is the distribution of the angle ϕ between the plane containing the e^+ and the e^- and the plane containing the π^+ and the π^- in the rest frame of the decaying K_L^0. The distribution contains pieces that vary as $\sin^2\phi$, $\cos^2\phi$, and $\sin\phi\cos\phi$, the last of which is CP violating. One mechanism that yields this final state is the CP-violating decay $K_L^0 \to \pi^+\pi^-$ followed by emission of a virtual photon that materializes as an electron–positron pair. To see CP violation requires interference and the dominant interfering process is the CP-conserving direct decay $K_L^0 \to \pi^+\pi^-\gamma$, with M1 photon emission. The asymmetry

$$A = \frac{N_{\sin\phi\cos\phi>0} - N_{\sin\phi\cos\phi<0}}{N_{\sin\phi\cos\phi>0} + N_{\sin\phi\cos\phi<0}} \tag{7.36}$$

is impressively large, 13.6%.

Because CP violation seems such a fundamental aspect of particle interactions, enormous efforts have been expended to measure the parameters η_{+-} and η_{00}. The values for the CP-violation parameters cited in the 2008 *Review of Particle Physics* are

$$|\eta_{+-}| = (2.233 \pm 0.012) \times 10^{-3}, \tag{7.37}$$

$$\phi_{+-} = \arg\eta_{+-} = 43.51° \pm 0.05°, \tag{7.38}$$

$$|\eta_{00}| = (2.222 \pm 0.012) \times 10^{-3}, \tag{7.39}$$

$$\phi_{00} = \arg\eta_{00} = 43.52° \pm 0.05°, \tag{7.40}$$

$$A_L = (3.32 \pm 0.06) \times 10^{-3}$$

The results indicate that η_{+-} and η_{00} are very nearly equal, or, equivalently, ϵ' is nearly zero. This could be explained if all the CP violation were due entirely to an interaction that changed strangeness by two units. All the CP violation then is in the K mass matrix and $\epsilon' = 0$. This is called the superweak model. Violation of CP in the decay of a neutral K, a $\Delta S = 1$ process, is called direct CP violation. The Standard Model of electroweak interactions, discussed in Chapter 12, predicts that ϵ'/ϵ is non-zero but small primarily because $|A_2/A_0| \approx 1/20$, a reflection of $\Delta I = 1/2$ rule. Any definitively non-zero result for ϵ'/ϵ suffices to rule out the superweak model. This can be accomplished by measuring $|\eta_{00}|^2$ and $|\eta_{+-}|^2$. Some uncertainties are reduced by taking the ratio

$$\left|\frac{\eta_{00}}{\eta_{+-}}\right|^2 \approx 1 - 6\mathrm{Re}\frac{\epsilon'}{\epsilon}, \qquad (7.41)$$

In fact, ϵ'/ϵ is known to be nearly real because the phases of ϵ' and ϵ are quite similar. Data from a 1988 experiment NA31 at CERN (Ref. 7.9) indicated a small, non-zero value, $\epsilon'/\epsilon = 0.0033 \pm 0.0011$, disfavoring the superweak model. Early results from experiment E731 at Fermilab did not confirm this non-zero value. Continued improvements in experiments at CERN (NA48, Ref. 7.10) and Fermilab (KTeV, Ref. 7.11) ultimately converged on small but distinctly non-zero values, with the average $\mathrm{Re}(\epsilon'/\epsilon) = (1.67 \pm 0.23) \times 10^{-3}$. The superweak model was more a straw man than a theory, but knocking it down took three decades.

Exercises

7.1 Derive the relation between the forward-scattering amplitude and the index of refraction by considering a plane wave of matter or light incident on a thin slab of material. Determine the shift in the phase of the wave passing through the material.

7.2 Show that the decay $\phi(1020) \to K_S^0 K_L^0$ is allowed but $\phi(1020) \to K_S^0 K_S^0$ and $\phi(1020) \to K_L^0 K_L^0$ are forbidden.

7.3 Verify the expression for the eigenstates of the neutral K system in matter. Estimate the size of the regeneration parameter in beryllium for a momentum of 1100 MeV, the conditions of the original CP-violation experiment. Estimate f_0 and $\bar{f}_0$ using the optical theorem and data for the K^+p and K^-p total cross sections.

7.4 A beam of K^0 is created at $t = 0$. Assuming CP conservation, what is the intensity of $\bar{K}^0$ in the beam as a function of the proper time? Plot the results for $|\Delta m|\tau_1 = 0, 1, 2, \infty$. See Camerini et al., Phys. Rev. **128**, 362 (1962).

7.5 Consider a neutral kaon beam that is purely K^0 at $t = 0$. Show that the rate of decay into $\pi^+\pi^-$ as a function of the proper time, τ, is proportional to

$$e^{-\Gamma_S \tau} + 2|\eta_{+-}|e^{-(\Gamma_S+\Gamma_L)\tau/2}\cos\left[\phi_{+-} - (m_L - m_S)\tau\right] + e^{-\Gamma_L \tau}|\eta_{+-}|^2.$$

Further Reading

The standard reference for the formalism is T. D. Lee and C. S. Wu *Ann. Rev. Nucl. Sci.*, **16**, 511 (1966).

A nice treatment of this material is given in *Weak Interactions of Leptons and Quarks*, by E. D. Commins and P. H. Bucksbaum, Cambridge University Press, Cambridge, 1983.

A comprehensive review is given in B. Winstein and L. Wolfenstein, *Rev. Mod. Phys.* **65**, 1113 (1993).

References

7.1 K. Lande et al., "Observation of Long Lived Neutral V Particles." *Phys. Rev.*, **103**, 1901 (1956).

7.2 W. F. Fry, J. Schneps, and M. S. Swami, "Evidence for Long-lived Neutral Unstable Particle." *Phys. Rev.*, **103**, 1904 (1956).

7.3 K. Lande, L. M. Lederman, and W. Chinowsky, "Report on Long Lived K^0 Mesons." *Phys. Rev.*, **104**, 1925 (1957).

7.4 F. Muller et al., "Regeneration and Mass Difference of Neutral K Mesons." *Phys. Rev. Lett.*, **4**, 418 (1960).

7.5 J. H. Christenson, J. W. Cronin, V. L. Fitch, and R. Turlay, "Evidence for the 2π Decay of the K_2^0 Meson." *Phys. Rev. Lett.*, **13**, 138 (1964).

7.6 S. Bennett et al., "Measurement of the Charge Asymmetry in the Decay $K_L^0 \to \pi^\pm + e^\mp + \nu$." *Phys. Rev. Lett.*, **19**, 993 (1967).

7.7 D. Dorfan et al., "Charge Asymmetry in the Muonic Decay of the K_2^0." *Phys. Rev. Lett.*, **19**, 987 (1967).

7.8 A. Alavi-Harati et al., KTeV Collaboration, "Observation of CP Violation in $K_L \to \pi^+\pi^- e^+ e^-$ Decays." *Phys. Rev. Lett.*, **84**, 408 (2000).

7.9 CERN–Dortmund–Edinburgh–Mainz–Orsay–Pisa–Siegen Collaboration, "First Evidence for Direct CP Violation." *Phys. Lett.*, **206B**, 169 (1988).

7.10 J. R. Batley et al., NA48 Collaboration, "A Precision Measurement of Direct CP Violation in the Decay of Neutral Kaons into Two Pions." *Phys. Lett.*, **544B**, 97 (2002).

7.11 A. Alavi-Harati et al., KTeV Collaboration, "Measurements of Direct CP Violation, CPT Symmetry, and Other Parameters in the Neutral Kaon System." *Phys. Rev.*, **D67**, 012005 (2003).

Observation of Long-Lived Neutral V Particles*

K. Lande, E. T. Booth, J. Impeduglia, and L. M. Lederman,
Columbia University, New York, New York

AND

W. Chinowsky, *Brookhaven National Laboratory, Upton, New York*
(Received July 30, 1956)

THE application of rigorous charge conjugation invariance to strange particle interactions has led to the prediction of rather startling properties for the θ^0-meson state.[1] Some of these are: (I) the existence of a second neutral particle, θ_2^0, for which two-pion decay is prohibited; (II) the consequent existence of a second

TABLE I. Data on V^0 events.

Event number	P_+ Mev/c	$I_+{}^a$	P_- Mev/c	$I_-{}^a$	θ^b	$Q^*_{\pi^-\pi^{\circ}}{}^c$ Mev	Comment
1	360±10	<2	206±37	<2	52°	96±17	Not $\tau^{0\,d}$
2	...		117±50	<2	151°	...	(−) track short
3	>100	<2	>100	<2	140°	...	Both tracks short
4	224±5	<2	58.5±4	<2	91.5°	66±2	Not τ^0; probable e^-
5	147±23	<2	197±6	<2	113.6°	121±12	Not τ^0
6	83±5	<2	137±3	<2	81.0°	40±2	$I_->I_+$, probable $\pi^-\mu^+$, π^-e^+, or μ^-e^+
7	142±13	<2	255±5	<2	124°	163±10	Not τ^0
8	197±25	<2	234±9	<2	97°	147±14	Not τ^0
9	241±33	<2	67±20	<2	142°	109±18	Not τ^0, probable e^-
10	194±8	<2	223±4	<2	140°	140±5	Not τ^0
11	111±4	<2	114±5	<2	77.5°	34±1.4	$I_->I_+$, like No. 6
12	249±5	<2	89±1	2–3	42.0°	38±1.2	Probable π^-
13	290±25	<2	86±25	<2	92°	103±12	μ^- or e^-, $\therefore$ not τ^0
14	183±5	<2	44±5	3–4	102.7	52±2	π^- or μ^-
15	>150	<2	62±7	<2	99°	...	3° deflection in +. $P_{sec}=287\pm30$. Possible $\pi-\mu$ decay. Probable e^-
16	508±18	<2	150±2	<2	65.9°	160±7	Not τ^0, $I_->I_+$
17	136±8	1.5–2	164±5	<2	118.6°	101±5	$I_+>I_-$. Probable π^+, not τ^0
18	251±15	<2	201±50	<2	65.9°	93±15	Not τ^0
19	327±15	<2	112±10	<2	38.3°	51±4	$\pi^+ \to \mu^+$, $P_\mu^* = 20\pm10$ Mev
20	167±3	<2	114±3	<2	65.5°	40±2	
21	152±5	<2	120±24	<2	112.5°	79±8	
22	283±10	<2	222±10	<2	50.1°	72±6	Coplanar, $P_\perp = 59\pm30$ Mev/c
23	89±1	<2	272±9	<3	128.5°	134±10	Not τ^0

[a] This is a visual estimate of the ionization, in units of minimum ionization as determined from nearby light tracks of $P<50$ Mev/c.
[b] Angle errors have not been computed. An average error of 3° has been used.
[c] $Q^*_{\pi\pi}$ for a normal θ^0 is 214 Mev.
[d] τ^0 is defined as $\to \pi^+ + \pi^- + \pi^0$ and is excluded by Q value or transverse momentum.

lifetime, considerably longer than that for two-pion decay of the θ_1^0 ($\sim 1\times 10^{-10}$ sec); (III) a complicated time dependence for the nuclear interaction properties.[2] The only additional assumption in this "particle mixture" theory is the nonidentity of θ^0 and its antiparticle.

These theoretical considerations have stimulated us to undertake a search for long-lived neutral particles. To this end, the Columbia 36-in. magnet cloud chamber was exposed to the neutral radiation emitted from a copper target at an angle of 68° to the 3-Bev external proton beam of the Brookhaven Cosmotron.[3] Charged particles are eliminated by the combination of a 4-ft long Pb collimator and a 4×10^5 gauss-inch sweeping magnet. The 6-meter flight path from target to chamber represents ~ 100 mean lives for the well-known Λ^0 and θ^0 particles which are produced at this energy.[4] To date twenty-six V^0 events have been observed. All of these events have anomalous Q values for two-pion decay, all but one are noncoplanar with the line of flight, and all but one demand at least one neutral secondary to balance transverse momentum.

The cloud chamber operates at a pressure of 0.91 atmos of He and 0.10 atmos of argon. The only additional matter in the direct path of the neutral radiation is the 1-cm thick Lucite chamber wall. A 1.5-in. thick lead filter was placed at the entrance to the collimator to reduce the γ-ray flux reaching the chamber. The aperture (5 in.$\times$1.5 in.) defined a solid angle of 0.002 steradian at 68° to the incident protons. The arrangement yielded readable photographs at a beam intensity of $\sim 10^8$ protons per pulse, although the flux through the chamber was estimated to be $\sim 10^4$ neutrons. The latter fact points up the virtue of the technique employed.

The relevant primary data on 23 measured V^0 events, found in a run of 1200 pictures, are listed in Table I. We have considered various background effects which could possibly simulate V^0 events:

(1) Production of meson pairs in the gas by neutrons or photons, the nuclear recoil track being too short to observe. However, the number of neutrons above meson production threshold energy at 68° was expected to be quite small. This was verified experimentally by the fact that no negative prongs (i.e., π^- mesons) were observed to emerge from 1218 neutron-induced stars in the gas.

(2) Decay of π^0 mesons, produced in the gas without recoil, into the alternate mode $e^+e^-\gamma$. This is ruled out kinematically for 16 of the events. The argument in (1) also applies.

(3) Production of large-angle electron pairs in the gas by photons.

(4) Bremsstrahlung or scattering of backward-moving particles with consequent large-angle deflections.

These possibilities lead to the prediction of thousands of smaller angle events and to the necessity for large fluxes of backward-moving particles. Neither of these is observed. These arguments will be detailed in a more complete report. They lead to the conclusion that

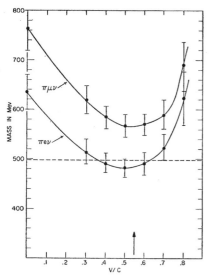

FIG. 1. Average calculated primary mass vs velocity of primary particle for assumed decay schemes $\pi^+e^-\nu^0$ and $\pi^+\mu^-\nu^0$. The arrow indicates the peak of the phase space spectra (reference 5). The vertical bars are average deviations of the mean.

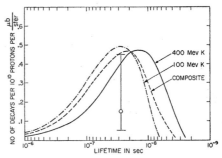

FIG. 2. Detection sensitivity for K mesons as function of lifetime. The composite curve is obtained with the spectra of reference 5. The point indicates the observed yield with a production cross section of ~ 20 μb/sterad.

the events listed in Table I are indeed examples of the disintegration of a long-lived neutral particle.

A preliminary analysis of the data yields some information on the properties of the new particle.

(1) All but three of the forty-six secondaries are determined to be lighter in mass than the K meson. None can be protons. We have assumed that all are pions, muons, or electrons. The identification of several of the decay products as pions or electrons is indicated in the table.

(2) We have considered various three-body decay schemes, motivated by the observed charged K-meson modes. In Fig. 1, we plot for assumed decay products $\pi^{\pm}e^{\mp}\nu^0$ and for $\pi^{\pm}\mu^{\mp}\nu^0$, the variation of the average computed mass (15 events were available) of the incoming primary as a function of its assumed velocity. Permutations of the relevant combinations of π's, μ's, e's, and ν's yield similar results. For example $\pi^+\mu^-\nu^0$, $\mu^+\pi^-\nu^0$, $\mu^+e^-\pi^0$, and $\pi^+e^-\pi^0$ are almost coincident. These graphs emphasize the conclusion that the resultant incoming velocity distribution is kinematically sensible only for primary masses near the K mass of 500 Mev.[5] One may also infer that, for a K mass primary, $\pi e\nu$ secondaries are more frequent than $\pi\mu\nu$ or, say, $\mu e\pi$.

(3) All but two of the events are kinematically inconsistent with a Λ^0-mass particle decaying into $\mu^{\pm}e^{\mp}N$ or $e^{\pm}e^{\mp}N$.[6]

(4) Figure 2 illustrates the detection sensitivity as a function of lifetime for a K-mass particle. Although the production cross section for K^0 mesons[7] has a large uncertainty, comparison with the observed yield serves to limit the lifetime to the range 10^{-6} sec$>\tau>3\times10^{-9}$ sec. The observed uniform distribution of events in the chamber, together with Fig. 1 also sets a lower limit: $\tau>1\times10^{-9}$ sec. If the lifetime is on the short side of above interval, then it is likely that many of the anomalous V^0's observed in cosmic rays are examples of this particle, and not alternate decay modes of the $\theta_1{}^0$.[8]

At the present stage of the investigation one may only conclude that Table I, Fig. 2, and Q^* plots are consistent with a K^0-type particle undergoing three-body decay. In this case the mode $\pi e\nu$ is probably prominent,[9] the mode $\pi\mu\nu$ and perhaps other combinations may exist but are more difficult to establish, and $\pi^+\pi^-\pi^0$ is relatively rare. Although the Gell-Mann–Pais predictions (I) and (II) have been confirmed, long lifetime and "anomalous" decay mode are not sufficient to identify the observed particle with $\theta_2{}^0$. In particular, a neutral τ meson, if three-pion decay has a small branching ratio, may have these properties. A much stronger test of particle mixtures must await the observation of nuclear interactions or of the striking interference effects which are also predicted by Pais and Piccioni,[2] Treiman and Sachs,[2] and Serber.[10]

The authors are indebted to Professor A. Pais whose elucidation of the theory directly stimulated this research. The effectiveness of Cosmotron staff collaboration is evidenced by the successful coincident operation of six magnets and the Cosmotron with the cloud chamber.

* Supported by the U. S. Atomic Energy Commission and the U. S. Atomic Energy Commission-Office of Naval Research Joint Program.
[1] M. Gell-Mann and A. Pais, Phys. Rev. 97, 1387 (1955).
[2] Further discussion of particle mixtures have been given by A. Pais and O. Piccioni, Phys. Rev. 100, 932 (1955); G. Snow, Phys. Rev. 103, 1111 (1956); S. Treiman and R. G. Sachs, Phys. Rev. 103, 1545 (1956); K. Case, Phys. Rev. 103, 1449 (1956).
[3] See Piccioni, Clark, Cool, Friedlander, and Kassner, Rev. Sci. Instr. 26, 232 (1955). The ejected beam is focused by a quadrupole magnet pair to a 3-in. diameter circle. Two bending magnets

were used to steer the beam onto the 1.5 in.×4 in.×5 in. long target.
 [4] Blumenfeld, Booth, Lederman, and Chinowsky, Phys. Rev. **102**, 1184 (1956).
 [5] We are grateful to R. Sternheimer for computing the energy spectrum of K mesons emitted at 68° under various assumptions as to the collision mechanism. These calculations yield similar spectra, all of which peak near 100 Mev. See Block, Harth, and Sternheimer, Phys. Rev. **100**, 324 (1956).
 [6] For example, one member of a Λ^0 parity doublet may have a long lifetime. See T. D. Lee and C. N. Yang, Phys. Rev. **102**, 290 (1956).
 [7] Collins, Fitch, and Sternheimer (private communication).
 [8] Kadyk, Trilling, Leighton, and Anderson, Bull. Am. Phys. Soc. Ser. II, **1**, 251 (1956). For a recent summary see Ballam, Grisaru, and Treiman, Phys. Rev. **101**, 1438 (1956).
 [9] Examples of this decay mode have been reported by Slaughter, Block, and Harth, Bull. Am. Phys. Soc. Ser. II, **1**, 186 (1956). A particularly clear event has been observed by the Ecole Polytechnique group. We are indebted to J. Tinlot and B. Gregory for this data and for helpful discussions on anomalous V^0's.
 [10] R. Serber (private communication).

Evidence for a Long-Lived Neutral Unstable Particle*

W. F. Fry, J. Schneps, and M. S. Swami

Department of Physics, University of Wisconsin, Madison, Wisconsin

(Received July 19, 1956)

DURING a systematic search for K^- mesons in a pellicle stack exposed to a channel of negative particles from the Berkeley Bevatron, four unusual events of the following nature were found: an unstable particle originated in the emulsion from a small star which was produced by a neutral particle. The channel was at 90° to a target bombarded by 6-Bev protons and defined a momentum of 280 Mev/c.

Since the events were of an unusual nature, each one will be described separately.

Event 1.—A $_\Lambda\text{He}^4$ hyperfragment originated from a star which also consisted of a π meson of 42 ± 12 Mev, two short recoil tracks, and a proton of 16.6 Mev. The star was produced by a neutral particle. The hyperfragment decayed from rest into a π^- meson, a proton, and a He^3 recoil. The binding of the Λ^0 particle was found to be 1.8 ± 0.6 Mev.[1] A drawing is shown in A of Fig. 1.

Event 2.—A negative K meson of 14 Mev was produced in a star which in addition had a low-energy electron and a nucleonic particle of 115 Mev. The K^- meson produced a star from rest consisting of a π meson, a hyperfragment which decayed nonmesonically, and nucleons. The nucleonic particle from the primary star left the stack and therefore the direction of its motion could not be established. However, the kinetic energy of this particle was not sufficient to produce a K^- meson if it were an incident Σ^- or Ξ^- hyperon; therefore we assume that it was an outgoing particle. This event is shown in B of Fig. 1.

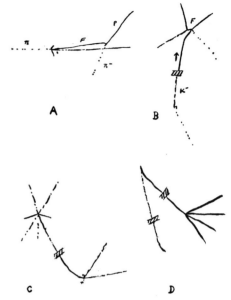

Fig. 1. Drawings of four events which were found in a pellicle stack as shown. A $_\Lambda\text{He}$ hyperfragment was produced in A, a K^- meson in B, probably a Σ^- hyperon in C, and a hyperfragment or a Σ^- hyperon in D.

Event 3.—A track, 730 microns long, from a seven-pronged star, has a two-pronged star associated with its ending. The primary star was produced by a neutral particle and has a visible kinetic energy of 160 Mev. The connecting particle had a charge of one and appeared to have stopped. The two nuclear particles from the secondary star had a total charge of two or three. In addition there are two tracks of low-energy electrons indicating that the connecting particle was captured. A mass measurement along the connecting track gave $(2200\pm750)m_e$. These facts indicate that the secondary star was produced by a Σ^- hyperon or possibly a Ξ^- hyperon or a K^- meson. A drawing of this event is shown in C of Fig. 1.

Event 4.—One particle from a small five-pronged star, which had no incoming particle, appears to have stopped and produced a one-prong star. The prong was most likely due to a proton of 9 Mev. The range of the connecting track is 230 microns. Gap counting showed that the secondary star was not the result of a scattering. Also it is possible to exclude the absorption of a slow π^- meson as its cause. The event is interpreted as a nonmesonic hydrogen hyperfragment decay or the capture of a negative hyperon or K meson. The event is shown in D of Fig. 1.

It is very improbable that these four events were produced by fast neutrons since the ratio of these events to energetic neutron stars in the stack is large ($\sim 1/10$). Also the flux of energetic neutrons at 90° to the target is expected to be small. They cannot be cosmic-ray events because their tracks can be followed through the stack, which was in a disassembled state both shortly before and after the machine exposure. These events can be interpreted as the nuclear interaction of a neutral unstable particle of the same "strangeness" as that of the K^- meson and the Λ and Σ hyperons.

The total energy in event 2 is too high to have been due to a neutral K from the charge exchange of a K^- meson in the beam. Furthermore, the expected number of such events, if due to charge exchange, is less than one per 20 cc of emulsion, whereas the density of these unusual events is probably greater than one per cc of emulsion.

Also it seems unlikely that the events could have been due to neutral θ's which were produced in the deflecting magnet because they would have had to have lived about 20 mean lifetimes.

These events can be explained by assuming that long-lived neutral K mesons were produced at the target with about the same frequency as the K^+ mesons. A small fraction of these neutral K mesons could have penetrated the shielding (about two feet of brass) between the plates and the target and then interacted in the pellicle stack. The lifetime of these particles must have been at least 10^{-8} sec. The existence of a long-lived neutral K meson was predicted by Gell-Mann and Pais.[2]

The authors are indebted to Dr. E. J. Lofgren for making possible the exposures to the Bevatron. Discussions with Dr. George Snow were stimulating and helpful.

---

* Supported in part by the U. S. Atomic Energy Commission and by the Graduate School from funds supplied by the Wisconsin Alumni Research Foundation.

[1] This hyperfragment has been described in detail (event 90) by Fry, Schneps, and Swami, Phys. Rev. **101**, 1526 (1956).

[2] M. Gell-Mann and A. Pais, Phys. Rev. **97**, 1387 (1955). Dr. G. A. Snow has pointed out the possibility of a long-lived τ^0 meson [Phys. Rev. **103**, 1111 (1956)].

REGENERATION AND MASS DIFFERENCE OF NEUTRAL K MESONS[*]

Francis Muller,[†] Robert W. Birge, William B. Fowler,[‡] Robert H. Good, Warner Hirsch,
Robert P. Matsen, Larry Oswald, Wilson M. Powell, and Howard S. White
Lawrence Radiation Laboratory, University of California, Berkeley, California

and

Oreste Piccioni
Brookhaven National Laboratory, Upton, New York
(Received March 29, 1960)

A very significant feature of the Gell-Mann–Pais particle mixture theory[1,2] is the regeneration of the $K1$ from the $K2$ neutral meson. We examine the three possible types of regeneration and give the results of an experiment that exhibits the expected transformations as demanded by the theory. The experiment also allows an estimate of the difference between the masses of $K1$ and $K2$.

One of the three types of regeneration has been described previously[3]: A plate inserted into a parallel beam of $K2$ particles produces a parallel beam of $K1$ particles. This phenomenon, which we will henceforth call transmission-regeneration, is in striking contrast with other known processes whereby a particle transforms into another one: a parallel beam of charged pions obviously cannot produce a parallel beam of neutral pions by interacting with a plate.

Here we point out another process that typically follows from the theory, namely the regeneration by diffraction. Because the $\overline{K}^0$ and the K^0 waves are diffracted by a nucleus with different amplitudes, the diffracted wave contains $K1$ as well as $K2$ particles. Thus $K1$ mesons are regenerated by a nucleus with a typical diffraction angular distribution.

Regeneration of $K1$ can also occur by interaction of $K2$ with single nucleons. The angular distribution of this nucleon-regeneration is broad, not essentially different from that obtained in K-nucleon scattering, and therefore it is not a crucial consequence of the particle mixture theory.

All three of these components will emerge from a plate traversed by a parallel beam of $K2$'s. The angular distribution should permit one to recognize each component separately.

Case[4] and Good[5] have shown that the intensity of the transmitted component is a very sensitive function of the mean life τ_1 of the $K1$ and of the difference δm between the masses of $K1$ and $K2$. The mass difference appears in the final expression because of the phase difference it introduces between the $K1$ and the $K2$ waves, an effect which was first noted by Serber[6] in connection with K^0 production. Moreover, Good pointed out that the intensities of both the transmitted and "scattered" component (in the forward direction) are proportional to $|f_{21}{}^0|^2$, $f_{21}{}^0$ being the amplitude of the regenerated $K1$, at zero angle, in a $K2$-nucleus collision. Good's "scattered" component must be identified with the diffracted component described above. Thus the intensity ratio of the transmitted wave to diffracted wave is a function only of δm and τ_1. We derive here in a more concise way the expression for this ratio.

The computation of the expected transmitted and diffracted intensities can be greatly simplified by neglecting, from the start, the regeneration of $K2$ from $K1$. As the number of $K1$'s is always less than one thousandth of the number of $K2$'s, this approximation is very good. We consider then a plane wave of $K2$ particles, of wavelength λ, crossing our plate, which contains N nuclei per cubic centimeter. If each nucleus produces $K1$'s with a forward amplitude $f_{21}{}^0$, an infinitesimal thickness dx of the plate at depth x

418

(x in the direction of the incoming $K2$ beam; $x = 0$ and $x = L$ denote the limits of the plate) produces a $K1$ wave amplitude $iN\lambda f_{21}{}^0 dx$ which arrives at the end of the plate with the amplitude

$$da_1 = iN\lambda f_{21}{}^0 dx \ \exp\left(-ik_2 x - ik_1(L-x) - \frac{L-x}{2v\gamma\tau_1} - \frac{L}{2u}\right).$$

Notice that the $K2$ wave has traveled to depth x before producing the $K1$ wave, which then travels from x to L; u is the collision mean free path, which is the same for $K1$ and $K2$, because both particles are a half-and-half mixture of K^0 and $\bar{K}^0$; v is the velocity of the particles; γ is the Lorentz factor; $\hbar k_1$ and $\hbar k_2$ are the momenta. Let us call $\Lambda = v\gamma\tau_1$ the decay mean free path of the $K1$'s and introduce the dimensionless quantities $l = L/\Lambda$ and $\delta = (m_2 - m_1)c^2/(\hbar/\tau_1)$. By integration with respect to x we obtain, for the transmitted intensity,

$$|a|^2 = \frac{4|f_{21}{}^0|^2 N^2 \Lambda^2 \lambda^2}{1+4\delta^2} |e^{-i\delta l} - e^{-l/2}|^2 e^{-L/u}. \quad (1)$$

On the other hand, the nuclei, incoherently from each other, regenerate $K1$'s by diffraction with a differential cross section $d\sigma_{21}/d\omega = |f_{21}|^2$. The number of diffraction-regenerated $K1$'s in the infinitesimal thickness dx at x, in the forward direction, surviving through the thickness $L-x$ is

$$d\left(\frac{dn_1}{d\omega}\right)^0 = |f_{21}{}^0|^2 N \exp\left(-\frac{L-x}{v\gamma\tau_1} - \frac{L}{u}\right) dx,$$

which is integrated to give

$$\left(\frac{dn_1}{d\omega}\right)^0 = |f_{21}{}^0|^2 N\Lambda(1-e^{-l})e^{-L/u}. \quad (2)$$

The ratio between (1) and (2) is

$$R = 4N\Lambda\lambda^2 |e^{-i\delta l} - e^{-l/2}|^2/[(1-e^{-l})(1+4\delta^2)]. \quad (3)$$

To observe these regeneration processes, we have inserted a plate in the Berkeley 30-inch propane chamber. The chamber was placed in a beam of $K3$ particles which traversed the instrument lengthwise and perpendicular to the plate. The experimental setup will be described in more detail in a later article. Here we give only a brief description.

A beam of 1.1-Bev/c negative pions impinged on a five-foot hydrogen target. The 670-Mev/c K^0 produced in the target travelled a distance of 22.5 feet before arriving at the 30-inch propane chamber, so that approximately one $K2$ crossed the chamber per 10^{11} protons in the Bevatron beam. About 200 000 pictures were taken, half of them with a 1.5-inch iron plate to enhance the diffracted wave relative to the transmitted wave, and the other half with a 6-inch iron plate which yields an intense transmitted wave.

We limited the analysis to those two-prong events in which the positive prong could be recognized to be a meson on the basis of ionization and momentum. We also required that the decay occur within two mean lives from the plate and that the primary momentum be equal, within the errors, to the beam momentum. The $Q(\pi,\pi)$ distribution of these selected events shows a marked peak around the expected value of 220 Mev, which fact proves the regeneration of $K1$. As a further selection, we keep only those events for which Q differs by no more than 1.4 standard deviations from the peak value.

By measuring the vector momenta of the two prongs, we determine the angle θ between the $K1$ and the incident $K2$ beams, within an error of about 2 degrees.

The angular distributions are shown in Fig. 1. The diffraction curve has been computed with the optical model method[7] using the known cross sections for K^+ and K^- with protons and nuclei.[8] The curve is quite close to the black-sphere distribution. Figure 1 clearly shows a diffraction component and a superimposed transmission peak. Most of the transmission peak is confined to angles smaller than 2.5 degrees ($\cos\theta > 0.999$), which is just what we expect from an infinitely narrow peak measured with our errors. The mere presence of such a large transmission peak is a proof that the mass difference is smaller than, say, $5\hbar/\tau_1$.

Referring to the data for the 6-inch plate, 29 events occur in the interval for $\cos\theta$ between 0.998 (3.5°) and 1, which should contain the total number T of the transmitted $K1$'s and a part D of the diffracted ones. Knowing the diffraction angular distribution we can compute D from the 31 particles in the 0.980 to 0.998 interval, assuming that that region contains only diffracted $K1$'s. A nucleonic background will actually be present in this interval, which we ignore for the moment. We thus obtain $D = 12$, which gives $T = 17$. For comparison with formula (3), we note that $(dn_1/d\omega) = 1.18(D/\omega)$, where the factor 1.18 is equal to the ratio of the intensity at zero degrees to the average intensity within ω, ω being the solid angle in the peak interval, that is, $2\pi \times 0.002$. It is convenient to compare $T/1.18D$

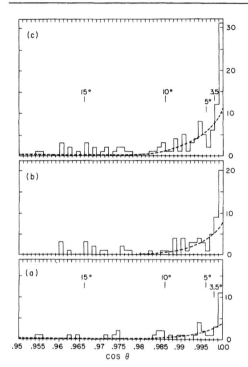

FIG. 1. Histograms of number of $K1$ decay events per 0.001 interval of $\cos\theta$ (θ is the angle between the direction of the primary $K2$ beam and the regenerated $K1$). (a) Data for the 1.5-inch plate; (b) data for the 6-inch plate; (c) combined data for the two plates. The curves are diffraction angular distributions normalized in the 0.980 to 0.998 interval for $\cos\theta$.

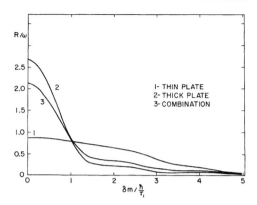

FIG. 2. Calculated intensity of the forward regenerated $K1$'s versus the $K1-K2$ mass difference. The ordinate R/ω is the ratio of the transmission-regenerated $K1$'s to the diffraction-regenerated $K1$'s in the interval $\cos\theta > 0.998$.

with R/ω, which is plotted in Fig. 2 versus the mass difference, in units of $\hbar/\tau_1$. We obtain $T/1.18D = 1.2 \pm 0.53$, which gives $\delta m = 0.85^{+0.4}_{-0.25}$. With a probability of 95%, $\delta m < 1.4$.

In the same way, we find from the thin plate alone $\delta m = 0$ ($\delta m < 4.5$ with 95% probability) and from the combination of thin and thick plate $\delta m = 0.85^{+0.3}_{-0.25}$ ($\delta m < 1.5$ with 95% probability).

If we correct for nucleonic or any other background, we would obtain a larger value for R, hence a smaller value for δm. For instance, assuming a uniform background from $\cos\theta = 0.96$ to $\cos\theta = 1$, the compounded data for both plates yield $\delta m = 0$ ($\delta m < 1.1$ with 95% probability). In view of a remark by Okun' and Pontecorvo,[9] this result indicates that decay rates for $\Delta S = 2$ are 10^5 times slower than for $\Delta S = 1$.

We are grateful to many people who generously contributed to this experiment, particularly Edward J. Lofgren, Myron L. Good, Richard L. Lander, Robert E. Lanou, Marian N. Whitehead, Roy Kerth, and Frank T. Solmitz. The pictures have been scanned by J. Peter Berge, Karl Brunstein, Layton Linch, Mrs. Glennette Anneson, Mrs. Rokalana Gamow, and Mrs. Ottilie Oldenbusch.

---

*Work performed under the auspices of the U. S. Atomic Energy Commission.

†On leave of absence from Ecole Polytechnique, Paris and Marine Nationale Francaise.

‡Now at Brookhaven National Laboratory, Upton, New York.

[1] M. Gell-Mann and A. Pais, Phys. Rev. 97, 1387 (1955).

[2] T. D. Lee, R. Oehme, and C. N. Yang, Phys. Rev. 106, 340 (1957).

[3] A. Pais and O. Piccioni, Phys. Rev. 100, 1487 (1955).

[4] K. M. Case, Phys. Rev. 103, 1449 (1956).

[5] M. L. Good, Phys. Rev. 106, 591 (1957); Phys. Rev. 110, 550 (1958).

[6] R. Serber (private communication), quoted in 3.

[7] S. Fernbach, R. Serber, and T. B. Taylor, Phys. Rev. 75, 1352 (1949).

[8] Total cross sections measured by H. C. Burrowes, D. O. Caldwell, D. H. Frisch, D. A. Hill, D. M. Ritson, and R. A. Schluter, Phys. Rev. Letters 2, 117 (1959). See also report by L. W. Alvarez at the

Ninth Annual International Conference on High-Energy Physics, Kiev, 1959 (unpublished). For K^+ we used the optical model potential of 28 Mev-i 19 Mev. We are indebted to B. Sechi Zorn and G. T. Zorn for communication of their unpublished results, and to L. S. Rodberg and R. M. Thaler for showing to us their unpublished article on K^+ scattering in emulsions.

[9]L. Okun' and B. Pontecorvo, J. Exptl. Theoret. Phys. (U.S.S.R.) 32, 1587 (1957) [translation: Soviet Phys. -JETP 5, 1297 (1957)].

EVIDENCE FOR THE 2π DECAY OF THE K_2^0 MESON*†

J. H. Christenson, J. W. Cronin,‡ V. L. Fitch,‡ and R. Turlay§
Princeton University, Princeton, New Jersey
(Received 10 July 1964)

This Letter reports the results of experimental studies designed to search for the 2π decay of the K_2^0 meson. Several previous experiments have served[1,2] to set an upper limit of $1/300$ for the fraction of K_2^0's which decay into two charged pions. The present experiment, using spark chamber techniques, proposed to extend this limit.

In this measurement, K_2^0 mesons were produced at the Brookhaven AGS in an internal Be target bombarded by 30-BeV protons. A neutral beam was defined at 30 degrees relative to the circulating protons by a $1\frac{1}{2}$-in. $\times$ $1\frac{1}{2}$-in. $\times$ 48-in. collimator at an average distance of 14.5 ft. from the internal target. This collimator was followed by a sweeping magnet of 512 kG-in. at ~20 ft. and a 6-in. $\times$ 6-in. $\times$ 48-in. collimator at 55 ft. A $1\frac{1}{2}$-in. thickness of Pb was placed in front of the first collimator to attenuate the gamma rays in the beam.

The experimental layout is shown in relation to the beam in Fig. 1. The detector for the decay products consisted of two spectrometers each composed of two spark chambers for track delineation separated by a magnetic field of 178 kG-in. The axis of each spectrometer was in the horizontal plane and each subtended an average solid angle of 0.7×10^{-2} steradians. The spark chambers were triggered on a coincidence between water Cherenkov and scintillation counters positioned immediately behind the spectrometers. When coherent K_1^0 regeneration in solid materials was being studied, an anticoincidence counter was placed immediately behind the regenerator. To minimize interactions K_2^0 decays were observed from a volume of He gas at nearly STP.

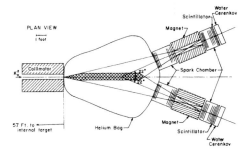

FIG. 1. Plan view of the detector arrangement.

The analysis program computed the vector momentum of each charged particle observed in the decay and the invariant mass, m^*, assuming each charged particle had the mass of the charged pion. In this detector the K_{e3} decay leads to a distribution in m^* ranging from 280 MeV to ~536 MeV; the $K_{\mu 3}$, from 280 to ~516; and the $K_{\pi 3}$, from 280 to 363 MeV. We emphasize that m^* equal to the K^0 mass is not a preferred result when the three-body decays are analyzed in this way. In addition, the vector sum of the two momenta and the angle, θ, between it and the direction of the K_2^0 beam were determined. This angle should be zero for two-body decay and is, in general, different from zero for three-body decays.

An important calibration of the apparatus and data reduction system was afforded by observing the decays of K_1^0 mesons produced by coherent regeneration in 43 gm/cm^2 of tungsten. Since the K_1^0 mesons produced by coherent regeneration have the same momentum and direction as the K_2^0 beam, the K_1^0 decay simulates the direct decay of the K_2^0 into two pions. The regenerator was successively placed at intervals of 11 in. along the region of the beam sensed by the detector to approximate the spatial distribution of the K_2^0's. The K_1^0 vector momenta peaked about the forward direction with a standard deviation of 3.4 ± 0.3 milliradians. The mass distribution of these events was fitted to a Gaussian with an average mass 498.1 ± 0.4 MeV and standard deviation of 3.6 ± 0.2 MeV. The mean momentum of the K_1^0 decays was found to be 1100 MeV/c. At this momentum the beam region sensed by the detector was 300 K_1^0 decay lengths from the target.

For the K_2^0 decays in He gas, the experimental distribution in m^* is shown in Fig. 2(a). It is compared in the figure with the results of a Monte Carlo calculation which takes into account the nature of the interaction and the form factors involved in the decay, coupled with the detection efficiency of the apparatus. The computed curve shown in Fig. 2(a) is for a vector interaction, form-factor ratio $f^-/f^+ = 0.5$, and relative abundance 0.47, 0.37, and 0.16 for the K_{e3}, $K_{\mu 3}$, and $K_{\pi 3}$, respectively.[3] The scalar interaction has been computed as well as the vector interaction

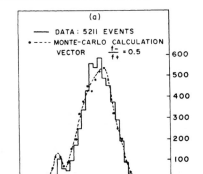

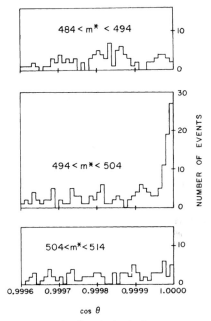

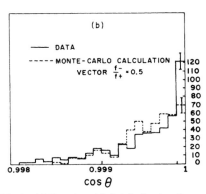

FIG. 2. (a) Experimental distribution in m^* compared with Monte Carlo calculation. The calculated distribution is normalized to the total number of observed events. (b) Angular distribution of those events in the range $490 < m^* < 510$ MeV. The calculated curve is normalized to the number of events in the complete sample.

FIG. 3. Angular distribution in three mass ranges for events with $\cos\theta > 0.9995$.

with a form-factor ratio $f^-/f^+ = -6.6$. The data are not sensitive to the choice of form factors but do discriminate against the scalar interaction.

Figure 2(b) shows the distribution in $\cos\theta$ for those events which fall in the mass range from 490 to 510 MeV together with the corresponding result from the Monte Carlo calculation. Those events within a restricted angular range ($\cos\theta > 0.9995$) were remeasured on a somewhat more precise measuring machine and recomputed using an independent computer program. The results of these two analyses are the same within the respective resolutions. Figure 3 shows the results from the more accurate measuring machine. The angular distribution from three mass ranges are shown; one above, one below, and one encompassing the mass of the neutral K meson.

The average of the distribution of masses of those events in Fig. 3 with $\cos\theta > 0.99999$ is found to be 499.1 ± 0.8 MeV. A corresponding calculation has been made for the tungsten data resulting in a mean mass of 498.1 ± 0.4. The difference is 1.0 ± 0.9 MeV. Alternately we may take the mass of the K^0 to be known and compute the mass of the secondaries for two-body decay. Again restricting our attention to those events with $\cos\theta > 0.99999$ and assuming one of the secondaries to be a pion, the mass of the other particle is determined to be 137.4 ± 1.8. Fitted to a Gaussian shape the forward peak in Fig. 3 has a standard deviation of 4.0 ± 0.7 milliradians to be compared with 3.4 ± 0.3 milliradians for the tungsten. The events from the He gas appear identical with those from the coherent regeneration in tungsten in both mass and angular spread.

The relative efficiency for detection of the three-body K_2^0 decays compared to that for decay to two pions is 0.23. We obtain 45 ± 9 events in

139

the forward peak after subtraction of background out of a total corrected sample of 22 700 K_2^0 decays.

Data taken with a hydrogen target in the beam also show evidence of a forward peak in the $\cos\theta$ distribution. After subtraction of background, 45 ± 10 events are observed in the forward peak at the K^0 mass. We estimate that ~10 events can be expected from coherent regeneration. The number of events remaining (35) is entirely consistent with the decay data when the relative target volumes and integrated beam intensities are taken into account. This number is substantially smaller (by more than a factor of 15) than one would expect on the basis of the data of Adair et al.[4]

We have examined many possibilities which might lead to a pronounced forward peak in the angular distribution at the K^0 mass. These include the following:

(i) K_1^0 coherent regeneration. In the He gas it is computed to be too small by a factor of ~10^6 to account for the effect observed, assuming reasonable scattering amplitudes. Anomalously large scattering amplitudes would presumably lead to exaggerated effects in liquid H_2 which are not observed. The walls of the He bag are outside the sensitive volume of the detector. The spatial distribution of the forward events is the same as that for the regular K_2^0 decays which eliminates the possibility of regeneration having occurred in the collimator.

(ii) $K_{\mu 3}$ or K_{e3} decay. A spectrum can be constructed to reproduce the observed data. It requires the preferential emission of the neutrino within a narrow band of energy, ± 4 MeV, centered at 17 ± 2 MeV ($K_{\mu 3}$) or 39 ± 2 MeV (K_{e3}). This must be coupled with an appropriate angular correlation to produce the forward peak. There appears to be no reasonable mechanism which can produce such a spectrum.

(iii) Decay into $\pi^+\pi^-\gamma$. To produce the highly singular behavior shown in Fig. 3 it would be necessary for the γ ray to have an average energy of less than 1 MeV with the available energy extending to 209 MeV. We know of no physical process which would accomplish this.

We would conclude therefore that K_2^0 decays to two pions with a branching ratio $R = (K_2 \to \pi^+ + \pi^-)/(K_2^0 \to \text{all charged modes}) = (2.0 \pm 0.4) \times 10^{-3}$ where the error is the standard deviation. As emphasized above, any alternate explanation of the effect requires highly nonphysical behavior of the three-body decays of the K_2^0. The presence of a two-pion decay mode implies that the K_2^0 meson is not a pure eigenstate of CP. Expressed as $K_2^0 = 2^{-1/2}[(K_0 - \bar{K}_0) + \epsilon(K_0 + \bar{K}_0)]$ then $|\epsilon|^2 \cong R_T \tau_1 \tau_2$ where τ_1 and τ_2 are the K_1^0 and K_2^0 mean lives and R_T is the branching ratio including decay to two π^0. Using $R_T = \tfrac{3}{2} R$ and the branching ratio quoted above, $|\epsilon| \cong 2.3 \times 10^{-3}$.

We are grateful for the full cooperation of the staff of the Brookhaven National Laboratory. We wish to thank Alan Clark for one of the computer analysis programs. R. Turlay wishes to thank the Elementary Particles Laboratory at Princeton University for its hospitality.

---

*Work supported by the U. S. Office of Naval Research.

†This work made use of computer facilities supported in part by National Science Foundation grant.

‡A. P. Sloan Foundation Fellow.

§On leave from Laboratoire de Physique Corpusculaire à Haute Energie, Centre d'Etudes Nucléaires, Saclay, France.

[1]M. Bardon, K. Lande, L. M. Lederman, and W. Chinowsky, Ann. Phys. (N.Y.) 5, 156 (1958).

[2]D. Neagu, E. O. Okonov, N. I. Petrov, A. M. Rosanova, and V. A. Rusakov, Phys. Rev. Letters 6, 552 (1961).

[3]D. Luers, I. S. Mittra, W. J. Willis, and S. S. Yamamoto, Phys. Rev. 133, B1276 (1964).

[4]R. Adair, W. Chinowsky, R. Crittenden, L. Leipuner, B. Musgrave, and F. Shively, Phys. Rev. 132, 2285 (1963).

8
The Structure of the Nucleon

Elastic and deep inelastic scattering from nucleons, 1956–1973.

Hadronic scattering experiments produced extensive and rich data revealing resonances and regularities of cross sections. While the quark model provided a firm basis for classifying the particles and resonances, the scattering cross sections were less easily interpreted. The early studies of strong interactions indicated that the couplings of the particles were large. This precluded the straightforward use of perturbation theory. While alternative approaches have yielded some important results, it is still true that even processes as basic as elastic proton–proton scattering are beyond our ability to explain in detail. In contradistinction, scattering of electrons by protons and neutrons is open to direct interpretation.

For the scattering of an electron by a proton it is a good approximation to assume that the interaction is due to the exchange of a single virtual photon. The small corrections to this approximation may be calculated if necessary. Each coupling of the photon gives a factor of e in the scattering amplitude, so a virtual photon's two couplings typically provides a factor $\alpha = e^2/4\pi \approx 1/137$. It is this small number that makes the approximation a good one.

The scattering of relativistic electrons ($E \gg m_e$) by a known charge distribution can be calculated using the standard methods of quantum mechanics. If the electron were spinless and scattered from a static point charge, the cross section would be given by the Rutherford formula:

$$\frac{d\sigma}{d\Omega} = \frac{\alpha^2}{4E^2 \sin^4 \tfrac{1}{2}\theta}, \tag{8.1}$$

where E is the energy of the incident relativistic electron and θ is its scattering angle in the laboratory. Taking into account the electron's spin gives the Mott cross section:

$$\frac{d\sigma}{d\Omega} = \frac{\alpha^2 \cos^2 \tfrac{1}{2}\theta}{4E^2 \sin^4 \tfrac{1}{2}\theta}. \tag{8.2}$$

If the electron is scattered by a static source, its final energy, E', is the same as the incident energy E, and the four-momentum transfer squared is $q^2 = -4E^2 \sin^2 \tfrac{1}{2}\theta$. If the target has finite mass, M, and thus recoils, then for elastic scattering

$$E' = \frac{E}{1 + \frac{2E}{M}\sin^2\tfrac{1}{2}\theta}, \tag{8.3}$$

$$q^2 = -4EE' \sin^2\tfrac{1}{2}\theta. \tag{8.4}$$

The elastic scattering of an electron by a pointlike Dirac particle of mass M has a cross section

$$\frac{d\sigma}{d\Omega} = \frac{\alpha^2 \cos^2\tfrac{1}{2}\theta}{4E^2 \sin^4\tfrac{1}{2}\theta} \cdot \frac{E'}{E}\left[1 - \frac{q^2}{2M^2}\tan^2\tfrac{1}{2}\theta\right], \tag{8.5}$$

which reduces to the Mott cross section as the target mass increases.

These simple results do not apply if the charge distribution of the target has some spatial extent. In the case of elastic scattering from a fixed charge distribution, $\rho(r)$, the scattering amplitude is modified by a form factor

$$F(q^2) = \int d^3 r\, e^{i\mathbf{q}\cdot\mathbf{r}} \rho(r), \tag{8.6}$$

so the Rutherford or Mott cross section would be multiplied by the factor $|F(q^2)|^2$. Since $\int d^3 r \rho(r) = 1$, the form factor reduces to unity for zero momentum transfer.

A relativistic treatment of the scattering of electrons by protons is obtained by writing the scattering amplitude as a product of three factors:

$$\mathcal{M} = \frac{4\pi\alpha}{q^2} J_\mu^{electron}(q) J^{\mu\, proton}(q), \tag{8.7}$$

where q is the four-momentum exchanged between the electron and the proton. The factor $1/q^2$ arises from the exchange of the virtual photon between the two. The current due to the electron is

$$J_\mu^{electron} = \bar{u}(k_f) \gamma_\mu u(k_i) \tag{8.8}$$

where k_i and k_f are the initial and final electron momenta and $\bar{u}$ and u are Dirac spinors as described in Chapter 6. The electromagnetic current for the proton involves two form factors,

$$J_\mu^{proton} = \bar{u}(p_f)\left[F_1(q^2)\gamma_\mu + i\frac{q^\nu \sigma_{\mu\nu} \kappa}{2M} F_2(q^2)\right] u(p_i). \tag{8.9}$$

Here p_i and p_f are the initial and final proton momenta and $q = k_i - k_f = p_f - p_i$ is the four-momentum transfer. The second term, proportional to the form factor $F_2(q^2)$, is the anomalous magnetic moment coupling and $\kappa = 1.79$ is the anomalous magnetic moment of the proton in units of the nuclear magneton, $e\hbar/(2Mc)$. The form factors, $F_1(q^2)$ and $F_2(q^2)$, are the analogs of $F(q^2)$ in the discussion above, and $F_1(0) = F_2(0) = 1$. If the proton were a pointlike Dirac particle like the electron, we would have instead $F_1(q^2) = 1$ and $\kappa F_2(q^2) = 0$. For a neutron, since the total charge is zero, $F_1(0) = 0$. The value of κ for the neutron is -1.91.

From these currents the differential cross section for elastic electron–proton scattering can be calculated in terms of the form factors. The result is known as the Rosenbluth formula:

$$\frac{d\sigma}{d\Omega} = \frac{\alpha^2 \cos^2 \frac{1}{2}\theta}{4E^2 \sin^4 \frac{1}{2}\theta} \cdot \frac{E'}{E} \cdot \left[\left(F_1^2 + \frac{\kappa^2 Q^2}{4M^2} F_2^2 \right) + \frac{Q^2}{2M^2} (F_1 + \kappa F_2)^2 \tan^2 \frac{1}{2}\theta \right], \quad (8.10)$$

where θ is the scattering angle of the electron in the laboratory and E is its initial energy. We have written Q^2 for $-q^2$, so Q^2 is positive.

The Rosenbluth formula follows from the assumption that a single photon is exchanged between the electron and the proton. All of our ignorance is subsumed in the two form factors, $F_1(Q^2)$ and $F_2(Q^2)$. The formula can be tested by multiplying the observed cross section by $(E^3/E') \sin^2 \frac{1}{2}\theta \tan^2 \frac{1}{2}\theta$ and plotting the result at fixed Q^2 as a function of $\tan^2 \frac{1}{2}\theta$. The result should be a straight line.

Elastic electron–proton scattering was measured by McAllister and Hofstadter using 188-MeV electrons **(Ref. 8.1)** produced by a linear accelerator at Stanford. The electrons scattered from a hydrogen target into a spectrometer that could be rotated around the interaction region.

The experiment was able to determine the root-mean-square charge radius of the proton by measuring the form factors at low momentum transfer. In this region, we can expand

$$F(q^2) = \int d^3 r \rho(r) \exp(i\mathbf{q} \cdot \mathbf{r})$$

$$= \int d^3 r \rho(r) [1 + i\mathbf{q} \cdot \mathbf{r} - (1/2)(\mathbf{q} \cdot \mathbf{r})^2 \cdots]$$

$$= 1 - \frac{q^2}{6} <r^2> \cdots \quad (8.11)$$

Assuming the same $<r^2>$ applied to both form factors, McAllister and Hofstadter found $<r^2>^{1/2} = 0.74 \pm 0.24$ fm.

Form factors exist as well for excitation processes like $ep \to e\Delta(1232)$. The number of form factors depends on the initial and final spins. The form factors are expected generally to decrease with momentum transfer, reflecting the spread in the charge and current distributions of the initial and final particles.

In the late 1960s, under the leadership of "Pief" Panofsky, the Stanford Linear Accelerator Center, SLAC, opened a vast new energy domain for exploration. The two-mile long accelerator produced electrons with energies up to about 18 GeV. The scattered electrons were detected and measured by very large magnetic spectrometers. At these high energies,

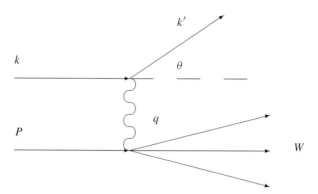

Figure 8.1. The kinematics of deep inelastic lepton–nucleon scattering. The incident lepton and proton have four-momenta k and P, respectively. The scattered lepton has four-momentum $k' = k - q$. The mass squared of the produced hadronic system is $W^2 = (P+q)^2$. The fundamental variables are $Q^2 = -q^2 = 4EE' \sin^2\frac{1}{2}\theta$ and $\nu = E - E'$, where E and E' are the initial and final lepton energies in the lab, and θ is the lab scattering angle of the lepton. The mass of the nucleon is M so $q \cdot P = (k - k') \cdot P = M\nu$, and $W^2 = M^2 + 2M\nu - Q^2$.

much of the scattering was inelastic, typically $ep \to ep\pi\pi...$ or $ep \to en\pi\pi...$. When the scattering is not elastic, the energy and direction of the scattered electron are independent variables, unlike the elastic scattering situation. From careful measurements of the direction, specified by a solid angle element $d\Omega$, and the energy E' of the scattered electron, the four-momentum transfer can be calculated. In this way, the differential cross section, $d\sigma/d\Omega dE'$ is determined as a function of E' and Q^2. The outgoing hadrons were generally not detected. The kinematics are shown in Figure 8.1.

A SLAC–MIT group (**Ref. 8.2**) scattered electrons from a hydrogen target and detected the outgoing electrons in a large magnetic spectrometer set at angles $\theta = 6°$ and $10°$. The scattered electrons' momenta were measured to 0.1%, and the spectrometer accepted a momentum interval $\Delta p/p = 3.5\%$. The potential background produced by charged pions entering the spectrometers was suppressed by observing the electron showers.

As expected, the data showed peaks when the mass W of the produced hadronic system corresponded to the mass of one of the resonances in the sequence N^* ($I = 1/2$ nonstrange baryons) or Δ ($I = 3/2$ nonstrange baryons). Each resonance showed the expected behavior as a function of Q^2. The production fell with increasing momentum transfer. What was surprising was that for W values beyond the resonances, the cross section did not fall with increasing Q^2.

Just as it is possible to write down a most general expression for the electromagnetic current of a proton for elastic scattering, it is possible to write down a general expression for the differential cross section measured in inelastic electron scattering when only the electron is measured in the final state. This expression depends on two functions, W_1 and W_2. These *structure functions* depend on two variables, ν, the energy lost by the electron in the laboratory, and Q^2. The full expression for the differential cross section is

$$\frac{d\sigma}{d\Omega dE'} = \frac{\alpha^2 \cos^2\frac{1}{2}\theta}{4E^2 \sin^4\frac{1}{2}\theta} \left[W_2 + 2W_1 \tan^2\frac{1}{2}\theta \right]. \tag{8.12}$$

This expression contains the Mott cross section as a factor and is analogous to the Rosenbluth formula. It follows from the assumption of single photon exchange and isolates the unknown physics in two functions, W_1 and W_2. Here, however, these are functions of two variables, ν and Q^2, not just one. In contrast, for elastic scattering, $(P+q)^2 = M^2$ so the two variables are not independent but rather are related by $Q^2 = 2M\nu$.

To determine W_1 and W_2 separately it is necessary to measure the differential cross section at two values of E' and θ that correspond to the same values of ν and Q^2. This is possible by varying the incident energy, E. At small values of θ, W_2 dominates, so it is most convenient to focus on this quantity.

The most important result of the experiment at SLAC was the discovery that νW_2 did not fall with increasing Q^2, but instead tended to a value that depended on the single variable $\omega = 2M\nu/Q^2$ (**Ref. 8.3**). This behavior, termed "scaling," had been anticipated first by Bjorken on the basis of a very complex study. By 1967, Bjorken was examining deep inelastic scattering by imagining the nucleon to be composed of pointlike quarks.

In an independent effort, Feynman had concluded from his analysis of hadronic collisions, that the proton ought to be composed of pointlike constituents, "partons" he called them. They shared the total momentum of the proton by taking up variable fractions, x, of that momentum. The probability of a parton carrying a fraction between x and $x + dx$ was written $f(x)dx$. The essential feature was that the function $f(x)$ was not to depend on the process at hand nor the energy of the proton, but was intrinsic to the proton so long as the proton had a large momentum. It was natural to assume that the partons were, in fact, quarks. There would not be just three quarks in a proton because in addition there could be many quark–antiquark pairs. The distribution functions for the various quarks were indicated by $u(x), d(x), \bar{u}(x)$, etc. Since the momenta had to add up to the proton's momentum, there was a constraint

$$\int dx\, x[u(x) + \bar{u}(x) + d(x) + \bar{d}(x) + ...] = 1. \tag{8.13}$$

As we shall see later, there is also a contribution from the uncharged constituents in the nucleon. In order for the quantum numbers of the proton to come out correctly, other conditions had to be satisfied:

$$\int dx[u(x) - \bar{u}(x)] = 2,$$

$$\int dx[d(x) - \bar{d}(x)] = 1,$$

$$\int dx[s(x) - \bar{s}(x)] = 0. \tag{8.14}$$

These replaced the statement that the proton was composed of two u quarks and a d quark. Thus in Feynman's model these "valence" quarks were supplemented by a "sea" of quark–antiquark pairs.

The combination of Bjorken's and Feynman's studies was a perfect explanation of "scaling," *i.e.* the dependence of νW_2 on the quantity ω alone. If the quark-partons were treated as real particles that had to be on-shell (that is, satisfied the relation $p^2 = E^2 - \mathbf{p}^2 = m^2$) both before and after being scattered by the virtual photon, then $p_f^2 = (p_i + q)^2 = (xP + q)^2 \approx 0$ if the masses of the quarks and the proton could be ignored, as seemed reasonable for very high energy collisions. From this followed

$$Q^2 = 2xP \cdot q = 2xM\nu. \tag{8.15}$$

This meant that the fraction x of the proton's momentum carried by the struck quark was simply the reciprocal of ω, the variable singled out in the experiment at SLAC. If the probability of there being a quark with momentum fraction x did not depend on the details of the event, scaling would follow, provided the scattering could be viewed as the incoherent sum of the scattering by the individual partons.

The precise connection between the parton distributions and the structure functions can be obtained by expressing the cross sections in terms of the Lorentz invariant variables $s = 2ME$, $x = Q^2/2M\nu$ and $y = \nu/E$. It is traditional to write $MW_1 = F_1$ and $\nu W_2 = F_2$. The dimensionless function F_1 and F_2, which must not be confused with the form factors of elastic scattering, are thus nominally functions of both x and Q^2. Substitution into the formula defining W_1 and W_2 gives

$$\frac{d\sigma}{dx\,dy} = \frac{4\pi\alpha^2 s}{Q^4} \left\{ \frac{1}{2}[1 + (1-y)^2]\, 2xF_1 + (1-y)(F_2 - 2xF_1) - \frac{M}{2E} xy F_2 \right\}. \tag{8.16}$$

This can be compared with the cross section for the scattering of an electron by a pointlike Dirac particle of unit charge carrying a fraction x of the proton's momentum. The cross section, which can be derived from the cross section given above for an electron on a pointlike Dirac particle, is

$$\frac{d\sigma}{dy} = \frac{4\pi\alpha^2 xs}{Q^4} \left\{ \frac{1}{2}[1 + (1-y)^2] - \frac{M}{2E} xy \right\}. \tag{8.17}$$

By comparing the results, we deduce the values of F_1 and F_2:

$$F_1 \equiv MW_1 = \frac{1}{2}\left[\frac{4}{9}u(x) + \frac{1}{9}d(x) + \frac{4}{9}\bar{u}(x) + \frac{1}{9}\bar{d}(x) + \cdots \right], \tag{8.18}$$

$$F_2 \equiv \nu W_2 = x\left[\frac{4}{9}u(x) + \frac{1}{9}d(x) + \frac{4}{9}\bar{u}(x) + \frac{1}{9}\bar{d}(x) + \cdots \right], \tag{8.19}$$

where the factors 4/9 and 1/9 arise as the squares of the quark charges. The connection $F_2 = 2xF_1$, known as the Callan–Gross relation, is a consequence of taking the partons to be pointlike Dirac particles. The absence of Q^2 dependence in F_1 and F_2 is the manifestation of scaling. With this stunningly simple formula, deep inelastic electron scattering becomes a powerful probe of the interior of the proton.

The simple parton picture was expected by Feynman to apply to very high energies. He reasoned that at high energies time dilation would cause the interactions between the partons to appear less frequent so that it would be a good approximation to ignore these interactions. Thus deep inelastic scattering could be regarded as the incoherent sum of the interactions with the individual partons.

A few years after these developments, important advances were made in understanding the theory of quantum chromodynamics (QCD). In this theory the interactions between quarks are the result of the exchange of vector particles called gluons. In many ways the theory is analogous to ordinary electrodynamics.

QCD finds very different behavior for quarks and gluons at short and long distances. Unlike the behavior of electric forces, the force between a quark and an antiquark does not decrease as their separation increases, but approaches a constant. Thus it takes an infinite amount of energy to separate them completely. Conversely, at short distances, the forces become weaker. It is the short-distance behavior that is probed in deep inelastic scattering, and thus QCD confirms Feynman's picture of non-interacting partons as the constituents of the proton.

Of course, the interactions between the quarks only decrease and do not disappear at short distances. As a result, the "kindergarten" parton model described above is only approximate. The quark and gluon distributions are weakly functions of Q^2 as well as x and scaling is only approximately satisfied.

This phenomenon can be understood by analogy with bremsstrahlung as described in Chapter 2. When an electron scatters from an electromagnetic field, it emits photons and the greater the scattering, the more bremsstrahlung there is. When a quark scatters, it emits gluons and some of its momentum is given to the gluons. As the momentum transfer is increased, the fraction of its momentum lost to gluons increases. Thus a quark with momentum fraction x at some low value of Q^2 becomes a quark with momentum fraction $x - x'$ and a gluon with momentum fraction x' at some higher value of Q^2. Thus for large values of x, $u(x, Q^2)$ falls with increasing Q^2. For low values of x, $u(x, Q^2)$ may increase because quarks with higher x may feed down quarks to it.

The parton model makes analogous predictions for deep inelastic neutrino scattering. Since the source of neutrino beams are the decays $\pi \to \mu\nu$ and $K \to \mu\nu$, ν_μ greatly dominate over ν_e (see Chapter 6). Thus in deep inelastic neutrino scattering by nucleons, one observes

$$\nu_\mu + \text{nucleon} \to \mu^- + \text{hadrons}$$

and

$$\overline{\nu_\mu} + \text{nucleon} \to \mu^+ + \text{hadrons}.$$

Because parity is not conserved in weak interactions, there are more structure functions for neutrino scattering than for electron scattering. Three structure functions contribute in the limit in which the lepton masses are ignored. If we use as variables $x = Q^2/2M\nu$ and $y = \nu/E$, the general forms are, in the context of the V-A theory,

$$\frac{d\sigma^\nu}{dx\,dy} = \frac{G_F^2 ME}{\pi}\left[(1-y)F_2^\nu + y^2 x F_1^\nu + (y - y^2/2)x F_3^\nu\right], \tag{8.20}$$

$$\frac{d\sigma^{\bar\nu}}{dx\,dy} = \frac{G_F^2 ME}{\pi}\left[(1-y)F_2^{\bar\nu} + y^2 x F_1^{\bar\nu} - (y - y^2/2)x F_3^{\bar\nu}\right]. \tag{8.21}$$

These forms are general (except that we have ignored the Cabibbo angle and corrections of order M/E) and F_1^ν, F_2^ν, and F_3^ν are functions of Q^2 and ν. In the Bjorken limit ($\nu \to \infty$, $Q^2 \to \infty$, $2M\nu/Q^2 = x$ finite), the F^ν's are nearly functions of x only.

The scattering of a neutrino by a pointlike fermion is much like the electromagnetic scattering of an electron by a pointlike fermion. In Chapter 6 we saw that the weak interaction current of the leptons has the V-A form, $\frac{1}{2}\gamma_\mu(1 - \gamma_5)$. For massless fermions, the quantity $\frac{1}{2}(1 - \gamma_5)$ projects out the left-handed piece of the fermion, while $\frac{1}{2}(1 + \gamma_5)$ projects out the right-handed piece. Now the coupling of the electromagnetic field to the fermion is governed by the current

$$\bar{u}(p')\gamma_\mu u(p). \tag{8.22}$$

If we consider an incident left-handed fermion we can write

$$\bar{u}(p')\gamma_\mu \frac{1}{2}(1 - \gamma_5)u(p) = \bar{u}(p')\frac{1}{2}(1 + \gamma_5)\gamma_\mu u(p) \tag{8.23}$$

$$= \left[\frac{1}{2}(1 - \gamma_5)u(p')\right]^\dagger \gamma_0\gamma_\mu u(p) \tag{8.24}$$

where, as usual the dagger indicates Hermitian conjugation. We see that the final fermion is also left-handed. Indeed, both vector and axial vector couplings have this property: the helicity (i.e. the projection of the spin along the direction of motion) of a massless fermion is unchanged by the interaction with an electromagnetic or weak current. It follows that we can consider the scattering as the incoherent sum of processes with specified helicities. We take as an example the electromagnetic process $e^-\mu^- \to e^-\mu^-$, ignoring the particle masses and using center-of-mass variables:

$$\frac{d\sigma}{d\Omega}(e_L^-\mu_L^- \to e_L^-\mu_L^-) = \frac{d\sigma}{d\Omega}(e_R^-\mu_R^- \to e_R^-\mu_R^-) = \frac{\alpha^2 s}{Q^4}, \tag{8.25}$$

$$\frac{d\sigma}{d\Omega}(e_L^-\mu_R^- \to e_L^-\mu_R^-) = \frac{d\sigma}{d\Omega}(e_R^-\mu_L^- \to e_R^-\mu_L^-) = \frac{\alpha^2 s}{Q^4}\frac{(1 + \cos\theta)^2}{4}. \tag{8.26}$$

The presence of the factor $(1 + \cos\theta)^2$ makes the last two cross sections vanish in the backward direction where $\cos\theta = -1$. This follows from the conservation of angular momentum. If the electron direction defines the z axis, the initial state $e_L^-\mu_R^-$ has $J_z = -1$

because the spins are antiparallel to the z axis and there is no orbital angular momentum along the direction of motion. For the final state $e_R^- \mu_L^-$ the same argument yields $J_z = +1$ if the scattering is at $180°$. Thus the scattering must vanish in this configuration.

The connection between the center-of-mass scattering angle and the invariant variables used above is $1 + \cos\theta = 2(1 - y)$. The addition of the four separate electromagnetic processes produces the characteristic $1 + (1 - y)^2$ behavior found in the deep inelastic electron scattering formulas.

The analogous weak cross sections follow the same pattern, except that only the left-handed parts of the fermions and the right-handed parts of the antifermions participate in charged-current processes, thus

$$\frac{d\sigma}{d\Omega}(\nu_\mu e_L^- \to \mu_L^- \nu_e) = \frac{G_F^2 s}{2\pi^2}, \qquad (8.27)$$

$$\frac{d\sigma}{d\Omega}(\nu_\mu \bar{\nu}_e \to \mu_L^- e_R^+) = \frac{G_F^2 s}{2\pi^2}\frac{(1+\cos\theta)^2}{4}. \qquad (8.28)$$

Using these simple formulas, we can determine the parton model values of the structure functions. Considering the scattering of a neutrino from a proton, we note that since the lepton loses charge ($\nu \to \mu^-$), the struck quark must gain charge. Thus it is only scattering from d quarks or $\bar{u}$ quarks that contributes. In this way we find for $\nu_\mu p \to \mu^- X$ and $\bar{\nu}_\mu p \to \mu^+ X$

$$\frac{d\sigma^\nu}{dx\,dy} = \frac{2MEG_F^2}{\pi}x\left[d(x) + (1-y)^2\bar{u}(x)\right], \qquad (8.29)$$

$$\frac{d\sigma^{\bar\nu}}{dx\,dy} = \frac{2MEG_F^2}{\pi}x\left[\bar{d}(x) + (1-y)^2 u(x)\right]. \qquad (8.30)$$

If the antiquarks, which are important only for rather small values of x, are ignored, the cross section for neutrino scattering is expected to be independent of y, while antineutrino scattering should vanish as $y \to 1$. To the extent to which the quark distributions are functions of x alone, the total cross section, σ, and the mean value of the momentum transfer squared, Q^2, are both proportional to E.

Comparing with the general formula for neutrino scattering, we deduce the structure functions for neutrino scattering in the parton model:

$$F_1^\nu = d(x) + \bar{u}(x), \qquad (8.31)$$
$$F_2^\nu = 2x[d(x) + \bar{u}(x)], \qquad (8.32)$$
$$F_3^\nu = 2[d(x) - \bar{u}(x)], \qquad (8.33)$$
$$F_1^{\bar\nu} = u(x) + \bar{d}(x), \qquad (8.34)$$
$$F_2^{\bar\nu} = 2x[u(x) + \bar{d}(x)], \qquad (8.35)$$
$$F_3^{\bar\nu} = 2[u(x) - \bar{d}(x)]. \qquad (8.36)$$

If the target is an equal mixture of u and d quarks, as is nearly the case for neutrino experiments, except with a hydrogen bubble chamber, each occurrence of u or d gets replaced by the average of u and d. Writing $q(x) = u(x) + d(x)$, $\bar{q}(x) = \bar{u}(x) + \bar{d}(x)$ we have

$$\frac{d\sigma^\nu}{dx\, dy} = \frac{MEG_F^2}{\pi} x \left[q(x) + (1-y)^2 \bar{q}(x) \right] \tag{8.37}$$

$$\frac{d\sigma^{\bar\nu}}{dx\, dy} = \frac{MEG_F^2}{\pi} x \left[\bar{q}(x) + (1-y)^2 q(x) \right] \tag{8.38}$$

Actually, we should include strange quarks as well. For energetic neutrino beams we have the processes $\nu_\mu s \to \mu^- c$ and $\bar\nu_\mu \bar{s} \to \mu^+ \bar{c}$. Here c is the charmed quark, to be discussed at length in Chapter 9. Our treatment has also been simplified by ignoring the Cabibbo angle.

The integrated cross sections are expressed in terms of $Q \equiv \int x\, dx\, q(x)$ and $\bar{Q} \equiv \int x\, dx\, \bar{q}(x)$, the momentum fractions carried by the quarks and the antiquarks.

$$\frac{d\sigma^\nu}{dy} = \frac{MEG_F^2}{\pi} \left[Q + (1-y)^2 \bar{Q} \right]; \qquad \sigma^\nu = \frac{MEG_F^2}{\pi} \left[Q + \frac{1}{3}\bar{Q} \right]; \tag{8.39}$$

$$\frac{d\sigma^{\bar\nu}}{dy} = \frac{MEG_F^2}{\pi} \left[\bar{Q} + (1-y)^2 Q \right]; \qquad \sigma^{\bar\nu} = \frac{MEG_F^2}{\pi} \left[\bar{Q} + \frac{1}{3} Q \right]. \tag{8.40}$$

Since we expect much more of the momentum in the proton to be carried by the quarks than the antiquarks, we anticipate

$$\frac{\sigma^{\bar\nu}}{\sigma^\nu} \approx \frac{1}{3}. \tag{8.41}$$

Inserting the values of the constants, we find

$$\frac{\sigma^\nu}{E} = 1.56 \left[Q + \frac{1}{3}\bar{Q} \right] 10^{-38} \text{ cm}^2 \text{ GeV}^{-1}. \tag{8.42}$$

The total cross sections were measured at CERN using a heavy liquid (freon) bubble chamber, Gargamelle, which had been constructed at Orsay, near Paris (**Ref. 8.4**). Separate neutrino and antineutrino beams were generated by the CERN Proton Synchroton (PS). Outgoing muons were identified by their failure to undergo hadronic interactions in the bubble chamber. The energy of the produced hadronic system was measured by adding the energy of the charged particles measured in a 20-kG magnetic field, to the energy of the neutral pions observed through conversion of photons in the heavy liquid. The neutrino flux was monitored by measuring the muon flux associated with it.

While the Gargamelle data covered very low energies, $E_\nu < 10$ GeV, the expected linear behavior of the cross section on the neutrino energy was observed, with the results $\sigma^\nu/E = 0.74 \pm 0.02 \times 10^{-38}$ cm^2 GeV^{-1}, $\sigma^{\bar\nu}/E = 0.28 \pm 0.01 \times 10^{-38}$ cm^2 GeV^{-1}. These results were in good accord with the expectations.

The Gargamelle results were severely limited by the low energy of the CERN PS. Later studies were carried out at Fermilab by the Harvard, Penn, Wisconsin, and Fermilab Collaboration (HPWF) and the Caltech, Columbia, Fermilab, Rochester, and Rockefeller Collaboration (CCFRR) and at the CERN SPS by the CERN, Dortmund, Heidelberg, and Saclay Collaboration (CDHS) and the CERN, Hamburg, Amsterdam, Rome, and Moscow Collaboration (CHARM). Bubble chamber studies have also been done with the 15-foot bubble chamber at Fermilab and the Big European Bubble Chamber (BEBC) at CERN. The counter detectors have active target regions, calorimetry, and a muon spectrometer. These experiments confirmed the linearity of the cross section as a function of the neutrino energy and also gave similar results for σ/E, about 0.67×10^{-38} cm^2 GeV^{-1} for neutrinos and 0.34×10^{-38} cm^2 GeV^{-1} for antineutrinos.

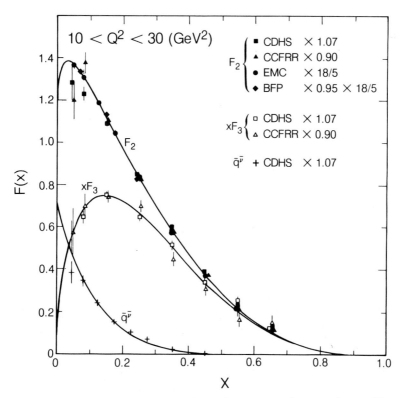

Figure 8.2. A compilation of data from neutrino and muon scattering experiments. The structure function F_2 is essentially proportional to the sum of the quark and antiquark distributions: $F_2(x) = x[q(x) + \bar{q}(x)]$. The structure function xF_3 is similarly related to the difference of the quark and antiquark distributions: $xF_3(x) = x[q(x) - \bar{q}(x)]$. The third combination shown is $\bar{q}^\nu(x) = x[\bar{u}(x) + \bar{d}(x) + 2\bar{s}(x)]$. The data shown are from the CDHS, CCFRR, EMC (European Muon Collaboration), and BFP (Berkeley, Fermilab, Princeton) groups [compilation taken from *Review of Particle Properties, Phys. Lett.*, **170B**, 79 (1986)]. The normalizations of the data sets have been modified as indicated to bring them into better agreement. A factor 18/5, the inverse of the average charge squared of a light quark, is applied to the muon data to compare them with the neutrino data.

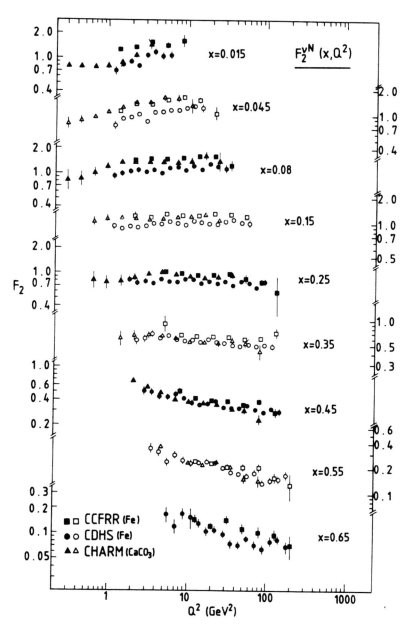

Figure 8.3. The structure functions F_2 for deep inelastic neutrino scattering as measured by the CDHS, CHARM and CCFRR collaborations. Scaling would require the structure functions to be independent of Q^2 at fixed x. The deviations seen from scaling are consistent with the predictions of QCD. From F. Dydak in *Proceedings of the 1983 International Lepton/Photon Symposium*, Cornell, 1983, p. 634.

The essence of the parton model is that the same quark distributions should work for all processes. For an isoscalar target, the electromagnetic structure function is

$$F_2 = \frac{5}{18}x(u + d + \bar{u} + \bar{d}) + \frac{1}{9}x(s + \bar{s}). \tag{8.43}$$

If the contribution from strange quarks is neglected, this is just 5/18 times the corresponding structure function for neutrinos on an isoscalar target. Neglecting the strange quarks is a good approximation for $x > 0.3$, where the antiquarks as well make a small contribution. The agreement between the electroproduction and neutrinoproduction data is satisfactory as is shown in Figure 8.2.

More detailed studies with electron, muon, and neutrino beams have demonstrated the Q^2 dependence predicted by QCD – the deviation from the scaling behavior of the "kindergarten" parton model. At high x, increasing Q^2 reduces the quark distribution because the quarks split into a quark and a gluon sharing the initial momentum, as described above. At low x, the structure functions increase as Q^2 increases because the momentum of high x quarks is degraded by the emission process. These features are seen in Figure 8.3 showing data for $d\sigma/dy$ from the CDHS and CHARM collaborations at CERN and the CCFRR collaboration at Fermilab. The deviations from scaling provide indirect evidence for the existence of gluons. Direct evidence awaited the development of high-energy e^+e^- colliding beam machines.

Exercises

8.1 Verify the curves in Figure 5 of McAllister and Hofstadter.

8.2 What static charge distributions would produce the form factors $F(\mathbf{q}^2) = 1/(1 + \mathbf{q}^2/m^2)$ and $F(\mathbf{q}^2) = 1/(1 + \mathbf{q}^2/m^2)^2$?

8.3 We can define cross sections in the lab frame for virtual photons with momentum q using polarization vectors ϵ_T and ϵ_L, where $\epsilon \cdot q = 0$. If $q = (\nu, 0, 0, \sqrt{\nu^2 + Q^2})$, where $Q^2 = -q^2$, let

$$\epsilon_T = (0, 1, 0, 0)$$

$$\epsilon_L = (\sqrt{\nu^2 + Q^2}, 0, 0, \nu)/Q$$

so $\epsilon_T \cdot \epsilon_T = -1$, $\epsilon_L \cdot \epsilon_L = 1$. Then

$$\frac{\sigma_L}{\sigma_T} = \frac{\epsilon_L^\mu \epsilon_L^\nu W_{\mu\nu}}{\epsilon_T^\mu \epsilon_T^\nu W_{\mu\nu}}.$$

Show that

$$\frac{\sigma_L}{\sigma_T} = \frac{W_2}{W_1}\left(1 + \frac{\nu^2}{Q^2}\right) - 1.$$

8.4 * The deep inelastic scattering process has an amplitude that can be represented as

$$\mathcal{M} = \frac{e^2}{q^2}\bar{u}(k')\gamma_\mu u(k) <F|J^\mu(0)|p>.$$

Here $q = k - k'$ is the four-momentum transfer, and k and k' are the initial and final lepton momenta, p is the initial nucleon momentum, and $|F>$ represents the final hadronic state. The cross section, summed over final states and averaged over initial lepton spins, is

$$d\sigma = \frac{(2\pi)^4}{4k\cdot p}\sum_F \delta^4(k+p-k'-p_F)\frac{d^3k'}{(2\pi)^3 2E'}$$

$$\times \prod_i \frac{d^3 p'_i}{(2\pi)^3 2E'_i}\left(\frac{4\pi\alpha}{q^2}\right)^2 \tfrac{1}{2}\mathrm{Tr}\,\slashed{k}'\gamma_\mu \slashed{k}\gamma_\nu <p|J^\nu(0)|F><F|J^\mu(0)|p>$$

where $\slashed{k} = k_\mu \gamma^\mu$ and where we treat the lepton as massless. The p'_i represent final state momenta of the produced hadrons. We define

$$W^{\mu\nu} = \frac{1}{2M}(2\pi)^3 \sum_F \int \prod_i \frac{d^3 p'_i}{(2\pi)^3 2E'_i}\delta^4(p+q-p_F)$$

$$\times <p|J^\mu(0)|F><F|J^\nu(0)|p>.$$

Current conservation requires that $q_\mu W^{\mu\nu} = q_\nu W^{\mu\nu} = 0$. The tensor $W^{\mu\nu}$ must be constructed from the vectors p and q. Show that the most general form for $W^{\mu\nu}$ may be written as

$$W^{\mu\nu} = \left(-g^{\mu\nu} + \frac{q^\mu q^\nu}{q^2}\right)W_1 + \left(p^\mu - \frac{p\cdot q\, q^\mu}{q^2}\right)\left(p^\nu - \frac{p\cdot q\, q^\nu}{q^2}\right)W_2/M^2.$$

Show that

$$\frac{d\sigma}{dE'd\Omega'} = \frac{4\alpha^2 E'^2}{Q^4}\left[2W_1 \sin^2\tfrac{1}{2}\theta + W_2 \cos^2\tfrac{1}{2}\theta\right]$$

where θ is the laboratory scattering angle of the lepton.

8.5 * If the sum defining $W_{\mu\nu}$ in Exercise 8.2 is restricted to elastic scattering, the Rosenbluth formula should be recovered. Demonstrate that this is so by taking

$$<F|J_\mu(0)|p> = \bar{u}(p')\left[F_1\gamma_\mu + iF_2\frac{\kappa q^\nu \sigma_{\mu\nu}}{2M}\right]u(p).$$

Further Reading

A unique insight into the structure of the nucleon is found in the seminal *Photon-Hadron Interactions*, by R. P. Feynman, W. A. Benjamin, 1972.

A more theoretical discussion is given by C. Quigg in *Gauge Theories of the Strong, Weak, and Electromagnetic Interactions*, Benjamin/Cummings, Menlo Park, CA, 1983.

References

8.1 R. W. McAllister and R. Hofstadter, "Elastic Scattering of 188-MeV Electrons from the Proton and Alpha Particles." *Phys. Rev.*, **102**, 851 (1956).

8.2 E. D. Bloom *et al.*, "High Energy Inelastic $e - p$ Scattering at $6°$ and $10°$." *Phys. Rev. Lett.*, **23**, 930 (1969).

8.3 M. Breidenbach *et al.*, "Observed Behavior of Highly Inelastic Electron Proton Scattering." *Phys. Rev. Lett.*, **23**, 935 (1969). For scattering off neutrons, see also A. Bodek *et al.*, Phys. Rev. **D20**, 1471 (1979).

8.4 T. Eichten *et al.*, "Measurement of the Neutrino-Nucleon and Anti Neutrino-Nucleon Total Cross Sections." *Phys. Lett.*, **46B**, 274 (1973).

Elastic Scattering of 188-Mev Electrons from the Proton and the Alpha Particle*†‡§||¶

R. W. McAllister and R. Hofstadter

Department of Physics and High-Energy Physics Laboratory, Stanford University, Stanford, California

(Received January 25, 1956)

The elastic scattering of 188-Mev electrons from gaseous targets of hydrogen and helium has been studied. Elastic profiles have been obtained at laboratory angles between 35° and 138°. The areas under such curves, within energy limits of ±1.5 Mev of the peak, have been measured and the results plotted against angle. In the case of hydrogen, a comparison has been made with the theoretical predictions of the Mott formula for elastic scattering and also with a modified Mott formula (due to Rosenbluth) taking into account both the anomalous magnetic moment of the proton and a finite size effect. The comparison shows that a finite size of the proton will account for the results and the present experiment fixes this size. The root-mean-square radii of charge and magnetic moment are each $(0.74\pm0.24)\times10^{-13}$ cm. In obtaining these results it is assumed that the usual laws of electromagnetic interaction and the Coulomb law are valid at distances less than 10^{-13} cm and that the charge and moment radii are equal. In helium, large effects of the finite size of the alpha-particle are observed and the rms radius of the alpha particle is found to be $(1.6\pm0.1)\times10^{-13}$ cm.

I. INTRODUCTION

IN principle, it is possible to discover the finite size and structure of nuclei by methods of elastic electron scattering at high energies.[1-3] It is even possible to determine the structure of the proton by these methods.[4] For the light nuclei the Born approximation is adequate to analyze the experimental data, while for heavier nuclei such as gold or even copper[5] Yennie *et al.* have shown that a more accurate phase shift analysis is required.

The proton, deuteron, and alpha particle are most interesting to study because they are among the simplest nuclear structures. Furthermore, nuclei are built up out of protons and neutrons and it is fascinating to think of what the proton itself is built. In this paper we shall examine the structure of the proton and alpha particle. In an earlier paper[6] the scattering from the deuteron was reported.

---

* The research reported in this document was supported jointly by the U. S. Navy (Office of Naval Research) and the U. S. Atomic Energy Commission, and by the U. S. Air Force through the Air Force Office of Scientific Research, Air Research and Development Command.

† Aided by a grant from the Research Corporation.

‡ These results were briefly reported at the Seattle Meeting of the American Physical Society, in July, 1954, but a comparison was not made at that time with the Rosenbluth results.

§ Miss Eva Wiener assisted in the early phases of this research. She was the victim of a fatal automobile accident in 1953.

|| Some of the material now reported was published earlier in brief form, *viz.*, R. Hofstadter and R. W. McAllister, Phys. Rev. **98**, 217 (1955).

¶ *Note added in proof.*—Results more recent than those reported in this paper and extending to 550 Mev, were presented at the New York meeting of the American Physical Society [Bull. Am. Phys. Soc. Ser. II, 1 (1956)] by Hofstadter, Chambers, and Blankenbeder. The newer experiments confirm in greater detail the results presented in this paper. The newer results are being submitted for publication in *The Physical Review*.

[1] Hofstadter, Fechter, and McIntyre, Phys. Rev. **91**, 422 (1953).
[2] Hofstadter, Fechter, and McIntyre, Phys. Rev. **92**, 978 (1953).
[3] Hofstadter, Hahn, Knudsen, and McIntyre, Phys. Rev. **95**, 512 (1954).
[4] R. Hofstadter and R. W. McAllister, Phys. Rev. **98**, 217 (1955).
[5] Yennie, Wilson, and Ravenhall, Phys. Rev. **92**, 1325 (1953).
[6] J. A. McIntyre and R. Hofstadter, Phys. Rev. **98**, 158 (1954).

II. EXPERIMENTAL METHODS

Many of the experimental procedures have been reported in earlier papers.[2,3] The only important new variation over earlier methods has been the substitution of a gaseous target for the previously used metallic foils. The gaseous target will now be described.

In Fig. 1, the basic design of the target assembly is given. The cylinder is made of 410 stainless steel and has been heat-treated to increase its strength. The end plates are made of 0.010-inch stainless steel and are deformed by the high-pressure gases into the approximate shape shown in the figure. The target cylinder is $3\frac{3}{8}$ inches long and $\frac{3}{4}$ inch in diameter. The end plates are sealed by means of O-rings shown in the figure. Pressures as high as 2000 pounds per square inch have been used successfully in this chamber over long periods of time.

The geometry of the scattering experiment using the gaseous target chamber is shown in Fig. 2. Because of the double-focusing characteristic of the magnetic spectrometer and because of the defining slits at the entrance and exit of the spectrometer, the effective target viewed by the spectrometer has the appearance indicated schematically in Fig. 2. It is evident that to a very good approximation the scattering yield at any given angle will be proportional to the cosecant of the angle of observation in the laboratory system since the

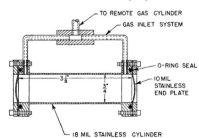

Fig. 1. Basic design of the gas chamber.

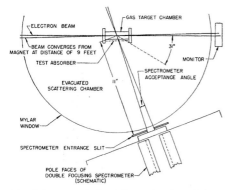

Fig. 2. Arrangement of parts in experiments on electron scattering from a gas target.

target thickness viewed by the spectrometer is proportional to this trigonometric function. Thus, to normalize the data to the same target thickness, the yield at any angle must be divided by the cosecant of the angle. In these experiments the effective target width viewed by the spectrometer is approximately $\frac{1}{2}$ inch at the gas chamber and this dimension is given by the vertical exit slit width (0.5 inch) imaged at the source.

The slit defining the acceptance angle in the plane common to the beam, scattering target, and entrance to the spectrometer, was $\frac{3}{4}$ inch wide and is indicated in Fig. 2. The exit collimator at the top of the magnet had a horizontal slit $\frac{5}{8}$ inch wide defining the energy band accepted by the Čerenkov detector, and a vertical slit $\frac{1}{2}$ inch wide. The vertical slit, together with the $\frac{3}{4}$ inch entrance slit, served to define the effective width of the target. In all the experiments herewith reported the incident beam was monochromatic within ± 1.0 Mev in 187 Mev.

At small angles, that is, angles less than 30°, it is possible for the spectrometer to view the end walls of the chamber and thus accept spurious electrons scattered by the target end plates. At 35° and 40°, a small residual effect of this type is present and is always subtracted from the yield furnished by the gas plus the target chamber. In other words, the scattering intensity is measured first with the gas in the chamber and then again with the gas removed from the target. The latter measurement gives the "background" due to the end wall effects. At all other angles, this effect is negligible.

Multiple scattering and radiation straggling from the 0.018-inch cylinder walls introduce only minor errors at the angles studied. This has been determined empirically by inserting a 0.010-inch stainless steel test absorber in the position so marked in Fig. 2. The test absorber was placed in the path of electrons scattered at angles 50°, 90°, and 130°. Elastic profiles were measured with the test absorber in and out of the path.

The peak of the elastic scattering profile was reduced 1 percent per mil of stainless steel in the direction of the scattered electrons, but the half-width of the curve was also increased by an amount such that the area under the elastic curve was the same, within 5 percent, whether the test absorber was in or out. This behavior may be understood as follows: The double focusing action of the spectrometer assures collection of all the electrons directed into the effective solid angle of the spectrometer, whether multiply-scattered or not and brings them to a focus beyond the energy slit (and from there into the Čerenkov detector). The only effects of multiple-scattering in the chamber walls are (a) to fuzz out the source of the scattered electrons in the gas, i.e., to increase or decrease the depth from which the scattered electrons appear to emerge from the target, (b) to reduce the angular resolution, and (c) to mix electrons scattered originally at different angles. Effect (a) may easily be seen to be of negligible importance. Effect (b) amounts to approximately $(\Delta\theta)_{rms} = \pm 1°$. Since the angular opening of the lower spectrometer slit is $\pm 2°$ and the multiple scattering is essentially Gaussian, the uncertainty in measuring the scattering angle is not appreciably increased by the effect of the side walls. The incoming end plate also contributes an uncertainty of $(\Delta\theta)_{rms} = \pm 0.7°$. The resulting uncertainty, combining all causes, is approximately $(\Delta\theta)_{rms} = \pm 2.4°$. The effect of multiple scattering in the hydrogen or helium gas volumes is of the order of 0.1° and hence negligible. In case (c), the error so introduced is of the order of tenths of a percent and is here neglected. In fact, plural scatterings are eliminated because of the energy selection of the spectrometer.

Radiation straggling of the electrons coming out through the walls of the chamber may be shown theoretically to contribute not more than a 5 percent relative correction between 50° and 90° and an equal figure between 90° and 130°, i.e., both the 50° yield and the 130° yield would each be lowered by something less than 5 percent with respect to the 90° yield. Our experiments with the test absorber have not demonstrated a consistent loss greater than 5 percent which could be attributed to straggling in the chamber walls. The statistical accuracy and drifts of the apparatus could have concealed an error of the order of 5 percent. Hence, we have not made a correction for straggling.

The lining-up procedure used a CsBr(Tl) crystal which could be moved remotely into or out of the beam. The crystal was placed along the beam axis just outside the scattering chamber. When it was desired to know the position of the beam, the crystal was moved into the beam and observed with a telescope. When the beam was lined up, say within $\pm \frac{1}{16}$ inch at the target, the crystal was withdrawn. Periodic checks showed whether or not the beam had moved. Very little beam motion was observed after an initial alignment.

An ion chamber was used in the early runs as a monitor of the incident beam, and a secondary electron emitter[7] in the later runs. The ion chamber showed a small amount of saturation at large beams and its runs were corrected by the empirically determined calibration of ion chamber *versus* secondary electron emitter. No correction so obtained was larger than 10 percent. If the correction had not been included, the proton size (see below) would have been a trifle larger.

The theoretical Schwinger radiation correction has not been applied since its angular dependence is very weak and well within the statistics of our experimental observations.

III. RESULTS

A. Hydrogen

Typical elastic profiles observed in a run with hydrogen at an incident energy of 185 Mev are shown in Fig. 3. Because of recoil of the struck proton the energy of the elastically deflected electron is a decreasing function of the angle of scattering. This may be observed by noting the variable position of the peaks in Fig. 3. Figure 4 shows the theoretical behavior of the energy of the scattered electron plotted against laboratory scattering angle for an incident energy of 187 Mev. The solid points show the positions of the peaks of the elastic scattering curves taken at the various angular positions during an experimental run at 187 Mev. The agreement is excellent except at extreme angles where small deviations are observed. The deviations are actually expected because of an increasing energy loss in the wall as the angle of entry becomes more and more oblique. The observed reduction in energy of the scattered electrons below the theoretical curve is in good agreement with the energy loss in the wall.

Because of the variation in energy of the scattered electrons we have been concerned that the solid angle effective in collecting electrons could have been smaller at small angles (high energies), where magnet saturation is important, than at large angles (smaller energies), where saturation is less important. To test this possibility we have measured the number of scattered electrons as a function of the entrance slit width at both

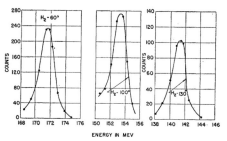

FIG. 3. Typical elastic profiles obtained with hydrogen gas at 185 Mev.

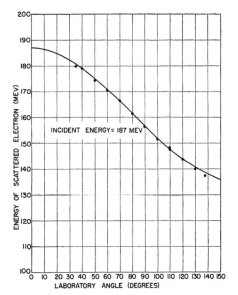

FIG. 4. The solid line gives the theoretical energy of the scattered electrons for an incident energy of 187 Mev. Relativistic kinematics are used to obtain the theoretical curve. The experimental points correspond to peak values of the elastic profiles and refer to experimental observations.

high (188 Mev) and low (139 Mev) energies. We have found that in both cases the number of scattered electrons for the ¾-inch entrance slit width is 15 percent below the number expected from the initial slope of the curve of number of scattered electrons *versus* slit width. The 15 percent reduction is due to the widest trajectories striking the magnet chamber walls. In the radial direction in the magnet no electrons are lost because of the small extent of the beam in this direction. In other words, the effective solid angle is the same at both low and high energies provided that the entrance slit width is not larger than ¾ inch. Hence correction for magnet saturation is not required.

Areas under the elastic peaks, such as those of Fig. 3, have been measured by numerical integration over a width of ±1.5 Mev about the peak. Such values have been plotted against laboratory angle as in Fig. 5. Areas over ±2 and ±2.5 Mev widths have also been obtained by numerical integration, but the relative results are essentially the same. Only the ±1.5 Mev results will be presented below.

Figure 5 presents a summary of all the data obtained over a period of several months. It may be noticed that the experimental spread of points is somewhat larger than the statistical errors might lead one to expect. The causes of the spread are probably connected with small, unnoticed horizontal shifts of beam, hysteresis in the spectrometer magnet, small changes in the bias of the Čerenkov counter detection equipment, variations

---

[7] G. W. Tautfest and H. R. Fechter, Phys. Rev. **96**, 35 (1954); Rev. Sci. Instr. **26**, 229 (1955).

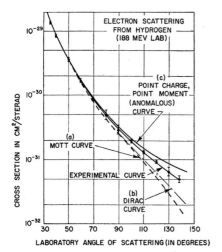

FIG. 5. Curve (a) shows the theoretical Mott curve for a spinless point proton. Curve (b) shows the theoretical curve for a point proton with the Dirac magnetic moment, curve (c) the theoretical curve for a point proton having the anomalous contribution in addition to the Dirac value of magnetic moment. The theoretical curves (b) and (c) are due to Rosenbluth.[8] The experimental curve falls between curves (b) and (c). This deviation from the theoretical curves represents the effect of a form factor for the proton and indicates structure within the proton, or alternatively, a breakdown of the Coulomb law. The best fit indicates a size of 0.70×10^{-13} cm.

in saturation of the ion chamber monitor response and in the integrating voltmeter, and perhaps other unknown items. In Fig. 5 we have drawn a curve, labeled "experimental curve," which is our best estimate of the accumulated data at 188 Mev. The limits of error represent the greatest variations we have observed in any runs. However all runs, not being absolute, are normalized to each other by "best fitting." The experimental curve is also normalized to the theoretical curve at small angles. Also plotted in Fig. 5 are (a) the theoretical Mott curve for a spinless point proton, (b) the theoretical curve for a point proton with the Dirac value of magnetic moment (gyromagnetic ratio 2.00), (c) the theoretical curve for a point proton with the anomalous value of the proton moment in addition to the Dirac moment (gyromagnetic ratio=5.58). The theoretical curves (b), (c) are obtained from calculations of Rosenbluth.[8] The experimental curve deviates from curves (a), (b), and (c) at the larger angles and is lower than the curve for a point proton with anomalous moment, but higher than the curve for a point proton with Dirac moment. This reduction at large angles below the curve for point charge represents the effect of a "structure factor" or a "form factor" for the proton and hence indicates the finite size of the proton. Since the usual electromagnetic relations and the Coulomb

[8] M. N. Rosenbluth, Phys. Rev. **79**, 615 (1950).

interaction have been used in Rosenbluth's calculation, we are here assuming the validity of these interactions at small distances ($<10^{-13}$ cm). Subject to this assumption, the experiment indicates the proton is not a point.

In order to carry out the form factor calculations, we have made use of Rosenbluth's formalism.[8] However we have given the charge and magnetic moment phenomenological interpretations in place of the meson theoretic interpretations originally presented by Rosenbluth.[9] We may write Rosenbluth's formulas as follows: for a point charge we have

$$\sigma = \sigma_{NS}\left\{1 + \frac{q^2}{4M^2}[2(1+\mu)^2 \tan^2(\theta/2) + \mu^2]\right\}, \quad (1)$$

where

$$\sigma_{NS} = \frac{e^4}{4E^2}\left(\frac{\cos^2(\theta/2)}{\sin^4(\theta/2)}\right)\frac{1}{1+(2E/M)\sin^2(\theta/2)}, \quad (2)$$

and where

$$q = \frac{(2/\lambda)\sin(\theta/2)}{[1+(2E/M)\sin^2(\theta/2)]^{\frac{1}{2}}}. \quad (3)$$

Here natural units, $\hbar = c = 1$, are used and the equations are written in terms of the laboratory coordinates; q is the invariant momentum transfer in the center-of-mass frame expressed in laboratory coordinates; E is the energy of the incident electrons; M the mass of the proton, and μ is the anomalous part of the proton's magnetic moment ($\mu = 1.79$). λ is the reduced de Broglie wavelength of the electron in the laboratory system.

For a diffuse proton we may write:

$$\sigma = \sigma_{NS}\left\{F_1^2 + \frac{q^2}{4M^2}[2(F_1+\mu F_2)^2 \tan^2(\theta/2) + \mu^2 F_2^2]\right\}, \quad (4)$$

where F_1 is the charge form factor (which also influences the intrinsic "Dirac" magnetic moment) and F_2 the anomalous magnetic moment form factor. In principle F_1 does not have to be the same as F_2. F_1 and F_2 may be written as functions of $\langle q\langle r\rangle\rangle$, where $\langle r\rangle$ is the root-mean-square radius of the appropriate charge, or moment distribution. F_1 and F_2 may also be identified with e'/e and $k'e'/k_0 e$ in Rosenbluth's article.

We have not made detailed analyses for different F_1 and F_2. Rather, as may be seen below, we have assumed $F_1 = F_2$. However, the data at all energies are quite consistent with this choice.

At the energies used in these experiments, the form factor (F_1 or F_2) is not appreciably shape dependent, i.e., one cannot distinguish between uniform, exponential, or Gaussian charge (or magnetic moment) distributions. All that can be determined is a mean square radius. Therefore we have tried to fit the experi-

[9] We are indebted to Dr. D. R. Yennie for formulation of Eqs. (1)–(4).

mental data with a phenomenological form factor corresponding to various values of the mean square radii up to values of $q\langle r\rangle \cong 1.0$. q is again the momentum transfer and $\langle r\rangle$ the root-mean-square radius of the charge or magnetic moment distributions. For simplicity, as stated above, we have assumed that $\langle r\rangle_{\text{charge}} = \langle r\rangle_{\text{anomalous magnetic moment}}$, although in principle this is not a necessary restriction. Hence we can expect only to obtain a first approximation to the structure and size of the proton.

When such form factors are applied to the point charge-point moment curve, the behavior of the experimental curve can be reproduced very well. In fact for $\langle r\rangle_{\text{charge}} = \langle r\rangle_{\text{magnetic moment}} = 0.70 \times 10^{-13}$ the theoretical curve cannot be distinguished from the experimental curve within the limits of error. A separate theoretical curve for 0.70×10^{-13} cm therefore has not been included in Fig. 5. The limits of error in the radius are conservatively estimated at $\pm 0.24 \times 10^{-13}$ cm.

A similar fitting procedure can be employed with data obtained with electrons at 236 Mev in the incident beam. In this case our measurements could be made only at angles larger than or equal to 90° since our magnetic spectrometer cannot bend electrons of energy higher than those scattered at 90° (or smaller angles): For an incident energy of 236 Mev the scattered electron at 90° has an energy of 189 Mev, the approximate limit of our apparatus.

Figure 6 shows the experimental points obtained in several runs at 236 Mev. The shape of the point charge-point moment curve is shown as well as the experimental points. No absolute values are known for the experimental points so that the best that can be done is to try to fit the shape of the experimental curve with Eq. (4) for various values of F_1 and F_2. Again the assumption $F_1 = F_2$ is made. Such attempts are shown in Fig. 6 and are labeled rms 6.2, 7.8, and 9.3×10^{-14} cm. The dotted curves corresponding to 6.2×10^{-14} cm and 9.3×10^{-14} cm may be shifted down or up respectively

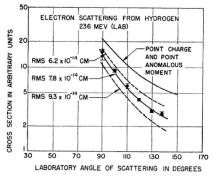

FIG. 6. This figure shows the experimental points at 236 Mev and the attempts to fit the shape of the experimental curve. The best fit lies near 0.78×10^{-13} cm.

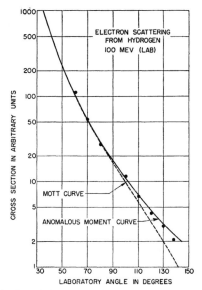

FIG. 7. Theoretical curves are shown for electrons of 100 Mev, along with the experimental observations at that energy.

to try to fit the experimental points, but neither curve will do so within the limits of error. Hence the data at 236 Mev support a "best" value of rms radius of $(0.78 \pm 0.20) \times 10^{-13}$ cm, conservatively speaking. This value is in good agreement with the best value $(0.70 \pm 0.24) \times 10^{-13}$ cm obtained above at 188 Mev.

In order to test some features of the apparatus, we have carried out a scattering experiment at an incident energy of 100 Mev. In this case, if our model of the proton is correct, the observed scattering should be quite close to the curve for a point charge and point moment because the $q\langle r\rangle$ value is small and $F^2 = 1.0$. Figure 7 shows that the agreement observed is highly satisfactory. At 100 Mev, the magnetic spectrometer is never saturated at any angle. Hence the "saturation" aspect and possible defocusing effects are not tested by this experiment. However, the 236-Mev and 188-Mev runs do test such possible effects since different energies correspond to different angular positions. The good agreement obtained between these latter two sets of data and the satisfactory behavior at 100 Mev is essentially what we have published earlier.[4]

These results may be summarized in the following way: If the proton can be assumed to (a) have distributions of charge and magnetic moment equal, or at least similar, in size and (b) if the Coulomb law and the usual electromagnetic laws are obeyed at distances of the order of 0.7×10^{-13} cm, then these experiments show that the proton has an rms radius of $(0.74 \pm 0.24) \times 10^{-13}$ cm. Of course, if the Coulomb law and the usual interactions are not valid, these findings could also be

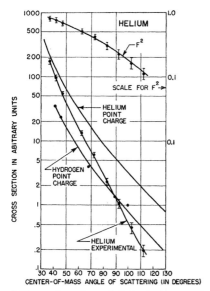

FIG. 8. The experimental curve for helium in the center-of-mass system, hydrogen normalizing points, and the helium point charge construction are shown. This figure also exhibits the square of the form factor as a function of angle. The best fit of theory to experiment corresponds to an rms radius of 1.60×10^{-13} cm.

interpreted in terms of a point charge and point moment. We suspect that the breakdown of the Coulomb law would have exhibited other consequences, possibly already recognized in the literature. Phenomenologically we cannot distinguish, at the present time with these experiments alone, between a finite size of the proton and a breakdown of the Coulomb law. Nevertheless, any meson theory would predict a finite size of the proton's magnetic moment and this is what we may have found in the proton.

B. Helium

The elastic peaks observed in helium are similar to those found in hydrogen, except that the recoil shifts are approximately four times smaller. To measure the form factor of the alpha particle with respect to electron scattering, we have made essentially simultaneous measurements of the scattering from helium and hydrogen and compared the results. The procedure involved carrying out the helium measurements, emptying the target chamber, and finally substitution of hydrogen for the helium. A series of measurements in hydrogen is thus made almost at the same time as the helium measurements. From a few representative hydrogen points, we can construct a point-charge *Mott curve* for hydrogen, say, between 35° and 90°. If we multiply this curve by four ($Z_{He}^2 = 2^2 = 4$) we obtain a theoretical point-charge curve for helium. Note that we use a Mott curve (spinless particle) since the alpha particle has no spin or magnetic moment. The ratio of the actually observed experimental curve in helium to the point charge curve for helium gives the square of the form factor. Thus the form factor can be compared with theoretical form factors for various size charge distributions.

Figure 8 shows the helium experimental curve in the center-of-mass system, the hydrogen normalizing points, and the helium point-charge construction. The incident energy was 188 Mev for these experiments. Corrections for the different energies in the center-of-mass system and for the different effective solid angles have been made. A glance at the figure shows that the elastic scattering from the alpha particle is considerably smaller at large angles (a factor of 10 at 110°) than that from a point charge.

Figure 8 also shows the ratio of the alpha-particle scattering to that of a point charge with $Z=2$. This curve represents the square of the "form factor." The scale is given in the upper right hand corner of Fig. 8. This curve is indistinguishable from a (form factor)^2 curve for an rms radius of $(1.60 \pm 0.10) \times 10^{-13}$ cm. For such a small nucleus and an energy 188 Mev, our analysis will not give more than an rms radius from these measurements. It is curious that the rms radius of the alpha particle is approximately twice that of the proton as determined from these scattering measurements. Allowing for the rms radius of each of the two protons in the alpha particle, as determined above, the rms radius of the alpha particle would be smaller. By subtracting mean squares, the rms radius to the charge centroid would be 1.41×10^{-13} cm. This approximate calculation probably overemphasizes the effect of the finite protonic size.

ACKNOWLEDGMENTS

We wish to acknowledge the valuable cooperation of Mr. E. E. Chambers in helping with some of the measurements. We wish to thank Dr. R. H. Dalitz, Dr. D. G. Ravenhall, Dr. L. I. Schiff, and Dr. D. R. Yennie for many interesting conversations. We acknowledge with much appreciation the valued help of the late Miss Eva Wiener.

HIGH-ENERGY INELASTIC e-p SCATTERING AT 6° AND 10°*

E. D. Bloom, D. H. Coward, H. DeStaebler, J. Drees, G. Miller, L. W. Mo, and R. E. Taylor
Stanford Linear Accelerator Center, Stanford University, Stanford, California 94305

and

M. Breidenbach, J. I. Friedman, G. C. Hartmann,† and H. W. Kendall
Department of Physics and Laboratory for Nuclear Science,‡
Massachusetts Institute of Technology, Cambridge, Massachusetts 02139
(Received 19 August 1969)

> Cross sections for inelastic scattering of electrons from hydrogen were measured for incident energies from 7 to 17 GeV at scattering angles of 6° to 10° covering a range of squared four-momentum transfers up to 7.4 $(GeV/c)^2$. For low center-of-mass energies of the final hadronic system the cross section shows prominent resonances at low momentum transfer and diminishes markedly at higher momentum transfer. For high excitations the cross section shows only a weak momentum-transfer dependence.

Inelastic electron-proton scattering at high four-momentum transfer and large electron-energy loss has been used to investigate the electromagnetic structure and interactions of the proton.[1] We have measured the double differential cross section $d^2\sigma(E, E', \theta)/d\Omega dE'$ for electrons on hydrogen in a new kinematic region made accessible by the Stanford linear accelerator. We report measurements made at 6° and 10° for several incident energies E, and for a range of scattered energies E', beginning at elastic scattering and ending at $E' \approx 3$ GeV. Only the scattered electron was detected. In this kind of measurement the two inelastic form factors[2] which describe the electromagnetic properties of the proton are functions of the squared four-momentum transfer, q^2, and the mass of the unobserved hadronic final state, W. We have measured several spectra at each angle to allow the calculation of model-independent radiative corrections[3] over a wide range of q^2 and W.

We observe the excitation of several nucleon resonances[4-7] whose cross sections fall rapidly with increasing q^2. The region beyond $W \approx 2$ GeV exhibits a surprisingly weak q^2 dependence. This Letter describes the experimental procedure and reports cross sections for $W \geq 2$ GeV. Discussion of the results and a detailed description of the resonance region will follow.[8]

The incident energies at $\theta = 10°$ were 17.7, 15.2, 13.5, 11, and 7 GeV, and at $\theta = 6°$ were 16, 13.5, 10, and 7 GeV. For fixed E and θ, along a spectrum of decreasing E', W increases and q^2 decreases. The maximum range of these variables over a single measured spectrum occurred at an incident energy of 17.7 GeV and an angle of 10°, where W varied from one proton mass to 5.2 GeV, and q^2 from 7.4 to 1.6 $(\text{GeV}/c)^2$. For each spectrum E' was changed in overlapping steps of 2% from elastic scattering, through the observed resonance region, to $W \approx 2$ GeV. Then steps corresponding to a change in W of 0.5 GeV were made.

The electron beam from the accelerator was momentum analyzed with values of $\Delta p/p$ between ± 0.1 and $\pm 0.25\%$ and then passed through a 7-cm liquid-hydrogen target. Two toroid charge monitors measured the integrated beam current with uncertainties of less than 0.5%. Electrons scattered in the target were momentum analyzed by a double-focusing magnetic spectrometer[9] capable of momentum analysis to 20 GeV/c. Particles selected by the spectrometer passed through a system of four hodoscopes to determine their trajectories and then into a pion-electron separation system based on the different cascade-shower properties of electrons and pions. This system considered of a 1-radiation-length slab of lead followed by three scintillation counters (dE/dx counters) to detect showers initiated in the lead. The showers were then further developed in a total-absorption counter consisting of sixteen 1-radiation-length lead slabs alternated with Lucite Cherenkov counters. The dE/dx counters increased the pion-electron separation efficiency by about a factor of 20 at lower E', but were not required for values of E' near the elastic peak. The electron-detection efficiency decreased with E' and was 88% at 5 GeV. The uncertainty in the electron-detection efficiency was $\pm 1.5\%$ above $E' = 5$ GeV and increased to $\pm 4\%$ at $E' = 3$ GeV.

The momentum acceptance of the spectrometer was $\Delta p/p = 3.5\%$ with momentum resolution of 0.1%. The angular acceptance was $\Delta\theta = 7$ mrad with a resolution of 0.3 mrad. The measured solid angle of the instrument was 6×10^{-5} sr with an uncertainty of $\pm 2\%$.

Extensive tests showed that there could be significant reductions in target density due to beam heating. In order to correct for changes in the density a second spectrometer[10] was simultaneously used to measure protons from elastic electron-proton scattering at low momentum transfer. The angle and momentum settings of this spectrometer remained fixed for each spectrum. Usually the density reductions were less than 4%, with the maximum value being 13%. An uncertainty of $\pm 1\%$ was assigned to the measured cross sections for this correction.

The main trigger for an event was provided by a logical "or" between the total-absorption counter and a coincidence of two scintillation trigger counters placed before and after the hodoscopes. The event information was buffered and written on magnetic tape by a SDS-9300 on-line computer, which also provided preliminary on-line data analysis.

The cross sections were determined from an event-by-event analysis of the hodoscopes and the electron-pion discrimination counters. Corrections were made for fast-electronics and computer dead times, hodoscope-counter inefficiencies, multiple tracks, inefficiencies of electron identification, and target-density fluctuations. Yields from an empty replica of the experimental target, typically 7%, were subtracted from the full-target measurements. Electrons originating from π^0 decay and pair production were measured by reversing the spectrometer polarity and measuring positron yields. This correction is important only for small E' and amounted to a maximum of 15%. The error associated with each point arose from counting statistics and uncertainties in electron-detection efficiencies, added in quadrature.

The data were analyzed separately at the Stanford Linear Accelerator Center (SLAC) and Massachusetts Institute of Technology (MIT) and averaged before each group began radiative corrections. The results of the analyses were in excellent agreement. For the results given in Table I, the two analyses differed from their mean by an average value of 0.35% with an rms deviation of 1.2%.

The radiative-correction procedures had two steps. The first was the subtraction of the calculated radiative tail of the elastic peak from each spectrum. Using the measured form factors

Table I. Measured cross sections for $W \geq 2.0$ GeV after all corrections. The errors are 1 standard deviation. The systematic error is not included in the table but is estimated at 5% for $E' > 5$ GeV, increasing to 10% at $E' \approx 3$ GeV.

θ (deg)	E (GeV)	E' (GeV)	q^2 (GeV/c)^2	W (GeV)	$d^2\sigma/d\Omega dE'$ (10^{-31} cm^2/sr-GeV)	θ (deg)	E (GeV)	E' (GeV)	q^2 (GeV/c)^2	W (GeV)	$d^2\sigma/d\Omega dE'$ (10^{-32} cm^2/sr-GeV)
6.000	7.000	5.130	.393	2.000	21.5 ± .49	10.000	10.988	7.915	2.643	2.001	5.66 ± .38
		4.586	.352	2.249	15.6 ± .40			6.879	2.297	2.509	6.17 ± .27
		3.750	.287	2.587	9.33 ± .64			5.634	1.881	3.008	5.59 ± .30
		3.250	.249	2.769	7.96 ± .73			4.163	1.390	3.507	5.19 ± .43
	10.005	7.886	.864	1.998	10.7 ± .23			3.000	1.001	3.856	5.38 ± .77
		7.349	.806	2.249	9.24 ± .20		13.534	9.737	4.004	2.000	1.80 ± .072
		6.745	.739	2.502	7.01 ± .27			9.270	3.812	2.252	2.20 ± .083
		5.361	.587	3.001	4.97 ± .24			8.737	3.593	2.508	2.62 ± .10
		3.724	.408	3.501	3.54 ± .30			7.534	3.098	3.007	3.03 ± .15
	13.529	11.00	1.630	1.999	4.26 ± .087			6.113	2.514	3.506	2.93 ± .15
		10.48	1.553	2.249	4.10 ± .093			4.473	1.839	4.005	2.98 ± .25
		9.936	1.473	2.480	3.85 ± .056			3.000	1.234	4.406	3.75 ± .54
		8.512	1.262	3.004	2.79 ± .085		15.201	10.86	5.016	2.002	.876 ± .058
		6.906	1.023	3.505	2.09 ± .11			9.868	4.558	2.516	1.57 ± .060
		5.054	.749	4.004	1.85 ± .11			8.691	4.014	3.014	1.94 ± .089
		3.394	.503	4.404	1.59 ± .27			7.300	3.372	3.512	2.08 ± .083
	16.049	13.16	2.314	1.998	2.19 ± .042			5.696	2.631	4.011	1.98 ± .094
		12.64	2.222	2.250	2.21 ± .043			4.258	1.967	4.410	2.26 ± .15
		12.03	2.116	2.510	2.16 ± .042			3.700	1.709	4.555	2.17 ± .22
		10.69	1.880	3.008	1.84 ± .042			3.000	1.386	4.732	2.46 ± .28
		9.109	1.602	3.507	1.59 ± .056		17.696	12.46	6.699	2.002	.336 ± .020
		7.282	1.280	4.006	1.24 ± .066			11.50	6.184	2.514	.617 ± .027
		5.644	.992	4.406	1.11 ± .074			10.36	5.571	3.012	.957 ± .042
		3.851	.677	4.805	1.13 ± .17			9.015	4.847	3.510	1.19 ± .057
10.000	7.010	4.802	1.023	2.000	2.82 ± .090			7.461	4.012	4.009	1.33 ± .073
		4.294	.915	2.250	2.34 ± .099			6.069	3.263	4.408	1.32 ± .091
		3.717	.792	2.504	2.05 ± .12			4.544	2.443	4.808	1.50 ± .15
								3.800	2.043	4.991	1.70 ± .21
								3.000	1.613	5.181	1.79 ± .37

for elastic electron-proton scattering, the radiative tail can be calculated to lowest order of α without using the peaking approximation.[11] Contributions from external and internal bremsstrahlung, including multiple photon emission, were calculated by two different methods. The differences between these methods were noticeable only for data with $E > 15$ GeV and $E' < 4$ GeV, and amounted to less than 2% in the corrected cross section. The maximum elastic-tail contributions to our data are 26% of the cross section at $E' = 3.8$ GeV in the 16-GeV, 6° spectrum and 21% at $E' = 3$ GeV in the 17.7-GeV, 10° spectrum.

The second step in the radiative correction procedure was a two-dimensional unfolding employing the peaking approximation.[11,12] All data at one angle were used to calculate the corrected cross sections. The computation involved no specific model for the cross section, but extensive interpolation and some extrapolation of the data were necessary. All experimental results having values of E' greater than the lowest value

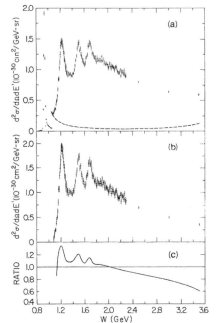

FIG. 1. The spectrum at $\theta = 6°$, $E = 10$ GeV (a) before and (b) after radiative corrections. In (a), the dashed line is the calculated elastic radiative tail which is subtracted before the two-dimensional unfolding is started. The elastic peak, but not the radiative tail, has been reduced by a factor of 6. (c) The ratio of the radiatively corrected to the uncorrected cross sections shown in (b) and (a). No systematic errors are shown. The radiative corrections increase the random errors.

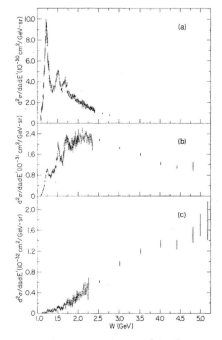

FIG. 2. Three representative radiatively corrected spectra at (a) $\theta = 6°$, $E = 7$ GeV; (b) $\theta = 6°$, $E = 16$ GeV; and (c) $\theta = 10°$, $E = 17.7$ GeV. The ranges of q^2 covered are (a) $0.2 \leq q^2 \leq 0.5$ (GeV/c)^2; (b) $0.7 \leq q^2 \leq 2.6$ (GeV/c)^2; and (c) $1.6 \leq q^2 \leq 7.3$ (GeV/c)^2. The elastic peaks are not shown.

of E (7 GeV) utilized no extrapolated data. A variety of numerical procedures involving different kinematic contours for interpolation and extrapolation have been studied.

The errors of the measured cross sections were propagated through the radiative unfolding procedure. Additional uncertainties resulted from the numerical procedures and the approximations made in the application of the radiative correction theories. From various studies we believe that uncertainties due to interpolation techniques are less than 1%, and uncertainties due to extrapolation procedures (most important at the lowest E spectrum at each angle) are less than 3%. Errors from theoretical approximations are more difficult to assess. They are small near the elastic peak and increase with decreasing E'; furthermore, they increase with increasing θ. We believe that for these data, errors due to theoretical approximations are on the order of 5% or less.

The results of the MIT and SLAC analyses, which involved different radiative correction procedures, differed typically from their mean by less than 1%, and nowhere by more than $\frac{1}{2}$ a standard deviation. The results have been averaged, and the differences have been included in the estimate of systematic error.

Figure 1 shows the 10-GeV, 6° spectrum. Figure 1(a) is the spectrum before radiative corrections. The dashed line is the calculated elastic radiative-tail contribution to this spectrum. Figure 1(b) shows this spectrum after complete radiative corrections, and Fig. 1(c) shows the ratio of the corrected to the measured data. Figure 2 shows three other corrected spectra with progressively increasing ranges of q^2. The q^2 dependence of the inelastic continuum at large W is clearly much weaker than that of the resonances.

Table I summarizes our results for $W \geq 2$ GeV. Data for $W \leq 2.3$ GeV are averages over a small number of neighboring data points, and all other data represent averages over the total spectrometer acceptance. The kinematic variables correspond to the central ray of the spectrometer. The errors are 1 standard deviation based on counting statistics and electron-detection efficiency, propagated through the radiative correction programs. Systematic errors are not included in Table I. Estimates of the combined systematic errors are 5% for $E' > 5$ GeV increasing to 10% at $E' \approx 3$ GeV. These data are in general agreement, to within the stated errors, with the preliminary data reported at Vienna.[7]

We are pleased to acknowledge the assistance of R. Cottrell, C. Jordan, J. Litt, and S. Loken. Professor W. K. H. Panofsky, Professor J. Pine, and Professor B. Barish participated in the initial planning of the experiment. We are indebted to E. Taylor and the Spectrometer Facilities Group, and to the Technical Division under R. B. Neal, especially the Research Area and Accelerator Operations Departments. We appreciate the help of Mrs. E. Miller during the analysis.

---

*Work supported by the U. S. Atomic Energy Commission.

†Now at Xerox Corp., Rochester, N. Y.

‡Work supported in part by the Atomic Energy Commission under Contract No. AT(30-1)2098.

[1]J. D. Bjorken, in Selected Topics in Particle Physics, Proceedings of the International School of Physics "Enrico Fermi," Course XLI, edited by J. Steinberger (Academic Press, Inc., New York, 1968).

[2]For example, the inelastic cross section may be represented in terms of the form factors W_1 and W_2 as

$$\frac{d^2\sigma}{d\Omega dE'} = \frac{e^4 \cos^2\tfrac{1}{2}\theta}{4E^2 \sin^4\tfrac{1}{2}\theta} [W_2(q^2, W) + 2W_1(q^2, W)\tan^2\tfrac{1}{2}\theta].$$

The squared four-momentum transfer is $q^2 = 4EE' \sin^2\tfrac{1}{2}\theta$. The mass of the final hadronic state is $W = [M_p^2 + 2M_p(E-E') - q^2]^{1/2}$. E and E' are the incident- and scattered-electron energies and θ is the scattering angle, all in the laboratory frame. M_p is the proton mass. For a discussion of the different form-factor notations see F. Gilman, Phys. Rev. 167, 1365 (1968).

[3]J. D. Bjorken, Ann. Phys. (N.Y.) 24, 201 (1963).

[4]A. A. Cone, K. W. Chen, J. R. Dunning, G. Hartwig, N. F. Ramsey, J. K. Walker, and R. Wilson, Phys. Rev. 156, 1490 (1967), and 163, 1854(E) (1967).

[5]F. W. Brasse, J. Engler, E. Ganssauge, and M. Schweizer, Nuovo Cimento 55A, 679 (1968).

[6]W. Bartel, B. Dudelzak, H. Krehbiel, J. McElroy, U. Meyer-Berkhout, W. Schmidt, V. Walther, and G. Weber, Phys. Letters 28B, 148 (1968).

[7]Preliminary results from the present experimental program are given in the report by W. K. H. Panofsky, in Proceedings of the Fourteenth International Conference on High Energy Physics, Vienna, Austria, September, 1968 (CERN Scientific Information Service, Geneva, Switzerland, 1968).

[8]See M. Breidenbach et al., following Letter [Phys. Rev. Letters 23, 935 (1969)] for a discussion of results. A description of the results obtained in the resonance region will be published.

[9]Brief descriptions of the spectrometer and the hodoscopes may be found in W. K. H. Panofsky, in Proceedings of the International Symposium on Electron and Photon Interactions at High Energies, Hamburg, Germany, 1965 (Springer-Verlag, Berlin, Germany, 1965), Vol. I; and A. M. Boyarski, F. Bulos, W. Busza, R. Diebold, S. D. Ecklund, G. E. Fischer, J. R. Rees, and B. Richter, Phys. Rev. Letters 20, 300 (1968).

[10]R. Anderson, D. Gustavson, R. Prepost, and D. Ritson, Nucl. Instr. Methods 66, 328 (1968).

[11]L. W. Mo and Y. S. Tsai, Rev. Mod. Phys. 41, 205 (1969).

[12]H. W. Kendall and D. Isabelle, Bull. Am. Phys. Soc. 9, 94 (1964).

OBSERVED BEHAVIOR OF HIGHLY INELASTIC ELECTRON-PROTON SCATTERING

M. Breidenbach, J. I. Friedman, and H. W. Kendall

Department of Physics and Laboratory for Nuclear Science,*
Massachusetts Institute of Technology, Cambridge, Massachusetts 02139

and

E. D. Bloom, D. H. Coward, H. DeStaebler, J. Drees, L. W. Mo, and R. E. Taylor

Stanford Linear Accelerator Center,† Stanford, California 94305

(Received 22 August 1969)

> Results of electron-proton inelastic scattering at 6° and 10° are discussed, and values of the structure function W_2 are estimated. If the interaction is dominated by transverse virtual photons, νW_2 can be expressed as a function of $\omega = 2M\nu/q^2$ within experimental errors for $q^2 > 1$ (GeV/c)^2 and $\omega > 4$, where ν is the invariant energy transfer and q^2 is the invariant momentum transfer of the electron. Various theoretical models and sum rules are briefly discussed.

In a previous Letter,[1] we have reported experimental results from a Stanford Linear Accelerator Center–Massachusetts Institute of Technology study of high-energy inelastic electron-proton scattering. Measurements of inelastic spectra, in which only the scattered electrons were detected, were made at scattering angles of 6° and 10° and with incident energies between 7 and 17 GeV. In this communication, we discuss some of the salient features of inelastic spectra in the deep continuum region.

One of the interesting features of the measurements is the weak momentum-transfer dependence of the inelastic cross sections for excitations well beyond the resonance region. This weak dependence is illustrated in Fig. 1. Here we have plotted the differential cross section divided by the Mott cross section, $(d^2\sigma/d\Omega dE')/(d\sigma/d\Omega)_{\text{Mott}}$, as a function of the square of the four-momentum transfer, $q^2 = 2EE'(1-\cos\theta)$, for constant values of the invariant mass of the recoiling target system, W, where $W^2 = 2M(E-E') + M^2 - q^2$. E is the energy of the incident electron, E' is the energy of the final electron, and θ is the scattering angle, all defined in the laboratory system; M is the mass of the proton. The cross section is divided by the Mott cross section

$$\left(\frac{d\sigma}{d\Omega}\right)_{\text{Mott}} = \frac{e^4}{4E^2}\frac{\cos^2\tfrac{1}{2}\theta}{\sin^4\tfrac{1}{2}\theta}$$

in order to remove the major part of the well-known four-momentum transfer dependence arising from the photon propagator. Results from both 6° and 10° are included in the figure for each value of W. As W increases, the q^2 dependence appears to decrease. The striking difference between the behavior of the inelastic and elastic cross sections is also illustrated in Fig. 1, where the elastic cross section, divided by the Mott cross section for $\theta = 10°$, is included. The q^2 dependence of the deep continuum is also consider-

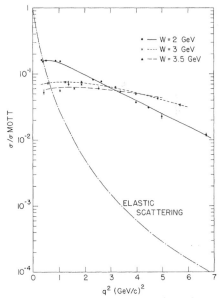

FIG. 1. $(d^2\sigma/d\Omega dE')/\sigma_{\text{Mott}}$, in GeV^{-1}, vs q^2 for W = 2, 3, and 3.5 GeV. The lines drawn through the data are meant to guide the eye. Also shown is the cross section for elastic e-p scattering divided by σ_{Mott}, $(d\sigma/d\Omega)/\sigma_{\text{Mott}}$, calculated for $\theta = 10°$, using the dipole form factor. The relatively slow variation with q^2 of the inelastic cross section compared with the elastic cross section is clearly shown.

ably weaker than that of the electroexcitation of the resonances,[2] which have a q^2 dependence similar to that of elastic scattering for $q^2 > 1$ (GeV/c)^2.

On the basis of general considerations, the differential cross section for inelastic electron scattering in which only the electron is detected can be represented by the following expression[3]:

$$\frac{d^2\sigma}{d\Omega dE'} = \left(\frac{d\sigma}{d\Omega}\right)_{\text{Mott}} (W_2 + 2W_1 \tan^2\tfrac{1}{2}\theta).$$

The form factors W_2 and W_1 depend on the properties of the target system, and can be represented as functions of q^2 and $\nu = E - E'$, the electron energy loss. The ratio W_2/W_1 is given by

$$\frac{W_2}{W_1} = \left(\frac{q^2}{\nu^2 + q^2}\right)(1+R), \quad R \geq 0,$$

where R is the ratio of the photoabsorption cross sections of longitudinal and transverse virtual photons, $R = \sigma_S/\sigma_T$.[4]

The objective of our investigations is to study the behavior of W_1 and W_2 to obtain information about the structure of the proton and its electromagnetic interactions at high energies. Since at present only cross-section measurements at small angles are available, we are unable to make separate determinations of W_2 and W_1. However, we can place limits on W_2 and study the behavior of these limits as a function of the invariants ν and q^2.

Bjorken[5] originally suggested that W_2 could have the form

$$W_2 = (1/\nu) F(\omega),$$

where

$$\omega = 2M\nu/q^2.$$

$F(\omega)$ is a universal function that is conjectured to be valid for large values of ν and q^2. This function is universal in the sense that it manifests scale invariance, that is, it depends only on the ratio ν/q^2. Since

$$\nu W_2 = \frac{\nu d^2\sigma/d\Omega\,dE'}{(d\sigma/d\Omega)_{\text{Mott}}} \left[1 + 2\frac{1}{1+R}\left(1 + \frac{\nu^2}{q^2}\right)\tan^2\tfrac{1}{2}\theta\right]^{-1},$$

the value of νW_2 for any given measurement clearly depends on the presently unknown value of R. It should be noted that the sensitivity to R is small when $2(1 + \nu^2/q^2)\tan^2\tfrac{1}{2}\theta \ll 1$. Experimental limits on νW_2 can be calculated on the basis of the extreme assumptions $R = 0$ and $R = \infty$. In Figs. 2(a) and 2(b) the experimental values of νW_2

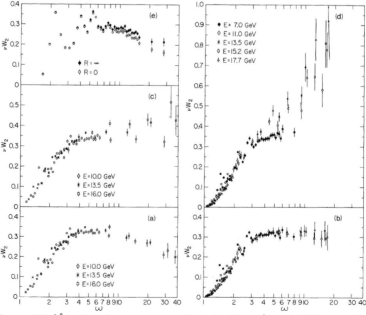

FIG. 2. νW_2 vs $\omega = 2M\nu/q^2$ is shown for various assumptions about $R = \sigma_S/\sigma_T$. (a) 6° data except for 7-GeV spectrum for $R = 0$. (b) 10° data for $R = 0$. (c) 6° data except for 7-GeV spectrum for $R = \infty$. (d) 10° data for $R = \infty$. (e) 6°, 7-GeV spectrum for $R = 0$ and $R = \infty$.

from the 6° and 10° data for $q^2 > 0.5$ (GeV/c)² are shown as a function of ω for the assumption that $R = 0$. Figures 2(c) and 2(d) show the experimental values of νW_2 calculated from the 6° and 10° data with $q^2 > 0.5$ (GeV/c)² under the assumption $R = \infty$. The 6°, 7-GeV results for νW_2, all of which have values of $q^2 \lesssim 0.5$ (GeV/c)², are shown for both assumptions in Fig. 2(e). The elastic peaks are not displayed in Fig. 2.

The results shown in these figures indicate the following:

(1) If $\sigma_T \gg \sigma_S$, the experimental results are consistent with a universal curve for $\omega \gtrsim 4$ and $q^2 \gtrsim 0.5$ (GeV/c)². Above these values, the measurements at 6° and 10° give the same results within the errors of measurements. The 6°, 7-GeV measurements of νW_2, all of which have values of $q^2 \lesssim 0.5$ (GeV/c)², are somewhat smaller than the results from the other spectra in the continuum region.

The values of νW_2 for $\omega \gtrsim 5$ show a gradual decrease as ω increases. In order to test the statistical significance of the observed slope, we have made linear least-squares fits to the values of νW_2 in the region $6 \leq \omega \lesssim 25$. These fits give $\nu W_2 = (0.351 \pm 0.023) - (0.00386 \pm 0.00088)\omega$ for data with $q^2 > 0.5$ (GeV/c)² and $\nu W_2 = (0.366 \pm 0.024) - (0.0045 \pm 0.0019)\omega$ for $q^2 > 1$ (GeV/c)². The quoted errors consist of the errors from the fit added in quadrature with estimates of systematic errors.

Since $\sigma_T + \sigma_S \simeq 4\pi^2 \alpha \nu W_2 / q^2$ for $\omega \gg 1$, our results can provide information about the behavior of σ_T if $\sigma_T \gg \sigma_S$. The scale invariance found in the measurements of νW_2 indicates that the q^2 dependence of σ_T is approximately $1/q^2$. The gradual decrease exhibited in νW_2 for large ω suggests that the photoabsorption cross section for virtual photons falls slowly at constant q^2 as the photon energy ν increases.

The measurements indicate that νW_2 has a broad maximum in the neighborhood of $\omega = 5$. The question of whether this maximum has any correspondence to a possible quasielastic peak[6] requires further investigation.

It should be emphasized that all of the above conclusions are based on the assumption that $\sigma_T \gg \sigma_S$.

(2) If $\sigma_S \gg \sigma_T$, the measurements of νW_2 do not follow a universal curve and have the general feature that at constant $2M\nu/q^2$, the value of νW_2 increases with q^2.

(3) For either assumption, νW_2 shows a threshold behavior in the range $1 \leq \omega \lesssim 4$. W_2 is constrained to be zero at inelastic threshold which corresponds to $\omega \simeq 1$ for large q^2. In the threshold region of νW_2, W_2 falls rapidly as q^2 increases at constant ν. This qualitatively different from the weak q^2 behavior for $\omega > 4$. For $q^2 \approx 1$ (GeV/c)², the threshold region contains the resonances excited in electroproduction. As q^2 increases, the variations due to these resonances damp out and the values of νW_2 do not appear to vary rapidly with q^2 at constant ω.

It can be seen from a comparison of Figs. 2(a) and 2(c) that the 6° data provide a measurement of νW_2 to within 10% up to a value of $\omega \approx 6$, irrespective of the values of R.

There have been a number of different theoretical approaches in the interpretation of the high-energy inelastic electron-scattering results. One class of models,[6-9] referred to as parton models, describes the electron as scattering incoherently from pointlike constituents within the proton. Such models lead to a universal form for νW_2, and the point charges assumed in specific models give the magnitude of νW_2 for $\omega > 2$ to within a factor of 2.[6] Another approach[10,11] relates the inelastic scattering to off-the-mass-shell Compton scattering which is described in terms of Regge exchange using the Pomeranchuk trajectory. Such models lead to a flat behavior of νW_2 as a function of ν but do not require the weak q^2 dependence observed and do not make any numerical predictions at this time. Perhaps the most detailed predictions made at present come from a vector-dominance model which primarily utilizes the ρ meson.[12] This model reproduces the gross behavior of the data and has the feature that νW_2 asymptotically approaches a function of ω as $q^2 \to \infty$. However, a comparison of this model with the data leads to statistically significant discrepancies. This can be seen by noting that the prediction for $d^2\sigma/d\Omega dE'$ contains a parameter ξ, the ratio of the cross sections for longitudinally and transversely polarized ρ mesons on protons, which is expected to be a function of W but which should be independent of q^2. For values of $W \geq 2$ GeV, the experimental values of ξ increase by about $(50 \pm 5)\%$ as q^2 increases from 1 to 4 (GeV/c)². This model predicts that

$$\sigma_S/\sigma_T = \xi(W)(q^2/m_\rho^2)[1 - q^2/2m\nu],$$

which will provide the most stringent test of this approach when a separation of W_1 and W_2 can be made.

The application of current algebra[13-17] and the use of current commutators leading to sum rules and sum-rule inequalities provide another way of comparing the measurements with theory. There have been some recent theoretical considerations[18-20] which have pointed to possible ambiguity in these calculations; however, it is still of considerable interest to compare them with experiment.

In general, W_2 and W_1 can be related to commutators of electromagnetic current densities.[6,16] The experimental value of the energy-weighted sum $\int_1^\infty (d\omega/\omega^2)(\nu W_2)$, which is related to the equal-time commutator of the current and its time derivative, is 0.16 ± 0.01 for $R = 0$ and 0.20 ± 0.03 for $R = \infty$. The integral has been evaluated with an upper limit $\omega = 20$. This integral is also important in parton theories where its value is the mean square charge per parton.

Gottfried[21] has calculated a constant-q^2 sum rule for inelastic electron-proton scattering based on a nonrelativistic quark model involving pointlike quarks. The resulting sum rule is

$$\int_1^\infty \frac{d\omega}{\omega}(\nu W_2) = \int_{q^2/2M}^\infty d\nu W_2$$

$$= 1 - \frac{G_{Ep}^2 + (q^2/4M^2)G_{Mp}}{1+q^2/4M^2},$$

where G_{Ep} and G_{Mp} are the electric and magnetic form factors of the proton. The experimental evaluation of this integral from our data is much more dependent on the assumption about R than the previous integral. We will thus use the 6° measurements of W_2 which are relatively insensitive to R. Our data for a value of $q^2 \simeq 1$ (GeV/$c)^2$, which extend to a value of ν of about 10 GeV, give a sum that is 0.72 ± 0.05 with the assumption that $R = 0$. For $R = \infty$, its value is 0.81 ± 0.06. An extrapolation of our measurements of νW_2 for each assumption suggests that the sum is saturated in the region $\nu \simeq 20\text{-}40$ GeV. Bjorken[13] has proposed a constant-q^2 sum-rule inequality for high-energy scattering from the proton and neutron derived on the basis of current algebra. His result states that

$$\int_1^\infty \frac{d\omega}{\omega}\nu(W_{2p}+W_{2n}) = \int_{q^2/2M}^\infty d\nu(W_{2p}+W_{2n}) \geq \tfrac{1}{2},$$

where the subscripts p and n refer to the proton and neutron, respectively. Since there are presently no electron-neutron inelastic scattering results available, we estimate W_{2n} in a model-dependent way. For a quark model[22] of the proton, $W_{2n} \simeq 0.8 W_{2p}$ whereas in the model[8] of Drell and co-workers, W_{2n} rapidly approaches W_{2p} as ν increases. Using our results, this inequality is just satisfied at $\omega \simeq 4.5$ for the quark model and at $\omega \simeq 4.0$ for the other model for either assumption about R. For example, this corresponds to a value of $\nu \simeq 4.5$ GeV for $q^2 = 2$ (GeV/$c)^2$. Bjorken[23] estimates that the experimental value of the sum is too small by about a factor of 2 for either model, but is should be noted that the q^2 dependence found in the data is consistent with the predictions of this calculation.

---

*Work supported in part through funds provided by the U. S. Atomic Energy Commission under Contract No. AT(30-1)2098.
†Work supported by the U. S. Atomic Energy Commission.
[1]E. Bloom et al., preceding Letter [Phys. Rev. Letters 23, 930 (1969)].
[2]Preliminary results from the present experimental program are given in the report by W. K. H. Panofsky, in Proceedings of the Fourteenth International Conference On High Energy Physics, Vienna, Austria, 1968 (CERN Scientific Information Service, Geneva, Switzerland, 1968), p. 23.
[3]R. von Gehlen, Phys. Rev. 118, 1455 (1960); J. D. Bjorken, 1960 (unpublished); M. Gourdin, Nuovo Cimento 21, 1094 (1961).
[4]See L. Hand, in Proceedings of the Third International Symposium on Electron and Photon Interactions at High Energies, Stanford Linear Accelerator Center, Stanford, California, 1967 (Clearing House of Federal Scientific and Technical Information, Washington, D. C., 1968), or F. J. Gilman, Phys. Rev. 167, 1365 (1968).
[5]J. D. Bjorken, Phys. Rev. 179, 1547 (1969).
[6]J. D. Bjorken and E. A. Paschos, Stanford Linear Accelerator Center, Report No. SLAC-PUB-572, 1969 (to be published).
[7]R. P. Feynman, private communication.
[8]S. D. Drell, D. J. Levy, and T. M. Yan, Phys. Rev. Letters 22, 744 (1969).
[9]K. Huang, in Argonne National Laboratory Report No. ANL-HEP 6909, 1968 (unpublished), p. 150.
[10]H. D. Abarbanel and M. L. Goldberger, Phys. Rev. Letters 22, 500 (1969).
[11]H. Harari, Phys. Rev. Letters 22, 1078 (1969).
[12]J. J. Sakurai, Phys. Rev. Letters 22, 981 (1969).
[13]J. D. Bjorken, Phys. Rev. Letters 16, 408 (1966).
[14]J. D. Bjorken, in Selected Topics in Particle Physics, Proceedings of the International School of Physics "Enrico Fermi," Course XLI, edited by J. Steinberger (Academic Press, Inc., New York, 1968).
[15]J. M. Cornwall and R. E. Norton, Phys. Rev. 177, 2584 (1969).
[16]C. G. Callan, Jr., and D. J. Gross, Phys. Rev. Letters 21, 311 (1968).
[17]C. G. Callan, Jr., and D. J. Gross, Phys. Rev. Letters 22, 156 (1969).

[18]R. Jackiw and G. Preparata, Phys. Rev. Letters **22**, 975 (1969).

[19]S. L. Adler and W.-K. Tung, Phys. Rev. Letters **22**, 978 (1969).

[20]H. Cheng and T. T. Wu, Phys. Rev. Letters **22**, 1409 (1969).

[21]K. Gottfried, Phys. Rev. Letters **18**, 1174 (1967).

[22]J. D. Bjorken, Stanford Linear Accelerator Center, Report No. SLAC-PUB-571, 1969 (unpublished).

[23]J. D. Bjorken, private communication.

MEASUREMENT OF THE NEUTRINO-NUCLEON AND ANTINEUTRINO-NUCLEON TOTAL CROSS SECTIONS

T. EICHTEN, H. FAISSNER, F.J. HASERT, S. KABE, W. KRENZ, J. Von KROGH,
J. MORFIN and K. SCHULTZE

III Physikalisches Institut der Technischen Hochschule, Aachen, Germany

G.H. BERTRAND-COREMANS, J. SACTON, W. Van DONINCK and P. VILAIN[*1]

Interuniversity Institute for High Energies, U L B , V U B , Brussels, Belgium

D.C. CUNDY, D. HAIDT, M. JAFFRE, S. NATALI[*2], P. MUSSET, J.B.M. PATTISON,
D.H. PERKINS[*3], A. PULLIA, A. ROUSSET, W. VENUS[*4] and H. WACHSMUTH

CERN, Geneva, Switzerland

V. BRISSON, B. DEGRANGE, M. HAGUENAUER, L. KLUBERG,
U. NGUYEN-KHAC and P. PETIAU

Laboratoire de Physique Nucléaire des Hautes Energies, Ecole Polytechnique, Paris, France

E. BELLOTTI, S. BONETTI, D. CAVALLI, C. CONTA[*5], E. FIORINI,
C. FRANZINETTI[*6] and M. ROLLIER

Istituto di Fisica dell'Università, Milano and I.N F.N., Milano, Italy

B. AUBERT, L.M. CHOUNET, P. HEUSSE, L. JAUNEAU,
A.M. LUTZ, C. PASCAUD and J.P. VIALLE

Laboratoire de l'Accélérateur Linéaire, Orsay, France

F.W. BULLOCK, M. DERRICK[*7], M.J. ESTEN, T.W. JONES, J. McKENZIE,
A.G. MICHETTE[*8], G. MYATT[*3] and W.G. SCOTT[*3,8]

University College, London, England

Received 10 August 1973

The ν and $\bar{\nu}$ nucleon total cross-sections have been determined as a function of energy using a sample of 2500 ν and 950 $\bar{\nu}$ event. The results are compared with predictions of scaling and charge symmetry hypotheses

Measurements of the total cross-sections for the two processes

$$\nu_\mu + \text{nucleon} \rightarrow \mu^- + \text{hadrons} \quad (1)$$

$$\bar{\nu}_\mu + \text{nucleon} \rightarrow \mu^+ + \text{hadrons} \quad (2)$$

[*1] Chercheur agréé de l'Institut Interuniversitaire des Sciences Nucléaires, Belgique.
[*2] Now at the University of Bari.
[*3] Also at the University of Oxford.
[*4] Now at the Rutherford High Energy Laboratory.
[*5] On leave of absence from the University and INFN-Pavia.
[*6] Also at the University of Turin.
[*7] Visiting research fellow on leave from Argonne National Laboratory.
[*8] Supported by the Science Research Council grant.

have been performed using the large heavy liquid bubble chamber Gargamelle exposed to the ν and $\bar{\nu}$ beams at the CERN PS. Gargamelle is a cylindrical chamber of length 4.8 m and diameter 1.85 m, placed in a magnetic field of 20 kG. The liquid filling was heavy freon CF_3Br with radiation length $X_0 = 11$ cm and interaction length $L_0 = 60$ cm. The analysis has been carried out using 95 000 and 174 000 pictures taken in the ν and $\bar{\nu}$ beams, respectively. Only those events located within a fiducial volume of 3 m^3 contained in the 7 m^3 visible volume of Gargamelle have been measured in the present work. This fiducial volume was chosen to provide a mean potential path length of 150 cm for the particles produced in the $\nu(\bar{\nu})$ interactions, allowing a reliable separation of the muons from the strongly interacting particles.

Muonic neutrino (antineutrino) induced reactions were characterised by the presence of a negative (positive) muon among the secondary products. Therefore, events containing at least one particle of negative (positive[+]) charge which did not undergo a strong interaction within the visible volume (absorption, nuclear interaction and/or large angle scattering $>30°$, or transverse momentum >100 MeV/c) were classified as neutrino (antineutrino) events. Of course, there exists a class of ambiguous events (hereafter called $\nu\bar{\nu}$ events) containing at least one "non-interacting" particle of each charge. Events without a muon-signature are called hadronic events [1] and are not analysed further in this paper

The energy of the neutrino (antineutrino) producing a given event was taken to be the total energy liberated in the interaction. The energy taken off by all charged and neutral secondaries must therefore be estimated. The momenta of charged particles were determined from track curvature or range measurements. The energy imparted to the neutrons and π^0 mesons was obtained by measuring the energy deposited by these particles in the visible volume of the chamber.

The muon energy is measured on average with a precision of about 8%, whilst the charged hadron energy is determined within 15%. Due to the short radiation length of heavy freon, γ-ray energies are only measured to a precision of about 30%.

Undetected neutrons constitute the main cause of missing energy. Only 75% of the emitted neutrons interact in the visible volume and produce a measurable neutron star. As on average only one third of the energy of a given neutron is deposited in the form of visible energy[++], the neutron energy was obtained by multiplying the visible energy of the neutron star by a factor of 3. Taking these facts into account the energy loss due to undetected neutrons was corrected for empirically by calculating the mean energy taken off by neutrons interactions with associated neutron stars, as a function of the total visible energy (E). The mean loss of energy per event was found to be:

for neutrinos $\quad 0.045 + 0.015\, E$ GeV

for antineutrinos $0.079 + 0.030\, E$ GeV.

A correction for undetected and unmeasurable γ-rays was also applied. On average, the energy imparted to an electron-positron pair was found to be 230 MeV, independent of the visible energy liberated in the neutrino interaction. This amount of energy was thus added to the events with an odd number of electron-positron pairs.

Some 7% of the tracks due to hadrons were unmeasurable because the particle interacts after a very short flight path so that any curvature measurement was meaningless. In these cases, either the hadron energy was estimated from an analysis of the secondary star, or if this was impossible, the particle was assigned an energy equal to the mean energy taken off by a hadron, i.e. 500 MeV.

Fast interacting protons and positive pions (kinetic energy $\gtrsim 1$ GeV) cannot be separated by a momentum-ionization analysis. However, as a result of a study of the distribution of δ-rays along their tracks, and the proton range spectrum, more than 80% of these interacting particles were found to be protons. Therefore, in the present study all these particles have been taken as protons.

The average energy correction per event due to all these causes amounts to $\sim (5 \pm 2.5)\%$ of the visible energy. Having applied these corrections, different cuts were imposed to select the final sample of events.

In order to reduce the background due to charged incoming particles interacting in the fiducial volume, cuts on the longitudinal momentum along the ν beam axis, P_L, and the total energy, E, and the four-momentum transfer q^2 were applied. Only those events with $P_L > 0.6$ GeV/c and $E \geqslant 1$ GeV and $q^2 \leqslant q^2_{max}$[+] were retained for further analysis.

Events have been discarded when the measurement error was greater than 30%. As a consequence, 4% of the events were rejected, the rejection rate being nearly independent of the value of the total energy

Details of the selected events in both the ν and $\bar{\nu}$ films are displayed in tables 1 and 2, respectively. The contamination of hadronic events among the ν, $\bar{\nu}$ and $\nu\bar{\nu}$ events has been estimated by calculating the prob-

---

[+] Stopping protons are easily identified at the scan table.
[++] This conclusion is obtained from the examination of all nucleon-nucleon data above 0.5 GeV/c

[+] q^2_{max} is the maximum allowed four-momentum transfer in a neutrino-nucleon collision: $q^2_{max} = 4E^2/(1+2E/M)$ where M is the nucleon mass

Table 1
Neutrino

E(GeV)	1–2	2–3	3–4	4–5	5–6	6–7	7–8	8–9	9–10	10
ν events	723	599	324	115	74	40	32	20	26	9
$\nu\bar{\nu}$ events	134	197	90	64	27	21	11	9	14	5
$\bar{\nu}$ events	44	24	10	9	2	3	–	1	1	1
Hadronic events	78	34	13	7	5	–	–	–	–	–
Estimated hadronic contamination in ν events	14.6	5 3	2.3	–	–	1 2	–	–	–	–
Final total of ν events	842	791	411	179	101	60	43	29	40	14

Table 2
Antineutrino

E(GeV)	1–2	2–3	3–4	4–5	5–6	6–7	7–8	8–9	9–10	10
$\bar{\nu}$ events	313	280	130	69	27	12	6	0	2	1
ν events	34	24	17	4	6	–	1	–	–	–
$\bar{\nu}\nu$ events	42	48	28	12	8	4	4	3	–	–
Estimated ν in $\bar{\nu}\nu$ events	5.8	8.7	4.1	2	1.9	–	0.24	–	–	–
Hadronic events	44	12	5	1	1	–	–	–	–	–
Estimated hadronic contamination in $\bar{\nu}$ events	13	5.5	0.3	1.5	–	–	–	–	–	–
Final total of $\bar{\nu}$ events	336	314	154	78	33	16	10	3	2	1

ability that a hadronic event can be assigned to any of these three categories. For neutrino events with $E < 2$ GeV, this contamination is found to be of the order of 2%; at higher energy, it becomes negligible.

Since in the ν beam the $\bar{\nu}$ flux is knwon to be small ($\sim 1\%$), all the $\nu\bar{\nu}$ events observed in the ν film have been considered as ν events. In fact all the so-called $\bar{\nu}$ events observed in the ν film are due to hadronic events in which one, or more positive hadrons leave the chamber without interacting. In the $\bar{\nu}$ film the number of ν events among the $\nu\bar{\nu}$ sample has been computed using the probability for a ν event to be ambiguous, as determined from the ν film.

The neutrino flux and energy spectrum were determined by measuring the muon radial flux distributions at different depths in the steel shielding [2]. The method was improved over the previous CERN neutrino experiments in that the muon flux was monitored continuously Furthermore the K to π production ratio has been measured extensively for 24 GeV protons incident on a Be target [2], i.e. at an energy near the one of the protons used in the present experiment (26 GeV).

The estimated error in the $\nu(\bar{\nu})$ spectra, 9% for energies ranging 2 to 6 GeV, is mainly due to the instabilities of the muon flux detectors, varying beam conditions and extrapolation from 24 to 26 GeV of the production data. Above 6 GeV, the error increases to approximately 12% due to uncertainties in the difference of the π and K meson absorption in the target. Below 2 GeV, as the neutrinos mainly come from pions for which the production data are not available and the corresponding muon flux cannot be measured, the neutrino flux can only be estimated by extrapolation.

It should be noted that the ratio of the ν and $\bar{\nu}$ spectra between 2 and 6 GeV is known to an accuracy of about 4%.

The total cross sections for ν and $\bar{\nu}$ are shown in fig. 1. The cross-sections for the quasi-elastic processes:

$$\nu_\mu + n \rightarrow \mu^- + p, \qquad \bar{\nu}_\mu + p \rightarrow \mu^+ + n$$

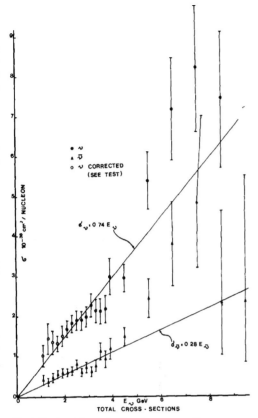

Fig. 1. Total neutrino and antineutrino cross-sections as a function of energy.

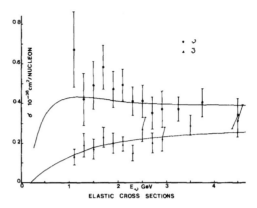

Fig. 2. Elastic neutrino and antineutrino cross-sections as a function of energy.

have also been determined, and are shown in fig. 2. For this analysis an elastic ν event is defined as one containing a single μ^- accompanied by not more than one proton of kinetic energy > 30 MeV.

An elastic antineutrino event is defined as one containing a single μ^+, no proton > 30 MeV and not more than one neutron > 30 MeV. A special scan assured that the scanning efficiency for this type of event was ~ 100%.

As stated above the flux between 1–2 GeV is not known with precision. Above 2 GeV the measured value of both the ν and $\bar\nu$ cross-sections are compatible with those expected using electromagnetic nucleon form-factors.

The curves shown in fig. 2 are the best fits for the elastic cross-section corrected for nuclear effects [3] for energies above 2 GeV.

From discrepancy of the neutrino elastic cross-section observed below 2 GeV, it is concluded that the extrapolation of the ν flux in this region is incorrect. Therefore the total cross-section measurements below 2 GeV have been corrected using the observed difference between the measured and expected elastic ν cross-section.

For the total ν and $\bar\nu$ cross-section data in fig. 1, best straight-lines have been fitted. In this fit account has been taken of the distortion expected for a linearly rising $\nu, \bar\nu$ cross section due to very rapidly falling $\nu, \bar\nu$ spectra and the measurement errors. The correction is of the order of ± 3%, except at 6 GeV where the cross-section is over-estimated by 10%. The best linear fits to the cross-sections are shown in table 3.

Table 3

		one parameter fit	two parameter fit
$E > 1$ GeV	ν	$(0.74\pm0.02)E$	$(0.70\pm0.07)E + (0.14\pm0.18)$
	$\bar\nu$	$(0.28\pm0.01)E$	$(0.26\pm0.04)E + (0.05\pm0.09)$
$E > 2$ GeV	ν	$(0.74\pm0.03)E$	$(0.77\pm0.09)E - (0.11\pm0.25)$
	$\bar\nu$	$(0.27\pm0.01)E$	$(0.32\pm0.06)E - (0.13\pm0.17)$

277

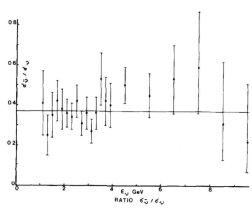

Fig. 3 The ratio $R = \sigma(\bar{\nu})/\sigma(\nu)$ of antineutrino to neutrino cross-sections as a function of energy.

To the extent that the ν and $\bar{\nu}$ cross-sections rise in an approximately linear fashion with energy, the data are consistent with the scaling hypothesis [4] which is well verified in the SLAC deep inelastic electron-nucleon scattering experiments [5].

Fig. 3 shows the ratio $R_1 = \sigma(\bar{\nu})/\sigma(\nu)$ of antineutrino to neutrino cross-sections as a function of energy E. Within the errors, R_1, is compatible with a constant independent of energy. The average are:

$R_1 = 0.37 \pm 0.02$, $E > 1$ GeV,

$R_1 = 0.38 \pm 0.02$, $E > 2$ GeV.

Scaling also implies a linear energy dependence of the mean value of q^2, called $\langle q^2 \rangle$. Fig. 4 shows the mean value $\langle q^2 \rangle$. Fig. 4 shows the mean value $\langle q^2 \rangle$ plotted against neutrino energy E, for events in the ν and $\bar{\nu}$ film. The results of linear fits are:

neutrino $\langle q^2 \rangle = 0.12 \pm 0.03 + (0.23 \pm 0.01)E$

antineu- $\qquad\qquad\qquad\qquad\qquad E > 1$ GeV
trino $\langle q^2 \rangle = 0.09 \pm 0.03 + (0.14 \pm 0.015)E$

neutrino $\langle q^2 \rangle = (0.22 \pm 0.06) + (0.21 \pm 0.02)E$

antineu- $\qquad\qquad\qquad\qquad\qquad E > 2$ GeV.
trino $\langle q^2 \rangle = (0.11 \pm 0.08) + (0.14 \pm 0.03)E$

The error on the slope depends very little on possible systematic errors in measuring the hadron energy in the events. For example, at $E = 4$ GeV, an 8% error in E results only in a 2% error in q^2 and therefore in the slope of the line.

The analysis of the energy transfer distributions (y) will be discussed in a forthcoming paper because it involves weighting procedures to take account of the π, μ ambiguities

The foregoing results are now discussed in terms of the hypotheses of scale invariance and charge symmetry However it should first be emphasized that most of the events have rather low energy, and are not in the "deep-inelastic" region, where $q^2 \gg M^2$ and $\nu \gg M$ (ν is the energy transfer, $E - E_\mu$), which are the conditions for scaling, as observed in the SLAC electron scattering experiments.

The scaling region cross-sections have the form:

$$\frac{d^2\sigma^{\nu,\bar{\nu}}}{dxdy} = \frac{G^2 ME}{\pi}[(1-y)F_2(x)$$

$$+ \frac{y^2}{2}[2xF_1(x)] \mp y(1-\frac{y}{2})xF_3(x)],$$

where $x = q^2/2M\nu$ and $y = \nu/E$ are dimensionless variables. For an equal number of neutron and proton targets, as is approximately true in the heavy liquid employed in the experiment, the hypothesis of charge symmetry of $\Delta S = 0$ weak processes results in the same values of the structure functions F_i for neutrinos and antineutrinos.

Experimentally, the cross-section ratio R_1 is close to the lowest bound allowed by scaling and charge symmetry i.e. 1/3. For this reason it is possible to obtain stringent bounds for ratios of the structure functions. In the case of $A = \int 2xF_1(x)\,dx/\int F_2(x)\,dx$ the bounds are:

$$(0.87 \pm 0.05) = \frac{3 - 3R_1}{1 + 3R_1} \leq A \leq 1.$$

This is in agreement with the Callan-Gross relation [6] as well as with the value found in the SLAC electroproduction experiments [5]. In terms of the parton model, a ratio equal to unity would imply partons of spin 1/2 only.

A similar bound can be obtained for the interference term, denoted by $B = -\int xF_3(x)\,dx/\int F_2(x)\,dx$, of

$$(0.90 \pm 0.04) = \frac{2(3 - 3R_1)}{3 + 3R_1} \geq B \geq \frac{(3 - 3R_1)}{1 + 3R_1}$$

$$= (0.87 \pm 0.05).$$

278

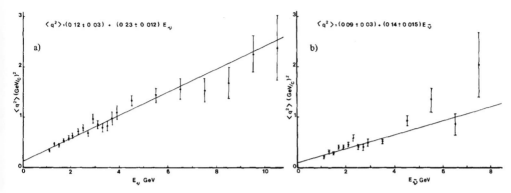

Fig. 4. (a) The mean value $\langle q^2 \rangle$ versus energy for neutrino interactions. (b) The mean value $\langle q^2 \rangle$ versus energy for antineutrino interactions.

The left limit is valid if one assumes the integrated Callan-Gross relation, i.e. $A = 1$. The fact that B is close to unity implies that parity is violated almost maximally.

Without making any additional assumptions scaling and charge symmetry can be tested by considering the average values $\langle xy \rangle_\nu$ and $\langle xy \rangle_{\bar\nu}$. The positivity conditions for the absorption of left, righthanded or scalar currents by the nucleon imply the following inequalities [7]:

$5(1+R_1) \geqslant 16(\langle xy \rangle_\nu - R_1 \langle xy \rangle_{\bar\nu})$

$7(1+R_1) \geqslant 16(\langle xy \rangle_\nu + R_1 \langle xy \rangle_{\bar\nu})$

$36 \langle xy \rangle_\nu \geqslant 6 R_1 \langle xy \rangle_{\bar\nu} \geqslant \langle xy \rangle_\nu$

$35(3 R_1 - 1) + 32 \langle xy \rangle_\nu \geqslant 192 R_1 \langle xy \rangle_{\bar\nu}$.

As $\langle q^2/E \rangle = 2M \langle xy \rangle$, the $\langle q^2 \rangle$ versus E plots give:

$\langle xy \rangle_\nu = 0.12 \pm 0.01$

$\langle xy \rangle_{\bar\nu} = 0.07 \pm 0.01$.

Hence, as the four above inequalities are satisfied by these experimental values, there is no indication of a violation of charge symmetry.

In addition assuming the local Callan-Gross relation $2xF_1(x) = F_2(x)$ it can be shown that

$R_2 \equiv \dfrac{\langle xy \rangle_{\bar\nu}}{\langle xy \rangle_\nu} = \dfrac{7 - 5B'}{R_1(7 + 5B')}$

where

$B' = -\int x (xF_3(x))\, dx \Big/ \int xF_2(x)\, dx.$

Experimentally

$B' = 0.87 \pm 0.08$.

The ratios B and B' are expected to be equal if $(xF_3(x))$ and $F_2(x)$ have the same functional x-dependence. The fact that B and B' are equal within the errors is compatible with this.

The value for the integral $\int F_2(x)\, dx$ can be determined from the relation:

$\sigma^\nu - \sigma^{\bar\nu} = \dfrac{G^2 ME}{\pi} \int F_2(x)\, dx \,(\tfrac{2}{3} B),$

to be

$0.49 \pm 0.03 \leqslant \int F_2(x)\, dx \leqslant 0.51 \pm 0.03$.

In the parton model this integral is interpreted as the relative momentum carried by isovector partons

$\int F_2(x)\, dx = \int x (U + D + \bar U + \bar D)\, dx,$

where $U, D, \bar U, \bar D$ are the isospin "up" and isospin "down" parton and antiparton momentum distribution functions in the proton. Therefore approximately 50% of the nucleon momentum is carried by either isoscalar partons or gluons.

The contribution of antipartons (isovector) can be estimated by means of the following relation [8]

$\dfrac{\int x(\bar U + \bar D)\, dx}{\int x(U + D)\, dx} \leqslant \tfrac{3}{8}(3 R_1 - 1) = 0.05 \pm 0.02$.

References

[1] F J. Hasert et al., Phys Lett. 46B (1973) 138.
[2] D. Bloess et al., Nuclear Instr. Methods 91 (1971) 605.
[3] J Løvseth, Nuovo Cim. 57A (1968) 382.
[4] J D Bjorken, Phys. Rev. 179 (1969) 1547.
[5] G Miller et al., Phys. Rev. D5 (1972) 528;
E D Bloom et al., Phys. Rev. Lett. 23 (1969) 930
[6] C.G Callan and D.G. Gross, Phys. Rev Lett. 22 (1969) 156.
[7] A. de Rujula and S L Glashow, Harvard Preprint (May 1973).
[8] E.A. Paschos, NAL Preprint-NAL Conf ,- 73/27 - THY (April 1973)

9

The J/ψ, the τ, and Charm

New forms of matter, 1974–1976.

In November 1974, Burton Richter at SLAC and Samuel Ting at Brookhaven were leading two very different experiments, one studying e^+e^- annihilation, the other the e^+e^- pairs produced in proton–beryllium collisions. Their simultaneous discovery of a new resonance with a mass of 3.1 GeV so profoundly altered particle physics that the period is often referred to as the "November Revolution." Word of the discoveries spread throughout the high energy physics community on November 11 and soon much of its research was directed towards the new particles.

Ting led a group from MIT and Brookhaven measuring the rate of production of e^+e^- pairs in collisions of protons on a beryllium target. The experiment was able to measure quite accurately the invariant mass of the e^+e^- pair. This made the experiment much more sensitive than an earlier one at Brookhaven led by Leon Lederman. That experiment differed in that $\mu^+\mu^-$ pairs were observed rather than e^+e^- pairs. Both these experiments investigated the Drell–Yan process whose motivation lay in the quark–parton model.

The Drell–Yan process is the production of e^+e^- or $\mu^+\mu^-$ pairs in hadronic collisions. Within the parton model, this can be understood as the annihilation of a quark from one hadron with an antiquark from the other to form a virtual photon. The virtual photon materializes some fraction of the time as a charged-lepton pair.

The e-pair and μ-pair approaches to measuring lepton-pair production each have advantages and disadvantages. Because high-energy muons are more penetrating than high-energy hadrons, muon pairs can be studied by placing absorbing material directly behind the interaction region. The absorbing material stops the strongly interacting π s, K s, and protons, but not the muons. This technique permits a very high counting rate since the muons can be separated from the hadrons over a large solid angle if enough absorber is available. The momenta of the muons can be determined by measuring their ranges. Together with the angle between the muons, this yields the invariant mass of the pair. Of course, the muons are subject to multiple Coulomb scattering in the absorber, so the resolution of the technique is limited by this effect. The spectrum observed by Lederman's group fell with increasing invariant mass of the lepton pair. There was, however, a shoulder in the spectrum between 3 and 4 GeV that attracted some notice, but whose real significance was obscured by the inadequate resolution.

By contrast, electrons can be separated from hadrons by the nature of the showers they cause or by measuring directly their velocity (using Cherenkov counters), which is much nearer the speed of light than that of a hadron of comparable momentum. The Cherenkov-counter approach is very effective in rejecting hadrons, but can be implemented easily only over a small solid angle. As a result, the counting rate is reduced. Ting's experiment used two magnetic spectrometers to measure separately the e^+ and e^-. The beryllium target was selected to minimize multiple Coulomb scattering. The achieved resolution was about 20 MeV for the e^+e^- pair, a great improvement over the earlier μ-pair experiment. The electrons and positrons were, in fact, identified using Cherenkov counters, time-of-flight information, and pulse height measurements.

In the early 1970s Richter, together with his co-workers, fulfilled his long-time ambition of constructing an e^+e^- ring, SPEAR, at SLAC to study collisions in the 2.5 to 7.5 GeV center-of-mass energy region. Lower energy machines had already been built at Novosibirsk, Orsay, Frascati, and Cambridge, Mass. Richter himself had worked as early as 1958 with Gerard O'Neill and others on the pioneering e^-e^- colliding-ring experiments at Stanford.

To exploit the new ring, SPEAR, the SLAC team, led by Richter and Martin Perl, and their LBL collaborators, led by William Chinowsky, Gerson Goldhaber, and George Trilling built a multipurpose large-solid-angle magnetic detector, the SLAC-LBL Mark I. The heart of this detector was a cylindrical magnetostrictive spark chamber inside a solenoidal magnet of 4.6 kG. This was surrounded by time-of-flight counters for particle velocity measurements, shower counters for photon detection and electron identification, and by proportional counters embedded in iron absorber slabs for muon identification.

What could the Mark I Collaboration expect to find in e^+e^- annihilations? In the quark–parton model, since interactions between the quarks are ignored, the process $e^+e^- \to q\bar{q}$ is precisely analogous to $e^+e^- \to \mu^+\mu^-$, except that the charge of the quarks is either 2/3 or $-1/3$ and that the quarks come in three colors, as more fully discussed in Chapter 10. Thus the ratio of the cross section for annihilation into hadrons to the cross section for the annihilation into muon pairs should simply be three times the sum of the squares of the charges of the quarks. This ratio, conventionally called R, was in 1974 expected to be $3[(-1/3)^2 + (2/3)^2 + (-1/3)^2] = 2$ counting the $u, d,$ and s quarks. In fact, measurements made at the Cambridge Electron Accelerator (CEA) found that R was not constant in the center-of-mass region to be studied at SPEAR, but instead seemed to grow to a rather large value, perhaps 6. The first results from the Mark I detector confirmed this puzzling result.

In 1974, Ting, Ulrich Becker, Min Chen and co-workers were taking data with their pair spectrometer at the Brookhaven AGS. By October of that year they found an e^+e^- spectrum consistent with expectations, except for a possible peak at 3.1 GeV. In view of the as-yet-untested nature of their new equipment, they proceeded to check and recheck this effect under a variety of experimental conditions and to collect more data.

During this same period, the Mark I experiment continued measurements of the annihilation cross section into hadrons with an energy scan with steps of 200 MeV. Since no abrupt structure was anticipated, these steps seemed small enough. The data confirming and extending the CEA results were presented at the London Conference in June 1974.

The data seemed to show a constant cross section rather than the $1/s$ behavior anticipated. (In the quark-parton model, there is no dimensionful constant, so the total cross section should vary as $1/s$ on dimensional grounds.) In addition, the value at center-of-mass energy 3.2 GeV appeared to be a little high. It was decided in June 1974 to check this by taking additional data at 3.1 and 3.3 GeV. Further irregularities at 3.1 GeV made it imperative in early November, 1974, before a cross section paper could be published, to remeasure this region. Scanning this region in very small energy steps revealed an enormous, narrow resonance. The increase in the cross section noticed at 3.2 GeV was due to the tail of the resonance and the anomalies at 3.1 GeV were caused by variations in the precise energy of the beam near the lower edge of the resonance, where the cross section was rising rapidly.

By Monday, November 11 (at which time the first draft of the ψ paper was already written) Richter learned from Sam Ting (who too had a draft of a paper announcing the new particle) about the MIT–BNL results on the resonance (named J by Ting), and *vice versa*. Clearly, both experiments had observed the same resonance. Word quickly reached Frascati, where Giorgio Bellettini and co-workers managed to push the storage ring beyond the designed maximum of 3 GeV and confirmed the discovery. Papers reporting the results at Brookhaven, SLAC, and Frascati all appeared in the same issue of *Physical Review Letters* (**Refs. 9.1, 9.2**, 9.3).

That the resonance was extremely narrow was apparent from the e^+e^- data, which showed an experimental width of 2 MeV. This was not the intrinsic width, but the result of the spread in energy of the electron and positron beams due to synchrotron radiation in the SPEAR ring. Additionally, the shape was spread asymmetrically by radiative corrections. If the natural width is much less than the beam spread, the area under the cross section curve

$$Area = \int dE\, \sigma \qquad (9.1)$$

is nearly the same as it would be in the absence of the beam spread and radiative corrections. The intrinsic resonance cross section is of the usual Breit–Wigner form given in Chapter 5

$$\sigma = \frac{2J+1}{(2S_1+1)(2S_2+1)} \frac{\pi}{p_{cm}^2} \frac{\Gamma_{in}\Gamma_{out}}{(E-E_0)^2 + \Gamma_{tot}^2/4}, \qquad (9.2)$$

where the incident particles have spin $S_1, S_2 = 1/2$ and momentum $p_{cm} \approx M_\psi/2 = E_0/2$. If the observed cross section is that for annihilation into hadrons, then $\Gamma_{out} = \Gamma_{had}$, the partial width for the resonance to decay into hadrons, while $\Gamma_{in} = \Gamma_{ee}$ is the electronic width. Assuming that the observed resonance has spin $J = 1$, we find by integrating the above,

$$Area = \frac{6\pi^2 \Gamma_{ee}\Gamma_{had}}{M_\psi^2 \Gamma_{tot}}. \qquad (9.3)$$

250   9. The J/ψ, the τ, and Charm

The area under the resonance curve measured at SPEAR is about 10 nb GeV. If we assume $\Gamma_{had} \approx \Gamma_{tot}$ and use the measured mass, $M_\psi = 3.1$ GeV, we find $\Gamma_{ee} \approx 4.2$ keV. The accepted value is 5.55 keV. Subsequent measurements of the branching ratio into electron pairs ($\approx 7\%$) led to a determination of the total width of between 60 and 70 keV, an astonishingly small value for a particle with a mass of 3 GeV.

Spurred by these results and theoretical predictions of a series of excited states like those in atomic physics, the SLAC–LBL Mark I group began a methodical search for other narrow states. It turned out to be feasible to modify the machine operation of SPEAR so that the energy could be stepped up by 1 MeV every minute. Ten days after the first discovery, a second narrow resonance was found (**Ref. 9.4**). The search continued, but no comparable resonances were found up to the maximum SPEAR energy of 7.4 GeV. The next such discovery had to wait until Lederman's group, this time at Fermilab and with much-improved resolution, continued their study of muon pairs into the 10 GeV region, as discussed in Chapter 11.

The discovery of the $\psi(3096)$ and its partner, ψ' or $\psi(3685)$ was the beginning of a period of intense spectroscopic work, which still continues. The spin and parity of the ψ s were established to be $J^P = 1^-$ by observing the interference between the ψ and the virtual photon intermediate states in $e^| e^- \to \mu^+ \mu^-$. The G-parity was found to be odd when the predominance of states with odd numbers of pions was demonstrated. Since C was known to be odd from the photon interference, the isospin had to be even and was shown to be nearly certainly $I = 0$. Two remarkable decays were observed quite soon after, $\psi' \to \psi \pi \pi$ and $\psi' \to \psi \eta$. Figure 9.1 shows a particularly clean $\psi' \to \psi \pi \pi$ decay with $\psi \to e^+ e^-$.

Prior to the announcement of the ψ, Tom Appelquist and David Politzer were investigating theoretically the binding of a charmed and an anticharmed quark, then hypothetical.

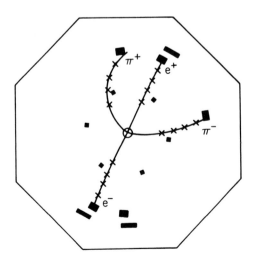

Figure 9.1. An example of the decay $\psi' \to \psi \pi^+ \pi^-$ observed by the SLAC–LBL Mark I Collaboration. The crosses indicate spark chamber hits. The outer dark rectangles show hits in the time-of-flight counters. Ref. 9.5.

They found that QCD predicted that there would be a series of bound states with very small widths, analogous to the e^+e^- bound states known as positronium. The $c\bar{c}$ bound states immediately became the leading explanation for the ψ and this interpretation was strengthened by the discovery of the ψ'. The ψ was seen as the lowest s-wave state with total spin equal to one. In spectroscopic notation it was the 1^3S_1. The ψ' was the next lowest spin-triplet, the s-wave state 2^3S_1.

The analogy between the $c\bar{c}$ bound states and positronium was striking. The two lowest energy states of positronium are the $^3S_1$ and the $^1S_0$. The former has $C = -1$ and the latter $C = +1$. It is this difference that first enabled Martin Deutsch to find experimental evidence for positronium in 1951. Because the triplet state has odd charge conjugation, it cannot decay into two photons like the charge-conjugation-even singlet state. As a consequence it decays into three photons and has a much longer lifetime. With detailed lifetime studies, Deutsch was able to find evidence for a long-lived species. QCD required that the triplet state of $c\bar{c}$ decay into three gluons, the quanta that bind the quarks together, while the singlet state could decay into two gluons. Again, this meant that the triplet state should be longer lived, that is, should have a narrow width.

In the nonrelativistic approximation, we can describe the $c\bar{c}$ system by a wave function, $\phi(r)$, satisfying a Schrödinger equation for some appropriate potential. The partial width, $\Gamma(\psi \to e^+e^-)$, is related to the wave function at zero separation, $\phi(0)$. The relation is obtained from the general prescription for a reaction rate, $\Gamma = \sigma\rho v$, where Γ is the reaction rate, σ the cross section, v is the relative velocity of the colliding particles and ρ is the target density. In this application $\rho = |\phi(0)|^2$. For the cross section we use the low energy limit of the process $c\bar{c} \to e^+e^-$,

$$\sigma = 3 \times \frac{2\pi\alpha^2 e_q^2}{\beta s}, \tag{9.4}$$

where α is the fine-structure constant ($\approx 1/137$), β is the velocity of the quark or antiquark in the center-of-mass frame, s is the center-of-mass energy squared ($\approx M_\psi^2$), and e_q is the charge of the quark measured in units of the proton's charge. A factor of 3 has been included to account for the three colors. The above cross section is averaged over the quark spins. The ψ is in fact a spin-triplet. The spin-singlet state has $C = +1$ and cannot annihilate through a virtual photon into e^+e^-. Since the cross section in the spin-singlet state is zero, the cross section in the spin-triplet state is actually 4/3 times the spin-averaged cross section. Noting that the relative velocity, v, is 2β, we have

$$\Gamma(\psi \to e^+e^-) = \frac{4}{3} \times 3 \times \frac{2\pi\alpha^2 e_q^2}{\beta M_\psi^2} \cdot 2\beta|\phi(0)|^2 \tag{9.5}$$

$$= \frac{16\pi\alpha^2 e_q^2}{M_\psi^2}|\phi(0)|^2. \tag{9.6}$$

The nonrelativistic model predicted that between the s-wave ψ and ψ' there would be a set of p-wave states. The spin-triplet states, ^3P, would have total angular momentum $J = 2, 1$, or 0. The spin-singlet state, ^1P, would have total angular momentum

$J = 1$. For a fermion–antifermion system the charge conjugation quantum number is $C = (-1)^{L+S}$, while the parity is $P = (-1)^{L+1}$. Thus the $^3P_{2,1,0}$ states would have $J^{PC} = 2^{++}, 1^{++}, 0^{++}$, while the $^1P_1$ state would have $J^{PC} = 1^{+-}$. The ψ' was expected to decay radiatively to the C-even states, which are now denoted χ (thus $\psi' \to \gamma\chi$). Such a transition was first observed at the PETRA storage ring at DESY in Hamburg by the Double Arm Spectrometer (DASP) group (**Ref. 9.6**). Evidence for all three χ states was then observed by the SLAC–LBL group with the Mark I detector, both by measuring the two photons in $\psi' \to \chi\gamma, \chi \to \psi\gamma$ and by detecting the first photon and a subsequent hadronic decay of the χ that was fully reconstructed.

The complete unraveling of these states took several years and was culminated in the definitive work of the Crystal Ball Collaboration, led by Elliott Bloom (**Ref 9.7**). Their detector was designed to provide high spatial and energy resolution for photons using 672 NaI crystals. A particularly difficult problem was the detection of the anticipated s-wave, spin-singlet states, $1\,^1S_0$ and $2\,^1S_0$ (denoted η_c and η_c') that were expected to lie just below the corresponding spin-triplet states, $1\,^3S_1$ and $2\,^3S_1$. Since these states have $C = +1$ and $J = 0$, they cannot be produced directly by e^+e^- annihilation through a virtual photon. Instead, they must be observed in the same way as the χ states, through radiative decays of the ψ and ψ'. The transitions are suppressed by kinematical and dynamical factors. They were identified only after a long effort.

In the simplest nonrelativistic model for the interaction between the charmed and anticharmed quarks, the potential is taken to be spin independent. In this approximation, the four p-states are degenerate, with identical radial wave functions. The $E1$ transitions, $\psi' \to \gamma\chi$ thus would occur with rates proportional to the statistical weights of the final states, $^3P_{0,1,2}$, i.e., $1 : 3 : 5$. In fact, as a result of spin-dependent forces, the splittings between the p-states are significant, so a better approximation is obtained by noting that the $E1$ rates are proportional to ω^3, where ω is the photon energy in the ψ' rest frame,

$$\omega = \frac{M_{\psi'}^2 - M_\chi^2}{2M_{\psi'}}. \tag{9.7}$$

If, for the masses of the ψ', χ_2, χ_1, χ_0 we take the measured values, 3.686, 3.556, 3.510, and 3.415 GeV, respectively, we find $\omega_2 = 0.128$ GeV, $\omega_1 = 0.172$ GeV, and $\omega_0 = 0.261$ GeV and the ratios

$$5 \times (0.128)^3 : 3 \times (0.172)^3 : 1 \times (0.261)^3 = 1 : 1.46 : 1.70. \tag{9.8}$$

The 2008 edition of the *Review of Particle Physics* gives branching ratios for $\psi' \to \gamma\chi_{2,1,0}$ of $8.3 \pm 0.4\%$, $8.8 \pm 0.8\%$, and $9.4 \pm 0.4\%$, in fair agreement with the above estimates.

It was during the exciting period of investigation of the ψ, ψ', and χ states that Martin Perl and co-workers of the SLAC–LBL group made a discovery nearly as dramatic as that of the ψ. Carefully sifting through 35,000 events, they found 24 with a μ and an opposite sign e, and no additional hadrons or photons. They interpreted these events as the pair production of a new lepton, τ, followed by its leptonic decay (**Ref. 9.8**). The

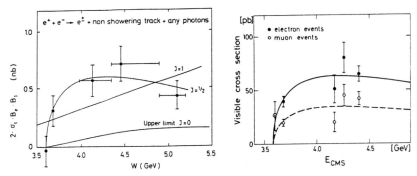

Figure 9.2. Left: The cross section from e^+e^- annihilation into candidates for τ leptons, as a function of center-of-mass energy, as measured by the DASP Collaboration. The threshold was determined to be very near 2×1800 MeV, that is, below the $\psi(3685)$ (Ref. 9.9). Right: Similar results from the DESY–Heidelberg group which give 1787^{+10}_{-18} MeV for the mass of the τ. The curves shown are for a spin-1/2 particle [W. Bartel et al., Phys. Lett. **B77**, 331 (1978)].

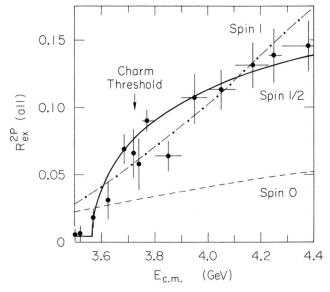

Figure 9.3. The production of anomalous two-prong events as a function of the center-of-mass energy, as determined by DELCO. These candidates for τ s yielded a threshold of 3564^{+4}_{-14} MeV, i.e. a mass of 1782^{+2}_{-7}. The threshold behavior confirmed the spin-1/2 assignment. (Ref. 9.10)

leptonic decays were $\tau \to e\nu\nu$ and $\tau \to \mu\nu\nu$. Figure 9.2 shows results obtained by the DASP Collaboration, using a double arm spectrometer, and by the DESY–Heidelberg Collaboration at the DORIS storage ring at DESY. Figure 9.3 show results from DELCO, the Direct Electron Counter at SPEAR. These established the spin and mass of the τ.

The decay $\tau \to e\nu\nu$ is exactly analogous to the decay $\mu \to e\nu\nu$. In both cases we can ignore the mass of the final state leptons. The decay rate for the μ is proportional to the square of the Fermi constant, G_F^2, which has dimension $[\text{mass}]^{-4}$. The decay rate for the

μ must then be proportional to m_μ^5. We expect then

$$\Gamma(\tau \to e\nu\bar{\nu}) = (m_\tau/m_\mu)^5 \Gamma(\mu \to e\nu\bar{\nu}) = 6 \times 10^{11}\,\text{s}^{-1}. \tag{9.9}$$

The measured lifetime of the τ is about 3.0×10^{-13} s and the branching ratio into $e\nu\nu$ is near 18%. Combining these gives a partial rate for $\tau \to e\nu\nu$ of roughly $6 \times 10^{11}\,\text{s}^{-1}$, in good agreement with the expectation.

Within a very short time, two new fundamental fermions had been discovered. The interpretation of the ψ as a bound state of a charmed quark and a charmed antiquark was backed by strong circumstantial evidence. What was lacking was proof that its constituents were indeed the charmed quarks first proposed by Bjorken and Glashow. As Glashow, Iliopoulos, and Maiani showed in 1970, charmed quarks were the simplest way to explain the absence of neutral strangeness-changing weak currents.

Until 1973 only weak currents that change charge had been observed. For example, in μ decay, the μ turns into ν_μ, and its charge changes by one unit. The neutral weak current, which can cause reactions like $\nu p \to \nu p$, as discussed in Chapter 12, does not change strangeness. If strangeness could be changed by a neutral current, then the decays $K^0 \to \mu^+\mu^-$ and $K^+ \to \pi^+ e^+ e^-$ would be possible. However, very stringent limits existed on these decays and others requiring strangeness-changing neutral weak currents. So restrictive were these limits that even second order weak processes would violate them in the usual Cabibbo scheme of weak interactions. Glashow, Iliopoulos, and Maiani showed that if in addition to the charged weak current changing an s quark into a u quark, there were another changing an s quark into a c quark, there would be a cancellation of the second order terms.

Consider the decay $K_L^0 \to \mu^+\mu^-$ for which the rate was known to be extremely small. The decay can proceed through the diagrams shown in Figure 9.4. Aside from other factors, the first diagram is proportional to $\sin\theta_C$ from the usW vertex and to $\cos\theta_C$ from the udW vertex. Here, W stands for the carrier of the weak interaction mentioned in Chapter 6 and discussed at length in Chapter 12.

The result given by this diagram alone would imply a decay rate that is not suppressed relative to normal K decay, in gross violation of the experimental facts. The proposal of Glashow, Iliopoulos, and Maiani was to add a fourth quark and correspondingly a second contribution to the charged weak current, which would become, symbolically,

$$\bar{u}(\cos\theta_C d + \sin\theta_C s) + \bar{c}(-\sin\theta_C d + \cos\theta_C s) = \begin{pmatrix} \bar{u} & \bar{c} \end{pmatrix} \begin{pmatrix} \cos\theta_C & \sin\theta_C \\ -\sin\theta_C & \cos\theta_C \end{pmatrix} \begin{pmatrix} d \\ s \end{pmatrix}. \tag{9.10}$$

Thus the Cabibbo angle would be simply a rotation, mixing the quarks d and s. Now when the $K_L^0 \to \mu^+\mu^-$ is calculated, there is a second diagram in which a c quark appears in place of the u quark. This amplitude has a term proportional to $-\sin\theta_C \cos\theta_C$, just cancelling the previous term. The surviving amplitude is higher order in G_F and does not conflict with experiment. The seminal quantitative treatment of this and related processes was given by M. K. Gaillard and B. W. Lee, who predicted the mass of the charmed quark to be about 1.5–2 GeV, in advance of the discovery of the ψ!

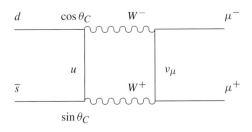

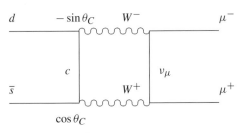

Figure 9.4. Two contributions to the decay $K_L^0 \to \mu^+\mu^-$ showing the factors present at the quark vertices. If only the upper contribution were present, the decay rate would be far in excess of the observed rate. The second contribution cancels most of the first. The cancellation would be exact if the c quark and u quark had the same mass. This cancellation is an example of the Glashow–Iliopoulos–Maiani mechanism.

As is described in Chapter 12, the discovery of strangeness non-changing neutral weak currents in 1973 made much more compelling the case for a unified theory of electromagnetism and weak interactions. The charmed quark was essential to this theoretical structure and the properties of the new quark were well specified by the theory. If the ψ was a bound state of a charmed quark and a charmed antiquark, there would have to be mesons with the composition $c\bar{u}$ and $c\bar{d}$, etc., that were stable against strong decays. The weak decay of a particle containing a c quark would yield an s quark. Thus the decay of a D^+ ($= c\bar{d}$) could produce a K^- ($= s\bar{u}$) but not a K^+ ($= \bar{s}u$).

There were a number of hints of charm already in the literature. K. Niu and collaborators working in Japan observed several cosmic-ray events in emulsion in which a secondary vertex was observed 10 to 100 μm from the primary vertex. These may have been decays of a particle with a lifetime in the 10^{-12} to 10^{-13} s range, just the lifetime expected for charmed particles. Nicolas Samios and Robert Palmer and co-workers, in a neutrino exposure of a hydrogen bubble chamber at Brookhaven, observed a single event that could have been a charmed baryon. See Figure 9.5. In other neutrino experiments, events with a pair of muons in the final state had been observed (Figure 9.6). These would be expected from processes in which the incident neutrino changed into a muon through the usual charged weak current and a strange quark was transformed into a charmed quark, again by the charged weak current. For that fraction in which the charmed particle decay produced a muon, two

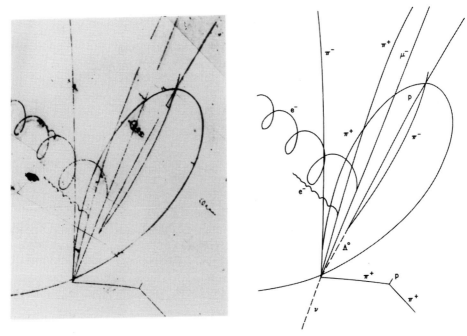

Figure 9.5. The event obtained in a neutrino exposure of the 7-ft hydrogen bubble chamber at Brookhaven that gave evidence for a charmed baryon. The overall reaction was most likely $\nu p \to \mu^- \Lambda^0 \pi^+ \pi^+ \pi^+ \pi^-$. The most probable assignments are shown in the sketch on the right. This violates the $\Delta S = \Delta Q$ rule. Such a violation can be understood if the process were really $\nu p \to \Sigma_c^{++} \mu^-$, followed by the strong decay $\Sigma_c^{++} \to \Lambda_c^+ \pi^+$. In the quark model $\Sigma_c^{++} = uuc$ and $\Lambda_c^+ = udc$. The decay of the Λ_c^+ to $\Lambda^0 \pi^+ \pi^+ \pi^-$ accounts for the violation of the $\Delta S = \Delta Q$ rule and is in accord with the pattern expected for charm decay. The mass of the Σ_c^{++} was measured to be 2426 ± 12 MeV. There were three possible choices for the pions to be joined to the Λ^0. Of these, one gave a mass splitting between the Σ_c^{++} and the Λ_c^+ of about 166 MeV, which agreed with the theoretical expectations [E. G. Cazzoli et al., Phys. Rev. Lett. **34**, 1125 (1975), figure courtesy N. Samios, Brookhaven National Laboratory].

muons would be observed in the final state, and they would have opposite charges. The evidence for a new phenomenon, perhaps charm, was accumulating.

The SLAC–LBL Mark I detector at SPEAR and the corresponding PLUTO and DASP at DESY were the leading candidates to produce convincing evidence for charmed particles. The rise in the e^+e^- annihilation cross section near a center-of-mass energy of 4 GeV strongly suggested that the threshold must be in that vicinity. The narrowness of the ψ' indicated that the threshold must be above that mass since the ψ' would be expected to decay rapidly into states like $c\bar{u}$ and $u\bar{c}$ if that were kinematically possible.

Despite advance knowledge of the approximate mass of the charmed particles and their likely decay characteristics, it took nearly two years before irrefutable evidence for them was obtained. The task turned out to be quite difficult because there were many different decay modes, with each having a branching ratio of just a few percent.

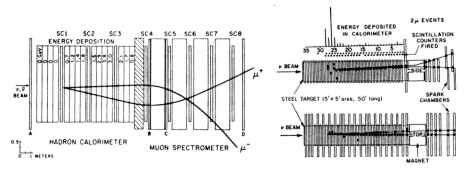

Figure 9.6. Early evidence for charm from opposite-sign dileptons observed in neutrino experiments at Fermilab. Left, one of fourteen events observed by the Harvard–Penn–Wisconsin Collaboration [A. Benvenuti et al., Phys. Rev. Lett. **34**, 419 (1975)]. Right, a similar event, one of eight seen by the Caltech–Fermilab Collaboration [B. C. Barish et al., Phys. Rev. Lett. **36**, 939 (1976)]. In addition, four events containing $\mu^- e^+ K_S^0$ were observed in the 15-ft bubble chamber at Fermilab [J. von Krogh et al., Phys. Rev. Lett. **36**, 710 (1976)] and two such events were seen in the Gargamelle bubble chamber at CERN [J. Blietschau et al., Phys. Lett. **60B**, 207 (1976)].

Ultimately, the SLAC–LBL Mark I group did succeed in isolating decays like $D^0 \to K^-\pi^+$ and $D^0 \to K^-\pi^-\pi^+\pi^+$ (**Ref. 9.11**), and soon after, $D^+ \to K^-\pi^+\pi^+$ (Ref. 9.12). See Figure 9.7. Overwhelming evidence was amassed identifying these new particles with the proposed charmed particles. Their masses were large enough to forbid the decay of the ψ' into a $D\bar{D}$ pair. The particles came in two doublets, (D^+, D^0) and $(\bar{D}^0, D^-)$, corresponding to $c\bar{d}, c\bar{u}$ and $\bar{c}u, \bar{c}d$. The decay mode $D^+ \to K^-\pi^+\pi^+$ was seen, but $D^+ \to K^+\pi^-\pi^+$ was not. It was possible to infer decay widths of less than 2 MeV, indicating that the decays were unlikely to be strong. The D's shared some properties of the K's. They were pair-produced with a particle of equal or greater mass, indicating the existence of a quantum number conserved in strong and electromagnetic interactions. In addition, their decays were shown to violate parity. Both nonleptonic and semileptonic decays were observed. The Cabibbo mixing in the four-quark model called for decays $c \to d$, suppressed by a factor roughly $\sin^2 \theta_C \approx 5\%$. These, too, were observed in $D^0 \to \pi^+\pi^-$ and $D^0 \to K^+K^-$. See Figure 9.8.

Further discoveries conformed to the charmed quark hypothesis. A set of partners about 140 MeV above the first states was found, with decays like $D^{*+} \to D^0\pi^+$ (Ref. 9.13). See Figure 9.9. These decays were strong, the analogs of $K^* \to K\pi$. Moreover, the spins of the D and D^* were consistent with the expected assignments, pseudoscalar and vector, respectively. Detailed studies of the charmed mesons were aided enormously by the discovery by the Lead Glass Wall collaboration of a resonance just above the charm threshold (Ref. 9.14), shown in Figure 9.10. The resonance, $\psi(3772)$, is primarily a d-wave bound state of $c\bar{c}$ with some mixture of $^3S_1$. The bound state decays entirely to $D\bar{D}$. The $\psi(3772)$ is thus a D-meson "factory" and has been the basis for a continuing study of charmed mesons.

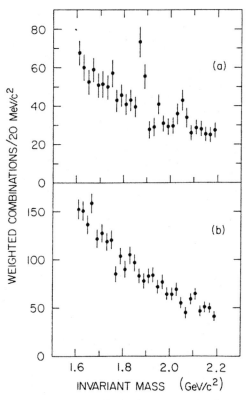

Figure 9.7. Invariant mass spectra for (a) $K^{\mp}\pi^{\pm}\pi^{\pm}$ and (b) $K^{\mp}\pi^{+}\pi^{-}$. Only the former figure shows a peak, in agreement with the prediction that D^+ decays to $K^-\pi^+\pi^+$, but not $K^+\pi^-\pi^+$. (Ref. 9.12)

The quark model requires that in addition to charmed mesons, there must be charmed baryons, in which one or more of the first three quarks are replaced by charmed quarks. Evidence for charmed baryons accumulated from a variety of experiments including neutrino bubble chamber experiments at Brookhaven and Fermilab, a photoproduction experiment at Fermilab, a spectrometer experiment at the CERN Intersecting Storage Ring (ISR), and the work of the Mark II group at SPEAR. The lowest mass charmed baryon has the composition udc and is denoted Λ_c^+. It has been identified in decays to $\Lambda\pi^+\pi^+\pi^-$, $\Lambda\pi^+$, pK_S^0, and $pK^-\pi^+$. In agreement with the results for meson decays, the decay of the charmed baryon yielded negative strangeness.

The strange-charmed meson with quark composition $c\bar{s}$ was even harder to find than the D. At first called the F^+ and now indicated D_s^+, it was observed by the CLEO detector at Cornell, by the ARGUS detector at DORIS (located at DESY), and by the TPC and HRS at PEP (located at SLAC). Evidence for this particle is shown in Figure 9.11. The F^* or D_s^* was also identified by TASSO at PETRA and the TPC, as well as the Mark III detector at SPEAR. It decays electromagnetically, $D_s^* \to D_s \gamma$. While the mass splitting

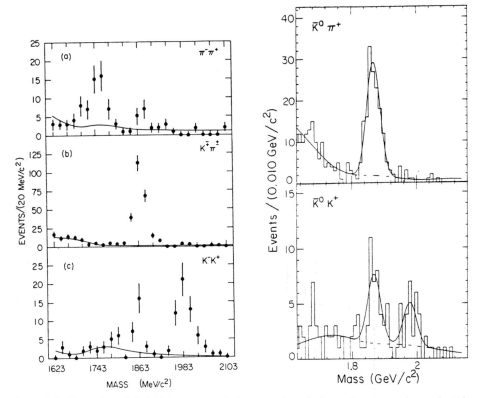

Figure 9.8. Examples of Cabibbo-suppressed decay modes of charmed mesons observed at the $\psi(3772)$. Left: $D^0 \to \pi^+\pi^-$ and $D^0 \to K^+K^-$ as well as the Cabibbo-allowed decay to $K^\mp\pi^\pm$. The data are from the Mark II experiment [G. S. Abrams et al., Phys. Rev. Lett. **43**, 481 (1979)]. Right: $D^+ \to \overline{K}^0 K^+$ as well as the Cabibbo-allowed mode $D^+ \to \overline{K}\pi^+$ from the Mark III experiment [R. M. Baltrusaitis et al., Phys. Rev. Lett. **55**, 150 (1985)]. For the suppressed modes, two peaks are observed. The one near 1865 MeV is the signal while the other is due to K/π misidentification.

is possibly large enough to permit $D_s^* \to D_s \pi^0$, this decay is suppressed by isospin conservation.

The lifetimes of the charmed mesons D^0, D^+, and D_s^+ as well as the charmed baryon Λ_c and the τ lepton are all in the region 10^{-13} s to 10^{-12} s and hence susceptible to direct measurement. The earliest measurements used photographic emulsions, with cosmic rays or beams at Fermilab or CERN providing the incident particles. This "ancient" technique is well suited to the few micron scale dictated by the small lifetimes. Studies were also conducted using special high resolution bubble chambers at CERN and SLAC. The required resolution was also achieved with electronic detectors at e^+e^- machines with the development of high precision vertex chambers pioneered by Mark II and later by MAC and DELCO at PEP, and TASSO, CELLO, and JADE at PETRA. The development returned the focus to hadronic machines where the production rate of charmed particles far exceeds that possible at e^+e^- machines. The detection with the requisite precision is achieved with

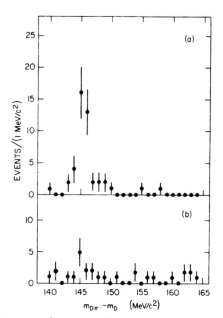

Figure 9.9. Data for $D^0\pi^+$ with $D^0 \to K^-\pi^+$. The abscissa is the difference between the $D\pi$ mass and the D mass. There is a clear enhancement near 145 MeV (G. J. Feldman *et al.* Ref. 9.13). The very small Q value for the D^{*+} decay, 5.88 ± 0.07 MeV, has become an important means of identifying the presence of a D^{*+} in high energy interactions. The data for $\overline{D}^0\pi^+$, a combination with the wrong quantum numbers to be a quark–antiquark state, show no enhancement.

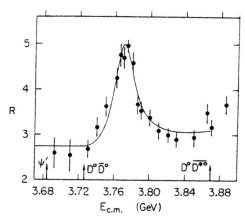

Figure 9.10. The $\psi(3772)$ resonance is broader than the $\psi(3096)$ and $\psi(3684)$ because it can decay into $D\overline{D}$. P. A. Rapidis *et al.*, (Ref. 9.14).

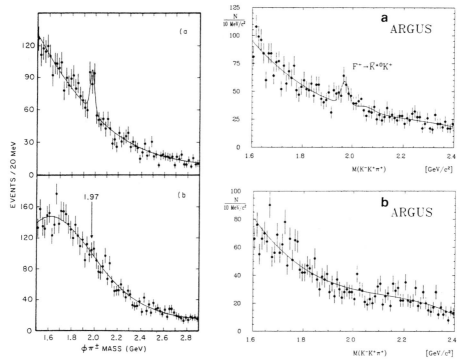

Figure 9.11. On the left, observation of the decay $D_s^+ \to \phi\pi^+$ by CLEO. In (a) only events in which the K^+K^- invariant mass is consistent with the mass of the ϕ are plotted. In (b) only $K^+K^-\pi$ events not containing a ϕ are shown [A. Chen et al., Phys. Rev. Lett., **51**, 634 (1983)]. On the right, observation of the decay $D_s^+ \to K^{*0}K^+$ by ARGUS. In (a) only events with $K^-\pi^+$ in the K^{*0} band are shown. In (b) only events without a K^{*0} are shown [ARGUS Collaboration, Phys. Lett. **179B**, 398 (1986)].

silicon microstrips. Experiments carried out at CERN and Fermilab achieved remarkable results, which required the analysis of 10^8 events in order to isolate several thousand charm decays.

Some of the lifetime measurements have relied on reconstructed vertices, others on impact parameters of individual tracks, as first employed in π^0 lifetime studies (Ref. 2.7). Figure 9.12 shows the photoproduction of a pair of charmed mesons from the SLAC Hybrid Facility Photon Collaboration. Both decay vertices are plainly visible. In the same figure a computer reconstruction of a digitized bubble chamber picture from LEBC at CERN, with an exaggerated transverse magnification, is shown. Again, pair production of charmed particles is demonstrated. Exponential decay distributions for charmed mesons obtained using a tagged photon beam at Fermilab are displayed in Figure 9.13.

The discoveries of the ψ, τ, and charm were pivotal events. They established the reality of the quark structure of matter and provided enormous circumstantial evidence for the theoretical view dubbed "The Standard Model," to be discussed in Chapter 12. The τ pointed the way to the third generation of matter, which is discussed in Chapter 11.

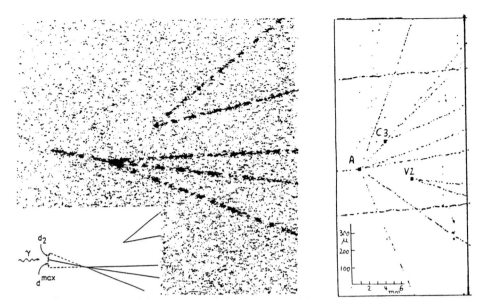

Figure 9.12. Left: A bubble chamber picture of the production and decay of a charged charmed particle and a neutral charmed particle. The charged particle decays into three tracks at 0.86 mm and the neutral decays after 1.8 mm. The quantities d_{max} and d_2, the largest and second largest impact distances were used in the lifetime calculations. The incident photon beam (E_{max} = 20 GeV) was obtained by Compton scattering of laser light off high energy electrons at SLAC [K. Abe *et al.*, *Phys. Rev. Lett.* **48**, 1526 (1982)]. Right: A computer reconstruction of a digitized bubble chamber picture. The transverse scale is exaggerated. The production vertex is at A. A charged charmed particle decays at $C3$ and a neutral charmed particle at $V2$. The picture was obtained with LEBC (Lexan Bubble Chamber) at CERN using a 360-GeV π^- beam [M. Aguilar-Benitez *et al.*, *Zeit. Phys.* **C31**, 491 (1986)].

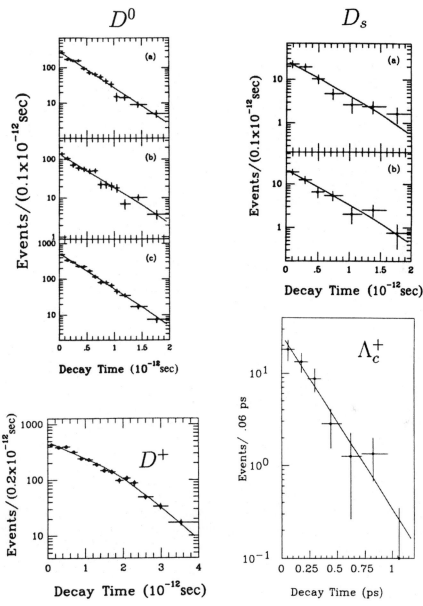

Figure 9.13. Proper time distributions for D^0, D^+, and D_s^+ mesons and Λ_c baryons from the Tagged Photon Spectrometer Collaboration at Fermilab, using silicon microstrip detectors [J. R. Raab et al., Phys. Rev. **D37**, 2391 (1988), J. C. Anjos et al., Phys. Rev. Lett. **60**, 1379 (1988)]. For the D^0, (a) corresponds to $D^{*+} \to D^0 \pi^+$, $D^0 \to K^-\pi^+$, (b) to $D^{*+} \to D^0\pi^+$, $D^0 \to K^-\pi^+\pi^+\pi^-$, and (c) to $D^0 \to K^-\pi^+$. For the D^+, the decay mode is $D^+ \to K^-\pi^+\pi^+$. For the D_s^+, (a) corresponds to $D_s^+ \to \phi\pi^+$ and (b) to $D_s^+ \to \overline{K}^{*0}K^+$, $\overline{K}^{*0} \to K^-\pi^+$. The lifetimes measured in these experiments were $\tau_{D^0} = (0.422 \pm 0.008 \pm 0.010) \times 10^{-12}$ s, $\tau_{D^+} = (1.090 \pm 0.030 \pm 0.025) \times 10^{-12}$ s, $\tau_{D_s} = (0.47 \pm 0.04 \pm 0.02) \times 10^{-12}$ s and $\tau_{\Lambda_c} = 0.22 \pm 0.03 \pm 0.02 \times 10^{-12}$ s.

Exercises

9.1 Estimate the lifetime of the D meson. Do you expect the neutral and charged D mesons to have the same lifetime? What do the data say?

9.2 Describe the baryons containing one or more charmed quarks that extend the lowest lying multiplets, the octet and decuplet. How many of these particles have been found? Compare with *Review of Practical Physics*. What do you expect their decay modes to be?

9.3 How have the most precise measurements of the mass of the ψ been made? See Ref. 9.15.

9.4 * Calculate the branching ratio for $\tau \to \pi \nu$. See Y. S. Tsai, *Phys. Rev.* **D4**, 2821 (1971); M. L. Perl, *Ann. Rev. Nucl. Part. Sci.* **30**, 229 (1980).

9.5 * Calculate the expected widths for $\psi' \to \gamma \chi_{2,1,0}$ in terms of the s- and p-state wave functions. Evaluate the results for a harmonic oscillator potential with the charmed quark mass set to 1.5 GeV and the spring constant adjusted to give the level splitting between the ψ and ψ' correctly. Calculate the partial width for $\psi \to \gamma \eta_c$. Why is the transition $\psi' \to \gamma \eta_c$ suppressed? Compare your results with the data given in the *Review of Particle Properties*. [See the lecture by J. D. Jackson listed in the Bibliography.]

9.6 * Show that the ψs produced in e^+e^- annihilation have their spins' components along the beam axis equal either to $+1$ or -1, but not 0. (Use the coupling of the ψ to $e^+e^-: \bar{e}\gamma_\mu e \psi^\mu$)

9.7 * What is the angular distribution of the γ's relative to the beam direction in $e^+e^- \to \psi' \to \gamma \chi_0$? What is the answer for χ_1 and χ_2 assuming that the transitions are pure $E1$? (See E. Eichten *et al.*, *Phys. Rev. Lett.* **34**, 369 (1975); G. J. Feldman and F. J. Gilman, *Phys. Rev.* **D12**, 2161 (1975); L. S. Brown and R. N. Cahn, *Phys. Rev.* **D13**, 1195 (1975).)

Further Reading

R. N. Cahn, ed., e^+e^- *Annihilation: New Quarks and Leptons*, Benjamin/Cummings, Menlo Park, CA, 1984. (A collection of articles from *Annual Review of Nuclear and Particle Science*.)

J. D. Jackson, "Lectures on the New Particles," in *Proc. of Summer Institute on Particle Physics, Stanford, CA, Aug. 2-13, 1976*, M. Zipf, ed.

G. J. Feldman and M. L. Perl, "Electron-Positron Annihilation above 2 GeV and the New Particles," *Phys. Rep.* **19**, 233 (1975) and **33**, 285 (1977).

G. H. Trilling, "The Properties of Charmed Particles," *Phys. Rep.* **75**, 57 (1981).

S. C. C. Ting, "Discovery of the J Particle: a Personal Recollection," *Rev. Mod. Phys.* **44(2)**, 235 (1977).

B. Richter, "From the Psi to Charm: the Experiments of 1975 and 1976," *Rev. Mod. Phys.* **44(2)**, 251 (1977).

A popular account of much of the historical material in the chapter is contained in contributions by S. C. C. Ting, G. Goldhaber, and B. Richter in *Adventures in Experimental Physics*, ϵ, B. Maglich, ed., World Science Education, Princeton, NJ, 1976. See also M. Riordan *The Hunting of the Quark*, Simon & Schuster, 1987.

References

9.1 J. J. Aubert et al., "Experimental Observation of a Heavy Particle J." *Phys. Rev. Lett.*, **33**, 1404 (1974).

9.2 J.-E. Augustin et al., "Discovery of a Narrow Resonance in e^+e^- Annihilation." *Phys. Rev. Lett.*, **33**, 1406 (1974).

9.3 C. Bacci et al., "Preliminary Result of Frascati (ADONE) on the Nature of a New 3.1 GeV Particle Produced in e^+e^- Annihilation." *Phys. Rev. Lett.*, **33**, 1408 (1974).

9.4 G. S. Abrams et al., "Discovery of a Second Narrow Resonance in e^+e^- Annihilation." *Phys. Rev. Lett.*, **33**, 1453 (1974).

9.5 G. S. Abrams et al., "Decay of $\psi(3684)$ into $\psi(3095)$." *Phys. Rev. Lett.*, **34**, 1181 (1974).

9.6 W. Braunschweig et al., "Observation of the Two Photon Cascade $3.7 \to 3.1 + \gamma\gamma$ via an Intermediate State P_c." *Phys. Lett.*, **B57**, 407 (1975).

9.7 R. Partridge et al., "Observation of an η_c Candidate State with Mass 2978 ± 9 MeV." *Phys. Rev. Lett.*, **45**, 1150 (1980); See also E. D.Bloom and C. W. Peck, "1983." *Ann. Rev. Nucl. Part. Sci.*, 30 (229). and J. E. Gaiser et al., "Charmonium Spectroscopy from Inclusive ψ' and J/ψ Radiative Decays." *Phys. Rev.*, **D34**, 711 (1986).

9.8 M. L. Perl et al., "Evidence for Anomalous Lepton Production in e^+e^- Annihilation." *Phys. Rev. Lett.*, **35**, 1489 (1975).

9.9 R. Brandelik et al., "Measurements of Tau Decay Modes and a Precise Determination of the Mass." *Phys. Lett.*, **73B**, 109 (1978).

9.10 W. Bacino et al., "Measurement of the Threshold Behavior of $\tau^+\tau^-$ Production in e^+e^- Annihilation." *Phys. Rev. Lett.*, **41**, 13 (1978).

9.11 G. Goldhaber et al., "Observation in e^+e^- Annihilation of a Narrow State at 1865 Mev/c^2 Decaying to $K\pi$ and $K\pi\pi\pi$." *Phys. Rev. Lett.*, **37**, 255 (1976).

9.12 I. Peruzzi et al., "Observation of a Narrow Charged State at 1876 MeV/c^2 Decaying to an Exotic Combination of $K\pi\pi$." *Phys. Rev. Lett.*, **37**, 569 (1976).

9.13 G. J. Feldman et al., "Observation of the Decay $D^{*+} \to D^0\pi^+$." *Phys. Rev. Lett.*, **38**, 1313 (1977).

9.14 P. A. Rapidis et al., "Observation of a Resonance in e^+e^- Annihilation Just Above Charm Threshold." *Phys. Rev. Lett.*, **39**, 526 (1977).

9.15 A. A. Zholentz et al., "High Precision Measurement of the ψ and ψ' Meson Masses." *Phys. Lett.*, **96B**, 214 (1980).

Experimental Observation of a Heavy Particle J†

J. J. Aubert, U. Becker, P. J. Biggs, J. Burger, M. Chen, G. Everhart, P. Goldhagen,
J. Leong, T. McCorriston, T. G. Rhoades, M. Rohde, Samuel C. C. Ting, and Sau Lan Wu
*Laboratory for Nuclear Science and Department of Physics, Massachusetts Institute of Technology,
Cambridge, Massachusetts 02139*

and

Y. Y. Lee
Brookhaven National Laboratory, Upton, New York 11973
(Received 12 November 1974)

> We report the observation of a heavy particle J, with mass $m = 3.1$ GeV and width approximately zero. The observation was made from the reaction $p + \text{Be} \rightarrow e^+ + e^- + x$ by measuring the e^+e^- mass spectrum with a precise pair spectrometer at the Brookhaven National Laboratory's 30-GeV alternating-gradient synchrotron.

This experiment is part of a large program to study the behavior of timelike photons in $p + p \rightarrow e^+ + e^- + x$ reactions[1] and to search for new particles which decay into e^+e^- and $\mu^+\mu^-$ pairs.

We use a slow extracted beam from the Brookhaven National Laboratory's alternating-gradient synchrotron. The beam intensity varies from 10^{10} to 2×10^{12} p/pulse. The beam is guided onto an extended target, normally nine pieces of 70-mil Be, to enable us to reject the pair accidentals by requiring the two tracks to come from the same origin. The beam intensity is monitored with a secondary emission counter, calibrated daily with a thin Al foil. The beam spot size is 3×6 mm², and is monitored with closed-circuit television. Figure 1(a) shows the simplified side view of one arm of the spectrometer. The two arms are placed at 14.6° with respect to the incident beam; bending (by $M1$, $M2$) is done vertically to decouple the angle (θ) and the momentum (p) of the particle.

The Cherenkov counter C_0 is filled with one atmosphere and C_e with 0.8 atmosphere of H_2. The counters C_0 and C_e are decoupled by magnets $M1$ and $M2$. This enables us to reject knock-on electrons from C_0. Extensive and repeated calibra-

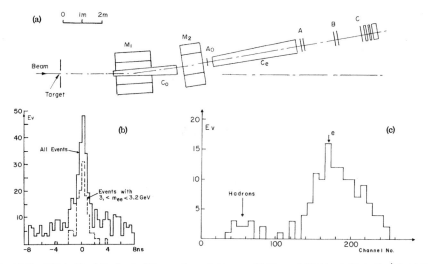

FIG. 1. (a) Simplified side view of one of the spectrometer arms. (b) Time-of-flight spectrum of e^+e^- pairs and of those events with $3.0 < m < 3.2$ GeV. (c) Pulse-height spectrum of e^- (same for e^+) of the e^+e^- pair.

tion of all the counters is done with approximately 6-GeV electrons produced with a lead converter target. There are eleven planes ($2 \times A_0$, $3 \times A$, $3 \times B$, $3 \times C$) of proportional chambers rotated approximately 20° with respect to each other to reduce multitrack confusion. To further reduce the problem of operating the chambers at high rate, eight vertical and eight horizontal hodoscope counters are placed behind chambers A and B. Behind the largest chamber C (1 m × 1 m) there are two banks of 25 lead glass counters of 3 radiation lengths each, followed by one bank of lead-Lucite counters to further reject hadrons from electrons and to improve track identification. During the experiment all the counters are monitored with a PDP 11-45 computer and all high voltages are checked every 30 min.

The magnets were measured with a three-dimensional Hall probe. A total of 10^5 points were mapped at various current settings. The acceptance of the spectrometer is $\Delta\theta = \pm 1°$, $\Delta\varphi = \pm 2°$, $\Delta m = 2$ GeV. Thus the spectrometer enables us to map the e^+e^- mass region from 1 to 5 GeV in three overlapping settings.

Figure 1(b) shows the time-of-flight spectrum between the e^+ and e^- arms in the mass region $2.5 < m < 3.5$ GeV. A clear peak of 1.5-nsec width is observed. This enables us to reject the accidentals easily. Track reconstruction between the two arms was made and again we have a clear-cut distinction between real pairs and accidentals. Figure 1(c) shows the shower and lead-glass pulse height spectrum for the events in the mass region $3.0 < m < 3.2$ GeV. They are again in agreement with the calibration made by the e beam.

Typical data are shown in Fig. 2. There is a clear sharp enhancement at $m = 3.1$ GeV. Without folding in the 10^5 mapped magnetic points and the radiative corrections, we estimate a mass resolution of 20 MeV. As seen from Fig. 2 the width of the particle is consistent with zero.

To ensure that the observed peak is indeed a real particle ($J \rightarrow e^+e^-$) many experimental checks were made. We list seven examples:

(1) When we decreased the magnet currents by 10%, the peak remained fixed at 3.1 GeV (see Fig. 2).

(2) To check second-order effects on the target, we increased the target thickness by a factor of 2. The yield increased by a factor of 2, not by 4.

(3) To check the pileup in the lead glass and shower counters, different runs with different voltage settings on the counters were made. No effect was observed on the yield of J.

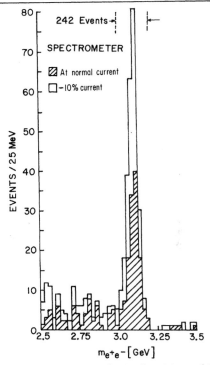

FIG. 2. Mass spectrum showing the existence of J. Results from two spectrometer settings are plotted showing that the peak is independent of spectrometer currents. The run at reduced current was taken two months later than the normal run.

(4) To ensure that the peak is not due to scattering from the sides of magnets, cuts were made in the data to reduce the effective aperture. No significant reduction in the J yield was found.

(5) To check the read-out system of the chambers and the triggering system of the hodoscopes, runs were made with a few planes of chambers deleted and with sections of the hodoscopes omitted from the trigger. No effect was observed on the J yield.

(6) Runs with different beam intensity were made and the yield did not change.

(7) To avoid systematic errors, half of the data were taken at each spectrometer polarity.

These and many other checks convinced us that we have observed a real massive particle $J \rightarrow ee$.

If we assume a production mechanism for J to be $d\sigma/dp_\perp \propto \exp(-6p_\perp)$ we obtain a yield of J of approximately 10^{-34} cm^2.

The most striking feature of J is the possibility that it may be one of the theoretically suggested charmed particles[2] or a's[3] or Z_0's,[4] etc. In order to study the real nature of J,[5] measurements are now underway on the various decay modes, e.g., an $e\pi\nu$ mode would imply that J is weakly interacting in nature.

It is also important to note the absence of an e^+e^- continuum, which contradicts the predictions of parton models.[6]

We wish to thank Dr. R. R. Rau and the alternating-gradient synchrotron staff who have done an outstanding job in setting up and maintaining this experiment. We thank especially Dr. F. Eppling, B. M. Bailey, and the staff of the Laboratory for Nuclear Science for their help and encouragement. We thank also Ms. I. Schulz, Ms. H. Feind, N. Feind, D. Osborne, G. Krey, J. Donahue, and E. D. Weiner for help and assistance. We thank also M. Deutsch, V. F. Weisskopf, T. T. Wu, S. Drell, and S. Glashow for many interesting conversations.

———

†Accepted without review under policy announced in Editorial of 20 July 1964 [Phys. Rev. Lett. 13, 79 (1964)].

[1]The first work on $p+p \to \mu^+ +\mu^- +x$ was done by L. M. Lederman et al., Phys. Rev. Lett. 25, 1523 (1970).

[2]S. L. Glashow, private communication.

[3]T. D. Lee, Phys. Rev. Lett. 26, 801 (1971).

[4]S. Weinberg, Phys. Rev. Lett. 19, 1264 (1967), and 27, 1688 (1971), and Phys. Rev. D 5, 1412, 1962 (1972).

[5]After completion of this paper, we learned of a similar result from SPEAR. B. Richter and W. Panofsky, private communication; J.-E. Augustin et al., following Letter [Phys. Rev. Lett. 33, 1404 (1974)].

[6]S. D. Drell and T. M. Yan, Phys. Rev. Lett. 25, 316 (1970). An improved version of the theory is not in contradiction with the data.

Discovery of a Narrow Resonance in e^+e^- Annihilation*

J.-E. Augustin,† A. M. Boyarski, M. Breidenbach, F. Bulos, J. T. Dakin, G. J. Feldman,
G. E. Fischer, D. Fryberger, G. Hanson, B. Jean-Marie,† R. R. Larsen, V. Lüth,
H. L. Lynch, D. Lyon, C. C. Morehouse, J. M. Paterson, M. L. Perl,
B. Richter, P. Rapidis, R. F. Schwitters, W. M. Tanenbaum,
and F. Vannucci‡

Stanford Linear Accelerator Center, Stanford University, Stanford, California 94305

and

G. S. Abrams, D. Briggs, W. Chinowsky, C. E. Friedberg, G. Goldhaber, R. J. Hollebeek,
J. A. Kadyk, B. Lulu, F. Pierre,§ G. H. Trilling, J. S. Whitaker,
J. Wiss, and J. E. Zipse

Lawrence Berkeley Laboratory and Department of Physics, University of California, Berkeley, California 94720

(Received 13 November 1974)

We have observed a very sharp peak in the cross section for $e^+e^- \to$ hadrons, e^+e^-, and possibly $\mu^+\mu^-$ at a center-of-mass energy of 3.105 ± 0.003 GeV. The upper limit to the full width at half-maximum is 1.3 MeV.

We have observed a very sharp peak in the cross section for $e^+e^- \to$ hadrons, e^+e^-, and possibly $\mu^+\mu^-$ in the Stanford Linear Accelerator Center (SLAC)–Lawrence Berkeley Laboratory magnetic detector[1] at the SLAC electron-positron storage ring SPEAR. The resonance has the parameters

$E = 3.105 \pm 0.003$ GeV,

$\Gamma \leq 1.3$ MeV

(full width at half-maximum), where the uncertainty in the energy of the resonance reflects the uncertainty in the absolute energy calibration of the storage ring. [We suggest naming this structure $\psi(3105)$.] The cross section for hadron production at the peak of the resonance is ≥ 2300 nb, an enhancement of about 100 times the cross section outside the resonance. The large mass, large cross section, and narrow width of this structure are entirely unexpected.

Our attention was first drawn to the possibility of structure in the $e^+e^- \to$ hadron cross section during a scan of the cross section carried out in 200-MeV steps. A 30% (6 nb) enhancement was observed at a c.m. energy of 3.2 GeV. Subsequently, we repeated the measurement at 3.2 GeV and also made measurements at 3.1 and 3.3 GeV. The 3.2-GeV results reproduced, the 3.3-GeV measurement showed no enhancement, but the 3.1-GeV measurements were internally inconsistent—six out of eight runs giving a low cross section and two runs giving a factor of 3 to 5 higher cross section. This pattern could have been caused by a very narrow resonance at an energy slightly larger than the nominal 3.1-GeV setting of the storage ring, the inconsistent 3.1-GeV cross sections then being caused by setting errors in the ring energy. The 3.2-GeV enhancement would arise from radiative corrections which give a high-energy tail to the structure.

We have now repeated the measurements using much finer energy steps and using a nuclear magnetic resonance magnetometer to monitor the ring energy. The magnetometer, coupled with measurements of the circulating beam position in the storage ring made at sixteen points around the orbit, allowed the relative energy to be determined to 1 part in 10^4. The determination of the absolute energy setting of the ring requires the knowledge of $\int B \, dl$ around the orbit and is accurate to $\pm 0.1\%$.

The data are shown in Fig. 1. All cross sections are normalized to Bhabha scattering at 20 mrad. The cross section for the production of hadrons is shown in Fig. 1(a). Hadronic events are required to have in the final state either ≥ 3 detected charged particles or 2 charged particles noncoplanar by $> 20°$.[2] The observed cross section rises sharply from a level of about 25 nb to a value of 2300 ± 200 nb at the peak[3] and then exhibits the long high-energy tail characteristic of radiative corrections in e^+e^- reactions. The detection efficiency for hadronic events is 45% over the region shown. The error quoted above includes both the statistical error and a 7% contribution from uncertainty in the detection efficiency.

Our mass resolution is determined by the energy spread in the colliding beams which arises from quantum fluctuations in the synchrotron radiation emitted by the beams. The expected Gaussian c.m. energy distribution ($\sigma = 0.56$ MeV), folded with the radiative processes,[4] is shown as the dashed curve in Fig. 1(a). The width of the resonance must be smaller than this spread; thus an upper limit to the full width at half-maximum is 1.3 MeV.

Figure 1(b) shows the cross section for e^+e^- final states. Outside the peak this cross section

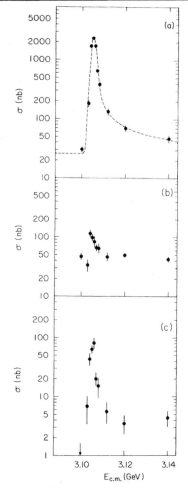

FIG. 1. Cross section versus energy for (a) multi-hadron final states, (b) e^+e^- final states, and (c) $\mu^+\mu^-$, $\pi^+\pi^-$, and K^+K° final states. The curve in (a) is the expected shape of a δ-function resonance folded with the Gaussian energy spread of the beams and including radiative processes. The cross sections shown in (b) and (c) are integrated over the detector acceptance. The total hadron cross section, (a), has been corrected for detection efficiency.

is equal to the Bhabha cross section integrated over the acceptance of the apparatus.[1]

Figure 1(c) shows the cross section for the production of collinear pairs of particles, excluding electrons. At present, our muon identifications system is not functioning and we therefore cannot separate muons from strongly interacting particles. However, outside the peak the data are consistent with our previously measured μ-pair cross section. Since a large $\pi\pi$ or KK branching ratio would be unexpected for a resonance this massive, the two-body enhancement observed is *probably* but not *conclusively* in the μ-pair channel.

The $e^+e^- \to$ hadron cross section is presumed to go through the one-photon intermediate state with angular momentum, parity, and charge conjugation quantum numbers $J^{PC} = 1^{--}$. It is difficult to understand how, without involving new quantum numbers or selection rules, a resonance in this state which decays to hadrons could be so narrow.

We wish to thank the SPEAR operations staff for providing the stable conditions of machine performance necessary for this experiment. Special monitoring and control techniques were developed on very short notice and performed excellently.

─────────

*Work supported by the U. S. Atomic Energy Commission.

†Present address: Laboratoire de l'Accélérateur Linéaire, Centre d'Orsay de l'Université de Paris, 91 Orsay, France.

‡Permanent address: Institut de Physique Nucléaire, Orsay, France.

§Permanent address: Centre d'Etudes Nucléaires de Saclay, Saclay, France.

[1]The apparatus is described by J.-E. Augustin *et al.*, to be published.

[2]The detection-efficiency determination will be described in a future publication.

[3]While preparing this manuscript we were informed that the Massachusetts Institute of Technology group studying the reaction $pp \to e^+e^- + x$ at Brookhaven National Laboratory has observed an enhancement in the e^+e^- mass distribution at about 3100 MeV. J. J. Aubert *et al.*, preceding Letter [Phys. Rev. Lett. **33**, 1402 (1974)].

[4]G. Bonneau and F. Martin, Nucl. Phys. **B27**, 381 (1971).

Discovery of a Second Narrow Resonance in e^+e^- Annihilation*†

G. S. Abrams, D. Briggs, W. Chinowsky, C. E. Friedberg, G. Goldhaber, R. J. Hollebeek,
J. A. Kadyk, A. Litke, B. Lulu, F. Pierre,‡ B. Sadoulet, G. H. Trilling, J. S. Whitaker,
J. Wiss, and J. E. Zipse

Lawrence Berkeley Laboratory and Department of Physics, University of California, Berkeley, California 94720

and

J.-E. Augustin,§ A. M. Boyarski, M. Breidenbach, F. Bulos, G. J. Feldman, G. E. Fischer,
D. Fryberger, G. Hanson, B. Jean-Marie,§ R. R. Larsen, V. Luth, H. L. Lynch, D. Lyon,
C. C. Morehouse, J. M. Paterson, M. L. Perl, B. Richter, P. Rapidis, R. F. Schwitters,
W. Tanenbaum, and F. Vannucci‖

Stanford Linear Accelerator Center, Stanford University, Stanford, California 94305
(Received 25 November 1974)

We have observed a second sharp peak in the cross section for $e^+e^- \to$ hadrons at a center-of-mass energy of 3.695 ± 0.004 GeV. The upper limit of the full width at half-maximum is 2.7 MeV.

The recent discovery of a very narrow resonant state coupled to leptons and hadrons[1-3] has raised the obvious question of the existence of other narrow resonances also coupled to leptons and hadrons. We therefore began a systematic search of the mass region accessible with the Stanford Linear Accelerator Center (SLAC) e^+e^- storage ring SPEAR and quickly found a second narrow resonance decaying to hadrons. The parameters of the new state [which we suggest calling $\psi(3695)$] are

$$M = 3.695 \pm 0.004 \text{ GeV}, \quad \Gamma < 2.7 \text{ MeV}$$

[full width at half-maximum (FWHM)], where the mass uncertainty reflects the uncertainty in the absolute energy calibration of the storage ring.

The $\psi(3695)$, like the $\psi(3105)$, was found using the SLAC–Lawrence Berkeley Laboratory magnetic detector at SPEAR.[4] The luminosity monitoring, event acceptance criteria, and storage-ring energy determination have been described previously.[1]

The new feature of this run is the search procedure used to hunt for narrow e^+e^- resonances. In the search mode the storage-ring energy is increased in about 1-MeV steps ($E_{c.m.} = 2 \times E_{beam}$)

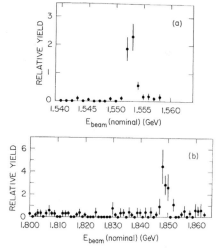

FIG. 1. Search-mode data (relative hadron yield) taken (a) in a 1-h calibration run over the $\psi(3105)$ (average luminosity of 2×10^{29} cm^{-2} sec^{-1}), and (b) during the run in which the $\psi(3695)$ was found (average luminosity of 5×10^{29} cm^{-2} sec^{-1}).

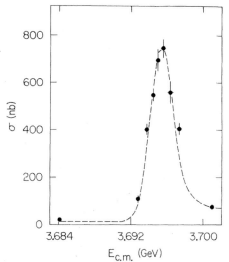

FIG. 2. Total cross section for $e^+e^- \to$ hadrons corrected for detection efficiency. The dashed curve is the expected resolution folded with the radiative corrections. The errors shown are statistical only.

every 3 min. The data taken during each step are analyzed in real time and the relative cross sections computed at the end of each step. Figure 1(a) shows the search-mode data taken during a calibration scan over the previously discovered $\psi(3105)$. Figure 1(b) shows the data taken during the first scan which began at a ring energy of 1.8 GeV. A clear indication of a narrow resonance with a mass of about 3.70 GeV is seen. It should be emphasized that we have not yet scanned any mass region other than that between 3.6 and 3.71 GeV.

On finding evidence of a resonance in the $e^+e^- \to$ hadron cross section, we switched to the normal SPEAR operating mode of longer runs at fixed energy. In this mode, smaller energy changes are possible than in the search mode. Figure 2 shows the cross section for $e^+e^- \to$ hadrons, corrected for the detection efficiency of about 55% over the energy region shown.

Our mass resolution is determined by the energy spread in the colliding beams, which depends on the energy of the beams. The expected Gaussian c.m. energy distribution ($\sigma = 1.2$ MeV) folded with the radiative processes[5] is shown as the dashed curve in Fig. 2. The width of the resonance must be smaller than this spread; thus,

an upper limit to the FWHM is 2.7 MeV.

In summary, the colliding-beam data now show two narrow resonances in the hadron production cross section. Our determination of the parameters of the resonance are as follows:

	Mass (GeV)	Γ (FWHM) (MeV)
$\psi(3105)$	3.105 ± 0.003	< 1.9 (Ref. 6)
$\psi(3695)$	3.695 ± 0.004	< 2.7

We are continuing the search for others.

We thank the SPEAR operations staff for the technological *tour de force* they accomplished whereby we are able to scan the machine energy in small, well-defined steps. We also acknowledge the cooperation of the Stanford Center for Information Processing in expediting the computation needs of this experiment.

---

*Work supported by the U. S. Atomic Energy Commission.

†Accepted without review under policy announced in Editorial of 20 July 1964 [Phys. Rev. Lett. 13, 79 (1964)].

‡Permanent address: Centre d'Etudes Nucléaires de Saclay, Saclay, France.

§Present address: Laboratoire de l'Accélérateur Linéaire, Centre d'Orsay de l'Université de Paris, 91 Orsay, France.
‖Permanent address: Institut de Physique Nucléaire, Orsay, France.

[1]J.-E. Augustin et al., Phys. Rev. Lett. 33, 1406 (1974).
[2]J. J. Aubert et al., Phys. Rev. Lett. 33, 1404 (1974).
[3]C. Bacci et al., Phys. Rev. Lett. 33, 1408 (1974).
[4]J.-E. Augustin et al., to be published.
[5]G. Bonneau and F. Martin, Nucl. Phys. B27, 321 (1971).
[6]In Ref. 1 a factor of $\sqrt{2}$ was omitted from the calculation of the experimental upper limit to the width of $\psi(3105)$. The correct value is 1.9 MeV rather than 1.3 MeV which appears in Ref. 1. However, the curve in Fig. 1(a) of Ref. 1 showing the calculated tail of the resonance is correct.

OBSERVATION OF THE TWO PHOTON CASCADE 3.7 → 3.1 + γγ VIA AN INTERMEDIATE STATE P_c

DASP-Collaboration

W. BRAUNSCHWEIG, H.-U. MARTYN, H.G. SANDER, D. SCHMITZ, W. STURM and W. WALLRAFF
I. Physikalisches Institut der RWTH Aachen, Germany

K. BERKELMAN*, D. CORDS, R. FELST, E. GADERMANN, G. GRINDHAMMER, H. HULTSCHIG,
P. JOOS, W. KOCH, U. KÖTZ, H. KREHBIEL, D. KREINICK, J. LUDWIG, K.-H. MESS,
K.C. MOFFEIT, A. PETERSEN, G. POELZ, J. RINGEL, K. SAUERBERG, P. SCHMÜSER,
G. VOGEL**, B.H. WIIK and G. WOLF
Deutsches Elektronen-Synchrotron DESY and II. Institut für Experimentalphysik der Universität Hamburg, Hamburg, Germany

G. BUSCHHORN, R. KOTTHAUS, U.E. KRUSE***, H. LIERL, H. OBERLACK, K. PRETZL, and M. SCHLIWA
Max-Planck-Institut für Physik und Astrophysik, München, Germany

S. ORITO, T. SUDA, Y. TOTSUKA and S. YAMADA
High Energy Physics Laboratory and Dept. of Physics, University of Tokyo, Tokyo, Japan

Received 22 July 1975

The two photon cascade decay of the 3.7 GeV resonance into the 3.1 GeV resonance has been observed in two nearly independent experiments. The clustering of the photon energies around 160 MeV and 420 MeV observed in the channel 3.7 → (3.1 → $\mu^+\mu^-$) + γγ indicates the existence of at least one intermediate state with even charge conjugation at a mass around 3.52 GeV or 3.26 GeV.

In studying the cascade transition of the 3.7 GeV resonance into the 3.1 GeV resonance we have observed the decay channel

$$3.7 \to 3.1 + \gamma\gamma \quad (1)$$

in two nearly independent experiments. The energies of the two photons cluster around 160 and 420 MeV suggesting the cascade decay to proceed via at least one new resonance with even charge conjugation.

The measurement has been performed at the DESY electron-positron storage rings DORIS using the double arm spectrometer DASP. The DASP detector is shown in fig. 1. It consists of two identical magnetic spectrometers arranged symmetrically with respect to the colliding beams. The details of this part of the spectrometer can be found elsewhere [1]. A large-aperture non magnetic detector is mounted between the two spectrometer arms. It consists of six sectors covering about 75% of 4π. To the top and bottom sectors, described already in previous publications [2], four similar sectors have been added during the course of the experiment. Each sector consists of a scintillation counter hodoscope, a 5 mm thick lead sheet and two or three layers of proportional tubes, all repeated four times and followed by a lead-scintillator shower detector of 7 radiation lengths. In addition the beam pipe is surrounded by a layer of 22 scintillation counters.

The two experiments were carried out as follows:

---
* On leave from Cornell University, Ithaca, N.Y.
** Now at Hochtemperatur Reaktorbau, Mannheim.
*** On leave from the University of Illinois, Urbana, Illinois.

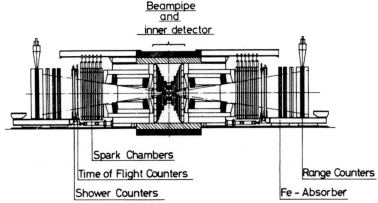

Fig. 1a. A schematic view of DASP showing the two magnetic spectrometer arms and the non magnetic detector (inner detector) mounted in the free space between the two spectrometer arms.

In the first experiment, the decay $3.7 \to 3.1 + \gamma\gamma$ was identified by measuring the two electrons from the decay of the 3.1 GeV resonance together with the two photons from the cascade, using the non magnetic detector only. The event was then completely reconstructed from the measurement of the emission angles of the four particles. In a second experiment, done concurrently with the first one, the reaction was identified by observing the decay of the 3.1 GeV resonance into a pair of muons with the magnetic spectrometers and selecting events with just two photons identified in the inner detector. The two experiments will be discussed separately.

A. $3.7 \to (3.1 \to e^+e^-) + \gamma\gamma$

Using the scanning criteria listed below events with just two electrons and two photons in the inner detector were selected.

For an electron we required:

(a) the appropriate beam pipe scintillation counter is fired.

(b) an electromagnetic shower is produced with energy greater than 700 MeV, as determined from the pulse height observed in the combined scintillator shower counter system (the loss of events due to this cut is at most a few percent).

Electron pairs in the top or bottom of the inner detector or in the horizontal part of the solid angle covered by the proportional chambers P1 and P2 were accepted. The angles of the electrons were determined to $\pm 1°$.

For a photon we required:

(a) a nonzero pulse height in the shower counter (with the threshold set at 0.3 of the pulse height for a minimum ionizing particle) or at least one proportional tube and one scintillation counter fired,

(b) the appropriate beam pipe counters and front counters in the inner detector module did not fire.

Photon showers which were within $15°$ of the axis of the electron shower were not accepted. The angular resolution for photons is $\pm 2°$ if they convert before a proportional tube chamber, and $\pm 8°$ in Φ and $\pm 20°$ in θ if they are detected by means of the counter hodoscope only.

Restricting the non-collinearity angle of the electrons to less than $45°$ we found 71 events with just two electrons and two photons as defined above. From the measured directions of the four particles and the known initial energy, a OC calculation was made yielding the momenta of the four particles. From the momenta so determined the effective mass distribution of the electron pairs was computed and is plotted in fig. 2a. The plot shows a clear peak centered at 3.1 GeV less than 200 MeV wide. This is consistent with the position and width expected for the decay:

$$3.7 \to (3.1 \to e^+e^-) + \gamma\gamma . \qquad (2)$$

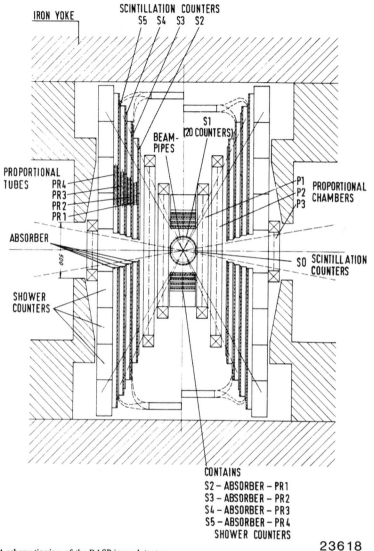

Fig. 1b. A schematic view of the DASP inner detector.

Here a Monte Carlo computation, including the measurement errors, predicts a peak at 3.1 GeV in the effective mass distribution of the e^+e^- pairs with a width of 130 MeV (FWHM).

The most serious background comes from the decay

$$3.7 \rightarrow (3.1 \rightarrow e^+e^-) + (\pi°\pi°), \qquad (3)$$

with only two of the four photons detected. The con-

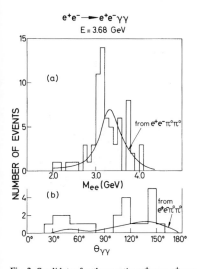

Fig. 2. Candidates for the reaction $e^+e^- \to e^+e^-\gamma\gamma$.
a) e^+e^- effective mass distribution
b) Distribution of the two-photon opening angle for events with $3.0 < M_{ee} < 3.2$ GeV.
The curves show the background from $e^+e^- \to (3.1 \to e^+e^-) \pi^0\pi^0$ as predicted by the Monte Carlo calculation.

tribution from this decay mode was calculated by a Monte Carlo calculation.

As an input to the Monte Carlo calculation the shape of the $\pi^0\pi^0$ mass distribution was taken to be identical to the mass distribution observed for the $\pi^+\pi^-$ pairs from the decay $3.7 \to (3.1 \to \mu^+\mu^-) + \pi^+\pi^-$ (see below). For the absolute prediction the branching ratio $\pi^0\pi^0/\pi^+\pi^-$ was assumed to be 0.5, as expected for a $\Delta I = 0$ transition. In the calculations the geometric acceptance and the measured photon detection efficiency were used. Also the measurement errors listed above were included in the computation. The shape and the magnitude of the effective e^+e^--mass distribution predicted by this calculation is plotted as the solid curve in fig. 2a. The predicted mass distribution of the e^+e^- pairs peaks at an effective mass of 3.3 GeV and is 600 MeV wide (FWHM) in disagreement with the narrow peak observed at 3.1 GeV. Neglecting the measurement errors changes neither the predicted width nor the position of the peak. We therefore conclude that the narrow peak observed at 3.1 GeV, cannot be explained by the background from the $\pi^0\pi^0$ decay mode from (3). The events outside the peak, however, are all consistent with being from $\pi^0\pi^0$ decay.

The decay

$$3.7 \to (3.1 \to e^+e^-) + (\eta \to \gamma\gamma, 3\pi^0) \qquad (4)$$

with two photons detected will also contribute to the background. A Monte Carlo calculation of $3.7 \to (3.1 \to e^+e^-) + (\eta \to \pi^0\pi^0\pi^0)$ leads to a wide mass distribution of the corresponding e^+e^- pairs. Since its absolute magnitude is very small compared to the background from (3) it is neglected. The contribution from $3.7 \to (3.1 \to e^+e^-) + (\eta \to \gamma\gamma)$ corresponds to electron pairs with a well defined mass centered at 3.1 GeV. However since the η is produced nearly at rest, the angle $\theta_{\gamma\gamma}$ between the two photons from its decay will be larger than 140°. The $\theta_{\gamma\gamma}$ distribution for events in the peak between 3.0 and 3.2 GeV is shown in fig. 2b together with the Monte Carlo prediction for the background from $e^+e^-\pi^0\pi^0$. It follows from fig. 2b that only 4 events in the narrow peak at 3.1 GeV can be due to $\eta \to \gamma\gamma$. To exclude these events and also reduce the background from the $\pi^0\pi^0$ decay mode only $\gamma\gamma$ events which fulfill the conditions $3.0 \text{ GeV} < M_{ee} < 3.2 \text{ GeV}$, and $\theta_{\gamma\gamma} < 120°$ were considered. There remains 14 events which satisfy these criteria. The Monte Carlo computation gives a $\pi^0\pi^0$ background of 3.7 events in this sample. From the data presented above we conclude that we have observed the cascade decay $3.7 \to (3.1 \to e^+e^-) + \gamma\gamma$.

B. $3.7 \to (3.1 \to \mu^+\mu^-) + \gamma\gamma$

The decay

$$3.7 \to (3.1 \to \mu^+\mu^-) + X \qquad (5)$$

was investigated by detecting the two muons with the two magnetic spectrometers of DASP and separating the various channels X using the information from the inner detector.

Muons were positively identified by requiring one or both of the oppositely charged particles, accepted by the magnet, to penetrate 70 cm of iron. The effective mass distribution of these pairs, plotted in fig. 3, shows a peak at 3.7 GeV from the direct decay of the 3.7 GeV resonance and the contribution from QED. A second peak, centered at a mass of 3.09 GeV, results from the cascade decay via the 3.1 GeV resonance. Accepting muon-pairs with an effective mass between

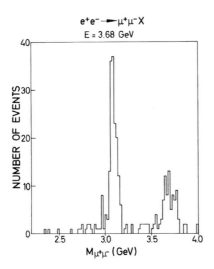

Fig. 3. Effective $\mu^+\mu^-$-mass distribution for the reaction $e^+e^- \to \mu^+\mu^- X$.

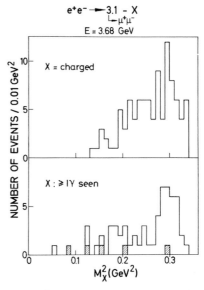

Fig. 4. The M_X^2-distribution for events of the type $e^+e^- \to \mu^+\mu^- X$ with 2.9 GeV $< M_{\mu^+\mu^-} <$ 3.2 GeV for
a) only charged particles observed in the inner detector
b) at least one photon detected in the inner detector. The shaded histogram shows the events of the reaction 3.7 $\to$ (3.1 $\to \mu^+\mu^-)\gamma\gamma$.

2.9 GeV and 3.2 GeV yields 164 events.

Using the information from the inner detector these events are split into two groups: The one group contains only charged particles, in the second group at least one photon must be observed. The distribution of the events as a function of their mass squared recoiling against the 3.1 GeV state was computed from the known muon momenta and is plotted for the two classes of events in fig. 4a and 4b. In this calculation the effective mass of the muon pair was constrained to be 3.09 GeV.

Candidates for the reaction

$$3.7 \to (3.1 \to \mu^+\mu^-) + \gamma\gamma \qquad (6)$$

were selected from the 164 events by requiring just two photons in the inner detector. Twelve such candidates were found. The energies of the two photons were calculated from the energy $E_{3.1}$ and momentum $P_{3.1}$ of the 3.1 GeV resonance and the measured direction of the two photons, where

$$P_{3.1} = P_{\mu_1} + P_{\mu_2} \text{ and } E_{3.1} = \sqrt{(3.09)^2 + P_{3.1}^2}.$$

For genuine events of reaction (6) the missing energy

$$\Delta E = E_{3.7} - E_{3.1} - E_{\gamma_1} - E_{\gamma_2}$$

and the angle $\Delta\theta$ between the momentum vector of the 3.1 GeV resonance and the plane spanned by the directions of the two photons should be zero, within experimental uncertainties. For the twelve candidates the errors in $\Delta\theta$ and ΔE were computed event by event. Five events were within 2σ of $\Delta\theta = \Delta E = 0$ and were taken to be genuine $\gamma\gamma$ events. The $\gamma\gamma$ effective mass spectrum for the five events is shown as the black squares in fig. 4b. One of the five events has a mass $M_X = (0.543 \pm 0.010)$ GeV and is therefore likely to be due to the decay:

$$3.7 \to (3.1 \to \mu^+\mu^-) + (\eta \to \gamma\gamma).$$

In the remainder we have demanded $M_X < 0.450$ GeV in order to eliminate a possible η contribution.

The background resulting from the $\pi^0\pi^0$ decay mode was evaluated by a similar Monte Carlo calculation as described above, with the e^+e^- detection efficiency replaced by the $\mu^+\mu^-$ detection efficiency. It was checked by comparing the predicted yield of the

$\pi^0\pi^0$ decay mode for 0, 1, 2, 3 or 4 of the photons detected in the inner detector with the experimental data. The results are listed in table 1.

Table 1

Number of photons detected:	0	1	2	3	4
Monte Carlo	: 8.8	12.7	8.2	2.2	0.2
Observed	: 5	19	8	4	0

The Monte Carlo calculation predicts a background distributed rather uniformly in $\Delta\theta$ and ΔE with a density consistent with that observed for the rejected events. The calculated background of 0.2 events is to be compared with the 4 events found within the same cuts on ΔE, $\Delta\theta$ and M_X. A 2C fit was made to the remaining four events in order to improve on the determination of the photon energies. The resulting photon energies and the effective masses of the photon pairs are listed in table 2. We observe that the events cluster in a narrow band of photon energies around 160 MeV and 420 MeV.

Table 2

Event#	E_{γ_1} (MeV)	E_{γ_2} (MeV)	$M_{\gamma\gamma}$ (MeV)
1	179 ± 53	408 ± 53	291
2	167 ± 16	404 ± 16	449
3	156 ± 23	406 ± 21	354
4	96 ± 22	463 ± 31	396

From the rate observed in $3.7 \to (3.1 \to e^+e^-) + \gamma\gamma$ we would expect to find 3 such events.

Conclusions: We have observed the decay $3.7 \to 3.1 + \gamma\gamma$ in two experiments. The observed tendency for the photon energies to cluster around 160 and 420 MeV strongly suggests that the $\gamma\gamma$ decay takes place in two stages via an intermediate particle, for which we suggest the name P_c:

$3.7 \to P_c + \gamma$

$P_c \to 3.1 + \gamma$.

The simultaneous emission of the two photons in a direct decay would produce photon pairs with a spectrum of energies, and would be expected to be very weak (of order α^2) relative to the strong mode $3.7 \to 3.1\ \pi^+\pi^-$. Since we have not determined whether the 160 or the 420 MeV photon is emitted first, the mass of the P_c is ambiguous:

$m_{P_c} = 3.52 \pm 0.05$ GeV

or

$m_{P_c} = 3.26 \pm 0.05$ GeV.

A very preliminary evaluation leads to a branching ratio $(3.7 \to P_c\gamma \to 3.1\ \gamma\gamma)/(3.7 \to \text{all})$ between 2% and 12%. Comparing the result of this experiment with the experimental limits on a monochromatic photon line [3], suggests the radiative decay mode $P_c \to 3.1 + \gamma$ to be a major decay mode of the new resonance.

We thank all engineers and technicians of the collaborating institutions who have participated in the construction of DASP. The invaluable cooperation with the technical support groups and the computer center at DESY is gratefully acknowledged. We are indebted to the DORIS machine group for their excellent support during the experiment. The non-DESY members of the collaboration thank the DESY Direktorium for the kind hospitality extended to them.

References

[1] DASP Collaboration, Phys. Letters 56B (1975) 491; DESY Report 75/14 (1975) and Phys. Lett. to be published.
[2] DASP Collaboration, Phys. Letters 53B (1974) 393, 491.
[3] J.W. Simpson et al., University of Stanford, HEPL-759 (1975) and Phys. Rev. Letters, to be published.

Observation of an η_c Candidate State with Mass 2978 ± 9 MeV

R. Partridge, C. Peck, and F. Porter
Physics Department, California Institute of Technology, Pasadena, California 91125

and

W. Kollmann,[a] M. Richardson, K. Strauch, and K. Wacker
Lyman Laboratory of Physics, Harvard University, Cambridge, Massachusetts 02138

and

D. Aschman, T. Burnett,[b] M. Cavalli-Sforza, D. Coyne, and H. Sadrozinski
Physics Department, Princeton University, Princeton, New Jersey 08544

and

R. Hofstadter, I. Kirkbride, H. Kolanoski,[c] K. Königsmann, A. Liberman,[d]
J. O'Reilly, and J. Tompkins
Physics Department and High Energy Physics Laboratory, Stanford University, Stanford, California 94305

and

E. Bloom, F. Bulos, R. Chestnut, J. Gaiser, G. Godfrey, C. Kiesling,[e] and M. Oreglia
Stanford Linear Accelerator Center, Stanford University, Stanford, California 94305
(Received 4 August 1980)

An η_c candidate state has been observed with a mass $M = 2978 \pm 9$ MeV and a natural line width $\Gamma < 20$ MeV (90% confidence level) using the Crystal Ball NaI(Tl) detector at SPEAR. Radiative transitions to this state are observed from ψ' (3684) and J/ψ (3095) in the inclusive photon spectra. The branching fraction to this state from the ψ' is (0.43 ± 0.08 ± 0.18)%, where the errors are statistical and systematic, respectively. In addition, evidence is presented for the decay of this new state into $\eta \pi^+\pi^-$ and an upper limit is presented on the decay into $\pi^0 K^+ K^-$.

PACS numbers: 14.40.Pe, 13.25.+n, 13.40.Hg

The properties of charmonium have been the subject of intense experimental and theoretical work since the J/ψ and ψ' were discovered more than five years ago. Of particular interest has been the question of the existence, mass, and width of the η_c, the 1^1S_0 partner of the 1^3S_1 charmonium state, the J/ψ. Potential models[1] and dispersion relation models[2] indicate that the η_c lies between 20 MeV above and 100 MeV below the J/ψ. In addition, a number of detailed calculations[3] have been made of the expected radiative rates to the η_c from J/ψ and ψ'.

From their first publication, these theoretical predictions have stimulated experimental searches for the η_c. An initial candidate for the η_c state, called $\chi(2830)$,[4] had a mass lower than theoretically expected. This state was not observed in inclusive photon spectra from the J/ψ; an upper limit[5] was set much below the expected branching fraction.[3] The $\chi(2830)$ was not confirmed by the Crystal Ball collaboration in a more sensitive experiment[6]; thus the question of the existence of the η_c has remained open.

We have observed a new state with mass $M = 2978 \pm 9$ MeV in radiative decays of the J/ψ and ψ', as well as the decay of this new state into $\eta \pi^+ \pi^-$. For the present we call this state the η_c candidate.

The data were obtained using the Crystal Ball NaI(Tl) detector[6,7] at the SPEAR e^+e^- storage ring of the Stanford Linear Accelerator Center (SLAC). The trigger efficiency for hadronic events is greater than 98%. Of the initial triggers at the J/ψ and ψ', roughly 30% and 20%, respectively, are hadronic events, the rest being cosmic rays, beam-gas interactions, and QED events. After removal of these three background sources by a series of software cuts, the efficiency of selection of hadronic decays is (93 ± 5)% with a residual background of less than 2%. The resulting final samples of 752×10^3 J/ψ and 775×10^3 ψ' hadronic decays, corresponding to integrated luminosities of 323 and 1630 nb^{-1}, respectively, were used in the present analysis.

A number of cuts were applied to obtain inclusive photon spectra[8,9]. Photon showers were re-

1150   © 1980 The American Physical Society

quired to be entirely contained in the detector and to be well separated from charged particles. Photon pairs that could be reconstructed to a π^0 were removed. Figure 1 shows the inclusive photon spectrum obtained from hadronic decays of the ψ'. The transitions[5,10] to the well-established χ states are indicated in the figure as are the cascade transitions.[10-12] Also clearly seen is a signal of greater than 5 standard deviations at $E_\gamma = 634 \pm 13$ MeV. The error in the photon energy is primarily systematic, resulting from a ±2% uncertainty in the absolute NaI(Tl) energy calibration. This signal corresponds to a transition to a state of mass $M = 2983 \pm 16$ MeV. Several systematic checks[9] were made to verify that the signal appears uniformly over the solid angle of the apparatus and in the data obtained in the earlier and later parts of the data collection period. To check the sensitivity of the detector to a small signal in the 630-MeV region,[9] we looked for the 617-MeV photon radiated in the reaction $e^+e^- \to \gamma J/\psi$ at the $\psi''(3770)$ resonance; this photon was seen at the expected level. In addition, to check that the signal is not an instrumental effect, the inclusive photon spectrum from hadronic decays of the J/ψ, shown in Fig. 2, was analyzed and no signal was found in the 630-MeV region.

If the signal from the ψ' corresponds to the hindered M1 transition[3] $\psi' \to \gamma \eta_c$, then we expect to observe the transition $J/\psi \to \gamma \eta_c$ at a photon energy of about 110 MeV. In the J/ψ inclusive photon spectrum, shown in Fig. 2, there appears to be an enhancement about a photon energy of 112 MeV, corresponding to a state of mass $M \sim 2981$ MeV. A simultaneous fit was therefore performed to the mass, M, and natural linewidth, Γ, of the η_c candidate for both the ψ' and J/ψ signal regions. The two observed signals were fit by a Breit-Wigner line shape convoluted with a Gaussian energy resolution; independent quadratic forms were used for the backgrounds. The Gaussian resolutions ($\sigma = 4.7$ MeV at $E_\gamma = 112$ MeV and $\sigma = 18.3$ MeV at $E_\gamma = 634$ MeV) were derived from other Crystal Ball measurements.[7]

Figures 3(a) and 3(b) show the best fit obtained, together with the data for the ψ' and J/ψ inclusive spectra, respectively, before and after background subtraction. The parameters from the best fit, excepting the primarily systematic error in M, are

$$M = 2981 \pm 15 \text{ MeV}, \quad \Gamma = 20^{+16}_{-11} \text{ MeV},$$
$$\chi^2 = 53 \text{ for 66 degrees of freedom.} \quad (1)$$

The signal obtained from the fit has a statistical significance of over 5 standard deviations. The systematic error in M arises mainly from the energy calibration uncertainty in the ψ' contribution to the fit, and uncertainty in the background shape in the J/ψ contribution; it dominates the ±2 MeV statistical error. The dependence of χ^2 on Γ exhibits a broad minimum in χ^2 centered at[13] $\Gamma = 20$ MeV, where the value of Γ is primarily determined from the J/ψ inclusive spectrum. The error in Γ, shown in (1), is essentially statistical; an additional uncertainty due to the choice of the functional form for the background to the J/ψ signal has not yet been evaluated.

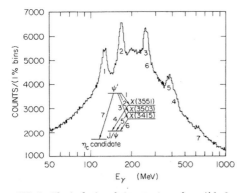

FIG. 1. The inclusive photon spectrum from ψ' hadronic decays. Counts are plotted in logarithmic bins since the resolution, $\Delta E/E$, is nearly constant in E for NaI(Tl).

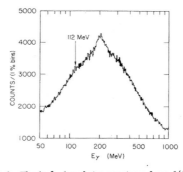

FIG. 2. The inclusive photon spectrum from J/ψ hadronic decays. The structure at $E_\gamma \sim 200$ MeV results from minimum ionizing charged particles which have been misidentified as photons (Refs. 8 and 9).

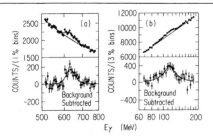
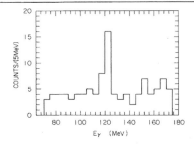

FIG. 3. Inclusive photon spectra from (a) ψ' and (b) J/ψ decays in the region of the η_c candidate signal, with fit results overplotted ($\Gamma = 20$ MeV). Unsubtracted and background subtracted spectra are shown; the background is determined from the fit described in the text.

FIG. 4. Fitted energy of the photon for events fitted to the hypothesis $J/\psi \to \gamma \eta \pi^+ \pi^-$.

The reconstruction efficiency for a single photon in an inclusive hadronic final state can at present be reliably estimated only for the ψ' decays, where the photon energy is 634 MeV. For the low-energy photons from the J/ψ, background sources are much more severe. Using a Monte Carlo estimate of the efficiency, assuming $J^P = 0^-$ for the η_c candidate, we obtain the inclusive branching ratio $R(\psi' \to \gamma \eta_c \text{ candidate}) = (0.43 \pm 0.08 \pm 0.18)\%$. The errors shown are statistical and systematic, respectively, with the latter dominated by the uncertainty in the photon reconstruction efficiency. The value for the branching ratio compares well with theoretical estimates[3] of $(0.2-0.4)\%$.

We have also looked for exclusive decays of the η_c candidate into hadrons by performing kinematic fits to exclusive final states with multiple photons and two charged hadrons.[14] The Crystal Ball measures both the energy and angle of electromagnetically showering particles; for charged hadrons (π, K) only the angles are measured well. Secondary interactions of the charged hadrons in the sodium iodide complicate the fitting of some events, but special pattern recognition algorithms have been developed to deal with this effect.

Events with a three-photon, two-charged-particle topology were selected from the sample of J/ψ hadronic decays and subjected to a three-constraint kinematic fit to the hypotheses

$$J/\psi \to \gamma \eta \pi^+ \pi^- \text{ and } \gamma \eta K^+ K^-, \quad \eta \to \gamma\gamma. \quad (2)$$

The energy spectrum for the low-energy radiated photon is shown in Fig. 4 for events which pass the fit with a probability of χ^2 greater than 0.10 for the $\eta \pi^+ \pi^-$ hypothesis. A clear signal is seen above background. No comparable signal is seen for the $\eta K^+ K^-$ hypothesis. A maximum likelihood fit gives the $\eta \pi^+ \pi^-$ mass corresponding to this signal as $2974 \pm 2 \pm 9$ MeV, where the first error is statistical and the second is an estimate of the systematic uncertainty. The mass agrees within errors with the value determined from the inclusive spectra.

The $\gamma \eta \pi^+ \pi^-$ data contain additional information on the width of the η_c candidate. Given the limited statistics of this measurement, we choose to combine the $\chi^2(\Gamma)$ function obtained from the likelihood fit to these data with the function obtained from the fit to the inclusive[13] data. The resulting function provides an upper limit on the natural line width of $\Gamma < 20$ MeV (90% C.L.).

The detection efficiency for the exclusive reaction $\gamma \eta \pi^+ \pi^-$ has been estimated by a Monte Carlo calculation where the η_c was assumed to have $J^P = 0^-$, and to decay with a phase-space distribution. The signal of 18 ± 6 events corresponds to a product branching ratio $R(J/\psi \to \gamma \eta_c \text{ candidate}) \cdot R(\eta_c \text{ candidate} \to \eta \pi^+ \pi^-) = (3.1 \pm 1.1 \pm 1.5) \times 10^{-4}$, where the errors are statistical and systematic, respectively.

The Mark II collaboration at SPEAR has observed an enhancement in $\psi' \to \gamma \pi^+ K^\mp K_s^0$ at $M(\pi KK) = 2980 \pm 8$ MeV.[15] We do not observe a signal in $J/\psi \to \gamma \pi^0 K^+ K^-$ in this mass range, with an upper limit of 1.5×10^{-4} (90% C.L.). Comparison of these two results awaits a reliable determination of $R(J/\psi \to \gamma \eta_c \text{ candidate})$.

In summary, an η_c candidate state is observed with mass 2978 ± 9 MeV. This estimate of the mass is obtained by averaging the masses determined from the inclusive and exclusive decays. The error shown is primarily systematic. The upper limit on the width of the state of $\Gamma < 20$ MeV

(90% C.L.) is consistent with the value of 5 MeV preferred by lowest-order quantum chromodynamics theory.[3] Final identification of this state as the pseudoscalar hyperfine partner of the J/ψ will depend on the determination of J^P and 0^-, and approximate agreement between the experimentally measured transition rate from the J/ψ and the value predicted by the charmonium model.[1]

We gratefully acknowledge the efforts of A. Baumgarten and J. Broeder (SLAC) and B. Beron, E. B. Hughes, and R. Parks (High Energy Physics Laboratory, Stanford University), as well as those of the linac and SPEAR staff at SLAC. This work was supported in part by the U. S. Department of Energy under Contracts No. DE-AC03-76SF00515, No. EY-76-C02-3064, and No. EY-76-C03-0068, and by the National Science Foundation under Contracts No. PHY78-00967, No. PHY78-07343, and No. PHY75-22980, and by the NATO Fellowships, the Chaim Weizmann Fellowship, and by the Sloan Foundation.

---

[a] Present address: Wentorfer Strasse 149, D-2050 Hamburg, Federal Republic of Germany.
[b] Present address: Physics Department, University of Washington, Seattle, Washington 98195.
[c] Present address: University of Bonn, D-5380 Bonn, Federal Republic of Germany.
[d] Present address: Schlumberger-Doll Research Center, Ridgefield, Connecticut 06877.
[e] Present address: Max Planck Institute for Physics and Astrophysics, D-8000 Munich 40, Federal Republic of Germany.

[1] T. Appelquist and H. D. Politzer, Phys. Rev. Lett. 34, 43 (1975); T. Appelquist, A. De Rújula, H. D. Politzer, and S. L. Glashow, Phys. Rev. Lett. 34, 365 (1975); H. J. Schnitzer, Phys. Rev. D 13, 74 (1976); R. Barbieri, R. Kögerler, Z. Kunszt, and R. Gatto, Nucl. Phys. B105, 125 (1976); H. J. Lipkin, H. R. Rubinstein, and N. Isgur, Phys. Lett. 78B, 295 (1978); P. Minkowski, Phys. Lett. 85B, 231 (1979); E. Eichten and F. L. Feinberg, Phys. Rev. Lett 43, 1205 (1979); C. Quigg, in *Proceedings of the Ninth International Symposium on Lepton and Photon Interactions at High Energies, Batavia, Illinois, 1979*, edited by T. B. W. Kirk and H. D. I. Abarbanel (Fermilab, Batavia, Ill., 1979), p. 247.

[2] V. A. Novikov, L. B. Okun, M. A. Shifman, A. I. Vainshtein, M. B. Voloshin, and V. I. Zakharov, Phys. Lett. 67B, 409 (1977); M. Shifman, A. Vainshtein, and V. Zakharov, Nucl. Phys. B147, 448 (1979).

[3] Two extensive reviews of the charmonium model and its comparison to experiment are T. Appelquist, R. M. Barnett, and K. D. Lane, Ann. Rev. Nucl. Part. Sci. 28, 387 (1978); E. Eichten, K. Gottfried, T. Kinoshita, K. D. Lane, and T.-M. Yan, Phys. Rev. D 21, 203 (1980).

[4] W. Braunschweig et al., Phys. Lett. 67B, 243 (1977); S. Yamada, in *Proceedings of the International Symposium on Lepton and Photon Interactions at High Energies, Hamburg, 1977*, edited by F. Gutbro (DESY, Hamburg, Germany, 1977), p. 69; W. D. Apel et al., Phys. Lett. 72B, 500 (1978).

[5] C. Biddick et al., Phys. Rev. Lett. 38, 1324 (1977).

[6] R. Partridge et al., Phys. Rev. Lett. 44, 712 (1980).

[7] E. D. Bloom, in *Proceedings of the Fourteenth Rencontre de Moriond, Les Arcs, France, 1979*, edited by J. Tran Than Van (Editions Frontieres, France, 1979); Y. Chan et al., IEEE Trans. Nucl. Sci. 25, 333 (1978); I. Kirkbride et al., IEEE Trans. Nucl. Sci. 26, 1535 (1979).

[8] E. D. Bloom, in *Proceedings of the Ninth International Symposium on Lepton and Photon Interactions at High Energies, Batavia, Illinois, 1979*, edited by T. B. W. Kirk and H. D. I. Abarbanel (Fermilab, Batavia, Ill., 1979), p. 92, and SLAC Report No. SLAC-PUB-2425, 1979 (to be published).

[9] C. W. Peck, in *Particles and Fields—1979*, edited by B. Margolis and D. G. Stairs, AIP Conference Proceedings No. 59 (American Institute of Physics, New York, 1980), and California Institute of Technology Report No. CALT 68-753, 1979 (unpublished).

[10] J. S. Whitaker et al., Phys. Rev. Lett. 37, 1596 (1976).

[11] W. Bartel et al., Phys. Lett. 79B, 492 (1978); W. Tanenbaum et al., Phys. Rev. D 17, 1731 (1978).

[12] M. Oreglia, in Proceedings of the Fifteenth Rencontre de Moriond, Les Arcs, France, 15–21 March 1980 (to be published), and SLAC Report No. SLAC-PUB-2529, 1980 (to be published).

[13] E. D. Bloom, in Proceedings of the Sixth International Conference on Experimental Meson Spectroscopy, Brookhaven National Laboratory, Upton, New York, 25–26 April, 1980 (to be published), and SLAC Report No. SLAC-PUB-2530, 1980 (to be published).

[14] D. G. Aschman, in Proceedings of the Fifteenth Rencontre de Moriond, Les Arcs, France, 15–21 March 1980 (to be published), and SLAC Report No. SLAC-PUB-2550, 1980 (to be published).

[15] T. M. Himel, Ph.D. thesis, 1979, SLAC Report No. SLAC-223 (unpublished); T. M. Himel et al., preceding Letter [Phys. Rev. Lett. 45, 1146 (1980)], and to be published.

Evidence for Anomalous Lepton Production in e^+-e^- Annihilation*

M. L. Perl, G. S. Abrams, A. M. Boyarski, M. Breidenbach, D. D. Briggs, F. Bulos, W. Chinowsky,
J. T. Dakin,† G. J. Feldman, C. E. Friedberg, D. Fryberger, G. Goldhaber, G. Hanson,
F. B. Heile, B. Jean-Marie, J. A. Kadyk, R. R. Larsen, A. M. Litke, D. Lüke,‡
B. A. Lulu, V. Lüth, D. Lyon, C. C. Morehouse, J. M. Paterson,
F. M. Pierre,§ T. P. Pun, P. A. Rapidis, B. Richter,
B. Sadoulet, R. F. Schwitters, W. Tanenbaum,
G. H. Trilling, F. Vannucci,‖ J. S. Whitaker,
F. C. Winkelmann, and J. E. Wiss

Lawrence Berkeley Laboratory and Department of Physics, University of California, Berkeley, California 94720, and Stanford Linear Accelerator Center, Stanford University, Stanford, California 94305
(Received 18 August 1975)

We have found events of the form $e^+ + e^- \rightarrow e^{\pm} + \mu^{\mp}$ + missing energy, in which no other charged particles or photons are detected. Most of these events are detected at or above a center-of-mass energy of 4 GeV. The missing-energy and missing-momentum spectra require that at least two additional particles be produced in each event. We have no conventional explanation for these events.

We have found 64 events of the form

$$e^+ + e^- \rightarrow e^{\pm} + \mu^{\mp} + \geq 2 \text{ undetected particles} \quad (1)$$

for which we have no conventional explanation. The undetected particles are charged particles or photons which escape the 2.6π sr solid angle of the detector, or particles very difficult to detect such as neutrons, K_L^0 mesons, or neutrinos. Most of these events are observed at center-of-mass energies at, or above, 4 GeV. These events were found using the Stanford Linear Accelerator Center–Lawrence Berkeley Laboratory (SLAC-

LBL) magnetic detector at the SLAC colliding-beams facility SPEAR.

Events corresponding to (1) are the signature for new types of particles or interactions. For example, pair production of heavy charged leptons[1-4] having the decay modes $l^- \to \nu_l + e^- + \bar{\nu}_e$, $l^+ \to \bar{\nu}_l + e^+ + \nu_e$, $l^- \to \nu_l + \mu^- + \bar{\nu}_\mu$, and $l^+ \to \bar{\nu}_l + \mu^+ + \nu_\mu$ would appear as such events. Another possibility is the pair production of charged bosons with decays $B^- \to e^- + \bar{\nu}_e$, $B^+ \to e^+ + \nu_e$, $B^- \to \mu^- + \bar{\nu}_\mu$, and $B^+ \to \mu^+ + \nu_\mu$. Charmed-quark theories[5,6] predict such bosons. Intermediate vector bosons which mediate the weak interactions would have similar decay modes, but the mass of such particles (if they exist at all) is probably too large[7] for the energies of this experiment.

The momentum-analysis and particle-identifier systems of the SLAC-LBL magnetic detector[8] cover the polar angles $50° \leq \theta \leq 130°$ and the full 2π azimuthal angle. Electrons, muons, and hadrons are identified using a cylindrical array of 24 lead-scintillator shower counters, the 20-cm-thick iron flux return of the magnet, and an array of magnetostrictive wire spark chambers situated outside the iron. Electrons are identified solely by requiring that the shower-counter pulse height be greater than that of a 0.5-GeV e. Incidently, the e's in the e-μ events thus selected give no signal in the muon chambers; and their shower-counter pulse-height distribution is that expected of electrons. Also the positions of the e's in the shower counters as determined from the relative pulse heights in the photomultiplier tubes at each end of the counters agree within measurement errors with the positions of the e tracks. Hence the e's in the e-μ events are not misidentified combinations of $\mu + \gamma$ or $\pi + \gamma$ in a single shower counter, except possibly for a few events already contained in the background estimates. Muons are identified by two requirements. The μ must be detected in one of the muon chambers after passing through the iron flux return and other material totaling 1.67 absorption lengths for pions. And the shower-counter pulse height of the μ must be small. All other charged particles are called hadrons. The shower counters also detect photons (γ). For γ energies above 200 MeV, the γ detection efficiency is about 95%.

To illustrate the method of searching for events corresponding to Reaction (1), we consider our data taken at a total energy ($\sqrt{s}$) of 4.8 GeV. This sample contains 9550 three-or-more-prong events and 25 300 two-prong events which include $e^+ + e^- \to e^+ + e^-$ events, $e^+ + e^- \to \mu^+ + \mu^-$ events, two-prong hadronic events, and the e-μ events described here. To study two-prong events we define a coplanarity angle

$$\cos\theta_{copl} = -(\vec{n}_1 \times \vec{n}_{e^+}) \cdot (\vec{n}_2 \times \vec{n}_{e^+})/ |\vec{n}_1 \times \vec{n}_{e^+}||\vec{n}_2 \times \vec{n}_{e^+}|, \quad (2)$$

where $\vec{n}_1$, $\vec{n}_2$, and $\vec{n}_{e^+}$ are unit vectors along the directions of particles 1, 2, and the e^+ beam. The contamination of events from the reactions $e^+ + e^- \to e^+ + e^-$ and $e^+ + e^- \to \mu^+ + \mu^-$ is greatly reduced if we require $\theta_{copl} > 20°$. Making this cut leaves 2493 two-prong events in the 4.8-GeV sample.

To obtain the most reliable e and μ identification[9] we require that each particle have a momentum greater than 0.65 GeV/c. This reduces the 2493 events to the 513 in Table I. The 24 e-μ events with no associated photons, called the signature events, are candidates for Reaction (1). The e-μ events can come conventionally from the two-virtual-photon process[10] $e^+ + e^- \to e^+ + e^- + \mu^+ + \mu^-$. Calculations indicate that this source is negligible, and the absence of e-μ events with charge 2 proves this point since the number of charge-2 e-μ events should equal the number of charge-0 e-μ events from this source.

We determine the background from hadron misidentification or decay by using the 9550 three-or-more-prong events and assuming that every particle called an e or a μ by the detector either was a misidentified hadron or came from the decay of a hadron. We use $P_{h \to l}$ to designate the sum of the probabilities for misidentification or decay causing a hadron h to be called a lepton l. Since the P's are momentum dependent[9] we use all the

TABLE I. Distribution of 513 two-prong events, obtained at $E_{c.m.} = 4.8$ GeV, which meet the criteria $|\vec{p}_1| > 0.65$ GeV/c, $|\vec{p}_2| > 0.65$ GeV/c, and $\theta_{copl} > 20°$. Events are classified according to the number N_γ of photons detected, the total charge, and the nature of the particles. All particles not identified as e or μ are called h for hadron.

N_γ Particles	0	1	>1	0	1	>1
	Total charge = 0			Total charge = ±2		
e-e	40	111	55	0	1	0
e-μ	24	8	8	0	0	3
μ-μ	16	15	6	0	0	0
e-h	20	21	32	2	3	3
μ-h	17	14	31	4	0	5
h-h	14	10	30	10	4	6

e-h, μ-h, and h-h events in column 1 of Table I to determine a "hadron" momentum spectrum, and weight the P's accordingly. We obtain the momentum-averaged probabilities $P_{h \to e} = 0.183 \pm 0.007$ and $P_{h \to \mu} = 0.198 \pm 0.007$. Collinear e-e and μ-μ events are used to determine $P_{e \to h} = 0.056 \pm 0.02$, $P_{e \to \mu} = 0.011 \pm 0.01$, $P_{\mu \to h} = 0.08 \pm 0.02$, and $P_{\mu \to e} < 0.01$.

Using these probabilities and assuming that all e-h and μ-h events in Table I result from particle misidentifications or particle decays, we calculate for column 1 the contamination of the e-μ sample to be 1.0 ± 1.0 event from misidentified e-e,[11] <0.3 event from misidentified μ-μ,[11] and 3.7 ± 0.6 events from h-h in which the hadrons were misidentified or decayed. The total e-μ background is then 4.7 ± 1.2 events.[12,13] The statistical probability of such a number yielding the 24 signature e-μ events is very small. The same analysis applied to columns 2 and 3 of Table I yields 5.6 ± 1.5 e-μ background events for column 2 and 8.6 ± 2.0 e-μ background events for column 3, both consistent with the observed number of e-μ events.

Figure 1(a) shows the momentum of the μ versus the momentum of the e for signature events.[14] Both p_μ and p_e extend up to 1.8 GeV/c, their average values being 1.2 and 1.3 GeV/c, respectively. Figure 1(b) shows the square of the invariant e-μ mass (M_i^2) versus the square of the missing mass (M_m^2) recoiling against the e-μ system. To explain Fig. 1(b) at least two particles must escape detection. Figure 1(c) shows the distribution in collinearity angle between the e and μ ($\cos\theta_{coll} = -\vec{p}_e \cdot \vec{p}_\mu / |\vec{p}_e||\vec{p}_\mu|$). The dip near $\cos\theta_{coll} = 1$ is a consequence of the coplanarity cut; however, the absence of events with large θ_{coll} has dynamical significance.

Figure 2 shows the *observed* cross section in the range of detector acceptance for signature e-μ events versus center-of-mass energy with the background subtracted at each energy as described above.[9] There are a total of 86 e-μ events summed over all energies, with a calculated background of 22 events.[12] The corrections to obtain the true cross section for the angle and momentum cuts used here depend on the hypothesis as to the origin of these e-μ events, and the corrected cross section can be many times larger than the observed cross section. While Fig. 2 shows an apparent threshold at around 4 GeV, the statistics are small and the correction fac-

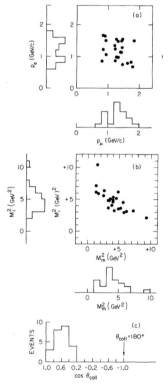

FIG. 1. Distribution for the 4.8-GeV e-μ signature events of (a) momenta of the e (p_e) and μ (p_μ); (b) square of the invariant mass (M_i^2) and square of the missing mass (M_m^2); and (c) $\cos\theta_{coll}$.

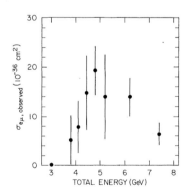

FIG. 2. The *observed* cross section for the signature e-μ events.

tors are largest for low $\sqrt{s}$. Thus, the apparent threshold may not be real.

We conclude that the signature e-μ events cannot be explained either by the production and decay of any presently known particles or as coming from any of the well-understood interactions which can conventionally lead to an e and a μ in the final state. A possible explanation for these events is the production and decay of a pair of new particles, each having a mass in the range of 1.6 to 2.0 GeV/c^2.

---

*Work supported by the U. S. Energy Research and Development Administration.
†Present address: Department of Physics and Astronomy, University of Massachusetts, Amherst, Mass. 01002.
‡Fellow of Deutsche Forschungsgemeinschaft.
§Centre d'Etudes Nucléaires de Saclay, Saclay, France.
‖Institut de Physique Nucléaire, Orsay, France.

[1]M. L. Perl and P. A. Rapidis, SLAC Report No. SLAC-PUB-1496, 1974 (unpublished).
[2]J. D. Bjorken and C. H. Llewellyn Smith, Phys. Rev. D 7, 887 (1973).
[3]Y. S. Tsai, Phys. Rev. D 4, 2821 (1971).
[4]M. A. B. Beg and A. Sirlin, Annu. Rev. Nucl. Sci. 24, 379 (1974).
[5]M. K. Gaillard, B. W. Lee, and J. L. Rosner, Rev. Mod. Phys. 47, 277 (1975).
[6]M. B. Einhorn and C. Quigg, Phys. Rev. D (to be published).
[7]B. C. Barish et al., Phys. Rev. Lett. 31, 180 (1973).
[8]J.-E. Augustin et al., Phys. Rev. Lett. 34, 233 (1975); G. J. Feldman and M. L. Perl, Phys. Rep. 19C, 233 (1975).
[9]See M. L. Perl, in Proceedings of the Canadian Institute of Particle Physics Summer School, Montreal, Quebec, Canada, 16–21 June 1975 (to be published).
[10]V. M. Budnev et al., Phys. Rep. 15C, 182 (1975); H. Terazawa, Rev. Mod. Phys. 45, 615 (1973).
[11]These contamination calculations do not depend upon the source of the e or μ; anomalous sources lead to overestimates of the contamination.
[12]Using *only* events in column 1 of Table I we find at 4.8 GeV $P_{h \to e} = 0.27 \pm 0.10$, $P_{h \to \mu} = 0.23 \pm 0.09$, and a total e-μ background of 7.9 ± 3.2 events. The same method yields a total e-μ background of 30 ± 6 events summed over all energies. This method of background calculation (Ref. 9) allows the hadron background in the two-prong, zero-photon events to be different from that in other types of events.
[13]Our studies of the two-prong and multiprong events show that there is *no* correlation between the misidentification or decay probabilities; hence the background is calculated using independent probabilities.
[14]Of the 24 events, thirteen are $e^+ + \mu^-$ and eleven are $e^- + \mu^+$.

Observation in e^+e^- Annihilation of a Narrow State at 1865 MeV/c^2 Decaying to $K\pi$ and $K\pi\pi\pi$ †

G. Goldhaber,* F. M. Pierre,‡ G. S. Abrams, M. S. Alam, A. M. Boyarski, M. Breidenbach,
W. C. Carithers, W. Chinowsky, S. C. Cooper, R. G. DeVoe, J. M. Dorfan, G. J. Feldman,
C. E. Friedberg, D. Fryberger, G. Hanson, J. Jaros, A. D. Johnson, J. A. Kadyk,
R. R. Larsen, D. Lüke,§ V. Lüth, H. L. Lynch, R. J. Madaras, C. C. Morehouse,‖
H. K. Nguyen,** J. M. Paterson, M. L. Perl, I. Peruzzi,†† M. Piccolo,††
T. P. Pun, P. Rapidis, B. Richter, B. Sadoulet, R. H. Schindler,
R. F. Schwitters, J. Siegrist, W. Tanenbaum, G. H. Trilling,
F. Vannucci,‡‡ J. S. Whitaker, and J. E. Wiss

*Lawrence Berkeley Laboratory and Department of Physics, University of California, Berkeley, California 94720,
and Stanford Linear Accelerator Center, Stanford University, Stanford, California 94305*
(Received 14 June 1975)

We present evidence, from a study of multihadronic final states produced in e^+e^- annihilation at center-of-mass energies between 3.90 and 4.60 GeV, for the production of a new neutral state with mass 1865 ± 15 MeV/c^2 and decay width less than 40 MeV/c^2 that decays to $K^{\pm}\pi^{\mp}$ and $K^{\pm}\pi^{\mp}\pi^{\pm}\pi^{\mp}$. The recoil-mass spectrum for this state suggests that it is produced only in association with systems of comparable or larger mass.

We have observed narrow peaks near 1.87 GeV/c^2 in the invariant-mass spectra for neutral combinations of the charged particles $K^{\pm}\pi^{\mp}$ ($K\pi$) and $K^{\pm}\pi^{\mp}\pi^{\pm}\pi^{\mp}$ ($K3\pi$) produced in e^+e^- annihilation. The agreement in mass, width, and recoil-mass spectrum for these peaks strongly suggests they represent different decay modes of the same object. The mass of this state is 1865 ± 15 MeV/c^2 and its decay width (full width at half-maximum) is less than 40 MeV/c^2 (90% confidence level). The state appears to be produced only in association with systems of comparable or higher mass.

Our results are based on studies of multihadronic events recorded by the Stanford Linear Accelerator Center–Lawrence Berkeley Laboratory magnetic detector operating at the colliding-beam facility SPEAR. Descriptions of the detector and event-selection procedures have been published.[1,2]

A new feature in our analysis is the use of time of flight (TOF) information to help identify hadrons. The TOF system includes 48 2.5 cm × 20 cm × 260 cm Pilot Y scintillation counters arranged in a cylindrical array immediately outside the tracking spark chambers at a radius of 1.5 m from the beam axis. Both ends of each counter are viewed by Amperex 56DVP photomultiplier tubes (PM); anode signals from each PM are sent to separate time-to-digital converters (TDC's), analog-to-digital converters, and latch-es. Pulse-height information is used to correct times given by the TDC's. The collision time is derived from a pickup electrode that senses the passage of the 0.2-nsec-long beam pulses; the period between successive collisions is 780 nsec. Run-to-run calibrations of the TOF system are performed with Bhabha scattering ($e^+e^- \to e^+e^-$) events. The rms resolution of the TOF system is 0.4 nsec.

Evidence for a new state in the $K\pi$ system was found among 29 000 hadronic events collected at center-of-mass (c.m.) energies between 3.90 and 4.60 GeV. As shown by the top row of Fig. 1, a

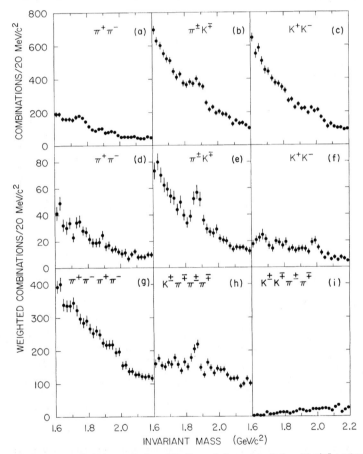

FIG. 1. Invariant-mass spectra for neutral combinations of charged particle. (a) $\pi^+\pi^-$ assigning π mass to all tracks, (b) $K^\mp\pi^\pm$ assigning K and π masses to all tracks, (c) K^+K^- assigning K mass to all tracks, (d) $\pi^+\pi^-$ weighted by $\pi\pi$ TOF probability, (e) $K^\pm\pi^\mp$ weighted by $K\pi$ TOF probability, (f) K^+K^- weighted by KK TOF probability, (g) $\pi^+\pi^-\pi^+\pi^-$ weighted by 4π TOF probability, (h) $K^\pm\pi^\mp\pi^\pm\pi^\mp$ weighted by $K3\pi$ TOF probability, (i) $K^+K^-\pi^+\pi^-$ weighted by $KK\pi\pi$ TOF probability.

significant signal[3] appears when we simply consider invariant-mass spectra for all possible neutral combinations of two charged particles *assuming* both π and K masses for the particles as was done in our previous search for the production of narrow peaks.[4] Through kinematic reflections, the signal appears near 1.74 GeV/c^2 for the $\pi^+\pi^-$ hypothesis [Fig. 1(a)], 1.87 GeV/c^2 in the case of $K^+\pi^-$ or $K^-\pi^+$ [Fig. 1(b)], and 1.98 GeV/c^2 for K^+K^- [Fig. 1(c)].

To establish the correct choice of final-state particles associated with these peaks, we use the TOF information. Because the typical time difference between a π and a K in the $K\pi$ signal is only about 0.5 nsec, we have used the following technique to extract maximal information on particle identity. First, tracks are required to have good timing information from both PM's, consistent with the extrapolated position of the track in the counter. Next, each track is assigned probabilities that it is a π or K; they are determined from the measured momentum and TOF assuming a Gaussian probability distribution with standard deviation 0.4 nsec. Tracks with net (π plus K) probability less than 1% are rejected.[5] Then, the relative π-K probabilities are renormalized so that their sum is unity, and two-particle combinations are weighted by the joint probability that the particles satisfy the particular π or K hypothesis assigned to them. In this way, the total weight assigned to all $\pi\pi$, $K\pi$, and KK combinations equals the number of two-body combinations and no double counting occurs.

Invariant-mass spectra weighted by the above procedure are presented in the second row of Fig. 1. We see that the $K\pi$ hypothesis [Fig. 1(e)] for the peak at the $K\pi$ mass 1.87 GeV/c^2 is clearly preferred over either $\pi^+\pi^-$ [Fig. 1(d)] or K^+K^- [Fig. 1(f)]. The areas under the small peaks remaining in the $\pi^+\pi^-$ and K^+K^- channels are consistent with the entire signal being $K\pi$ and the resulting misidentification of true $K\pi$ events expected for our TOF system. From consideration of possible residual uncertainties in the TOF calibration, we estimate that the confidence level for this signal to arise only from $\pi^+\pi^-$ or K^+K^- is less than 1%. Assuming the entire signal in Figs. 1(d)-1(f) to be in the $K\pi$ channel, we find a total of 110 ± 25 decays of the new state; the significance of the peak in Fig. 1(e) is greater than 5 standard deviations. No signal occurs in the corresponding doubly charged channels.

Evidence for the decay of this state to neutral combinations of a charged K and three charged π's is presented in the third row of Fig. 1. Again, we employ the TOF weighting technique discussed above; the hadron event sample is the same as that used for the $K\pi$ study. Four-body mass combinations are weighted by their joint π-K probabilities. In order to recover tracks when an extra particle is present in the TOF counter, or when they miss a counter, all tracks failing the timing quality criteria are called π.

As can be seen in Fig. 1(h), a clear signal is obtained in the $K3\pi$ system at a mass near 1.86 GeV/c^2. No corresponding signal is evident at this mass or the appropriate kinematically reflected mass for either the $\pi^+\pi^-\pi^+\pi^-$ or $K^+K^-\pi^+\pi^-$ systems. We estimate the number of $K3\pi$ decays in the 1.86-GeV/c^2 peak to be 124 ± 21, an effect of more than 5 standard deviations. Again, there is no signal in the corresponding doubly charged channel.

To determine the masses and widths of the peaks in the $K\pi$ and $K3\pi$ mass spectra, we have fitted the data represented by Fig. 1 with a Gaussian for the peak and linear and quadratic background terms under various conditions of bin size, event-selection criteria, and kinematic cuts. Masses for the $K\pi$ signal center at 1870 MeV/c^2; those for the $K3\pi$ signal center at 1860 MeV/c^2. The spread in central-mass values for the various fits is ± 5 MeV/c^2. Within the statistical errors of ± 3 to 4 MeV/c^2, the widths obtained by these fits agree with those expected from experimental resolution alone. From Monte Carlo calculations we expect a rms mass resolution of 25 MeV/c^2 for the $K\pi$ system and 13 MeV/c^2 for the $K3\pi$ system. Systematic errors in momentum measurement are estimated to contribute a ± 10-MeV/c^2 uncertainty in the absolute mass determination, and can account for the 10-MeV/c^2 mass difference observed between the $K\pi$ and $K3\pi$ systems. Thus, both signals are consistent with being decays of the same state and, from our mass resolution, we deduce a 90%-confidence-level upper limit of 40 MeV/c^2 for the decay width of this state.

In Fig. 2, we show the spectra of masses recoiling against neutral $K\pi$ and $K3\pi$ systems in the signal region. The entries are weighted by the TOF likelihood as discussed above. Background estimates are obtained by plotting smooth curves corresponding to the recoil spectra for $K\pi$ and $K3\pi$ invariant-mass combinations in bands on either side of the signal region. The normalizations of these curves are fixed by the areas of the respective control regions.

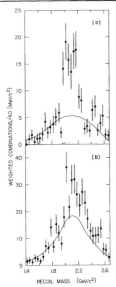

FIG. 2. Recoil-mass spectra for combinations in the $K\pi$ and $K3\pi$ peaks. Smooth curves are estimates of the background obtained from combinations whose invariant masses are on either side of the peak mass region. (a) $K^\pm \pi^\mp$, peak mass region of 1.84 to 1.90 GeV/c^2 and background mass regions of 1.70 to 1.82 GeV/c^2 and 1.92 to 2.04 GeV/c^2. (b) $K^\pm \pi^\mp \pi^+ \pi^-$, peak mass region of 1.84 to 1.88 GeV/c^2 and background mass regions of 1.74 to 1.82 GeV/c^2 and 1.90 to 1.98 GeV/c^2.

From Fig. 2 we find no evidence for the production of recoil systems having masses less than or equal to 1.87 GeV/c^2 in either spectrum. The $K\pi$ data of Fig. 2(a) show a large signal for recoil masses in the range 1.96 to 2.20 GeV/c^2 with contributions up to 2.5 GeV/c^2. The $K3\pi$ recoil-mass spectrum [Fig. 2(b)] has more background, but appears to be consistent with the $K\pi$ spectrum. These spectra suggest that the $K\pi$ and $K3\pi$ systems are produced with thresholds occuring above 3.7-GeV c.m. energy; more detailed interpretations of Fig. 2 are made difficult by the broad range of c.m. energies over which this data sample was collected.

As a further test of this apparent threshold behavior, we have examined 150 000 multihadronic events collected at the ψ mass ($E_{c.m.}$ = 3.1 GeV) and 350 000 events at the ψ' mass ($E_{c.m.}$ = 3.7 GeV) for $K\pi$ and $K3\pi$ signals near 1.87 GeV/c^2. Because of the large cascade decay rate[6] of ψ' to ψ and the large second-order electromagnetic decay rate[7] of the ψ, the resonance events contain 72 000 examples of hadron production by a virtual photon of c.m. energy 3.1 GeV. From fits to invariant-mass spectra (with the signal mass near 1.87 GeV/c^2) we find no $K\pi$ signal larger than 0.3 standard deviations and no $K3\pi$ signal larger than 1.2 standard deviations in this large sample of events. The upper limits (90% confidence level) are 60 events for the $K\pi$ signal and 200 events for the $K3\pi$ signal.

The threshold behavior noted above as well as the narrow widths argue against the interpretation of the structure in Fig. 1 as being a conventional K^*, e.g., the strange counterpart of the $g(1680)$.

Preliminary Monte Carlo calculations to estimate detection efficiencies for two modes have been performed; present systematic uncertainties in these detection efficiencies could be as large as ± 50%. Our estimate of the cross section times branching ratio σB (errors quoted are statistical) averaged over our 3.9–4.6-GeV c.m. energy data is 0.20± 0.05 nb for the $K\pi$ mode and 0.67± 0.11 nb for the $K3\pi$ mode. These are to be compared with the average total hadronic cross section σ_T in this energy region[8] of 27± 3 nb. We have also searched for these signals in the events at higher c.m. energies. In our previous search for the production of narrow peaks[4] at 4.8 GeV, there was a small $K\pi$ signal at 1.87 GeV/c^2 corresponding to a σB of 0.10± 0.07 nb. This signal set the upper limit quoted in the paper (σB < 0.18 nb for the $K\pi$ system of mass between 1.85 and 2.40 GeV/c^2) but lacked the statistical significance necessary to be considered a convincing peak. The value of σ_T at 4.8 GeV is 18± 2 nb.[2] In the c.m. energy range 6.3 to 7.8 GeV the $K\pi$ σB is 0.04± 0.03 nb and the average σ_T is 10± 2 nb.

In summary, we have observed significant peaks in the invariant-mass spectra of $K^\pm \pi^\mp$ and $K^\pm \pi^\mp \pi^+ \pi^-$ that we associate with the decay of a state of mass 1865± 15 MeV/c^2 and width less than 40 MeV/c^2. The recoil-mass spectra indicate that this state is produced in association with systems of comparable or larger mass.

We find it significant that the threshold energy for pair-producing this state lies in the small interval between the very narrow ψ' and the broader structures present in e^+e^- annihilation near 4 GeV.[8] In addition, the narrow width of this state, its production in association with systems of even greater mass, and the fact that the decays we observe involve kaons form a pattern of observation that would be expected for a state possessing the proposed new quantum number charm.[9,10]

[†]Work supported by the U. S. Energy Research and Development Administration.
[*]Miller Institute for Basic Research in Science, Berkeley, Calif. (1975–1976).
[‡]Permanent address: Centre d'Etudes Nucléaires de Saclay, Saclay, France.
[§]Fellow of Deutsche Forschungsgemeinschaft.
[∥]Permanent address: Varian Associates, Palo Alto, California.
[**]Permanent address: Laboratoire de Physique Nucléaire et Haute Energie, Université Paris VI, Paris, France.
[††]Permanent address: Laboratori Nazionali, Frascatti, Rome, Italy.
[‡‡]Permanent address: Institut de Physique Nucléaire, Orsay, France.

[1]J.-E. Augustin et al., Phys. Rev. Lett. 34, 233 (1975).
[2]J.-E. Augustin et al., Phys. Rev. Lett. 34, 764 (1975).
[3]The only other feature in the $K\pi$ system that we observe in this data sample is the $K^*(890)$.
[4]A. M. Boyarski et al., Phys. Rev. Lett. 35, 196 (1975).
[5]This cut rejects most nucleons (p and $\bar{p}$) as well as tracks accompanied by extra particles in the TOF counter.
[6]G. S. Abrams et al., Phys. Rev. Lett. 34, 1181 (1975).
[7]A. M. Boyarski et al., Phys. Rev. Lett. 34, 1357 (1975).
[8]J. Siegrist et al., Phys. Rev. Lett. 36, 700 (1976).
[9]S. L. Glashow, J. Iliopoulos, and L. Maiani, Phys. Rev. D 2, 1285 (1970); S. L. Glashow, in *Experimental Meson Spectroscopy—1974*, AIP Conference Proceedings No. 21, edited by D. A. Garelick (American Institute of Physics, New York, 1974), p. 387.
[10]Other indications of possible charmed-particle production have come from experiments involving neutrino interactions. See, for example, A. Benvenuti et al., Phys. Rev. Lett. 34, 419 (1975); E. G. Cazzoli et al., Phys. Rev. Lett. 34, 1125 (1975); J. Bleitschau et al., Phys. Lett. 60B, 207 (1976); J. von Krogh et al., Phys. Rev. Lett. 36, 710 (1976); B. C. Barish et al., Phys. Rev. Lett. 36, 939 (1976).

10
Quarks, Gluons, and Jets

The quanta of quantum chromodynamics, 1974–1982.

The striking success of the parton model in describing deep inelastic–nucleon and neutrino–nucleon scattering provided strong circumstantial evidence for the Feynman–Bjorken picture and for its complete elaboration as quantum chromodynamics (QCD). QCD describes all strong interactions as resulting from the interactions of spin-1/2 quarks and spin-1 gluons. The fundamental coupling is of the gluon to the quarks, in a fashion analogous to the coupling of a photon to electrons. In addition, the gluons couple directly to each other. $SU(3)$ plays a central role. Just as in the Gell-Mann–Zweig model of hadrons, there are three basic constituents. The u quark, for example, comes in three versions, say, red, blue, and green. Similarly, every other kind of quark comes in these three versions or "colors." Often it is convenient to refer to u, d, s, and c as "flavors" of quarks, to contrast with the three colors in which every flavor comes. While the $SU(3)$ of flavor is an approximate symmetry, the $SU(3)$ of color is an exact symmetry, thus the three colors of the u quark are exactly degenerate in mass, while the u, d, and s quarks are not degenerate.

The rules of $SU(3)$ state that if we combine a 3 (a quark) with a 3* (an antiquark), we get $1 + 8$, a singlet and an octet. In terms of mesons, this explains that combining the three quark flavors (u, d, s) with the three antiquark flavors yields an $SU(3)$ singlet (η') and an octet (the pseudoscalar octet of π, K, η). $SU(3)$ color works the same way. Suppose we take red, blue, and green u quarks and combine them with antired, antiblue, and antigreen $\bar{d}$ quarks. We get nine combinations, each of which is $u\bar{d}$. One linear combination is a color singlet and the eight others form a color octet. The gluons are octets of color. From the rule $3 \times 3^* = 1 + 8$ we learn that a quark (3) and an antiquark (3*) of the same flavor can combine to make a gluon (8).

It was an initial postulate of QCD that only color singlet objects could appear as physical particles. Thus the π^+ would be the color singlet combination of $u\bar{d}$, while the remaining eight combinations would not correspond to physically observed states. Combining three quarks is described by the $SU(3)$ relation $3 \times 3 \times 3 = 1 + 8 + 8 + 10$. When applied to the $SU(3)$ of u, d, and s, this means that baryons should come in 1-, 8-, and 10-dimensional representations. Indeed, these are the representations observed, while mesons are not observed in 10-dimensional representations. When applied to color $SU(3)$, the relation shows that there is one way to combine three colors to make a color singlet. This single way corresponds to the antisymmetric combination of the three elements, $rbg - rgb + grb - gbr + bgr - brg$. The combinations producing nonsinglet states do not correspond to physical particles. Indeed an initial impetus for introducing three colors was to explain how the $\Delta^{++}(1232)$ could be a low-lying state. Since it is a uuu and presumably entirely s-wave (as are all lowest-lying states), its wave function is apparently symmetric under interchange of any two quarks. This is not allowed for fermions. A solution to this puzzle was proposed in 1964, before the development of QCD by O. W. Greenberg who added to the other quantum numbers of the quark an index that could take on three values. This index is equivalent to the color quantum number. The color singlet combination of three colors is completely antisymmetric thus making the overall wave function satisfy the Pauli principle.

A single quark cannot be a color singlet and thus should not occur as a physical particle. This property is called "confinement." The quarks are confined inside physical hadrons, which are always color singlets.

The e^+e^- annihilation process produces a virtual photon which according to the quark–parton model couples to the various quarks according to their electric charges. It couples to each color of quark equally. Suppose that the virtual photon produces a $u\bar{u}$ pair that is red–antired. These quarks will be receding from each other rapidly if the energy of the collision is large. Why do they not emerge as isolated quarks? According to QCD, the force between the quarks becomes a constant for large separation. Thus the potential energy is proportional to the separation. When this is large enough, it is energetically favorable to produce a new quark–antiquark pair out of the vacuum, thus reducing the separation between the quark and the antiquark. Suppose this new pair is located so that its antiquark is near the original quark. These may bind to form a meson. The unpaired new quark is still receding from the initial antiquark so it may become favorable to create another new pair. This continues until all the quarks and antiquarks are paired. A similar mechanism permits the creation of baryons.

If the quarks are never free, how can they be observed? Of course they were observed indirectly in deep-inelastic scattering. However, the parton model and QCD indicated that more direct evidence should be obtained by studying certain reactions, the simplest being e^+e^- annihilation. While the produced quarks could not be seen, the initial quarks should materialize into jets of hadrons moving nearly along the directions of the quarks. In a very high energy collision, the hadrons would lie nearly along this single axis, with momenta transverse to it of a few hundred MeV. This estimate was derived from the observation that in most hadronic collisions at high energy, the transverse momentum of the secondaries rarely exceeded this amount.

In an idealized picture, the annihilation of the electron and positron would occur into $\mu^+\mu^-$ pairs and quark–antiquark pairs with frequencies proportional to the squares of the final particle charges. The hadronic final states would come from u, d, and s quarks with probabilities proportional to $3(2/3)^2$, $3(-1/3)^2$, $3(-1/3)^2$, relative to 1 for the muons. The factor 3 arises from the three possible colors. The ratio of the hadronic to muonic final states is called R and is thus predicted to be 2 if there are three quarks and three colors. This prediction failed in a spectacular way, as described in the previous chapter. Ultimately, the prediction for R was verified at energies away from the ψ resonances and provided one of the best pieces of evidence for the correctness of QCD. See Fig. 10.1. A second prediction is that the angular distribution of the muons and the quarks should be $1+\cos^2\theta$, relative to the direction of the electron and positron beams. Of course, the direction of the quarks cannot be measured since the quarks are never seen. However, there is an axis for each event, defined by the initial quark direction. This axis is obscured by the transverse momentum acquired by the final-state particles in the "hadronization," in which the initial quarks become hadrons. At sufficiently high energy the axis is clear, but at low energy, the momentum of the final-state particles is not much more than the few hundred MeV anticipated for transverse momentum.

Evidence for jets arising from quarks was first obtained by comparing data taken at various center-of-mass energies (**Ref. 10.1**), using the SLAC–LBL Mark I detector at the SPEAR storage ring located at SLAC. Since the jets could not be discerned by simply looking at the pattern of outgoing tracks, it was necessary to define an algorithm for defining the jet axis. The one selected was that originally proposed by Bjorken and Brodsky. The axis was taken to be the direction such that the sum of the squares of the momenta transverse to the axis was a minimum. For each event, such an axis could be found. Each event was assigned a value of the "sphericity" defined to be

$$S = \frac{3 \sum_i \mathbf{p}_{\perp i}^2}{2 \sum_i \mathbf{p}_i^2}, \qquad (10.1)$$

where $\mathbf{p}_{\perp i}$ is the momentum of the ith particle perpendicular to the sphericity axis. A completely jetlike event with outgoing particles aligned precisely with the axis would have $S = 0$. An isotropic event would have $S \approx 1$. An alternative variable that characterizes e^+e^- events is "thrust." Events with two, well-defined, back-to-back jets have thrust near 1. Spherical events have thrust near 0.

There are two predictions that can be made. First, as the energy increases, the events should become more jetlike so the sphericity should decrease. More importantly, the jet axis should have an angular distribution identical to that for muons. To test the first prediction the sphericity measured at SPEAR was compared at 3.0, 6.2, and 7.4 GeV center-of-mass energy to the predictions of two models, one using an isotropic phase space distribution and one simulating the parton model, with limited transverse momentum relative to the event axis. At 3.0 GeV both models adequately described the sphericity distribution, but at the higher energies only the jetlike parton model succeeded.

Because the Mark I detector was limited in its acceptance in the polar angle, high statistical accuracy was required to test directly the prediction $d\sigma/d\Omega \propto 1 + \cos^2\theta$. However, since the beams at SPEAR were polarized at 7.4 GeV, with electron polarization parallel to the magnetic field responsible for the bending of the beams, another approach was available. If the beams were completely polarized, the angular distribution in $e^+e^- \to \mu^+\mu^-$

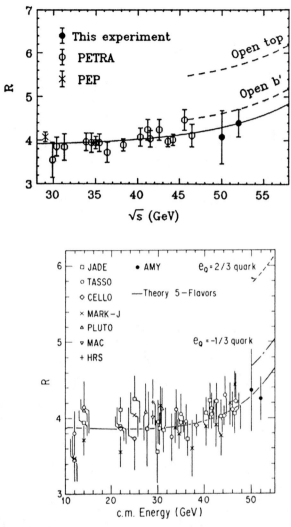

Figure 10.1. Data for e^+e^- annihilation into hadrons as a function of the c.m. energy, including results at $\sqrt{s}$ = 50 GeV and 52 GeV from the TRISTAN storage ring located at the KEK Laboratory in Japan: above, results from the TOPAZ Collaboration, [I. Adachi et al., Phys. Rev. Lett. **60**, 97 (1988)]; below, results from the AMY Collaboration [H. Sagawa et al., Phys. Rev. Lett. **60**, 93 (1988)]. Also shown are results obtained at lower energy machines. The basic prediction of the quark–parton model, including the b-quark discussed in the next Chapter, is $R = 11/3$. QCD radiative corrections and contributions from the Z (discussed in Chapter 12) increase this, and account for the rising prediction at higher c.m. energies. If there were a sixth quark with low enough mass to be pair-produced in this energy region, the value of R would jump as indicated by the curves. Extensive searches at PETRA found no evidence for a sixth quark up to $\sqrt{s} = 46$ GeV. The searches at TRISTAN also show no evidence of a new quark.

would be

$$d\sigma/d\Omega \propto 1 + \cos^2\theta + \sin^2\theta \cos 2\phi, \tag{10.2}$$

where ϕ is the azimuthal angle measured from the plane of the storage ring. If the degree of polarization of each beam is P, then

$$d\sigma/d\Omega \propto (1 - P^2)(1 + \cos^2\theta) + P^2(1 + \cos^2\theta + \sin^2\theta \cos 2\phi)$$
$$\propto 1 + \cos^2\theta + P^2 \sin^2\theta \cos 2\phi. \tag{10.3}$$

This behavior had been confirmed in earlier measurements of the $\mu^+\mu^-$ final state by Mark I at SPEAR. The angular distribution for the hadronic jets would be expected to be the same if the quarks could be regarded as nearly massless spin-1/2 objects with purely pointlike (Dirac) couplings. If, on the other hand, the partons were spin-0, the expected distribution would be

$$d\sigma/d\Omega \propto 1 - \cos^2\theta - P^2 \sin^2\theta \cos 2\phi. \tag{10.4}$$

These two cases are the extremes. The Dirac coupling of relativistic spin-1/2 particles to the photon produces a "transverse cross section" in that the electromagnetic current matrix element is perpendicular to the outgoing quark direction, while the coupling of the spin-0 particles to the photon produces a "longitudinal cross section" with the current parallel to the outgoing parton direction. The most general form is

$$d\sigma/d\Omega \propto 1 + \alpha \cos^2\theta + P^2 \alpha \sin^2\theta \cos 2\phi, \tag{10.5}$$

where $-1 \leq \alpha \leq 1$. The square of the polarization was measured to be $P^2 = 0.47 \pm 0.05$ at 7.4 GeV using the $e^+e^- \to \mu^+\mu^-$ process. The hadronic jets gave an angular distribution with $\alpha = 0.45 \pm 0.07$. After correcting for detector effects, this became $\alpha = 0.78 \pm 0.12$ at 7.4 GeV, near the value $\alpha = 1$ predicted for the purely spin-1/2 case. Previously, the Mark I collaboration had measured the angular distribution of produced hadrons, rather than the distribution of the sphericity axis, relative to the beam (Ref. 10.2). There too, the azimuthal dependence indicated that the underlying partons that coupled to the virtual photon produced in e^+e^- annihilation had spin 1/2.

QCD not only encompasses the quark model, it predicts deviations from the simplest form of that model, as discussed in Chapter 8. Deviations from scaling in deep inelastic lepton scattering were predicted using "asymptotic freedom," a property of the theory that states that at high momentum transfer, the coupling between the quarks and the gluons becomes small. This means that in this regime, predictions can be made on the basis of perturbation theory, just as they are in quantum electrodynamics (QED). There are two primary differences. Instead of $\alpha \approx 1/137$, the coupling is $\alpha_s(Q^2)$, a function of the momentum transfer, Q^2. Typically, in the region where perturbation theory applies, $\alpha_s(Q^2) \approx 0.1 - 0.2$. Secondly, unlike photons, gluons can couple to themselves.

Actually, the α used in QED can also be thought of as a function of the momentum transfer. Because of vacuum polarization, the force between two point charges with separation r is not just α/r^2, but is more accurately $\alpha[1+\alpha f(r)]/r^2$, where $f(r)$ represents the effect of vacuum polarization and is important for r less than the Compton wavelength of the electron. The vacuum polarization in QED increases the force between charges as the distance between them decreases, or equivalently, as the momentum transfer increases. In QCD, the behavior is just the opposite. The coupling gets weaker as the momentum transfer increases. The leading behavior can be expressed as

$$\alpha_s(Q^2) = \frac{4\pi}{(33 - 2n_f)\ln(Q^2/\Lambda^2)}, \tag{10.6}$$

where n_f is the number of quark flavors (u, d, s, etc.) with mass less than $Q/2$ and Λ is a parameter to be determined experimentally, and is typically found to be about 200 MeV.

The basic process in e^+e^- annihilation into hadrons is, according to the quark–parton model, $e^+e^- \to \gamma^* \to q\bar{q}$. In addition, there are corrections that produce $e^+e^- \to \gamma^* \to q\bar{q}g$, where g is a gluon. The cross section for this is of order α_s relative to the process in which no gluon is produced. It is conventional to define scaled variables $x_i = E_i/E$, where the energies of the $q, \bar{q}$, and g are E_1, E_2, and E_3, and the electron and positron beam energies are E, so that $x_1 + x_2 + x_3 = 2$. If σ_0 represents the cross section for $e^+e^- \to q\bar{q}$, then

$$\frac{1}{\sigma_0}\frac{d\sigma_{q\bar{q}g}}{dx_1\,dx_2} = \frac{2\alpha_s}{3\pi}\frac{x_1^2 + x_2^2}{(1-x_1)(1-x_2)}. \tag{10.7}$$

The cross section is seen to diverge if $x_1 \to 1$ or $x_2 \to 1$. These limits obtain when the gluon is parallel to either the quark or antiquark, or if x_3 goes to zero. If the gluon and the quark are moving in nearly the same direction, it becomes difficult to discern that the gluon is present: the $q\bar{q}g$ state merges into the $q\bar{q}$ state.

While the $q\bar{q}g$ state could be produced at the energy available at SPEAR or DORIS (an e^+e^- collider at DESY with an energy similar to that at SPEAR), we have already seen that the jets in $q\bar{q}$ could just barely be distinguished there. To identify $q\bar{q}g$ states required higher energy. This was achieved first at PETRA, an e^+e^- collider located at DESY, which was able to reach more than 30 GeV total center-of-mass energy.

PETRA had four intersection regions. These were initially occupied by the TASSO, PLUTO, MARK J, and JADE detectors. All found evidence for the $q\bar{q}g$ final state (**Refs. 10.3**, 10.4, 10.5, 10.6). Some data from MARK J, PLUTO, and JADE are shown in Figures 10.2, 10.3, and 10.4. The TASSO collaboration defined three orthogonal axes, $\mathbf{n}_1$, $\mathbf{n}_2$, and $\mathbf{n}_3$, for each event. The direction $\mathbf{n}_3$ was the sphericity axis, the one relative to which the sum of the squares of the transverse momenta was minimal. The direction $\mathbf{n}_1$ maximized the sum of the squares of the transverse momenta. The remaining axis was orthogonal to the other two. The $\mathbf{n}_2$–$\mathbf{n}_3$ plane was thus such that the sum of the squares

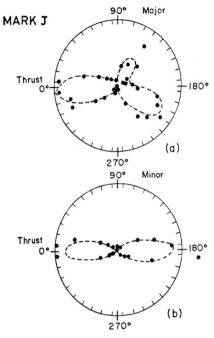

Figure 10.2. Data from the MARK J detector showing the energy flow distribution projected onto a plane. Events showing the typical two-jet distribution are not included. The distance from the center to the data point is proportional to the energy deposited. The dashed line represents the expectation of a $q\bar{q}g$ model. (a) Projection on the plane of the thrust and major axes. (b) Projection on the plane of the thrust and minor axes. The thrust axis is similar to the sphericity axis while the major and minor axes are analogous to the directions $\mathbf{n}_2$ and $\mathbf{n}_1$ defined in the text. (Ref. 10.4)

of the momenta out of it was a minimum. This plane could be viewed as the event plane. Components perpendicular to the primary axis, in and out of the plane were defined:

$$< p_\perp^2 >_{out} = \frac{1}{N} \sum_{j=1}^{N} (\mathbf{p}_j \cdot \mathbf{n}_1)^2 \tag{10.8}$$

$$< p_\perp^2 >_{in} = \frac{1}{N} \sum_{j=1}^{N} (\mathbf{p}_j \cdot \mathbf{n}_2)^2 \tag{10.9}$$

The experiment sought to distinguish between two possibilities. The first was that all $e^+e^- \to$ hadron events were basically of the form $e^+e^- \to q\bar{q}$, but as the jet energy increased, the jets became "fatter," i.e. had more transverse momentum relative to the jet axis. The second was that as energy increased, more and more events were due to $e^+e^- \to q\bar{q}g$. The data showed that at high energies, there were events with $< p_\perp^2 >_{in} \gg < p_\perp^2 >_{out}$. This could be understood as the result of $q\bar{q}g$ final states, but not from $q\bar{q}$ final states. Some of the events displayed very clean three-jet topology, providing visual evidence for the existence of the gluon.

300   10. Quarks, Gluons, and Jets

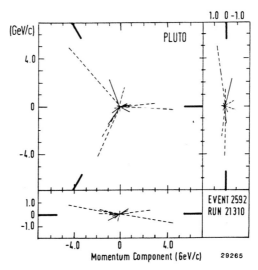

Figure 10.3. Track momentum vectors for a single event observed by the PLUTO collaboration, shown in three projections. The solid lines represent charged particles; the dashed lines, neutral particles. The dark bars show the inferred directions of the three jets. The upper left projection is onto a plane analogous to the $\mathbf{n}_2$–$\mathbf{n}_3$ plane. The bottom projection corresponds to the $\mathbf{n}_1$–$\mathbf{n}_2$ plane and the right projection to the $\mathbf{n}_1$–$\mathbf{n}_3$ plane. (Ref. 10.5)

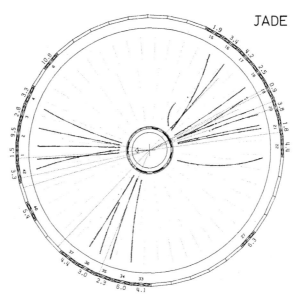

Figure 10.4. A three-jet event measured by the JADE Collaboration, viewed along the beam axis. [P. Söding and G. Wolf, *Ann. Rev. Nucl. Part. Sci.*, **31**, 231 (1981).]

According to QCD, the fundamental interactions of quarks are due to the exchange of gluons. However, these interactions are obscured because the coupling of the gluons to the quarks is large when the momentum transfer is small. Thus, many gluons are emitted and absorbed in low energy processes. In contrast, at high energies, when the momentum transfer is great, the coupling is small and a single exchange of a gluon may dominate the process.

Experiments at Fermilab, using its 400 GeV proton beam, and the ISR, a proton–proton colliding beam machine at CERN capable of reaching about 60 GeV in the center of mass, sought to identify jets of particles with large transverse momentum. These could arise from the scattering of a quark from the incident proton by a quark from the target proton. The fundamental interaction is the exchange of a gluon between the quarks. This process is entirely analogous to the scattering of alpha particles observed by Geiger and Marsden in 1909. Rutherford inferred from the large angle scattering the existence of a compact, hard nucleus inside the atom. Hadronic jets would support the evidence from electroproduction that inside the nucleon are more fundamental partons, the quarks and gluons. The difficulty was to identify the outgoing jets of particles.

There is no a priori definition that specifies which outgoing particles should be grouped together in identifying a jet. Inevitably, the least energetic particles in a jet merge into the particles not associated with the jet. It is necessary in each experiment to set out an algorithm that defines a jet. This is an especially serious problem at lower energies where jet structures are not clear. Despite years of determined effort, the results from Fermilab and the ISR were not conclusive but only suggestive of jets.

With the operation of the Sp$\bar{\text{p}}$S Collider at CERN, the energies available increased enormously, to $\sqrt{s} = 540$ GeV. Two large detectors, UA-1 and UA-2, were prepared to measure the anticipated high transverse momentum events with highly segmented calorimeters.

Early results from the UA-2 detector showed unambiguous evidence for large transverse momentum jets (**Ref. 10.7**). The UA-2 detector featured one set of calorimeters covering from 40° to 140° in polar angle and a second set covering from 20° to 37.5° and from 142.5° to 160°. The azimuthal coverage in the central region was 300° and consisted of 200 cells.

Since transverse momentum is the signal of interest, the energy measurements are converted to "transverse energy," $E_T = E \sin\theta$, where θ is the polar angle between the beam direction and the jet, and E is the energy deposited into some portion of the detector. In lower energy experiments events with large total transverse energy ΣE_T were observed, but often the transverse energy was not localized into two distinct directions representing two jets, but rather was spread over a large portion of the total solid angle. The UA-2 collaboration was able to provide evidence for well-defined jets at the high energy offered by the Sp$\bar{\text{p}}$S Collider.

To give an operational definition of a jet, the UA-2 collaboration defined a "cluster" of calorimeter cells as a set of contiguous cells each showing an energy deposit greater than 400 MeV. It was then found that as ΣE_T increased, a larger and larger fraction of the total was contained in the two clusters having the largest E_T. This was quite clear evidence for the long-sought-for jets. Some individual events showed strikingly clear evidence for the jets, which could be displayed in "Lego" plots showing the energy deposited in the various calorimeter cells. A series of comparisons showed qualitative agreement with the two-jet picture.

The measured distributions for high-transverse-momentum jets was in reasonable agreement with predictions made from QCD-based models. These models used quark and gluon distributions derived from deep inelastic scattering, together with cross sections calculated from perturbative QCD for the processes like $q\bar{q} \to q\bar{q}$ and $gq \to gq$. The fastest partons in a proton are quarks, so the very high transverse momentum events should arise from the $q\bar{q} \to q\bar{q}$ process. However, the cross sections for these events are small. At more modest transverse momenta, where there are more events, it is actually $gg \to gg$ that is expected to dominate. This is so because of the large number of gluons in the structure functions at high Q^2 and at not too large x, and because the coupling of gluons to other gluons is stronger than the coupling of gluons to quarks.

While high-precision tests were lacking, the qualitative features of the jets found at SPEAR, PETRA, and the Sp$\bar{p}$S Collider confirmed the general predictions of QCD and established its applicability in both leptonically and hadronically induced processes.

Exercises

10.1 Using numerical methods, determine the fraction of e^+e^- events that produce $q\bar{q}g$, where $x_1, x_2, x_3 < 0.9$. Suppose it is also required that $E_1, E_2, E_3 > 5$ GeV. What fraction of $e^+e^- \to$ hadrons events at $E_{cm} = 30, 60, 90$ GeV satisfy this condition as well? Take $\alpha_s = 0.1$.

10.2 Consider the cross section for qq scattering if the quarks are of different flavors (e.g. u and d). The gluon coupling to quarks is completely analogous to the photon coupling with electrons, except that there is a matrix specifying the color interaction:

$$g_s \bar{q}_a \frac{1}{2} \lambda_i^{ab} \gamma_\mu q_b,$$

where $a, b = 1, 2, 3$. The λ_is, $i = 1, \ldots, 8$ are 3×3 traceless matrices satisfying

$$\text{Tr} \lambda_i \lambda_j = 2\delta_{ij} \qquad i, j = 1, \ldots 8,$$

and $g_s^2 = 4\pi \alpha_s$. Find $d\sigma/d\Omega$ for the elastic scattering relative to what it would be without color factors, remembering to average over initial states and sum over final states.

10.3 Suppose the color gauge group were $SO(3)$ (the rotation group) instead of $SU(3)$ and suppose that the quarks came in three colors corresponding to the three-dimensional (vector) representation of $SO(3)$. Assume that hadrons must still be color singlets. Why would this not produce just the usual mesons and baryons?

10.4 * Verify that if the electrons and positrons are completely polarized with their spins perpendicular to the plane of the ring (antiparallel to each other), the angular distribution for $e^+e^- \to \mu^+\mu^-$ is

$$d\sigma/d\Omega \propto 1 + \cos^2\theta + \sin^2\theta \cos 2\phi$$

where θ measures the polar angle away from the beam direction and ϕ the azimuthal angle from the plane of the ring. [Consider the matrix element for producing the virtual photon, $\mathcal{M} \propto \bar{v}\epsilon_\mu \gamma^\mu u$, where ϵ is the polarization vector of the virtual photon and show that if the electron and positron spins are perpendicular to the plane of the ring, so is ϵ. Then consider the matrix element for the decay into massless fermions, $\mathcal{M} \propto \bar{u}(k)\epsilon_\mu \gamma^\mu v(k')$ and calculate $|\mathcal{M}|^2$, summing over final-state spins to find the angular distribution.] Do the same for the final state with two spin-0 particles. The decay matrix element is proportional to $(k-k')_\mu \epsilon^\mu$, where k and k' are the final-state momenta.

Further Reading

Extensive coverage of QCD is given in C. Quigg, *Gauge Theories of the Strong, Weak, and Electromagnetic Interactions*, Benjamin/Cummings, Menlo Park, CA, 1983, and Westview Press, 1997. See especially Chapter 8.

A comprehensive treatment is given in R. K. Ellis, W. J. Stirling, and B. R. Webber, *QCD and Collider Physics*, Cambridge University Press, 1996.

The polarization of electron and positron beams in storage rings caused by spin-flip emission of synchrotron radiation is beautifully explained in J. D. Jackson, *Rev. Mod. Phys.* **48**, 417 (1976).

References

10.1 G. Hanson et al., "Evidence for Jet Structure in Hadron Production by e^+e^- Annihilation." *Phys. Rev. Lett.*, **35**, 1609 (1975).

10.2 R. F. Schwitters et al., "Azimuthal Asymmetry in Inclusive Hadron Production by e^+e^- Annihilation." *Phys. Rev. Lett.*, **35**, 1320 (1975).

10.3 TASSO Collaboration, R. Brandelik et al., "Evidence for Planar Events in e^+e^- Annihilation at High Energies." *Phys. Lett.*, **86B**, 243 (1979).

10.4 MARK J Collaboration, D. P. Barber et al., "Discovery of Three-jet Events and a Test of Quantum Chromodynamics at PETRA." *Phys. Rev. Lett.*, **43**, 830 (1979).

10.5 PLUTO Collaboration, Ch. Berger et al., "Evidence for Gluon Bremsstrahlung in e^+e^- Annihilation at High Energies." *Phys. Lett.*, **86B**, 418 (1979).

10.6 JADE Collaboration, W. Bartel et al., "Observation of Planar Three-jet Events in e^+e^- Annihilation and Evidence for Gluon Bremsstrahlung." *Phys. Lett.*, **91B**, 142 (1980).

10.7 UA-2 Collaboration, M. Banner et al., "Observation of Very Large Transverse Momentum Jets at the CERN $\bar{p}p$ Collider." *Phys. Lett.*, **118B**, 203 (1982).

Evidence for Jet Structure in Hadron Production by e^+e^- Annihilation*

G. Hanson, G. S. Abrams, A. M. Boyarski, M. Breidenbach, F. Bulos,
W. Chinowsky, G. J. Feldman, C. E. Friedberg, D. Fryberger, G. Goldhaber,
D. L. Hartill,† B. Jean-Marie, J. A. Kadyk, R. R. Larsen, A. M. Litke,
D. Lüke,‡ B. A. Lulu, V. Lüth, H. L. Lynch, C. C. Morehouse,
J. M. Paterson, M. L. Perl, F. M. Pierre,§ T. P. Pun, P. A. Rapidis,
B. Richter, B. Sadoulet, R. F. Schwitters, W. Tanenbaum,
G. H. Trilling, F. Vannucci, ‖ J. S. Whitaker,
F. C. Winkelmann, and J. E. Wiss

*Lawrence Berkeley Laboratory and Department of Physics, University of California, Berkeley, California 94720,
and Stanford Linear Accelerator Center, Stanford University, Stanford, California 94305*
(Received 8 October 1975)

We have found evidence for jet structure in $e^+e^- \to$ hadrons at center-of-mass energies of 6.2 and 7.4 GeV. At 7.4 GeV the jet-axis angular distribution integrated over azimuthal angle was determined to be proportional to $1 + (0.78 \pm 0.12)\cos^2\theta$.

In quark-parton constituent models of elementary particles, hadron production in e^+e^- annihilation reactions proceeds through the annihilation of the e^+ and e^- into a virtual photon which subsequently produces a quark-parton pair, each member of which decays into hadrons. At sufficiently high energy the limited transverse-momentum distribution of the hadrons with respect to the original parton production direction, characteristic of all strong interactions, results in oppositely directed jets of hadrons.[1-4] The spins of the constituents can, in principle, be determined from the angular distribution of the jets.

In this Letter we report the evidence for the existence of jets and the angular distribution of the jet axis.

The data were taken with the Stanford Linear Accelerator Center–Lawrence Berkeley Laboratory magnetic detector at the SPEAR storage ring of the Stanford Linear Accelerator Center. Hadron production, muon pair production, and Bhabha scattering data were recorded simultaneously. The detector and the selection of events have been described previously.[5,6] The detector subtended $0.65 \times 4\pi$ sr with full acceptance in azimuthal angle and acceptance in polar angle from

50° to 130°. We have used the large blocks of data at center-of-mass energies ($E_{c.m.}$) of 3.0, 3.8, 4.8, 6.2, and 7.4 GeV. We included only those hadronic events in which three or more particles were detected in order to avoid background contamination in events with only two charged tracks due to beam-gas interactions and photon-photon processes.

To search for jets we find for each event that direction which minimizes the sum of squares of transverse momenta.[7] For each event we calculate the tensor

$$T^{\alpha\beta} = \sum_i (\delta^{\alpha\beta} \vec{p}_i^{\,2} - p_i^{\alpha} p_i^{\beta}), \quad (1)$$

where the summation is over all detected particles and α and β refer to the three spatial components of each particle momentum $\vec{p}_i$. We diagonalize $T^{\alpha\beta}$ to obtain the eigenvalues λ_1, λ_2, and λ_3 which are the sums of squares of transverse momenta with respect to the three eigenvector directions. The smallest eigenvalue (λ_3) is the minimum sum of squares of transverse momenta. The eigenvector associated with λ_3 is defined to be the reconstructed jet axis. In order to determine how jetlike an event is, we calculate a quantity which we call the sphericity (S):

$$S = \frac{3\lambda_3}{\lambda_1 + \lambda_2 + \lambda_3} = \frac{3(\sum_i p_{\perp i}^{\,2})_{\min}}{2\sum_i \vec{p}_i^{\,2}}. \quad (2)$$

S approaches 0 for events with bounded transverse momenta and approaches 1 for events with large multiplicity and isotropic phase-space particle distributions.

The data at each energy were compared to Monte Carlo simulations which were based on either an isotropic phase-space (PS) model or a jet model. In both models only pions (charged and neutral) were produced. The total multiplicity was given by a Poisson distribution. The jet model modified phase space according to the square of a matrix element of the form

$$M^2 = \exp(-\sum_i p_{\perp i}^{\,2}/2b^2), \quad (3)$$

where $p_\perp$ is the momentum perpendicular to the jet axis.

The angular distribution for the jet axis is expected to have the form

$$d\sigma/d\Omega \propto 1 + \alpha \cos^2\theta + P^2 \alpha \sin^2\theta \cos(2\varphi), \quad (4)$$

where θ is the polar angle of the jet axis with respect to the incident positron direction, φ is the azimuthal angle with respect to the plane of the storage ring, $\alpha = (\sigma_T - \sigma_L)/(\sigma_T + \sigma_L)$ with σ_T and σ_L the transverse and longitudinal production cross sections, and P is the polarization of each beam. (The polarization term will be discussed later.) The angular distribution given by Eq. (4) was used in the jet-model simulation. The simulations included the geometric acceptance, the trigger efficiency, and all other known characteristics of the detector. The total multiplicity and the charged-neutral multiplicity ratio for both models were obtained by fitting to the observed charged-particle mean multiplicity and mean momentum at each energy. In the jet model the parameter b was determined by fitting to the observed mean S at the highest energy (7.4 GeV). For lower energies the value of b was determined by requiring that the mean $p_\perp$ in the jet model be the same (315 MeV/c) as at 7.4 GeV.

Figure 1 shows the observed mean S and the model predictions. Both models are consistent with the data in the 3–4-GeV region. At higher energies the data have significantly lower mean S than the PS model and agree with the jet model. Figure 2 shows the S distributions at several energies. At 3.0 GeV the data agree with either the PS or the jet model [Fig. 2(a)]. At 6.2 and 7.4 GeV the data are peaked toward low S, favoring

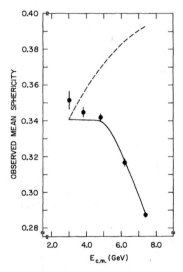

FIG. 1. Observed mean sphericity versus center-of-mass energy $E_{c.m.}$ for data, jet model with $\langle p_\perp \rangle = 315$ MeV/c (solid curve), and phase-space model (dashed curve).

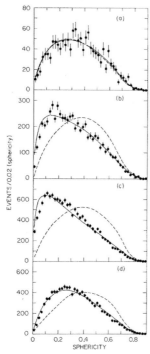

FIG. 2. Observed sphericity distributions for data, jet model with $\langle p_\perp \rangle = 315$ MeV/c (solid curves), and phase-space model (dashed curves) for (a) $E_{c.m.} = 3.0$ GeV; (b) $E_{c.m.} = 6.2$ GeV; (c) $E_{c.m.} = 7.4$ GeV; and (d) $E_{c.m.} = 7.4$ GeV, events with largest $x < 0.4$. The distributions for the Monte Carlo models are normalized to the number of events in the data.

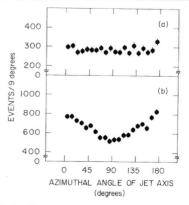

FIG. 3. Observed distributions of jet-axis azimuthal angles from the plane of the storage ring for jet axes with $|\cos\theta| \leq 0.6$ for (a) $E_{c.m.} = 6.2$ GeV and (b) $E_{c.m.} = 7.4$ GeV.

the jet model [Figs. 2(b) and 2(c)]. At the highest two energies, the PS model poorly reproduces the single-particle momentum spectra, having fewer particles with $x > 0.4$ ($x = 2p/E_{c.m.}$ and p is the particle momentum) than the data.[8] The jet-model x distributions are in better agreement. For $x < 0.4$ the x distributions for both models agree with the data. Therefore, we show in Fig. 2(d) the S distributions at 7.4 GeV for those events in which *no* particle has $x > 0.4$. The jet model is still preferred.

At $E_{c.m.} = 7.4$ GeV the electron and positron beams in the SPEAR ring are transversely polarized, and the hadron inclusive distributions show an azimuthal asymmetry.[9] The φ distributions of the jet axis for jet axes with $|\cos\theta| \leq 0.6$ are shown in Fig. 3 for 6.2 and 7.4 GeV.[10] At 6.2 GeV, the beams are unpolarized[9] and the φ distribution is flat, as expected. At 7.4 GeV, the φ distribution of the jet axis shows an asymmetry with maxima and minima at the same values of φ as for $e^+e^- \to \mu^+\mu^-$.

The φ distribution shown in Fig. 3(b) and the value for P^2 (0.47 ± 0.05) measured simultaneously by the reaction[9] $e^+e^- \to \mu^+\mu^-$ were used to determine the parameter α of Eq. (4). The value obtained for the *observed* jet axis is $\alpha = 0.45 \pm 0.07$. This observed value of α will be less than the true value which describes the production of the jets because of the incomplete acceptance of the detector, the loss of neutral particles, and our method of reconstructing the jet axis. We have used the jet-model Monte Carlo simulation to estimate the ratio of observed to produced values of α and find this ratio to be 0.58 at 74 GeV. Thus the value of α describing the *produced* jet-axis angular distribution is $\alpha = 0.78 \pm 0.12$ at $E_{c.m.} = 7.4$ GeV. The error in α is statistical only; we estimate that the systematic errors in the observed α can be neglected. However, we have not studied the model dependence of the correction factor relating observed to produced values of α.

The sphericity and the value of α as determined above are properties of whole events. The simple jet model used for the sphericity analysis can also be used to predict the single-particle inclusive angular distributions for all values of the secondary particle momentum. In Fig.

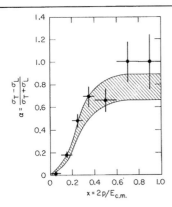

FIG. 4. Observed inclusive α versus x (from Ref. 9) for particles with $|\cos\theta| \leq 0.6$ in hadronic events at $E_{c.m.} = 7.4$ GeV. The prediction of the jet-model Monte Carlo simulation for jet-axis angular distribution with $\alpha = 0.78 \pm 0.12$ is represented by the shaded band.

4 values for the inclusive hadron α as a function of x at 7.4 GeV[9] are compared with the jet-model calculation. The model assumed the value $\alpha = 0.78 \pm 0.12$ for the jet-axis angular distribution. The prediction agrees well with the data for all values of x.

We conclude that the data strongly support the jet hypothesis for hadron production in e^+e^- annihilation. The data show a decreasing mean sphericity with increasing $E_{c.m.}$ and the sphericity distributions peak more strongly at low values as $E_{c.m.}$ increases. Both of these trends agree with a jet model and disagree with an isotropic PS model. The mean transverse momentum relative to the jet axis obtained using the jet-model Monte Carlo simulation was found to be 315 ± 2 MeV/c. At $E_{c.m.} = 7.4$ GeV the coefficient α for the jet-axis angular distribution in Eq. (4) has been found to be nearly $+1$ giving a value for σ_L/σ_T of 0.13 ± 0.07. The jet model also reproduces well the inclusive hadron α versus x. All of this indicates not only that there are jets but also that the helicity along the jet axis is ± 1. In the framework of the quark-parton model, the partons must must have spin $\frac{1}{2}$ rather than spin 0.

---

*Work supported by the U. S. Energy Research and Development Administration.
†Alfred P. Sloan Fellow.
‡Fellow of Deutsche Forschungsgemeinschaft.
§Permanent address: Centre d'Etudes Nucléaires de Saclay, Saclay, France
∥Permanent address: Institut de Physique Nucléaire, Orsay, France.

[1]S. D. Drell, D. J. Levy, and T. M. Yan, Phys. Rev. 187, 2159 (1969), and Phys. Rev. D 1, 1617 (1970).
[2]N. Cabibbo, G. Parisi, and M. Testa, Lett. Nuovo Cimento 4, 35 (1970).
[3]J. D. Bjorken and S. J. Brodsky, Phys. Rev. D 1, 1416 (1970).
[4]R. P. Feymann, *Photon-Hadron Interactions* (Benjamin, Reading, Mass., 1972), p. 166.
[5]J.-E. Augustin et al., Phys. Rev. Lett. 34, 233 (1975).
[6]J.-E. Augustin et al., Phys. Rev. Lett. 34, 764 (1975).
[7]It is impossible to determine the jet axis exactly, even with perfect detection efficiency; the method described here, which was suggested in Ref. 3, is the best approximation known to us.
[8]The momentum distributions will be discussed in a subsequent paper.
[9]R. F. Schwitters et al., Phys. Rev. Lett. 35, 1320 (1975).
[10]Since the jet axis is a symmetry axis, the azimuthal angle $\varphi + 180°$ is equivalent to the azimuthal angle φ.

EVIDENCE FOR PLANAR EVENTS IN e^+e^- ANNIHILATION AT HIGH ENERGIES

TASSO Collaboration

R. BRANDELIK, W. BRAUNSCHWEIG, K. GATHER, V. KADANSKY, K. LÜBELSMEYER,
P. MÄTTIG, H.-U. MARTYN, G. PEISE, J. RIMKUS, H.G. SANDER, D. SCHMITZ,
A. SCHULTZ von DRATZIG, D. TRINES and W. WALLRAFF
I. Physikalisches Institut der RWTH Aachen, Germany [5]

H. BOERNER, H.M. FISCHER, H. HARTMANN, E. HILGER, W. HILLEN, G. KNOP,
W. KORBACH, P. LEU, B. LÖHR, F. ROTH [1], W. RÜHMER, R. WEDEMEYER, N. WERMES
and M. WOLLSTADT
Physikalisches Institut der Universität Bonn, Germany [5]

R. BÜHRING, R. FOHRMANN, D. HEYLAND, H. HULTSCHIG, P. JOOS, W. KOCH,
U. KÖTZ, H. KOWALSKI, A. LADAGE, D. LÜKE, H.L. LYNCH, G. MIKENBERG [2],
D. NOTZ, J. PYRLIK, R. RIETHMÜLLER, M. SCHLIWA, P. SÖDING, B.H. WIIK and
G. WOLF
Deutsches Elektronen-Synchrotron DESY, Hamburg, Germany

M. HOLDER, G. POELZ, J. RINGEL, O. RÖMER, R. RÜSCH and P. SCHMÜSER
II. Institut für Experimentalphysik der Universität Hamburg, Germany [5]

D.M. BINNIE, P.J. DORNAN, N.A. DOWNIE, D.A. GARBUTT, W.G. JONES, S.L. LLOYD,
D. PANDOULAS, A. PEVSNER [3], J. SEDGEBEER, S. YARKER and C. YOUNGMAN
Department of Physics, Imperial College, London, England [6]

R.J. BARLOW, R.J. CASHMORE, J. ILLINGWORTH, M. OGG and G.L. SALMON
Department of Nuclear Physics, Oxford University, England [6]

K.W. BELL, W. CHINOWSKY [4], B. FOSTER, J.C. HART, J. PROUDFOOT, D.R. QUARRIE,
D.H. SAXON and P.L. WOODWORTH
Rutherford Laboratory, Chilton, England [6]

Y. EISENBERG, U. KARSHON, E. KOGAN, D. REVEL, E. RONAT and A. SHAPIRA
Weizmann Institute, Rehovot, Israel [7]

J. FREEMAN, P. LECOMTE, T. MEYER, SAU LAN WU and G. ZOBERNIG
Department of Physics, University of Wisconsin, Madison, WI, USA [8]

Received 29 August 1979

---

[1] Now at University Kiel, Germany. [2] On leave from Weizmann Institute, Rehovot, Israel. [3] On leave from Johns Hopkins University, Baltimore, MD, USA. [4] On leave from University of California, Berkeley, USA. [5] Supported by the Deutsches Bundesministerium für Forschung und Technologie. [6] Supported by the UK Science Research Council. [7] Supported by the Minerva Gesellschaft für die Forschung mbH, Munich, Germany. [8] Supported in part by the US Department of Energy, contract EY-76-C-02-0881.

Hadron jets produced in e^+e^- annihilation between 13 GeV and 31.6 GeV in c.m. at PETRA are analyzed. The transverse momentum of the jets is found to increase strongly with c.m. energy. The broadening of the jets is not uniform in azimuthal angle around the quark direction but tends to yield planar events with large and growing transverse momenta in the plane and smaller transverse momenta normal to the plane. The simple $q\bar{q}$ collinear jet picture is ruled out. The observation of planar events shows that there are three basic particles in the final state. Indeed, several events with three well-separated jets of hadrons are observed at the highest energies. This occurs naturally when the outgoing quark radiates a hard noncollinear gluon, i.e., $e^+e^- \to q\bar{q}g$ with the quarks and the gluons fragmenting into hadrons with limited transverse momenta.

It has been conjectured that hadron production in e^+e^- annihilation proceeds by quark pair production with the quarks fragmenting into two nearly collinear jets of hadrons [1–4]. The transverse momentum of the hadrons is limited at high energies such that narrow jets are predicted in the PETRA energy range. Data from SPEAR and DORIS support this picture [5,6]. In this paper we examine our data on e^+e^- annihilation at high energies for deviations from the simple quark parton picture [+1]. Such deviations have been observed [8] in deep inelastic lepton scattering experiments and are expected [9] to occur in any field theory of strong interactions. Furthermore, final states observed [10] in the hadronic decay of the $\Upsilon(9.46)$ disagree with a naive $q\bar{q}$ picture.

This analysis is based on data collected at the DESY e^+e^- storage ring PETRA using the TASSO detector. A description of the experimental setup and of the data analysis has already been published [11]. We list here only the main points. The detector, which measures charged secondaries over approximately 87% of 4π, was triggered by demanding at least three tracks with a transverse momentum of more than 0.32 GeV/c with respect to the beam axis, at total energies $W = 2E_{\text{beam}} = 13$ and 17 GeV. At least four tracks were required at $W = 27.4, 27.7, 30,$ and 31.6 GeV. The hadronic events were separated from the beam gas background by demanding that the event vertex was in the interaction volume, and that the sum of the detected momenta were above 3 GeV/c at 13 GeV energy, above 4 GeV/c at 17 GeV energy, and above 9 GeV/c at the higher energies. Further cuts on the event topology essentially eliminated $\gamma\gamma$ and $\tau^+\tau^-$ events. The resulting detection efficiency varied with energy between 75% and 78%. The final data sample consists of 75 events at 13 GeV, 40 events at 17 GeV, 118 events at 27.4 and 27.7 GeV, 135 events at 30 GeV, and 40 events at 31.6 GeV. For the following analysis only tracks that reached at least the sixth zero degree layer in the drift chamber ($|\cos\theta| < 0.87$) and had transverse momentum relative to the beam axis of at least 0.10 GeV/c, are used. Hadrons resulting from kaon decays were not removed while electrons from pair conversion or Dalitz decays were removed.

The effects of the selection criteria described above, as well as the effects of measuring errors and the efficiencies of the pattern recognition programs and the whole analysis chain, were checked by propagating Monte Carlo events through the simulated detector and the analysis programs. It was thus ascertained that the results presented in this paper are not subject to any significant bias due to these sources.

The quark parton picture of the process $e^+e^- \to q\bar{q}$ is depicted in fig. 1a. It will produce back-to-back jets of hadrons with typical transverse momenta of 0.30 GeV/c. In field theories of the strong interactions this picture will be modified [9,12] to include other processes including the lowest-order diagrams shown in fig. 1b. The radiated field quanta (gluons) are also expected to evolve into jets of hadrons. This has clear experimental implications [12–16]: The p_T distribution of the final hadrons will broaden with increasing energy. If the coupling constant is less than one there will be a tendency for only one of the jets to be broadened. The $q\bar{q}g$ state is necessarily planar; this should be reflected in the final hadron configuration which should retain the planarity with small transverse momenta with respect to the plane and large transverse momentum in the plane. If the gluon is radiated with a transverse momentum that is large compared to the typical transverse momentum of 0.30 GeV/c, then the event will have a three-jet topology [12].

Since in e^+e^- interactions the direction of the jet axis is not known from the initial state, it has to be

---
[+1] Evidence for planar events in e^+e^- annihilation has been reported earlier, see ref. [7]. The present results are based on a five-fold increase of statistics.

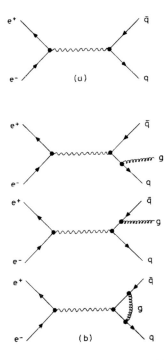

Fig. 1. Quark pair production and gluon bremsstrahlung to lowest order.

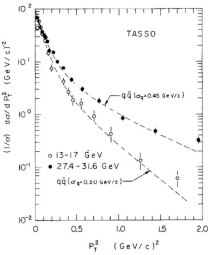

Fig. 2. $\sigma^{-1} d\sigma/dp_T^2$ at 13 and 17 GeV combined (o) and for 27.4 GeV, 27.7 GeV, 30.0 GeV and 31.6 GeV combined (•) as a function of p_T^2. The curves are fits to the data for $p_T^2 < 1.0$ GeV/c using the Field–Feynman quark–parton model including u, d, s, c and b quarks with σ_q as a free parameter.

determined event by event from the final state hadrons. We have done this in two ways: Minimizing Σp_T^2 (which gives the sphericity axis) [3,5] or maximizing $\Sigma |p_{\parallel}|$ (which gives the thrust axis) [13,17]. The summations are over all observed hadrons. The results presented in this paper were generally found not to depend to any significant degree on the choice between these two possibilities [*2]. Studies with Monte Carlo $q\bar{q}$ jets [18] [*3] further showed that the jet axes so determined deviate by an average of less than 5° from the true jet axis at our higher energies.

The normalized transverse momentum distribution $\sigma^{-1} d\sigma/dp_T^2$ evaluated with respect to the sphericity

---

[*2] An exception is the "seagull effect" shown in fig. 3. Here only the thrust axis is useful since the sphericity weights fast particles too heavily and thus biases the particles of high $z = p_h/p_{beam}$ towards small values of p_T.
[*3] For comparison with the data we used the version given by Field and Feynman [18].

axis is plotted in fig. 2 versus p_T^2. The data at 13 GeV and 17 GeV are identical within statistics and are averaged; similarly the higher energy data between 27.4 GeV and 31.6 GeV are combined. The data at both energies are in reasonable agreement for $p_T^2 < 0.2$ (GeV/c)^2 but the high-energy data are well above the low-energy data for larger values of p_T^2. The average value of p_T^2 increases from 0.15 ± 0.02 (GeV/c)^2 at 13 and 17 GeV, to 0.27 ± 0.02 (GeV/c)^2 for the combined high-energy data. The low-energy data have been fitted for $p_T^2 < 1.0$ (GeV/c)^2 with the jet model [18] [*3] of Field and Feynman, extended to include c and b quarks [*4]. In this model, the parameter σ_q determining the width of the p_T distribution was varied from the original value $\sigma_q = 0.25$ GeV/c to $\sigma_q = 0.30$ GeV/c to obtain a fit to our data. This is shown by the curve in fig. 2. To fit the higher-energy data with the Field–Feynman model, σ_q must be increased to 0.45 GeV/c. This is in contradiction to the naive quark parton model which assumes the quark to frag-

---

[*4] The branching ratios for B meson decay were taken from ref. [19].

ment into hadrons with an energy-independent transverse momentum distribution. On the other hand, field theories of the strong interactions naturally predict the transverse momentum to increase with energy due to gluon bremsstrahlung by one of the outgoing quarks. The production of a new quark flavour will also lead to an increase in the average value of p_T. We do not find any evidence for the production of a new flavour in agreement [20] with other groups at PETRA.

If hard noncollinear gluon emission is a rare process, then there should usually be only one such gluon in these events. Dividing each event into two halves by a plane perpendicular to the jet axis and determining $\langle p_T^2 \rangle$ separately for the two sides, the "narrow" side should rarely have a noncollinear hard gluon. Thus $\langle p_T^2 \rangle_{\text{narrow}}$ will increase with energy less rapidly than $\langle p_T^2 \rangle_{\text{wide}}$. This is observed as shown in fig. 3 where $\langle p_T^2 \rangle$ is plotted as a function of $z = p/p_{\text{beam}}$ for the wide and the narrow jet separately. The $q\bar{q}$ model with $\sigma_q = 0.30$ GeV/c is also plotted in fig. 3a. It fits the data rather well at low energy; the narrow–wide asymmetry is due to statistical fluctuations. The model fails to describe the data at high energies (fig. 3b); however, increasing σ_q to 0.45 GeV/c, obtained from the fit to the p_T^2 distribution, approximately reproduces the observed z distribution. Therefore we can fit, within the statistical uncertainties, both the p_T^2 distributions and the seagull plot at both energies by increasing the value of σ_q with energy.

Regardless of the value of $\langle p_T \rangle$ in the naive quark jet picture hadrons resulting from the fragmentation of the quark must be on the average uniformly distributed in azimuthal angle around the quark axis. Therefore, apart from statistical fluctuations, the two-jet process $e^+e^- \to q\bar{q}$ will not lead to planar events whereas the radiation of a hard gluon, $e^+e^- \to q\bar{q}g$, will result in an approximately planar configuration of hadrons with large transverse momenta in the plane and small transverse momenta with respect to the plane. Thus the observation of such planar events at a rate significantly above the rate expected from statistical fluctuations of the $q\bar{q}$ jets shows in a model-independent way that there must be a third particle in the final state, which might be identified with a gluon.

The shape of the events is evaluated using the following method. For each event we construct the second-rank tensor [3,5] from the hadron momenta

$$M_{\alpha\beta} = \sum_{j=1}^{N} p_{j\alpha} p_{j\beta} \quad (\alpha, \beta = x, y, z),$$

summing over all N observed charged particles. Let $\hat{n}_1, \hat{n}_2$ and $\hat{n}_3$ be the unit eigenvectors of this tensor associated with the smallest, intermediate, and largest eigenvalues Λ_1, Λ_2 and Λ_3, respectively. The principal jet axis is then the $\hat{n}_3$ direction, the event plane is the $\hat{n}_2$–$\hat{n}_3$ plane, and $\hat{n}_1$ defines the direction in which the sum of the squared hadron momenta components is minimized [21]. We compare in fig. 4 the distribution of

$$\langle p_T^2 \rangle_{\text{out}} = \frac{1}{N} \sum_{j=1}^{N} (p_j \cdot \hat{n}_1)^2$$

(the momentum component normal to the event plane squared) with that of

$$\langle p_T^2 \rangle_{\text{in}} = \frac{1}{N} \sum_{j=1}^{N} (p_j \cdot \hat{n}_2)^2$$

(the momentum component in the event plane per-

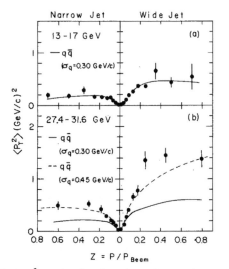

Fig. 3. $\langle p_T^2 \rangle$ as a function of $z = p/p_{\text{beam}}$ for the wide and narrow jet separately, for the low-energy (a) and the high-energy (b) data. The curves show the prediction from the $q\bar{q}$ model [18] [+3] with $\sigma_q = 0.30$ GeV/c (solid curve) and $\sigma_q = 0.45$ GeV/c (dotted curve). The model includes u, d, s, c and b quarks.

246

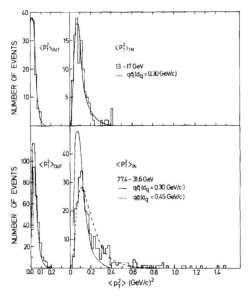

Fig. 4. The mean transverse momentum squared normal to the event plane $\langle p_T^2 \rangle_{out}$ and in the event plane $\langle p_T^2 \rangle_{in}$ per event for the low-energy and the high-energy data. The predictions from the quark model [18] ‡3 are shown assuming $\sigma_q = 0.30$ GeV/c (solid curves) and $\sigma_q = 0.45$ GeV/c (dotted curve). The model includes u, d, s, c and b quarks.

pendicular to the jet axis). The data show only little increase in $\langle p_T^2 \rangle_{out}$ between the low-energy and the high-energy point. The distribution of $\langle p_T^2 \rangle_{in}$, however, becomes much wider at high energies, and in particular there is a long tail of events with a large value of $\langle p_T^2 \rangle_{in}$ not observed at lower energies. The curves show the expectations from the Monte Carlo qq̄ jets. Hadrons resulting from pure qq̄ events will on the average be distributed uniformly around the jet axis, however, some asymmetry between $\langle p_T^2 \rangle_{in}$ and $\langle p_T^2 \rangle_{out}$ is caused by statistical fluctuations. Fair agreement with the qq̄ model is found both for $\langle p_T^2 \rangle_{in}$ and $\langle p_T^2 \rangle_{out}$ at the low-energy point. Thus the asymmetry observed at this energy can be explained by statistical fluctuations only.

At the high energy, we find fair agreement between $\langle p_T^2 \rangle_{out}$ and the qq̄ model with $\sigma_q = 0.30$ GeV/c, however, the observed long tail of the $\langle p_T^2 \rangle_{in}$ distribution is not reproduced by the model. This discrepancy can-

not be removed by increasing the mean transverse momentum of the qq̄ jets. The result from the model with $\sigma_q = 0.45$ GeV/c is also plotted in fig. 4. The agreement is poor. We therefore conclude that the data include a number of planar events not reproduced by the qq̄ model, independent of the assumption on the average p_T in that model.

The normalized eigenvalues,

$$Q_k = \Lambda_k \Big/ \sum_{j=1}^{N} p_j^2 ,$$

may be used for a more detailed study of the shape of the events. These normalized eigenvalues Q_k satisfy $Q_1 + Q_2 + Q_3 = 1$ and are arranged such that $0 \leq Q_1 \leq Q_2 \leq Q_3$. We express our data in terms of two variables, aplanarity A and sphericity S:

$$A = \tfrac{3}{2} Q_1 = \tfrac{3}{2} \langle p_T^2 \rangle_{out}/\langle p^2 \rangle ,$$

$$S = \tfrac{3}{2} (Q_1 + Q_2) = \tfrac{3}{2} \langle p_T^2 \rangle/\langle p^2 \rangle .$$

All the events are then inside a triangle shown in fig. 5, where S is the hypotenuse of the triangle. The event distribution in A and S is shown in fig. 5, for the low-energy and high-energy data separately.

Collinear two-jet events lie in the left-hand corner

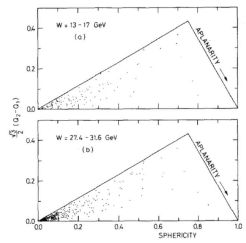

Fig. 5. Distribution of the events as a function of aplanarity $A = \tfrac{3}{2} Q_1 = \tfrac{3}{2} \langle p_T^2 \rangle_{out}/\langle p^2 \rangle$ and sphericity $S = \tfrac{3}{2}(Q_1 + Q_2) = \tfrac{3}{2} \langle p_T^2 \rangle/\langle p^2 \rangle$ for the low (a) and high (b) energy data.

(A, S small), uniform disk shaped events in the upper corner (A small, S large), and spherical events in the lower right-hand corner (A, S large), while coplanar events will occupy a band along the larger of the two small sides of the triangle in fig. 5.

Collinear two-jet events are seen to dominate at all energies, the collinearity being most pronounced at the highest energy. We exclude these events and select the candidates for planar events by requiring that $A < 0.04$ and $S > 0.25$. At 13 and 17 GeV we observe six events in this region compared to 3.5 events predicted by the $q\bar{q}$ model with $\sigma_q = 0.30$ GeV/c. At the higher energies we find 18 events compared to 4.5 events predicted by the $q\bar{q}$ model, independent of σ_q between 0.30 and 0.45 GeV/c. As an independent test of the planar structure, a randomization procedure [*5] was applied to the data to destroy any natural correlations. This estimate of accidentally planar events yields six events in the 13–17 GeV data and four events in the higher-energy data. Thus at the higher energies there is an excess of planar events well above the level predicted from statistical fluctuations of the $q\bar{q}$ jets. This shows that $e^+e^- \to$ hadrons proceeds via the creation and decay of at least three primary particles that subsequently fragment into hadrons. Field theories of the strong interactions predict such a topology resulting from the radiation of a field quantum (gluon) by one of the quarks, i.e., $e^+e^- \to q\bar{q}g$.

If this is the correct explanation and the gluon materializes as a jet of hadrons with limited transverse momentum then a small fraction of the events should display a three-jet structure. The events were analyzed for a three-jet structure as described in ref. [21]. All the coplanar events gave a good fit to the three-jet hypothesis. We further determined the transverse momenta of the hadrons with respect to the axis to which they were assigned. For the 18 events defined above we find an average transverse momentum of about 0.30 GeV/c, close to the mean p_T observed in two-jet events at lower energies.

To compare this new class of three-jet events with the predominant class of two-jet events, fig. 6 shows a characteristic event of each type in momentum

---

[*5] The sphericity axis was chosen as a reference, and all tracks were rotated by a random azimuthal angle around the jet direction; this preserves both p_T and $p_\parallel$. Then at random the sign of $p_\parallel$ was changed.

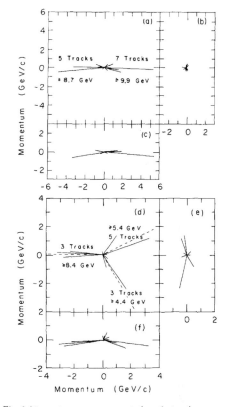

Fig. 6. Momentum space representation of a two-jet event (a)–(c) and a three-jet event (d)–(f) in each of three projections. (a), (d) $\hat{n}_2$–$\hat{n}_3$ plane; (b), (e) $\hat{n}_1$–$\hat{n}_2$ plane; (c), (f) $\hat{n}_1$–$\hat{n}_3$ plane.

space in all three projections. Figs. 6a and 6d show a two-jet and a three-jet event, respectively, in the $\hat{n}_2$–$\hat{n}_3$ plane; this is the plane containing the largest components of momenta. The first event shows two clearly delineated jets. The three-jet event, on the other hand, shows a much broader distribution of momenta transverse to the $\hat{n}_3$ axis. Figs. 6b and 6e show the projection on the plane perpendicular to the jet direction ($\hat{n}_3$). Here one clearly sees the small transverse momenta for the two-jet event and the tendency of the large transverse momentum to lie along the $\hat{n}_2$ direction for the three-jet event. Finally figs. 6c and 6f show the remaining projection on the $\hat{n}_1$–$\hat{n}_3$ plane.

In summary, we have studied $e^+e^- \to$ hadrons for values of Q^2 between 170 (GeV/c)^2 and 1000 (GeV/c)^2. We observe a change in the p_T^2 distribution and a strong increase of $\langle p_T^2 \rangle$ with increasing energy. This increase occurs predominantly in only one of the two jets. The distribution of the transverse momentum perpendicular to the "event plane" does not show a pronounced energy dependence while a strong broadening takes place in the event plane at the highest values of Q^2. We observe planar events at a rate which is well above the rate computed for statistical fluctuations of the q$\bar{\text{q}}$ jets. The planar events exhibit three axes, the average transverse momentum of the hadrons with respect to these axes being 0.30 GeV/c. This establishes in a model-independent way that a small fraction of the e^+e^- annihilation events proceeds via ma- emission of three primary particles, each of which materializes as a jet of hadrons in the final state. The data are most naturally explained by hard noncollinear bremsstrahlung $e^+e^- \to$ q$\bar{\text{q}}$g. Indeed, the data are in agreement with predictions based on first-order perturbative QCD as will be discussed in a forthcoming paper.

We thank the DESY directorate for its strong support of this experiment. We are grateful for the diligent efforts of the PETRA machine group which made this experiment possible. We are indebted to all the engineers and technicians of the collaborating institutions who have participated in the construction and maintenance of TASSO. The excellent cooperation with the technical support groups at DESY and at the Rutherford Lab. is gratefully acknowledged. The Wisconsin group thanks the Physics Department of the University of Wisconsin for support, especially the High Energy Group and the Graduate School Research Committee.

References

[1] S.D. Drell, D.J. Levy and T.M. Yan, Phys. Rev. 187 (1969) 2159; D1 (1970) 1617.
[2] N. Cabibbo, G. Parisi and M. Testa, Lett. Nuovo Cimento 4 (1970) 35.
[3] J.D. Bjorken and S.J. Brodsky, Phys. Rev. D1 (1970) 1416.
[4] R.P. Feynman, Photon–hadron interactions (Benjamin, Reading, MA, 1972) p. 166.
[5] G. Hanson et al., Phys. Rev. Lett. 35 (1975) 1609.
[6] Ch. Berger et al., Phys. Lett. 78B (1978) 176.
[7] B.H. Wiik, Proc. Intern. Neutrino Conf. (Bergen, Norway, 18–22 June 1979);
R. Cashmore, P. Söding and G. Wolf, Proc. EPS Intern. Conf. on High energy physics (Geneva, Switzerland, 27 June–4 July 1979).
[8] E.M. Riordan et al., SLAC preprint PUB-1639 (1975); D.J. Fox et al., Phys. Rev. Lett. 37 (1974) 1504; H.L. Anderson et al., Phys. Rev. Lett. 37 (1976) 4; P.C. Bosetti et al., Nucl. Phys. B142 (1978) 1; J.G.H. de Groot et al., Particles and Fields (Z. Phys. C) 1 (1979) 143; Phys. Lett. 82B (1979) 292, 456.
[9] J. Kogut and L. Susskind, Phys. Rev. D9 (1974) 697, 3391;
A.M. Polyakov, Proc. 1975 Intern. Symp. on Lepton and photon interactions at high energies (Stanford), ed. W.T. Kirk (SLAC, Stanford, 1975) p. 855.
[10] PLUTO Collaboration, Ch. Berger et al., Phys. Lett. 82B (1979) 449;
S. Brandt, Paper presented at the Intern. Conf. on High energy physics (Geneva, 27 June–4 July 1979).
[11] TASSO Collaboration, R. Brandelik et al., Phys. Lett. 83B (1979) 201.
[12] J. Ellis, M.K. Gaillard and G.G. Ross, Nucl. Phys. B111 (1976) 253; Erratum B130 (1977) 516.
[13] A. de Rujula, J. Ellis, E.G. Floratos and M.K. Gaillard, Nucl. Phys. B138 (1978) 387.
[14] T.A. DeGrand, Yee Jack Ng and S.-H.H. Tye, Phys. Rev. D16 (1977) 3251.
[15] P. Hoyer, P. Osland, H.G. Sander, T.F. Walsh and P.M. Zerwas, DESY 78/21, to be published.
[16] G. Kramer and G. Schierholz, Phys. Lett. 82B (1979) 102;
G. Kramer, G. Schierholz and J. Willrodt, Phys. Lett. 79B (1978) 249;
G. Curci, M. Greco and Y. Srivastava, CERN-Report 2632-1979.
[17] E. Fahri, Phys. Rev. Lett. 39 (1977) 1587; S. Brandt and H.D. Dahmen, Particles and Fields (Z. Phys. C) 1 (1979) 61.
[18] B. Anderson, G. Gustafson and C. Peterson, Nucl. Phys. B135 (1978) 273;
R.D. Field and R.P. Feynman, Nucl. Phys. B136 (1978) 1;
[19] A. Ali, J.G. Körner, J. Willrodt and G. Kramer, Particles and Fields (Z. Phys. C) 1 (1979) 269; 2 (1979) 33.
[20] See papers contributed by the JADE, MARK J, PLUTO and TASSO Collaborations to the 1979 Intern. Symp. on Lepton and photon interactions at high energies (Batavia, IL, USA, August 23–29, 1979).
[21] S.L. Wu and G. Zobernig, Particles and Fields (Z. Phys. C) 2 (1979) 107.

OBSERVATION OF VERY LARGE TRANSVERSE MOMENTUM JETS AT THE CERN $\bar{p}p$ COLLIDER

The UA2 Collaboration

M. BANNER [f], Ph. BLOCH [f], F. BONAUDI [b], K. BORER [a], M. BORGHINI [b], J.-C. CHOLLET [d],
A.G. CLARK [b], C. CONTA [e], P. DARRIULAT [b], L. Di LELLA [b], J. DINES-HANSEN [c], P.-A. DORSAZ [b],
L. FAYARD [d], M. FRATERNALI [e], D. FROIDEVAUX [b,d], J.-M. GAILLARD [d], O. GILDEMEISTER [b],
V.G. GOGGI [e,1], H. GROTE [b], B. HAHN [a], H. HÄNNI [a], J.R. HANSEN [b], P. HANSEN [c], T. HIMEL [b],
V. HUNGERBÜHLER [b], P. JENNI [b], O. KOFOED-HANSEN [c], M. LIVAN [e], S. LOUCATOS [f],
B. MADSEN [c], B. MANSOULIÉ [f], G.C. MANTOVANI [e,2], L. MAPELLI [b], B. MERKEL [d],
M. MERMIKIDES [b], R. MØLLERUD [c], B. NILSSON [c], C. ONIONS [b], G. PARROUR [b,d], F. PASTORE [b],
H. PLOTHOW-BESCH [d], J.-P. REPELLIN [d], J. RINGEL [b], A. ROTHENBERG [b], A. ROUSSARIE [f],
G. SAUVAGE [d], J. SCHACHER [a], J.L. SIEGRIST [b], F. STOCKER [a], J. TEIGER [f], V. VERCESI [e],
H.H. WILLIAMS [b], H. ZACCONE [f] and W. ZELLER [a]

[a] *Laboratorium für Hochenergiephysik, Universität Bern, Sidlerstrasse 5, Bern, Switzerland*
[b] *CERN, 1211 Geneva 23, Switzerland*
[c] *Niels Bohr Institute, Blegdamsvej 17, Copenhagen, Denmark*
[d] *Laboratoire de l'Accélérateur Linéaire, Université de Paris-Sud, Orsay, France*
[e] *Istituto di Fisica Nucleare, Università di Pavia and INFN, Sezione di Pavia, Via Bassi 6, Pavia, Italy*
[f] *Centre d'Etudes Nucléaires de Saclay, Gif sur Yvette, France*

Received 25 August 1982

The distribution of total transverse energy ΣE_T over the pseudorapidity interval $-1 < \eta < 1$ and an azimuthal range $\Delta\phi = 300°$ has been measured in the UA2 experiment at the CERN $\bar{p}p$ collider ($\sqrt{s}$ = 540 GeV) using a highly segmented total absorption calorimeter. In the events with very large ΣE_T ($\Sigma E_T \geq 60$ GeV) most of the transverse energy is found to be contained in small angular regions as expected for high transverse momentum hadron jets. We discuss the properties of a sample of two-jet events with invariant two-jet masses up to 140 GeV/c^2 and we measure the cross section for inclusive jet production in the range of jet transverse momenta between 15 and 60 GeV/c.

1. Introduction. The suggestion that hard scattering of hadron constituents should result in two jets with the same momenta as the scattered partons [1] has motivated an intense experimental effort [2]. Earlier ISR experiments [3] have reported observations of such double-jet structures. However these jets were not as clearly identified as they are in the hadronic final states of high-energy e^+e^- annihilations [4], because in hadronic collisions the jets carry only a fraction of the total energy available. As a consequence,

jets are accompanied by several soft hadrons which may make their identification more difficult and in general they are not collinear.

The recent successful operation of the CERN $\bar{p}p$ collider [5] has opened a new possibility to observe high transverse momentum hadron jets. At $\sqrt{s}$ = 540 GeV the yield of jets with $E_T > 20$ GeV is expected to increase by about four orders of magnitude with respect to the top ISR energy [6] whereas the average particle density in the central region for an ordinary collision has increased by less than a factor of 2 [7].

We report here on results from the UA2 experiment at the CERN $\bar{p}p$ collider. This experiment uses a large

---

[1] Now also at Istituto di Fisica, Università di Udine, Italy.
[2] Now also at Istituto di Fisica, Università di Perugia, Italy.

0 031-9163/82/0000–0000/$02.75 © 1982 North-Holland

solid angle total absorption hadron calorimeter subdivided in small cells, a device well suited to the detection of hadron jets.

2. Apparatus.
A cross section of the UA2 detector is shown in fig. 1. At the centre of the apparatus a vertex detector consisting of cylindrical proportional and drift chambers measures particle trajectories in a region without magnetic field.

The vertex detector is surrounded by a highly segmented electromagnetic and hadronic calorimeter (the central calorimeter) which covers the pseudorapidity interval $-1 < \eta < 1$ (polar angle $40° < \theta < 140°$) and an azimuthal range of $300°$. In the present stage of the experiment the remaining azimuthal interval ($\pm 30°$ around the horizontal plane) is covered by a single arm spectrometer to measure charged and neutral particle production [8].

The forward and backward regions ($20° < \theta < 37.5°$ and $142.5° < \theta < 160°$, respectively), are each instrumented by twelve toroidal magnet sectors followed by drift chambers, multitube proportional chambers and electromagnetic calorimeters.

The central calorimeter is segmented into 200 cells, each covering $15°$ in ϕ and $10°$ in θ and built in a tower structure pointing to the centre of the interaction region. The cells are segmented longitudinally into a 17 radiation length thick electromagnetic compartment (lead-scintillator) followed by two hadronic compartments (iron-scintillator) of two absorption lengths each. The light from each compartment is collected by two BBQ-doped light guide plates on opposite sides of the cell.

All calorimeters, including the forward–backward modules, have been calibrated in a 10 GeV/c beam from the CERN PS using incident electrons and muons. The calibration has since been tracked with a Xe light flasher system. In addition, the response of the electromagnetic compartments is checked regularly by accurately positioning a Co^{60} source in front of each cell and measuring the direct current from each photomultiplier. The systematic uncertainty in the energy calibration for the data discussed here is less than ±2% for the electromagnetic calorimeter and less than ±3% for the hadronic one.

The response of the calorimeter to electrons, single hadrons and multi-hadrons (produced in a target located in front of the calorimeter) has been measured at the CERN PS and SPS machines using beams from 1 to 70 GeV. In particular, we have studied the longi-

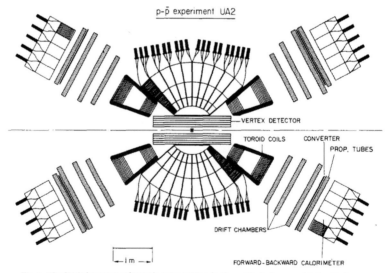

Fig. 1. The UA2 detector: schematic cross section in the vertical plane containing the beam.

tudinal and transverse shower development and the effect of particles impinging near the cell boundaries.

The energy resolution for electrons is measured to be $\sigma_E/E = 0.14/\sqrt{E}$ (E in GeV). In the case of hadrons, σ_E/E varies from 32% at 1 GeV to 11% at 70 GeV. The resolution for multi-hadron systems of more than 20 GeV is similar to that of single hadrons.

Details of the construction and performance of the calorimeter are reported elsewhere [9].

3. *Data taking.* The data discussed in this paper were recorded using a trigger sensitive to events with large transverse energy in the central calorimeter [10]. The gains of the photomultipliers were adjusted so that their signals were proportional to transverse energy. The signals were linearly added and their sum was required to exceed a given threshold (typically set at 20 GeV).

In order to suppress background from sources other than $\bar{p}p$ collisions, we required a coincidence with two additional signals obtained from two scintillator arrays surrounding the vacuum chamber 10.3 metres downstream of the interaction point and covering an interval $\Delta\eta = 1.1$ around $\eta = \pm 4.7$.

A sample of "minimum bias" data was recorded simultaneously by using only the coincidence of the signals from the two scintillator arrays. The rate of such coincidences provided also a measurement of the luminosity [8].

Furthermore the electronics were enabled between $\bar{p}p$ crossings and the cosmic ray background was found to be negligible.

4. *Data reduction.* Data have been recorded for an integrated luminosity of 79 μb^{-1}. The full data sample has been used in the following analysis for $\Sigma E_T > 30$ GeV. Partial samples are used for $\Sigma E_T < 30$ GeV and for minimum bias events.

The background events from beam halo particles and the accidental overlap of beam halo particles with minimum bias events exhibit a characteristic pattern in the detector different from that for events from $\bar{p}p$ collisions. A series of selection criteria is applied to remove background events:
- A fast filter requires that the pattern of hits in the vertex detector be consistent with charged particles coming from a common vertex.
- The ratios of the energies in each calorimeter layer (electromagnetic, first and second hadronic) to the total observed energy must be physically sensible.
- The local patterns of energy deposition in depth in the central calorimeter have to be consistent with that expected for particles emerging from a $\bar{p}p$ collision.
- Finally, for the highest ΣE_T events we reconstruct the event vertex and verify that it is within the collision region.

The combination of these requirements reduces strongly the number of background events. We find that the ratio of signal to background triggers in the initial event sample is ΣE_T-dependent, varying from ~ 10 at $\Sigma E_T \lesssim 10$ GeV to ~ 1 at 30 GeV and ~ 0.25 at $\Sigma E_T > 60$ GeV. After applying all of the above cuts we estimate from a visual scan of events that the final event sample contains less than 10% of background events independently of ΣE_T.

To study the loss of good events introduced by the cuts described above we use test beam data, Monte Carlo simulations and we investigate the effect of varying the cuts on the data. We estimate that the loss of good events is $\lesssim 15\%$, independent of ΣE_T.

5. *Transverse energy distribution.* The total hadronic energy in a cell is measured as the sum of the energies in the three compartments (at least one compartment must have 150 MeV, well above pedestal fluctuations).

The distribution of events as a function of the transverse energy ΣE_T in the interval $-1 < \eta < +1$, $30° < \phi < 330°$ is presented in fig. 2a; it is observed to fall off exponentially as has been observed in experiments [11] with a similar solid angle coverage at the SPS nd at Fermilab ($\sqrt{s} = 24$ GeV). There are 10 events having $\Sigma E_T > 60$ GeV.

We estimate that the uncertainty in the energy scale due to systematic effects (150 MeV minimum cell energy, variation in response of the electromagnetic compartment for charged and neutral pions, and calibration errors [+1]) to be $\pm 5\%$. In the lower energy experiments, events with large ΣE_T (the largest accessible value is approximately 20 GeV) have predominantly high multiplicity, cylindrically symmetrical

---

[+1] The response to hadrons has been measured to be linear to within 2% over the whole energy range, that of the electromagnetic compartment differs by 17% between charged and neutral pions.

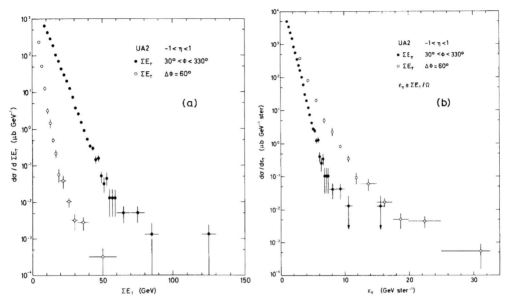

Fig. 2. Transverse energy (a) and its average density (b): observed distributions over the whole azimuthal acceptance (full dots) and over a restricted azimuthal region, $\Delta\phi = 60°$ (open circles).

configurations. It is however commonly expected [12] that at the present collider energy, $\sqrt{s}$ = 540 GeV, sufficiently large values of ΣE_T can be reached that two-jet configurations resulting from hard scattering will dominate. An indication of this is already provided by the fact that the two highest energy points in fig. 2a lie significantly above the exponential.

A more sensitive test is provided by studying the ΣE_T distribution into a solid angle which is substantially smaller though still sufficiently large to contain most of the energy for hadron jets in the E_T region of interest ($E_T > 15$ GeV). The $d\sigma/d\Sigma E_T$ distribution for $-1 < \eta < 1$, $\Delta\phi = 60°$ is shown in fig. 2a; it demonstrates a clear departure from an exponential. In fig. 2b, the same data samples are plotted as a function of the transverse energy density, $\epsilon_T = \Sigma E_T/\Delta\Omega$. Figs. 2a,b illustrate that regions of high energy density are confined to within a small solid angle.

Direct evidence for an increase of the energy clustering towards higher values of ΣE_T is obtained by constructing clusters of adjacent cells, each cell containing more than a fraction f of the total energy deposited in the central calorimeter. When ΣE_T increases from 30 to >60 GeV, the number of such clusters decreases from $\simeq 15$ to $\simeq 6$ on the average for $f = 1\%$, while it increases from $\simeq 0.7$ to $\simeq 2$ for $f = 8\%$: events with a high value of ΣE_T are made of relatively less numerous but more energetic clusters.

The same fact is illustrated by determining for each event the two non-overlapping 45° × 40° solid angles which together contain the largest fraction, g, of the total transverse energy, and studying g as a function of ΣE_T; the average value of g increases from $\simeq 40\%$ at $\Sigma E_T = 30$ GeV to $\simeq 60\%$ for $\Sigma E_T > 60$ GeV.

To study these high ΣE_T events in detail we use a clustering algorithm which joins into a cluster all cells which share a common side and contain an energy $E_{cell} > E_{cell}^{min}$; E_{cell}^{min} is normally chosen to be 400 MeV, though the results obtained are relatively insensitive to the exact value chosen. Clusters having two or more local maxima separated by a valley deeper than 2 GeV are then split. On the average, for $\Sigma E_T > 30$ GeV, we obtain $\simeq 4.4$ clusters having $E_T > 2$ GeV per event, each cluster consisting of typically 4 cells.

The cluster (resp. the two clusters) with the largest transverse energy in an event accounts for a fraction

206

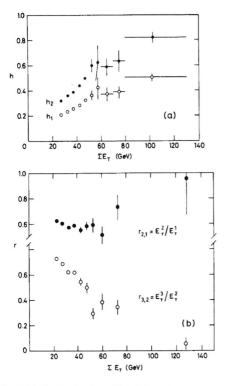

Fig. 3. (a) The fractions h_1 and h_2 of the total transverse energy ΣE_T contained in the cluster, and respectively the two clusters, having the largest E_T are displayed versus ΣE_T. (b) Dependence upon ΣE_T of the ratios $r_{21} = E_T^2/E_T^1$ and $r_{32} = E_T^3/E_T^2$ (see text).

h_1 (resp. h_2) of the total transverse energy ΣE_T measured in the central calorimeter. The dependence upon ΣE_T of h_1 and h_2 is illustrated in fig. 3a. In particular all events with ΣE_T in excess of 60 GeV have an average of 67% of the transverse energy contained in only two clusters.

6. *Two-jet events.* Fig. 4 shows the configuration of the event with the largest value of ΣE_T, 127 GeV. It exhibits striking features: energy is concentrated within two small regions separated in azimuth by $\Delta\phi \simeq 180°$ and towards which several collimated tracks are observed to point. In addition the transverse energies of the two clusters are approximately equal (57 and 60 GeV).

We now investigate to which extent such characteristics are also present in the other events having a large value of ΣE_T.

In each event we rank the central calorimeter clusters (C1, C2, ...) in order of decreasing transverse energies ($E_T^1 > E_T^2 > ...$). Fig. 5a shows the azimuthal separation $\Delta\phi$ between C1 and C2 for $E_T^{1,2} > 10$ GeV and $E_T^{1,2} > 14$ GeV. It peaks strongly near $\Delta\phi \simeq 180°$. If however we release the constraint that both E_T^1 and E_T^2 must exceed a high threshold, we find events in which E_T^2 may be only a small fraction of E_T^1: for example, in the sample of 59 events having $E_T^1 > 20$ GeV, 41 have $E_T^2/E_T^1 < 0.4$. For those $\Delta\phi$ is observed to have a uniform distribution between 0 and 180° while it always exceeds 140° in the rest of the sample. It is then natural to consider the possibility that a large localised transverse energy be produced outside the central calorimeter acceptance. Indeed, of 385 events having $E_T^1 > 10$ GeV and $E_T^2/E_T^1 < 0.4$, 22 deposit more than 5 GeV transverse energy in a sector of one of the forward/backward electromagnetic calorimeters: in such cases the azimuthal separation $\Delta\phi$ between this sector and C1 is nearing 180° (fig. 5b). These observations suggest an interpretation of the events in terms of a hard parton collision, a characteristic feature of which is the coplanarity of the scattered partons with the incident beams.

Further evidence is obtained from the fact that C1 and C2 are associated with jets of particles. We select a sample S_{JJ} (55 events) by requiring $E_T^{1,2} > 10$ GeV and $\Delta\phi > 140°$. We make the following observations:

(i) several tracks, measured in the vertex detector, are observed to aim towards the cluster centers (to within 13° on the average for $E_T > 20$ GeV);

(ii) longitudinal shower developments, measured from the contributions of electromagnetic and hadronic compartments, are inconsistent with that of a single particle but consistent with that of a jet fragmenting into charged and neutral pions;

(iii) cluster diameters are $\approx 80\%$ larger than expected from the transverse extension of showers induced by single particles but are consistent with fragmenting jets.

We use the same event sample S_{JJ} to investigate a third property of hard parton collisions: the approximate equality of the transverse momenta of the scattered partons. We calculate the transverse momentum P_T and the invariant mass M associated with the pair

207

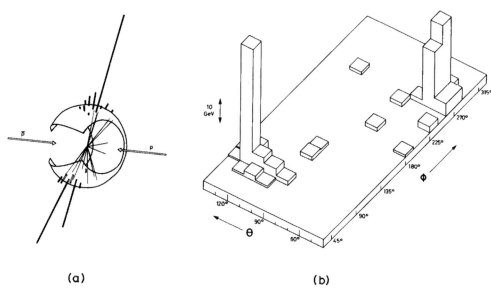

(a)   (b)

Fig. 4. Configuration of the event with the largest value of ΣE_T, 127 GeV (M = 140 GeV): (a) charged tracks pointing to the inner face of the central calorimeter are shown together with cell energies (indicated by heavy lines with lengths proportional to cell energies). (b) the cell energy distribution as a function of polar angle θ and azimuth ϕ.

(C1, C2) in each event (we assign to each cluster a four-momentum (Eu, E), E being the cluster energy and u the unit vector pointing from the event vertex to the cluster center). We measure P_T to be 6 GeV/c on the average, of which at least 3 GeV/c are of instrumental nature (non-inclusion of large angle frag-

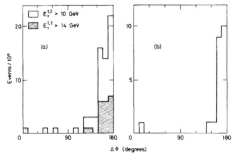

Fig. 5. (a) Azimuthal separation between C1 and C2 (see text) for $E_T^{1,2} >$ 10 and 14 GeV. (b) Azimuthal separation between C1 and the forward/backward sector having $E_T >$ 5 GeV for $E_T^1 >$ 10 GeV and $E_T^2/E_T^1 <$ 0.4 (see text).

ments in the cluster, energy resolution, edge effects, etc.).

The above observations support the interpretation of S_{JJ} as a sample of two-jet events resulting from a hard parton collision. We remark however that the spectacular configuration illustrated in fig. 4 is not representative of the whole sample. As shown in fig. 3a the two-jet system accounts for only a fraction of ΣE_T. The rest of the transverse energy in the event, $\widetilde{E}_T$, is distributed among clusters, of which typically 2 to 3 are in excess of 1 GeV. Their detailed study is beyond the scope of the present report. We simply remark that they are only weakly correlated with the jet directions and that their multiplicity and transverse energy distributions are the same as in events having $\Sigma E_T = \widetilde{E}_T$.

Given the presence of relatively abundant and hard clusters accompanying the two-jet system, we further ascertain the emergence of a two-jet (as opposed to multi-jet) structure by measuring the dependence upon ΣE_T of the ratios $r_{21} = E_T^2/E_T^1$ and $r_{32} = E_T^3/E_T^2$. As ΣE_T increases, r_{21} increases and r_{32} decreases (fig. 3b), again illustrating the dominance of two-jet events for ΣE_T exceeding $\simeq$60 GeV.

7. *Inclusive jet production.* There are 59 events containing at least one cluster with $E_T > 20$ GeV. The evaluation of the inclusive jet production cross section from these events requires the knowledge of the detector acceptance and luminosity.

The detector acceptance is obtained from a Monte Carlo simulation that generates jets with the E_T-distribution given in ref. [6], superimposed on a system of soft hadrons accounting for the remaining fraction of $\sqrt{s}$. The jets fragment into hadrons with an average transverse momentum of 0.45 GeV/c with respect to the jet axis according to a fragmentation function of the form $(1 - x)^2/x$ (x is the fractional momentum of the fragment along the jet axis). All of these hadrons (assumed to be charged and neutral pions only) are then followed into the calorimeters to generate a pattern of energy depositions. Both the longitudinal and lateral shower developments as well as the energy resolution are taken into account.

The data generated by the Monte Carlo simulation undergo the same analysis chain as the real data. In particular, we find that the distribution of cluster size in the Monte Carlo data is very similar to that of the real data, indicating that both hadronic fragmentation and shower developments are correctly described in the simulation program.

The comparison of the E_T-distribution of the Monte Carlo data with that used as an input provides the correction function $\alpha(E_T)$ by which the observed cross section must be divided to obtain the jet inclusive cross section. We have checked that varying some of the analysis parameters, in particular those related to the cluster definition, changes both the observed E_T distribution and $\alpha(E_T)$ but the correct cross section always varies by less than 10%. The function $\alpha(E_T)$ varies by less than a factor of 2 over the range $20 < E_T < 60$ GeV.

The integrated luminosity is obtained by counting the total number of minimum bias events which occurred during data taking. From the fluctuations measured during different running conditions we assign an uncertainty of ±17% to its value. An additional uncertainty results from the fact that, as already mentioned, the trigger to record large-E_T events required a coincidence with a pair of small angle charged secondaries. This requirement introduces a bias which may affect both the absolute magnitude of the cross section and its E_T-dependence.

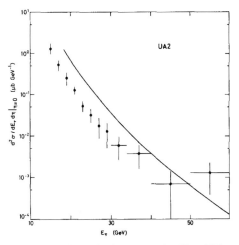

Fig. 6. Inclusive jet production cross section. The solid line (ref. [6]) uses $\Lambda = 0.5$ GeV while $\Lambda = 0.15$ GeV would bring the calculated rates in better agreement with the data. However various uncertainties preclude a determination of Λ from the data [13].

Fig. 6 shows the cross section for inclusive jet production as a function of the jet transverse energy E_T. The errors shown are only statistical. There is an overall uncertainty of ±20% in the vertical scale reflecting the uncertainties in the knowledge of the total luminosity and in the Monte Carlo calculated acceptance. An uncertainty of ±2.5% in the E_T-scale, reflecting the calorimeter energy calibration uncertainties, results in an additional vertical uncertainty of ±20%. From a visual scan of the events the contribution from sources other than $\bar{p}p$ collisions is estimated to be <10%, independent of E_T.

Our measured cross section is at a level comparable with the QCD calculation of Horgan and Jacob [6], which is also shown in fig. 6. In the framework of this model, inclusive jet production is dominated by gluon–gluon scattering in the kinematical region of this experiment.

We finally note that the possible merging of two high E_T clusters produced with a small angular separation may increase the measured cluster energy by as much as 2 GeV on the average. This effect is not accounted for in ref. [6] where jet fragmentation is ignored.

8. Conclusion. We have observed that events with a large transverse energy (in excess of $\simeq 60$ GeV) in a rapidity interval of two units around 90° have a dominant two-jet structure at $\sqrt{s} = 540$ GeV. This is in strong contrast with the situation at $\sqrt{s} = 24$ GeV [11]. The inclusive jet production cross section is measured at a level similar to that predicted by QCD calculations [6]. A sample of two-jet events has been studied and observed to feature properties characteristic of hard scattering of partons. This observation provides the first evidence for highly collimated jets produced in hadron collisions.

We deeply acknowledge the help of the technical staffs of the Institutes collaborating in UA2 and particularly that of the CERN staff who assembled the central calorimeter and its associated electronics: F. Bourgeois and his team, L. Bonnefoy, J.-M. Chapuis, Y. Cholley, G. Gurrieri and A. Sigrist. Financial support from the Danish Natural Science Research Council to the Niels Bohr Institute group and from the Schweizerischer National fonds zur Förderung der wissenschaftlichen Forschung to the Bern group are acknowledged.

References

[1] S.M. Berman, J.D. Bjorken and J.B. Kogut, Phys. Rev. 4D (1971) 3388.
[2] For reviews see: K. Hansen and P. Hoyer, eds., Jets in high energy collisions, Phys. Scr. 19 (1979);
P. Darriulat, Ann. Rev. Nucl. Part. Sci. 30 (1980) 159.
[3] M.G. Albrow et al., Nucl. Phys. B160 (1979) 1;
A.L.S. Angelis et al., Phys. Scr. 19 (1979) 116;
A.G. Clark et al., Nucl. Phys. B160 (1979) 397;
D. Drijard et al., Nucl. Phys. B166 (1980) 233.
[4] For a review see: P. Söding and G. Wolf, Ann. Rev. Nucl. Part. Sci. 31 (1981) 231.
[5] The Staff of the CERN $\bar{p}p$ project, Phys. Lett. 107B (1981) 306.
[6] R. Horgan and M. Jacob, Nucl. Phys. B179 (1981) 441.
[7] K. Alpgård et al., Phys. Lett. 107B (1981) 310.
[8] M. Banner et al., Phys. Lett. 115B (1982) 59;
The UA2 Collab., First results from the UA2 experiment at the SPS $p\bar{p}$ collider, presented XIIIth Intern. Symp. on Multiparticle Dynamics (Volendam, The Netherlands, June 1982).
[9] A.G. Clark, Proc. Intern. Conf. on Instrumentation for Colliding beam physics (SLAC, 1982);
The UA2 Collab., Status and first results from the UA2 experiment, presented 2nd Intern. Conf. on Physics in Collission (Stockholm, Sweden, June 1982);
F. Bonaudi et al., in preparation.
[10] V. Hungerbühler, Proc. Topical Conf. on the Applications of microprocessors to high energy physics experiments, CERN 81-07 (1981) p. 46.
[11] C. de Marzo et al., Phys. Lett. 112B (1982) 173;
M. Arenton et al., Evidence for jets from a transverse energy triggered calorimeter experiment at Fermilab, Contrib. XXIst Intern. Conf. on High Energy Physics (Paris, France, July 1982);
B. Brown et al., Properties of high transverse energy hadronic events, NAL preprint Fermilab-Conf. − 82/34 − Exp. (1982).
[12] M. Jacob, in: Proton−antiproton collider physics (1981), AIP Conf. Proc. nr. 85, p. 651;
H.U. Bengtsson and T. Åkesson, Transverse energy distribution and hard constituent scattering in hadronic collisions, contribution XXIst Intern. Conf. on High Energy Physics (Paris, July 1982).
[13] M. Jacob, private communication.

11

The Fifth Quark

Discovery of the Υ and the B meson, 1977–1987.

The discovery of the J/ψ and charmed quark seemed to complete a family of fermions, (c, s, ν_μ, μ), entirely analogous to (u, d, ν_e, e). If this pattern were indicative, then the τ and its neutrino presaged a new pair of quarks. Both e^+e^- annihilation and hadronic production of lepton pairs, the techniques that had uncovered the charmed quark, were extended in the search for the next quark.

Leon Lederman and his co-workers (**Ref. 11.1**) pressed the search for peaks in the $\mu^+\mu^-$ spectrum to high energies by studying the collisions of 400 GeV protons on nuclear targets at Fermilab. Their apparatus was a double-arm spectrometer set to measure $\mu^+\mu^-$ pairs with invariant masses above 5 GeV with a resolution of 2%. Hadrons were eliminated by using long beryllium filters in each arm. In mid 1977, a clear, statistically significant $\mu^+\mu^-$ peak was observed in the 9.5 GeV region with an observed width of about 1.2 GeV. A more detailed analysis showed better agreement with two peaks at 9.44 and 10.17 GeV, respectively, which were given the names Υ and Υ'. It soon became evident that this was a repetition of the J/ψ and ψ' story.

With the help of an energy upgrade, in May 1978 two groups at the DORIS e^+e^- storage ring at DESY were able to observe the Υ in the PLUTO and DASP II detectors. The results of the experiments are reproduced here (Refs. **11.2**, 11.3) and in Figure 11.1. The determination of the mass of the resonance was greatly improved with the result $M_\Upsilon = 9.46 \pm 0.01$ GeV. Moreover, the observed width was limited only by the energy spread of the beams, so that it was less than 1/100 as much as that observed in hadronic production. Just as for the J/ψ, it was possible to infer the partial width for $\Upsilon \to e^+e^-$ from the area under the resonance curve, with the result $\Gamma_{e^+e^-}(\Upsilon) = 1.3 \pm 0.4$ keV. Using nonrelativistic potential models derived from the ψ system and the assumption that the potential was independent of the quark type, it was possible to predict the wave function at the origin and thus $\Gamma_{e^+e^-}(\Upsilon)$ for the cases of charge $-1/3$ and $+2/3$. The comparison indicated that the new quark had charge $-1/3$ rather than $+2/3$. The new quark was dubbed the b for "bottom," reflecting the practice of writing the quark pairs (u, d) and (c, s) with the charge $-1/3$ beneath the charge $2/3$ quark. Thus the sixth quark is referred to as t or "top."

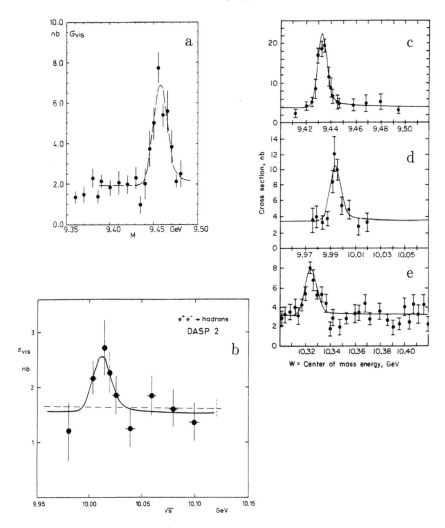

Figure 11.1. Measurements of the e^+e^- cross section at the lower Υ states. Measurements from the DASP II experiment at DORIS show (a) the Υ and (b) the Υ' (Ref. 11.3). Measurements by the CLEO group at CESR show (c) the Υ, (d) the Υ', and (e) the Υ''. Discrepancies between the mass measurements by the two groups were later resolved (Ref. 11.5).

After additional cavities were added to increase the energy of the DORIS ring, DASP II and the DESY–Heidelberg sodium-iodide and lead-glass detector were able to observe the Υ' (Refs. 11.3, **11.4**). The $\Upsilon' - \Upsilon$ splitting was found to be very nearly the same as that for $\psi' - \psi$.

By 1980, the Cornell Electron Storage Ring (CESR) with its two detectors, CLEO and CUSB, became operational. They both observed the Υ and Υ', and additional resonances, Υ'' and Υ''' (Refs. 11.5, 11.6, 11.7, **11.8**). The first three states, with masses 9.460, 10.023, and 10.355 GeV, are narrow, with observed widths consistent with the beam spread of the machine. They are analogous to the ψ and ψ', and correspond to 1^3S_1, 2^3S_1, and 3^3S_1 states of a $b\bar{b}$ system. Figure 11.1 shows the Υ and Υ' as observed by DASP II and the Υ, Υ', and Υ'' as observed by CLEO. The Υ''' at 10.577 GeV is a broader state, like the $\psi(3772) = \psi''$, and is interpreted as the 4^3S_1 state, lying above the threshold for $B\bar{B}$ production, where B represents a meson containing a $\bar{b}$ quark and a u or d quark. Thus $B^+ = \bar{b}u$, $B^0 = \bar{b}d$, $B^- = b\bar{u}$, $\bar{B}^0 = b\bar{d}$.

The existence of a series of s-wave bound states required that there be p-wave states as well. These were observed through radiative transitions from the s-wave states, $\Upsilon' \to \chi_b \gamma$, where χ_b represents a $C = +1$, $P = +1$ p-wave state. Evidence was obtained from the inclusive photon spectrum, $\Upsilon' \to \gamma +$ anything, and from the cascade $\Upsilon' \to \gamma \chi_b$, $\chi_b \to \gamma \Upsilon$, $\Upsilon \to l^+l^-$, where l represents e or μ. Measurements were carried out by CUSB and CLEO at CESR and by the Crystal Ball at DORIS II after the detector was shipped from Stanford to Hamburg. In Figure 11.2 some results from the Crystal Ball are shown.

What is the role of the b quark in weak interactions? Beta decay is described at the quark level by the transition $d \to u e^- \bar{\nu}$. Positron emission is the result of $u \to d e^+ \nu$. The strangeness-changing semileptonic weak decays (e.g. $\Lambda \to p e \bar{\nu}$) are described by $s \to u e^- \bar{\nu}$ whose inverse is $u \to s e^+ \nu$. The Cabibbo hypothesis is that the weak current is really $u \to (\cos\theta_c d + \sin\theta_c s)$. As discussed in Chapter 9, the introduction of a fourth

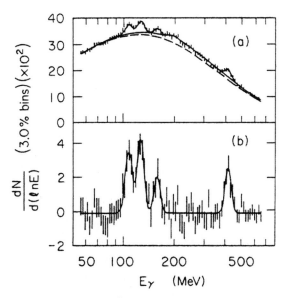

Figure 11.2. The photon spectrum from Υ' decays obtained by the Crystal Ball Collaboration at DORIS II. A triplet of lines corresponding to $\Upsilon' \to \gamma \chi_b(^3P_{2,1,0})$ is seen between 100 and 200 MeV. The decays $\chi_b \to \gamma \Upsilon$ produce the unresolved signal between 400 and 500 MeV [R. Nernst et al., Phys. Rev. Lett. **54**, 2195 (1985)].

quark makes the Cabibbo angle into a rotation, with the current described by

$$(\bar{u} \quad \bar{c}) \begin{pmatrix} \cos\theta_c & \sin\theta_c \\ -\sin\theta_c & \cos\theta_c \end{pmatrix} \begin{pmatrix} d \\ s \end{pmatrix}. \quad (11.1)$$

The V-A structure $\gamma_\mu(1 - \gamma_5)$ has been suppressed for clarity. The 2 × 2 matrix can be viewed either as a rotation of the charge $-1/3$ quarks or of the charge $+2/3$ quarks, though by convention it is usually the charge $-1/3$ quarks that are subjected to rotation.

With the discovery of the b quark it was apparent that the Cabibbo matrix would have to be expanded to a 3×3 matrix. Indeed this possibility had been anticipated by M. Kobayashi and T. Maskawa before the discovery of even the charmed quark. They observed that if there were a third generation, that is a third pair like (u, d) and (c, s), the 3×3 mixing matrix would allow for CP violation.

In order to provide for CP violation, we need a complex term in the interaction $J_\mu^\dagger J^\mu$ where $J_\mu = \overline{U}\gamma_\mu V(1-\gamma_5)D$ is the weak current. If there are n families, U represents the column of n charge $+2/3$ quarks and D the column of n charge $-1/3$ quarks. The matrix V is unitary and has n^2 complex or $2n^2$ real parameters. Unitarity imposes the conditions $V_{ij}V_{kj}^* = \delta_{ik}$, which give $n(n-1)/2$ complex constraints for $i \neq k$ and n real constraints for $i = k$. Altogether there are n^2 remaining free parameters in V.

It is possible to eliminate some of the complex phases in V by redefining the phases of the $2n$ quark fields. Changing all of the fields by the same phase changes nothing so $2n - 1$ phases from V can be eliminated in this way. Thus the number of real parameters characterizing V is $n^2 - 2n + 1 = (n-1)^2$. For two families this gives just one parameter, which is the Cabibbo angle. For three families there are four parameters. Now if V were purely real it would be a 3×3 rotation matrix, which is determined by three real parameters. Thus the fourth parameter of V must necessarily introduce a complex component into V, one that cannot be absorbed into a redefinition of the quark fields.

We can represent the Cabibbo–Kobayashi–Maskawa (CKM) matrix by

$$\begin{bmatrix} \bar{u} & \bar{c} & \bar{t} \end{bmatrix} \begin{bmatrix} V_{ud} & V_{us} & V_{ub} \\ V_{cd} & V_{cs} & V_{cb} \\ V_{td} & V_{ts} & V_{tb} \end{bmatrix} \begin{bmatrix} d \\ s \\ b \end{bmatrix}. \quad (11.2)$$

In principle, the squares of the various matrix elements can be measured by observing a variety of weak decays. The comparison of nuclear beta decay and muon decay indicates $|V_{ud}| \approx 0.97$, while the strangeness-changing decays give $|V_{us}| \approx 0.22$. These two are just $\cos\theta_c$ and $\sin\theta_c$ in the Cabibbo scheme. The production of charmed particles in neutrino (or antineutrino) nucleus scattering is proportional to $|V_{cd}|^2$. Data from the CDHS Collaboration led to a value $|V_{cd}| = 0.21 \pm 0.03$. The decay of mesons containing b quarks is controlled by V_{ub} and V_{cb}. The relative size of these elements determines the ratio of the semileptonic decays $\Gamma(b \to ul\nu)/\Gamma(b \to cl\nu)$. Because of the greater phase space available, the $b \to ul\nu$ decay produces leptons with higher momentum than does $b \to cl\nu$.

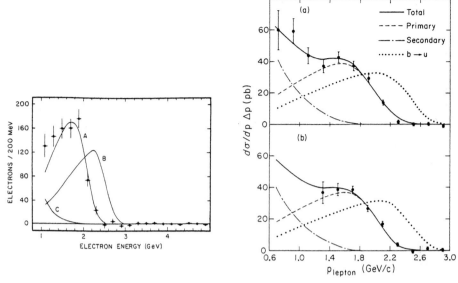

Figure 11.3. Lepton spectra for semileptonic B meson decays. Left, CUSB data from CESR together with the curves expected for (A) $b \to ce\nu$, (B) $b \to ue\nu$, and (C) $b \to cX, c \to se\nu$ [C. Klopfenstein et al., Phys. Lett. **130B**, 444 (1983)]. Right, data from CLEO, also taken at CESR. The upper figure is for electrons and the lower for muons. The solid curves are predictions without any $b \to ul\nu$ while the dotted curves are predictions for purely $b \to ul\nu$ [A. Chen et al., Phys. Rev. Lett. **52**, 1084, (1984)]. All the figures indicate that $\Gamma(b \to cl\nu) \gg \Gamma(b \to ul\nu)$. This Cabibbo–Kobayashi–Maskawa suppression is analogous to the Cabibbo suppression observed in the decays of charmed particles to states without strangeness.

In Figure 11.3 we show data from CUSB and CLEO for the lepton spectra. The evidence overwhelmingly supports $b \to cl\nu$ as the dominant mode. Data indicated that

$$\frac{\Gamma(b \to ul\nu)}{\Gamma(b \to cl\nu)} < 0.08. \qquad (11.3)$$

Correcting for the difference in phase space available for the two modes gives

$$|V_{ub}/V_{cb}| < 0.22. \qquad (11.4)$$

Actual identification of B meson decays promised to be a formidable task, even though some lessons had been learned from the study of charm. By focusing on the 4^3S_1 Υ''' it was possible to obtain a good sample of $\Upsilon(4S) \to B\bar{B}$ events **(Ref. 11.9)**. The technique used was to identify candidates for D^0 s and D^{*+} s, using only entirely charged decay modes, and combine these with either one or two charged pions. In analogy with the fundamental decay $c \to s$ leading to $D^+ \to K^-\pi^+\pi^+$, the transition $b \to c$ produces $B^- \to D^+\pi^-\pi^-$, $D^{*+}\pi^-\pi^-$. The combinations $B \to D\pi$ and $B \to D\pi\pi$ were required to produce B s with energy equal to the beam energy since the decay is $\Upsilon(4S) \to B\bar{B}$. An accumulation of events for mass near 5.275 GeV suggested the observation of exclusive B

decays. Ultimately the branching fractions for these modes were determined to be an order of magnitude or more smaller than in this first report.

The CUSB Collaboration observed photons of energy about 50 MeV associated with $B\bar{B}$ production at energies above the $\Upsilon(4S)$, which they ascribed to the production of $B^*\bar{B}$ and the subsequent decay $B^* \to B\gamma$. The splitting between the spin-1 B^* and the pseudoscalar B was determined to be $52 \pm 2 \pm 4$ MeV. Some of the CUSB data are shown in Figure 11.4. Precision measurements later by CUSB and CLEO and by the LEP experiments, to be discussed in Chapter 12, refined the mass difference to 45.78 ± 0.35 MeV.

Semileptonic decays were exploited in several experiments to obtain B-enriched samples of events for B lifetime measurements. The Mark II Collaboration at PEP built a vertex detector using a precision drift chamber located close to the interaction point, which allowed measurements of the distance of closest approach of the lepton tracks to the beam–beam collision region. The experiment found a B lifetime of $1.2^{+0.45}_{-0.36} \pm 0.30$ ps.

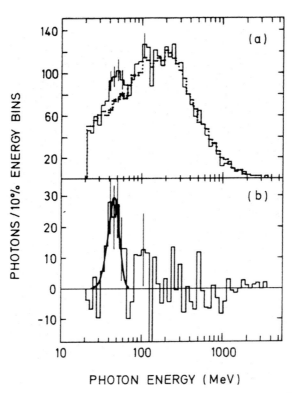

Figure 11.4. The photon energy spectrum obtained by the CUSB Collaboration for events with high energy leptons and thrust less than 0.88 (indicative of events more spherical than the ordinary two-jet events produced in e^+e^- annihilation). These criteria signal the presence of B mesons. The e^+e^- c.m. energy for the solid histogram in (a) is 10.62 – 25 GeV, above the $B^*\bar{B}$ threshold. The dotted histogram in (a) was taken at the $\Upsilon(4S)$, below the $B^*\bar{B}$ threshold. In (b) the spectrum with the background subtracted shows a line near 50 MeV, ascribed to $B^* \to B\gamma$ [K. Han et al., Phys. Rev. Lett. **55**, 36 (1985)].

A contemporaneous measurement by the MAC experiment at PEP found a similar result: $1.8\pm0.6\pm0.4$ ps. The surprisingly large value was confirmed subsequently by the DELCO experiment at PEP and TASSO and JADE experiments at PEP and PETRA. This unexpectedly long lifetime indicated that the V_{cb} matrix element is quite small, between 0.030 and 0.062. From the unitarity of the KM matrix, we conclude that $|V_{cd}|^2+|V_{cs}|^2+|V_{cb}|^2 = 1$, so $|V_{cs}| \approx 0.97$, assuming there are just three generations.

The pattern of decreasing CKM matrix elements – diagonal, first generation to second generation, second generation to third generation, first generation to third generation – led Lincoln Wolfenstein to propose a particularly convenient representation for the the matrix:

$$V = \begin{bmatrix} 1 - \lambda^2/2 & \lambda & \lambda^3 A(\rho - i\eta) \\ -\lambda & 1 - \lambda^2/2 & \lambda^2 A \\ \lambda^3 A(1 - \rho - i\eta) & -\lambda^2 A & 1 \end{bmatrix} \quad (11.5)$$

where λ, ρ, η, and A are real parameters. This satisfies unitarity up to corrections of order λ^4 in the imaginary part and of order λ^5 in the real part.

The CKM picture predicts that the dominant B meson decays are due to $b \to c\bar{u}d$ and $b \to c\bar{c}s$. The CLEO Collaboration at CESR and ARGUS Collaboration at the DORIS storage ring at DESY identified decay modes modes of both categories. Each of the numerous final states of the form $D\pi$, $D^*\pi$, $D^*\pi\pi$, etc. accounts for less than 2% of the total, but together they constitute the vast majority of decays. The final state like $c\bar{d}\bar{c}s$ can appear as $D^+D_s^-$ or alternatively as $J/\psi K^0$, and similar states with additional pions. The decay to $J/\psi K_S^0$ with a branching fraction of about 0.45×10^{-3} has a spectacular signal when the J/ψ decays to e^+e^- or $\mu^+\mu^-$ and was destined to play a central role in future studies.

An upgraded detector, CLEO II, featured greatly improved tracking and particle identification. As the number of accumulated events increased, it became possible to search for rarer decays. Of particular importance was the observation in 1993 of $B^0 \to K^{*0}\gamma$ and $B^+ \to K^{*+}\gamma$ (**Ref. 11.10**). The decay $B \to K\gamma$ is forbidden since the final state necessarily has a component of angular momentum along the γ direction of ± 1. The signal was isolated by requiring the photon to have a laboratory energy between 2.1 GeV and 2.9 GeV and demanding that there be a $K\pi$ with an invariant mass consistent with that of the K^*. The dominant background, from non $B - \bar{B}$ events, was suppressed by excluding events with a two-jet appearance, characteristic of the production of lower mass quarks. The final signals were 6.6 ± 2.8 events in the $B^0 \to K^{*0}\gamma$ channel and 4.1 ± 2.3 in the $B^+ \to K^{*+}\gamma$ channel, resulting in branching fractions $(4.0 \pm 1.7 \pm 0.8) \times 10^{-5}$ and $(5.7 \pm 3.1 \pm 1.1) \times 10^{-5}$. The significance of these decays is that they establish the existence of the whimsically named "penguin" process, shown in Fig. 11.5.

The discovery of the b quark provided an excellent opportunity to test the models proposed to explain the phenomena associated with the charmed quark. These tests have been quite successful in a qualitative and semiquantitative way. The general spacing of bound-state levels in the two systems can be understood from a single potential. The systematics of the fine structure (the splitting of the p-wave states) is in accord with expectations. The rates for radiative decays are in general agreement with the nonrelativistic model. The b

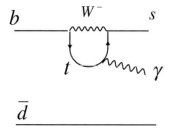

Figure 11.5. A penguin process, in which the decaying quark emits a W boson, then reabsorbs it, is responsible for the class of decays $b \to s\gamma$, and in particular for $B \to K^*\gamma$. Analogous decays occur with the γ replaced by a gluon. Since the gluon carries no isospin, the penguin process $s \to dg$ is purely $\Delta I = 1/2$ and explains, in part, the $\Delta I = 1/2$ rule. All charge 2/3 quarks contribute to the loop, but the t quark makes the dominant contribution because of its large mass.

quark provided, as well, a possible explanation for CP violation through the complex phase in the Cabibbo–Kobayashi–Maskawa matrix.

Exercises

11.1 Suppose the quark–antiquark potential obeys the power law $V(r) = ar^\nu$. Show that the binding energies vary with the quark mass as $E \propto m^{-\nu/(\nu+2)}$ and that the density at the origin $|\phi(0)|^2$ for an s-wave state varies as $m^{3/(\nu+2)}$. Given that the splitting $\Upsilon' - \Upsilon$ is nearly identical to that for $\psi' - \psi$, predict $\Gamma(\Upsilon \to e^+e^-)$ from $\Gamma(\psi \to e^+e^-)$ if the charge of the new quark is $-1/3$ or is $+2/3$.

11.2 Show that $|\phi(0)|^2$, the s-wave wave function at the origin squared, is related to the average force by

$$|\phi(0)|^2 = \frac{m}{2\pi} <F> .$$

Hint: write the Schrödinger wave equation for the radial wave function, u and multiply by u'. Integrate the result from $r = 0$ to $r = \infty$.

11.3 Use the results of Exercise 6.5 to determine the e^- spectrum in the decays $b \to ce^-\bar{\nu}$, $b \to ue^-\bar{\nu}$. Take $m_b = 5$ GeV, $m_c = 1.5$ GeV, $m_u = 0.3$ GeV.

11.4 * Suppose that the $b\bar{b}$ or $c\bar{c}$ interaction can be represented approximately by a non-relativistic Schrödinger equation:

$$\left[-\frac{1}{2m_{red}}\nabla^2 + V(r)\right]\psi = E\psi$$

where $m_{red} = m/2$ is the reduced mass. Then the energy levels are spin-independent so $^3S_1$ and $^1S_0$ are degenerate, as are $^3P_{2,1,0}$ and $^1P_1$, etc. Now consider as perturbations the spin dependent forces

$$\mathbf{L}\cdot\mathbf{S}\,V_{so}(r)$$

$$\sigma_1\cdot\sigma_2\,V_{spin-spin}(r)$$

$$\frac{3\,\sigma_1\cdot\mathbf{r}\,\sigma_2\cdot\mathbf{r}-\sigma_1\cdot\sigma_2 r^2}{r^2}V_{tensor}(r)\equiv S_{12}V_{tensor}(r).$$

Here $\mathbf{S}=\frac{1}{2}(\sigma_1+\sigma_2)$ is the total spin, and σ_1 and σ_2 are the quark and antiquark spin operators.

a. Which degenerate states are split by each of the interactions? Which nondegenerate state are mixed by the interactions?

b. Use the relations (try to prove them, too)

$$<J,L=J+1,\ S=1,M|S_{12}|J,L=J+1,\ S=1,M>=-2\frac{L+1}{2L-1}$$
$$<J,L=J,\ S=1,M|S_{12}|J,L=J,\ S=1,M>=+2$$
$$<J,L=J-1,\ S=1,M|S_{12}|J,L=J-1,\ S=1,M>=-2\frac{L}{2L+3}$$

to analyze the observed splittings of the ^3P states in the Υ and ψ systems. Here $|J,L,S,M>$ is a state with total angular momentum J, orbital angular momentum L, total spin angular momentum S, and $J_Z=M$. [See J. D. Jackson, "Lectures on the New Particles" in *Proc. of Summer Institute on Particle Physics*, Stanford, CA, Aug. 2-13, 1976, M. Zipf, ed.]

11.5 * The relation between the standard relativistic Lorentz invariant amplitude, $\mathcal{M}$, (the usual Feynman rules generate $-i\mathcal{M}$) and the conventional scattering amplitude of potential theory is

$$f=-\frac{1}{8\pi\sqrt{s}}\mathcal{M}$$

where s is the square of the center-of-mass energy. The center-of-mass differential cross section is $d\sigma/d\Omega=|f|^2$. In potential theory, the Born value for f is

$$f=-\frac{m}{2\pi}\int d^3r\,e^{-i\mathbf{p}'\cdot\mathbf{r}}V(r)e^{i\mathbf{p}\cdot\mathbf{r}}$$

where $\mathbf{p}$ and $\mathbf{p}'$ are the initial and final momenta. Two body scattering can be treated analogously with the modification $m\to m_{reduced}$. If the particles have spin 1/2, we generalize the wave function to $\psi(r)\chi_1\chi_2$, where χ_1 and χ_2 are two-component spinors. Thus

$$f=-\frac{m_{red}}{2\pi}\int d^3r\,\chi_2'^{\dagger}\chi_1'^{\dagger}e^{-i\mathbf{p}'\cdot\mathbf{r}}V(r)e^{i\mathbf{p}\cdot\mathbf{r}}\chi_1\chi_2.$$

Suppose $\mathcal{M}$ has the form of vector exchange but with some more general dependence on momentum transfer:

$$\mathcal{M} = \bar{u}(p_4)\gamma_\mu u(p_2)\, \bar{u}(p_3)\gamma^\mu u(p_1)\, \tilde{V}(p_1 - p_3).$$

Show that the spin dependent potential is, to leading order

$$V(r) = V_0(r) + \frac{3}{2m^2}\frac{1}{r}\frac{dV_0}{dr}\mathbf{L}\cdot\mathbf{S} + \frac{1}{12m^2}S_{12}\left(\frac{1}{r}\frac{dV_0}{dr} - \frac{d^2V_0}{dr^2}\right) + \frac{\boldsymbol{\sigma}_1\cdot\boldsymbol{\sigma}_2}{6m^2}\nabla^2 V_0$$

where $\mathbf{S} = \frac{1}{2}(\boldsymbol{\sigma}_1 + \boldsymbol{\sigma}_2)$ is the total spin and

$$\int d^3r\, V_0(r) e^{i\mathbf{q}\cdot\mathbf{r}} = \tilde{V}(q).$$

Further Reading

For a discussion of the Cabibbo–Kobayashi–Maskawa matrix see the article in the current edition of the *Review of Particle Physics*.

References

11.1 S. W. Herb et al., "Observation of a Dimuon Resonance at 9.5 GeV in 400 GeV Proton Nucleus Collisions." *Phys. Rev. Lett.*, **39**, 252 (1977).
11.2 PLUTO Collaboration, Ch. Berger et al., "Observation of a Narrow Resonance Formed in e^+e^- Annihilation at 9.46 GeV." *Phys. Lett.*, **76B**, 243 (1978).
11.3 C. W. Darden et al., "Observation of a Narrow Resonance at 9.46 GeV in Electron-Positron Annihilation." *Phys. Lett.*, **76B**, 246 (1978). Also *Phys. Lett.* **78B**, 364 (1978).
11.4 J. K. Bienlein et al., "Observation of a Narrow Resonance at 10.02 GeV in e^+e^- Annihilation." *Phys. Lett.*, **78B**, 360 (1978).
11.5 D. Andrews et al., "Observation of Three Upsilon States." *Phys. Rev. Lett.*, **44**, 1108 (1980).
11.6 T. Böhringer et al., "Observation of Υ, Υ', and Υ'' at the Cornell Electron Storage Ring." *Phys. Rev. Lett.*, **44**, 1111 (1980).
11.7 D. Andrews et al., "Observation of a Fourth Upsilon State in e^+e^- Annihilation." *Phys. Rev. Lett.*, **45**, 219 (1980).
11.8 G. Finocchiaro et al., "Observation of the Υ''' at the Cornell Electron Storage Ring." *Phys. Rev. Lett.*, **45**, 222 (1980).
11.9 S. Behrends et al., "Observation of Exclusive Decay Modes of b Flavored Mesons." *Phys. Rev. Lett.*, **50**, 881 (1983).
11.10 R. Ammar et al., "Evidence for Penguin-Diagram Decays: First Observation $B \to K^*\gamma$." *Phys. Rev. Lett.*, **71**, 674 (1993).

Observation of a Dimuon Resonance at 9.5 GeV in 400-GeV Proton-Nucleus Collisions

S. W. Herb, D. C. Hom, L. M. Lederman, J. C. Sens,[a] H. D. Snyder, and J. K. Yoh
Columbia University, New York, New York 10027

and

J. A. Appel, B. C. Brown, C. N. Brown, W. R. Innes, K. Ueno, and T. Yamanouchi
Fermi National Accelerator Laboratory, Batavia, Illinois 60510

and

A. S. Ito, H. Jöstlein, D. M. Kaplan, and R. D. Kephart
State University of New York at Stony Brook, Stony Brook, New York 11974
(Received 1 July 1977)

Accepted without review at the request of Edwin L. Goldwasser under policy announced 26 April 1976

> Dimuon production is studied in 400-GeV proton-nucleus collisions. A strong enhancement is observed at 9.5 GeV mass in a sample of 9000 dimuon events with a mass $m_{\mu^+\mu^-} > 5$ GeV.

We have observed a strong enhancement at 9.5 GeV in the mass spectrum of dimuons produced in 400-GeV proton-nucleus collisions. Our conclusions are based upon an analysis of 9000 dimuon events with a reconstructed mass $m_{\mu^+\mu^-}$ greater than 5 GeV corresponding to 1.6×10^{16} protons incident on Cu and Pt targets:

$$p + (\text{Cu, Pt}) \to \mu^+ + \mu^- + \text{anything}.$$

The produced muons are analyzed in a double-arm magnetic-spectrometer system with a mass resolution $\Delta m/m$ (rms) $\approx 2\%$.

The experimental configuration (Fig. 1) is a modification of an earlier dilepton experiment in the Fermilab Proton Center Laboratory.[1-3] Narrow targets (~ 0.7 mm) with lengths corresponding to 30% of an interaction length are employed.

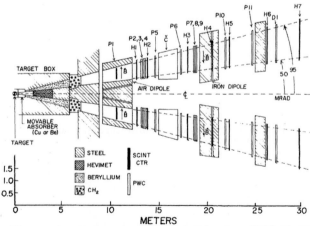

FIG. 1. Plan view of the apparatus. Each spectrometer arm includes eleven PWC's P1–P11, seven scintillation counter hodoscopes H1–H7, a drift chamber D1 and a gas-filled threshold Čerenkov counter Č. Each arm is up/down symmetric and hence accepts both positive and negative muons.

TABLE I. Yield of muon pairs for various running conditions.

Analyzing-magnet current (A)	Target	First 30 cm absorber	No. of incident protons	No. of events $m_{\mu^+\mu^-} \geq 5$ GeV
1500	Cu	Cu	8.1×10^{15}	4093
1500	Pt	Cu	4.1×10^{15}	2076
1250	Cu	Cu	2.5×10^{15}	1891
1250	Pt	Be	1.6×10^{15}	911

Beryllium (18 interaction lengths) is used as a hadron filter, covering the 50–95-mrad (70–110° c.m.) horizontal and ± 10-mrad vertical aperture in each arm. The Be is closely packed against steel and tungsten which minimize particle leakage from outside the aperture, especially from the tungsten beam dump located 2.2 m downstream of the target. Polyethylene (1.5 m) and a 2.2-m steel collimator complete the shielding. The first 30 cm of beryllium (starting 13 cm downstream of the target center) can be remotely exchanged for 30 cm of copper.

The spectrometer dipoles deflect vertically, decoupling the production angle of each muon from its momentum determination. At full excitation (1500 A), the magnets provide a transverse momentum kick $p_t \approx 1.2$ GeV. In order to maximize the usable luminosity, no detectors are placed upstream of the magnet. Conventional proportional wire chambers (PWC's) and scintillation hodoscopes serve to define the muon trajectory downstream of the air dipole. Following the PWC's is a solid iron magnet (1.8 m long, energized to 20 kG) used to refocus partially the muons vertically and to redetermine the muon momentum to ± 15%. A threshold Čerenkov counter on each arm also helps prevent possible low-momentum muon triggers. The apparatus is ar-

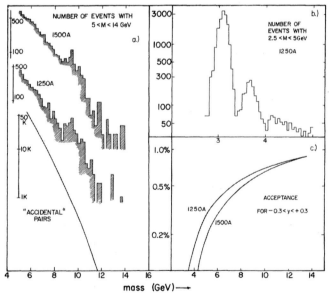

FIG. 2. (a) Dimuon yield at 1500 and 1250 A; the data with Cu and Pt targets have been combined. Also shown is the mass spectrum generated by combining two muons from different events. (b) Excess of opposite-sign over equal-sign muon pairs in the ψ, ψ' region. (c) Dimuon mass acceptance for the two excitations of the air dipole.

ranged symmetrically with respect to the horizontal median plane in order to detect both μ^+ and μ^- in each arm.

The data sets presented here are listed in Table I. Low-current runs produced ~ 15 000 J/ψ and 1000 ψ' particles which provide a test of resolution, normalization, and uniformity of response over various parts of the detector. Figure 2(b) shows the 1250-A J/ψ and ψ' data. The yields are in reasonable agreement with our earlier measurements.[2]

High-mass data (1250 and 1500 A) were collected at a rate of 20 events/h for $m_{\mu^+\mu^-} > 5$ GeV using $(1.5-3)\times 10^{11}$ incident protons per accelerator cycle. The proton intensity is limited by the requirement that the singles rate at any detector plane not exceed 10^7 counts/sec. The copper section of the hadron filter has the effect of lowering the singles rates by a factor of 2, permitting a corresponding increase in protons on target. The penalty is an ~ 15% worsening of the resolution at 10 GeV mass. Figure 2(a) shows the yield of muon pairs obtained in this work.

At the present stage of the analysis, the following conclusions may be drawn from the data [Fig. 3(a)]:

(1) A statistically significant enhancement is observed at 9.5-GeV $\mu^+\mu^-$ mass.

(2) By exclusion of the 8.8–10.6-GeV region, the continuum of $\mu^+\mu^-$ pairs falls smoothly with mass. A simple functional form,

$$[d\sigma/dmdy]_{y=0} = Ae^{-bm},$$

with $A = (1.89 \pm 0.23)\times 10^{-33}$ cm^2/GeV/nucleon and $b = 0.98 \pm 0.02$ GeV^{-1}, gives a good fit to the data for 6 GeV $< m_{\mu^+\mu^-} < 12$ GeV ($\chi^2 = 21$ for 19 degrees of freedom).[4,5]

(3) In the excluded mass region, the continuum fit predicts 350 events. The data contain 770 events.

(4) The observed width of the enhancement is greater than our apparatus resolution of a full width at half-maximum (FWHM) of 0.5 ± 0.1 GeV. Fitting the data minus the continuum fit [Fig. 3(b)] with a simple Gaussian of variable width yields the following parameters (B is the branching ratio to two muons):

Mass = 9.54 ± 0.04 GeV,

$[Bd\sigma/dy]_{y=0} = (3.4 \pm 0.3)\times 10^{-37}$ cm^2/nucleon,

with FWHM = 1.16 ± 0.09 GeV and $\chi^2 = 52$ for 27

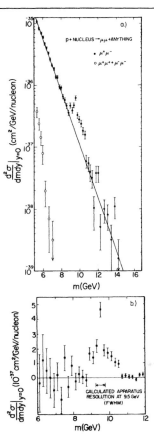

FIG. 3. (a) Measured dimuon production cross sections as a function of the invariant mass of the muon pair. The solid line is the continuum fit outlined in the text. The equal-sign–dimuon cross section is also shown. (b) The same cross sections as in (a) with the smooth exponential continuum fit subtracted in order to reveal the 9–10-GeV region in more detail.

degrees of freedom (Ref. 5). An alternative fit with two Gaussians whose widths are fixed at the resolution of the apparatus yields

Mass = 9.44 ± 0.03 and 10.17 ± 0.05 GeV,

$[Bd\sigma/dy]_{y=0} = (2.3 \pm 0.2)$ and (0.9 ± 0.1)

$\times 10^{-37}$ cm^2/nucleon,

with $\chi^2 = 41$ for 26 degrees of freedom (Ref. 5).

The Monte Carlo program used to calculate the acceptance [see Fig. 2(c)] and resolution of the apparatus assumed a mass, p_t, rapidity and decay angular distribution of the $\mu^+\mu^-$ pair consistent with these data and previously published dilepton searches. It also included all multiple-scattering effects in the hadron absorber and detector resolutions. The conclusions stated above are insensitive to these assumptions. In particular we note that the acceptance is relatively flat in the 9–10-GeV region.

The following checks have been made to verify the validity of the conclusions reached above:

(1) The spectrum of $\mu^+\mu^+$ and $\mu^-\mu^-$ events in Fig. 3(a) constitute an upper limit on the combined effects of accidental coincidences and hadronic decays. Misidentified $\psi \to \mu^+ + \mu^-$ decays are prevented from producing background at high mass by the remeasurement of the muon momenta both downstream by the second magnet and also by the PWC at the center of the first magnet. This is confirmed by the clean separation of the ψ and ψ' peaks in Fig. 2(a). Their widths agree with the calculated apparatus resolution.

(2) Various subsets of the data were studied in order to search for possible apparatus bias. In addition to the subsets shown in Table I, data were studied as a function of magnet polarity and magnetic bend direction. All fits showed enhancements consistent with the values quoted above.

(3) To check our analysis software (and as a further check of the apparatus), we mixed muons from different events, yielding the smooth mass spectrum shown in Fig. 2(a). The geometrical distribution of events in the 9–10-GeV region at the various detector planes in the apparatus is consistent with that of events in neighboring mass bins.

(4) The longitudinal distribution of muon-pair vertices at the target (FWHM = 16 cm) is cleanly separated from events generated in the beam dump, 220 cm downstream. A separate target-out run with 6×10^{14} incident protons produced no acceptable $\mu^+\mu^-$ candidates above 6 GeV (an equivalent run with a Cu or Pt target would have yielded about 200 events with 25 of these in the 9–10-GeV region).

In conclusion, the measured spectrum of $\mu^+\mu^-$ pairs produced in proton-nucleus collisions shows significant structure[6] in the 9–10-GeV region on an exponentially falling continuum. The structure is wider than the apparatus resolution. The 9.5-GeV enhancement and the continuum are in agreement with our previous measurements.[7]

We owe much to Ken Gray, Karen Kephart, Frank Pearsall, and S. Jack Upton for technical assistance, and to F. William Sippach for our electronic systems design. We also thank Brad Cox, William Thomas, and the staffs of the Fermilab Proton Department and Accelerator Division for their efforts. The work at Columbia University and at the State University of New York at Stony Brook was supported by the National Science Foundation, and that at the Fermi National Accelerator Laboratory by the U. S. Energy Research and Development Administration.

---

[a] Permanent address: Foundation for Fundamental Research on Matter, The Netherlands.

[1] D. C. Hom et al., Phys. Rev. Lett. 36, 1236 (1976).

[2] H. D. Snyder et al., Phys. Rev. Lett. 36, 1415 (1976).

[3] D. C. Hom et al., Phys. Rev. Lett. 37, 1374 (1976).

[4] Cu and Pt yields were reduced to cross sections per nucleon by assuming an atomic-number dependence of $A^{1.0}$.

[5] The errors quoted on the magnitude of the continuum and resonance cross sections and the resonance masses are statistical only. Systematic normalization effects are probably less than 25% and do not affect the conclusions drawn here. Systematic errors on the mass calibration are probably less than 1%.

[6] Following Ref. 1, a reasonable designation for this enhancement is $\Upsilon(9.5)$.

[7] We note that the 9–10-GeV mass bin in the e^+e^- and $\mu^+\mu^-$ spectra previously published by this group (Ref. 3) shows an excess of events, consistent with the statistically more significant results here. If we add our preliminary unpublished e^+e^- data to our published e^+e^- yield (Ref. 1), the spectrum contains a cluster of 6 e^+e^- events near 9.5 GeV where ~ 5 events would be expected on the assumption of μ-e universality.

OBSERVATION OF A NARROW RESONANCE FORMED IN e^+e^- ANNIHILATION AT 9.46 GeV

PLUTO Collaboration

Ch. BERGER, W. LACKAS, F. RAUPACH, W. WAGNER
I. Physikalisches Institut der RWTH Aachen, FRG

G. ALEXANDER[1], L. CRIEGEE, H.C. DEHNE, K. DERIKUM, R. DEVENISH, G. FLÜGGE, G. FRANKE
Ch. GERKE, E. HACKMACK, P. HARMS, G. HORLITZ, Th. KAHL[2], G. KNIES, E. LEHMANN,
B. NEUMANN, R.L. THOMPSON[3], U. TIMM, P. WALOSCHEK, G.G. WINTER,
S. WOLFF, W. ZIMMERMANN
Deutsches Elektronen-Synchrotron DESY, Hamburg, FRG

O. ACHTERBERG, V. BLOBEL, L. BOESTEN, H. DAUMANN, A.F. GARFINKEL[4], H. KAPITZA,
B. KOPPITZ, W. LÜHRSEN, R. MASCHUW, H. SPITZER, R. van STAA, G. WETJEN
II. Institut für Experimentalphysik der Universität Hamburg, FRG

A. BÄCKER, J. BÜRGER, C. GRUPEN, H.J. MEYER, G. ZECH
Gesamthochschule Siegen, FRG

H.J. DAUM, H. MEYER, O. MEYER, M. RÖSSLER, K. WACKER
Gesamthochschule Wuppertal, FRG

Received 9 May 1978

> An experiment using the PLUTO detector has observed the formation of a narrow, high mass, resonance in e^+e^- annihilations at the DORIS storage ring. The mass is determined to be 9.46 ± 0.01 GeV which is consistent with that of the Upsilon. The gaussian width σ is observed as 8 ± 1 MeV and is equal to the DORIS energy resolution. This suggests that the resonance is a bound state of a new heavy quark-antiquark pair. An electronic width Γ_{ee} = 1.3 ± 0.4 keV was obtained. In standard theoretical models, this favors a quark charge assignment of −1/3.

The existence of the Upsilon particle recently discovered in proton interactions by Herb et al. [1] is of considerable interest. Not only is it exceptionally massive (9.4 GeV), but its observed decay to muon pairs implies that it is probably quite narrow. The observed width of about 500 MeV is consistent with their experimental resolution. The situation is analogous to that of the J/ψ and suggests that the Upsilon is a bound state of a new heavy quark-antiquark pair [2]. We report on an experiment, which has observed the formation of the Upsilon, using the PLUTO detector at the electron positron storage ring DORIS. It sets a substantially lower upper limit on the total width. From our measurement of the Upsilon production cross section we obtain Γ_{ee} — its electronic partial width. Γ_{ee} in turn may be used to determine the charge of the constituent quark-antiquark pair.

The energy of the storage ring DORIS has recently

---

[1] On leave from Tel-Aviv University, Israel.
[2] Now at Max-Planck-Institut für Physik und Astrophysik, Munchen, Germany.
[3] On leave from Humboldt University, Arcata, CA., USA.
[4] On leave from Purdue University, W. Lafayette In., USA.

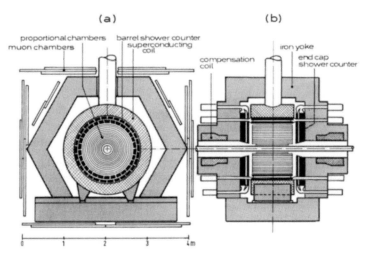

Fig. 1. Schematic views of the PLUTO detector. Sections (a) perpendicular to and (b) containing the beam axis.

been upgraded to make possible measurements for center of mass energies up to 10 GeV. In order to achieve this goal the machine was modified to operate in a single ring, single bunch mode. The detector PLUTO [3], see fig. 1, has also been improved by installing a cylindrical array (barrel) of shower counters (8.6 radiation lengths) and proportional tubes and by covering the ends of the cylindrical detector by a second set of shower counters (10.5 radiation lengths) and proportional wire chambers. These improvements bring the total gamma and electron coverage to approximately 94% of 4π. Cylindrical proportional wire chambers covering 92% of 4π and operating in a field of 1.69 T are used for track recognition and momentum measurement. The detector was triggered either by the presence of tracks in the wire chambers, by sufficient detected energy in the shower counters, or by a combination of the two. The luminosity was monitored by a set of shower counter telescopes which record Bhabha scatters at an angle to the beam of about 7°. We find good agreement between this monitor and the rate of large angle Bhabhas observed in the barrel shower counters. Reference points were taken at center of mass energies of 9.20 GeV and 9.30 GeV. The resonance search was made from 9.35 GeV upwards in steps of either 5 or 10 MeV. The average integrated luminosity per 10 MeV was $\sim$20 nb^{-1}.

In order to obtain substantial background reduction and fast feedback during the energy scan we analysed in the first off-line pass only those events with energy equivalent in the shower counters greater than 2 GeV. The bulk of the data reported here are subject to this restriction. Cosmic ray background was further reduced by making use of the bunch crossing time.

To obtain the total cross section for hadron production, we selected events with at least two charged tracks having a vertex within the prescribed interaction region. QED events were removed by the combined usage of a coplanarity cut and a shower recognition algorithm. To remove beam gas interactions cuts were imposed on the total visible energy and the missing mass in the final state. Remaining beam gas events were removed if they showed a substantial excess of positive charge. The effect of the energy cut was studied by analysing a subsample completely. We estimate that 51% of the hadronic events are being detected and analysed in the first off-line step.

Fig. 2 shows the total cross section for hadron production as a function of the center of mass energy. In addition to the statistical errors, shown in the figure, there is an overall estimated systematic uncertainty of 20%. Radiative corrections have not been applied. A resonance is seen in the region of 9.5 GeV. Fitting a gaussian to the peak on a 1/s background we obtain a

244

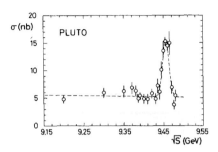

Fig. 2. Total cross section for hadron production in e^+e^--annihilation as a function of center of mass energy. There is an additional systematic error (not shown) of 20%. Contributions from the heavy lepton are included.

mass of 9.46 ± 0.01 GeV. The error in mass comes from the 0.1% uncertainty in the absolute calibration of DORIS energy. Our mass value is in agreement with the value of 9.40 GeV measured by Innes et al. [1], when their quoted systematic uncertainty of less than 1% is considered [4]. The gaussian width σ is 7.8 ± 0.9 MeV which is consistent with the theoretically estimated energy spread [5] of 8 MeV in the storage ring. The actual resonance width is therefore less than 18 MeV (FWHM). This adds weight to the supposition that the Upsilon is a bound state of a new quark-antiquark pair.

The relation between the integral over the cross section for resonant hadronic production σ_h, the resonance mass M_R, the electronic width Γ_{ee}, the hadronic width Γ_h, and the total width Γ_{tot} is given by

$$\int \sigma_h \, dM = \frac{6\pi^2}{M_R^2} \frac{\Gamma_{ee} \Gamma_h}{\Gamma_{tot}}.$$

On the standard assumption that the hadronic width dominates the total width, ($\Gamma_h \approx \Gamma_{tot}$) one obtains Γ_{ee} directly from the measured integral and the mass. Our result is Γ_{ee} = 0.8 ± 0.2 keV. An estimate of the radiative corrections raises this to 1.3 ± 0.4 keV. Models for quark binding in nonrelativistic potentials relate Γ_{ee} to the charge of the constituent quarks. In the standard model [6] our measurement implies a charge of $-1/3$ for the new heavy quark.

In summary we have observed the formation of a high mass, narrow resonance in e^+e^- annihilation at DORIS. We determine the mass to be (9.46 ± 0.01) GeV and therefore associate it with the Upsilon resonance seen by Herb et al. [1]. We observe a gaussian width of 7.8 MeV which is consistent with the energy resolution of the storage ring and supports the interpretation that the Upsilon is a bound state of a new heavy quark-antiquark pair. Our determination of Γ_{ee} as 1.3 ± 0.4 keV favors an assignment of $-1/3$ for the charge of the new quark.

We are grateful to D. Degèle and his colleagues at DORIS, H. Gerke, K. Holm, R.D. Kohaupt, G. Mühlhaupt, H. Nesemann, S. Pätzold, A. Piwinski, R. Rossmanith, K. Wille and A. Wrulich, whose outstanding efforts made these measurements possible. We also wish to thank C. Brown, W. Innes and L. Lederman for additional information about systematics of their Upsilon mass determination. We are indebted to all the service groups which supported the experiment, namely the computer center, the synchrotron staff, the gas supply group and the vacuum group. Our special thanks go to our technicians, those from DESY, Hamburg University and Siegen who have constructed most of the detector parts and took care of it during running times, and those from the cryogenic group who have maintained the superconducting operation during all the years. The non-DESY members of the PLUTO group want to thank the directorium for support and hospitality extended to them.

References

[1] S.W. Herb et al., Phys. Rev. Lett. 39 (1977) 252;
    W.R. Innes et al., Phys. Rev. Lett. 39 (1977) 1240.
[2] T. Appelquist and H.D. Politzer, Phys. Rev. Lett. 34 (1975) 43.
[3] PLUTO Collaboration, J. Burmester et al., Phys. Lett. 66B (1977) 395;
    A. Bäcker, Thesis, Internal Report DESY F33-77/03, December 1978, unpublished.
[4] Their latest estimate for the most probable mass is now 9.45 GeV. B.C. Brown, W.R. Innes and L.M. Ledermann, private communication.
[5] D. Degèle, private communication.
[6] K. Gottfried, Proc. Intern. Symposium on Lepton and photon interactions at high energies, Hamburg (1977) p. 667.

OBSERVATION OF A NARROW RESONANCE AT 10.02 GeV IN e^+e^- ANNIHILATIONS

J.K. BIENLEIN, E. HÖRBER [1], M. LEISSNER, B. NICZYPORUK [2], C. RIPPICH, M. SCHMITZ and H. VOGEL [3]
DESY, Hamburg, Germany

U. GLAWE, F.H. HEIMLICH, P. LEZOCH and U. STROHBUSCH
I. Institut für Experimentalphysik, Hamburg, Germany

P. BOCK, G. HEINZELMANN and B. PIETRZYK
Physikalisches Institut der Universität, Heidelberg, Germany

and

G. BLANAR, W. BLUM, H. DIETL, E. LORENZ and R. RICHTER
Max-Planck-Institut für Physik, Munich, Germany

Received 5 September 1978

The Υ' state has been observed as a narrow resonance at $M(\Upsilon') = 10.02 \pm 0.02$ GeV in e^+e^- annihilations, using a NaI and lead-glass detector in the DORIS storage ring at DESY. The ratio $\Gamma_{ee}\Gamma_{had}/\Gamma_{tot}$ of electronic, hadronic, and total widths has been measured to be 0.32 ± 0.13 keV. The parameters of the Υ particle have also been determined to be $M(\Upsilon) = 9.46 \pm 0.01$ and $\Gamma_{ee}\Gamma_{had}/\Gamma_{tot} = 1.04 \pm 0.28$ keV. The mass difference is $M(\Upsilon') - M(\Upsilon) = 0.56 \pm 0.01$ GeV.

The two massive particles $\Upsilon(9.4)$ and $\Upsilon'(10.0)$ discovered by Herb et al. [1] were produced in 400 GeV proton nucleus collisions and were observed in their $\mu^+\mu^-$ decay with a mass resolution of about 200 MeV (rms). It was thought that they could be bound states of a quark–antiquark pair in analogy to the $J/\Psi(3.1)$ and $\Psi'(3.7)$ states, but composed of a new type of quark. In this context it is essential to establish a narrow width, and to measure the mass difference accurately. In the framework of such quarkonium models [2], the charge of the new quark is related to the electronic partial widths Γ_{ee} which are, therefore, of considerable interest. These quantities can most conveniently be determined if the Υ resonances are formed in e^+e^- annihilations, a process expected to occur since they decayed into $\mu^+\mu^-$.

This had prompted the efforts of the DORIS machine group to extend the energy range of the storage ring beyond its original limits. DORIS was modified to operate as a single-ring single-bunch machine, and in April 1978 it had reached energies up to 9.6 GeV. Using this modified machine the existence of the Υ resonance was confirmed by the PLUTO [3] and DASP II [4] groups. The mass of the resonance was found to be 9.46 ± 0.01 GeV, and $\Gamma_{ee}\Gamma_{had}/\Gamma_{tot}$ was measured to be 1.3 ± 0.4 keV. In July 1978 additional cavities were installed in the DORIS ring, which allowed a search for the Υ' resonance expected above 10 GeV.

Here we present the results of measurements performed with the DESY–Heidelberg NaI and lead-glass detector in June and August 1978. During the first period the Υ resonance was measured, whereas during the second period the 10 GeV region was scanned and the Υ' resonance observed.

---

[1] On leave from: Physikalisches Institut der Universität, Würzburg, Germany.
[2] On leave from: Institute of Nuclear Physics, Cracow, Poland.
[3] On leave from: Physikalisches Institut der Universität, Erlangen, Germany.

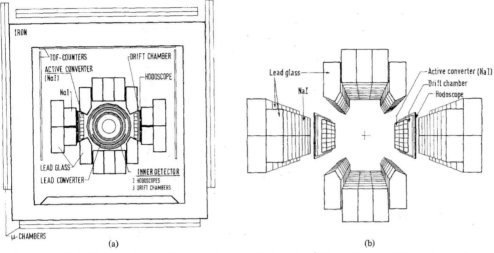

Fig. 1. Layout of the NaI lead-glass detector: (a) cross section perpendicular to the e^+e^- beam directions; (b) "exploded view" of the arrangement of energy detectors.

The layout of the apparatus is shown in fig. 1. This non-magnetic detector was described in detail in ref. [5]. For the measurements presented here, use has been made of the following parts: an inner detector measures the directions of charged particles; it has three cylindrical double-drift chambers and two scintillation hodoscopes. An outer box of energy counters (NaI and lead glass) determines the energy and direction of photons and electrons originating from the interaction point. The side walls have a total thickness of 15.2 X_0 (radiation lengths). The lead glass at top and bottom has 12.7 X_0. The lead converter between the second and third drift chamber of the inner detector has a thickness of 1 X_0. Minimum ionizing particles deposit about 200 MeV energy. The inner detector, as well as the energy counters, cover a solid angle of 86% of 4π.

The trigger consisted of several combinations of charged track multiplicities and a minimum total deposited energy as described in ref. [5]. An on-line filter eliminated beam–gas interactions that were easily recognized by their oblique incidence in the chambers. In the data reduction the following cuts were applied in order to isolate events of the type $e^+e^- \to$ hadrons: more than 1.8 GeV seen in the energy detector; at least three charged tracks recognized; at least 10% of the energy seen in the energy detector correlated with charged tracks. In a hand-scan, all events were eliminated which had drift chamber timing information inconsistent with the geometry of beam-beam interactions. Table 1 shows the breakdown of the events as they pass through the filtering procedure.

The "visible cross section" σ_{vis} was obtained by dividing the number of observed hadronic events by the time-integrated luminosity measured with the large-angle ($\alpha > 36°$) Bhabha events in the same apparatus for each energy point. The luminosity was also measured in a set of four counter telescopes for Bhabha scattering under 7°. Both measurements agreed within ±10%.

Fig. 2 shows how σ_{vis} depends on the centre-of-mass energy $\sqrt{s}$. The Υ is seen near 9.46 GeV above a

Table 1
Number of events in the filtering procedure and luminosity.

	Υ region	Υ' region
Triggers	3.4×10^6	5.1×10^6
Left after on-line filter	4.6×10^5	9.1×10^5
Left after off-line filter	2500	770
Left after hand-scan	1200	420
Integrated luminosity (nb^{-1})	173	120

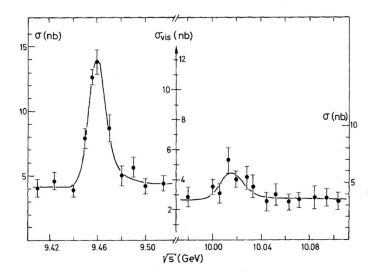

Fig. 2. Observed cross section $\sigma_{vis}(e^+e^- \to$ hadrons) in the Υ and Υ' regions (centre scale common to both measurements). The two outer scales represent the data after normalization to the expected level of continuum based on $R = 4.7$ (see text). The two outer scales are different by $\approx 10\%$, reflecting the systematic errors between the independent normalizations of the two measurements.

nearly constant continuum of 3.7 nb, and the Υ' resonance is seen for the first time to be formed in e^+e^- annihilations, above a continuum of 2.7 nb. We note that the Υ' signal is also very clearly visible in the event sample defined by the off-line filter.

The energy dependence of σ_{vis} was fitted around each resonance by a radiatively corrected gaussian resolution function (according to ref. [6]) over a background proportional to $1/s$. At the present time we are not able to calculate the exact amount of losses of hadronic events due to our filtering procedure. They are probably between 10% and 30% and are roughly independent of s. The visible cross section was normalized to 4.16 nb at 9.4 GeV and to 3.68 nb at 10.0 GeV to correspond to a value of $R = \sigma_{had}/\sigma_{\mu\mu}$ of 4.2 derived from $R(\sqrt{s} = 5$ GeV$) = 4.7$ [7] minus 0.5 units for undetected heavy-lepton decays due to our event selection. R was assumed to be energy independent between 5 and 10 GeV.

We foudn the resonances at masses of $M(\Upsilon) = 9.46 \pm 0.01$ GeV, $M(\Upsilon') = 10.02 \pm 0.02$ GeV; $\Delta M = 0.56 \pm 0.01$ GeV where the errors are from the machine [8]. The mass difference is only 29 ± 10 MeV less than the one between J/Ψ and Ψ', contrary to expectations based on universal potentials, "Coulomb + linear" [2], for the charmed and the new quarks.

The observed widths were found to be $\sigma(\Upsilon) = (7.1 \pm 0.8)$ MeV and $\sigma(\Upsilon') = (12 \pm 4)$ MeV. These are consistent with the energy resolution of the storage ring calculated to be $\sigma(9.5) = (7.8 \pm 0.8)$ MeV at the Υ and $\sigma(10.0) = (8.7 \pm 0.9)$ MeV at the Υ'. The machine value was used in the following determination of the area under the Υ'.

The areas under the normalized resonance curves were $A(\Upsilon) = 208 \pm 25$ MeV nb and $A(\Upsilon') = 59 \pm 15$ MeV nb, which turned into $A_0(\Upsilon) = 267 \pm 32$ MeV nb and $A_0(\Upsilon') = 74 \pm 19$ MeV nb after radiative corrections. We assume the Υ and Υ' are $J^P = 1^-$ objects, for which the ratio of electronic, hadronic, and total widths is

$$\Gamma_{ee}\Gamma_{had}/\Gamma_{tot} = (E_{res}^2/6\pi^2)A_0 \approx \Gamma_{ee}.$$

We obtained:

$$\Gamma_{ee}\Gamma_{had}/\Gamma_{tot} = (1.04 \pm 0.28) \text{ keV for the } \Upsilon,$$
$$(0.32 \pm 0.13) \text{ keV for the } \Upsilon',$$

thus confirming the earlier measurement of the Υ [3,4], whereas the Υ'-value is new. The errors contain a statistical contribution of ±0.13 keV (Υ) and ±0.08 keV (Υ'), and a common systematic contribution of ±15%. The ratio 0.32/1.04 is, therefore, known with a precision of ±28%. In the quarkonium picture [2] such values of Γ_{ee} favour the assignment of charge 1/3 to the new quark. Our small $\Gamma_{ee}(\Upsilon')$ seems to exclude a charge of 2/3 [9].

This work would not have been possible without the successful operation of the DORIS storage ring, which the machine group was able to run far beyond its original specifications. We are very grateful to the old DESY—Heidelberg group, who built the experiment and let us use their software. We are indebted to all the service groups who supported the experiment, i.e. the computer centre, the synchrotron staff, and the vacuum group as well as our technicians. The non-DESY members of our collaboration want to thank the DESY directorate for their hospitality.

References

[1] S.W. Herb et al., Phys. Rev. Lett. 39 (1977) 252;
    W.R. Innes et al., Phys. Rev. Lett. 39 (1977) 1240.
[2] K. Gottfried, Proc. Intern. Symp. on Lepton and photon interactions at high energies (Hamburg, 1977) ed. F. Gutbrod (DESY, Hamburg, 1977) p. 667.
[3] Ch. Berger et al., Phys. Lett. 76B (1978) 243.
[4] C.W. Darden et al., Phys. Lett. 76B (1978) 246.
[5] W. Bartel et al., Phys. Lett. 66B (1976) 483; 77B (1978) 331.
[6] J.D. Jackson and D.L. Scharre, Nucl. Instr. Meth. 128 (1975) 13.
[7] G. Knies, Proc. Intern. Symp. on Lepton and photon interactions at high energies (Hamburg, 1977) ed. F. Gutbrod (DESY, Hamburg, 1977) p. 93.
[8] D. Degèle of DORIS, private communication.
[9] J.L. Rosner et al., Phys. Lett. 74B (1978) 350.

Observation of the Υ''' at the Cornell Electron Storage Ring

G. Finocchiaro, G. Giannini, J. Lee-Franzini, R. D. Schamberger, Jr., M. Sivertz,
L. J. Spencer, and P. M. Tuts
The State University of New York at Stony Brook, Stony Brook, New York 11794

and

T. Böhringer, F. Costantini,[a] J. Dobbins, P. Franzini, K. Han, S. W. Herb, D. M. Kaplan,
L. M. Lederman,[b] G. Mageras, D. Peterson, E. Rice, and J. K. Yoh
Columbia University, New York, New York 10027

and

G. Levman
Louisiana State University, Baton Rouge, Louisiana 70803
(Received 21 April 1980)

During an energy scan at the Cornell Electron Storage Ring, with use of the Columbia University–Stony Brook NaI detector, an enhancement in $\sigma(e^+e^- \to \text{hadrons})$ is observed at center-of-mass energy ~ 10.55 GeV. The mass and leptonic width of this state (Υ''') suggest that it is the 4^3S_1 bound state of the b quark and its antiquark. After applying to the data a cut in a (pseudo) thrust variable, the natural width is measured to be $\Gamma = 12.6 \pm 6.0$ MeV, indicating that the Υ''' is above the threshold for $B\bar{B}$ production.

PACS numbers: 13.65.+i, 14.40.Pe

In the quarkonium model,[1] vector mesons are considered to be triplet S states of a quark-antiquark system in a "Coulombic-plus-confining" potential, with the number of quasistable radial excited states increasing with the mass of the quark. In particular, for the b quark ($M \sim 5$ GeV) the 1^3S_1, 2^3S_1, and 3^3S_1 states (Υ, Υ', and Υ'') have been observed as narrow enhancements in $\sigma(e^+e^- \to \text{hadrons})$,[2-4] as well as in proton-nucleon scattering.[5] A 4^3S_1 state should also exist, with an excitation energy of ~ 1.15 GeV. The 4^3S_1 state is expected to be close to the threshold for $B\bar{B}$ production,[6] where B is a pseudoscalar bound system of b and $\bar{u}$ or $\bar{d}$ quarks. If the 4^3S_1 state lies below the $B\bar{B}$ threshold, its natural width would be well below 1 whereas if it lies above the $B\bar{B}$ threshold, the opening up of decay channels would result in a natural width which increases rapidly with $M(4^3S_1) - 2M(B)$.[7]

We report here on the production of a new state in the Υ family, the Υ''', which we identify, from a measurement of its leptonic width Γ_{ee}, with the 4^3S_1 $b\bar{b}$ state. In addition, we observe a total width $\Gamma_{\text{obs}}(4^3S_1) \sim 19$ MeV; unfolding the contribution ($\Gamma \sim 10.8$ MeV) due to the Cornell Electron Storage Ring (CESR) beam-energy spread gives a natural width $\Gamma \sim 12.6$ MeV. This suggests that this state is above the $B\bar{B}$ threshold. We also obtain the (pseudo) thrust distribution for this new state and compare it with the corresponding distribution obtained for the continuum and for the $\Upsilon(1^3S_1)$.

These results were obtained in a twenty-day run covering the energy range 10.46 to 10.60 GeV, with an integrated luminosity of 1100 nb^{-1}. We had seen preliminary evidence for the new state during a run covering 10.55 to 10.80 GeV with a total integrated luminosity of 400 nb^{-1}. During the present run, we also collected data at the $1^3S_1(\Upsilon)$ with 300 nb^{-1} and at the $3^3S_1(\Upsilon'')$ with 150 nb^{-1}.

The principle of the Columbia University–Stony Brook layered NaI detector has been described in our previous Letter,[4] in which we reported the first measurements of the Υ, Υ', and Υ'' at CESR. For the present run, the complete NaI array was available, and for half of the run, in three of the four quadrants, drift chambers between the beam pipe and the NaI array were in operation. The NaI array consists of five radial layers, each subdivided into two polar halves and 32 azimuthal sectors (see Fig. 1). In contrast to its earlier configuration, our detector now covers the polar angle range $45° < \theta < 135°$ and a solid angle of approximately two-thirds of 4π, with each of the 64 sectors covering $\sim 1\%$ of 4π. In the following, we consider the detector as being composed of eight octants, four with $90° < \theta < 135°$ (West) and four with $45° < \theta < 90°$ (East), each spanning $\Delta\varphi$ intervals of $90°$.

The operation of the detector was also described in Ref. 4. All NaI signals are integrated every

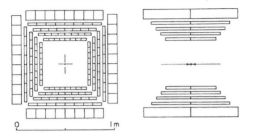

FIG. 1. A front view and a cut-away side view of the NaI array. An error bar shows the length of the interaction region.

machine cycle (2.56 μs) but are digitized only if a trigger is present. Only a single total-energy trigger is used, requiring ≥ 700 MeV to be deposited in the outer three layers of the NaI array. Hadronic and large-angle Bhabha-scattering yields in the detector are monitored on line by counting on scalers events satisfying appropriate criteria. The hadronic criterion requires that the total energy deposited in the outer three NaI layers, E_{NaI}, be in the range 1.2 GeV $< E_{NaI} <$ 3.6 GeV, that two or more octants in each half have ≥ 100 MeV deposited in the outer four layers, and that two such octants be collinear with the interaction point. The hadron monitor sensitivity is such that the Υ and Υ'' signals are clearly visible in a few hours' run (~ 15 nb^{-1}); this monitor also provided an on-line indication of the Υ'''. The Bhabha-scattering criterion requires that $E_{NaI} >$ 5.4 GeV, and that two collinear octants each have ≥ 100 MeV in the outer four layers. These on-line "Bhabhas" comprise approximately 80% of those found later in off-line analysis and serve as an on-line consistency check of the beam luminosity measurement obtained from the luminosity counters (which detect small-angle Bhabha scatterings). Additional electronic criteria were used to monitor and veto poor beam conditions. Typically we write 150 events on tape per inverse nanobarn of integrated luminosity, of which, at the Υ, about 15% are hadronic events and 10% are large-angle Bhabha scattering events. The live time of our detection system is 99.0%.

"$e^+e^- \to$ hadrons" events are recognized clearly in our detector. Photons from π^0 and η decay are detected as electromagnetic showers. Charged particles which do not shower electromagnetically nevertheless leave a clear signature since the five layers of NaI give five independent measurements of dE/dx. Recognition of "tracks" left in the NaI by minimum-ionizing particles is central to our hadronic-event-selection algorithms. Events containing at least one "track" pointing towards the beam and some additional energy deposition are hadronic candidates. Additional criteria imposed at succeeding levels of analysis include requirements that additional "tracks" or showers be present, and that energy be deposited in both the East and West halves of the detector.

Several independent hadronic-event-selection computer algorithms were developed. To guide us in this process, one or more physicists have examined over 95% of the hadronic-event candidates in the Υ''' energy scan. The various algorithms have efficiencies for continuum events of from 60 to 75% and background contaminations between 1 and 5%. The inefficiencies include loss of events due to detector solid angle. The background estimates are obtained from single-beam runs, and from reconstructed vertex position for those events having drift-chamber information. While the numbers of events found by the various algorithms and the on-line hadron monitor differ, the determinations of masses, widths, and relative cross sections of the resonances are in good agreement.

The present run yielded 5000 hadronic events at the Υ, 450 at the Υ'', and 3000 in the region around the Υ''', for a total 1550 nb^{-1} of integrated luminosity. The data are shown in Fig. 2(a). They have been normalized with use of the measured small-angle and large-angle Bhabha yields. Comparison of our observed cross sections with those measured at DORIS[2] and by CLEO[3] indicates that the product of acceptance by efficiency for our detector is approximately 73% for the continuum and 82% for the Υ.

A peak in the cross section is visible at a mass of 10.55 GeV. As has been shown in the DORIS experiments,[8] the spatial distributions of the continuum and Υ decay events are different. We choose here to test this difference using a simplified thrust variable, T', defined as the maximum of $\sum |\vec{E}_{NaI} \cdot \hat{n}|/\sum E_{NaI}$ over all possible $\hat{n}$ perpendicular to the beam axis. The distributions of T' for Υ and continuum events are given in Fig. 3. The difference between the two cases is quite evident and agrees with the conjecture that continuum events have a two-jet-like structure while resonance decays have a more spherical distribution. A cut at $T' <$ 0.85 has been made for all the data and the result is shown in Fig. 2(b).

223

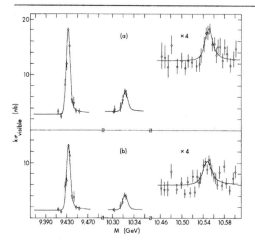
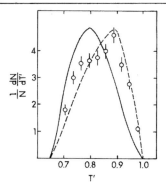

FIG. 2. (a) The observed cross section for $e^+e^- \to$ hadrons multiplied by $k = (M/M_\Upsilon)^2$. M is the e^+e^- invariant mass. (b) The same cross section after removing events with $T \geq 0.85$. The lines are fitted to the data including machine energy spread and radiative corrections. See text for explanation.

FIG. 3. Solid line, pseudothrust (see text) distribution for events in the Υ region; dashed line, distribution for continuum events. Data points are for events from the Υ''' region, showing contributions from both distributions.

The cut removes 52% of the continuum events but only 26% of resonance events.

We determine the parameters of the Υ''' by fitting the data sample with $T' < 0.85$; the uncut data give similar results but with less statistical significance. A fit to a constant continuum plus a Gaussian with radiative corrections gives an apparent machine energy spread of $\Gamma = 19 \pm 4$ MeV. We have also fitted the parameters of the Υ using our data from this running period, cut on $T' < 0.85$. This gives a full width at half maximum (FWHM) machine energy spread at the Υ of 8.7 ± 0.7 MeV. Scaling this by the expected $(E_{\text{beam}})^2$ dependence yields a FWHM at the Υ''' of $\Gamma = 10.8 \pm 0.9$ MeV. This is inconsistent with the observed result by 2 standard deviations. We therefore assume a Breit-Wigner, rather than Gaussian, resonance shape for the enhancement, fold in the machine energy spread and radiative corrections, and fit the Υ''' data with the resulting curve. The mass values are calculated from the CESR energy calibration, which gives a mass for the Υ 0.3% below the DORIS values.[2-4] The mass difference is $M(\Upsilon''') - M(\Upsilon) = 1114 \pm 2$ MeV with systematic uncertainty of 5 MeV. The ratio of leptonic widths calculated from the fitted areas is $\Gamma_{ee}(\Upsilon''')/\Gamma_{ee}(\Upsilon) = 0.25 \pm 0.07$. Both the mass difference and the ratio of leptonic widths are in excellent agreement with many phenomenological calculations for the 4^3S_1 state of $b\bar{b}$.[1,6,9] Therefore, we conclude that the enhancement observed at $M = 10.547$ GeV is most likely that state.

The natural width of the enhancement is also of great interest. Our fit gives a natural width $\Gamma = 12.6 \pm 6.0$ MeV. If we constrain the natural width to be much smaller than the machine energy spread, χ^2 increased by 8.3, from 40.3 for thirty degrees of freedom to 48.6 for 29 degrees of freedom. Thus, our value for the natural width is inconsistent with the expected width of less than 1 MeV for a resonance below $B\bar{B}$ threshold. A similar result has been obtained by the CLEO collaboration at CESR.[10] This implies that Υ''' is above threshold and that the mass of the B is less than 5.275 GeV. It also implies a production rate for B mesons which is greatly enhanced above the level in the neighboring continuum. If this is confirmed, the study of the Υ''' events should contribute enormously toward our understanding of the B meson.

We gratefully acknowledge the assistance of the CESR operating staff and again thank the people who assisted in the design, construction, and installation of our detector. One of us (S.W.H.) acknowledges the support of the Sloan Foundation. This research was supported in part by the National Science Foundation, and in part by the U. S. Department of Energy.

---

[a] On leave from University of Pisa, I-56100 Pisa,

Italy, and Istituto Nazionale di Fisica Nucleare, I-56100, Pisa, Italy.
[(b)]Also at Fermilab, Batavia, Ill. 60510.

[1]C. Quigg and J. L. Rosner, Phys. Rep. **56**, 169 (1979).

[2]Ch. Berger et al., Phys. Lett. **76B**, 243 (1978); C. W. Darden et al., Phys. Lett. **76B**, 246 (1978); J. K. Bienlein et al., Phys. Lett. **78B**, 360 (1978); C. W. Darden et al., Phys. Lett. **78B**, 364 (1978).

[3]D. Andrews et al., Phys. Rev. Lett. **44**, 1108 (1980).

[4]T. Böhringer et al., Phys. Rev. Lett. **44**, 1111 (1980).

[5]S. W. Herb et al., Phys. Rev. Lett. **39**, 252 (1977); W. R. Innes et al., Phys. Rev. Lett. **39**, 1240, 1640(E) (1977); K. Ueno et al., Phys. Rev. Lett. **42**, 486 (1979).

[6]E. Eichten et al., Phys. Rev. D **17**, 3090 (1978), and **21**, 203 (1980).

[7]An example is the $\psi(3770)$. See P. A. Rapidis et al., Phys. Rev. Lett. **39**, 526 (1977).

[8]Ch. Berger et al., Phys. Lett. **78B**, 176 (1978), and **82B**, 449 (1979); F. H. Heimlich et al., Phys. Lett. **86B**, 399 (1979).

[9]G. Bhanot and S. Rudaz, Phys. Lett. **78B**, 119 (1978).

[10]D. Andrews et al., preceding Letter [Phys. Rev. Lett. **45**, 219 (1980)].

Observation of Exclusive Decay Modes of b-Flavored Mesons

S. Behrends, K. Chadwick, J. Chauveau,[a] P. Ganci, T. Gentile, Jan M. Guida, Joan A. Guida,
R. Kass, A. C. Melissinos, S. L. Olsen, G. Parkhurst, D. Peterson, R. Poling,
C. Rosenfeld, G. Rucinski, and E. H. Thorndike
University of Rochester, Rochester, New York 14627

and

J. Green, R. G. Hicks, F. Sannes, P. Skubic,[b] A. Snyder, and R. Stone
Rutgers University, New Brunswick, New Jersey 08854

and

A. Chen, M. Goldberg, N. Horwitz, A. Jawahery, M. Jibaly, P. Lipari, G. C. Moneti,
C. G. Trahern, and H. van Hecke
Syracuse University, Syracuse, New York 13210

and

M. S. Alam, S. E. Csorna, L. Garren, M. D. Mestayer, and R. S. Panvini
Vanderbilt University, Nashville, Tennessee 37235

and

D. Andrews,[c] P. Avery, C. Bebek, K. Berkelman, D. G. Cassel, J. W. DeWire, R. Ehrlich,
T. Ferguson, R. Galik, M. G. D. Gilchriese, B. Gittelman, M. Halling, D. L. Hartill,
D. Herrup,[d] S. Holzner, M. Ito, J. Kandaswamy, V. Kistiakowsky,[e]
D. L. Kreinick, Y. Kubota, N. B. Mistry, F. Morrow, E. Nordberg,
M. Ogg, R. Perchonok, R. Plunkett,[f] A. Silverman, P. C. Stein,
S. Stone, R. Talman, D. Weber, and R. Wilcke
Cornell University, Ithaca, New York 14853

and

A. J. Sadoff
Ithaca College, Ithaca, New York 14850

and

R. Giles, J. Hassard, M. Hempstead, J. M. Izen,[g] K. Kinoshita, W. W. MacKay,
F. M. Pipkin, J. Rohlf, and Richard Wilson
Harvard University, Cambridge, Massachusetts 02138

and

H. Kagan
Ohio State University, Columbus, Ohio 43210
(Received 24 January 1983)

B-meson decays to final states consisting of a D^0 or $D^{*\pm}$ and one or two charged pions have been observed. The charged-B mass is $5270.8 \pm 2.3 \pm 2.0$ MeV and the neutral-B mass is $5274.2 \pm 1.9 \pm 2.0$ MeV.

PACS numbers: 14.40.Jz, 13.25.+m, 13.65.+i

The upsilon states[1] are interpreted as q-$\bar{q}$ resonances of a new quark, the b quark. The first three resonances are narrow,[2,3] implying bound b flavor and a suppressed strong decay. The large width of the $\Upsilon(4S)$ resonance discovered at the e^+e^- storage ring CESR[4] (Cornell Electron Storage Ring) and the observation that the decay products from the $\Upsilon(4S)$ include high-momentum leptons[5] imply that the $\Upsilon(4S)$ decays strongly into $B\bar{B}$ meson pairs, which then decay weakly. Until now, however, the b-flavored mesons themselves had not been found. Here we report that discovery.

The b quark has been shown to decay predominantly to the c quark.[6] Thus the principal decay mode of the B meson will be to a charmed meson plus pions. Since the high multiplicity in $\Upsilon(4S)$ decay[7] leads to large combinatorial background, we have restricted our search to low-multiplicity decay modes, D^0 or $D^{*\pm}$ plus one or two charged pions.

The data sample used is 40.7 pb^{-1} of $\Upsilon(4S)$ data and 19.6 pb^{-1} of continuum data taken with the CLEO detector at CESR. The $\Upsilon(4S)$ cross section is a 1.0-nb enhancement above a 2.5-nb continuum contribution. The detector has been described in detail elsewhere.[8] In this work we have used the cylindrical drift chamber inside a 1.0-T solenoid magnet to determine momenta of charged particles. In addition we have used the dE/dx-measuring wire proportional chambers and the time-of-flight scintillation counters located outside the solenoid magnet to identify charged kaons over a momentum range from 0.45 to 1.0 GeV/c.

The two-body decay modes, $D^0 \to K^-\pi^+$ and its charge conjugate, were used to find D^0 mesons. Identified kaons were paired with each oppositely charged particle in the event (assumed to be a pion). The combination was kept only if its momentum was below 2.6 GeV/c, since D^0's from B decay cannot exceed this momentum. The resulting $K^\pm \pi^\mp$ mass distribution is shown in Fig. 1(a). Mass combinations within ± 40 MeV of the D^0 mass were kept as D^0 candidates.

We looked for charged D^* mesons through the cascade $D^{*+} \to D^0 \pi^+$, $D^0 \to K^-\pi^+$ and its charge conjugate. We did not require that the charged kaon be identified as such. Rather we first formed mass combinations of all pairs of oppositely charged particles in an event, assuming that each particle in turn is a kaon. We then added an additional particle (assumed to be a pion) of charge opposite to that of the assumed kaon. We kept as D^* candidates the combinations for which the $[K\pi\pi, K\pi]$ mass difference was within ± 3.0 MeV of the $[D^{*\pm}, D^0]$ mass difference of 145.4 MeV,[9] and $K\pi$ masses within ± 80 MeV of the D^0 mass. We further required that the D^* candidate have momentum below 2.6 GeV/c, eliminating the high-momentum D^* contribution from the continuum. With these requirements the $D^{*\pm}$ signal is hidden under considerable background. By demanding a more restrictive set of conditions [see Fig. 1(b)], we demonstrate that the $\Upsilon(4S)$ decays contain a $D^{*\pm}$ signal. Because these latter restrictions lower

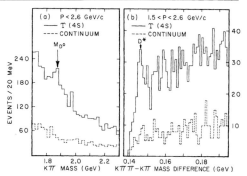

FIG. 1. (a) Mass distribution of $K^\pm \pi^\mp$, for $K\pi$ momenta below 2.6 GeV/c, using identified kaons. The solid line shows data from 40.7 pb^{-1} of $\Upsilon(4S)$ running; the dashed line is from 19.6 pb^{-1} of continuum running at energies just below the $\Upsilon(4S)$. A D^0 signal at 1.86 GeV is evident in the 4S data. (b) $(K^\pm \pi^\mp \pi^\pm) - (K^\pm \pi^\mp)$ mass-difference distribution, for $K\pi\pi$ momenta between 1.5 and 2.6 GeV/c and $K\pi$ masses within 20 MeV of the D^0 mass. Kaons were not directly identified. Curves are as in (a). The signal at the $D^{*\pm} - D^0$ mass difference (145.4 MeV) is evidence of $D^{*\pm}$ production in $\Upsilon(4S)$ decay.

the $D^{*\pm}$ detection efficiency they were not used in the search for B mesons.

Each event containing a D^0 or $D^{*\pm}$ candidate was fitted to the following hypotheses[10] (or their charge conjugates):

$$B^- \to D^0 \pi^-, \qquad (1)$$

$$\bar{B}^0 \to D^0 \pi^+ \pi^-, \qquad (2)$$

$$\bar{B}^0 \to D^{*+} \pi^-, \qquad (3)$$

$$B^- \to D^{*+} \pi^- \pi^-. \qquad (4)$$

We considered only these charge combinations, since they preserve the quark decay scheme $b \to c \to s$. In making the fit, we constrained the B-meson energy to the beam energy and constrained the D^0 or $D^{*\pm}$ decay products to the known D^0, $D^{*\pm}$ masses, respectively. This fitting procedure measures the B mass relative to the CESR beam energy, which is scaled to agree with the VEPP4 measurement of the $\Upsilon(1S)$ mass.[11] Since the threshold for $B\bar{B}$ production is known to lie between the $\Upsilon(3S)$ and $\Upsilon(4S)$ resonances, we considered B-meson mass combinations between half the $\Upsilon(3S)$ and $\Upsilon(4S)$ resonance masses (i.e., 5180 and 5290 MeV). Candidate fits were required to have a χ^2 value less than 14. If an event had two acceptable fits in this mass interval we took the hypothesis with the lower χ^2. Each successful fit was examined visually to reject B candidates involving incorrectly fitted drift-chamber tracks (a 15% rejection).

The B masses for all successful fits are shown in Fig. 2. The 18 events in the peak near 5275 MeV are divided 2, 5, 5, and 6 for reactions (1)–(4), respectively. The width of the mass peak is consistent with the resolution expected from Monte Carlo studies. We have estimated the background under the mass peak in several ways. (1) We changed our selection criteria to accept $K^{\mp}\pi^{\pm}$ mass combinations that differed from the D^0 mass by ± 200 MeV. The spectrum of reconstructed "B-meson" masses for this "sideband" search is shown in Fig. 3(a). (2) We considered wrong charge combinations, which corresponded to doubly charged B's, or corresponded to decay sequences other than $b \to c \to s$. The mass spectrum for wrong charge combinations is shown in Fig. 3(b). Both distributions in Fig. 3 have been normalized so that the vertical scales are directly comparable with Fig. 2. (3) We performed Monte Carlo studies of background from $B\bar{B}$ events and from continuum events, to determine how the background in Fig. 2 should be extrapolated from lower masses to the peak region. (4) We searched 19.6 pb^{-1} of data accumulated just below the $\Upsilon(4S)$ for apparent B's, finding two in the region of the mass peak. These studies lead to estimates of the background under the peak at 5275 MeV which lie between 4 and 7 events.

If a B decay contains a low-energy particle that escapes detection, the remaining particles from that B may still be consistent with the beam-energy constraint and give an acceptable fit. We frequently cannot distinguish reactions (1) and (2) from similar reactions with the D^0 replaced by D^{*0}, where $D^{*0} \to D^0\pi^0$ or $D^0\gamma$. Similarly, the decay $\bar{B}^0 \to D^{*+}\pi^-$, $D^{*+} \to D^0\pi^+$ (π^+ not detected), can masquerade as $B^- \to D^0\pi^-$, causing us to assign an incorrect charge to the B. Monte Carlo studies show that the reconstructed mass is shifted down a few megaelectronvolts from the true B mass, and the mass resolution is worsened slightly. The problem of missed low-energy particles is not important for reactions (3) and (4), and therefore we use only these to determine the B mass.

We find a mass of $5274.2 \pm 1.9 \pm 2.0$ MeV for the neutral B, and $5270.8 \pm 2.3 \pm 2.0$ MeV for the charged B, where the first error is statistical and the second error systematic. The $[\bar{B}^0, B^-]$ mass difference is $3.4 \pm 3.0 \pm 2.0$ MeV, consistent

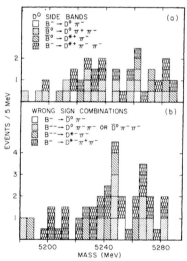

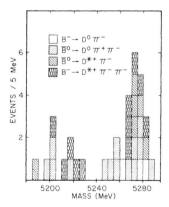

FIG. 2. Mass distribution of B-meson candidates. The $B \to$ final-state decay labels should be interpreted as including the charge-conjugate reaction.

FIG. 3. Mass distribution for two estimates of the background to the B-meson candidates of Fig. 2. (a) D^0's chosen from sidebands. The events shown are plotted with a weight of $\frac{1}{2}$ event, since there are approximately twice as many events in the sidebands as in the D region. (b) Wrong charge combinations. The events shown in this distribution have been scaled to account for the difference in the number of combinations leading to a wrong-sign B compared to those leading to the correct-sign charged B.

with the theoretical prediction[12] of 4.4 MeV. The average of charged and neutral B masses is 5272.3± 1.5± 2.0 MeV. This corresponds to a mass difference ΔM of 32.4± 3.0± 4.0 MeV between the mass of the $\Upsilon(4S)$ and twice the B-meson mass. If the $[B^*, B]$ mass difference is ~ 50 MeV as expected theoretically,[12] the $\Upsilon(4S)$ must decay exclusively to $B\bar{B}$, with no contribution from $B^*\bar{B}$. Previous experimental information on ΔM comes from the fact that Schamberger et al. do not observe monochromatic photons from $B^* \to \gamma B$ decay.[13] Their experiment sets an upper limit of 50 MeV on ΔM. Theoretical calculations of ΔM using the width and the height of the $\Upsilon(4S)$ fall either above[12] or below[14] our result. Using our measured value for ΔM and the theoretical value of 4.4 MeV for the $[\bar{B}^0, B^-]$ mass difference, we obtain the branching fractions $B(\Upsilon(4S) \to B^+B^-) = 0.60 \pm 0.02$ and $B(\Upsilon(4S) \to B^0\bar{B}^0) = 0.40 \pm 0.02$. We estimate branching ratios of 4.2± 4.2%, 13± 9%, 2.6± 1.9%, and 4.8± 3.0% for reactions (1)–(4), respectively.[15,16]

In conclusion, we have explicitly demonstrated the existence of the B meson through its decay into exclusive final states and have measured its mass.

We gratefully acknowledge the efforts of the CESR staff who made this work possible. We thank H. Tye for useful discussions. This work was supported in part by the National Science Foundation and the U. S. Department of Energy.

---

[a] Permanent address: Laboratoire de Physique Corpusculaire, College de France, F-75231 Paris, France.
[b] Present address: University of Oklahoma, Norman, Okla. 73019.
[c] Present address: AIRCO Superconductors, Carteret, N.J. 07008.
[d] Present address: Lawrence Berkeley Laboratory, Berkeley, California 94720.
[e] Permanent address: Massachusetts Institute of Technology, Cambridge, Mass. 02139.
[f] Present address: CERN, EP Division, CH-1211 Geneva 23, Switzerland.
[g] Present address: University of Wisconsin, Madison, Wisc. 53706.

[1] S. W. Herb et al., Phys. Rev. Lett. 39, 252 (1977).
[2] C. Berger et al., Phys. Lett. 76B, 243 (1978); C. W. Darden et al., Phys. Lett. 76B, 246 (1978), and 78B, 364 (1978); J. Bienlein et al., Phys. Lett. 78B, 360 (1978).
[3] D. Andrews et al., Phys. Rev. Lett. 44, 1108 (1980); T. Bohringer et al., Phys. Rev. Lett. 44, 222 (1980).
[4] D. Andrews et al., Phys. Rev. Lett. 45, 219 (1980); G. Finocchiaro et al., Phys. Rev. Lett. 45, 222 (1980).
[5] C. Bebek et al., Phys. Rev. Lett. 46, 84 (1981); K. Chadwick et al., Phys. Rev. Lett. 46, 88 (1981); L. J. Spencer et al., Phys. Rev. Lett. 47, 771 (1981).
[6] A. Brody et al., Phys. Rev. Lett. 48, 1070 (1982); D. Andrews et al., Cornell University Report No. CLNS 82/547, 1982 (to be published), p. 14; Spencer et al., Ref. 5.
[7] M. S. Alam et al., Phys. Rev. Lett. 49, 357 (1982).
[8] D. Andrews et al., in Cornell University Report No. CLNS 82/538 (to be published).
[9] M. Roos et al. (Particle Data Group), Phys. Lett. 111B, 1 (1982).
[10] Our notation for neutral B's follows the convention used for neutral kaons, i.e., $\bar{B}^0 = (b\bar{d})$, $B^0 = (\bar{b}d)$.
[11] A. S. Artamonov et al., Institute of Nuclear Physics, Academy of Science of the U.S.S.R., Novosibirsk Report No. 82-94 (to be published).
[12] E. Eichten, Phys. Rev. D 22, 1819 (1980).
[13] R. D. Schamberger et al., Phys. Rev. D 26, 720 (1982).
[14] M. Gronau et al., Phys. Rev. D 25, 3100 (1982); I. I. Bigi and S. Ono, Nucl. Phys. B189, 229 (1981).
[15] D^0 branching ratios are from R. H. Schindler et al., Phys. Rev. D 24, 78 (1981).
[16] As noted earlier, the branching ratio for reaction (1) includes a contamination from $B^- \to D^{*0}\pi^-$ and $\bar{B}^0 \to D^{*+}\pi^-$, while reaction (2) includes $\bar{B}^0 \to D^{*0}\pi^+\pi^-$.

Evidence for Penguin-Diagram Decays: First Observation of $B \to K^*(892)\gamma$

R. Ammar,[1] S. Ball,[1] P. Baringer,[1] D. Coppage,[1] N. Copty,[1] R. Davis,[1] N. Hancock,[1] M. Kelly,[1] N. Kwak,[1] H. Lam,[1] Y. Kubota,[2] M. Lattery,[2] J. K. Nelson,[2] S. Patton,[2] D. Perticone,[2] R. Poling,[2] V. Savinov,[2] S. Schrenk,[2] R. Wang,[2] M. S. Alam,[3] I. J. Kim,[3] B. Nemati,[3] J. J. O'Neill,[3] H. Severini,[3] C. R. Sun,[3] M. M. Zoeller,[3] G. Crawford,[4] M. Daubenmeir,[4] R. Fulton,[4] D. Fujino,[4] K. K. Gan,[4] K. Honscheid,[4] H. Kagan,[4] R. Kass,[4] J. Lee,[4] R. Malchow,[4] F. Morrow,[4] Y. Skovpen,[4,*] M. Sung,[4] C. White,[4] J. Whitmore,[4] P. Wilson,[4] F. Butler,[5] X. Fu,[5] G. Kalbfleisch,[5] M. Lambrecht,[5] W. R. Ross,[5] P. Skubic,[5] J. Snow,[5] P. L. Wang,[5] M. Wood,[5] D. Bortoletto,[6] D. N. Brown,[6] J. Fast,[6] R. L. McIlwain,[6] T. Miao,[6] D. H. Miller,[6] M. Modesitt,[6] S. F. Schaffner,[6] E. I. Shibata,[6] I. P. J. Shipsey,[6] P. N. Wang,[6] M. Battle,[7] J. Ernst,[7] H. Kroha,[7] S. Roberts,[7] K. Sparks,[7] E. H. Thorndike,[7] C. H. Wang,[7] J. Dominick,[8] S. Sanghera,[8] T. Skwarnicki,[8] R. Stroynowski,[8] M. Artuso,[9] D. He,[9] M. Goldberg,[9] N. Horwitz,[9] R. Kennett,[9] G. C. Moneti,[9] F. Muheim,[9] Y. Mukhin,[9] S. Playfer,[9] Y. Rozen,[9] S. Stone,[9] M. Thulasidas,[9] G. Vasseur,[9] G. Zhu,[9] J. Bartelt,[10] S. E. Csorna,[10] Z. Egyed,[10] V. Jain,[10] P. Sheldon,[10] D. S. Akerib,[11] B. Barish,[11] M. Chadha,[11] S. Chan,[11] D. F. Cowen,[11] G. Eigen,[11] J. S. Miller,[11] C. O'Grady,[11] J. Urheim,[11] A. J. Weinstein,[11] D. Acosta,[12] M. Athanas,[12] G. Masek,[12] B. Ong,[12] H. Paar,[12] M. Sivertz,[12] A. Bean,[13] J. Gronberg,[13] R. Kutschke,[13] S. Menary,[13] R. J. Morrison,[13] S. Nakanishi,[13] H. N. Nelson,[13] T. K. Nelson,[13] J. D. Richman,[13] A. Ryd,[13] H. Tajima,[13] D. Schmidt,[13] D. Sperka,[13] M. S. Witherell,[13] M. Procario,[14] S. Yang,[14] R. Balest,[15] K. Cho,[15] M. Daoudi,[15] W. T. Ford,[15] D. R. Johnson,[15] K. Lingel,[15] M. Lohner,[15] P. Rankin,[15] J. G. Smith,[15] J. P. Alexander,[16] C. Bebek,[16] K. Berkelman,[16] D. Besson,[16] T. E. Browder,[16] D. G. Cassel,[16] H. A. Cho,[16] D. M. Coffman,[16] P. S. Drell,[16] R. Ehrlich,[16] M. Garcia-Sciveres,[16] B. Geiser,[16] B. Gittelman,[16] S. W. Gray,[16] D. L. Hartill,[16] B. K. Heltsley,[16] C. D. Jones,[16] S. L. Jones,[16] J. Kandaswamy,[16] N. Katayama,[16] P. C. Kim,[16] D. L. Kreinick,[16] G. S. Ludwig,[16] J. Masui,[16] J. Mevissen,[16] N. B. Mistry,[16] C. R. Ng,[16] E. Nordberg,[16] M. Ogg,[16,†] J. R. Patterson,[16] D. Peterson,[16] D. Riley,[16] S. Salman,[16] M. Sapper,[16] H. Worden,[16] F. Würthwein,[16] P. Avery,[17] A. Freyberger,[17] J. Rodriguez,[17] R. Stephens,[17] J. Yelton,[17] D. Cinabro,[18] S. Henderson,[18] K. Kinoshita,[18] T. Liu,[18] M. Saulnier,[18] F. Shen,[18] R. Wilson,[18] H. Yamamoto,[18] M. Selen,[19] and A. J. Sadoff[20]

(CLEO Collaboration)

[1] *University of Kansas, Lawrence, Kansas 66045*
[2] *University of Minnesota, Minneapolis, Minnesota 55455*
[3] *State University of New York at Albany, Albany, New York 12222*
[4] *Ohio State University, Columbus, Ohio 43210*
[5] *University of Oklahoma, Norman, Oklahoma 73019*
[6] *Purdue University, West Lafayette, Indiana 47907*
[7] *University of Rochester, Rochester, New York 14627*
[8] *Southern Methodist University, Dallas, Texas 75275*
[9] *Syracuse University, Syracuse, New York 13244*
[10] *Vanderbilt University, Nashville, Tennessee 37235*
[11] *California Institute of Technology, Pasadena, California 91125*
[12] *University of California, San Diego, La Jolla, California 92093*
[13] *University of California, Santa Barbara, California 93106*
[14] *Carnegie-Mellon University, Pittsburgh, Pennsylvania 15213*
[15] *University of Colorado, Boulder, Colorado 80309-0390*
[16] *Cornell University, Ithaca, New York 14853*
[17] *University of Florida, Gainesville, Florida 32611*
[18] *Harvard University, Cambridge, Massachusetts 02138*
[19] *University of Illinois, Champaign-Urbana, Illinois, 61801*
[20] *Ithaca College, Ithaca, New York 14850*

(Received 24 May 1993)

We have observed the decays $B^0 \to K^*(892)^0 \gamma$ and $B^- \to K^*(892)^- \gamma$, which are evidence for the quark-level process $b \to s\gamma$. The average branching fraction is $(4.5 \pm 1.5 \pm 0.9) \times 10^{-5}$. This value is consistent with standard model predictions from electromagnetic penguin diagrams.

PACS numbers: 13.40.Hq, 14.40.Jz

One-loop, flavor-changing neutral current diagrams, known as penguins, were originally introduced into the theory of weak decays to explain the $\Delta I = \frac{1}{2}$ rule in K meson decays [1]. They were later identified as a possible source of direct CP violation in kaon decay, and hence as a contribution to ϵ'/ϵ [2]. Their importance in B meson decays has also been noted [3].

One of the clearest signatures for penguin diagrams is the radiative process $b \to s\gamma$ (Fig. 1). There are many calculations of the rate for this process, which depends on the as yet unknown mass of the top quark. After including substantial QCD corrections, the branching ratio for $b \to s\gamma$ is expected to be in the range $(2-4) \times 10^{-4}$ [4]. Other standard model contributions have been considered, and found to be at least an order of magnitude smaller than the penguin contribution [5]. Observation of a rate much larger than 4×10^{-4} would be evidence for nonstandard-model contributions. In the recent literature possible contributions to $b \to s\gamma$ from supersymmetry, a fourth generation, and a charged Higgs boson have been discussed in some detail [6].

The fraction of $b \to s\gamma$ decays that hadronize to any exclusive final state is much less reliably predicted than the inclusive rate. Estimates for the fraction of $B \to K^*(892)\gamma$ range from 5% to 40% [7]. In this Letter, we report observation of the decay $B \to K^*\gamma$ [8], in both $B^0 \to K^{*0}\gamma$ and $B^- \to K^{*-}\gamma$ modes, at a rate that is consistent with the predictions from the penguin diagram.

The data sample used in this study was collected with the CLEO-II detector [9] at the Cornell Electron Storage Ring (CESR). It consists of 1377 pb^{-1} of integrated luminosity on the $\Upsilon(4S)$ resonance (1.39×10^6 $B\bar{B}$ events) and 633 pb^{-1} at a center-of-mass energy 55 MeV below the resonance. Charged particles are tracked using three nested cylindrical wire chambers operated in a 1.5 T magnetic field. The outer tracking chamber also measures specific ionization (dE/dx), which is used for particle identification. The tracking chambers are followed by time-of-flight (TOF) counters, which provide additional particle identification information. Beyond these counters but still inside the superconducting magnet coil is an electromagnetic calorimeter consisting of 7800 CsI crystals. Outside the coil is an iron yoke for field return, with chambers interspersed for muon identification.

The dominant experimental problem in identifying $B \to K^*\gamma$ is the large background from continuum (non-$B\bar{B}$) processes. We suppressed this background with a series of cuts determined from Monte Carlo studies. For $B^0 \to K^{*0}\gamma$, we selected events with at least 3 charged tracks and a visible energy of at least 30% of the center-of-mass energy. Our high energy photon candidates are clusters with energy E_γ satisfying $2.1 < E_\gamma < 2.9$ GeV, with polar angles θ (relative to the beam axis) satisfying $|\cos\theta| < 0.7$, not matched to charged tracks, and with shower shapes consistent with single photons. Photon candidates are rejected if they form π^0's (η's) when combined with another photon of energy greater than 30 (200) MeV.

There are two main sources of high energy photons from the continuum: initial state radiation (ISR) and fragmentation of non-$b\bar{b}$ quarks ($q\bar{q}$). Most continuum processes have a two-jet topology, which is used to distinguish them from the more spherical $B\bar{B}$ events with B mesons decaying almost at rest. We suppress $q\bar{q}$ backgrounds by applying cuts on the shape variables R_2, the normalized second Fox-Wolfram moment [10], and $S_\perp$, a measure of the momentum transverse to the photon direction [11]. We require $R_2 < 0.5$ and $0.25 < S_\perp < 0.60$ (the upper restriction on $S_\perp$ provides rejection of ISR). To further suppress the ISR background we use variables evaluated in the rest frame of the e^+e^- following the radiation of the high energy photon (the primed frame). In this frame we require $R_2' < 0.3$, where R_2' is evaluated excluding the photon, and $|\cos\theta'| > 0.5$, where θ' is the angle between the photon and the thrust axis of the rest of the event.

We look for K^{*0} candidates in the decay mode $K^{*0} \to K^+\pi^-$. Each charged track must pass standard track quality cuts, and must have a value of dE/dx and/or TOF which is within 2.5 standard deviations (σ) of that expected for the mass hypothesis. Particles lacking both TOF and dE/dx information are considered to be pions, but not kaons. A $K^+\pi^-$ pair must have an invariant mass $M_{K\pi}$ satisfying $821 < M_{K\pi} < 971$ MeV, and a decay helicity angle $\theta_{K\pi}$ satisfying $|\cos\theta_{K\pi}| < 0.8$, since K^*'s from $B \to K^*\gamma$ decay with a $\sin^2\theta_{K\pi}$ distribution, whereas the background is expected to have a flat distribution in $\cos\theta_{K\pi}$.

We combined the high energy photon and K^* candidates to form candidates for $B^0 \to K^{*0}\gamma$. Having made all particle assignments, we imposed two further cuts: (1) the angle θ_{thr} between the high energy photon direction and the thrust axis of the particles not from the B candidate, $|\cos\theta_{thr}| < 0.7$ (expected to be flat for signal, peaked at 1.0 for continuum background); (2) the production polar angle of the B, $|\cos\theta_B| < 0.85$ (expected to be $\sin^2\theta_B$ for signal, flat for background). Finally, we required that the B candidates have an energy close to the beam energy, $|\Delta E| < 90$ MeV (2.2σ), where $\Delta E = E_{\text{beam}} - E_{K^*\gamma}$. For candidates passing this cut, we scaled the photon energy to obtain $\Delta E = 0$, and computed $M_{K^*\gamma}$. The $M_{K^*\gamma}$ resolution of 2.8 MeV rms is dominated by the beam energy spread. The mass distribution for the B candidates is shown in Fig. 2. There are 8 events in the signal region, the mass interval 5.274–5.286 GeV.

The background is still mostly from the continuum, even after the continuum suppression cuts have been optimized. This background varies smoothly as a function

FIG. 1. Penguin diagram for $b \to s\gamma$. The photon may be radiated from any of the four lines.

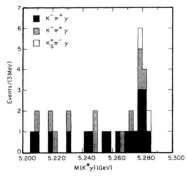

FIG. 2. The $K^*\gamma$ mass distributions for $B^0 \to K^{*0}\gamma$; $B^- \to K^{*-}\gamma$, $K^{*-} \to K^0_S\pi^-$; and $B^- \to K^{*-}\gamma$, $K^{*-} \to K^-\pi^0$ candidates.

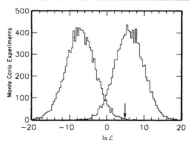

FIG. 3. The $\ln\mathcal{L}$ distributions for 10 000 "experiments" of 8 events each, drawn from $B^0 \to K^{*0}\gamma$ Monte Carlo (right curve) or continuum Monte Carlo (left curve) simulations. The value for the 8 $B^0 \to K^{*0}\gamma$ candidate events is indicated by an arrow.

of ΔE and $M_{K^*\gamma}$, so the amount of background in the signal region of the on-resonance data can be reliably estimated by scaling the events observed in sideband regions by an appropriate factor. For $B^0 \to K^{*0}\gamma$ the sideband was chosen to be $|\Delta E| < 280$ MeV and $M_{K^*\gamma} > 5.2$ GeV, excluding $|\Delta E| < 100$ MeV and $M_{K^*\gamma} > 5.274$ GeV (the signal region plus a narrow boundary in $|\Delta E|$). The relative population of the sideband and signal regions depends on the momentum distributions of the photon and two charged particles making up the $K^*\gamma$ candidate, and on their transverse momentum distributions relative to a common axis. Using Monte Carlo tuned to match the off-resonance data we determined the population ratio to be 25.4:1 with an error of $\pm$ 8%. For the actual background determination, we counted events in the sideband regions of the on- and off-resonance data samples, and the signal region of the off-resonance sample, and scaled the total of 41 events by a factor of 37.6 [12] to obtain a background estimate of 1.1 $\pm$ 0.2. The binomial probability [13] that 8 + 41 events in signal plus sideband regions would distribute themselves such that 8 or more were in the signal region, given that the intrinsic relative populations are 1:37.6, is 3.5 $\times 10^{-5}$.

For $B^- \to K^{*-}\gamma$, we looked for K^{*-} candidates in both the $K^0_S\pi^-$ and the $K^-\pi^0$ modes. A $K^0_S \to \pi^+\pi^-$ decay is required to have a vertex more than 5 mm from the beam axis, a good χ^2 for the vertex fit, and a $\pi^+\pi^-$ mass within 10 MeV (2σ) of the K^0_S mass. The π^0's are selected from pairs of photons with an invariant mass within 15 MeV (2.5σ) of the π^0 mass. The photons are selected from showers in the calorimeter that are not matched to charged tracks, have shower shapes consistent with isolated photons, and have energies above 30 (50) MeV in the barrel (end-cap) regions. Other cuts [14] are similar to those described for $B^0 \to K^{*0}\gamma$. The $K^{*-}\gamma$ mass distributions for the two modes are shown in Fig. 2, and the numbers of signal events and estimated backgrounds are given in Table I. The combined probability of both K^{*-} results being fluctuations is 7.0×10^{-4}.

We obtain additional evidence that the signal events are not all continuum background by examining the distributions inside the cuts of the variables $M_{K^*\gamma}$, ΔE, $\cos\theta_B$, $\cos\theta_{K\pi}$, $M_{K\pi}$, R_2, and $\cos\theta_{\text{thr}}$. We do this with a likelihood ratio test [15], which reduces the information contained in several variables to a single number. In Fig. 3 we show the distribution in log likelihood ratio ($\ln\mathcal{L}$) for two groups of 10 000 simulated experiments, one drawing 8 events from a sample of Monte Carlo $B^0 \to K^{*0}\gamma$ events, the other drawing from a sample of Monte Carlo

TABLE I. Summary of results for $B \to K^*\gamma$.

	$B^0 \to K^{*0}\gamma$ $K^{*0} \to K^+\pi^-$	$B^- \to K^{*-}\gamma$	
		$K^{*-} \to K^0_S\pi^-$	$K^{*-} \to K^-\pi^0$
Signal events	8	2	3
Sideband events	41	2	10
Sideband scale factor	37.6	40	12
Sideband background	1.1$\pm$0.2	0.05$\pm$0.03	0.8$\pm$0.3
Binomial probability	3.5$\times 10^{-5}$	3.7$\times 10^{-3}$	7.3$\times 10^{-2}$
Residual $B\overline{B}$ background	0.30$\pm$0.15	0.01$\pm$0.01	0.10$\pm$0.05
Efficiency	(11.9$\pm$1.8)%	(2.0$\pm$0.3)%	(3.1$\pm$0.5)%
Branching ratio	(4.0$\pm$1.7$\pm$0.8)$\times 10^{-5}$	(5.7$\pm$3.1$\pm$1.1)$\times 10^{-5}$	

continuum events. Only 0.11% of the 8-event continuum samples have values of $\ln \mathcal{L}$ as large as that of the 8 $B^0 \to K^{*0}\gamma$ candidate events, while about half of the signal samples do. Allowing for systematic and statistical errors, the probability of observing a value of $\ln \mathcal{L}$ this large from a sample of 8 continuum events is less than 1%. Similar studies for $B^- \to K^{*-}\gamma$ support the interpretation of a $K^{*-} \to K_S^0 \pi^-$ signal, and are not inconsistent with a $K^{*-} \to K^- \pi^0$ signal.

We have assessed sources of background to $B \to K^*\gamma$ from other B decays. First, using a fast Monte Carlo that did not include a full detector simulation, we considered $b \to c$, $b \to u$, and $b \to sg$ decays, with particular attention to the $D^{*0}\gamma$, $\rho\pi^0$, $K^*\pi^0$, and $K^*\rho^-$ channels, which we anticipated might be troublesome. These sources accounted for < 0.11, < 0.06, and < 0.02 events as background to $K^-\pi^+\gamma$, $K^-\pi^0\gamma$, and $K_S^0\pi^-\gamma$, respectively. Then using techniques largely based on data, we studied possible sources of false 2-3 GeV photons such as random overlaps of uncorrelated photons, merging of photons from π^0's, clusters caused by K_L^0's, and clusters caused by antineutrons. These sources accounted for < 0.25, < 0.06, and < 0.02 events as background to $K^-\pi^+\gamma$, $K^-\pi^0\gamma$, and $K_S^0\pi^-\gamma$, respectively. A final contribution, from feeddown into $B \to K^*\gamma$ from other $b \to s\gamma$ decays, is estimated using the theoretical prediction of an inclusive rate of 4×10^{-4}, and a model for the hadronization. We find backgrounds of 0.4 and 0.1 events in $K^-\pi^+\gamma$ and $K^-\pi^0\gamma$, respectively. The feeddown into the $K_S^0\pi^-\gamma$ mode is negligible. Approximately half of the $B\overline{B}$ backgrounds are included in the background estimated from the sidebands. The residual backgrounds from $B\overline{B}$, not included in the sideband subtraction, are listed in Table I [16].

Table I summarizes our results. The net yield of $B^0 \to K^{*0}\gamma$ events is 6.6 ± 2.8. The efficiency for $B^0 \to K^{*0}\gamma$ decays, including K^* branching ratio, is $(11.9 \pm 1.8)\%$. The data sample contains 1.39×10^6 $B\overline{B}$ decays, which we assume to be half charged, half neutral. From this we obtain a branching ratio for $B^0 \to K^{*0}\gamma$ of $(4.0 \pm 1.7 \pm 0.8) \times 10^{-5}$, where the first error is statistical and the second is a $\pm 20\%$ systematic error to account for uncertainties in efficiency and background. The net yield of $B^- \to K^{*-}\gamma$ events is 4.1 ± 2.3. The efficiencies for the $K_S^0\pi^-$ and $K^-\pi^0$ modes, including K^* and K^0 branching ratios, are 2.0% and 3.1%, respectively, giving an overall efficiency for the $B^- \to K^{*-}\gamma$ decay of 5.1%. From this we obtain a branching ratio for $B^- \to K^{*-}\gamma$ of $(5.7 \pm 3.1 \pm 1.1) \times 10^{-5}$.

In conclusion, we have obtained compelling evidence for the existence of the decay $B^0 \to K^{*0}\gamma$, and supporting evidence for the existence of the closely related decay $B^- \to K^{*-}\gamma$. If one makes the reasonable assumption that the branching ratios for these two decays are equal, then an average branching ratio of $(4.5 \pm 1.5 \pm 0.9) \times 10^{-5}$ is obtained. This is entirely consistent with the theoretical predictions from the penguin diagram which are in the range $(1-15) \times 10^{-5}$ [4,7]. Our result does not require any contributions beyond the standard model, but it is an order of magnitude larger than would be expected [5] if the penguin diagram were not present. This is strong evidence for the existence of the penguin diagram.

We gratefully acknowledge the effort of the CESR staff in providing us with excellent luminosity and running conditions. J.P.A. and P.S.D. thank the PYI program of the NSF, I.P.J.S. thanks the YI program of the NSF, K.H. thanks the Alexander von Humboldt Stiftung, G.E. thanks the Heisenberg Foundation, K.K.G. and A.J.W. thank the SSC Fellowship program of TNRLC, K.K.G., H.N.N., J.D.R., and H.Y. thank the OJI program of DOE, and R.P. and P.R. thank the A.P. Sloan Foundation for support. This work was supported by the National Science Foundation and the U.S. Department of Energy.

---

* Permanent address: INP, Novosibirsk, Russia.
† Permanent address: Carleton University, Ottawa, Canada K1S 5B6.

[1] M.K. Gaillard and B.W. Lee, Phys. Rev. Lett. **33**, 108 (1974); G. Altarelli and L. Maiani, Phys. Lett. **52B**, 351 (1974); J. Ellis, M.K. Gaillard, and D.V. Nanopoulos, Nucl. Phys. **B100**, 313 (1975); A.I. Vainshtein, V.I. Zakharov, and M.A. Shifman, Pis'ma Zh. Eksp. Teor. Fiz. **22**, 123 (1975) [JETP Lett. **22**, 55 (1975)].
[2] F. Gilman and M.B. Wise, Phys. Lett. **83B**, 83 (1979).
[3] B.A. Campbell and P.J. O'Donnell, Phys. Rev. D **25**, 1989 (1982); M. Bander, D. Silverman, and A. Soni, Phys. Rev. Lett. **43**, 242 (1979). For a review, see J. Rosner, Enrico Fermi Institute Report No. EFI 90-63, 1990 (unpublished).
[4] S. Bertolini, F. Borzumati, and A. Masiero, Phys. Rev. Lett. **59**, 180 (1987); N.G. Deshpande et al., Phys. Rev. Lett. **59**, 183 (1987); B. Grinstein, R. Springer, and M.B. Wise, Phys. Lett. B **202**, 138 (1988); R. Grigjanis et al., Phys. Lett. B **213**, 355 (1988); A. Ali and C. Greub, Z. Phys. C **49**, 431 (1991).
[5] N.G. Deshpande et al., Phys. Lett. B **214**, 467 (1988); P.Colangelo et al., Z. Phys. C **45**, 575 (1990).
[6] S. Bertolini et al., Nucl. Phys. **B353**, 591 (1991); J.L. Hewett, Phys. Rev. Lett. **70**, 1045, (1993); V. Barger et al., Phys. Rev. Lett. **70**, 1368 (1993).
[7] T. Altomari, Phys. Rev. D **37**, 677 (1988); C.A. Dominguez et al., Phys. Lett. B **214**, 459 (1988); N.G. Deshpande et al., Z. Phys. C **40**, 369 (1988); T.M. Aliev et al., Phys. Lett. B **237**, 569 (1990); P.J. O'Donnell and H. Tung, Phys. Rev. D **44**, 741 (1991); A. Ali and C. Greub, Phys. Lett. B **259**, 182 (1991); A. Ali, T. Ohl, and T. Mannel, Phys. Lett. B **298**, 195 (1993).
[8] We use K^* to denote $K^*(892)$ from here on.
[9] Y. Kubota et al., Nucl. Instrum. Methods Phys. Res., Sect. A **320**, 66 (1992).
[10] G. Fox and S. Wolfram, Phys. Rev. Lett. **41**, 1581 (1978).
[11] $S_\perp$ is defined as the sum of the magnitudes of the momenta transverse to the photon direction, summed over all particles more than 45° away from the photon axis, divided by the sum of the magnitudes of the momenta of all particles except the photon.

[12] This is $25.4 \times 1.464 + 0.464$, where 0.464 is the scale factor between off-resonance and on-resonance data samples.

[13] The Poisson probability that a mean of 1.1 gives rise to 8 or more events is 2.0×10^{-5}, but this estimate of significance does not consider the statistical error on the background, while the procedure with binomial probability does.

[14] Cuts that differ are at least 5 tracks, $|\cos\theta_{K\pi}| < 0.85$, $|\cos\theta_{\mathrm{thr}}| < 0.85$, $|\cos\theta_B| < 0.80$, $|\Delta E| < 0.75$, $R_2 < 0.35$, and no cuts on $S_\perp$, R'_2, or $|\cos\theta'|$.

[15] The likelihood analysis is described in Cornell Laboratory of Nuclear Studies Report No. CLNS 93/1212 (unpublished).

[16] The binomial probabilities for continuum plus $B\overline{B}$ backgrounds fluctuating up to the numbers of observed signal events are larger than the probabilities for continuum background alone by at most a factor of 4 for the $K^{*0}\gamma$ mode, and a factor of 1.4 for the $K^{*-}\gamma$ modes.

12

From Neutral Currents to Weak Vector Bosons

The unification of weak and electromagnetic interactions, 1973–1987.

Fermi's theory of weak interactions survived nearly unaltered over the years. Its basic structure was slightly modified by the addition of Gamow–Teller terms and finally by the determination of the V-A form, but its essence as a four fermion interaction remained. Fermi's original insight was based on the analogy with electromagnetism; from the start it was clear that there might be vector particles transmitting the weak force the way the photon transmits the electromagnetic force. Since the weak interaction was of short range, the vector particle would have to be heavy, and since beta decay changed nuclear charge, the particle would have to carry charge. The weak (or W) boson was the object of many searches. No evidence of the W boson was found in the mass region up to 20 GeV.

The V-A theory, which was equivalent to a theory with a very heavy W, was a satisfactory description of all weak interaction data. Nevertheless, it was clear that the theory was not complete. As described in Chapter 6, it predicted cross sections at very high energies that violated unitarity, the basic principle that says that the probability for an individual process to occur must be less than or equal to unity. A consequence of unitarity is that the total cross section for a process with angular momentum J can never exceed $4\pi(2J+1)/p_{cm}^2$. However, we have seen that neutrino cross sections grow linearly with increasing center-of-mass energy. When the energy exceeds about 300 GeV, there would be a contradiction.

It might be hoped that the theory could be calculated more completely, to a higher order in the Fermi coupling constant. In a complete theory, these corrections could bring the predictions back into the allowed range. Unfortunately, the Fermi theory cannot be calculated to higher order because the results are infinite. Infinities arise in calculating quantum electrodynamics (QED) to higher order, as well. In QED, it is possible to absorb these infinities so that none appears in the physical results. This is impossible in the Fermi theory. Writing the Fermi theory in terms of the W bosons enhances the similarity with QED, but the infinities remain.

The first step in the solution to this problem came from C. N. Yang and R. Mills, who in 1954 developed a theory of massless interacting vector particles. This theory could accommodate particles like the photon, W^+, and W^- that would interact with one another, but it required them to be massless. The infinities in the model could be reabsorbed (the model was "renormalizable"). An important advance was made by Peter Higgs, who in 1964 showed how a theory initially containing a massless photon and two scalar particles could turn into a theory with a massive vector particle and one scalar. This "Higgs mechanism" was a key ingredient in the final model.

The Standard Model of electroweak interactions, developed largely by Glashow, Weinberg, and Salam begins with massless Yang–Mills particles. These are denoted W^+, W^-, W^0, and B (not to be confused with the B meson of the previous chapter, which plays no role here). The W's form a triplet of a new symmetry, "weak isospin," while the B is an isosinglet. The Higgs mechanism is invoked to give mass to the W bosons. At the same time, the two neutral particles, W^0 and B mix to produce two physical particles, the photon (represented by the field A) and the Z. The photon, of course, is massless. The Z acquires a mass comparable to that of the W.

The Fermi theory is equivalent to the exchange of only charged weak bosons. This allows for processes like $\nu_\mu e^- \to \mu^- \nu_e$, which may be viewed as emission of a W^+ by the initial neutrino, which turns into a muon and its absorption by the electron, which turns into an electron-neutrino. When the W is emitted or absorbed, the charges of the interacting particles are changed. The currents to which the W attaches, for example $\bar{e}\gamma_\mu(1-\gamma_5)\nu_e$, are called charged currents. The process $\nu_\mu e^- \to \nu_\mu e^-$ cannot proceed in the Fermi theory because the charged current can change ν_μ only to μ^-, not to e^-, as was shown by the two-neutrino experiment discussed in Chapter 7. The Z boson adds new interactions, ones with neutral currents. The ν_μ can emit a Z which is absorbed by the electron, thus permitting the process $\nu_\mu e^- \to \nu_\mu e^-$. No charge is transferred. The existence of weak neutral currents is a dramatic prediction of the model.

In fact, neutral-current processes had been searched for in decays like $K^+ \to \pi^+ e^- e^+$ and $K_L^0 \to \mu^+ \mu^-$ (where the $e^+ e^-$ or $\mu^+ \mu^-$ would be viewed as coming from a virtual Z) and found to be very rare or nonexistent. These searches had been limited invariably to strangeness-changing neutral currents, for example the current that transformed a K^+ into a π^+. The reason for this limitation was simple. In most instances where there is no change of strangeness, if a Z can be exchanged, so can a photon. Thus the effect of the Z, and hence of the neutral weak current, was always masked by a much larger electromagnetic effect. One way to avoid this was to look for scattering initiated by a neutrino that emitted a Z that subsequently interacted with a nuclear target. This process could not occur electromagnetically since the neutrino does not couple to photons. The signature of such a process was the absence of a charged lepton in the final state.

Although neutral currents were predicted in the model of Glashow, Weinberg, and Salam, the intensity of the search for them increased dramatically in the early 1970s when, through the work of G. 't Hooft and others, the theory was shown to be renormalizable. Weinberg and Salam had conjectured that the theory was renormalizable, but there was no proof initially.

The discovery of neutral-weak-current interactions was made in mid 1973 by A. Lagarrigue, P. Musset, D. H. Perkins, A. Rousset, and co-workers using the Gargamelle bubble chamber at CERN (**Ref. 12.1**). The experiment used separate neutrino and antineutrino beams. The beams were overwhelmingly muon-neutrinos, so the task was to demonstrate the occurrence of events without a final-state muon. Muons could be distinguished from hadrons in the bubble chamber because it was filled with a rather dense material, freon, in which most of the produced hadrons would either interact or range out. The muons, then, were signaled by the particles exiting from the chamber without undergoing a hadronic interaction.

The background with the greatest potential to obscure the results was due to neutrino interactions occurring in the shielding before the bubble chamber. Neutrons produced in these interactions could enter the bubble chamber without leaving a track and cause an event from which, of course, no muon would emerge. The Gargamelle team was able to control this background by studying a related class of events. Some ordinary charged-current events occurring within the bubble chamber yielded neutrons that subsequently had hadronic collisions inside the bubble chamber. These events were quite analogous to the background events in which the initial neutrino interaction took place in the shielding. By studying the events in which the neutron's source was apparent, it was possible to place limits on the neutron background arising outside the chamber. In addition, the neutral-current events had another characteristic that indicated they were due to neutrinos. They were evenly distributed along the length of the bubble chamber. If they had come from neutrons there would have been more of them at the front and fewer at the back as a consequence of the depletion of the neutrons traveling through the freon. The neutrinos have such a small cross section that there is no measurable attenuation.

Not only did the experiment find convincing evidence for the neutral-current events, it measured the ratio of neutral-current to charged-current events both for neutrinos and antineutrinos. This was especially important because it provided a means of measuring the value of the neutral weak charge to which the Z boson coupled.

The electroweak theory contains three fundamental parameters aside from the masses of the particles and the mixing angles in the Kobayashi–Maskawa matrix. Once these are determined, all purely electroweak processes can be predicted. To determine the three parameters, it is necessary to measure three fundamental quantities. There is, however, a great deal of freedom in choosing these experimental quantities. It is natural to take two of them to be $\alpha_{em} \approx 1/137$ and $G_F \approx 1.166 \cdot 10^{-5}$ GeV^{-2} since these are quite well measured. The third quantity must involve some new feature introduced by the electroweak model. The strength of the neutral weak currents is such a quantity. The result is often expressed in terms of the weak mixing angle θ_W that indicates the degree of mixing of the W^0 and B bosons that generates the photon and Z:

$$A = \sin\theta_W W^0 + \cos\theta_W B; \qquad W^0 = \sin\theta_W A + \cos\theta_W Z; \qquad (12.1)$$
$$Z = \cos\theta_W W^0 - \sin\theta_W B; \qquad B = \cos\theta_W A - \sin\theta_W Z. \qquad (12.2)$$

The photon couples to particles according to their charges. We can represent the coupling to a fermion f by

$$\overline{f}\gamma_\mu e Q f A^\mu \qquad (12.3)$$

Table 12.1. *The weak interaction quantum numbers of quarks and leptons in the Standard Model. The subscripts indicate left-handed and right-handed components.*

	e_L	e_R	ν_L	u_L	u_R	d_L	d_R
Q	-1	-1	0	$2/3$	$2/3$	$-1/3$	$-1/3$
T_3	$-1/2$	0	$1/2$	$1/2$	0	$-1/2$	0
$\frac{1}{2}Y$	$-1/2$	-1	$-1/2$	$1/6$	$2/3$	$1/6$	$-1/3$

or, in shorthand, eQA, where Q measures the charge of a particle in units of the proton charge, e, and A is the electromagnetic vector potential.

The absorption of a W^+ boson changes an electron into a neutrino. This action can be represented by the isospin operator T_+ if the neutrino and electron form a doublet with the neutrino being the $T_3 = 1/2$ component. Of course, we already know that it is only the left-handed component of the electron that participates, so we assign zero weak isospin to the right-handed part of the electron. The quarks are treated analogously, with the absorption of a W^+ changing a left-handed d into a left-handed u.

The B boson couples to fermions according to another new quantum number, the "weak hypercharge," Y. These new quantum numbers satisfy an analog of the Gell-Mann–Nishijima relation $Q = T_3 + Y/2$ as shown in Table 12.1.

After the mixing of the B and W^0 that produces the photon and the Z, the coupling of the photon to fermions is given by eQ and that of the Z by

$$\frac{e}{\sin\theta_W \cos\theta_W} \left[T_3 - Q \sin^2\theta_W \right] \tag{12.4}$$

where T_3 has an implicit $(1-\gamma_5)/2$ included to project out the left-hand part of the fermion. This is explained in greater detail below. Because the Z couples differently to left-handed and right-handed fermions, its interactions are parity violating. By comparing the couplings of the Z to that of the W, it is possible to derive a relation for the ratio of neutral-current events to charged-current events in deep inelastic neutrino scattering, NC/CC, using the parton model discussed in Chapter 8. Although the parton model is expected to work best at very high energies, the early Gargamelle results on charged currents showed that the model worked well even at the low energies available to Gargamelle using the CERN Proton Synchrotron. If the scattering of the neutrinos from antiquarks is ignored (a 10–20% correction), the predictions are

$$R_\nu = \left(\frac{NC}{CC} \right)_\nu = \frac{1}{2} - \sin^2\theta_W + \frac{20}{27} \sin^4\theta_W, \tag{12.5}$$

$$R_{\bar{\nu}} = \left(\frac{NC}{CC}\right)_{\bar{\nu}} = \frac{1}{2} - \sin^2\theta_W + \frac{20}{9}\sin^4\theta_W. \tag{12.6}$$

In these relations, it is assumed that

$$m_Z^2 = m_W^2/\cos^2\theta_W, \tag{12.7}$$

a prediction of the simplest version of the Standard Model of electroweak interactions as discussed below. The Gargamelle results indicated that $\sin^2\theta_W$ was in the range 0.3 to 0.4.

These results were followed by confirmation from other laboratories. The neutral-current events were not rare. They were easy to find. The problem was to demonstrate that they were not due to any of the various backgrounds. The Harvard–Penn–Wisconsin (HPW) experiment at Fermilab did verify the result, but only after some considerable difficulty in determining their efficiency for identifying muons (Ref. 12.2). The HPW experiment was a counter experiment. The target and detector were combined into a segmented unit. This was followed by a muon spectrometer. A diagram of the apparatus and the appearance of an event in the detector are shown in Figure 12.1. Inevitably there was the problem of determining how many muons failed, for geometrical reasons, to enter the muon spectrometer.

Another Fermilab experiment, a Caltech–Fermilab collaboration, also confirmed the existence of neutral currents (Ref. 12.3). A good measurement of $\sin^2\theta_W$, however, had to await the results of the CERN experiments, carried out by the CDHS, CHARM, and BEBC collaborations mentioned in Chapter 8. The CERN experiments used a beam from the Super Proton Synchrotron (SPS). The values obtained were about 0.30 for R_ν and 0.38 for $R_{\bar{\nu}}$. Later analyses of the neutral-current data found a value $\sin^2\theta_W = 0.23$.

The existence of the neutral currents was important circumstantial evidence for the electroweak model. The neutral-current to charged-current ratios lay close to the curve required by the model. Very impressive evidence came from a different kind of neutral-current experiment performed at SLAC. This experiment measured the interference between an electromagnetic amplitude and one due to neutral weak currents.

The experiment of Prescott and co-workers (**Ref. 12.4**) measured the scattering of longitudinally polarized electrons from a deuterium target. A dependence of the cross section on the value of $\sigma_e \cdot \mathbf{p}_e$, where σ_e is the electron's spin, is necessarily a parity violation since this is a pseudoscalar quantity. The experiment actually measured the asymmetry

$$A = \frac{\sigma_R - \sigma_L}{\sigma_R + \sigma_L}. \tag{12.8}$$

where the subscript on the cross section indicates a right-handed or left-handed electron incident.

The right-handed and left-handed electron beams were produced by a source using a laser shining on a GaAs crystal. A Pockels cell allowed linearly polarized laser light to be changed into circularly polarized light, with the polarization changed pulse to pulse in a

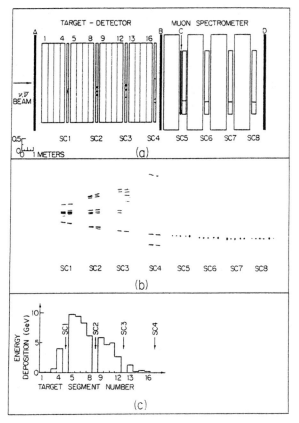

Figure 12.1. Diagram (a) of the HPW neutrino detector used at Fermilab (Ref. 12.2). There were eight spark chambers (SC) and sixteen liquid scintillator segments. The muon spectrometer contained four magnetized iron toroids. Additional scintillator counters are labeled A, B, C, D. An event is seen in the spark chambers and the same event is shown enlarged in (b). A muon track is visible in the muon spectrometer so this is a charged current event. The energy deposition for the event is displayed in (c).

random way, which was recorded. The polarized photons ejected polarized electrons from the crystal, with an average polarization of 37%.

On the basis of very general considerations, it was possible to see that the weak–electromagnetic interference effect should give A a value of order $G_F Q^2/\alpha$ where Q^2 is the momentum transferred squared of the electron (and is not to be confused with the charge operator!). A more complete calculation shows the effect ought to be about one-tenth this size, or near 10^{-4} for Q^2 of about 1 GeV^2. In order that such an effect not be masked by statistical fluctuations, about 10^{10} events are needed. This was achieved by integrating outputs of phototubes rather than counting individual events.

The scattered electrons were collected in a magnetic spectrometer like that used in the pioneering deep inelastic scattering experiments carried out 10 years before. Measurements were made for several beam energies. Because the beam was bent through an angle of 24.5°

before scattering, the polarized electrons precessed. This provided an additional check of the measurements.

The asymmetry can be predicted within the standard electroweak model. The result is a function of $y = (E - E')/E$, the fraction of the incident electron's energy that is lost:

$$A = \frac{-G_F Q^2}{2\sqrt{2}\pi\alpha} \frac{9}{10} \left\{ 1 - \frac{20}{9} \sin^2\theta_W + (1 - 4\sin^2\theta_W) \left[\frac{1 - (1-y)^2}{1 + (1-y)^2} \right] \right\}. \quad (12.9)$$

The result of the experiment, $A/Q^2 = (-9.5 \pm 1.6) \times 10^{-5}$ GeV^{-2}, was in good agreement with the Standard Model for a value $\sin^2\theta_W = 0.20 \pm 0.03$.

The measurement of the weak mixing angle in the neutral-current experiments made it possible to predict the masses of the W and Z. Masses arise in the Standard Model from the Higgs mechanism, which is due to hypothetical scalar particles, known as Higgs particles. The field corresponding to a neutral Higgs particle obtains a vacuum expectation value that is non-zero because this minimizes the energy of the vacuum. The various massless particles in the theory obtain masses by interacting with this ubiquitous non-zero field. The coupling of the vector (gauge) particles is governed by the analog of the usual minimal coupling of electrodynamics:

$$D_\mu = \partial_\mu - ieQA_\mu. \quad (12.10)$$

In the conventional model, the Higgs particle is part of a complex isodoublet. This is analogous to the kaon multiplet. There are four states, two charged and two neutral. We can represent this as a two component vector

$$\begin{pmatrix} \phi^+ \\ \phi^0 \end{pmatrix} \quad (12.11)$$

and its complex conjugate. In the vacuum, the field ϕ^0 is non-zero: $<\phi^0> = v/\sqrt{2}$. The analog of the minimal coupling is

$$D_\mu = \partial_\mu - ig\mathbf{T} \cdot \mathbf{W}_\mu - ig'(Y/2)B_\mu \quad (12.12)$$

where the three components of $\mathbf{T}$ are the generators of the weak isospin and where g and g' are two coupling constants, one for $SU(2)$ and one for $U(1)$. Rewritten in terms of the physical particles, this is

$$D_\mu = \partial_\mu - ieQA_\mu - ig(T_+ W_\mu^+ + T_- W_\mu^-)/\sqrt{2} - ig(T_3 - \sin^2\theta_W Q)Z_\mu/\cos\theta_W. \quad (12.13)$$

The relations between e, g, g', and θ_W are

$$\tan\theta_W = \frac{g'}{g}, \quad 1/g^2 + 1/g'^2 = 1/e^2. \quad (12.14)$$

A comparison with the usual V-A theory shows that

$$\frac{G_F}{\sqrt{2}} = \frac{g^2}{8m_W^2}. \tag{12.15}$$

This determines the W mass:

$$m_W^2 = \frac{\pi \alpha}{\sqrt{2} \sin^2 \theta_W G_F}. \tag{12.16}$$

In fact, a more precise result is obtained by using the electromagnetic coupling measured not at zero momentum transfer but rather at a momentum squared equal to m_W^2, $\alpha(m_W^2) \approx 1/129$, a value that takes into account vacuum polarization corrections. Inserting the vacuum expectation value of the Higgs field we find mass terms from

$$(D_\mu \phi)^\dagger D_\mu \phi \rightarrow \frac{1}{2} \left[\frac{g^2 v^2}{4} (W_\mu^+ W^{-\mu} + W_\mu^- W^{+\mu}) + \frac{g^2 v^2}{4 \cos^2 \theta_W} Z_\mu Z^\mu \right]. \tag{12.17}$$

This gives for the Z mass

$$m_Z^2 = m_W^2 / \cos^2 \theta_W \tag{12.18}$$

and with $\sin^2 \theta_W = 0.23$, $m_W = 80$ GeV, $m_Z = 91$ GeV.

With a promising theory and a good measurement of $\sin^2 \theta_W$, the search for the W and Z now took a different character. The masses could be predicted from the results of the neutral-current measurements of $\sin^2 \theta_W$ and lay outside the range of existing machines. Following a proposal by D. Cline, C. Rubbia, and P. McIntyre, a major effort at CERN, led by C. Rubbia and S. van der Meer, transformed the SPS into a colliding beam machine, the Sp$\bar{\text{p}}$S. The regular proton beam was used to create antiprotons, which were captured and stored. The antiprotons then re-entered the SPS, but moving in the opposite direction. A particularly difficult problem was to compress the beam of antiprotons so that it would be dense enough to cause many collisions when the protons moving the other way passed through it.

If a u quark from a proton and a $\bar{d}$ quark from an antiproton collided, a W^+ could be created if the energy of the pair were near the mass of the W. The W^+ would decay into $e^+ \nu$ about 8% of the time. The cross section for this process was calculated to be a fraction of a nanobarn (10^{-33}cm^2). A more spectacular signal could be obtained from Z's that decayed into $e^+ e^-$ or $\mu^+ \mu^-$.

The W and Z bosons were discovered by the two large collaborations, UA-1 (**Ref. 12.5**) and UA-2 (Ref. 12.6), working at the Sp$\bar{\text{p}}$S Collider. The UA-1 detector used a uniform magnetic field of 0.7 T (7 kG) perpendicular to the beam. Inside the field was a high quality drift chamber. External to the drift chamber was extensive coverage by electromagnetic and hadronic calorimeters. The critical capability of discriminating between electrons and hadrons was achieved using many radiation lengths of material, segmented into layers.

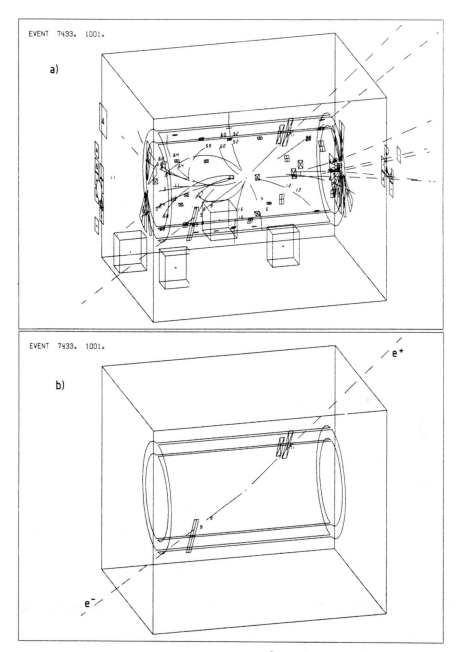

Figure 12.2. A UA-1 event display for a candidate for $Z^0 \to e^+e^-$. (a) Display of reconstructed tracks and calorimeter hits. (b) Display of tracks with $p_T > 2$ GeV/c and calorimeter hits with $E_T > 2$ GeV. The electron pair emerges cleanly from the event (Ref. 12.8).

By covering nearly all of the full 4π steradians with calorimetry, it was possible to check momentum balance in the plane perpendicular to the beam. This, in effect, provided a neutrino detector for those neutrinos with transverse momentum above 15 GeV or so.

In colliding-beam machines like SPEAR, DORIS, PETRA, PEP, ISR, and the $S\bar{p}pS$ Collider, the event rate is related to the cross section by

$$\text{Rate} = \mathcal{L}\sigma \qquad (12.19)$$

where $\mathcal{L}$ is the luminosity and is measured in cm^{-2} s^{-1}. The luminosity depends on the density of the intersecting beams and their degree of overlap. The total number of events is $\sigma \int \mathcal{L} dt$. For the results reported by UA-1, $\int \mathcal{L} dt = 18$ nb^{-1} at an energy of $\sqrt{s} = 540$ GeV. The total event rate was high so various triggers were used to choose the small subset of events to be recorded.

Events with electron candidates that had high transverse momentum detected in the central part of the calorimeter and that were well-separated from any other high transverse momentum particles were selected. This class contained 39 events. Five of these contained no hadronic jets and thus had a significant transverse momentum imbalance, as would be expected for decays $W \to e\nu$. An alternative search through the same recorded events sought those with large momentum imbalance. The same five events were ultimately isolated, together with two additional events that were candidates for $W \to \tau\nu$.

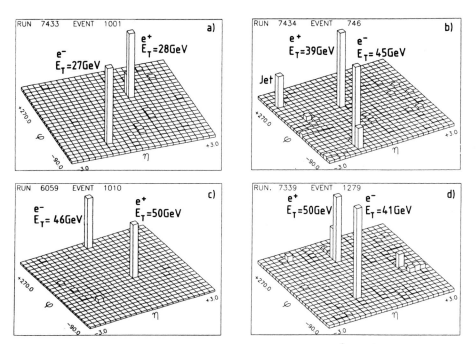

Figure 12.3. Lego plots for four UA-1 events that were candidates for $Z^0 \to e^+e^-$. The plots show the location of energy deposition in ϕ, the azimuthal angle, and $\eta = -\ln\tan(\theta/2)$, the pseudorapidity. The isolated towers of energy indicate the cleanliness of the events (Ref. 12.8).

The mass of the W could be estimated from the observed transverse momenta. The result was $m_W = 81 \pm 5$ GeV, in good agreement with predictions of the Standard Model using the weak mixing angle as measured in neutral-current experiments. Similar results were obtained by the UA-2 collaboration and also by observing the muonic decay of the W (Ref. 12.7).

Later, the two collaborations detected the Z through its decays $Z \to e^+e^-$ (Refs. 12.8, 12.9) and $Z \to \mu^+\mu^-$ (Ref. 12.10). This discovery took longer because the cross section for Z production is somewhat smaller than that for W's and because the branching ratios $Z \to e^+e^-$ and $Z \to \mu^+\mu^-$ are expected to be only 3% each, while $W \to e\nu$ and $W \to \mu\nu$ should be 8% each. However, the signature of two leptons with large invariant mass was unmistakable, and only a few events were necessary to establish the existence of the Z with a mass consistent with the theoretical expectation. An event that is a Z^0 candidate measured by the UA-1 Collaboration is shown in Figure 12.2. The "lego" plots for four UA-1 Z^0 candidates are shown in Figure 12.3. An event measured by the UA-2 Collaboration is shown in Figure 12.4, together with its lego plot. During running at an increased center-of-mass energy of 630 GeV additional data were accumulated. Results for the decay $Z \to e^+e^-$ obtained by the UA-1 and UA-2 Collaborations are shown in Figure 12.5.

The discovery of the W and the Z dramatically confirmed the basic features of the electroweak theory. Its unification of the seemingly unrelated phenomena of nuclear beta decay and electromagnetism is one of the major achievements of twentieth-century physics. With elegance and simplicity, it subsumes the phenomenological V-A theory, extends that theory to include neutral current phenomena and meets the theoretical demand of renormalizability. The unification of electromagnetism and weak interactions remarkably confirms Fermi's prescient observation that the fundamental process of beta decay, $n \to pe\nu$ might be viewed as the interactions of two currents. While the Fermi theory worked only in lowest order, the new theory predicted higher order radiative corrections. Just as the Lamb shift and $g - 2$ provided crucial test of QED, the real test of the electroweak theory was still to come in higher precision measurements.

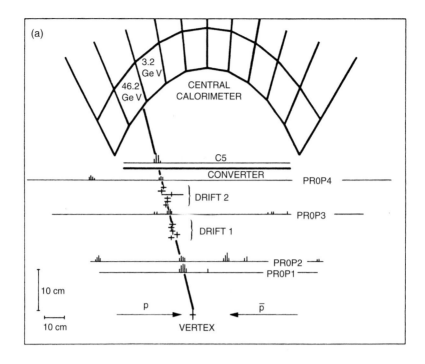

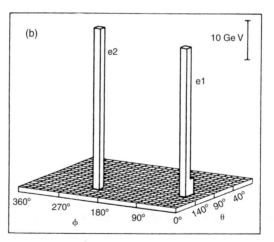

Figure 12.4. A UA-2 candidate for $Z^0 \to e^+e^-$. The upper diagram shows a track detected by a series of proportional chambers and a chamber following a tungsten converter. The calorimeter cells indicate energy measured by the electromagnetic calorimeter. The lego plot for the event shows two isolated depositions of electromagnetic energy, indicative of an e^+e^- pair (Ref. 12.9).

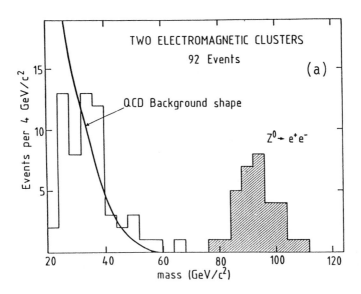
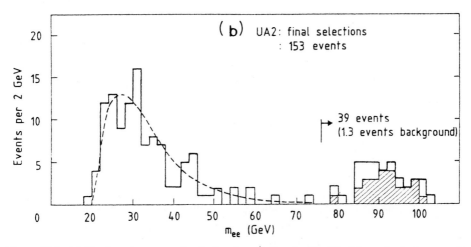

Figure 12.5. (a) The invariant mass distribution for e^+e^- pairs identified through electromagnetic calorimetry in the UA-1 detector. (Figure supplied by UA-1 Collaboration) (b) The analogous plot for the UA-2 data (Ref. 12.12). In both data sets, the Z appears well-separated from the lower mass background.

Exercises

12.1 Make a graph with $(NC/CC)_\nu$ as abscissa and $(NC/CC)_{\bar\nu}$ as ordinate. Draw the curve of values allowed by the Standard Model ignoring contributions from antiquarks. Plot the results quoted in the text. What value of $\sin^2\theta_W$ do you find?

12.2 Derive the predictions for $(NC/CC)_\nu$ and $(NC/CC)_{\bar\nu}$ by comparing the couplings of the W:

$$gT_+/\sqrt{2} = eT_+/\sin\theta_W\sqrt{2}$$

and the Z:

$$e(T_3 - Q\sin^2\theta_W)/\sin\theta_W\cos\theta_W.$$

Use an isoscalar target. For $|q|^2 << m_W^2, m_Z^2$, the vector boson propagator is essentially $1/m_W^2$ or $1/m_Z^2$.

12.3 Use relations analogous to those in Chapter 8 for $e_L^-\mu_L^- \to e_L^-\mu_L^-, e_R^-\mu_L^-$, etc. to derive the expression for the asymmetry, A, in polarized-electron–deuteron scattering.

12.4 The classical equation for the motion of a charged particle with mass m, charge e, and g-factor, g, in a plane perpendicular to a uniform field **B** is

$$\frac{d\beta}{dt} = \frac{e}{m\gamma}\beta\times\mathbf{B}$$

where β is the velocity vector (J. D. Jackson, *Classical Electrodynamics*, 2nd Edition, Wiley, New York, 1975. p. 559). If the direction of the spin is denoted **s**, then

$$\frac{d\mathbf{s}}{dt} = \frac{e}{m}\mathbf{s}\times\left(\frac{g}{2} - 1 + \frac{1}{\gamma}\right)\mathbf{B}$$

Use these equations to verify the precession equation, Eq. (4) of Ref. 12.4.

12.5 * Assume that the W production at the Sp$\bar{\text{p}}$S Collider is due to the annihilation of a quark from the proton and an antiquark from the antiproton. Show that if the proton direction defines the z axis, the produced W's have $J_z = -1$. Show that in the W rest frame the outgoing negative leptons from $W^- \to l^-\bar\nu$ have the angular distribution $(1+\cos\theta^*)^2$, while the positive leptons from $W^+ \to l^+\nu$ have the angular distribution $(1-\cos\theta^*)^2$, where θ^* is measured from the z (proton) direction. What is expected for Z decay? Compare with available data, e.g. S. Geer, in *Proceedings of the XXIII International Conference on High Energy Physics, Berkeley, 1986*, S. C. Loken ed., World Scientific, Singapore, 1987, p. 982.

Further Reading

A more theoretical, but non-technical presentation is given by C. Quigg *Gauge Theories of the Strong, Weak, and Electromagnetic Interactions*, Benjamin/Cummings, Menlo Park, CA, 1983.

The Standard Model is covered in E. D. Commins and P. H. Bucksbaum, *Weak Interactions of Leptons and Quarks*, Cambridge University Press, Cambridge, 1983.

A semi-popular account of the W and Z discoveries is given by P. Watkins, *Story of the W and Z*, Cambridge University Press, New York, 1986.

References

12.1 F. J. Hasert *et al.*, "Observation of Neutrino-like Interactions without Muon or Electron in the Gargamelle Neutrino Experiment." *Phys. Lett.*, **46B**, 138 (1973).

12.2 A. Benvenuti *et al.*, "Observation of Muonless Neutrino-Induced Inelastic Interactions." *Phys. Rev. Lett.*, **32**, 800 (1974).

12.3 B. C. Barish *et al.*, "Neutral Currents in High Energy Neutrino Collisions: An Experimental Search." *Phys. Rev. Lett.*, **34**, 538 (1975).

12.4 C. Y. Prescott *et al.*, "Parity Non-Conservation in Inelastic Electron Scattering." *Phys. Lett.*, **77B**, 347 (1978).

12.5 UA 1 Collaboration, "Experimental Observation of Isolated Large Transverse Energy Electrons with Associated Missing Energy at $\sqrt{s} = 540$ GeV." *Phys. Lett.*, **122B**, 103 (1983).

12.6 UA 2 Collaboration, "Observation of Single Isolated Electrons of High Transverse Momentum in Events with Missing Transverse Energy at the CERN $\bar{p}p$ Collider." *Phys. Lett.*, **122B**, 476 (1983).

12.7 UA 1 Collaboration, "Observation of the Muonic Decay of the Charged Intermediate Vector Boson." *Phys. Lett.*, **134B**, 469 (1984).

12.8 UA 1 Collaboration, "Experimental Observation of Lepton Pairs of Invariant Mass around 95 GeV/c^2 at the CERN SPS Collider." *Phys. Lett.*, **126B**, 398 (1983).

12.9 UA 2 Collaboration, "Evidence for $Z^0 \to e^+e^-$ at the CERN $\bar{p}$ p Collider." *Phys. Lett.*, **129B**, 130 (1983).

12.10 UA 1 Collaboration, "Observation of Muonic Z^0 Decay at the $\bar{p}p$ Collider." *Phys. Lett.*, **147B**, 241 (1984).

12.11 UA 1 Collaboration, "Recent Results on Intermediate Vector Boson Properties at the CERN Super Proton Synchrotron Collider." *Phys. Lett.*, **166B**, 484 (1986).

12.12 UA 2 Collaboration, "Measurements of the Standard Model Parameters from a Study of W and Z Bosons." *Phys. Lett.*, **186B**, 440 (1987).

OBSERVATION OF NEUTRINO-LIKE INTERACTIONS WITHOUT MUON OR ELECTRON IN THE GARGAMELLE NEUTRINO EXPERIMENT

F.J. HASERT, S. KABE, W. KRENZ, J. Von KROGH, D. LANSKE, J. MORFIN,
K. SCHULTZE and H. WEERTS

III. Physikalisches Institut der Technischen Hochschule, Aachen, Germany

G.H. BERTRAND-COREMANS, J. SACTON, W. Van DONINCK and P. VILAIN[*1]

Interuniversity Institute for High Energies, U.L.B., V.U.B. Brussels, Belgium

U. CAMERINI[*2], D.C. CUNDY, R. BALDI, I. DANILCHENKO[*3], W.F. FRY[*2], D. HAIDT,
S. NATALI[*4], P. MUSSET, B. OSCULATI, R. PALMER[*4], J.B.M. PATTISON,
D.H. PERKINS[*6], A. PULLIA, A. ROUSSET, W. VENUS[*7] and H. WACHSMUTH

CERN, Geneva, Switzerland

V. BRISSON, B. DEGRANGE, M. HAGUENAUER, L. KLUBERG,
U. NGUYEN-KHAC and P. PETIAU

Laboratoire de Physique Nucléaire des Hautes Energies, Ecole Polytechnique, Paris, France

E. BELOTTI, S. BONETTI, D. CAVALLI, C. CONTA[*8], E. FIORINI and M. ROLLIER

Istituto di Fisica dell'Università, Milano and I.N.F.N. Milano, Italy

B. AUBERT, D. BLUM, L.M. CHOUNET, P. HEUSSE, A. LAGARRIGUE,
A.M. LUTZ, A. ORKIN-LECOURTOIS and J.P. VIALLE

Laboratoire de l'Accélérateur Linéaire, Orsay, France

F.W. BULLOCK, M.J. ESTEN, T.W. JONES, J. McKENZIE, A.G. MICHETTE[*9]
G. MYATT[*] and W.G. SCOTT[*6,*9]

University College, London, England

Received 25 July 1973

Events induced by neutral particles and producing hadrons, but no muon or electron, have been observed in the CERN neutrino experiment. These events behave as expected if they arise from neutral current induced processes. The rates relative to the corresponding charged current processes are evaluated.

We have searched for the neutral current (NC) and charged current (CC) reactions:

NC $\nu_\mu/\bar{\nu}_\mu + N \rightarrow \nu_\mu/\bar{\nu}_\mu$ + hadrons, (1)

CC $\nu_\mu/\bar{\nu}_\mu + N \rightarrow \mu^-/\mu^+$ + hadrons (2)

which are distinguished respectively by the absence of any possible muon, or the presence of one, and only one, possible muon. A small contamination of $\nu_e/\bar{\nu}_e$ exists in the $\nu_\mu/\bar{\nu}_\mu$ beams giving some CC events which are easily recognised by the e^-/e^+ signature. The analysis is based on 83 000 ν pictures and 207 000 $\bar{\nu}$ pictures taken at CERN in the Gargamelle bubble chamber filled with freon of density 1.5×10^3 kg/m³ [‡]. The dimensions of this chamber are such that most

[*1] Chercheur agréé de L'Institut Interuniversitaire des Sciences Nucléaires, Belgique.
[*2] Also at Physics Department, University of Wisconsin.
[*3] Now at Serpukhov.
[*4] Now at University of Bari.
[*5] Now at Brookhaven National Laboratory.
[*6] Also at University of Oxford.
[*7] Now at Rutherford High Energy Laboratory.
[*8] On leave of absence from University and INFN-Pavia.
[*9] Supported by Science Research Council grant.

[‡] A more detailed account of the analysis of this experiment appears in a paper to be submitted to Nuclear Physics.

138

hadrons are unambiguously identified by interaction or by range-momentum and ionisation. Any track which could possibly be due to a muon has consigned the event to reaction (2).

Analysis of the signal. To estimate the background of neutral hadrons coming from neutrino interactions in the shielding and simulating reaction (1), events where a visible charged current interaction produces an identified neutron star in the chamber (associated, AS, events) were also studied. To obtain a good estimate of the true neutral hadron direction from the direction of the observed total momentum a cut in visible total energy of > 1 GeV was applied to the NC and AS events, as well as to the hadronic part of the CC events.

We have observed, in a fiducial volume of 3 m^3, 102 NC, 428 CC and 15 AS in the ν run and 64 NC, 148 CC and 12 AS in the $\bar{\nu}$ run. Using these numbers without background substraction the ratios NC/CC are then 0.24 for ν and 0.42 for $\bar{\nu}$, whilst the NC/AS ratios are 6.8 and 5.3 respectively.

The spatial distributions of the NC events have been compared to those of the CC events and found to be similar. In particular, the distribution along the beam direction of NC (fig. 1) has the same shape as the CC distribution. In contrast the observed distribution of low energy neutral stars shows a typical exponential attenuation as expected for neutron background. The distributions of radial position, hadron total energy, and angle between measured hadron total momentum and beam direction are also indistinguishable for NC and CC.

Using the direction of measured total momentum of the hadrons in NC and CC events, a Bartlett method has been used to evaluate the apparent interaction mean free paths, λ_a, for NC and CC which are found to be compatible with infinity. For the NC events we find $\lambda_a > 2.6$ m at 90% CL; this corresponds to 3.5 times the neutron interaction length for high energy (> 1 GeV) inelastic collisions in freon.

Evaluation of the background. Since the outgoing neutrinos cannot be detected in reaction (1), the NC events may be simulated by neutral hadrons coming from the ν beam or elsewhere.

As a check for cosmic ray origin, the up-down asymmetries of NC events in vertical position and momenta have been measured and found to be $(3 \pm 8)\%$ and $(-8 \pm 8)\%$ respectively. In addition, a cosmic ray

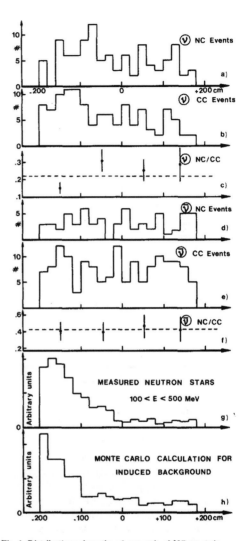

Fig. 1. Distributions along the ν-beam axis. a) NC events in ν. b) CC events in ν (this distribution is based on a reference sample of $\sim 1/4$ of the total ν film). c) Ratio NC/CC in ν (normalized). d) NC in $\bar{\nu}$. e) CC events in $\bar{\nu}$. f) Ratio NC/CC in $\bar{\nu}$. g) Measured neutron stars with $100 < E < 500$ MeV having protons only. h) Computed distribution of the background events from the Monte-Carlo.

exposure of 15 000 pictures shows no NC type event satisfying the selection criteria. We conclude that the cosmic background is negligible.

The low energy muons (< 100 MeV/c) captured at rest in the ν run could be mistaken as protons. A study of the observed muon spectrum in CC events, as well as a theoretical estimate of the low end of this spectrum shows that the correction to be applied is 0 ± 5 events.

Interactions of neutral hadrons produced by the primary protons up to and including the target should produce events at an equal rate in ν and $\bar{\nu}$ runs. On the contrary, we observe an absolute rate 4 times larger in the ν run than in the $\bar{\nu}$ run. If the neutral hadrons are due to defocussed secondary pions and kaons, the disagreement is larger since we expect $1-2$ times more events in $\bar{\nu}$ than in ν. Since the whole installation is shielded from below by earth we should again expect up-down asymmetries in the NC events. This is not observed.

The most important source of background is the interaction of neutral hadrons produced by the undetected neutrino interactions in the shielding. The high elasticity (0.7) of the neutrons causes a cascade effect in propagation through the shielding. The neutron energy spectrum at production can, in principle, be obtained from the AS events together with available nucleon-nucleus data. Due to the limited statistics in the AS events we make the extreme assumption that all the NC events are neutron produced an use their observed energy spectrum to calculate the neutron spectrum from neutrino interactions. This gives an energy dependence described by E^{-2}. The effective interaction length λ_e of neutrons in the shielding is then found to be 2.5 times the inelastic interaction length, λ_i. A smaller effective interaction length is found for K_L^0 although the background from this source must be negligible since we find no examples of Λ^0 hyperon production among the NC events.

From the absolute value of the number of AS events, we can calculate the number of background events. This has been done by Monte-Carlo generation of events in the shielding surrounding the fiducial volume according to the radial intensity distribution of the beam. The ratio of background events (B) to AS events is found to be B/AS = 0.7 for $\lambda_e = 2.5 \lambda_i$.

If the NC sample has to be explained as being entirely due to neutral hadrons, the Monte-Carlo requires $\lambda_e/\lambda_i > 10$, instead of the best estimate of 2.5. Both ratios would predict distributions along the beam direction in the chamber in strong disagreement with those observed.

Another evaluation of this type of background has been made using the simple assumption that an equilibrium of neutral hadrons with neutrinos exists throughout the entire chamber/shielding assembly. For a radially uniform ν flux it gives B/AS < 1.0 which confirms the Monte-Carlo prediction.

Conclusion. We have observed events without secondary muon or electron, induced by neutral penetrating particles. We are not able to explain the bulk of the signal by any known source of background, unless the effective interaction length of neutrons and K_L^0 is at least 10 times the inelastic interaction length. These events behave similarly to the hadronic part of the charged current events. They could be attributed to neutral current induced reactions, other penetrating particles than ν_μ and ν_e, heavy leptons decaying mainly into hadrons, or by penetrating particles produced by neutrinos and in equilibrium with the ν beam.

On subtraction of the best estimate of the neutral hadron background, and taking into account the $\nu(\bar{\nu})$ contamination in the $\bar{\nu}$ (ν) beam, our best estimates of the NC/CC ratios are

$$(NC/CC)_\nu = 0.21 \pm 0.03$$

$$(NC/CC)_{\bar{\nu}} = 0.45 \pm 0.09$$

where the stated errors are statistical only. If the events are due to neutral currents, these two results are compatible with the same value of Weinberg parameter, $\sin^2 \theta_W$ [1-3] in the range 0.3 to 0.4.

References

[1] S. Weinberg, Phys. Rev. D5 (1972) 1412.
[2] A. Pais and S.B. Treiman, Phys. Rev. D6 (1972) 2700.
[3] E.A. Paschos and L. Wolfenstein, Phys. Rev. D7 (1973) 91.

140

PARITY NON-CONSERVATION IN INELASTIC ELECTRON SCATTERING ☆

C.Y. PRESCOTT, W.B. ATWOOD, R.L.A. COTTRELL, H. DeSTAEBLER, Edward L. GARWIN,
A. GONIDEC [1], R.H. MILLER, L.S. ROCHESTER, T. SATO [2], D.J. SHERDEN, C.K. SINCLAIR,
S. STEIN and R.E. TAYLOR
Stanford Linear Accelerator Center, Stanford University, Stanford, CA 94305, USA

J.E. CLENDENIN, V.W. HUGHES, N. SASAO [3] and K.P. SCHÜLER
Yale University, New Haven, CT 06520, USA

M.G. BORGHINI
CERN, Geneva, Switzerland

K. LÜBELSMEYER
Technische Hochschule Aachen, Aachen, West Germany

and

W. JENTSCHKE
II. Institut für Experimentalphysik, Universität Hamburg, Hamburg, West Germany

Received 14 July 1978

> We have measured parity violating asymmetries in the inelastic scattering of longitudinally polarized electrons from deuterium and hydrogen. For deuterium near $Q^2 = 1.6$ (GeV/c)^2 the asymmetry is $(-9.5 \times 10^{-5})Q^2$ with statistical and systematic uncertainties each about 10%.

We have observed a parity non-conserving asymmetry in the inelastic scattering of longitudinally polarized electrons from an unpolarized deuterium target. In this experiment a polarized electron beam of energy between 16.2 and 22.2 GeV was incident upon a liquid deuterium target. Inelastically scattered electrons from the reaction

$$e(\text{polarized}) + d \rightarrow e' + X, \quad (1)$$

☆ Work supported by the Dept. of Energy.
[1] Permanent address: Annecy (LAPP), 74019 Annecy-le-Vieux, France.
[2] Permanent address: National Laboratory for High Energy Physics, Tsukuba, Japan.
[3] Present address: Department of Physics, Kyoto University, Kyoto, Japan.

were momentum analyzed in a magnetic spectrometer at 4° and detected in a counter system instrumented to measure the electron flux, rather than to count individual scattered electrons. The momentum transfer, Q^2, to the recoiling hadronic system varied between 1 and 1.9 (GeV/c)^2 (see table 1).

Parity violating effects may arise from the interference between the weak and electromagnetic amplitudes. Calculations of the expected effects in deep inelastic experiments have been reported by several authors [1–7], and asymmetries at the level of $10^{-4} Q^2$ are predicted for the kinematics of our experiment. Previous experiments with muons [8] and electrons [9,10] have not achieved sufficient accuracy to observe such small effects. This same interference of amplitudes may also give rise to measurable effects in

Table 1
Kinematic conditions at which data were taken. The average Q^2 and y values were calculated for the shower counter using a Monte Carlo program.

Beam energy E_0 (GeV)	$g-2$ precession angle θ_{prec} (rad)	Spectrometer setting E' (GeV)	Kinematic quantities averaged over spectrometer	
			Q^2 (GeV/c)^2	y
16.18	5.0π	12.5	1.05	0.18
17.80	5.5π	13.5	1.25	0.19
19.42	6.0π	14.5	1.46	0.21
22.20	6.9π	17.0	1.91	0.21

atomic spectra; experiments on transitions in the spectrum of bismuth have already been reported [11–13].

Of crucial importance to this experiment was the development of an intense source of longitudinally polarized electrons. The source consisted of a gallium arsenide crystal mounted in a structure similar to a regular SLAC gun with the GaAs replacing the usual thermionic cathode. The polarized electrons were produced by optical pumping with circularly polarized photons between the valence and conduction bands in the GaAs, which had been treated to assure a surface with negative electron affinity [14,15]. The light source was a dye laser operated at 710 nm and pulsed to match the linac (1.5 μs pulses at 120 pulses per second). Linearly polarized light from the laser was converted to circularly polarized light by a Pockels cell, a crystal with birefringence proportional to the applied electric field. The plane of polarization of the light incident on the Pockels cell could be varied by rotating a calcite prism. Reversing the sign of the high voltage pulse driving the Pockels cell reversed the helicity of the photons which in turn reversed the helicity of the electrons. This reversal was done randomly on a pulse to pulse basis. The rapid reversals minimized the effects of drifts in the experiment, and the randomization avoided changing the helicity synchronously with periodic changes in experimental parameters. Pulsed beam currents of several hundred milliamperes were achieved, with intensity fluctuations of a few percent.

The longitudinally polarized electrons were accelerated with negligible depolarization as confirmed by earlier tests [16] [+1]. Both the sign and the magnitude of the polarization of the beam at the target were measured periodically by observing the asymmetry in Møller (elastic electron–electron) scattering from a magnetized iron foil [16]. The polarization, $|P_e|$, averaged 0.37. Each measurement had a statistical error less than 0.01; we estimate an overall systematic uncertainty of 0.02. The beam intensity at the target varied between 1 and 4×10^{11} electrons per pulse.

A schematic of the apparatus is shown in fig. 1. The target was a 30 cm cell of liquid deuterium. The spectrometer consisted of a dipole magnet, followed by a single quadrupole and a second dipole. The scattering angle was 4° and the momentum setting was about 20% below the beam energy (see table 1 for the kinematic settings). The acceptance was ±7.4 mrad in scattering angle, ±16.6 mrad in azimuth and about ±30% in momentum, as determined from a Monte Carlo model of the spectrometer.

Two separate electron detectors intercepted electrons analyzed by the spectrometer. The first was a nitrogen-filled Cerenkov counter operated at atmospheric pressure. The second was a lead-glass shower counter with a thickness of nine radiation lengths (the TA counter). Approximately 1000 scattered electrons per pulse entered the counters.

The high rates were handled by integrating the outputs of each phototube rather than by counting individual particles. For each pulse, i, the integrated output of each phototube, N_i, was divided by the integrated beam intensity (charge), Q_i, to form the yield for that pulse, $Y_i = N_i/Q_i$. For the distributions of the Y_i we verified experimentally that the (charge weighted) means of the distributions, $\langle Y \rangle$, were independent of Q, within errors of about ±0.3%, and that the (charge weighted) standard deviations, ΔY, were consistent with the statistical fluctuations expected from the number of scattered electrons per pulse. For a run with n beam pulses the statistical uncertainty on $|Y|$ was given by $\Delta Y/\sqrt{n}$.

As a check on our procedures we measured the asymmetry for a series of runs using the unpolarized beam from the regular SLAC gun for which the asymmetry should be zero. For a given run the experimental asymmetry was given by:

$$A_{\exp} = [\langle Y(+) \rangle - \langle Y(-) \rangle] / [\langle Y(+) \rangle + \langle Y(-) \rangle], \qquad (2)$$

[+1] The present experiment used the same target as ref. [16], but used a different spectrometer and detectors.

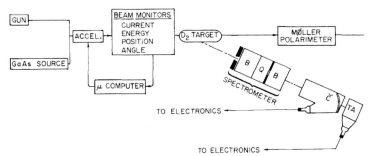

Fig. 1. Schematic layout of the experiment. Electrons from the GaAs source or the regular gun are accelerated by the linac. After momentum analysis in the beam transport system the beam passes through a liquid deuterium target. Particles scattered at 4° are analyzed in the spectrometer (bend-quad-bend) and detected in two separate counters (a gas Cerenkov counter, and a lead-glass shower counter). A beam monitoring system and a polarization analyzer are only indicated, but they provide important information in the experiment.

where + and − were assigned by the same random number generator that determined the sign of the voltage applied to the Pockels cell. For the shower counter we obtained a value of $(-2.5 \pm 2.2) \times 10^{-5}$ for A_{exp} divided by 0.37, the average value of $|P_e|$ for polarized beams from the GaAs source. The individual values were distributed about zero consistent with the calculated statistical errors. We conclude that asymmetries can be measured in this apparatus to a level of about 10^{-5}.

The same procedures were next applied to a similar series of runs using polarized beams. The helicity of the electrons coming from the source depended on the orientation of the linearly polarizing prism as well as on the sign of the voltage on the Pockels cell. Rotation of the plane of polarization by rotating the calcite prism through an angle ϕ_p caused the net electron helicity to vary as $\cos(2\phi_p)$. We chose three operating conditions:

(a) prism orientation at 0°, producing + (−) helicity electrons for + (−) Pockels cell voltage;

(b) prism orientation at 45°, producing unpolarized electrons for either sign of Pockels cell voltage; and

(c) prism orientation at 90°, producing − (+) helicity electrons for + (−) Pockels cell voltage.

Positive helicity indicates that the spin is parallel to the direction of motion. As the prism is rotated by 90°, A_{exp} should change sign since it is defined only with respect to the sign of the voltage on the Pockels cell. We may define a physics asymmetry, A, whose sign depends on the helicity of the beam at the target

$$A_{exp} = |P_e| A \cos(2\phi_p), \qquad (3)$$

where ϕ_p is the angle of orientation of the calcite prism.

Fig. 2 shows the results at 19.4 GeV for $A_{exp}/|P_e|$. For the 45° point we used a value of 0.37 for $|P_e|$. These data are in satisfactory agreement with expecta-

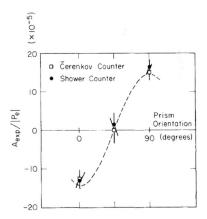

Fig. 2. The experimental asymmetry shows the expected variation (dashed line) as the beam helicity changes due to the change in orientation of the calcite prism. The data are for 19.4 GeV and deuterium. Since the same scattered particles strike both counters, they are not statistically independent. No systematic errors are shown. No corrections have been made for helicity dependent differences in beam parameters.

349

tions, and serve to separate effects due to the helicity of the beam from possible systematic effects associated with the reversal of the Pockels cell voltage. Only statistical errors are shown. The results at 45° are consistent with zero and indicate that other sources of error in A_{exp} must be small. Furthermore, the asymmetries measured at 0° and 90° are equal and opposite, within errors, as expected. Fig. 2 shows data from both the Cerenkov counter and the shower counter. Although these two separate counters were not statistically independent, they were analyzed with independent electronics and responded quite differently to potential backgrounds. The consistency between these counters serves as a check that such backgrounds are small.

At 19.4 GeV with the prism at 0° the helicity at the target was positive for positive Pockels cell voltage. However, this helicity depended on beam energy, owing to the $g - 2$ precession of the spin in the transport magnets which deflected the beam through 24.5° before reaching the target. Because of the anomalous magnetic moment of the electron, the electron spin direction precessed relative to the momentum direction by an angle

$$\theta_{\text{prec}} = \frac{E_0}{m_e c^2} \frac{g-2}{2} \theta_{\text{bend}} = \frac{E_0 \text{(GeV)}}{3.237} \pi \text{ rad}, \quad (4)$$

where m_e is the mass and g the gyromagnetic ratio of the electron. Thus we expect

$$A_{\text{exp}} = |P_e| A \cos[(E_0 \text{(GeV)}/3.237)\pi], \quad (5)$$

where the signs of values of A_{exp} for the prism at 90° have been reversed before combining with values for the prism at 0°. Fig. 3 shows the results for the kinematic points in table 1 as a function of beam energy. At each point Q^2 is different. Since we expect A to be proportional to Q^2, we divide A_{exp} by Q^2 [‡2]. Fig. 3 also shows the expected curve normalized to the point at 19.4 GeV. The data clearly follow the $g - 2$ modulation of the helicity. At 17.8 GeV the spin is transverse; any effects from transverse components of the spin are expected to be negligible, in agreement with our data.

We conclude from figs. 2 and 3 that the observed asymmetries are due to electron helicity. Nevertheless,

---

[‡2] This fact is true in all models. It arises because the electromagnetic amplitude has a $1/Q^2$ dependence, giving an asymmetry proportional to Q^2.

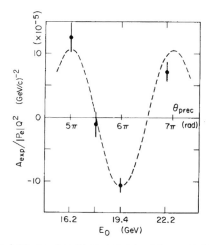

Fig. 3. The experimental asymmetry shows the expected variation (dashed line) as the beam helicity changes as a function of beam energy due to the $g - 2$ precession in the beam transport system. The data are for the shower counter and the deuterium target. No systematic errors are shown. No corrections have been made for helicity dependent differences in beam parameters.

it is essential to search for and set limits on asymmetries due to effects other than helicity. Systematic effects due to slow drifts in phototube gains, magnet currents, etc., were minimized by the rapid, random reversals of polarization, and had negligible effects on A_{exp}. Effects due to random fluctuations in the beam parameters were small compared to the 3% pulse to pulse fluctuations due to counting statistics in the detectors. This was verified experimentally by measuring A_{exp} with unpolarized beams from the regular SLAC gun, and also by generating "fake" asymmetries using pulses of the same helicity from the polarized data runs themselves.

A more serious source of potential error came from small systematic differences between the beam parameters for the two helicities. Small changes in position, angle, current or energy of the beam can influence the measured yields. If these changes are correlated with reversals of the beam helicity, they may cause apparent parity violating asymmetries. Using an extensive beam monitoring system based on microwave cavities, measurements were made for each beam pulse of the average energy and position [17]. Angles were determined from cavities 50 m apart. The beam charge was determined using the standard toroid monitors [18]. The resolutions per pulse were about 10 μm in position, 0.3 μrad in angle, 0.01% in energy, and 0.02% in beam intensity. A microcomputer driven feedback system used position and energy signals to stabilize the average beam position, angle, and energy. Using the measured pulse to pulse beam information together with the measured sensitivities of the yield to each of the beam parameters, we made corrections to the asymmetries for helicity dependent differences in beam parameters. For these corrections, we have assigned a systematic error equal to the correction itself. The most significant imbalance was less than one part per million in E_0 which contributed -0.26×10^{-5} to A/Q^2.

We combine the values of A/Q^2 from the shower counter for the two highest energy points to obtain

$$A/Q^2 = (-9.5 \pm 1.6) \times 10^{-5} \ (\text{GeV}/c)^{-2} \ (\text{deuterium}). \quad (6)$$

We do not include the point at 16.2 GeV because it contains fairly strong elastic and resonance contributions. The sign implies a greater yield from electrons with spin antiparallel to momentum. For this combined point the average value of $y = 1 - E'/E_0$ is 0.21 and the average value of Q^2 is 1.6 $(\text{GeV}/c)^2$. The quoted error, based on preliminary analysis, is derived from a statistical error of $\pm 0.86 \times 10^{-5}$ added linearly to estimated systematic uncertainties of 5% in the value of $|P_e|$, and of 3.3% from asymmetries in beam parameters. We determined experimentally that the π^- background contributed less than 0.1×10^{-5} to A/Q^2. The result in eq. (6) includes normalization corrections of 2% for the π^- background, and 3% for radiative corrections.

Any observation of non-conservation of parity in interactions involving electrons adds new information on the nature of neutral currents and gauge theories. Certain classes of gauge theory models predict no observable parity violations in experiments such as ours. Among these are those left–right symmetric models in which the difference between neutral current neutrino and anti-neutrino scattering cross sections is explained as a consequence of the handedness of the neutrino and anti-neutrino, while the underlying dynamics are parity conserving. Such models are incompatible with the results presented here.

The simplest gauge theories are based on the gauge

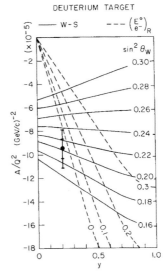

Fig. 4. Comparison of our result for deuterium with two SU(2) × U(1) predictions using the simple quark-parton model for nucleons. The outer error bars correspond to the error quoted in the text (eq. (6)). The inner error bars correspond to the statistical error. The y-dependence of A/Q^2 for various values of $\sin^2\theta_W$ is shown for two models: Weinberg–Salam (solid lines) and the hybrid model (dashed line).

group SU(2) × U(1). Within this framework the original Weinberg–Salam (W–S) model makes specific weak isospin assignments: the left-handed electron and quarks are in doublets, the right-handed electron and quarks are singlets [19]. Other assignments are possible, however. In particular, the "hybrid" or "mixed" model that assigns the right-handed electron to a doublet and the right-handed quarks to singlets has not been ruled out by neutrino experiments.

To make specific predictions for parity violation in inelastic electron scattering, it is necessary to have a model for the nucleon, and the customary one is the simple quark-parton model. The predicted asymmetries depend on the kinematic variable y as well as on the weak isospin assignments and on $\sin^2\theta_W$, where θ_W is the Weinberg angle. Fig. 4 compares our result for two SU(2) × U(1) models. The simplest model (W–S) is in good agreement with our measurement for $\sin^2\theta_W = 0.20 \pm 0.03$ which is consistent with the values obtained in neutrino experiments. The hybrid model is consistent with our data only for values of $\sin^2\theta_W \lesssim 0.1$.

We took a limited amount of data at 19.4 GeV using a liquid hydrogen target with the result

$$A/Q^2 = (-9.7 \pm 2.7) \times 10^{-5} \, (\text{GeV}/c)^{-2} \text{ (hydrogen)}, \tag{7}$$

where the error contains both statistical and systematic uncertainties. A proton target provides a different mix of quarks and is expected to give a slightly smaller asymmetry than deuterium [7]. Our results are not inconsistent with this expectation.

It is a pleasure to acknowledge the support we received from many people at SLAC. In particular we would like to thank M.J. Browne, G.J. Collet, R.L. Eisele, Z.D. Farkas, H.A. Hogg, C.A. Logg and H.L. Martin for especially significant contributions.

References

[1] A. Love et al., Nucl. Phys. B49 (1972) 513.
[2] E. Derman, Phys. Rev. D7 (1973) 2755.
[3] W.W. Wilson, Phys. Rev. D10 (1974) 218.
[4] S.M. Berman and J.R. Primack, Phys. Rev. D9 (1974) 2171; D10 (1974) 3895 (erratum).
[5] M.A.B. Beg and G. Feinberg, Phys. Rev. Lett. 33 (1974) 606.
[6] S.M. Bilenkii et al., Sov. J. Nucl. Phys. 21 (1975) 657.
[7] R.N. Cahn and F.J. Gilman, Phys. Rev. D17 (1978) 1313; further references to the theory may be found in this reference.
[8] Y.B. Bushnin et al., Sov. J. Nucl. Phys. 24 (1976) 279.
[9] M.J. Alguard et al., Phys. Rev. Lett. 37 (1976) 1258, 1261; 41 (1978) 70.
[10] W.B. Atwood et al., SLAC preprint SLAC-PUB-2123 (1978).
[11] L.L. Lewis et al., Phys. Rev. Lett. 39 (1977) 795.
[12] P.E.G. Baird et al., Phys. Rev. Lett. 39 (1977) 798.
[13] L.M. Barkov and M.S. Zolotorev, Zh. Eskp. Teor. Fiz. Pis'ma 26 (1978) 379.
[14] E.L. Garwin, D.T. Pierce and H.C. Siegmann, Swiss Physical Society Meeting (1974), Helv. Phys. Acta 47 (1974) 393 (abstract only); the full paper is available as SLAC-PUB-1576 (1975) (unpublished).
[15] D.T. Pierce et al., Phys. Lett. 51A (1975) 465; Appl. Phys. Lett. 26 (1975) 670.
[16] P.S. Cooper et al., Phys. Rev. Lett. 34 (1975) 1589.
[17] Z.D. Farkas et al., SLAC-PUB-1823 (1976).
[18] R.S. Larsen and D. Horelick, in: Proc. Symp. on Beam intensity measurement, DNPL/R1, Daresbury Nuclear Physics Laboratory (1968); their contribution is available as SLAC-PUB-398.
[19] S. Weinberg, Phys. Rev. Lett. 19 (1967) 1264; A. Salam, in: Elementary particle theory: relativistic groups and analyticity, Nobel Symp. No. 8, ed. N. Svartholm (Almqvist and Wiksell, Stockholm, 1968) p. 367.

EXPERIMENTAL OBSERVATION OF ISOLATED LARGE TRANSVERSE ENERGY ELECTRONS WITH ASSOCIATED MISSING ENERGY AT $\sqrt{s}$ = 540 GeV

UA1 Collaboration, CERN, Geneva, Switzerland

G. ARNISON [j], A. ASTBURY [j], B. AUBERT [b], C. BACCI [i], G. BAUER [1], A. BÉZAGUET [d], R. BÖCK [d], T.J.V. BOWCOCK [f], M. CALVETTI [d], T. CARROLL [d], P. CATZ [b], P. CENNINI [d], S. CENTRO [d], F. CERADINI [d], S. CITTOLIN [d], D. CLINE [1], C. COCHET [k], J. COLAS [b], M. CORDEN [c], D. DALLMAN [d], M. DeBEER [k], M. DELLA NEGRA [b], M. DEMOULIN [d], D. DENEGRI [k], A. Di CIACCIO [i], D. DiBITONTO [d], L. DOBRZYNSKI [g], J.D. DOWELL [c], M. EDWARDS [c], K. EGGERT [a], E. EISENHANDLER [f], N. ELLIS [d], P. ERHARD [a], H. FAISSNER [a], G. FONTAINE [g], R. FREY [h], R. FRÜHWIRTH [1], J. GARVEY [c], S. GEER [g], C. GHESQUIÈRE [g], P. GHEZ [b], K.L. GIBONI [a], W.R. GIBSON [f], Y. GIRAUD-HÉRAUD [g], A. GIVERNAUD [k], A. GONIDEC [b], G. GRAYER [j], P. GUTIERREZ [h], T. HANSL-KOZANECKA [a], W.J. HAYNES [j], L.O. HERTZBERGER [2], C. HODGES [h], D. HOFFMANN [a], H. HOFFMANN [d], D.J. HOLTHUIZEN [2], R.J. HOMER [c], A. HONMA [f], W. JANK [d], G. JORAT [d], P.I.P. KALMUS [f], V. KARIMÄKI [e], R. KEELER [f], I. KENYON [c], A. KERNAN [h], R. KINNUNEN [e], H. KOWALSKI [d], W. KOZANECKI [h], D. KRYN [d], F. LACAVA [d], J.-P. LAUGIER [k], J.-P. LEES [b], H. LEHMANN [a], K. LEUCHS [a], A. LÉVÊQUE [k], D. LINGLIN [b], E. LOCCI [k], M. LORET [k], J.-J. MALOSSE [k], T. MARKIEWICZ [d], G. MAURIN [d], T. McMAHON [c], J.-P. MENDIBURU [g], M.-N. MINARD [b], M. MORICCA [i], H. MUIRHEAD [d], F. MULLER [d], A.K. NANDI [j], L. NAUMANN [d], A. NORTON [d], A. ORKIN-LECOURTOIS [g], L. PAOLUZI [i], G. PETRUCCI [d], G. PIANO MORTARI [i], M. PIMIÄ [e], A. PLACCI [d], E. RADERMACHER [a], J. RANSDELL [h], H. REITHLER [a], J.-P. REVOL [d], J. RICH [k], M. RIJSSENBEEK [d], C. ROBERTS [j], J. ROHLF [d], P. ROSSI [d], C. RUBBIA [d], B. SADOULET [d], G. SAJOT [g], G. SALVI [f], G. SALVINI [i], J. SASS [k], J. SAUDRAIX [k], A. SAVOY-NAVARRO [k], D. SCHINZEL [f], W. SCOTT [j], T.P. SHAH [j], M. SPIRO [k], J. STRAUSS [1], K. SUMOROK [c], F. SZONCSO [1], D. SMITH [h], C. TAO [d], G. THOMPSON [f], J. TIMMER [d], E. TSCHESLOG [a], J. TUOMINIEMI [e], S. Van der MEER [d], J.-P. VIALLE [d], J. VRANA [g], V. VUILLEMIN [d], H.D. WAHL [1], P. WATKINS [c], J. WILSON [c], Y.G. XIE [d], M. YVERT [b] and E. ZURFLUH [d]

Aachen [a] – Annecy (LAPP) [b] – Birmingham [c] – CERN [d] – Helsinki [e] – Queen Mary College, London [f] – Paris (Coll. de France) [g] – Riverside [h] – Rome [i] – Rutherford Appleton Lab. [j] – Saclay (CEN) [k] – Vienna [1] Collaboration

Received 23 January 1983

We report the results of two searches made on data recorded at the CERN SPS Proton–Antiproton Collider: one for isolated large-E_T electrons, the other for large-E_T neutrinos using the technique of missing transverse energy. Both searches converge to the same events, which have the signature of a two-body decay of a particle of mass ~80 GeV/c^2. The topology as well as the number of events fits well the hypothesis that they are produced by the process $\bar{p} + p \rightarrow W^\pm + X$, with $W^\pm \rightarrow e^\pm + \nu$; where $W^\pm$ is the Intermediate Vector Boson postulated by the unified theory of weak and electromagnetic interactions.

---

[1] University of Wisconsin, Madison, WI, USA.
[2] NIKHEF, Amsterdam, The Netherlands.

1. Introduction.

It is generally postulated that the beta decay, namely (quark) → (quark) + $e^{\pm}$ + ν is mediated by one of two charged Intermediate Vector Bosons (IVBs), W^+ and W^- of very large masses. If these particles exist, an enhancement of the cross section for the process (quark) + (antiquark) → $e^{\pm}$ + ν should occur at centre-of-mass energies in the vicinity of the IVB mass (pole), where direct experimental observation and a study of the properties of such particles become possible. The CERN Super Proton Synchrotron (SPS) Collider, in which proton and antiproton collisions at $\sqrt{s}$ = 540 GeV provide a rich sample of quark–antiquark events, has been designed with this search as the primary goal [1].

Properties of IVBs become better specified within the theoretical frame of the unified weak and electromagnetic theory and of the Weinberg–Salam model [2]. The mass of the IVB is precisely predicted [3]:

$$M_{W^{\pm}} = (82 \pm 2.4) \text{ GeV}/c^2$$

for the presently preferred [4] experimental value of the Weinberg angle $\sin^2\theta_W$ = 0.23 ± 0.01. The cross section for production is also reasonably well anticipated [5]

$$\sigma(p\bar{p} \to W^{\pm} \to e^{\pm} + \nu) \simeq 0.4 \times 10^{-33} k \text{ cm}^2 ,$$

where k is an enhancement factor of ~1.5, which can be related to a similar well-known effect in the Drell–Yan production of lepton pairs. It arises from additional QCD diagrams in the production reaction with emission of gluons. In our search we have reduced the value of k by accepting only those events which show no evidence for associated jet structure in the detector.

2. The detector.

The UA1 apparatus has already been extensively described elsewhere [6]. Here we concentrate on those aspects of the detector which are relevant to the present investigation.

The detector is a transverse dipole magnet which produces a uniform field of 0.7 T over a volume of 7 × 3.5 × 3.5 m³. The interaction point is surrounded by the central detector (CD): a cylindrical drift chamber volume, 5.8 m long and 2.3 m in diameter, which yields a bubble-chamber quality picture of each $p\bar{p}$ interaction in addition to measuring momentum and specific ionization of all charged tracks.

Momentum precision for high-momentum particles is dominated by a localization error inherent to the system (≤100 μm) and the diffusion of electrons drifting in the gas (proportional to $\sqrt{l}$ and about 350 μm after l = 22 cm maximum drift length). This results in a typical relative accuracy of ±20% for a 1 m long track at p = 40 GeV/c, and in the plane normal to the magnetic field. The precision, of course, improves considerably for longer tracks. The ionization of tracks can be measured by the classical method of the truncated mean of the 60% lowest readings to an accuracy of 10%. This allows an unambiguous identification of narrow, high-energy particle bundles (e^+e^- pairs or pencil jets) which cannot be resolved by the drift chamber digitizings.

The central section of electromagnetic and hadronic calorimetry has been used in the present investigation to identify electrons over a pseudorapidity interval $|\eta| < 3$ with full azimuthal coverage. Additional calorimetry, both electromagnetic and hadronic, extends to the forward regions of the experiment, down to 0.2° (for details, see table 1).

The central electromagnetic calorimeters consist of two different parts:

(i) 48 semicylindrical modules of alternate layers of scintillator and lead (gondolas), arranged in two cylindrical half-shells, one on either side of the beam axis with an inner radius of 1.36 m. Each module extends over approximately 180° in azimuth and measures 22.5 cm in the beam direction. The light produced in each of the four separate segmentations in depth is seen by wavelength shifter plates on each side of the counter, in turn connected to four photomultipliers (PMs), two at the top and two at the bottom. Light attenuation is exploited in order to further improve the calorimetric information: the comparison of the pulse heights of the top and bottom PM of each segment gives a measurement of the azimuthal angle ϕ for localized energy depositions, $\Delta\phi$ (rad) = 0.3/$[E(\text{GeV})]^{1/2}$. A similar localization along the beam direction is possible using the complementary pairing of PMs. The energy resolution for electrons using all four PMs is $\Delta E/E = 0.15/[E(\text{GeV})]^{1/2}$.

(ii) 64 petals of end-cap electromagnetic shower counters (bouchons), segmented four times in depth, on both sides of the central detector at 3 m distance from the beam crossing point. The position of each shower is measured with a position detector located inside the calorimeter at a depth of 11 radiation lengths, i.e. after the first two segments. It consists of

104

Table 1
Calorimetry.

Calorimeter		Angular coverage θ (deg)	Thickness		Cell size		Sampling step	Segmentation in depth	Resolution
			No. rad. lengths	No. abs. lengths	$\Delta\theta$ (deg)	$\Delta\phi$ (deg)			
barrel EM:	gondolas	25 −155	26.4/sin θ	1.1/sin θ	5	180	1.2 mm Pb 1.5 mm scint.	3.3/6.5/10.1/6.5 X_0	$0.15/\sqrt{E}$
hadr.:	c's	25 −155	–	5.0/sin θ	15	18	50 mm Fe 10 mm scint.	2.5/2.5 λ	$0.8/\sqrt{E}$
end-caps EM:	bouchons	5 − 25	27/cos θ	1.1/cos θ	20	11	4 mm Pb 6 mm scint.	4/7/9/7 X_0	$0.12/\sqrt{E_T}$
hadr.:	I's	155 −175	–	7.1/cos θ	5	10	50 mm Fe 10 mm scint.	3.5/3.5 λ	$0.8/\sqrt{E}$
calcom	EM	0.7− 5	30	1.2	4	45	3 mm Pb 3 mm scint.	4 × 7.5 X_0	$0.15/\sqrt{E}$
	hadr.	175 −179.3	–	10.2	–	–	40 mm Fe 8 mm scint.	6 × 1.7 λ	$0.8/\sqrt{E}$
very forward	EM	0.2− 0.7	24.5	1.0	0.5	90	3 mm Pb 6 mm scint.	5.7/5.3/5.8/7.7 X_0	$0.15/\sqrt{E}$
	hadr.	179.3−179.8	–	5.7	0.5	90	40 mm Fe 10 mm scint.	5 × 1.25 λ	$0.8/\sqrt{E}$

two planes of orthogonal proportional tubes of 2 × 2 cm^2 cross section and it locates the centre of gravity of energetic electromagnetic showers to ±2 mm in space. The attenuation length of the scintillator has been chosen to match the variation of sin θ over the radius of the calorimeters, so as to directly measure in first approximation $E_T = E \sin\theta$ rather than the true energy deposition E, which can, however, be determined later, using the information from the position detector. This technique permits us to read out directly from the end-cap detectors the amount of transverse energy deposited, without reconstruction of the event topology.

3. Electron identification. Electromagnetic showers are identified by their characteristic transition curve, and in particular by the lack of penetration in the hadron calorimeter behind them. The performance of the detectors with respect to hadrons and electrons has been studied extensively in a test beam as a function of the energy, the angle of incidence, and the location of impact. The fraction of hadrons (pions) delivering an energy deposition E_c below a given threshold in the hadron calorimeter is a rapidly falling function of energy, amounting to about 0.3% for $p \simeq 40$ GeV/c

and $E_c <$ 200 MeV. Under these conditions, 98% of the electrons are detected.

4. Neutrino identification. The emission of one (or more) neutrinos can be signalled only by an apparent visible energy imbalance of the event (missing energy). In order to permit such a measurement, calorimeters have been made completely hermetic down to angles of 0.2° with respect to the direction of the beams. (In practice, 97% of the mass of the magnet is calorimetrized.) It is possible to define an energy flow vector ΔE, adding vectorially the observed energy depositions over the whole solid angle. Neglecting particle masses and with an ideal calorimeter response and solid-angle coverage, momentum conservation requires $\Delta E = 0$. We have tested this technique on minimum bias and jet-enriched events for which neutrino emission ordinarily does not occur. The transverse components ΔE_y and ΔE_z exhibit small residuals centred on zero with an rms deviation well described by the law $\Delta E_{y,z} = 0.4(\Sigma_i |E_T^i|)^{1/2}$, where all units are in GeV and the quantity under the square root is the scalar sum of all transverse energy contributions recorded in the event (fig. 1). The distributions have gaussian shape and no prominent tails. The longitudinal component

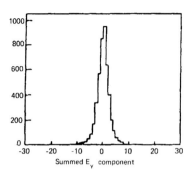
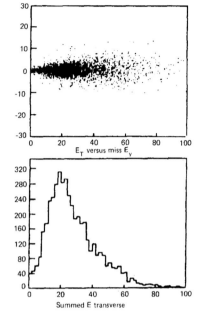

Fig. 1. The missing transverse energy in the y direction [ΔE_y (GeV)] plotted versus the scalar sum of missing transverse energy [E_T (GeV)] for minimum bias triggers. The y-axis is pointing up vertically.

of energy ΔE_x is affected by the energy flow escaping through the 0° singularity of the collider's beam pipe and it cannot be of much practical use. We remark that, like neutrinos, high-energy muons easily penetrate the calorimeter and leak out substantial amounts of energy. A muon detector, consisting of stacks of eight planes of drift chambers, surrounds the whole apparatus and has been used to identify such processes, which are occurring at the level of 1 event per nanobarn for $\Delta E_{y,z} \geqslant 10$ GeV.

5. Data-taking and initial event selections. The present work is based on data recorded in a 30-day period during November and December 1982. The integrated luminosity after subtraction of dead-time and other instrumental inefficiencies was 18 nb^{-1}, corresponding to about 10^9 collisions between protons and antiprotons at $\sqrt{s}$ = 540 GeV.

For each beam–beam collision detected by scintillator hodoscopes, the energy depositions in all calorimeter cells after fast digitization were processed, in the time prior to the occurrence of the next beam–beam crossing, by a fast arithmetic processor in order to recognize the presence of a localized electromagnetic energy deposition, namely of at least 10 GeV of transverse energy either in two gondola elements or in two bouchon petals. In addition, we have simultaneously operated three other trigger conditions: (i) a jet trigger, with $\geqslant 15$ GeV of transverse energy in a localized cluster [‡1] of electromagnetic and hadron calorimeters; (ii) a global E_T trigger, with >40 GeV of total transverse energy from all calorimeters with $|\eta| < 1.4$; and (iii) a muon trigger, namely at least one penetrating track with $|\eta| < 1.3$ pointing to the diamond.

The electron trigger rate was about 0.2 event per second at the (peak) luminosity $L = 5 \times 10^{28}$ cm^{-2}s^{-1}. Collisions with residual gas or with vacuum chamber walls were completely negligible, and the apparatus in normal machine conditions yielded an almost pure sample of beam–beam collisions. In total, 9.75×10^5 triggers were collected, of which 1.4×10^5 were char-

---

[‡1] We define a cluster as: (i) a group of eight gondolas and the two hadron calorimeter elements immediately behind; or (ii) a quadrant of bouchon elements (8) with the corresponding hadron calorimeters.

acterized by an electron trigger flag.

Event filtering by calorimetric information was further perfected by off-line selection of 28 000 events with $E_T > 15$ GeV in two gondolas, or $E_T > 15$ GeV in two bouchon petals with valid position-detector information. These events were finally processed with the central detector reconstruction. Of these events there are 2125 with a good quality, vertex-associated charged track of $p_T > 7$ GeV/c. This sample will be used for the subsequent analysis of events in the gondolas.

6. *Search for electron candidates.* We now require three conditions in succession in order to ensure that the track is isolated, namely to reject the debris of jets:

(i) The fast track ($p_T > 7$ GeV/c) as recorded by the central detector must hit a pair of adjacent gondolas with transverse energy $E_T > 15$ GeV (1106 events).

(ii) Other charged tracks, entering the same pair of gondolas, must not add up to more than 2 GeV/c of transverse momenta (276 events).

(iii) The ϕ information from pulse division from gondola phototubes must agree within 3σ with the impact of the track (167 events).

Next we introduce two simple conditions to enhance its electromagnetic nature:

(iv) The energy deposition E_c in the hadronic calorimeters aimed at by the track must not exceed 600 MeV (72 events).

(v) The energy deposited in the gondolas E_{gon} must match the measurement of the momentum of the track p_{CD}, namely $|1/p_{CD} - 1/E_{gon}| < 3\sigma$.

At this point only 39 events are left, which were individually examined by physicists on the visual scanning and interactive facility Megatek. The surviving events break up cleanly into three classes, namely 5 events with no jet activity [+2], 11 with a jet opposite

to the track within a 30° angle in ϕ, and 23 with two jets (one of which contains the electron candidate) or clear e^+e^- conversion pairs. A similar analysis performed on the bouchon has led to another event with no jets. The classes of events have striking differences. We find that whilst events with jet activity have essen-

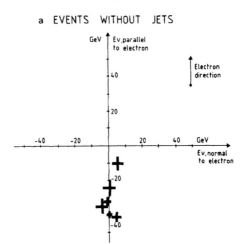

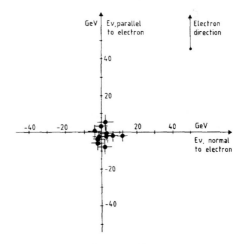

Fig. 2. The missing transverse energy (E_ν) is plotted vectorially against the electron direction for the events yielded by the electron search: (a) without jets, (b) with jets.

[+2] The definition of a jet is based on the UA1 standard algorithm, applied separately on the calorimetry and on the central detector data. Positive results on either set are taken as evidence for a jet. In the calorimetry a four-vector (k_i, E_i) pointing to the interaction vertex is associated with each struck cell. Working in the transverse plane, all vectors with $k_T > 2.5$ GeV are ordered and are used as potential jet initiators. They are combined if their separation in phase space satisfies the cut $\Delta R = [(\Delta\eta)^2 + (\Delta\phi)^2]^{1/2} < 1$ (with ϕ in radians). The remaining soft particles are added to the nearest jet in $\Delta\eta$ and $\Delta\phi$, provided the relative p_T is < 1 GeV and $\Delta\theta < 45°$. A jet is considered valid if $E_T^{jet} > 10$ GeV. This same procedure is used for central detector tracks with appropriately adjusted parameters.

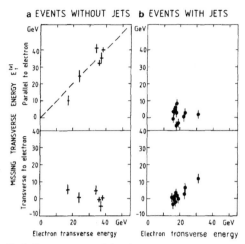

Fig. 3. The components of the missing energy parallel and perpendicular to the electron momentum plotted versus the electron energy for the events found in the electron search: (a) without jets, (b) with jets.

tially no missing energy (fig. 2b) [+3], the ones with no jets show evidence of a missing transverse energy of the same magnitude as the transverse electron energy (fig. 3a), with the vector momenta almost exactly balanced back-to-back (fig. 2a). In order to assess how significant the effect is, we proceed to an alternative analysis based exclusively on the presence of missing transverse energy.

7. *Search for events with energetic neutrinos.* We start again with the initial sample of 2125 events with a charged track of $p_T > 7$ GeV/c. We now move to pick up validated events with a high missing transverse energy and with the candidate track not part of a jet:

(i) The track must point to a pair of gondolas with deposition in excess of $E_T > 15$ GeV and no other track with $p_T > 2$ GeV/c in a 20° cone (911 events).

(ii) Missing transverse energy imbalance in excess of 15 GeV.

Only 70 events survive these simple cuts, as shown in fig. 4. The previously found 5 jetless events of the gondolas are clearly visible. At this point, as for the

---

[+3] The 11 events with an electron and a jet exhibit a p_T^{-4} spectrum with the highest event at $p_T = 32$ GeV/c.

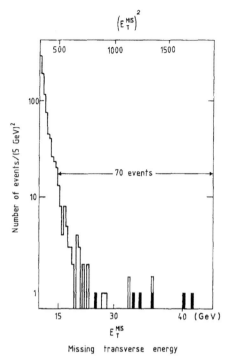

Fig. 4. The distribution of the square of the missing transverse energy for those events which survive the cuts requiring association of the central detector isolated track and a struck gondola in the missing-energy search. The five jetless events from the electron search are indicated.

electron analysis, we process the events at the interactive facility Megatek:

(iii) The missing transverse energy is validated, removing those events in which jets are pointing to where the detector response is limited, i.e. corners, light-pipe ducts going up and down. Some very evident, big secondary interactions in the beam pipe are also removed. We are left with 31 events, of which 21 have $E_c > 0.01\ E_{gon}$ and 10 events in which $E_c < 0.01\ E_{gon}$.

(iv) We require that the candidate track be well isolated, that there is no track with $p_T > 1.5$ GeV in a cone of 30°, and that $E_T < 4$ GeV for neutrals in neighbouring gondolas at similar ϕ angle. Eighteen events survive: ten with $E_c \neq 0$ and eight with $E_c = 0$.

The events once again divide naturally into the two classes: 11 events with jet activity in the azimuth op-

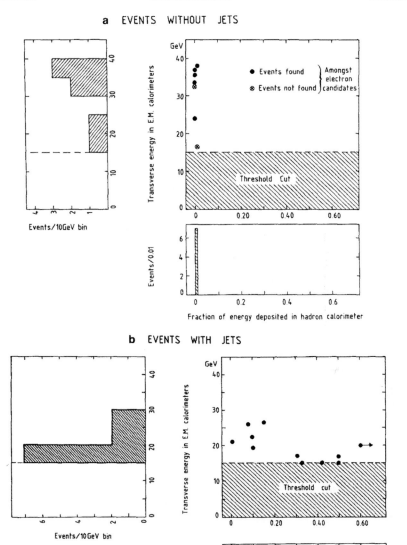

Fig. 5. A plot of the transverse energy in the EM calorimeters versus the fraction of energy deposited in the hadron calorimeters for events which survive the missing-energy search: (a) without jets, (b) with jets.

Table 2
Main parameters of electron events with a large missing transverse energy.

Run, event	Properties of the electron track										Calorimeter information						General event topology			
	E_T (GeV)	E (GeV)	p (GeV/c)	Δp a)	Q	dE/dx I/I_0	y b)	Track No.	Length (m)	Sagitta (mm)	Electromagnetic energy deposition				E_{had} (GeV)	E_{tot} (GeV)	Missing E_T (GeV)	Δφ c) (deg.)	Charged tracks	$\Sigma\|E_T\|$ (GeV)
											Sample 1 (GeV)	Sample 2 (GeV)	Sample 3 (GeV)	Sample 4 (GeV)						
A 2958 1279	26	42	33.8	+6.3 −4.6	−	1.22 ±0.2	+1.1	36	1.36	1.7	4	35	3	0.2	0	278	24.4 ± 4.6	179	65	81
B 3522 214	17	46	47.5	+8.2 −6.1	−	1.37 ±0.16	+1.7	18	1.64	1.5	2	32	10	0.5	0	296	10.9 ± 4.0	219	49	60
C 3524 197	34	45	21.6	+21.8 −7.2	−	1.37 ±0.3	−0.8	26	1.25	2.11	1	30	14	0.2	0	367	41.3 ± 3.6	187	21	68
D 3610 760	38	40	33.4	+33.0 −11.1	−	1.64 ±0.34	+0.3	9	0.98	0.75	3	9	26	2.2	0.4	111	40.0 ± 2.0	181	10	47
E 3701 305	37	37	56.2	+121.3 −22.8	+	1.54 ±0.28	−0.1	12	0.95	0.4	1	18	17	0.9	0	363	35.5 ± 4.3	173	39	87
F 4017 838	37	70	53.1	+6.6 −5.3	−	1.30 ±0.26	+1.4	3	2.01	2.0	19	48	3	0.3	0	177	32.3 ± 2.4	179	14	49
G 3262 1108	40	40	6.7	+1.9 −1.2	−	1.23 ±0.28	0.0	21	0.85	3.0	2	22	15	0.9	0	218	33.4 ±2.9	172	21	63

a) Including 200 μm systematic error. b) y is defined as positive in the direction of the outgoing p̄.
c) Angle between electron and missing energy (neutrino).

posite to the track, and 7 events without detectable jet structure. If we now examine E_c, we see that these two classes are strikingly different, with large E_c for the events with jets (fig. 5b) and negligible E_c for the jetless ones (fig. 5a). We conclude that whilst the first ones are most likely to be hadrons, the latter constitute an electron sample.

We now compare the present result with the candidates of the previous analysis based on electron signature. We remark that five out of the seven events constitute the previous final sample (fig. 5a). Two new events have been added, eliminated previously by the test on energy matching between the central detector and the gondolas. Clearly the same physical process that provided us with the large-p_T electron delivers also high-energy neutrinos. The selectivity of our apparatus is sufficient to isolate such a process from either its electron or its neutrino features individually. If (ν_e, e) pairs and (ν_τ, τ) pairs are both produced at comparable rates, the two additional new events can readily be explained since missing energy can arise equally well from ν_e and ν_τ. Indeed, closer inspection of these events shows them to be compatible with the τ hypothesis, for instance, $\tau^- \to \pi^- \pi^0 \nu_\tau$ with leading π^0. However, our isolation requirements on the charged track strongly biases against most of the τ decay modes.

8. *Detailed description of the electron–neutrino events.* The main properties of the final sample of six events (five gondolas, one bouchon) are given in table 2 and marked A through F. The event G is a τ candidate. One can remark that both charges of the electrons are represented. The successive energy depositions in the gondola samples are consistent with test beam findings. All but event D have no energy deposition in the hadron calorimeter; event D has a 400 MeV visible, 1% leakage beyond 26.4 radiation lengths. Test beam measurements show that this is a possible fluctuation. Multiplicity of the events is widely different: event F (fig. 6b, fig. 7b) has a small charged multiplicity (14), whilst event A (fig. 6a, fig. 7a) is very rich in particles (65). Event B is the bouchon event, and it has a number of features which must be mentioned. A 100 MeV/c track emerges from the vacuum chamber near the exit point of the electron track, which might form a part of an asymmetric electron pair with the candidate. The initial angle between the two tracks would then be 11°, not incompatible with this hypothesis once Coulomb scattering and measurement errors of the two tracks are taken into account. There is also some activity in the muon detector opposite to the electron candidate; the muon track is unmeasurable in the central detector. For these reasons we prefer to limit our final analysis to the events in the gondolas, although we believe that everything is still consistent with event B being a good event.

9. *Background evaluations.* We first consider possible backgrounds to the electron signature for events with no jets. Missing energy (neutrino signature) is not yet advocated. We have taken the following into consideration:

(1) A high-p_T charged pion (hadron) misidentified as an electron, or a high-p_T charged pion (hadron) overlapping with one or more π^0.

The central detector measurement obviously gives only the momentum p of the charged pion. In addition, the electromagnetic detectors can accumulate an arbitrary amount of electromagnetic energy from π^0's, which would simulate the electron behaviour. Since gondolas are thick enough to absorb the electromagnetic cascade, the energy deposition in the hadron calorimeter is dominated by the punch-through of the charged pion of momentum p measured in the central detector, for which rejection tables exist from test beam results. In our 18 nb^{-1} sample we have searched for single-track events with $p_T > 20$ GeV/c, no associated jet, $E_c > 600$ MeV to ensure hadronic signature, and a reasonable energy balance (within 3 SD) between the charged track momentum measurement and the sum of hadronic and electromagnetic energy depositions. We have found no such event. Once the measured pion rejection table is folded in, this background is entirely negligible. A further test against pile-up is given by the matching in the x-direction between the charged track of the central detector and the centroid of the energy depositions in the gondolas, and which is very good for all events.

(2) High-p_T π^0, η^0, or γ internally (Dalitz) or externally converted to an e$^+$e$^-$ pair with one leg missed. The number of isolated EM conversions (π^0, η, γ, etc.) per unit of rapidity has been directly measured as a function of E_T in the bouchons, using the position detectors over the interval 10–40 GeV. From this spectrum, the Bethe–Heitler formula for pair creation, and the Kroll–Wada formula for Dalitz pairs [7], the ex-

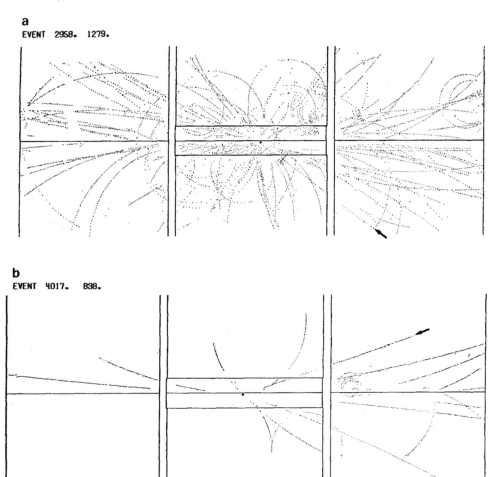

Fig. 6. The digitization from the central detector for the tracks in two of the events which have an identified, isolated, well-measured high-p_T electron: (a) high-multiplicity, 65 associated tracks; (b) low-multiplicity, 14 associated tracks.

pected number of events with a "single" $e^\pm$ with $p_T > 20$ GeV/c is $0.2\,p_0$ (GeV), largely independent of the composition of the EM component; p_0 is the effective momentum below which the low-energy leg of the pair becomes undetectable. Very conservatively, we can take $p_0 = 200$ MeV/c (curvature radius 1.2 m) and conclude that this background is negligible.

(3) Heavy quark associated production, followed by pathological fragmentation and decay configuration, such that $Q_1 \to e(\nu X)$ with the electron leading and the rest undetected, and $Q_2 \to \nu(\ell X)$, with the neutrino leading and the rest undetected. In 5 nb^{-1} we have observed one event in which there is a muon and an electron in separate jets, with $p_T^{(\mu)} = 4.4$ GeV/c and

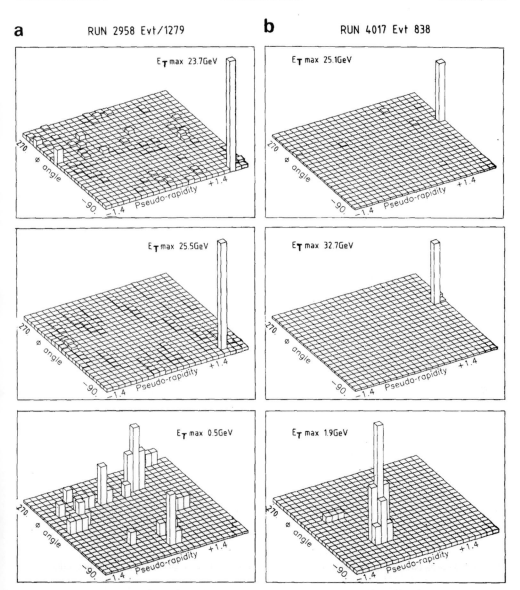

Fig. 7. The energy deposited in the cells of the central calorimetry and the equivalent plot for track momenta in the central detector for the two events of fig. 6. The top diagram shows the electromagnetic cells, the middle shows the central detector tracks, and the bottom plot, with a very much increased sensitivity, shows the energy in the hadron calorimeter. The plots reveal no hadronic energy behind the electron and no jet structure; (a) high-multiplicity; (b) low-multiplicity.

$p_T^{(e)} = 13.3$ GeV/c. Requiring (i) extrapolation to the energy of the events, (ii) fragmentation functions for leading lepton, and (iii) a detection hole for all remaining particles, makes the rate of these background events negligible.

In conclusion, we have been unable to find a background process capable of simulating the observed high-energy electrons. Thus we are led to the conclusion that they are electrons. Likewise we have searched for backgrounds capable of simulating large-E_T neutrino events. Again, none of the processes considered appear to be even near to becoming competitive.

10. Comparison between events and expectations from W decays.

The simultaneous presence of an electron and (one) neutrino of approximately equal and opposite momenta in the transverse direction (fig. 8) suggests the presence of a two-body decay, $W \to e + \nu_e$. The main kinematical quantities of the events are given in table 3. A lower, model-independent bound to the W mass m_W can be obtained from the transverse mass, $m_T^2 = 2 p_T^{(e)} p_T^{(\nu)} (1 - \cos \phi_{\nu e})$, remarking that $m_W \geq m_T$ (fig. 9). We conclude that:

$m_W > 73$ GeV/c^2 (90% confidence level).

A better accuracy can be obtained from the data if one assumes W decay kinematics and standard V − A couplings. The transverse momentum distribution of the W at production also plays a role. We can either (i) extract it from the events (table 3); or, (ii) use theoretical predictions [8].

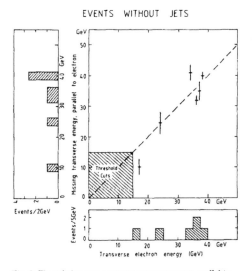

Fig. 8. The missing transverse energy component parallel to the electron, plotted versus the transverse electron energy for the final six electron events without jets (5 gondolas, 1 bouchon) All the events in the gondolas appear well above the threshold cuts used in the searches.

As one can see from fig. 10, there is good agreement between two extreme assumptions of a theoretical model [8] and our observations. By requiring no associated jet, we may have actually biased our sample towards the narrower first-order curve. Fitting of the in-

Table 3
Transverse mass and transverse momentum of a W decaying into an electron and a neutrino computed from the events of table 2.

Run, event	$p_T^{(e)}$ of electron (GeV/c)	$p_T^{(\nu)} =$ missing E_T (GeV)	Transverse mass (GeV/c)^2	$p_T^{(W)} = \|p_T^{(e)} + p_T^{(\nu)}\|$ (GeV)
A 2958 1279	24 ± 0.6	24.4 ± 4.6	48.4 ± 4.6	0.6 ± 4.6
B 3522 214	17 ± 0.4	10.9 ± 4.0	26.5 ± 4.6	10.8 ± 4.0
C 3524 197	34 ± 0.8	41.3 ± 3.6	74.8 ± 3.4	8.6 ± 3.7
D 3610 760	38 ± 1.0	40.0 ± 2.0	78.0 ± 2.2	2.1 ± 2.2
E 3701 305	37 ± 1.0	35.5 ± 4.3	72.4 ± 4.5	4.7 ± 4.4
F 4017 838	36 ± 0.7	32.3 ± 2.4	68.2 ± 2.6	3.8 ± 2.5

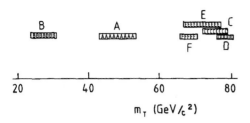

Fig. 9. The distribution of the transverse mass derived from the measured electron and neutrino vectors of the six electron events.

clusive electron spectrum and using full QCD smearing gives $m_W = (74^{+4}_{-4})$ GeV/c^2. The method finally used is the one of correcting, on an event-to-event basis, for the transverse W motion from the $(E_\nu - E_e)$ imbalance, and using the Drell–Yan predictions with no smearing. The result of a fit on electron angle and energy and neutrino transverse energy with allowance for systematic errors, is

$m_W = (81^{+5}_{-5})$ GeV/c^2 ,

in excellent agreement with the expectation of the Weinberg–Salam model [2].

We find that the number of observed events, once detection efficiencies are taken into account, is in agreement with the cross-section estimates based on structure functions, scaling violations, and the Weinberg–Salam parameters for the W particle [5].

We gratefully acknowledge J.B. Adams and L. Van Hove, CERN Directors-General during the initial phase of the project and without whose enthusiasm and support our work would have been impossible. The success of the collider run depended critically upon the superlative performance of the whole of the CERN accelerator complex, which was magnificently operated by its staff.

We are thankful to the management and staff of CERN and of all participating Institutes who have vigorously supported the experiment.

The following funding Agencies have contributed to this programme:
Fonds zur Förderung der Wissenschaftlichen Forschung, Austria.
Valtion luonnontieteellinen toimikunta, Finland.
Institut National de Physique Nucléaire et de Physique des Particules and Institut de Recherche Fondamentale (CEA), France.
Bundesministerium für Forschung und Technologie, Germany.
Istituto Nazionale di Fisica Nucleare, Italy.
Science and Engineering Research Council, United Kingdom.
Department of Energy, USA.

Thanks are also due to the following people who have worked with the collaboration in the preparation and data collection on the runs described here,
F. Bernasconi, F. Cataneo, A.-M. Cnops, L. Dumps, J.-P. Fournier, A. Micolon, S. Palanque, P. Quéru, P. Skimming, G. Stefanini, M. Steuer, J.C. Thevenin, H. Verweij and R. Wilson.

References

[1] C. Rubbia, P. McIntyre and D. Cline, Proc. Intern. Neutrino Conf. (Aachen, 1976) (Vieweg, Braunschweig, 1977) p. 683;
Study Group, Design study of a proton–antiproton colliding beam facility, CERN/PS/AA 78-3 (1978), reprinted in Proc. Workshop on Producing high-luminosity, high-energy proton–antiproton collisions (Berkeley, 1978) report LBL-7574, UC34, p. 189;
The staff of the CERN proton–antiproton project, Phys. Lett. 107B (1981) 306.

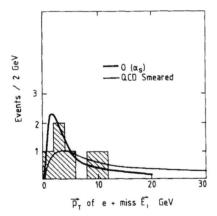

Fig. 10. The transverse momentum distribution of the W derived from our events, using the electron and missing-energy vectors. This is compared with the theoretical predictions of Halzen et al. [8] for W production without [O(α_s)] and with QCD smearing.

[2] S. Weinberg, Phys. Rev. Lett. 19 (1967) 1264;
A. Salam, Proc. 8th Nobel Symp. (Aspenäsgården, 1968) (Almqvist and Wiksell, Stockholm, 1968) p. 367.
[3] A. Sirlin, Phys. Rev. D22 (1980) 971;
W.J. Marciano and A. Sirlin, Phys. Rev. D22 (1980) 2695;
C.H. Llewellyn Smith and J.A. Wheater, Phys. Lett. 105B (1981) 486.
[4] For a review, see: M. Davier, Proc. 21st Intern. Conf. on High-energy physics (Paris, 1982), J. Phys. (Paris) 43 (1982) C3-471.
[5] F.E. Paige, Proc. Topical Conf. on the Production of new particles at super-high energies (University of Wisconsin, Madison, 1979);
L.B. Okun and M.B. Voloshin, Nucl. Phys. B120 (1977) 459;
C. Quigg, Rev. Mod. Phys. 94 (1977) 297;
J. Kogut and J. Shigemitsu, Nucl. Phys. B129 (1977) 461;
R. Horgan and M. Jacob, Proc. CERN School of Physics (Malente, Fed. Rep. Germany, 1980), CERN 81-04, p. 65;
R.F. Peierls, T. Trueman and L.L. Wang, Phys. Rev. D16 (1977) 1397.
[6] UA1 proposal, A 4π solid-angle detector for the SPS used as a proton–antiproton collider at a centre-of-mass energy of 540 GeV, CERN/SPSC 78-06 (1978);
M. Barranco Luque et al., Nucl. Instrum. Methods 176 (1980) 175;
M. Calvetti et al., Nucl. Instrum. Methods 176 (1980) 255;
K. Eggert et al., Nucl. Instrum. Methods 176 (1980) 217, 233;
A. Astbury, Phys. Scr. 23 (1981) 397.
[7] H.M. Kroll and W. Wada, Phys. Rev. 98 (1955) 1355.
[8] F. Halzen and D.M. Scott, Phys. Lett. 78B (1978) 318;
P. Aurenche and F. Lindfors, Nucl. Phys. B185 (1981) 301;
F. Halzen, A.D. Martin, D.M. Scott and M. Dechantsreiter, Phys. Lett. 106B (1981) 147;
M. Chaichian, M. Hayashi and K. Yamagishi, Phys. Rev. D25 (1982) 130;
A. Martin, Proc. Conf. on Antiproton–proton collider physics (Madison, 1981) AIP Proc. No. 85 (American Institute of Physics, New York, 1982) p. 216;
F. Halzen, A.D. Martin and D.M. Scott, Phys. Rev. D25 (1982) 754;
V. Barger and R.J.N. Phillips, University of Wisconsin preprint MAD/PH/78 (1982).

13

Testing the Standard Model

Precision Measurements of the Z and W; Search for the Higgs.

The ψ and Υ resonances were startling and largely unanticipated. By contrast, it was apparent far in advance that the Z would be spectacular in e^+e^- annihilation. Indeed, within the Standard Model nearly every aspect of the Z could be predicted to the extent that $\sin^2\theta_W$ was known. Despite this, the study of the Z in e^+e^- annihilation was a singular achievement in particle physics.

After initial planning as early as 1976, CERN began construction of the Large Electron Positron collider in 1983. Because ultrarelativistic electrons lose energy rapidly through synchrotron radiation, whose intensity varies as E^4/ρ, where ρ is the radius of curvature, LEP was designed with a large circumference, 26.67 km. The first collisions occurred on August 13, 1989.

In a daring move, SLAC aimed to reach the Z before LEP by colliding electron and positron beams generated with its linear accelerator. At the Stanford Linear Collider each bunch would be lost after colliding with the opposing bunch. While the Mark II detector, which had seen service at PEP, was refurbished, four new detectors – ALEPH, DELPHI, L3, and OPAL – were built at CERN.

SLC indeed got to the Z first (Ref. 13.1), but with a disappointing luminosity. In July 1989, Mark II reported for the Z a mass of 91.11 ± 0.23 GeV and a width of $1.61^{+0.60}_{-0.43}$ GeV, based on 106 events.

These results were soon surpassed by measurements at Fermilab. The original accelerator at Fermilab began operation in 1972 with an energy of 200 GeV. At the time of the discovery of the Υ in 1977, it was operating at 400 GeV. Fermilab pioneered the use of superconducting magnets, which increased the operating field to 4 T, allowing the beam energy to be doubled to 800 GeV. Following the lead of the SPS at CERN, Fermilab also constructed a ring in which antiprotons could be accumulated. The Tevatron Collider brought together protons and antiprotons inside the main ring. Through this series of improvements, the operating c.m. energy of the machine increased from about $\sqrt{s} = 20$ GeV to $\sqrt{s} = 1.6$ TeV, from which it was subsequently raised to 1.8 TeV.

The first detector at the Tevatron Collider was CDF, the Collider Detector Facility. A descendant of UA1 and UA2, CDF featured cylindrical geometry, tracking with a drift

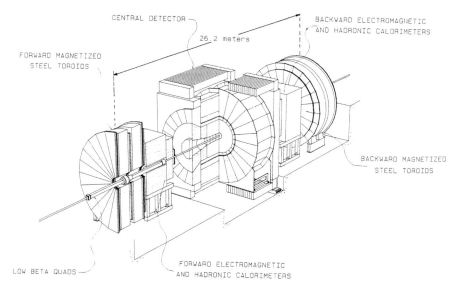

Figure 13.1. The CDF detector circa 1988. From the inside out, the major sections are the inner detector, the electromagnetic and hadronic calorimeters, and finally the magnetized steel toroids for muon identification and measurement. [Courtesy Fermilab and CDF Collaboration.]

chamber inside an axial magnetic field of 1.4 T, and both electromagnetic and hadronic calorimetry outside the magnet. The final layer provided for muon detection and measurement. During the 1988/89 run, a total of 4 pb^{-1} was accumulated.

A second detector at the Tevatron Collider, D0, was completed in 1992. It complemented CDF by optimizing calorimetry at the cost of tracking. In particular, it had no magnetic field in its tracking region. D0's advantage lay in measuring jets at high transverse momentum and in detecting missing transverse momentum, a sign of neutrinos or other non-interacting particles. The energies of electrons and muons could be measured using electromagnetic calorimetry for the former and magnetized absorbers in the outermost layers for the latter.

CDF, pursuing the hadron collider path set by UA-1 and UA-2, found a Z mass of $90.9 \pm 0.3 \pm 0.2$ GeV and a width $3.8 \pm 0.8 \pm 1.0$ GeV (Ref. 13.2) from 188 events. Back at SLAC, Mark II announced new results in October 1989, based on 480 events: $m_Z = 91.14 \pm 0.12$ GeV, $\Gamma_Z = 2.42^{+0.45}_{-0.35}$ GeV.

The high precision measurement of initial interest was the full line shape of the Z because it would reveal the total number of light neutrinos that couple to the Z. While the apparent number was simply three – ν_e, ν_μ, ν_τ – additional generations would appear if their neutrinos were light even if their charged leptons and quarks were too heavy to be produced.

The shape of the Z resonance is determined primarily by the Breit–Wigner form discussed in Chapters 5 and 9. A relativistic version for e^+e^- annihilation through the Z to produce the final state f at cm energy $\sqrt{s}$ is

$$\sigma_f(s) = \frac{12\pi}{m_Z^2} \frac{\Gamma_e \Gamma_f}{\Gamma^2} \frac{s\Gamma^2}{(s-m_Z^2)^2 + s^2\Gamma^2/m_Z^2}. \tag{13.1}$$

Here Γ represents the full width of the Z including its decays to neutrinos, while Γ_f represents the partial width into some final state f and in particular Γ_e is the partial width into e^+e^-. Because the light electrons and positrons can emit photons before annihilating, there is an important radiative correction. This reduces the height at the peak and makes the shape asymmetric. The cross section is higher above the peak than below it because the higher energy electrons and positrons can lose energy and move closer to the resonance.

From the fit to the line shape, the full width Γ could be determined. The peak cross section (with radiative corrections removed) is

$$\sigma_{peak} = \frac{12\pi}{m_Z^2} \text{BR}(\ell)\, \text{BR(had)} \tag{13.2}$$

where BR(had) is the branching ratio for Z into hadrons and BR(ℓ) is the branching ratio for the Z into one of the three charged leptons, assuming the three to be equal. The relative frequency of the charged lepton and hadronic final states, $R_\ell = \text{BR(had)}/\text{BR}(\ell)$, could be measured as well. From Γ_Z, σ_{peak}, and R_ℓ, the partial widths Γ_ℓ and $\Gamma_{hadrons}$ could be deduced. If the remainder is assumed to be due to N_ν species of neutrinos, we can write

$$\Gamma = \Gamma_{hadrons} + 3\Gamma_\ell + N_\nu \Gamma_\nu \tag{13.3}$$

where Γ_ν is the partial width of the Z into a single neutrino species. If the Standard Model prediction is used for this quantity, then the number of neutrino species can be derived. The original Mark II data gave $N_\nu = 3.8 \pm 1.4$. With 480 events, the result was $N_\nu = 2.8 \pm 0.6$, with $N_\nu = 3.9$ excluded at 95% CL.

In November 1989, the LEP experiments reported their first results, each with a few thousand events (Refs. 13.3, **13.4**, 13.5, 13.6). The masses clustered near 91.1 GeV with uncertainties less than 100 MeV. The widths were all near 2.5 GeV, with uncertainties typically 150 MeV. The number of neutrino generations was found to be near three, with each experiment having an uncertainty of about 0.5. Together, the evidence was overwhelmingly for precisely three neutrino generations.

LEP studied the Z from 1989 to 1995 and tested the Standard Model in exquisite detail. The LEP detectors followed the conventional scheme of a generally cylindrical design, with charged-particle tracking close to the interaction point, followed by electromagnetic calorimetry, hadronic calorimetry, and finally by muon identification and measurement. Still, each detector had its own character. ALEPH and DELPHI both used large time projection chambers for tracking, with axial magnetic fields of 1.5 T and 1.2 T respectively. See Figure 13.2. The OPAL and L3 detectors used magnetic fields of 0.5 T. The magnet for L3 was outside the rest of the detector, providing an enormous volume over which muons could be tracked to give excellent measurements of their momenta.

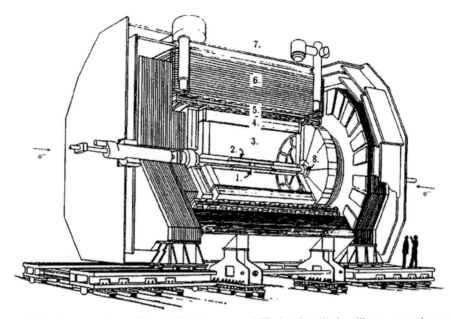

Figure 13.2. Cut-away view of the ALEPH detector at LEP showing (1) the silicon vertex detector, (2) inner trigger chamber, (3) time projection chamber, (4) electromagnetic calorimeter, (5) superconducting coil, (6) hadron calorimeter, (7) muon chambers, (8) luminosity monitors. Figure taken from M. Martinez et al., Rev. Mod. Phys. **71**, 575 (1999).

The tremendous number of events accumulated by the LEP detectors did not guarantee high precision results. Critical to this goal were accurate measurements of the luminosity and the beam energy. Cross sections could be measured only as well as luminosities and the Z mass only as well as the beam energy. Each detector monitored the luminosity by measuring Bhabha scattering, whose cross section is well known and whose rate is so large that statistics were basically unlimited. Ultimately, with very careful measurements of the luminosity monitor geometries, uncertainties were reduced below one part in a thousand.

The beam energy at LEP was measured with extreme accuracy by using the technique of resonant depolarization. This technique, developed at Novosibirsk where it was used to measure the mass of the J/ψ to high precision, resulted in a measurement of the beam energy to approximately 1 MeV once effects from the Earth's tides and the Geneva train system were fully understood.

The thousands of events grew to 16 million, shared between the four detectors. The most precise results were ultimately obtained by combining the data from ALEPH, DELPHI, L3 and OPAL, with the results $m_Z = 91.1876 \pm 0.0021$ GeV and $\Gamma_Z = 2.4952 \pm 0.0023$ GeV. The high precision measurement of the mass of the Z is especially important because it, together with $\alpha = 1/137.03599911 \pm 0.00000046$, and $G_F = 1.16637 \pm 0.00001 \times 10^{-5}$ GeV^{-2} can be taken as the three inputs that define the fundamental constants of the Standard Model. The peak cross section was found to be 41.540 ± 0.037 nb and the ratio of the hadronic to leptonic width was given by $R_\ell = 20.767 \pm 0.025$.

13. Testing the Standard Model

The Standard Model, described in Chapter 12 is a theory rather than a model in that it gives complete predictions, not just approximations. Every prediction can be expressed in terms of the three fundamental physical quantities, α, G_F, and m_Z. Other parameters of the Standard Model, like the quark and lepton masses can enter, as well. In practice, all the quark masses are small compared to the scale m_Z except for the mass of the top quark, to be discussed in Chapter 14. The mass of the Higgs boson, M_H, plays a role, too, but the dependence in radiative corrections turns out to be on $\ln M_H^2$ rather than on M_H^2 directly. Two kinds of radiative corrections turn out to be dominant: those involving m_t and the shift from using α evaluated as the static constant, $\alpha = 1/137.036...$, and α evaluated at the short distance given by the Compton wavelength of the Z. Because we are interested in processes at the energy scale m_Z, the expressions are simplest when written in terms of $\alpha(m_Z) \approx 1/129$.

The LEP program was to measure branching ratios, asymmetries, and polarizations, which could be compared to Standard Model results, looking for possible discrepancies that could signal new particles or forces.

The Standard Model makes very explicit predictions for the branching ratios of the Z. Using the relations given in Chapter 12, we find that for a decay to a left-handed fermion (and a right-handed antifermion),

$$\Gamma(Z \to f_L \bar{f}_R) = \frac{\sqrt{2} G_F m_Z^3}{6\pi} (T_3 - Q \sin^2 \theta_W)^2 \qquad (13.4)$$

where Q is the charge of the fermion, T_3 is its third component of weak isospin (1/2 for u, c, $-1/2$ for d, s, and b) and θ_W is the weak mixing angle. If the fermion is a quark rather than a lepton, we must multiply by a color factor of three. For right-handed fermions (and left-handed antifermions), we have similarly,

$$\Gamma(Z \to f_R \bar{f}_L) = \frac{\sqrt{2} G_F m_Z^3}{6\pi} (Q \sin^2 \theta_W)^2. \qquad (13.5)$$

There is a correction from QCD for the width to quark pairs, which in lowest order is a factor $1 + \frac{\alpha_s}{\pi} \approx 1.03$.

The angular dependence of the production of the various fermion pairs is governed by the simple expressions analogous to those given in Chapter 8, which reflect angular momentum conservation. Because the Z has only vector and axial vector couplings to fermions a left-handed electron can annihilate only a right-handed positron. If the electron's direction is the z-axis, the pair annihilates into a Z with $J_z = -1$. If the final fermion f is left-handed, then the antifermion is right-handed and angular momentum conservation prevents the fermion from coming out in the negative z direction. Thus we find

$$\frac{d\sigma}{d\Omega}(e_L^- e_R^+ \to Z \to f_L \bar{f}_R) \propto (1 + \cos\theta)^2 \qquad (13.6)$$

$$\frac{d\sigma}{d\Omega}(e_L^- e_R^+ \to Z \to f_R \bar{f}_L) \propto (1 - \cos\theta)^2 \qquad (13.7)$$

$$\frac{d\sigma}{d\Omega}(e_R^- e_L^+ \to Z \to f_L \bar{f}_R) \propto (1 - \cos\theta)^2 \qquad (13.8)$$

$$\frac{d\sigma}{d\Omega}(e_R^- e_L^+ \to Z \to f_R \bar{f}_L) \propto (1+\cos\theta)^2 \tag{13.9}$$

Since the cross sections are proportional to $\Gamma_e \Gamma_f$ we have for unpolarized scattering

$$\frac{d\sigma}{d\Omega}(e^-e^+ \to Z \to f\bar{f})$$
$$\propto [\Gamma_{e_L} + \Gamma_{e_R}][\Gamma_{f_L} + \Gamma_{f_R}](1+\cos^2\theta) + 2[\Gamma_{e_L} - \Gamma_{e_R}][\Gamma_{f_L} - \Gamma_{f_R}]\cos\theta. \tag{13.10}$$

An asymmetry can be formed by comparing the number of events F in which the fermion f goes forward, that is, into the hemisphere in the electron's direction to the number B in which f goes into the backward hemisphere. We find

$$A_{FB}^f \equiv \frac{F-B}{F+B} = \frac{3}{4}\frac{[\Gamma_{e_L} - \Gamma_{e_R}][\Gamma_{f_L} - \Gamma_{f_R}]}{[\Gamma_{e_L} + \Gamma_{e_R}][\Gamma_{f_L} + \Gamma_{f_R}]} \equiv \frac{3}{4}\mathcal{A}_e \mathcal{A}_f \tag{13.11}$$

where $\mathcal{A}_f = (\Gamma_{f_L} - \Gamma_{f_R})/(\Gamma_{f_L} + \Gamma_{f_R})$. The measurement of the forward–backward asymmetry in $e^+e^- \to Z \to \mu^+\mu^-$, for example, provides a clean measurement of $\sin^2\theta_W$ since we have

$$\mathcal{A}_\ell = \frac{1 - 4\sin^2\theta_W}{(1 - 2\sin^2\theta_W)^2 + 4\sin^4\theta_W}. \tag{13.12}$$

The combined LEP result was $A_{FB}^\ell = 0.0169 \pm 0.0013$.

The SLC's luminosity improved over the years, though it never rivaled that at LEP. Still SLC did have a capability that made it competitive for this class of measurements: beam polarization. Using the same technique that was used in the measurement of the left–right asymmetry in deep inelastic scattering of electrons off protons discussed in Chapter 12, left-handed and right-handed electrons were injected into the SLAC linac. It was not necessary to polarize the positrons since the coupling only allows annihilation of pairs with parallel spins.

An asymmetry can be formed for left-handed and right-handed electrons producing any final state, f. That asymmetry is simply equal to $\mathcal{A}_e$. If the degree of polarization of the beams is P, then $\mathcal{A}_e$ is simply given by $1/P$ times the observed asymmetry. Ultimately, an electron polarization of about 80% was achieved. The careful measurement of the polarization by scattering a polarized beam from the polarized electron beam was essential to the measurement. The result reported in 1997 by the SLD Collaboration (Ref. 13.7) was $\mathcal{A}_e = 0.151 \pm 0.011$, equivalent to $A_{FB}^\ell = 0.0171 \pm 0.0025$. The final analysis of the full data set gave an improved result, $\mathcal{A}_e = 0.1516 \pm 0.0021$, equivalent to $A_{FB}^\ell = 0.0171 \pm 0.0005$, consistent with the LEP result, but more precise.

With the measurement of the Z mass pinned down, the third fundamental parameter of the Standard Model, the measurement of the W mass became a critical test. The basic prediction for the W mass is

$$m_W^2 = \frac{\pi \alpha}{\sqrt{2} G_F \sin^2 \theta_W} \quad (13.13)$$

where $\sin^2 \theta_W$ itself depends on m_W:

$$\sin^2 \theta_W = 1 - \frac{m_W^2}{m_Z^2}. \quad (13.14)$$

This is modified by radiative corrections. However, the dominant correction is simply to replace the usual fine structure constant $\alpha(0)$ by $\alpha(m_Z^2)$. Additional corrections depend on m_t^2 and $\ln(m_H/m_Z)$. See Problem 13.5. Thus a precision measurement of the W mass could predict the mass of the top quark, with only a weak dependence on the unknown mass of the Higgs boson.

While e^+e^- annihilation provided an unbeatable method for studying the Z, LEP was not suited for studying the W. The original measurements of the W mass by UA-1 and UA-2 had uncertainties of several GeV. In 1990, CDF reported on 1722 events combining results from the $W \to e\nu$ and $W \to \mu\nu$ channels. CDF found $m_W = 79.91 \pm 0.39$ GeV. By 1992, UA-2 had reduced the error by accumulating more than 2000 events of the decay $W \to e\nu$. For the ratio m_W/m_Z they found $0.8813 \pm 0.0036 \pm 0.0019$. The ratio could be determined more precisely than either value separately because some of the uncertainties were common to the two measurements. At the time, the mass of the Z had already been measured to ± 20 MeV at LEP, giving a combined result of $m_W = 80.35 \pm 0.33$(stat.) $\pm$ 0.017(syst.) GeV.

In Run I at the Fermilab Tevatron Collider, from 1992 to 1995, CDF and D0 both accumulated large numbers of W's and Z's. The errors for each experiment were reduced to near 100 MeV, with a combined result of 80.450 ± 0.063 GeV, reported in 1999.

An entirely new approach to measuring the W mass became possible once the energy at LEP was increased above the WW threshold in June, 1996. The W pair cross section rises gradually rather than abruptly because the substantial width of the W makes it possible to produce one real and one virtual W. While one can measure the W mass through careful determination of the threshold rise, in fact the method found more effective at LEP-II was to reconstruct the mass from final states in $W \to q\bar{q}$, $W \to q\bar{q}$ and $W \to q\bar{q}$, $W \to \ell\nu$ events.

In 1997, more than 50 pb^{-1} of data were accumulated near $\sqrt{s} = 180$ GeV. The mass of the W could be determined with a statistical uncertainty of about 130 MeV by each experiment. Combining the experiments gave $80.38 \pm 0.07 \pm 0.03 \pm 0.02$ GeV, with the uncertainties arising from the experiment itself, from theoretical issues, and from the LEP beam energy. Further measurements were made as the c.m. energy was increased up to 206 GeV. The combined LEP result was $m_W = 80.376 \pm 0.033$ GeV. An upgraded CDF detector, running at the Tevatron Collider's Run II, remeasured the W with greatly increased statistics and found a result in 2007 completely compatible with CERN's, $m_W = 80.413 \pm 0.048$ GeV.

Even before the discovery of the top quark in 1995, the W mass measurements were accurate enough to predict m_t to be around 180 GeV, assuming the Higgs mass was in the range of 100–1000 GeV.

The Higgs boson is the least constrained part of the Standard Model. Indeed, there is no a priori limit on its mass. If the mass is sufficiently large, more than say 1.5 TeV, the width of the Higgs boson becomes comparable to its mass and it is hard to justify calling it a particle at all. On the other hand, there is no reason to suppose that there is just a single Higgs boson. Indeed some models, like supersymmetry, require that there be more than one neutral Higgs boson. Because the Higgs boson couples feebly to light particles (that is why they are light!), it is best sought in conjunction with heavy particles. LEP II offered an ideal approach: $e^+e^- \to ZH$. The electron–positron pair annihilate into a virtual Z, which then decays to a real Z and the Higgs boson. In this way, a Higgs boson could be found up to very near the kinematic limit, $m_H = \sqrt{s} - m_Z$.

The Higgs boson couples to fermion pairs according to their masses, making $H \to b\bar{b}$ and $H \to \tau^+\tau^-$ the best targets. The accompanying Z can be detected in any of its decay channels. One vexing background comes from the ZZ final state, when one Z decays to $b\bar{b}$. With data taken at a center-of-mass energy of 189 GeV, three of the LEP experiments were able to set lower limits of about 95 GeV on a Standard Model Higgs boson, while the limit from ALEPH, the remaining experiment, was about 90 GeV.

Still there was more to be wrung out of LEP. Between 1995 and 1999 one after another upgrade was carried out to raise the energy higher and higher, opening each time a new window in which the Higgs boson might appear. The enormous effort this entailed was justified because detailed fits, which depended on $\ln m_H^2$, of the electroweak data from the Z pointed to a low value of the Higgs mass, around 100 GeV. The center-of-mass energy leapt to 204 GeV, then in a series of small steps to 209.2 GeV. No sign of a Higgs boson was seen until the data at 206 GeV were analyzed.

In the fall of 2000, ALEPH reported events above the background expected, consistent with a Higgs boson with a mass of 115 GeV. Some confirmation came from L3, but none from DELPHI or OPAL. Combining the data from all events in November 2000, the signal had a 2.9 σ significance. Luciano Maiani, the Director General of CERN faced a dilemma. Should he continue to raise the energy of LEP2 and accept a delay in CERN's next big project, the Large Hadron Collider, which was to use the LEP tunnel? The decision was made to terminate LEP2. Further analysis of the data in the summer of 2001 showed that the effect was somewhat smaller, 2.2 σ, but whether there is a 115-GeV Higgs boson will be settled by a hadron collider.

Exercises

13.1 Use the final LEP values for the width of the Z, σ_{peak}, and R_ℓ to determine N_ν. For Γ_ν / Γ_ℓ use the Standard Model value of 1.99.

13.2 Determine the expression for the left–right forward–backward asymmetry for the production of a fermion-antifermion pair at the Z when the initial electron polarization is P. How well can $\mathcal{A}_\mu$ be measured with N events of $e^+e^- \to \mu^+\mu^-$? Assume $\mathcal{A}_e$ is known from measuring the total cross section for left- and right-polarized electrons. Take $P = 0.75$. How much is the measurement of $\mathcal{A}_\mu$ improved by using polarized beams? Compare your estimate with SLC Collaboration, K. Abe et al., *Phys. Rev. Lett.* **86**, 1162 (2001).

13.3 If a τ at rest decays, the angular distribution of the pion is $dN/d\cos\theta \propto 1 + \cos\theta$, where θ is the angle between the pion's direction and the spin of the τ and the mass of the pion is neglected. Show that this is consistent with the V-A nature of weak interactions. If a high energy τ decays to $\pi\nu$, what is the expected distribution of its visible energy, i.e. the pion's energy, if the τ is left-handed? Consider $Z \to \tau^+\tau^-$ and let $x = E_\pi/E_\tau$ be the fraction of τ's energy that is given to the π. Find the joint distribution in θ, the polar angle relative to the e^- direction and x, in terms of $\mathcal{A}_e$ and $\mathcal{A}_\tau$. See, ALEPH Collaboration, A. Heister et al., *Eur. Phys. J.* **C20**, 401 (2001).

13.4 The stored LEP electron beam develops a polarization perpendicular to the plane of the ring. As described in Problem 12.4, the electron's spin makes $\nu_0 = \gamma a_e = (E_{beam}/m_e)a_e$ cycles around its polarization for each circuit of the ring, where $a_e \approx \alpha/2\pi$ is the anomalous magnetic moment of the electron in Bohr magnetons. Determine the value of ν_0 when LEP ran at the Z using the more precise value $a_e = 0.0115965$. At a single spot, the electron's spin will seem to advance only by $[\nu_0]$, the non-integer part of ν_0. If a radial magnetic field is applied with a frequency $[\nu_0]$ times the frequency of the electron's revolution around the ring, electron spins will flip, destroying or reversing the polarization. At LEP, the frequency of the depolarizing resonance was measured to 2 Hz. What uncertainty in the mass of the Z would this cause? See L. Arnaudon et al., *Zeit. f. Phys.* **C66**, 45 (1995).

13.5 The W mass can be predicted from the Z mass using the formula

$$m_W^2 = \frac{1}{2}\left[1 + \sqrt{1 - \frac{4\pi\alpha(1+\Delta r)}{\sqrt{2}m_Z^2 G)F}}\right]m_Z^2$$

where Δr incorporates the radiative corrections, including the shift of α from its static value to the value at the scale m_Z. The radiative corrections depend on the value of m_t and m_H. An adequate representation [A. Ferroglila et al., *Phys. Rev.* **D65**, 113002 (2002)] is

$$m_W(\text{GeV}) = 80.387 - 0.572\ln(m_H/100\text{ GeV}) - 0.0090\,[\ln(m_H/100\text{ GeV})]^2$$
$$+ 0.540\,[(m_t/174.3\text{ GeV})^2 - 1].$$

Compare the current measurements of m_t and m_W. What does this indicate about the mass of the Higgs boson? Compare with the direct information from LEP II.

13.6 A value of $\sin^2\theta_W$ can be inferred from measurements of the forward–backward asymmetry at LEP. Within the Standard Model, it can be predicted in terms of the three basic parameters, α, G_F, and m_Z if m_t and m_H are known. The latter two occur through radiative corrections. An adequate representation is

$$\sin^2 \theta_{eff}^{lept} = 0.2314 + 4.9 \times 10^{-4} \ln(m_H/100 \text{ GeV})$$
$$+ 3.41 \times 10^{-5} [\ln(m_H/100 \text{ GeV})]^2$$
$$- 2.7 \times 10^3 [(m_t/174.3 \text{ GeV})^2 - 1].$$

The results from LEP for the forward–backward asymmetry for leptonic final states gave $\sin^2 \theta_{eff}^{lept} = 0.23113(21)$ while for hadronic final states the result was $\sin^2 \theta_{eff}^{lept} = 0.23220(29)$. What do these results suggest about the mass of the Higgs? Compare with the results of Exercise 13.5.

Further Reading

ALEPH, DELPHI, L3, OPAL, and SLD Collaborations, LEP Electroweak Working Group, SLD Electroweak and Heavy Flavor Working Groups, "Precision Electroweak Measurements at the Z Resonance," *Phys. Rep.* **427**, 257 (2006).

References

13.1 G. A. Abrams *et al.*, "Initial Measurements of Z-boson Resonance Parameters in e^+e^- Annihilation." *Phys. Rev. Lett.*, **63**, 724 (1989).
13.2 F. Abe *et al.*, "Measurement of the Mass and Width of the Z^0 Boson at the Fermilab Tevatron." *Phys. Rev. Lett.*, **63**, 720 (1989).
13.3 L3, "A Determination of the Properties of the Neutral Intermediate Vector Boson Z0." *Phys. Lett.*, **B 231**, 509 (1989).
13.4 ALEPH , "The Number of Light Neutrino Species." *Phys. Lett.*, **B 231**, 519 (1989).
13.5 OPAL, "Measurement of the Z^0 Mass and Width with the OPAL Detector at LEP." *Phys. Lett.*, **B 231**, 530 (1989).
13.6 DELPHI, "Measurement of the Mass and Width of the Z^0-particle from Multi-hadronic Final States Produced in e^+e^- Annihilations." *Phys. Lett.*, **B 231**, 539 (1989).
13.7 SLD, "Direct Measurement of Leptonic Coupling Asymmetries with Polarized Z Bosons." *Phys. Rev. Lett.*, **79**, 804 (1997).

DETERMINATION OF THE NUMBER OF LIGHT NEUTRINO SPECIES

ALEPH Collaboration

D. DECAMP, B. DESCHIZEAUX, J.-P. LEES, M.-N. MINARD
Laboratoire de Physique des Particules (LAPP), F-74019 Annecy-le-Vieux Cedex, France

J.M. CRESPO, M. DELFINO, E. FERNANDEZ [1], M. MARTINEZ, R. MIQUEL, M.L. MIR,
S. ORTEU, A. PACHECO, J.A. PERLAS, E. TUBAU
Laboratorio de Fisica de Altas Energias, Universidad Autonoma de Barcelona, E-08193 Bellaterra (Barcelona), Spain [2]

M.G. CATANESI, M. DE PALMA, A. FARILLA, G. IASELLI, G. MAGGI, A. MASTROGIACOMO,
S. NATALI, S. NUZZO, A. RANIERI, G. RASO, F. ROMANO, F. RUGGIERI, G. SELVAGGI,
L. SILVESTRIS, P. TEMPESTA, G. ZITO
INFN, Sezione di Bari e Dipartimento di Fisica dell' Università, I-70126 Bari, Italy

Y. CHEN, D. HUANG, J. LIN, T. RUAN, T. WANG, W. WU, Y. XIE, D. XU, R. XU, J. ZHANG,
W. ZHAO
Institute of High-Energy Physics, Academia Sinica, Beijing, P.R. China

H. ALBRECHT [3], F. BIRD, E. BLUCHER, T. CHARITY, H. DREVERMANN, Ll. GARRIDO,
C. GRAB, R. HAGELBERG, S. HAYWOOD, B. JOST, M. KASEMANN, G. KELLNER,
J. KNOBLOCH, A. LACOURT, I. LEHRAUS, T. LOHSE, D. LÜKE [3], A. MARCHIORO, P. MATO,
J. MAY, V. MERTENS, A. MINTEN, A. MIOTTO, P. PALAZZI, M. PEPE-ALTARELLI,
F. RANJARD, J. RICHSTEIN [4], A. ROTH, J. ROTHBERG [5], H. ROTSCHEIDT, W. VON RÜDEN,
D. SCHLATTER, R. ST.DENIS, M. TAKASHIMA, M. TALBY, H. TAUREG, W. TEJESSY,
H. WACHSMUTH, S. WHEELER, W. WIEDENMANN, W. WITZELING, J. WOTSCHACK
European Organization for Nuclear Research (CERN), CH-1211 Geneva 23, Switzerland

Z. AJALTOUNI, M. BARDADIN-OTWINOWSKA, A. FALVARD, P. GAY, P. HENRARD,
J. JOUSSET, B. MICHEL, J-C. MONTRET, D. PALLIN, P. PERRET, J. PRAT, J. PRORIOL,
F. PRULHIÈRE
Laboratoire de Physique Corpusculaire, Université Blaise Pascal, Clermont-Ferrand, F-63177 Aubière, France

H. BERTELSEN, F. HANSEN, J.R. HANSEN, J.D. HANSEN, P.H. HANSEN, A. LINDAHL,
B. MADSEN, R. MØLLERUD, B.S. NILSSON, G. PETERSEN
Niels Bohr Institute, DK-2100 Copenhagen, Denmark [6]

E. SIMOPOULOU, A. VAYAKI
Nuclear Research Center Demokritos (NRCD), Athens, Greece

J. BADIER, D. BERNARD, A. BLONDEL, G. BONNEAUD, J. BOUROTTE, F. BRAEMS,
J.C. BRIENT, M.A. CIOCCI, G. FOUQUE, R. GUIRLET, P. MINÉ, A. ROUGÉ, M. RUMPF,
H. VIDEAU, I. VIDEAU [1], D. ZWIERSKI
Laboratoire de Physique Nucléaire Hautes Energies, École Polytechnique, F-91128 Palaiseau Cedex, France

0370-2693/89/$ 03.50 © Elsevier Science Publishers B.V.
(North-Holland Physics Publishing Division)

D.J. CANDLIN
Department of Physics, University of Edinburgh, Edinburgh EH9 3JZ, UK [7]

A. CONTI, G. PARRINI
Dipartimento di Fisica, Università di Firenze, I-50125 Florence, Italy

M. CORDEN, C. GEORGIOPOULOS, J.H. GOLDMAN, M. IKEDA, D. LEVINTHAL [8],
J. LANNUTTI, M. MERMIKIDES, L. SAWYER
High-Energy Particle Physics Laboratory, Florida State University, Tallahassee, FL 32306, USA [9,10,11]

A. ANTONELLI, R. BALDINI, G. BENCIVENNI, G. BOLOGNA, F. BOSSI, P. CAMPANA,
G. CAPON, V. CHIARELLA, G. DE NINNO, B. D'ETTORRE-PIAZZOLI, G. FELICI, P. LAURELLI,
G. MANNOCCHI, F. MURTAS, G.P. MURTAS, G. NICOLETTI, P. PICCHI, P. ZOGRAFOU
Laboratori Nazionali dell' INFN (LNF-INFN), I-00044 Frascati, Italy

B. ALTOON, O. BOYLE, A.J. FLAVELL, A.W. HALLEY, I. TEN HAVE, J.A. HEARNS, I.S. HUGHES,
J.G. LYNCH, D.J. MARTIN, R. O'NEILL, C. RAINE, J.M. SCARR, K. SMITH [1], A.S. THOMPSON
Department of Natural Philosophy, University of Glasgow, Glasgow G12 8QQ, UK [7]

B. BRANDL, O. BRAUN, R. GEIGES, C. GEWENIGER [1], P. HANKE, V. HEPP, E.E. KLUGE,
Y. MAUMARY, M. PANTER, A. PUTZER, B. RENSCH, A. STAHL, K. TITTEL, M. WUNSCH
Institut für Hochenergiephysik, Universität Heidelberg, D-6900 Heidelberg, FRG [12]

G.J. BARBER, A.T. BELK, R. BEUSELINCK, D.M. BINNIE, W. CAMERON [1], M. CATTANEO,
P.J. DORNAN, S. DUGEAY, R.W. FORTY, D.N. GENTRY, J.F. HASSARD, D.G. MILLER,
D.R. PRICE, J.K. SEDGBEER, I.R. TOMALIN, G. TAYLOR
Department of Physics, Imperial College, London SW7 2BZ, UK [7]

P. GIRTLER, D. KUHN, G. RUDOLPH
Institut für Experimentalphysik, Universität Innsbruck, A-6020 Innsbruck, Austria

T.J. BRODBECK, C. BOWDERY [1], A.J. FINCH, F. FOSTER, G. HUGHES, N.R. KEEMER,
M. NUTTALL, B.S. ROWLINGSON, T. SLOAN, S.W. SNOW
Department of Physics, University of Lancaster, Lancaster LA1 4YB, UK [7]

T. BARCZEWSKI, L.A.T. BAUERDICK, K. KLEINKNECHT, D. POLLMANN [13], B. RENK,
S. ROEHN, H.-G. SANDER, M. SCHMELLING, F. STEEG
Institut für Physik, Universität Mainz, D-6500 Mainz, FRG [12]

J.-P. ALBANESE, J.-J. AUBERT, C. BENCHOUK, A. BONISSENT, F. ETIENNE, R. NACASCH,
P. PAYRE, B. PIETRZYK [1], Z. QIAN
Centre de Physique des Particules, Faculté des Sciences de Luminy, F-13288 Marseille, France

W. BLUM, P. CATTANEO, M. COMIN, B. DEHNING, G. COWAN, H. DIETL,
M. FERNANDEZ-BOSMAN, D. HAUFF, A. JAHN, E. LANGE, G. LÜTJENS, G. LUTZ,
W. MÄNNER, H.-G. MOSER, Y. PAN, R. RICHTER, A. SCHWARZ, R. SETTLES, U. STIEGLER,
U. STIERLIN, G. STIMPFL [14], J. THOMAS, G. WALTERMANN
Max-Planck-Institut für Physik und Astrophysik, Werner-Heisenberg-Institut für Physik, D-8000 Munich, FRG [12]

J. BOUCROT, O. CALLOT, A. CORDIER, M. DAVIER, G. DE BOUARD, G. GANIS, J.-F. GRIVAZ,
Ph. HEUSSE, P. JANOT, V. JOURNÉ, D.W. KIM, J. LEFRANÇOIS, D. LLOYD-OWEN,
A.-M. LUTZ, P. MAROTTE, J.-J. VEILLET
Laboratoire de l'Accélérateur Linéaire, Université de Paris-Sud, F-91405 Orsay Cedex, France

S.R. AMENDOLIA, G. BAGLIESI, G. BATIGNANI, L. BOSISIO, U. BOTTIGLI, C. BRADASCHIA,
I. FERRANTE, F. FIDECARO, L. FOÀ [1], E. FOCARDI, F. FORTI, A. GIASSI, M.A. GIORGI,
F. LIGABUE, A. LUSIANI, E.B. MANNELLI, P.S. MARROCCHESI, A. MESSINEO, F. PALLA,
G. SANGUINETTI, S. SCAPELLATO, J. STEINBERGER, R. TENCHINI, G. TONELLI,
G. TRIGGIANI
Dipartimento di Fisica dell' Università, INFN Sezione di Pisa, e Scuola Normale Superiore, I-56010 Pisa, Italy

J.M. CARTER, M.G. GREEN, A.K. McKEMEY, P.V. MARCH, T. MEDCALF, M.R. SAICH,
J. STRONG [1], R.M. THOMAS, T. WILDISH
Department of Physics, Royal Holloway & Bedford New College, University of London, Surrey TW20 0EX, UK [7]

D.R. BOTTERILL, R.W. CLIFFT, T.R. EDGECOCK, M. EDWARDS, S.M. FISHER, D.L. HILL,
T.J. JONES, G. McPHERSON, M. MORRISSEY, P.R. NORTON, D.P. SALMON, G.J. TAPPERN,
J.C. THOMPSON, J. HARVEY
High-Energy Physics Division, Rutherford Appleton Laboratory, Chilton, Didcot, Oxon OX11 0QX, UK [7]

B. BLOCH-DEVAUX, P. COLAS, C. KLOPFENSTEIN, E. LANÇON, E. LOCCI, S. LOUCATOS,
L. MIRABITO, E. MONNIER, P. PEREZ, F. PERRIER, B. PIGNARD, J. RANDER, J.-F. RENARDY,
A. ROUSSARIE, J.-P. SCHULLER, R. TURLAY
Département de Physique des Particules Elémentaires, CEN-Saclay, F-91191 Gif-sur-Yvette Cedex, France

J.G. ASHMAN, C.N. BOOTH, F. COMBLEY, M. DINSDALE, J. MARTIN, D. PARKER,
L.F. THOMPSON
Department of Physics, University of Sheffield, Sheffield S3 7RH, UK [7]

S. BRANDT, H. BURKHARDT, C. GRUPEN, H. MEINHARD, E. NEUGEBAUER, U. SCHÄFER,
H. SEYWERD, K. STUPPERICH
Fachbereich Physik, Universität Siegen, D-5900 Siegen, FRG [12]

B. GOBBO, F. LIELLO, E. MILOTTI, F. RAGUSA [15], L. ROLANDI [1]
Dipartimento di Fisica, Università di Trieste e INFN Sezione di Trieste, I-34127 Trieste, Italy

L. BELLANTONI, J.F. BOUDREAU, D. CINABRO, J.S. CONWAY, D.F. COWEN, Z. FENG,
J.L. HARTON, J. HILGART, R.C. JARED [16], R.P. JOHNSON, B.W. LECLAIRE, Y.B. PAN,
T. PARKER, J.R. PATER, Y. SAADI, V. SHARMA, J.A. WEAR, F.V. WEBER, SAU LAN WU,
S.T. XUE and G. ZOBERNIG
Department of Physics, University of Wisconsin, Madison, WI 53706, USA [17]

Received 12 October 1989

The cross-section for $e^+e^- \to$ hadrons in the vicinity of the Z boson peak has been measured with the ALEPH detector at the CERN Large Electron Positron collider, LEP. Measurements of the Z mass, $M_Z = (91.174 \pm 0.070)$ GeV, the Z width $\Gamma_Z = (2.68 \pm 0.15)$ GeV, and of the peak hadronic cross-section, $\sigma_{had}^{peak} = (29.3 \pm 1.2)$ nb, are presented. Within the constraints of the standard electroweak model, the number of light neutrino species is found to be $N_\nu = 3.27 \pm 0.30$. This result rules out the possibility of a fourth type of light neutrino at 98% CL.

1. Introduction

The Z boson was discovered in 1983 at the pp̄ collider at CERN [1,2]. Detailed studies of its properties can now be performed in e^+e^- collisions at LEP. In the standard electroweak model [3], the Z boson is expected to decay with comparable probability into all species of fermions that are kinematically allowed. The decay rate of the Z into light, neutral, penetrating particles such as neutrinos, that would otherwise escape detection, can be measured through an increase in the total width Γ_Z. Detection of additional neutrino species would put in evidence additional fermion families, even if the masses of their charged partners are inaccessible at presently available energies. One additional species of neutrino would result in an increase of 6.6% in Γ_Z. Furthermore the peak cross-section to any detectable final state f is very sensitive to this change and would decrease by 13% for one more neutrino species. This cross-section can be expressed in terms of Γ_Z and of the partial widths, Γ_{ee}, Γ_ν, Γ_f, of the Z into e^+e^-, neutrinos, and the final state f as

$$\sigma_f^{peak} = \frac{12\pi}{M_Z^2} \frac{\Gamma_{ee}\Gamma_f}{\Gamma_Z^2}(1-\delta_{rad}) \equiv \sigma_f^0(1-\delta_{rad}), \quad (1)$$

with

$$\Gamma_Z = N_\nu \Gamma_\nu + 3\Gamma_{ee} + \Gamma_{had}, \quad (2)$$

where N_ν is the number of light neutrino species. In the particular case where f comprised mostly hadronic final states, uncertainties in the overall scale of partial widths, such as those related to the lack of knowledge of the top quark mass and more generally to electroweak radiative effects, as well as uncertainties in the ratio of hadronic to leptonic partial widths, largely cancel in this formula. The QED initial state radiative correction, δ_{rad}, is quite large, but has been calculated to an accuracy believed to be better than 0.5% by several authors [4].

Cross-sections are measured by taking the ratio of the number of selected Z decays, hereafter referred to as hadronic events, to the number of small angle e^+e^- events from the well calculable Bhabha scattering process, hereafter referred to as luminosity events.

2. Description of the ALEPH detector

The data presented here have been collected with the ALEPH detector during the first three weeks of running at LEP, from 20 September to 9 October 1989. Guided by earlier measurements of the Z mass by the CDF [5] and MarkII [6] Collaborations, data were collected mainly at the Z peak and in the near vicinity of it. Integrated luminosities of 64 nb^{-1} at the peak and 88 nb^{-1} on the sides of the peak were recorded.

A detailed description of the ALEPH detector is in preparation [7]. The principal components relevant for this measurement are:
- The Inner Tracking Chamber, ITC, and 8-layer cylindrical drift chamber with sense wires parallel to the beam axis from 13 cm to 29 cm in radius. Tracks with polar angles from 14° to 166° traverse all 8 layers.
- The large cylindrical Time Projection Chamber, TPC, extending from an inner radius of 31 cm to an outer radius of 180 cm over a length of 4.4 m. Up to 21 space coordinates are recorded for tracks with polar angles from 47° to 133°. Requiring 4 coordinates, tracks are reconstructed down to 15°.
- The Electromagnetic Calorimeter, ECAL, a lead wire-chamber sandwich. The cathode readout is subdivided into a total of 73 728 projective towers. Each

---

[1] Present address: CERN, CH-1211 Geneva 23, Switzerland.
[2] Supported by CAICYT, Spain.
[3] Permanent address: DESY, D-2000 Hamburg 52, FRG.
[4] Present address: Lecroy, Geneva, Switzerland.
[5] On leave of absence from University of Washington, Seattle, WA 98195, USA.
[6] Supported by the Danish Natural Science Research Council.
[7] Supported by the UK Science and Engineering Research Council.
[8] Supported by SLOAN fellowship, contract BR 2703.
[9] Supported by the US Department of Energy, contract DE-FG05-87ER40319.
[10] Supported by the NSF, contract PHY-8451274.
[11] Supported by the US Department of Energy, contract DE-FC0S-85ER250000.
[12] Supported by the Bundesministerium für Forschung und Technologie.
[13] Permanent address: Institut für Physik, Universität Dortmund, D-4600 Dortmund 50, FRG.
[14] Present address: FSU, Tallahassee, FL 32306, USA.
[15] Present address: INFN, Sezione di Milano, I-20133 Milan, Italy.
[16] Permanent address: LBL, Berkeley, CA 94720, USA.
[17] Supported by the US Department of Energy, contract DE-AC02-76ER00881.

tower of about 1° × 1° solid angle is read out in three stacks of 10, 23 and 12 layers (respectively 4, 9 and 9 radiation lengths). The signals from the 45 wire planes of each of the 36 modules are also read out. The two endcaps cover polar angles from 11° to 40° and 140° to 169° and the barrel covers polar angles from 40° to 140°.
- The superconducting, solenoidal coil, providing a magnetic field of 1.5 T.
- The Hadron Calorimeter, HCAL, comprised of 23 layers of streamer tubes interleaved in the iron of the magnet return yoke, read out by a total of 4608 projective towers. Signals from each of the tubes are also read out. The modules are rotated in azimuth by ~2° with respect to the electromagnetic calorimeter so that inactive zones do not align in the two calorimeters. The two endcaps and the barrel of the hadron calorimeter cover polar angles down to 6°.
- The Small Angle Tracking chamber, SATR, with 9 planes of drift tubes, covering angles from 40 to 90 mrads, for precise measurement of small angle electron tracks.
- The Luminosity Calorimeter, LCAL, similar in its construction and read-out to ECAL, extending from 50 to 180 mrads, providing energy and position measurement of the showers produced by luminosity events.

Hadronic events were triggered by two independent first level triggers: (i) an ECAL-based trigger, requiring a total energy of 6 GeV deposited in the ECAL barrel or 3 GeV in either of the ECAL endcaps or 1 GeV in both in coincidence; (ii) an ITC–HCAL coincidence, for penetrating charged particles, requiring 6 ITC wire planes and 4 to 8 planes of HCAL tubes in the same azimuthal region.

Luminosity events were also triggered in two ways: (i) a coincidence of 15 GeV deposited in LCAL on one side with 10 GeV deposited on the other side, without requiring azimuthal correlation; (ii) a single arm requirement of 32 GeV deposited in either side of LCAL. In addition, prescaled single arm triggers with 10 and 15 GeV thresholds were recorded to provide an estimate of the beam related background.

All types of events were processed simultaneously through the same trigger system (with the same deadtime), through the same data acquisition and reconstruction programs. In order to ensure that the events were counted during the same life-time, the list of enabled triggers and the status of each of the relevant subdetectors was recorded with each event. Events were accepted only when both ECAL and LCAL were running and when both ECAL and LCAL triggers were enabled. A possible bias could come from the few data acquisition failures that occurred during the run. A careful investigation of all the events before, during, and after each failure revealed a possible but small loss of 0.4% of the hadronic events. This check could be done for the events that were mishandled because their trigger pattern was always recorded and the trigger patterns of hadronic and luminosity events are unique enough to allow an estimate of the losses in each category.

3. Event selection

To provide a check, two independent event selections were used. The first one selected hadronic Z decays only, and was based on TPC tracks. The second one selected decays of the Z into hadrons as well as τ pairs, and was based on calorimetric energy. The event samples overlapped to the extent of 95%. The efficiencies of both methods were very close to unity and systematic uncertainties in these efficiencies were less than 1%. Altogether 3112 events were retained in the track selection and 3320 events in the calorimetric selection.

3.1. Selection with TPC tracks

Hadronic Z decays (a typical event is shown in fig. 1) were selected on the basis of charged tracks only, requiring at least 5 charged tracks. The energy sum of all charged tracks was required to be at least 10% of the centre-of-mass energy. Tracks were required to have a polar angle larger than 18.2°, to be reconstructed from at least 4 TPC coordinates, and to originate from a 2 cm radius 20 cm long cylinder around the nominal beam position. The performance of the TPC was in remarkable agreement with expectations. In order to estimate the acceptance, a complete simulation of the $e^+e^- \to$ hadrons process was performed, including initial state radiation effects and hadronization. The properties that are relevant for acceptance calculation, total charged-particle energy, track multiplicity, sphericity distribution and polar

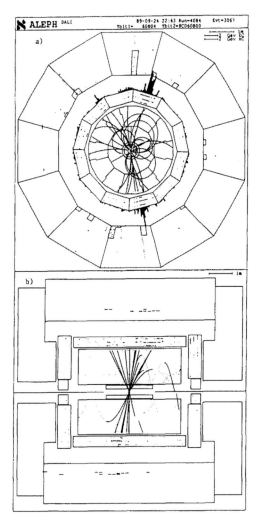

Fig. 1. A hadronic Z decay in the ALEPH detector: (a) x-y view; (b) r-z view.

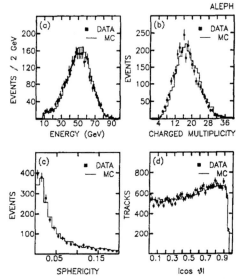

Fig. 2. Properties of charged tracks in hadronic events and comparison with simulations. In each plot, the solid points represent data and the lines represent the simulation normalized to the data: (a) distribution of the charged-track energy sum per event; (b) charged-track multiplicity distribution; (c) sphericity distribution for events where the sphericity axis had a polar angle such that $|\cos \vartheta_{sph}| < 0.8$; and (d) polar angle distribution for charged tracks.

angle distribution, are in good agreement with the simulation, as shown in fig. 2. The efficiency of this selection method is 0.975 ± 0.006, on the peak; the error corresponds to assigning a conservative 20% energy scale uncertainty near the cut. Uncertainties in hadronization models were reduced to a very small level by using the measured sphericity distribution for the acceptance calculation. Due to initial state radiation, this efficiency varies slightly on the side of the Z peak by up to -0.003.

Contamination of the sample of hadronic events by τ pairs from Z decays was estimated to be (5.1 ± 1.5) events, and in fact three events compatible with that hypothesis were found in the sample. Contamination by beam–gas interactions was estimated from the number of events found passing the selection cuts except for the longitudinal vertex position: about one event is expected. Finally the background from $e^+e^- \rightarrow e^+e^- +$ hadrons ("two-photon" events) was calculated to be about 15 pb, representing a contamination of 0.5×10^{-3} to the peak cross-section.

3.2. Selection using the calorimeters

The aim of this method was to select hadronic and τ events. The basic requirement was that the total calorimetric energy be above 20 GeV, as well as either ≥ 6 GeV in the ECAL barrel or at least 1.5 GeV in each ECAL endcap. These requirements reduce both two-photon events and muon-pair events to a negligible level. Large-angle e^+e^- events were rejected on the basis of their characteristic tight energy clusters in the electromagnetic calorimeter. For the few events with no tracks at all, cuts were applied to eliminate cosmic rays: a timing cut and a cut on the minimal number (2) of clusters above 3 GeV in ECAL. On the basis of Monte Carlo simulation as well as the scanning of events the resulting selection efficiency is 0.994 ± 0.005 for hadronic events and 0.60 ± 0.05 for τ events, giving a combined $\sigma_{had} + \sigma_\tau$ efficiency of 0.974 ± 0.006. Contamination by events other than hadronic or τ was estimated to be less than 0.4%.

The two event samples were compared event by event. Differences were well understood given the different characteristics of the two selections.

3.3. Trigger efficiency

The trigger efficiency was measured by counting events where one or both of the ECAL and ITC–HCAL triggers occurred; it was found to be 100% for the ECAL trigger and 87% for the ITC–HCAL trigger, giving an overall efficiency of 100%.

4. Determination of the luminosity

Luminosity events were selected on the basis of the energy deposited in the LCAL towers. Neighboring towers containing more than 50 MeV were joined into clusters, giving energy and position of the shower. Events were required to have a shower reconstructed on each side of LCAL.

In order to minimize the dependence of the acceptance upon precise knowledge of the beam parameters, an asymmetric selection was performed: on one side (e.g., the e^+ side), showers were required to have more than half of their energy deposited in a fiducial volume (fig. 3) excluding the towers situated at the edge of the detector; the total energy deposit on that

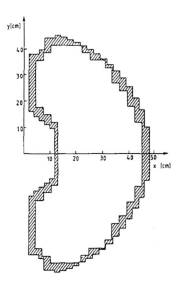

Fig. 3. End view of the Luminosity Calorimeter, showing the tower limits; the shaded area represent the towers excluded from the fiducial area.

side was also required to be larger than 55% of the beam energy; on the other side (e^- side), only a total energy deposition of more than 44% of beam energy was required. The respective roles of the e^+ and e^- sides were interchanged in every other event. Finally, the difference in azimuth, $\Delta\phi$, between the e^+ and the e^- was required to be larger than 170°.

The accepted cross-section was calculated using a first order event generator [8] with a full simulation of the detector. The value obtained for the cross-section for Bhabha scattering into the region within these cuts was found to be $(31.12 \pm 0.45_{exp} \pm 0.31_{th})$ nb, at a centre-of-mass energy of 91.0 GeV and for a Z mass of 91.0 GeV. the first error represents the effect of uncertainties in the simulation, calibration, and positioning of the apparatus; the second one represents the possible error due to neglecting higher order radiative effects, and the uncertainty in the photon vacuum polarization [9]. Properties of the luminosity events are shown in fig. 4, and compared with the simulation. The optimum energy resolution has not yet been obtained, but the acceptance is quite insensitive to this. Other properties are in good agreement with expectations. The displacements of the beam

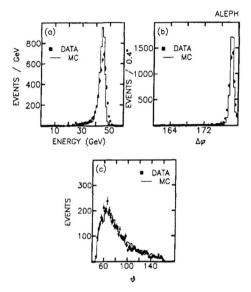

Fig. 4. Properties of luminosity events passing selection criteria and comparison with simulations. In each plot, the solid points represent data and the lines represent the simulation normalized to the data: (a) shower energy distribution for events passing the tight fiducial cut; (b) azimuthal separation $\Delta\phi$ of the two opposite clusters; (c) polar angle distribution.

Table 1
Summary of systematic errors in the luminosity measurement.

transverse and longitudinal shower profile	±0.005
energy scale	±0.002
energy resolution and cell-to-cell calibration	±0.007
external alignment and beam parameters	±0.002
internal alignment and inner radius	±0.010
description of material	±0.005
higher order radiative effects	±0.01
total uncertainty:	±0.02

were measured using luminosity events themselves and corrected for.

The beam-related background contamination was estimated from single arm, prescaled triggers. These triggers were combined into artificial double arm events. The number of such combinations passing the selection cuts was normalized to the number of real coincidences with $\Delta\phi < 90°$. This background subtraction was performed for each run and was of the order of 1% or smaller.

The contamination by physics sources such as $e^+e^- \to \gamma\gamma$ or $e^+e^- \to e^+e^- f\bar{f}$ has been estimated to be less than 2×10^{-3}. The interference of the Z exchange diagram with the purely QED contribution has been taken into account when determining the resonance parameters.

The trigger efficiency was measured for each data taking period by comparing the number of events passing the selection criteria that set the single arm trigger, the coincidence trigger, or both. Inefficiencies in the trigger were traced down to faulty electronic channels. Efficiencies vary with the run, ranging from 0.98 to 1.00 with an average value of 0.997 ± 0.002.

A summary of the luminosity systematic errors is given in table 1. A relatively small normalization error was possible mainly as a result of the excellent spatial resolution of the calorimeter (~ 300 μm for electrons near the tower boundaries), the good background conditions delivered by the machine, the availability of a redundant set of triggers and progress in the theoretical calculations [10]. The relative normalization uncertainty of cross-section measurements at different energies comes mostly from differences in background conditions and trigger efficiency. These uncertainties were taken into account in the statistical error.

5. Determination of the Z resonance parameters

The number of Z events, of luminosity events, together with the cross-sections are given in table 2 for the two event selection methods.

Two different fits were performed to the data. In the first fit, the basic parameters of the Z resonance, its mass M_Z, width Γ_Z, and QED corrected peak cross-section

$$\sigma_f^0 \equiv \frac{12\pi}{M_Z^2} \frac{\Gamma_{ee}\Gamma_f}{\Gamma_Z^2}$$

are extracted with little model dependence. In the second fit, the constraints from the standard model are applied to determine M_Z and N_ν.

The three parameter fit was performed using com-

Table 2
Event numbers and cross-section as a function of centre-of-mass energy. The overall systematic error of ± 2% in the cross-sections is not included.

Energy GeV	Selection from TPC tracks			Selection by calorimeters		
	N_{had}	N_{lumi}	σ_{had} (nb)	$N_{had}+N_\tau$	N_{lumi}	$\sigma_{had}+\sigma_\tau$ (nb)
89.263	120	443	9.00 ± 0.92	134	450	9.89 ± 0.97
90.265	406	715	18.43 ± 1.14	445	736	19.62 ± 1.17
91.020	656	668	31.29 ± 1.72	678	669	32.28 ± 1.76
91.266	1156	1295	28.16 ± 1.14	1243	1309	29.96 ± 1.19
92.260	258	377	21.11 ± 1.72	268	374	22.10 ± 1.78
92.519	125	247	15.52 ± 1.71	142	260	16.75 ± 1.76
93.264	391	883	13.36 ± 0.82	410	889	13.92 ± 0.84

puter programs by Burgers [11] and Borelli et al. [12], folding a Breit–Wigner resonance (with s-dependent width) with the second order exponentiated initial state radiation spectrum. Results from fitting with the two programs were in good agreement with each other. These approximate programs agree adequately with complete electroweak calculations (see D.Y. Bardin et al., in ref. [4] and ref. [13]) for centre-of-mass energies within ± 2.5 GeV of the peak. These fits yield the values for the mass and width of the Z boson shown in table 3.

The data points and the result of the fit are shown in fig. 5 for the track selected events.

The error in M_Z does not yet include the uncertainty in the mean e^+e^- collision energy. This error was determined by the LEP division [14] on the basis of measurements of uncertainties in the magnetic field integrals and in the orbit positions to be 5×10^{-4} or 45 MeV. Including this uncertainty, we find

$$M_Z = (91.174 \pm 0.055_{exp} \pm 0.045_{LEP})\text{ GeV}. \quad (3)$$

The value agrees with the two previous best measurements, refs [5,6], but the uncertainty is smaller by a factor of 2.

M_Z is effectively uncorrelated with the other two parameters. The correlation between Γ_Z and σ^0 for the track selected data sample is shown in fig. 6.

In the standard model, Γ_{ee}, Γ_{had} and Γ_ν are calculable with a small uncertainty of about ± 1% due to (i) electroweak radiative effects involving unknown

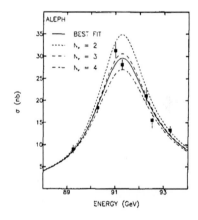

Fig. 5. The cross-section for $e^+e^-\to$ hadrons as a function of centre-of-mass energy and result of the three parameter fit.

Table 3

	M_Z (GeV)	Γ_Z (GeV)	σ^0 (nb)	σ^{peak} (nb)
hadronic events	91.178 ± 0.055	2.66 ± 0.16	39.1 ± 1.6	29.3 ± 1.2
hadronic + τ events	91.170 ± 0.054	2.70 ± 0.15	40.9 ± 1.7	30.5 ± 1.3
combined	91.174 ± 0.054	2.68 ± 0.15		

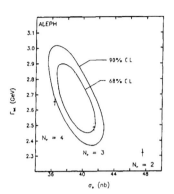

Fig. 6. The total width versus the peak hadronic cross-section, with 68% and 90% CL experimental contours. The standard model prediction for 2, 3 or 4 species of neutrinos is also shown, with its theoretical error.

Table 4
Standard model partial widths of the Z in MeV for the measured value of M_Z; $\alpha_s = 0.12 \pm 0.02$ and $\sin^2\theta_w = 0.230 \pm 0.006$ have been used as input.

Γ_{ee} (MeV)	83.5 ± 0.5
Γ_ν (MeV)	166.5 ± 1.0
Γ_{had} (MeV)	1737 ± 22

particles, such as the top quark; (ii) the value of the strong coupling constant α_s. The standard model predictions for the partial widths are given in table 4. The standard model predictions for σ^0 and Γ_Z assuming 2, 3, and 4 species of light neutrinos are shown in fig. 6. The value $N_\nu = 3$ is preferred. More precise information on N_ν is contained in the peak cross-section σ^0.

If the partial widths are taken from standard model predictions, a two parameter fit can be performed, leaving M_Z and N_ν as only free parameters. This fit, performed with the programs of refs. [12,15] on the two data samples, leaves the value of M_Z unchanged. The result for N_ν is

$$N_\nu = 3.27 \pm 0.24_{stat} \pm 0.16_{sys} \pm 0.05_{th}, \quad (4)$$

where the errors coming from statistics, experimental systematics and theoretical uncertainty are shown separately. Theoretical uncertainties due to electroweak radiative effects consist mostly in a change of the overall scale of the partial widths, and largely cancel in σ^0. The uncertainty in Γ_{had}, related to the QCD correction, cancels in part in σ^0; the resulting uncertainty $\Delta\sigma^0/\sigma^0 = 0.4\Delta\Gamma_{had}/\Gamma_Z = 0.003$ is much smaller than the effect produced by a single neutrino family, $\Delta\sigma^0/\sigma^0 = -0.13$. Combining these errors in quadrature one finds

$$N_\nu = 3.27 \pm 0.30. \quad (5)$$

The hypothesis $N_\nu = 4$ is ruled out at 98% confidence level. This measurement improves in a decisive way upon previous determinations of the number of neutrino species from the UA1 [16] and UA2 [17] experiments, from PEP [18] and PETRA [19], from cosmological [20] or astrophysical [21] arguments, as well as from a similar determination at the Z peak [22].

The demonstration that there is a third neutrino confirms that the τ neutrino is distinct from the e and μ neutrinos. The absence of a fourth light neutrino indicates that the quark–lepton families are closed with the three which are already known, except for the possibility that higher order families have neutrinos with masses in excess of ~ 30 GeV.

Acknowledgement

We would like to express our gratitude and admiration to our colleagues of the LEP division for the timely and beautiful operation of the machine. We thank the Technical Coordinator, Pierre Lazeyras, and the technical staff of the ALEPH Collaboration for their excellent work. We would like to dedicate this paper to the memory of those who died during LEP construction: Luigi Barito, Miloud Ferras, Luigi Filippi, Frédéric Mouly, Noël Piccini and François Pierrus. Those of us from non-member countries thank CERN for its hospitality.

References

[1] UA1 Collab., G. Arnison et al., Phys. Lett. B 126 (1983) 398.
[2] UA2 Collab., P. Bagnaia et al., Phys. Lett. B 129 (1983) 130.

[3] S.L. Glashow, Nucl. Phys. 22 (1961) 579;
S. Weinberg, Phys. Rev. Lett. 19 (1967) 1264;
A. Salam, Elementary particle theory, ed. N. Svartholm (Almquist and Wiksell, Stockholm, 1968) p. 367.

[4] For references see D.Y. Bardin et al., Z-line-shape group, in: Proc. Workshop of Z physics at LEP, CERN report 89-08;
R.N. Cahn, Phys. Rev. D 36 (1987) 2666;
O. Nicrosini and L. Trentadue, Phys. Lett. B 196 (1987) 551;
F.A. Berends, G. Burgers, W. Hollik and W.L. van Neerven, Phys. Lett. B 203 (1988) 177;
G. Burgers, in: Polarization at LEP, preprint CERN 88-06 (1988);
D.C. Kennedy et al., Nucl. Phys. B 321 (1989) 83.

[5] F. Abe et al., Phys. Rev. Lett. 63 (1989) 720.

[6] G.S. Abrams et al., Phys. Rev. Lett. 63 (1989) 724.

[7] ALEPH – a detector for electron–positron annihilation at LEP, Nucl. Instrum. Methods, to be published.

[8] F.A. Berends and R. Kleiss, Nucl. Phys. B 228 (1983) 737;
M. Böhm, A. Denner and W. Hollik, Nucl. Phys. B 304 (1988) 687;
F.A. Berends, R. Kleiss and W. Hollik, Nucl. Phys. B 304 (1988) 712.

[9] H. Burkhardt, F. Jegerlehner, G. Penso and C. Verzegnassi, Z. Phys. C 43 (1989) 497.

[10] D.Y. Bardin et al., Monte Carlo working group, Proc. Workshop of Z physics at LEP, CERN report 89-08.

[11] Computer program ZAPP, courtesy of G. Burgers.

[12] A. Borelli, M. Consoli, L. Maiani and R. Sisto, preprint CERN-TH-5441 (1989).

[13] Computer program ZHADRO, courtesy of G. Burgers.

[14] S. Myers, private communication; and LEP note, to appear.

[15] Computer program ZAPPH, courtesy of G. Burgers.

[16] UA1 Collab., C. Albajar et al., Phys. Lett. B 185 (1987) 241; B 198 (1987) 271.

[17] UA2 Collab., R. Ansari et al., Phys. Lett. B 186 (1987) 440.

[18] MAC Collab.. W.T. Ford et al., Phys. Rev. D 19 (1986) 3472.
ASP Collab., C. Hearty et al., Phys. Rev. Lett. 58 (1987) 1711.

[19] CELLO Collab., H.J. Behrend et al., Phys. Lett. B 215 (1988) 186.

[20] G. Steigman, K.A. Olive, D.N. Schramm and M.S. Turner, Phys. Lett. B 176 (1986) 33;
J. Ellis, K. Enqvist, D.V. Nanopoulos and S. Sarkar, Phys. Lett. B 167 (1986) 457.

[21] J. Ellis and K.A. Olive, Phys. Lett. B 193 (1987) 525;
R. Schaeffer, Y. Declais and S. Jullian, Nature 330 (1987) 142;
L.M. Krauss, Nature 329 (1987) 689.

[22] MarkII Collab., J.M. Dorfan, Intern. Europhysics Conf. on High energy physics (Madrid, Spain, September 1989).

14
The Top Quark

Completing the Third Generation.

No one could doubt that there would be a sixth quark, the top or t, but it was equally certain that initially no one knew where it would be found. With the b quark near 5 GeV, 15 GeV or so seemed reasonable for the mass of the top quark. Every new accelerator that came on line had the potential to make the discovery and every one of them came up empty handed. Particularly disappointing were the cases of TRISTAN, an e^+e^- collider at KEK, which reached c.m. energy of 61.4 GeV and set a lower bound of 30.2 GeV and the $Sp\bar{p}S$ collider at CERN, which found the W and Z. Even SLC and LEP searched to no avail, setting limits at half the mass of the Z. This left the search to hadron colliders.

In 1984 and 1985 CERN's $Sp\bar{p}S$ collider reigned as the world's highest energy machine, with $\sqrt{s} = 630$ GeV. Having already discovered the W and Z, it was positioned to look for the top quark through the decay $W \to t\bar{b}$ and early results from UA-1 gave evidence for a top quark with a mass of 40 ± 10 GeV. However, additional running and further analysis did not confirm the result but instead produced a bound of 55 GeV.

With further running at the $Sp\bar{p}S$ in 1988/9 both UA-1 and UA-2 improved this limit. Using signals from muons and jets, UA-1 ruled out a top quark below 60 GeV, while UA-2, which looked in the electron plus jets channel excluded masses below 69 GeV.

At the same time, at CDF the lower limit on the top quark mass was raised to 77 GeV. Adding additional channels moved the limit higher, to 85 GeV, then to 91 GeV.

The Standard Model gave no direct information on the mass of the top quark, for all the quark masses are simply arbitrary parameters. However, using detailed electroweak measurements it was possible to make inferences about the mass of the top quark. Of all the particles in the Standard Model, only the t and the Higgs remained to be discovered. The prediction of the W mass in terms of the Z mass in lowest order is

$$m_W^2 = \frac{1}{2}m_Z^2\left(1 + \sqrt{1 - \frac{4\pi\alpha}{\sqrt{2}G_F m_Z^2}}\right). \tag{14.1}$$

The W and Z can undergo virtual transitions, the W to $t\bar{b}$ and the Z to $t\bar{t}$ or $b\bar{b}$. These result in small radiative corrections to the relation between their masses. It is also possible for the W and Z to make virtual transitions by emitting and reabsorbing a Higgs particle. It turns out that the mass of the Higgs boson enters these effects only as $\ln m_H^2$ while there are corrections to the W and Z masses squared proportional to m_t^2, as discussed in Chapter 13. A good measurement of the W mass together with a rough guess for the Higgs mass was enough to make a reasonable prediction of the mass of the top quark. By 1994, these estimates centered on values around 180 GeV.

Though the mass of the top was uncertain, its behavior was completely predicted by the Standard Model. Once the limits on the top mass exceeded the mass of the W it was clear that its decay would be $t \to W^+ b$, whose width is

$$\Gamma_t = \frac{G_F m_t^3}{8\pi \sqrt{2}} \left(1 - \frac{m_W^2}{m_t^2}\right)^2 \left(1 + 2\frac{m_W^2}{m_t^2}\right). \tag{14.2}$$

If the t quark were well above the bW threshold, this width would be on the order of a GeV, meaning that any narrow bound states $t\bar{t}$ would be completely obscured. More picturesquely, the t would decay before it could bind.

The W^+ could decay leptonically to $e^+ \nu_e$, $\mu^+ \nu_\mu$, or $\tau^+ \nu_\tau$, or nonleptonically, primarily to $u\bar{d}$ or $c\bar{s}$. Since the t was pair produced, there were four general forms for the events:

$$\begin{array}{llll}
t \to & b(\ell^+ \nu) & \bar{t} \to & \bar{b}(\ell^- \bar{\nu}) \\
t \to & b(q\bar{q}) & \bar{t} \to & \bar{b}(\ell^- \bar{\nu}) \\
t \to & b(\ell^+ \nu) & \bar{t} \to & \bar{b}(q\bar{q}) \\
t \to & b(q\bar{q}) & \bar{t} \to & \bar{b}(q\bar{q})
\end{array}$$

The last of these would be particularly hard to isolate since these events would be masked by much more common events in which jets were produced by ordinary QCD interactions. Leptons were thus the key signature for the t quark. The b and q and $\bar{q}$ quarks would appear as jets while the neutrinos would result in large missing "transverse energy," i.e. a transverse momentum imbalance.

The Tevatron Collider resumed running in 1992 and collected data during Run Ia (1992/3) and Run Ib (1994/5). Using Run Ia data, D0 raised the limit on the top quark mass to 131 GeV. By May 1994, CDF had enough events to declare that they had "evidence for the top quark," though they stopped short of announcing its discovery (Ref. 14.1). Two events were found that contained both an e and a μ, and which had both two additional jets (presumably from the b quarks) and missing transverse energy. Events in which a single lepton was found faced more severe backgrounds and additional requirements had to be imposed. Only events in which there were three or more jets were considered. In addition, there had to be evidence that at least one jet came from a b. This evidence was obtained in two ways. A lepton too "soft" (i.e. with not very high transverse momentum) to indicate a W was circumstantial evidence for the semileptonic decay of a b. Alternatively, the presence of the b could be demonstrated by finding a sign of the B decay itself.

The silicon vertex detector (SVX), the innermost part of CDF's tracking system, could measure tracks with a precision of tens of microns. This was good enough to identify B

decays, for which a typical decay length would be $c\tau = 450$ μm times the boost due to the motion of the B. Six events with an apparent separated B vertex were found, with a background of 2.3 ± 0.3. The soft lepton tag found seven with a background of 3.1 ± 0.3. Three of the seven had SVX tags as well.

The mass of the t quark could be obtained from the events with a single lepton (and thus just one missing neutrino). Seven of the ten events had four or more jets and those with the highest transverse momentum were used to fit the hypothesis $(b\ell\nu)(bq\bar{q})$ where the b's and q's would appear as jets. This kinematical fit gave a mass determination (though no discovery was claimed!) of $m_t = 174 \pm 10 ^{+10}_{-12}$ GeV.

In February, 1995, both CDF (**Ref. 14.2**) and D0 (**Ref. 14.3**) were ready to declare the top quark found. CDF had 48 pb^{-1} of new data to add to their previous 19 pb^{-1}. Moreover, improvements in the silicon vertex detector increased its efficiency for finding b vertices in top events by a factor of two, to about 40%. There were 21 events in the data sample in which the SVX found vertices that were candidates for b decays. In six of the events, two jets were tagged. Additional candidates with soft lepton tags together with jets were identified. Six dilepton events were recorded. The refined mass measurement, $m_t = 176 \pm 8 \pm 10$ GeV, was quite close to that in the earlier CDF paper.

The D0 Collaboration had to overcome handicaps from the design of their detector, which was less suited for the task than was CDF, lacking both a magnetic field and a high precision silicon tracking device. See Figure 14.1. The basic strategy was the same as for CDF: identify leptons as candidates for decays of W's and jets as candidates for both the b quark-jets and products of nonleptonic W decays. To compensate for the limitations of the detector, D0 developed effective cuts that reduced background, in particular a cut on the total transverse energy. In the dilepton channels, two jets were required as well as missing transverse energy. In the single lepton channels, at least three jets were required. This was increased to four for events in which there was no lepton tag that would signal a b quark. Combining seven channels, D0 found 17 events with an anticipated background of 3.8. With their sample, D0 was not able to determine the mass with as much precision as CDF. Their result, $m_t = 199 ^{+19}_{-21}$ GeV, was consistent, however, with the CDF result. Subsequent running at the Tevatron Collider enabled both experiments to observe additional top quark events and to reduce the uncertainty in the mass measurement. Their measurements in both dilepton and single lepton channels were in good agreement. The combined result from the two experiments for Run 1, $m_t = 174.3 \pm 5.1$ GeV, had the smallest fractional error of any quark mass determination.

The $t\bar{t}$ pairs are produced in two ways: $q\bar{q} \rightarrow t\bar{t}$ and $gg \rightarrow t\bar{t}$. Calculations show that at the Tevatron Collider, the former dominates. It is also possible to produce a single top quark through processes like $u\bar{d} \rightarrow W^+ \rightarrow t\bar{b}$, but these should not have passed the cuts imposed by the CDF and D0 experiments. The predicted cross section for $t\bar{t}$ was about 5–6 pb at the Tevatron Collider with $\sqrt{s} = 1.8$ GeV. The cross sections measured by CDF and D0 in Run 1 were near this, $6.5 ^{+1.7}_{-1.4}$ pb and 5.9 ± 1.7 pb.

The energy and luminosity of the Tevatron Collider were increased for Run 2, which began in 2001. By early 2008 more than 3 fb^{-1} had been collected and results were

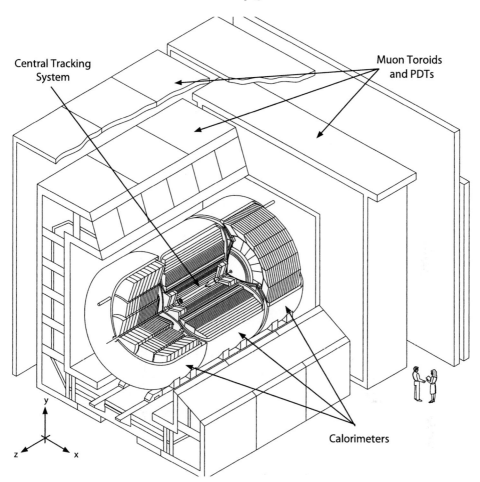

Figure 14.1. The D0 detector at the Fermilab Collider. Optimized for calorimetric measurements, D0 nonetheless was able to observe the t quark. [Courtesy D0 Collaboration.]

available for about 2 fb^{-1}, to be compared with 100 pb^{-1} in Run 1. The combined CDF and D0 result for m_t stood at 172.6 ± 1.4 GeV.

While the general agreement between the expected and measured cross sections and the conformity of the event structure to that anticipated from the Standard Model provides evidence that we do understand these processes, more exacting tests are needed to exclude exotic alternatives. The top quarks might, for example, be decay products of more massive particles rather than directly produced themselves. Absent such a surprise, the t may seem the most mundane of all quarks. Because of its rapid decay it doesn't produce stable hadrons as do all other quarks. In the t the quark concept is reduced to its most fundamental. Its interactions are for the most part described by perturbative QCD.

But this may be an illusion. Does the very large mass of the t quark point to a special role? Is it an indication of some new interactions not enjoyed by the lighter quarks?

Exercises

14.1 A t quark decays into a b quark, whose momentum is measured, and W, which decays to $\mu\nu$. The momentum of the μ is measured and the momentum of the ν transverse to the beam direction, p_ν, is inferred from the missing transverse momentum in the event. If the transverse and longitudinal components of the muon momentum are $p_{\mu\perp}$ and $p_{\mu\|}$, find the two possible values of the longitudinal momentum of the ν. When is there no solution for the longitudinal momentum of the ν? Do not assume that the mass of the t quark is known.

14.2 The coupling of the t, b, and W is described by

$$\frac{g}{\sqrt{2}} \bar{b}_L \gamma \cdot W t$$

where $b_L = \frac{1}{2}(1-\gamma_5)b$. Here b and t stand for the corresponding spinor fields and W for its field. The square of the decay matrix element can be shown (perhaps by the reader) to be

$$\frac{g^2}{2}[2 p_b \cdot \epsilon \, p_t \cdot \epsilon - p_b \cdot p_t \, \epsilon \cdot \epsilon].$$

The polarization of the W is ϵ, which obeys

$$\epsilon \cdot \epsilon = -1; \; \epsilon \cdot p_W = 0.$$

The three polarizations of the W are given by two choices of three-vectors perpendicular to the momentum of the W (transverse polarization) and one choice with both a time component and a space component parallel to the momentum of the W. (In writing the square of the matrix element the polarization vector was assumed real so the tranverse polarizations must be linear.) Using the two-body decay formula

$$d\Gamma = \frac{1}{32\pi^2}|\mathcal{M}|^2 \frac{p_{cm}}{M^2} d\Omega$$

and ignoring the mass of the b quark, confirm the formula in the text for the decay rate of the t. Show that the ratio of longitudinal to transverse W's is $\frac{1}{2}(m_t^2/m_W^2)$.

Further Reading

D. Chakraborty, J. Konigsberg, and D. Rainwater, *Ann. Rev. Nucl. Part. Sci.* **53**, 301 (2003).

References

14.1 F. Abe et al., "Evidence for Top Quark Production in $\bar{p}p$ Collisions at $\sqrt{s} = 1.8$ TeV." *Phys. Rev. Lett.*, **73**, 225 (1994).

14.2 F. Abe *et al.*, "Observation of Top Quark Production in $\bar{p}p$ Collisions with the Collider Detector at Fermilab." *Phys. Rev. Lett.*, **74**, 2626 (1995).

14.3 S. Abachi *et al.*, "Observation of the Top Quark." *Phys. Rev. Lett.*, **74**, 2632 (1995).

Observation of Top Quark Production in $\bar{p}p$ Collisions with the Collider Detector at Fermilab

F. Abe,[14] H. Akimoto,[32] A. Akopian,[27] M. G. Albrow,[7] S. R. Amendolia,[24] D. Amidei,[17] J. Antos,[29] C. Anway-Wiese,[4] S. Aota,[32] G. Apollinari,[27] T. Asakawa,[32] W. Ashmanskas,[15] M. Atac,[7] P. Auchincloss,[26] F. Azfar,[22] P. Azzi-Bacchetta,[21] N. Bacchetta,[21] W. Badgett,[17] S. Bagdasarov,[27] M. W. Bailey,[19] J. Bao,[35] P. de Barbaro,[26] A. Barbaro-Galtieri,[15] V. E. Barnes,[25] B. A. Barnett,[13] P. Bartalini,[24] G. Bauer,[16] T. Baumann,[9] F. Bedeschi,[24] S. Behrends,[3] S. Belforte,[24] G. Bellettini,[24] J. Bellinger,[34] D. Benjamin,[31] J. Benlloch,[16] J. Bensinger,[3] D. Benton,[22] A. Beretvas,[7] J. P. Berge,[7] S. Bertolucci,[8] A. Bhatti,[27] K. Biery,[12] M. Binkley,[7] D. Bisello,[21] R. E. Blair,[1] C. Blocker,[3] A. Bodek,[26] W. Bokhari,[16] V. Bolognesi,[24] D. Bortoletto,[25] J. Boudreau,[23] G. Brandenburg,[9] L. Breccia,[2] C. Bromberg,[18] E. Buckley-Geer,[7] H. S. Budd,[26] K. Burkett,[17] G. Busetto,[21] A. Byon-Wagner,[7] K. L. Byrum,[1] J. Cammerata,[13] C. Campagnari,[7] M. Campbell,[17] A. Caner,[7] W. Carithers,[15] D. Carlsmith,[34] A. Castro,[21] G. Cauz,[24] Y. Cen,[26] F. Cervelli,[24] H. Y. Chao,[29] J. Chapman,[17] M.-T. Cheng,[29] G. Chiarelli,[24] T. Chikamatsu,[32] C. N. Chiou,[29] L. Christofek,[11] S. Cihangir,[7] A. G. Clark,[24] M. Cobal,[24] M. Contreras,[5] J. Conway,[28] J. Cooper,[7] M. Cordelli,[8] C. Couyoumtzelis,[24] D. Crane,[1] D. Cronin-Hennessy,[6] R. Culbertson,[5] J. D. Cunningham,[3] T. Daniels,[16] F. DeJongh,[7] S. Delchamps,[7] S. Dell'Agnello,[24] M. Dell'Orso,[24] L. Demortier,[27] B. Denby,[24] M. Deninno,[2] P. F. Derwent,[17] T. Devlin,[28] M. Dickson,[26] J. R. Dittmann,[5] S. Donati,[24] R. B. Drucker,[15] A. Dunn,[17] N. Eddy,[17] K. Einsweiler,[15] J. E. Elias,[7] R. Ely,[15] E. Engels, Jr.,[23] D. Errede,[11] S. Errede,[11] Q. Fan,[26] I. Fiori,[2] B. Flaugher,[7] G. W. Foster,[7] M. Franklin,[9] M. Frautschi,[19] J. Freeman,[7] J. Friedman,[16] H. Frisch,[5] T. A. Fuess,[1] Y. Fukui,[14] S. Funaki,[32] G. Gagliardi,[23] S. Galeotti,[24] M. Gallinaro,[21] M. Garcia-Sciveres,[15] A. F. Garfinkel,[25] C. Gay,[9] S. Geer,[7] D. W. Gerdes,[17] P. Giannetti,[24] N. Giokaris,[27] P. Giromini,[8] L. Gladney,[22] D. Glenzinski,[13] M. Gold,[19] J. Gonzalez,[22] A. Gordon,[9] A. T. Goshaw,[6] K. Goulianos,[27] H. Grassmann,[7,*] L. Groer,[28] C. Grosso-Pilcher,[5] G. Guillian,[17] R. S. Guo,[29] C. Haber,[15] S. R. Hahn,[7] R. Hamilton,[9] R. Handler,[34] R. M. Hans,[35] K. Hara,[32] B. Harral,[22] R. M. Harris,[7] S. A. Hauger,[6] J. Hauser,[4] C. Hawk,[28] E. Hayashi,[32] J. Heinrich,[22] M. Hohlmann,[1,5] C. Holck,[22] R. Hollebeek,[22] L. Holloway,[11] A. Hölscher,[12] S. Hong,[17] G. Houk,[22] P. Hu,[23] B. T. Huffman,[23] R. Hughes,[26] J. Huston,[18] J. Huth,[9] J. Hylen,[7] H. Ikeda,[32] M. Incagli,[24] J. Incandela,[7] J. Iwai,[32] Y. Iwata,[10] H. Jensen,[7] U. Joshi,[7] R. W. Kadel,[15] E. Kajfasz,[7,*] T. Kamon,[30] T. Kaneko,[32] K. Karr,[33] H. Kasha,[35] Y. Kato,[20] L. Keeble,[8] K. Kelley,[16] R. D. Kennedy,[28] R. Kephart,[7] P. Kesten,[15] D. Kestenbaum,[9] R. M. Keup,[11] H. Keutelian,[7] F. Keyvan,[4] B. J. Kim,[26] D. H. Kim,[7,*] H. S. Kim,[12] S. B. Kim,[17] S. H. Kim,[32] Y. K. Kim,[15] L. Kirsch,[3] P. Koehn,[26] K. Kondo,[32] J. Konigsberg,[9] S. Kopp,[5] K. Kordas,[12] W. Koska,[7] E. Kovacs,[7,*] W. Kowald,[6] M. Krasberg,[17] J. Kroll,[7] M. Kruse,[25] T. Kuwabara,[32] S. E. Kuhlmann,[1] E. Kuns,[28] A. T. Laasanen,[25] N. Labanca,[24] S. Lammel,[7] J. I. Lamoureux,[3] T. LeCompte,[11] S. Leone,[24] J. D. Lewis,[7] P. Limon,[7] M. Lindgren,[4] T. M. Liss,[11] N. Lockyer,[22] O. Long,[22] C. Loomis,[28] M. Loreti,[21] J. Lu,[30] D. Lucchesi,[24] P. Lukens,[7] S. Lusin,[34] J. Lys,[15] K. Maeshima,[7] A. Maghakian,[27] P. Maksimovic,[16] M. Mangano,[24] J. Mansour,[18] M. Mariotti,[21] J. P. Marriner,[7] A. Martin,[11] J. A. J. Matthews,[19] R. Mattingly,[16] P. McIntyre,[30] P. Melese,[27] A. Menzione,[24] E. Meschi,[24] S. Metzler,[22] C. Miao,[17] G. Michail,[9] S. Mikamo,[14] R. Miller,[18] H. Minato,[32] S. Miscetti,[8] M. Mishina,[14] H. Mitsushio,[32] T. Miyamoto,[32] S. Miyashita,[32] Y. Morita,[14] J. Mueller,[23] A. Mukherjee,[7] T. Muller,[4] P. Murat,[24] H. Nakada,[32] I. Nakano,[32] C. Nelson,[7] D. Neuberger,[4] C. Newman-Holmes,[7] M. Ninomiya,[32] L. Nodulman,[1] S. Ogawa,[32] S. H. Oh,[6] K. E. Ohl,[35] T. Ohmoto,[10] T. Ohsugi,[10] R. Oishi,[32] M. Okabe,[32] T. Okusawa,[20] R. Oliver,[22] J. Olsen,[34] C. Pagliarone,[2] R. Paoletti,[24] V. Papadimitriou,[31] S. P. Pappas,[35] S. Park,[7] J. Patrick,[7] G. Pauletta,[24] M. Paulini,[15] L. Pescara,[21] M. D. Peters,[15] T. J. Phillips,[6] G. Piacentino,[2] M. Pillai,[26] K. T. Pitts,[7] R. Plunkett,[7] L. Pondrom,[34] J. Proudfoot,[1] F. Ptohos,[9] G. Punzi,[24] K. Ragan,[12] A. Ribon,[21] F. Rimondi,[2] L. Ristori,[24] W. J. Robertson,[6] T. Rodrigo,[7,*] J. Romano,[5] L. Rosenson,[16] R. Roser,[11] W. K. Sakumoto,[26] D. Saltzberg,[5] A. Sansoni,[8] L. Santi,[24] H. Sato,[32] V. Scarpine,[30] P. Schlabach,[9] E. E. Schmidt,[7] M. P. Schmidt,[35] G. F. Sciacca,[24] A. Scribano,[24] S. Segler,[7] S. Seidel,[19] Y. Seiya,[32] G. Sganos,[12] A. Sgolacchia,[2] M. D. Shapiro,[15] N. M. Shaw,[25] Q. Shen,[25] P. F. Shepard,[23] M. Shimojima,[32] M. Shochet,[5] J. Siegrist,[15] A. Sill,[31] P. Sinervo,[12] P. Singh,[23] J. Skarha,[13] K. Sliwa,[33] D. A. Smith,[24] F. D. Snider,[13] T. Song,[17] J. Spalding,[7] T. Speer,[24] P. Sphicas,[16] L. Spiegel,[7] A. Spies,[13] L. Stanco,[21] J. Steele,[34] A. Stefanini,[24] K. Strahl,[12] J. Strait,[7] D. Stuart,[7] G. Sullivan,[5] A. Soumarokov,[29] K. Sumorok,[16] J. Suzuki,[32] T. Takada,[32] T. Takahashi,[20] T. Takano,[32] K. Takikawa,[32] N. Tamura,[10] F. Tartarelli,[24] W. Taylor,[12] P. K. Teng,[29] Y. Teramoto,[20] S. Tether,[16] D. Theriot,[7] T. L. Thomas,[19] R. Thun,[17] M. Timko,[33] P. Tipton,[26] A. Titov,[27] S. Tkaczyk,[7] D. Toback,[5] K. Tollefson,[26] A. Tollestrup,[7] J. Tonnison,[25] J. F. de Troconiz,[9] S. Truitt,[17] J. Tseng,[13] N. Turini,[24] T. Uchida,[32] N. Uemura,[32] F. Ukegawa,[22] G. Unal,[22] S. C. van den Brink,[23] S. Vejcik III,[17] G. Velev,[24] R. Vidal,[7] M. Vondracek,[11] D. Vucinic,[16] R. G. Wagner,[1] R. L. Wagner,[7] J. Wahl,[5] R. C. Walker,[26] C. Wang,[6] C. H. Wang,[29] G. Wang,[24] J. Wang,[5] M. J. Wang,[29] Q. F. Wang,[27] A. Warburton,[12] G. Watts,[26] T. Watts,[28] R. Webb,[30] C. Wei,[6] C. Wendt,[34] H. Wenzel,[15] W. C. Wester III,[7] A. B. Wicklund,[1] E. Wicklund,[7] R. Wilkinson,[22] H. H. Williams,[22] P. Wilson,[5] B. L. Winer,[26] D. Wolinski,[17] J. Wolinski,[30] X. Wu,[24] J. Wyss,[21] A. Yagil,[7] W. Yao,[15]

K. Yasuoka,[32] Y. Ye,[12] G. P. Yeh,[7] P. Yeh,[29] M. Yin,[6] J. Yoh,[7] C. Yosef,[18] T. Yoshida,[20] D. Yovanovitch,[7] I. Yu,[35] J. C. Yun,[7] A. Zanetti,[24] F. Zetti,[24] L. Zhang,[34] W. Zhang,[22] and S. Zucchelli[2]

(CDF Collaboration)

[1] *Argonne National Laboratory, Argonne, Illinois 60439*
[2] *Istituto Nazionale di Fisica Nucleare, University of Bologna, I-40126 Bologna, Italy*
[3] *Brandeis University, Waltham, Massachusetts 02254*
[4] *University of California at Los Angeles, Los Angeles, California 90024*
[5] *University of Chicago, Chicago, Illinois 60637*
[6] *Duke University, Durham, North Carolina 27708*
[7] *Fermi National Accelerator Laboratory, Batavia, Illinois 60510*
[8] *Laboratori Nazionali di Frascati, Istituto Nazionale di Fisica Nucleare, I-00044 Frascati, Italy*
[9] *Harvard University, Cambridge, Massachusetts 02138*
[10] *Hiroshima University, Higashi-Hiroshima 724, Japan*
[11] *University of Illinois, Urbana, Illinois 61801*
[12] *Institute of Particle Physics, McGill University, Montreal, Canada H3A 2T8
and University of Toronto, Toronto, Canada M5S 1A7*
[13] *The Johns Hopkins University, Baltimore, Maryland 21218*
[14] *National Laboratory for High Energy Physics (KEK), Tsukuba, Ibaraki 305, Japan*
[15] *Lawrence Berkeley Laboratory, Berkeley, California 94720*
[16] *Massachusetts Institute of Technology, Cambridge, Massachusetts 02139*
[17] *University of Michigan, Ann Arbor, Michigan 48109*
[18] *Michigan State University, East Lansing, Michigan 48824*
[19] *University of New Mexico, Albuquerque, New Mexico 87131*
[20] *Osaka City University, Osaka 588, Japan*
[21] *Università di Padova, Instituto Nazionale di Fisica Nucleare, Sezione di Padova, I-35131 Padova, Italy*
[22] *University of Pennsylvania, Philadelphia, Pennsylvania 19104*
[23] *University of Pittsburgh, Pittsburgh, Pennsylvania 15260*
[24] *Istituto Nazionale di Fisica Nucleare, University and Scuola Normale Superiore of Pisa, I-56100 Pisa, Italy*
[25] *Purdue University, West Lafayette, Indiana 47907*
[26] *University of Rochester, Rochester, New York 14627*
[27] *Rockefeller University, New York, New York 10021*
[28] *Rutgers University, Piscataway, New Jersey 08854*
[29] *Academia Sinica, Taipei, Taiwan 11529, Republic of China*
[30] *Texas A&M University, College Station, Texas 77843*
[31] *Texas Tech University, Lubbock, Texas 79409*
[32] *University of Tsukuba, Tsukuba, Ibaraki 305, Japan*
[33] *Tufts University, Medford, Massachusetts 02155*
[34] *University of Wisconsin, Madison, Wisconsin 53706*
[35] *Yale University, New Haven, Connecticut 06511*

(Received 24 February 1995)

We establish the existence of the top quark using a 67 pb^{-1} data sample of $\bar{p}p$ collisions at $\sqrt{s} = 1.8$ TeV collected with the Collider Detector at Fermilab (CDF). Employing techniques similar to those we previously published, we observe a signal consistent with $t\bar{t}$ decay to $WWb\bar{b}$, but inconsistent with the background prediction by 4.8σ. Additional evidence for the top quark is provided by a peak in the reconstructed mass distribution. We measure the top quark mass to be $176 \pm 8(\text{stat}) \pm 10(\text{syst})$ GeV/c^2, and the $t\bar{t}$ production cross section to be $6.8^{+3.6}_{-2.4}$ pb.

PACS numbers: 14.65.Ha, 13.85.Qk, 13.85.Ni

Recently the Collider Detector at Fermilab (CDF) Collaboration presented the first direct evidence for the top quark [1], the weak isodoublet partner of the b quark required in the standard model. We searched for $t\bar{t}$ pair production with the subsequent decay $t\bar{t} \to Wb W\bar{b}$. The observed topology in such events is determined by the decay mode of the two W bosons. Dilepton events ($e\mu$, ee, and $\mu\mu$) are produced primarily when both W bosons decay into $e\nu$ or $\mu\nu$. Events in the lepton + jets channel (e, μ + jets) occur when one W boson decays into leptons and the other decays into quarks. To suppress background in the lepton + jets mode, we identify b quarks by reconstructing secondary vertices from b decay (SVX tag) and by finding additional leptons from b semileptonic decay (SLT tag). In Ref. [1] we found a 2.8σ excess of signal over the expectation from background. The interpretation of the excess as top quark production was supported by a peak in the mass distribution for fully reconstructed events. Additional evidence was found in the jet energy distributions in lepton + jets events [2]. An upper limit on the $t\bar{t}$ production cross section has been published by the D0 Collaboration [3].

We report here on a data sample containing 19 pb^{-1} used in Ref. [1] and 48 pb^{-1} from the current Fermilab Collider run, which began early in 1994 and is expected to continue until the end of 1995.

The CDF consists of a magnetic spectrometer surrounded by calorimeters and muon chambers [4]. A new low-noise, radiation-hard, four-layer silicon vertex detector, located immediately outside the beampipe, provides precise track reconstruction in the plane transverse to the beam and is used to identify secondary vertices from b and c quark decays [5]. The momenta of charged particles are measured in the central tracking chamber (CTC), which is in a 1.4 T superconducting solenoidal magnet. Outside the CTC, electromagnetic and hadronic calorimeters cover the pseudorapidity region $|\eta| < 4.2$ [6] and are used to identify jets and electron candidates. The calorimeters are also used to measure the missing transverse energy $\slashed{E}_T$, which can indicate the presence of undetected energetic neutrinos. Outside the calorimeters, drift chambers in the region $|\eta| < 1.0$ provide muon identification. A three-level trigger selects the inclusive electron and muon events used in this analysis. To improve the $t\bar{t}$ detection efficiency, triggers based on $\slashed{E}_T$ are added to the lepton triggers used in Ref. [1].

The data samples for both the dilepton and lepton + jets analyses are subsets of a sample of high-P_T inclusive lepton events that contain an isolated electron with $E_T > 20$ GeV or an isolated muon with $P_T > 20$ GeV/c in the central region ($|\eta| < 1.0$). Events which contain a second lepton candidate are removed as possible Z bosons if an ee or $\mu\mu$ invariant mass is between 75 and 105 GeV/c^2. For the lepton + jets analysis, an inclusive W boson sample is made by requiring $\slashed{E}_T > 20$ GeV. Table I classifies the W events by the number of jets with observed $E_T > 15$ GeV and $|\eta| < 2.0$. The dilepton sample consists of inclusive lepton events that also have a second lepton with $P_T > 20$ GeV/c, satisfying looser lepton identification requirements. The two leptons must have opposite electric charge.

The primary method for finding top quarks in the lepton + jets channel is to search for secondary vertices from b quark decay (SVX tagging). The vertex-finding efficiency is significantly larger now than previously due to an improved vertex-finding algorithm and the performance of the new vertex detector. The previous vertex-finding algorithm searched for a secondary vertex with two or more tracks. The new algorithm first searches for vertices with three or more tracks with looser track requirements, and if that fails, searches for two-track vertices using more stringent track and vertex quality criteria. The efficiency for tagging a b quark is measured in inclusive electron and muon samples which are enriched in b decays. The ratio of the measured efficiency to the prediction of a detailed Monte Carlo simulation is 0.96 ± 0.07, with good agreement ($\pm 2\%$) between the electron and muon samples. The efficiency for tagging at least one b quark in a $t\bar{t}$ event with ≥ 3 jets is determined from Monte Carlo simulation to be $(42 \pm 5)\%$ in the current run, compared to the $(22 \pm 6)\%$ reported in the previous publication [7]. In this Letter we apply the new vertex-finding algorithm to the data from the previous and the current runs.

In Ref. [1], we presented two methods for estimating the background to the top quark signal. In method 1, the observed tag rate in inclusive jet samples is used to calculate the background from mistags and QCD-produced heavy quark pairs ($b\bar{b}$ and $c\bar{c}$) recoiling against a W boson. This is an overestimate of the background because there are sources of heavy quarks in an inclusive jet sample that are not present in W + jet events. In method 2, the mistag rate is again measured with inclusive jets, while the fraction of W + jet events that are $Wb\bar{b}$ and $Wc\bar{c}$ is estimated from a Monte Carlo sample, using measured tagging efficiencies. In the present analysis, we use method 2 as the best estimate of the SVX-tag background. The improved performance of the new vertex detector, our ability to simulate its behavior accurately, and the agreement between the prediction and data in the W + 1-jet and W + 2-jet samples make this the natural choice. The calculated background, including the small contributions from non-W background, Wc production, and vector boson pair production, is given in Table I.

The numbers of SVX tags in the 1-jet and 2-jet samples are consistent with the expected background plus a small $t\bar{t}$ contribution (Table I and Fig. 1). However, for the $W + \geq 3$-jet signal region, 27 tags are observed compared to a predicted background of 6.7 ± 2.1 tags [8]. The probability of the background fluctuating to ≥ 27 is calculated to be 2×10^{-5} (see Table II) using the procedure outlined in Ref. [1] (see [9]). The 27 tagged jets are in 21 events; the six events with two tagged jets can be compared with four expected for the top + background hypothesis and ≤ 1 for background alone. Figure 1 also shows the decay lifetime distribution for the SVX tags in $W + \geq 3$-jet events. It is consistent with the distribution predicted for b decay from the $t\bar{t}$ Monte Carlo simulation. From the number of SVX-tagged events, the estimated background, the calculated

TABLE I. Number of lepton + jet events in the 67 pb^{-1} data sample along with the numbers of SVX tags observed and the estimated background. Based on the excess number of tags in events with ≥ 3 jets, we expect an additional 0.5 and 5 tags from $t\bar{t}$ decay in the 1- and 2-jet bins, respectively.

N_{jet}	Observed events	Observed SVX tags	Background tags expected
1	6578	40	50 ± 12
2	1026	34	21.2 ± 6.5
3	164	17	5.2 ± 1.7
≥ 4	39	10	1.5 ± 0.4

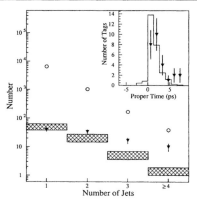

FIG. 1. Number of events before SVX tagging (circles), number of tags observed (triangles), and expected number of background tags (hatched) versus jet multiplicity. Based on the excess number of tags in events with ≥3 jets, we expect an additional 0.5 and 5 tags from $t\bar{t}$ decay in the 1- and 2-jet bins, respectively. The inset shows the secondary vertex proper time distribution for the 27 tagged jets in the $W+ \geq$3-jet data (triangles) compared to the expectation for b quark jets from $t\bar{t}$ decay.

$t\bar{t}$ acceptance, and the integrated luminosity of the data sample, we calculate the $t\bar{t}$ production cross section to be $6.8^{+3.6}_{-2.4}$ pb, where the uncertainty includes both statistical and systematic effects. This differs from the cross section given in Ref. [1] by 6.9 ± 5.9 pb.

The second technique for tagging b quarks (SLT tagging) is to search for an additional lepton from semileptonic b decay. Electrons and muons are found by matching CTC tracks with electromagnetic energy clusters or tracks in the muon chambers. To maintain acceptance for leptons coming directly from b decay and from the daughter c quark, the P_T threshold is kept low (2 GeV/c). The only significant change to the selection algorithm compared to Ref. [1] is that the fiducial region for SLT muons has been increased from $|\eta| < 0.6$ to $|\eta| < 1.0$, resulting in an increase of the SLT total acceptance and background by a factor of 1.2.

The major backgrounds in the SLT analysis are hadrons that are misidentified as leptons, and electrons from unidentified photon conversions. These rates and the smaller $Wb\bar{b}$ and $Wc\bar{c}$ backgrounds are determined directly from inclusive jet data. The remaining backgrounds are much smaller and are calculated using the techniques discussed in Ref. [1]. The efficiency of the algorithm is measured with photon conversion and $J/\psi \to \mu\mu$ data. The probability of finding an additional e or μ in a $t\bar{t}$ event with ≥3 jets is (20 ± 2)%. Table II shows the background and number of observed tags for the signal region ($W+ \geq$3 jets). There are 23 tags in 22 events, with 15.4 ± 2.0 tags expected from background. Six events contain both an SVX and SLT tag, compared to the expected four for top + background and one for background alone.

The dilepton analysis is very similar to that previously reported [1], with slight modifications to the lepton identification requirements to make them the same as those used in the single lepton analysis. The dilepton data sample, described above, is reduced by additional requirements on $\not{E}_T$ and the number of jets. In order to suppress background from Drell-Yan lepton pairs, which have little or no true $\not{E}_T$, the $\not{E}_T$ is corrected to account for jet energy mismeasurement [1]. The magnitude of the corrected $\not{E}_T$ is required to be at least 25 GeV and, if $\not{E}_T$ is less than 50 GeV, the azimuthal angle between the $\not{E}_T$ vector and the nearest lepton or jet must be greater than 20°. Finally, all events are required to have at least two jets with observed $E_T > 10$ GeV and $|\eta| < 2.0$.

The major backgrounds are Drell-Yan lepton pairs, $Z \to \tau\tau$, hadrons misidentified as leptons, WW, and $b\bar{b}$ production. We calculate the first three from data and the last two with Monte Carlo simulation [1]. As is shown in Table II, the total background expected is 1.3 ± 0.3 events. We observe a total of seven events, 5 $e\mu$ and 2 $\mu\mu$. The relative numbers are consistent with our dilepton acceptance, 60% of which is in the $e\mu$ channel. Although we estimated the expected background from radiative Z decay to be small (0.04 event), one of the $\mu\mu$ events contains an energetic photon with a $\mu\mu\gamma$ invariant mass of 86 GeV/c^2. To be conservative, we removed that event from the final sample, which thus contains six events. Three of these events contain a total of five b tags, compared with an expected 0.5 if the events are background. We would expect 3.6 tags if the events are from $t\bar{t}$ decay. When the requirement that the leptons have opposite charge is relaxed, we find one same-sign dilepton event ($e\mu$) that passes all the other event selection criteria. The expected number of same-sign events is 0.5, of which 0.3 is due to background and 0.2 to $t\bar{t}$ decay.

In summary, we find 37 b-tagged $W+ \geq$3-jet events [10] that contain 27 SVX tags compared to 6.7 ± 2.1 expected from background and 23 SLT tags with an estimated background of 15.4 ± 2.0. There are six dilepton events compared to 1.3 ± 0.3 events expected from background. We have taken the product (P) of the three probabilities in Table II and calculated the likelihood that a fluctuation of the background alone would yield a value of P no larger than that which we observe. The result

TABLE II. The number of tags or events observed in the three channels along with the expected background and the probability that the background would fluctuate to the observed number or more.

Channel	SVX	SLT	Dilepton
Observed	27 tags	23 tags	6 events
Expected background	6.7 ± 2.1	15.4 ± 2.0	1.3 ± 0.3
Background probability	2×10^{-5}	6×10^{-2}	3×10^{-3}

is 1×10^{-6}, which is equivalent to a 4.8σ deviation in a Gaussian distribution [11]. Based on the excess number of SVX-tagged events, we expect an excess of 7.8 SLT tags and 3.5 dilepton events from $t\bar{t}$ production, in good agreement with the observed numbers.

We performed a number of checks of this analysis. A good control sample for b tagging is Z + jet events, where no top contribution is expected. We observe 15, 3, and 2 tags (SVX and SLT) in the Z + 1-jet, 2-jet, and $\geq$3-jet samples, respectively, compared with the background predictions of 17.5, 4.2, and 1.5. The excess over background that was seen in Ref. [1] is no longer present. In addition, there is no discrepancy between the measured and predicted W + 4-jet background, in contrast to a small deficit described in Ref. [1] (see [12]).

Single-lepton events with four or more jets can be kinematically reconstructed to the $t\bar{t} \to Wb\bar{W}b$ hypothesis, yielding for each event an estimate of the top quark mass [1]. The lepton, neutrino ($\not{E}_T$), and the four highest-E_T jets are assumed to be the $t\bar{t}$ daughters [13]. There are multiple solutions, due to both the quadratic ambiguity in determining the longitudinal momentum of the neutrino and the assignment of jets to the parent W's and b's. For each event, the solution with the lowest fit χ^2 is chosen. Starting with the 203 events with $\geq$3 jets, we require each event to have a fourth jet with $E_T > 8$ GeV and $|\eta| < 2.4$. This yields a sample of 99 events, of which 88 pass a loose χ^2 requirement on the fit. The mass distribution for these events is shown in Fig. 2. The distribution is consistent with the predicted mix of approximately 30% $t\bar{t}$ signal and 70% W + jets background. The Monte Carlo background shape agrees well with that meaured in a limited-statistics sample of Z + 4-jet events as well as in a QCD sample selected to approximate non-W background. After requiring an SVX or SLT b tag, 19 of the events remain, of which $6.9^{+2.5}_{-1.9}$ are expected to be background. For these events, only solutions in which the tagged jet is assigned to one of the b quarks are considered. Figure 3 shows the mass distribution for the tagged events. The mass distribution in the current run is very similar to that from the previous run. Furthermore, we employed several mass fitting techniques which give nearly identical results.

To find the most likely top mass, we fit the mass distribution to a sum of the expected distributions from the W + jets background and a top quark of mass M_{top} [1]. The $-\ln(\text{liklihood})$ distribution from the fit is shown in the Fig. 3 inset. The best fit mass is 176 GeV/c^2 with a ± 8 GeV/c^2 statistical uncertainty. We make a conservative extrapolation of the systematic uncertainty from our previous publication, giving $M_{\text{top}} = 176 \pm 8 \pm 10$ GeV/c^2. Further studies of systematic uncertainties are in progress.

The shape of the mass peak in Fig. 3 provides additional evidence for top quark production, since the number of observed b tags is independent of the observed mass distribution. After including systematic effects in the predicted background shape, we find a 2×10^{-2} probability that the observed mass distribution is consistent with the background (Kolmogorov-Smirnov test). This is a conservative measure because it does not explicitly take into account the observed narrow mass peak.

In conclusion, additional data confirm the top quark evidence presented in Ref. [1]. There is now a large

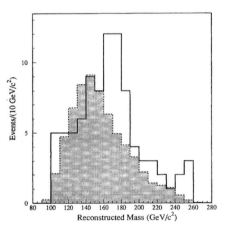

FIG. 2. Reconstructed mass distribution for the $W + \geq$4-jet sample prior to b tagging (solid). Also shown is the background distribution (shaded) with the normalization constrained to the calculated value.

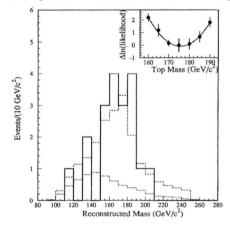

FIG. 3. Reconstructed mass distribution for the b-tagged $W+ \geq$4-jet events (solid). Also shown are the background shape (dotted) and the sum of background plus $t\bar{t}$ Monte Carlo simulations for $M_{\text{top}} = 175$ GeV/c^2 (dashed), with the background constrained to the calculated value, $6.9^{+2.5}_{-1.9}$ events. The inset shows the likelihood fit used to determine the top mass.

excess in the signal that is inconsistent with the background prediction by 4.8σ, and a mass distribution with a 2×10^{-2} probability of being consistent with the background shape. When combined, the signal size and mass distribution have a 3.7×10^{-7} probability of satisfying the background hypothesis (5.0σ). In addition, a substantial fraction of the jets in the dilepton events are b tagged. This establishes the existence of the top quark. The preliminary mass and cross section measurements yield $M_{\text{top}} = 176 \pm 8 \pm 10$ GeV/c^2 and $\sigma_{t\bar{t}} = 6.8^{+3.6}_{-2.4}$ pb.

This work would not have been possible without the skill and hard work of the Fermilab staff. We thank the staffs of our institutions for their many contributions to the construction of the detector. This work is supported by the U.S. Department of Energy, the National Science Foundation, the Natural Sciences and Engineering Research Council of Canada, the Istituto Nazionale di Fisica Nucleare of Italy, the Ministry of Education, Science and Culture of Japan, the National Science Council of the Republic of China, and the A. P. Sloan Foundation.

---

*Visitor.

[1] F. Abe et al., Phys. Rev. D **50**, 2966 (1994); Phys. Rev. Lett. **73**, 225 (1994).
[2] F. Abe et al., Phys. Rev. D (to be published).
[3] S. Abachi et al., Phys. Rev. Lett. **72**, 2138 (1994); following Letter, Phys. Rev. Lett. **74**, 2632 (1995).
[4] F. Abe et al., Nucl. Instrum. Methods Phys. Res., Sect. A **271**, 387 (1988).
[5] P. Azzi et al., Fermilab Report No. FERMILAB-CONF-94/205-E, 1994 (unpublished). Our previous silicon vertex detector is described in D. Amidei et al., Nucl. Instrum. Methods Phys. Res., Sect. A **350**, 73 (1994).
[6] In the CDF coordinate system, θ is the polar angle with respect to the proton beam direction. The pseudorapidity η is defined as $-\ln \tan(\theta/2)$. The transverse momentum of a particle is $P_T = P \sin\theta$. If the magnitude of this vector is obtained using the calorimeter energy rather than the spectrometer momentum, it becomes the transverse energy E_T. The difference between the vector sum of all the transverse energies in an event and zero is the missing transverse energy ($\not{E}_T$).
[7] A factor of 1.65 increase comes from the improvements noted. The remaining factor of 1.15 results from correcting an error in the b baryon lifetime used in the simulation of $t\bar{t}$ decay in Ref. [1].
[8] For comparison we note that if we had used both the tagging algorithm and background calculation (method 1) presented in Ref. [1], we would have 24 observed tags with a predicted background of 8.8 ± 0.6 tags.
[9] We get essentially the same probability if we use method 1 for the SVX-tag background because of its smaller systematic uncertainty.
[10] There are 21 events with SVX tags and 22 events with SLT tags. Six of these events have both SVX and SLT tags.
[11] This technique is chosen because we are combining channels with very different expected background rates. For comparison, if we apply the method used in Ref. [1] to the SVX and dilepton channels, the two low background modes, we obtain a probability of 1.5×10^{-6}.
[12] The improved agreement is due to the smaller $t\bar{t}$ production cross section obtained in this analysis as well as correcting an overestimate in Ref. [1] in the Monte Carlo background prediction.
[13] The jet energies used in the mass fitting have been corrected for instrumental and fragmentation effects.

Observation of the Top Quark

S. Abachi,[12] B. Abbott,[33] M. Abolins,[23] B. S. Acharya,[40] I. Adam,[10] D. L. Adams,[34] M. Adams,[15] S. Ahn,[12] H. Aihara,[20] J. Alitti,[36] G. Álvarez,[16] G. A. Alves,[8] E. Amidi,[27] N. Amos,[22] E. W. Anderson,[17] S. H. Aronson,[3] R. Astur,[38] R. E. Avery,[29] A. Baden,[21] V. Balamurali,[30] J. Balderston,[14] B. Baldin,[12] J. Bantly,[4] J. F. Bartlett,[12] K. Bazizi,[7] J. Bendich,[20] S. B. Beri,[31] I. Bertram,[34] V. A. Bezzubov,[32] P. C. Bhat,[12] V. Bhatnagar,[31] M. Bhattacharjee,[11] A. Bischoff,[7] N. Biswas,[30] G. Blazey,[12] S. Blessing,[13] A. Boehnlein,[12] N. I. Bojko,[32] F. Borcherding,[12] J. Borders,[35] C. Boswell,[7] A. Brandt,[12] R. Brock,[23] A. Bross,[12] D. Buchholz,[29] V. S. Burtovoi,[32] J. M. Butler,[12] D. Casey,[35] H. Castilla-Valdez,[9] D. Chakraborty,[38] S.-M. Chang,[27] S. V. Chekulaev,[32] L.-P. Chen,[20] W. Chen,[38] L. Chevalier,[36] S. Chopra,[31] B. C. Choudhary,[7] J. H. Christenson,[12] M. Chung,[15] D. Claes,[38] A. R. Clark,[20] W. G. Cobau,[21] J. Cochran,[7] W. E. Cooper,[12] C. Cretsinger,[35] D. Cullen-Vidal,[4] M. Cummings,[14] D. Cutts,[4] O. I. Dahl,[20] K. De,[41] M. Demarteau,[12] R. Demina,[27] K. Denisenko,[12] N. Denisenko,[12] D. Denisov,[12] S. P. Denisov,[32] W. Dharmaratna,[13] H. T. Diehl,[12] M. Diesburg,[12] G. Di Loreto,[23] R. Dixon,[12] P. Draper,[41] J. Drinkard,[6] Y. Ducros,[36] S. R. Dugad,[40] S. Durston-Johnson,[35] D. Edmunds,[23] A. O. Efimov,[32] J. Ellison,[7] V. D. Elvira,[12,*] R. Engelmann,[38] S. Eno,[21] G. Eppley,[34] P. Ermolov,[24] O. V. Eroshin,[32] V. N. Evdokimov,[32] S. Fahey,[23] T. Fahland,[4] M. Fatyga,[3] M. K. Fatyga,[35] J. Featherly,[3] S. Feher,[38] D. Fein,[2] T. Ferbel,[35] G. Finocchiaro,[38] H. E. Fisk,[12] Yu. Fisyak,[24] E. Flattum,[23] G. E. Forden,[2] M. Fortner,[28] K. C. Frame,[23] P. Franzini,[10] S. Fredriksen,[39] S. Fuess,[12] A. N. Galjaev,[32] E. Gallas,[41] C. S. Gao,[12,†] S. Gao,[12,†] T. L. Geld,[23] R. J. Genik II,[23] K. Genser,[12] C. E. Gerber,[12,‡] B. Gibbard,[3] M. Glaubman,[27] V. Glebov,[35] S. Glenn,[5] J. F. Glicenstein,[36] B. Gobbi,[29] M. Goforth,[13] A. Goldschmidt,[20] B. Gomez,[1] P. I. Goncharov,[32] H. Gordon,[3] L. T. Goss,[42] N. Graf,[3] P. D. Grannis,[38] D. R. Green,[12] J. Green,[28] H. Greenlee,[12] G. Griffin,[6] N. Grossman,[12] P. Grudberg,[20] S. Grünendahl,[35] J. A. Guida,[38] J. M. Guida,[3] W. Guryn,[3] S. N. Gurzhiev,[32] Y. E. Gutnikov,[32] N. J. Hadley,[21] H. Haggerty,[12] S. Hagopian,[13] V. Hagopian,[13] K. S. Hahn,[35] R. E. Hall,[6] S. Hansen,[12] R. Hatcher,[23] J. M. Hauptman,[17] D. Hedin,[28] A. P. Heinson,[7] U. Heintz,[12] R. Hernandez-Montoya,[9] T. Heuring,[13] J. D. Hobbs,[12] B. Hoeneisen,[1,§] J. S. Hoftun,[4] F. Hsieh,[22] Ting Hu,[38] Tong Hu,[16] T. Huehn,[7] S. Igarashi,[12] A. S. Ito,[12] E. James,[2] J. Jaques,[30] S. A. Jerger,[23] J. Z.-Y. Jiang,[38] T. Joffe-Minor,[29] H. Johari,[27] K. Johns,[2] M. Johnson,[12] H. Johnstad,[39] A. Jonckheere,[12] M. Jöstlein,[12] S. Y. Jun,[29] C. K. Jung,[38] S. Kahn,[3] J. S. Kang,[18] R. Kehoe,[30] M. Kelly,[30] A. Kernan,[7] L. Kerth,[20] C. L. Kim,[18] S. K. Kim,[37] A. Klatchko,[13] B. Klima,[12] B. I. Klochkov,[32] C. Klopfenstein,[38] V. I. Klyukhin,[32] V. I. Kochetkov,[32] J. M. Kohli,[31] D. Koltick,[33] A. V. Kostritskiy,[32] J. Kotcher,[3] J. Kourlas,[26] A. V. Kozelov,[32] E. A. Kozlovski,[32] M. R. Krishnaswamy,[40] S. Krzywdzinski,[12] S. Kunori,[21] S. Lami,[38] G. Landsberg,[38] R. E. Lanou,[4] J-F. Lebrat,[36] J. Lee-Franzini,[38] A. Leflat,[24] H. Li,[38] J. Li,[41] Y. K. Li,[29] Q. Z. Li-Demarteau,[12] J. G. R. Lima,[8] D. Lincoln,[22] S. L. Linn,[13] J. Linnemann,[23] R. Lipton,[12] Y. C. Liu,[29] F. Lobkowicz,[35] S. C. Loken,[20] S. Lökös,[38] L. Lueking,[12] A. L. Lyon,[21] A. K. A. Maciel,[8] R. J. Madaras,[20] R. Madden,[13] I. V. Mandrichenko,[32] Ph. Mangeot,[36] S. Mani,[5] B. Mansoulié,[36] H. S. Mao,[12,†] S. Margulies,[15] R. Markeloff,[28] L. Markosky,[2] T. Marshall,[16] M. I. Martin,[12] M. Marx,[38] B. May,[29] A. A. Mayorov,[32] R. McCarthy,[38] T. McKibben,[15] J. McKinley,[23] H. L. Melanson,[12] J. R. T. de Mello Neto,[8] K. W. Merritt,[12] H. Miettinen,[34] A. Milder,[2] C. Milner,[39] A. Mincer,[26] J. M. de Miranda,[8] C. S. Mishra,[12] M. Mohammadi-Baarmand,[38] N. Mokhov,[12] N. K. Mondal,[40] H. E. Montgomery,[12] P. Mooney,[1] M. Mudan,[26] C. Murphy,[16] C. T. Murphy,[12] F. Nang,[4] M. Narain,[12] V. S. Narasimham,[40] A. Narayanan,[2] H. A. Neal,[22] J. P. Negret,[1] E. Neis,[22] P. Nemethy,[26] D. Nešić,[4] D. Norman,[42] L. Oesch,[21] V. Oguri,[8] E. Oltman,[20] N. Oshima,[12] D. Owen,[23] P. Padley,[34] M. Pang,[17] A. Para,[12] C. H. Park,[12] Y. M. Park,[19] R. Partridge,[4] N. Parua,[40] M. Paterno,[35] J. Perkins,[12] M. Peters,[14] H. Piekarz,[13] Y. Pischalnikov,[33] A. Pluquet,[36] V. M. Podstavkov,[32] B. G. Pope,[23] H. B. Prosper,[13] S. Protopopescu,[3] D. Pušeljić,[20] J. Qian,[22] P. Z. Quintas,[12] R. Raja,[12] S. Rajagopalan,[38] O. Ramirez,[15] M. V. S. Rao,[40] P. A. Rapidis,[12] L. Rasmussen,[38] A. L. Read,[12] S. Reucroft,[27] M. Rijssenbeek,[38] T. Rockwell,[23] N. A. Roe,[20] J. M. R. Roldan,[1] P. Rubinov,[38] R. Ruchti,[30] S. Rusin,[24] J. Rutherfoord,[2] A. Santoro,[8] L. Sawyer,[41] R. D. Schamberger,[38] H. Schellman,[29] D. Schmid,[39] J. Sculli,[26] E. Shabalina,[24] C. Shaffer,[13] H. C. Shankar,[40] R. K. Shivpuri,[11] M. Shupe,[2] J. B. Singh,[31] V. Sirotenko,[28] W. Smart,[12] A. Smith,[2] R. P. Smith,[12] R. Snihur,[29] G. R. Snow,[25] S. Snyder,[3] J. Solomon,[15] P. M. Sood,[31] M. Sosebee,[41] M. Souza,[8] A. L. Spadafora,[20] R. W. Stephens,[41] M. L. Stevenson,[20] D. Stewart,[22] F. Stocker,[39] D. A. Stoianova,[32] D. Stoker,[6] K. Streets,[26] M. Strovink,[20] A. Taketani,[12] P. Tamburello,[21] J. Tarazi,[6] M. Tartaglia,[12] T. L. Taylor,[29] J. Teiger,[36] J. Thompson,[21] T. G. Trippe,[20] P. M. Tuts,[10] N. Varelas,[23] E. W. Varnes,[20] P. R. G. Virador,[20] D. Vititoe,[2] A. A. Volkov,[32] E. von Goeler,[27] A. P. Vorobiev,[32] H. D. Wahl,[13] J. Wang,[12,†] L. Z. Wang,[12,†] J. Warchol,[30] M. Wayne,[30] H. Weerts,[23] W. A. Wenzel,[20] A. White,[41] J. T. White,[42] J. A. Wightman,[17] J. Wilcox,[27] S. Willis,[28] S. J. Wimpenny,[7]

0031-9007/95/74(14)/2632(6)$06.00 © 1995 The American Physical Society

J. V. D. Wirjawan,[42] Z. Wolf,[39] J. Womersley,[12] E. Won,[35] D. R. Wood,[12] H. Xu,[4] R. Yamada,[12] P. Yamin,[3] C. Yanagisawa,[38] J. Yang,[26] T. Yasuda,[27] C. Yoshikawa,[14] S. Youssef,[13] J. Yu,[35] Y. Yu,[37] Y. Zhang,[12,†] Y. H. Zhou,[12,†] Q. Zhu,[26] Y. S. Zhu,[12,†] Z. H. Zhu,[35] D. Zieminska,[16] A. Zieminski,[16] A. Zinchenko,[17] and A. Zylberstejn[36]

(D0 Collaboration)

[1]*Universidad de los Andes, Bogota, Colombia*
[2]*University of Arizona, Tucson, Arizona 85721*
[3]*Brookhaven National Laboratory, Upton, New York 11973*
[4]*Brown University, Providence, Rhode Island 02912*
[5]*University of California, Davis, California 95616*
[6]*University of California, Irvine, California 92717*
[7]*University of California, Riverside, California 92521*
[8]*LAFEX, Centro Brasileiro de Pesquisas Físicas, Rio de Janeiro, Brazil*
[9]*Centro de Investigacion y de Estudios Avanzados, Mexico City, Mexico*
[10]*Columbia University, New York, New York 10027*
[11]*Delhi University, Delhi, India 110007*
[12]*Fermi National Accelerator Laboratory, Batavia, Illinois 60510*
[13]*Florida State University, Tallahassee, Florida 32306*
[14]*University of Hawaii, Honolulu, Hawaii 96822*
[15]*University of Illinois, Chicago, Illinois 60680*
[16]*Indiana University, Bloomington, Indiana 47405*
[17]*Iowa State University, Ames, Iowa 50011*
[18]*Korea University, Seoul, Korea*
[19]*Kyungsung University, Pusan, Korea*
[20]*Lawrence Berkeley Laboratory, Berkeley, California 94720*
[21]*University of Maryland, College Park, Maryland 20742*
[22]*University of Michigan, Ann Arbor, Michigan 48109*
[23]*Michigan State University, East Lansing, Michigan 48824*
[24]*Moscow State University, Moscow, Russia*
[25]*University of Nebraska, Lincoln, Nebraska 68588*
[26]*New York University, New York, New York 10003*
[27]*Northeastern University, Boston, Massachusetts 02115*
[28]*Northern Illinois University, DeKalb, Illinois 60115*
[29]*Northwestern University, Evanston, Illinois 60208*
[30]*University of Notre Dame, Notre Dame, Indiana 46556*
[31]*University of Panjab, Chandigarh 16-00-14, India*
[32]*Institute for High Energy Physics, 142-284 Protvino, Russia*
[33]*Purdue University, West Lafayette, Indiana 47907*
[34]*Rice University, Houston, Texas 77251*
[35]*University of Rochester, Rochester, New York 14627*
[36]*Commissariat à l'Energie Atomique, DAPNIA/Service de Physique des Particules,
Centre d'Etudes de Saclay, Saclay, France*
[37]*Seoul National University, Seoul, Korea*
[38]*State University of New York, Stony Brook, New York 11794*
[39]*Superconducting Super Collider Laboratory, Dallas, Texas 75237*
[40]*Tata Institute of Fundamental Research, Colaba, Bombay 400005, India*
[41]*University of Texas, Arlington, Texas 76019*
[42]*Texas A&M University, College Station, Texas 77843*
(Received 24 February 1995)

The D0 Collaboration reports on a search for the standard model top quark in $p\bar{p}$ collisions at $\sqrt{s} = 1.8$ TeV at the Fermilab Tevatron with an integrated luminosity of approximately 50 pb^{-1}. We have searched for $t\bar{t}$ production in the dilepton and single-lepton decay channels with and without tagging of b-quark jets. We observed 17 events with an expected background of 3.8 ± 0.6 events. The probability for an upward fluctuation of the background to produce the observed signal is 2×10^{-6} (equivalent to 4.6 standard deviations). The kinematic properties of the excess events are consistent with top quark decay. We conclude that we have observed the top quark and measured its mass to be 199^{+19}_{-21} (stat) ±22 (syst) GeV/c^2 and its production cross section to be 6.4 ± 2.2 pb.

PACS numbers: 14.65.Ha, 13.85.Qk, 13.85.Ni

In the standard model (SM), the top quark is the weak isospin partner of the b quark. The D0 Collaboration published a lower limit on the mass of the top quark of 131 GeV/c^2, at a confidence level (C.L.) of 95%, based on an integrated luminosity of 13.5 pb^{-1} [1]. A subsequent publication [2] reported the top quark production cross section as a function of the assumed top quark mass. In that analysis, we found nine events with an expected background of 3.8 ± 0.9 events (statistical significance 1.9 standard deviations) corresponding to a production cross section of 8.2 ± 5.1 pb for an assumed top quark mass of 180 GeV/c^2. The Collider Detector at Fermilab (CDF) Collaboration published evidence for top quark production with a statistical significance of 2.8 standard deviations, a top quark of mass 174 ± 10$^{+13}_{-12}$ GeV/c^2, and a production cross section of 13.9$^{+6.1}_{-4.8}$ pb [3]. Precision electroweak measurements predict a SM top quark mass of approximately 150–210 GeV/c^2, depending on the mass of the Higgs boson [4]. In the present Letter, we report new results from the D0 experiment that firmly establish the existence of the top quark.

We assume that the top quark is pair produced and decays according to the minimal SM (i.e., $t\bar{t} \to W^+W^-b\bar{b}$). We searched for the top quark in channels where both W bosons decayed leptonically ($e\mu$ + jets, ee + jets, and $\mu\mu$ + jets) and in channels where just one W boson decayed leptonically (e + jets and μ + jets). The single-lepton channels were subdivided into b-tagged and untagged channels according to whether or not a muon was observed consistent with $b \to \mu + X$. The muon-tagged channels are denoted e + jets/μ and μ + jets/μ.

Here we present an analysis based on data collected at the Fermilab Tevatron at $\sqrt{s} = 1.8$ TeV with an integrated luminosity of 44–56 pb^{-1}, depending on the channel. In the present analysis, the signal-to-background ratio for a high-mass top quark was substantially improved relative to Ref. [2]. An optimization of the selection criteria was carried out using Monte Carlo top quark events for signal and our standard background estimates. The result of this procedure was a factor of 3.7 better background rejection while retaining 70% of the acceptance for 180 GeV/c^2 top quarks. This corresponds to a signal-to-background ratio of 1:1 for a top quark mass of 200 GeV/c^2, assuming the expected SM top cross section [5]. The improved rejection arises primarily by requiring events to have a larger total transverse energy.

The D0 detector and data collection systems are described in Ref. [6]. The triggers and reconstruction algorithms for jets, electrons, muons, and neutrinos were the same as those used in our previous top quark searches [1,2].

The signature for the dilepton channels was defined as two isolated leptons, at least two jets, and large missing transverse energy $\not{E}_T$. The signature for the single-lepton channels was defined as one isolated lepton, large $\not{E}_T$, and a minimum of three jets (with muon tag) or four jets (without tag). The minimum transverse momentum p_T of tagging muons was 4 GeV/c. Requirements pertaining to the magnitude and direction of the $\not{E}_T$, the aplanarity of the jets $\mathcal{A}$, and the allowed ranges of pseudorapidity η were similar to Ref. [2]. Muons were restricted to $|\eta| <$ 1 for the last 70% of the data because of forward muon chamber aging. Events in the $\mu\mu$ + jets and μ + jets/μ channels were required to be inconsistent with the Z + jets hypothesis, based on a global kinematic fit. The principal difference between the present analysis and the analysis of Ref. [2] was the imposition of a minimum requirement in all channels on a quantity H_T, which we defined as the scalar sum of the transverse energies E_T of the jets (for the single-lepton and $\mu\mu$ + jets channels) or the scalar sum of the E_T's of the leading electron and the jets (for the $e\mu$ + jets and ee + jets channels). The kinematic requirements for our standard event selection for all seven channels are summarized in Table I. In addition to the standard selection, we defined a set of loose event selection requirements, which differed from the standard set by the removal of the H_T requirement and by the relaxation of the aplanarity requirement for e + jets and μ + jets from $\mathcal{A} > 0.05$ to $\mathcal{A} > 0.03$.

For the dilepton channels, the main backgrounds were from Z and continuum Drell-Yan production ($Z, \gamma^* \to ee, \mu\mu$, and $\tau\tau$), vector boson pairs (WW, WZ), heavy flavor ($b\bar{b}$ and $c\bar{c}$) production, and backgrounds with jets misidentified as leptons. For the single-lepton channels, the main backgrounds were from W + jets, Z + jets, and multijet production with a jet misidentified as a lepton. The method for estimating these backgrounds was the same as in our previously published analyses [1,2].

H_T is a powerful discriminator between background and high-mass top quark production. Figure 1 shows a comparison of the shapes of the H_T distributions expected from background and 200 GeV/c^2 top quarks in the channels (a) $e\mu$ + jets and (b) untagged single-lepton + jets. We tested our understanding of background H_T distributions by comparing data and calculated background in background-dominated channels such as electron + two jets and electron + three jets (Fig. 2). The observed H_T distribution agrees with the background calculation, which includes contributions from both W + jets as calcu-

TABLE I. Minimum kinematic requirements for the standard event selection (energy in GeV).

Channel	Leptons		Jets				
	$E_T(e)$	$p_T(\mu)$	N_{jets}	E_T	$\not{E}_T$	H_T	$\mathcal{A}$
$e\mu$ + jets	15	12	2	15	20	120	...
ee + jets	20		2	15	25	120	...
$\mu\mu$ + jets		15	2	15	...	100	...
e + jets	20		4	15	25	200	0.05
μ + jets		15	4	15	20	200	0.05
e + jets/μ	20		3	20	20	140	...
μ + jets/μ		15	3	20	20	140	...

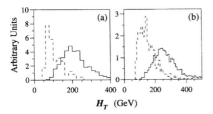

FIG. 1. Shape of H_T distributions expected for the principal backgrounds (dashed line) and 200 GeV/c^2 top quarks (solid line) for (a) $e\mu$ + jets and (b) untagged single-lepton + jets.

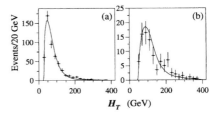

FIG. 2. Observed H_T distributions (points) compared to the distributions expected from background (line) for $\rlap{/}{E}_T >$ 25 GeV/c and (a) $e+$ $\geq$2 jets and (b) $e+$ $\geq$3 jets.

lated by the VECBOS Monte Carlo program [7] and multijet events.

The acceptance for $t\bar{t}$ events was calculated using the ISAJET event generator [8] and a detector simulation based on the GEANT program [9]. As a check, the acceptance was also calculated using the HERWIG event generator [10]. The difference between ISAJET and HERWIG was included in the systematic error.

From all seven channels, we observed 17 events with an expected background of 3.8 ± 0.6 events (see Table II). The probability of an upward fluctuation of the background to 17 or more events is 2×10^{-6}, which corresponds to 4.6 standard deviations for a Gaussian probability distribution. Our measured cross section as a function of the top quark mass hypothesis is shown in Fig. 3. Assuming a top quark mass of 200 GeV/c^2, the production cross section is 6.3 ± 2.2 pb. The error in the cross section includes an overall 12% uncertainty in the luminosity. The cross section determined from the loose selection criteria is in good agreement with this value, demonstrating that the backgrounds are well understood. We calculated the probability for our observed distribution of excess events among the seven channels and find that our results are consistent with top quark branching fractions at the 53% C.L. Thus,

we observe a statistically significant excess of events and the distribution of events among the seven channels is consistent with top quark production.

Additional confirmation that our observed excess contains a high-mass object comes from the invariant masses of jet combinations in single-lepton + jets events. For this analysis, we selected single-lepton + four-jet events using the loose event selection requirements (27 events). An invariant mass analysis was performed, based on the hypothesis $t\bar{t} \rightarrow W^+W^-b\bar{b} \rightarrow \ell\nu q\bar{q}b\bar{b}$. One jet was assigned to the semileptonically decaying top quark, and three jets were assigned to the hadronically decaying top quark. The jet assignment algorithm attempted to assign one of the two highest-E_T jets to the semileptonically decaying top quark and to minimize the difference between the masses of the two top quarks. The invariant mass of the three jets assigned to the hadronically decaying top quark is denoted by m_{3j}. Among the three possible jet pairs from the hadronically decaying top quark, the smallest invariant mass is denoted by m_{2j}. Figure 4 shows the distribution of m_{3j} vs m_{2j} for (a) background (W + jets and multijet) (b) 200 GeV/c^2 top Monte Carlo simulation, and (c) data. The data are peaked at higher invariant mass, in both dimensions, than the background. Based

TABLE II. Efficiency × branching fraction ($\varepsilon \times \mathcal{B}$) using standard event selection and the expected number of top quark events ($\langle N \rangle$) in the seven channels, based on the central theoretical $t\bar{t}$ production cross section of Ref. [5], for four top masses. Also given are the expected background, integrated luminosity, and the number of observed events in each channel.

m_t (GeV/c^2)		$e\mu$ + jets	ee + jets	$\mu\mu$ + jets	e + jets	μ + jets	e + jets/μ	μ + jets/μ	All
140	$\varepsilon \times \mathcal{B}$ (%)	0.17 ± 0.02	0.11 ± 0.02	0.06 ± 0.01	0.50 ± 0.10	0.33 ± 0.08	0.36 ± 0.07	0.20 ± 0.05	
	$\langle N \rangle$	1.36 ± 0.21	1.04 ± 0.19	0.46 ± 0.08	4.05 ± 0.94	2.47 ± 0.68	2.93 ± 0.68	1.48 ± 0.42	13.80 ± 2.07
160	$\varepsilon \times \mathcal{B}$ (%)	0.24 ± 0.02	0.15 ± 0.02	0.09 ± 0.02	0.80 ± 0.10	0.57 ± 0.13	0.50 ± 0.08	0.25 ± 0.06	
	$\langle N \rangle$	0.94 ± 0.13	0.69 ± 0.12	0.34 ± 0.07	3.13 ± 0.54	2.04 ± 0.53	1.95 ± 0.39	0.92 ± 0.24	10.01 ± 1.41
180	$\varepsilon \times \mathcal{B}$ (%)	0.28 ± 0.02	0.17 ± 0.02	0.10 ± 0.02	1.20 ± 0.30	0.76 ± 0.17	0.56 ± 0.09	0.35 ± 0.08	
	$\langle N \rangle$	0.57 ± 0.07	0.40 ± 0.07	0.19 ± 0.04	2.42 ± 0.67	1.41 ± 0.36	1.14 ± 0.22	0.64 ± 0.16	6.77 ± 1.09
200	$\varepsilon \times \mathcal{B}$ (%)	0.31 ± 0.02	0.20 ± 0.03	0.11 ± 0.02	1.70 ± 0.20	0.96 ± 0.21	0.74 ± 0.11	0.41 ± 0.08	
	$\langle N \rangle$	0.34 ± 0.04	0.25 ± 0.05	0.11 ± 0.02	1.84 ± 0.31	0.95 ± 0.24	0.81 ± 0.16	0.41 ± 0.10	4.71 ± 0.66
Background		0.12 ± 0.03	0.28 ± 0.14	0.25 ± 0.04	1.22 ± 0.42	0.71 ± 0.28	0.85 ± 0.14	0.36 ± 0.08	3.79 ± 0.55
$\int \mathcal{L} dt$ (pb^{-1})		47.9 ± 5.7	55.7 ± 6.7	44.2 ± 5.3	47.9 ± 5.7	44.2 ± 5.3	47.9 ± 5.7	44.2 ± 5.3	
Data		2	0	1	5	3	3	3	17

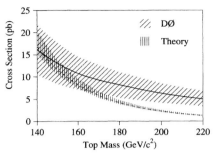

FIG. 3. D0 measured $t\bar{t}$ production cross section (solid line with one standard deviation error band) as a function of assumed top quark mass. Also shown is the theoretical cross section curve (dashed line) [5].

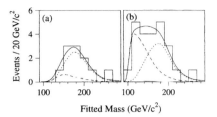

FIG. 5. Fitted mass distribution for candidate events (histogram) with the expected mass distribution for 199 GeV/c^2 top quark events (dotted curve), background (dashed curve), and the sum of top and background (solid curve) for (a) standard and (b) loose event selection.

only on the shapes of the distributions, the hypothesis that the data are a combination of top quark and background events (60% C.L.) is favored over the pure background hypothesis (3% C.L.).

To measure the top quark mass, single-lepton + four-jet events were subjected to two-constraint kinematic fits to the hypothesis $t\bar{t} \rightarrow W^+W^-b\bar{b} \rightarrow \ell\nu q\bar{q}b\bar{b}$. Kinematic fits were performed on all permutations of the jet assignments of the four highest-E_T jets, with the provision that muon-tagged jets were always assigned to a b quark in the fit. A maximum of three permutations with $\chi^2 < 7$ (two degrees of freedom) were retained, and a single χ^2-probability-weighted average mass ("fitted mass") was calculated for each event. Monte Carlo studies using the ISAJET and HERWIG event generators showed that the fitted mass was strongly correlated with the top quark mass. Gluon radiation, jet assignment combinatorics, and the event selection procedure introduced a shift in the fitted mass (approximately -20 GeV/c^2 for 200 GeV/c^2 top quarks), which was taken into account in the final mass determination.

Eleven of the 14 single-lepton + jets candidate events selected using the standard cuts were fitted successfully. Figure 5(a) shows the fitted mass distribution. An unbinned likelihood fit, incorporating top quark and background contributions, with the top quark mass allowed to

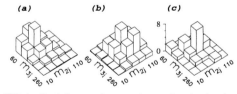

FIG. 4. Single-lepton + jets, two-jet vs three-jet invariant mass distribution for (a) background, (b) 200 GeV/c^2 top Monte Carlo simulation (ISAJET), and (c) data.

vary, was performed on the fitted mass distribution. The top quark contribution was modeled using ISAJET. The background contributions were constrained to be consistent with our background estimates. The likelihood fit yielded a top quark mass of 199^{+31}_{-25} (stat) GeV/c^2 and described the data well.

To increase the statistics available for the mass fit, and to remove any bias from the standard H_T requirement, we repeated the mass analysis on events selected using the loose requirements. Of 27 single-lepton + four-jet events, 24 were fitted successfully. The removal of the H_T requirement introduced a substantial background contribution at lower mass in addition to the top signal, as shown in Fig. 5(b). A likelihood fit to the mass distribution resulted in a top quark mass of 199^{+19}_{-21} (stat) GeV/c^2, consistent with the result obtained from the standard event selection. The result of the likelihood fit did not depend significantly on whether the normalization of the background was constrained. Using HERWIG to model the top quark contribution resulted in a mass 4 GeV/c^2 below that found using ISAJET. This effect was included in the systematic error. The total systematic error in the top quark mass is 22 GeV/c^2, which is dominated by the uncertainty in the jet energy scale (10%).

In conclusion, we report the observation of the top quark. We measure the top quark mass to be 199^{+19}_{-21} (stat) ± 22 (syst) GeV/c^2 and measure a production cross section of 6.4 ± 2.2 pb at our central mass.

We thank the Fermilab Accelerator, Computing, and Research Divisions, and the support staffs at the collaborating institutions for their contributions to the success of this work. We also acknowledge the support of the U.S. Department of Energy, the U.S. National Science Foundation, the Commissariat à L'Energie Atomique in France, the Ministry for Atomic Energy and the Ministry of Science and Technology Policy in Russia, CNPq in Brazil, the Departments of Atomic Energy and Science and Education in India, Colciencias in Colombia, CONA-CyT in Mexico, the Ministry of Education, Research Foundation and KOSEF in Korea, and the A. P. Sloan Foundation.

*Visitor from CONICET, Argentina.
†Visitor from IHEP, Beijing, China.
‡Visitor from Universidad de Buenos Aires, Argentina.
§Visitor from University San Francisco de Quito, Ecuador.

[1] D0 Collaboration, S. Abachi et al., Phys. Rev. Lett. **72**, 2138 (1994).
[2] D0 Collaboration, S. Abachi et al., Phys. Rev. Lett. **74**, 2422 (1995); Fermilab Report No. FERMILAB-PUB-95/020-E, 1995 (to be published).
[3] CDF Collaboration, F. Abe et al., Phys. Rev. D **50**, 2966 (1994); Phys. Rev. Lett. **73**, 225 (1994).
[4] D. Schaile, in Proceedings of the 27th International Conference on High Energy Physics, Glasgow, 1994 (CERN Report No. CERN-PPE/94-162) (unpublished).
[5] E. Laenen, J. Smith, and W. van Neerven, Phys. Lett. B **321**, 254 (1994).
[6] D0 Collaboration, S. Abachi et al., Nucl. Instrum. Methods Phys. Res., Sect. A **338**, 185 (1994).
[7] F. A. Berends, H. Kuijf, B. Tausk, and W. T. Giele, Nucl. Phys. **B357**, 32 (1991).
[8] F. Paige and S. Protopopescu, BNL Report No. BNL38034, 1986 (unpublished), release v 6.49.
[9] F. Carminati et al., "GEANT Users Guide," CERN Program Library, 1991 (unpublished).
[10] G. Marchesini et al., Comput. Phys. Commun. **67**, 465 (1992).

15

Mixing and CP Violation in Heavy Quark Mesons

Testing the Standard Model with B, B_s, and D.

Just as for K^0 and $\overline{K}^0$, there can be mixing between the B^0 and $\overline{B}^0$ mesons. In fact, this is possible for two distinct systems, the non-strange $B_d^0 = B^0 = \bar{b}d$ and the strange $B_s^0 = \bar{b}s$. If a $B^0\overline{B}^0$ pair is created and both mesons decay semileptonically, the B^0 would be expected to give a positive lepton ($\bar{b} \to \bar{c}l^+\nu$) and the $\overline{B}^0$ a negative lepton. If there is B^0–$\overline{B}^0$ mixing, it is possible that both leptons will have the same sign. An unfortunate background arises from the chain $b \to c \to sl\nu$ since the semileptonic decay of the c would give a lepton of the sign opposite that expected from a b decay. While some evidence for $B^0 - \overline{B}^0$ mixing was found by UA-1 at the Sp$\bar{\text{p}}$S in the same-sign dilepton signal, clear, convincing evidence was first obtained in an e^+e^- experiment.

In 1987, the ARGUS Collaboration working at the $\Upsilon(4S)$ (**Ref. 15.1**) found one example of $\Upsilon(4S) \to B_d^0 B_d^0$, as demonstrated by specific semileptonic decays, each with a positive muon, and both described by $B^0 \to D^{*-}\mu^+\nu$. Additional evidence for mixing was obtained by measuring the inclusive like-sign dilepton signal. A third independent measurement came from identifying complete B^0 decays and observing semileptonic decays of the accompanying meson. Finding a positive lepton opposite an identified B^0 is evidence for mixing. Combining the results of these measurements determined the ratio of wrong-sign decays to right-sign decays: $r_d = 0.21 \pm 0.08$.

Because so many channels are open for B meson decay, it is reasonable to assume that the two eigenstates will have very similar lifetimes, quite unlike the situation for neutral K mesons. The result from Chapter 7 for the ratio of "wrong-sign" decays to "right-sign" decays then becomes

$$r_d = \frac{x_d^2}{2 + x_d^2}, \tag{15.1}$$

where $x_d = \Delta m/\Gamma$ and the d specifies B_d mesons. The ARGUS Collaboration revisited this measurement in 1992 with a data sample more than twice the size of the original one. Using much the same techniques, they confirmed the result with a refined determination: $r_d = 0.206 \pm 0.070$ or $x = 0.72 \pm 0.15$.

15.1 Mixing and the CKM Model

The mixing of B^0 and $\overline{B}^0$ is analogous to the mixing of K^0 and $\overline{K}^0$ and the mass eigenstates can be found by diagonalizing a matrix just like that considered in Chapter 7:

$$\begin{pmatrix} M - i\dfrac{\Gamma}{2} & M_{12} - i\dfrac{\Gamma_{12}}{2} \\ M_{12}^* - i\dfrac{\Gamma_{12}^*}{2} & M - i\dfrac{\Gamma}{2} \end{pmatrix}. \tag{15.2}$$

Quite generally, the lighter and heavier mass eigenstates can be written

$$|B_L\rangle = p|B^0\rangle + q|\overline{B}^0\rangle,$$
$$|B_H\rangle = p|B^0\rangle - q|\overline{B}^0\rangle, \tag{15.3}$$

where

$$\left(\frac{q}{p}\right)^2 = \frac{M_{12}^* - i\Gamma_{12}^*/2 - M_{12}^* - \frac{i}{2}\Gamma 12^8}{M_{12} - i\Gamma_{12}/2 - M_{12} - \frac{i}{2}\Gamma 12}. \tag{15.4}$$

The mixing depends on the existence of common states to which both B^0 and $\overline{B}^0$ can couple. The B^0 favors decays to states like $\overline{D}\pi$, while the $\overline{B}^0$ prefers $D\pi$. Both however, can decay to $D\overline{D}$, or to any state composed of $c\,\overline{d}\,\overline{c}\,d$, albeit as a CKM-suppressed decay. Similarly, they both can make virtual transitions to states containing $t\overline{t}$. Mixing arises both from real and virtual transitions.

Mixing of neutral mesons depends on both the quark masses and the Cabibbo–Kobayashi–Maskawa matrix. If the u, c, and t quarks were degenerate in mass, we could redefine them so that the d quark connected only to u, s only to c, and b only to t. Then the CKM matrix would be the unit matrix and there would be no intermediate quark states possible in Figure 15.1. The same would be true if the masses of the d, s, and b were degenerate. Mixing, then, depends critically on quark mass differences, emphasizing the importance of the heavy quarks.

The measured values of the CKM matrix show that jumping from the first generation to the second is suppressed in amplitude by roughly $\lambda = 0.22$, where λ is the parameter introduced in Wolfenstein parameterization of the CKM matrix, Eq. (11.5), equivalent to the sine of the Cabibbo angle. A second-to-third generation amplitude is reduced by $\lambda^2 \approx 0.05$, while a first-to-third transition is suppressed by roughly $\lambda^3 \approx 0.01$.

The most conspicuous feature of K^0–$\overline{K}^0$ mixing is the disparate lifetimes of the two mass eigenstates. For B mesons the lifetime difference is quite small and it is the oscillations reflecting Δm that are predominant. The lifetime difference and thus Γ_{12} can be neglected for B_d (but not for B_s). For B_d mixing then we find that the light and heavy eigenmasses are

$$\mu_L = M - i\frac{\Gamma}{2} - |M_{12}|,$$
$$\mu_H = M - i\frac{\Gamma}{2} + |M_{12}|, \tag{15.5}$$

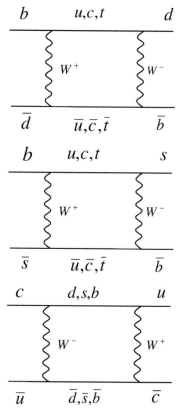

Figure 15.1. Box diagrams showing how mixing occurs at the quark level for B, B_s, and D mesons. Each diagram contributes in two ways: once with quarks as intermediate states and once with W bosons as intermediate states. If the u, c, and t quarks had identical masses, we could redefine the states so b coupled only to t, s only to c, and d only to u. There would then be no mixing. The value of the mixing diagrams thus depends on the differences of the quark masses and on the size of the CKM matrix elements that couple the quarks to the W bosons. For B mesons, the t quark contributions dominate. For D mesons, the couplings to the b quarks are small and the d and s quark contributions dominate.

so that the mass splitting is $\Delta m = 2|M_{12}|$. The eigenstates are

$$|B_L\rangle = \frac{1}{\sqrt{2}}\left(|B^0\rangle - \frac{|M_{12}|}{M_{12}}|\bar{B}^0\rangle\right),$$

$$|B_H\rangle = \frac{1}{\sqrt{2}}\left(|B^0\rangle + \frac{|M_{12}|}{M_{12}}|\bar{B}^0\rangle\right), \quad (15.6)$$

and they evolve simply

$$|B_L(t)\rangle = e^{-i(M-\frac{\Delta m}{2}-i\frac{\Gamma}{2})t}|B_L\rangle,$$

$$|B_H(t)\rangle = e^{-i(M+\frac{\Delta m}{2}-i\frac{\Gamma}{2})t}|B_H\rangle. \quad (15.7)$$

These states are analogous to K_L and K_S, except that the lifetime difference is ignored. In Chapter 7 we saw that CP violation in mixing was due to the imaginary part of Γ_{12}/M_{12}. Since we have neglected Γ_{12} in this approximation there is no CP violation in mixing itself, though mixing will contribute to visible CP violation in time-dependent decay rates. A state that at $t = 0$ is purely B^0, will oscillate into $\overline{B}^0$:

$$|B^0_{phys}(t)\rangle = e^{-i(M-i\Gamma/2)t}\left[\cos\frac{\Delta m}{2}t\,|B^0\rangle - i\frac{|M_{12}|}{M_{12}}\sin\frac{\Delta m}{2}t\,|\overline{B}^0\rangle\right], \quad (15.8)$$

while its counterpart behaves as

$$|\overline{B}^0_{phys}(t)\rangle = e^{-i(M-i\Gamma/2)t}\left[\cos\frac{\Delta m}{2}t\,|\overline{B}^0\rangle - i\frac{M_{12}}{|M_{12}|}\sin\frac{\Delta m}{2}t\,|B^0\rangle\right]. \quad (15.9)$$

A state that begins as a B^0 will produce semileptonic decays exponentially damped by $e^{-\Gamma t}$, with the "right" sign modulated by $\cos^2\frac{1}{2}\Delta mt$ and with the "wrong" sign modulated by $\sin^2\frac{1}{2}\Delta mt$.

15.2 CP Violation

The mixing of B^0 and $\overline{B}^0$ provides an opportunity to explore CP violation just as the analogous mixing in the neutral K system does. While it is also possible to measure CP violation by showing an inequality between the rate for B^+ decays to some state and B^- decays to the CP conjugate, decays of neutral B's can be analyzed more incisively.

The presence of phases in the CKM matrix is the source of CP violation in the Standard Model. These phases enter into decay matrix elements and into the mixing described by M_{12}. In the Wolfenstein parameterization, Eq. (11.5), phases occur only in transitions between quarks of the first and third generation. One way to represent the CKM matrix is with the "unitarity triangle," shown in Figure 15.2. The three angles of this triangle at the vertices 0, 1, and $\rho + i\eta$ are traditionally called α, β, and γ. The matrix element V_{ub} corresponds to the transition $b \to u$, while the Figure shows that $V^*_{ub} \propto \rho + i\eta$ and has the phase γ. Thus the $b \to u$ transition picks up the phase $-\gamma$. Similarly, the transition $d \to t$ picks up the phase of $V_{td} \propto 1 - \rho - i\eta$, which is $-\beta$.

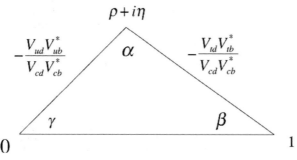

Figure 15.2. The unitarity triangle for B decays expresses the relation $V_{ud}V^*_{ub} + V_{cd}V^*_{cb} + V_{td}V^*_{tb} = 0$ in the complex plane. The angles α, β, and γ can be measured in the time dependence of B decays.

If a state that is initially B^0 decays at a later time into a final state f, there will be interference between the decay of the piece that has remained B^0 and the piece that has become $\overline{B}^0$. The phase between the two interfering amplitudes will depend on the relative phases of $\langle f|\mathcal{H}|B^0\rangle$ and $\langle f|\mathcal{H}|\overline{B}^0\rangle$ and on the phase of M_{12}.

Oscillations in the decay of a B to a CP eigenstate are especially interesting because $\langle f|\mathcal{H}|B^0\rangle$ is then related to $\langle f|\mathcal{H}|\overline{B}^0\rangle$ in a simple way. The weak interaction Hamiltonian is made up of many pieces $\mathcal{H}_j$: strangeness increasing, strangeness decreasing, charm increasing, charm decreasing, etc. Altogether the Hamiltonian must be Hermitian so that the theory will be unitary (conserving probability). If CP is conserved, the Hamiltonian takes the form

$$\mathcal{H} = \sum_j \mathcal{H}_j + \sum_j \mathcal{H}_j^\dagger \tag{15.10}$$

where $CP\mathcal{H}_j CP = \mathcal{H}_j^\dagger$. On the other hand, if CP is violated, the CKM matrix introduces phases into the currents that make up the weak interaction. The current that raises one quantum number has a phase opposite that of the current that lowers that quantum number. The Hamiltonian then takes the form

$$\mathcal{H} = \sum_j e^{i\phi_j}\mathcal{H}_j + \sum_j e^{-i\phi_j}\mathcal{H}_j^\dagger \tag{15.11}$$

where each piece acquires its phase from a particular combination of CKM matrix elements. The result then is that while $CP\mathcal{H}_j CP = \mathcal{H}_j^\dagger$, we see that, in general, $CP\mathcal{H}CP \neq \mathcal{H}$.

If one single part $\mathcal{H}_j$ of the weak Hamiltonian is responsible for the decay $B^0 \to f$ then

$$\begin{aligned}\langle f|\mathcal{H}|B^0\rangle &= \langle f|e^{i\phi_j}\mathcal{H}_j|B^0\rangle &&= \langle f|e^{i\phi_j}CP\mathcal{H}_j^\dagger CP|B^0\rangle \\ &= \eta_f e^{2i\phi_j}\langle f|e^{-i\phi_j}\mathcal{H}_j^\dagger|\overline{B}^0\rangle &&= \eta_f e^{2i\phi_j}\langle f|\mathcal{H}|\overline{B}^0\rangle,\end{aligned} \tag{15.12}$$

where η_f is the value of CP for the state f.

Interference in the decay of a neutral B depends on the weak phases ϕ_j, which come from the CKM matrix, and on the phase introduced by M_{12}. Mixing results from the processes shown in Figure 15.1. For M_{12} itself, the dominant diagram has t-quark intermediates and $M_{12} \propto (V_{tb}V_{td}^*)^2$ with a negative coefficient of proportionality with our convention $CP|B^0\rangle = |\overline{B}^0\rangle$. It follows that $|M_{12}|/M_{12} = -e^{-2i\beta}$. Combining all these results we find

$$\langle f|\mathcal{H}|B^0_{phys}(t)\rangle \propto e^{-\Gamma t/2} A_f \left[\cos\frac{\Delta m}{2}t + i\lambda_f \sin\frac{\Delta m}{2}t\right], \tag{15.13}$$

$$\langle f|\mathcal{H}|\overline{B}^0_{phys}(t)\rangle \propto e^{-\Gamma t/2} \overline{A}_f \left[\cos\frac{\Delta m}{2}t + i\frac{1}{\lambda_f}\sin\frac{\Delta m}{2}t\right], \tag{15.14}$$

where
$$A_f = \langle f|\mathcal{H}|B^0\rangle; \qquad \overline{A}_f = \langle f|\mathcal{H}|\overline{B}^0\rangle, \qquad (15.15)$$

and where
$$\lambda_f = -\frac{|M_{12}|}{M_{12}}\frac{\overline{A}_f}{A_f}$$
$$= \eta_f e^{-2i\beta} e^{-2i\phi_{wk}}. \qquad (15.16)$$

Here ϕ_{wk} is the single weak phase in the amplitude for $B^0 \to f$. We see that $|\lambda| = 1$, a consequence of our assumptions that only one mechanism contributes to the decay and that $\Delta\Gamma$ can be ignored for B_d. The decay rate is then governed by

$$|\langle f|\mathcal{H}|B^0_{phys}(t)\rangle|^2 \propto e^{-\Gamma t}\left[1 + \eta_f \sin 2(\beta + \phi_{wk})\sin \Delta mt\right], \qquad (15.17)$$
$$|\langle f|\mathcal{H}|\overline{B}^0_{phys}(t)\rangle|^2 \propto e^{-\Gamma t}\left[1 - \eta_f \sin 2(\beta + \phi_{wk})\sin \Delta mt\right]. \qquad (15.18)$$

What is remarkable here is that there are no unknown matrix elements involving hadrons: when just a single weak phase occurs, the hadronic uncertainty disappears.

15.3 $B \to J/\psi$ and $\sin 2\beta$

A particularly important example is the decay $B \to J/\psi K_S$. Since the J/ψ with $CP = +1$ and the K_S with $CP = +1$ must be combined in a p-wave ($CP = -1$) to make the spin-zero B, we have $\eta_f = -1$. Here the underlying transition is $\overline{b} \to \overline{c}c\overline{s}$. Because this involves only second and third generation quarks, no weak phase is introduced ($\phi_{wk} = 0$). This process, then, measures the phase of M_{12}, which is predicted by the Standard Model to be 2β. See Figure 15.3.

In Run I at the Tevatron Collider, which lasted from 1991 to 1996, CDF demonstrated that such measurements can be made in the intense environment of a hadron collider. The task was not just to reconstruct $B \to J/\psi K_S$ decays and determine the time elapsed from the production of the B meson until its decay: it was necessary to infer whether the B had begun as a B^0 or $\overline{B}^0$. The method used initially by CDF was to look at the particles accompanying the B meson that decayed to $J/\psi K_S$. A $B^0 = \overline{b}d$, is more likely to have a $\overline{d}u = \pi^+$ nearby than a $\overline{u}d = \pi^-$. This preference can be measured quantitatively by observing $B^0 - \overline{B}^0$ oscillations.

Any means used to determine whether the B observed as $J/\psi K_S$ began as a B^0 or $\overline{B}^0$ will be imperfect. If it is wrong a fraction w of the time, a distribution that should be $1 - A \sin \Delta mt$ will instead appear as $(1-w)(1 - A\sin \Delta mt) + w(1 + A\sin \Delta mt) = 1 - DA\sin \Delta mt$, where the dilution D is just $1 - 2w$. A figure of merit for an experiment is $Q = \sum \epsilon_i D_i^2$, where the ith tagging category captures a fraction ϵ_i of the neutral B events and has a dilution D_i. A collection of N events with efficiency ϵ_i and dilutions D_i in each category i has the statisical power of NQ perfectly tagged events. In 1998, CDF

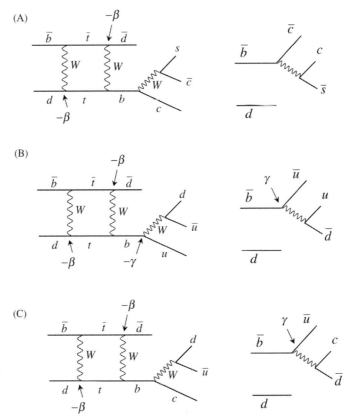

Figure 15.3. Oscillations and decay provide a means of measuring CP violation in neutral B meson decay. The sine of the relative phase between the diagram on the right and the corresponding diagram on the left gives the coefficient of $\sin \Delta mt$ in the time-dependent decay distribution. In the Wolfenstein phase convention only transitions between the first and third generations give significant CP-violating phases: $t \to d$ gives β and $u \to b$ gives γ. (A) In decays $\bar{b} \to \bar{c}cs$ ($B^0 \to J/\psi K_S$) no weak phase aside from that in mixing occurs and the measured asymmetry is proportional to $\sin 2\beta$. Note that K^0–$\overline{K}^0$ mixing is essential here for there to be interference. (B) In decays like $B \to \pi\pi$ the relative weak phase is $2\beta + 2\gamma = 2\pi - 2\alpha$. Because of contributions from penguin diagrams with different phases, this decay does not give a direct measurement of $\sin 2\alpha$. (C) The decay $\overline{B}^0 \to D^+\pi^-$ is CKM-favored and interferes with the CKM-suppressed $B^0 \to D^+\pi^-$ through mixing. The relative weak phase is $2\beta + \gamma$, but a relative strong phase enters as well.

reported (Ref. 15.2) a value $\sin 2\beta = 1.8 \pm 1.1 \pm 0.3$ taking only events in which both the J/ψ-decay muons were seen by the SVX. The dilution was about 17%, with an efficiency of 65%, so Q was about 0.02.

In 2000, CDF added two more means of tagging the initial B meson by looking for indications of the other B in the event. A semileptonic decay would be decisive: a positive lepton would indicate $\bar{b} \to \bar{c}\ell^+\nu$ and thus the decay of a B, not $\overline{B}$ meson. Even just measuring the charges of particles likely to be part of the B meson decay provided some evidence for its nature. In these ways CDF achieved a total Q of about 6%. The data sample included about 400 events in which the J/ψ was seen in its $\mu^+\mu^-$ decay mode and the K_S^0 was seen in $\pi^+\pi^-$. In about half of the events, the muons were measured by the silicon vertex detector (SVX) providing precise information on the distance traveled before the decay to $J/\psi K_S^0$. With this much larger dataset, an improved result (Ref. 15.3), $\sin 2\beta = 0.79^{+0.41}_{-0.44}$, was reported.

The $\Upsilon(4S)$, which provided such an excellent source of B mesons at CESR, can be used to study CP violation as well. However, in contrast to the production of $B\overline{B}$ pairs at a hadron collider, which can be regarded as incoherent, the production of $B\overline{B}$ pairs at the $\Upsilon(4S)$ is completely coherent. If at some instant, say $t = 0$, one B is known to be a B^0, then at the same time the other must be a $\overline{B}^0$. This follows from Bose statistics, which requires that the odd spatial wave function (for angular momentum one) must be balanced by a wave function odd under particle interchange.

At hadron colliders, where the initial B and $\overline{B}$ are produced incoherently, t measures the time since their simultaneous production and is necessarily positive. At e^+e^- colliders running at the $\Upsilon(4S)$, since there are no particles produced aside from the B and $\overline{B}$, tagging can only be done by observing features of the "other" B meson, the one not being fully reconstructed. The pair of neutral B mesons is produced coherently and t measures time from the decay of the B that is tagged as a B^0 or $\overline{B}^0$ to the time of the decay of the other neutral B meson. If the decay of the tagging B occurs before the fully observed decay, t is positive, but if the decay of the tagging B comes later, then t is negative. At the $\Upsilon(4S)$, the time dependence for the decay to a CP eigenstate f of a state known to be a B^0 at $t = 0$ is

$$|\langle f|\mathcal{H}|B^0_{phys}(t)\rangle|^2 \propto e^{-\Gamma|t|}\left[1 + \eta_f \sin 2(\beta + \phi_{wk}) \sin \Delta mt\right]. \tag{15.19}$$

Integrating over all t, positive and negative, cancels the asymmetry. To measure the asymmetry, then, the actual time dependence must be seen. This is hardly possible in a collider like CESR. There, the $\Upsilon(4S)$ is produced at rest and the B mesons it yields go about 30 μm on average before decaying. Such decay lengths are too short to be measured with sufficient accuracy to see the oscillations.

15.4 Asymmetric B Factories

To overcome this, asymmetric-energy e^+e^- colliders were built at SLAC and at the Japanese high energy physics facility, KEK, following the original proposal of Pier Oddone. The general features of the accelerators and detectors at the two locations were quite similar. At SLAC, the energy of the electron beam was about 9 GeV and that of the positron beam was near 3 GeV. This produces an $\Upsilon(4S)$ resonance with a relativistic factor $\beta\gamma = 0.56$. At KEK, the energy asymmetry was less, with $\beta\gamma = 0.42$. The typical B path length at SLAC was 250 μm. Such distances can be measured reliably with a silicon vertex detector.

In both the Belle detector at KEK and the BaBar detector at SLAC, particle identification relied on Cherenkov radiation, either as a threshold device or with imaging to reconstruct the angle of the Cherenkov cone, and on measurements of dE/dx. Crystals of CsI provided electromagnetic calorimetry with the requisite precision.

The new asymmetric-energy colliders at KEK and SLAC reached luminosities of order 10^{33} cm^{-2} s^{-1} remarkably quickly and over the years increased this another factor of ten. This enabled the Belle and BaBar experiments to improve substantially the fundamental measurements of Δm_d and τ_B and of the CKM matrix elements V_{cb} and V_{ub}. The mass splitting uncertainty was decreased from the 5% achieved at LEP and the Tevatron Collider to 1% ($\Delta m_d = 0.507 \pm 0.005$ ps^{-1}) and the uncertainty in the B^0 lifetime went from 2% to 0.7% ($\tau_B^0 = 1.530 \pm 0.009$ ps). Detailed studies of the semileptonic decays from $b \to c \ell^- \bar{\nu}$ yielded $|V_{cb}| = (41.6 \pm 0.6) \times 10^{-3}$, while those from the CKM-suppressed decays $b \to u \ell^- \bar{\nu}$ gave $|V_{ub}| = (4.13 \pm 0.30) \times 10^{-3}$.

The real power of the asymmetric-energy B factories lay in their ability to measure time-dependent quantities. The oscillations between B^0 and $\overline{B}^0$ were apparent in comparisons of events with a B^0 and a $\overline{B}^0$ with events having either two B^0s or two $\overline{B}^0$s. See Figure 15.4.

By March 2001 both the Belle and BaBar Collaborations reported new values for $\sin 2\beta$ or $\sin 2\phi_1$, as it is called by Belle. The Belle result (**Ref. 15.4**) was $0.58^{+0.32}_{-0.34}$(stat)$^{+0.09}_{-0.10}$(sys) while that from BaBar (**Ref. 15.5**) was $0.34 \pm 0.20 \pm 0.05$. Combining the CDF, Belle, and BaBar results gave 0.49 ± 0.16, strongly indicating a non-zero result, but still too limited

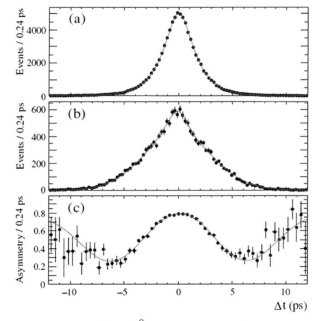

Figure 15.4. Oscillations between B^0 and $\overline{B}^0$ seen in events with two charged leptons. Top: events with two opposite-sign leptons. Middle: events with same-sign leptons. Bottom: the asymmetry. Because both neutral and charged B mesons are included, the idealized prediction would be an asymmetry $(1 + \cos \Delta mt)/2$, assuming equal numbers of charged and neutral B meson pairs. Figure from the BaBar Collaboration, *Phys. Rev. Lett.* **88**, 221803 (2002).

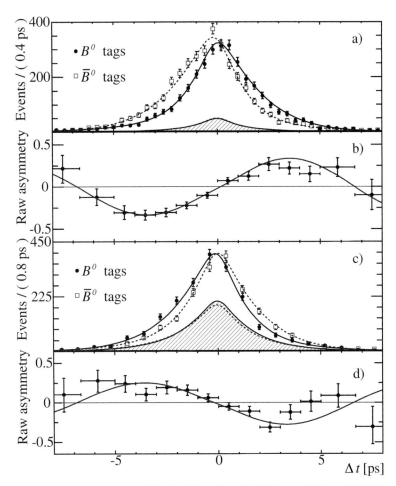

Figure 15.5. The upper panel shows the time-dependent distributions of $B \to J/\psi K_S$ and analogous processes with J/ψ replaced by $\psi(2S)$, χ_{c1}, and η_c. Data points with open circles are for events where the other B meson was tagged as a $\overline{B}^0$ while filled circles are for events tagged with a B^0. Directly below is the raw asymmetry, $(N_{B^0} - N_{\overline{B}^0})/(N_{B^0} + N_{\overline{B}^0})$. The lower two panels show the analogous distributions for $B \to J/\psi K_L$. The asymmetry has the opposite sign as expected from the replacement of the (mostly) CP-even K_S by the K_L. Figure from the BaBar Collaboration, B. Aubert et al., Phys. Rev. Lett. **99**, 171803(2007).

by statistics to provide a sharp test of the Standard Model. A few months later, the BaBar Collaboration announced an updated result (Ref. 15.6) $\sin 2\beta_1 = 0.59 \pm 0.14 \pm 0.05$, which taken alone was enough to establish CP violation in the B system. In subsequent years the luminosity of both machines increased beyond 10^{34} cm^{-2} s^{-1} and hundreds of inverse femtobarns of data, and thus hundreds of millions of $B\overline{B}$ pairs, were accumulated. By the end of 2007, the results from the two teams for $\sin 2\beta$ converged on a value 0.680 ± 0.025, or $\alpha = 21.45 \pm 1.0$ degrees. Figure 15.5 shows representative data for $B \to J/\psi K_S$ and related channels.

15.5 α and γ

While the time dependence of $B \to J/\psi K_S$ by itself provided an effective path to measuring β, the angles α and γ were more challenging and required a variety of channels. The decay $B \to \pi\pi$ would appear to provide direct access to the angle α. This decay will result from the rather suppressed process $b \to \bar{u}ud$, which introduces the CKM matrix element V_{ub}^* and thus the phase γ. If this were the only contribution, the decay's time dependence would be

$$|\langle \pi^+\pi^-|\mathcal{H}|B^0_{phys}(t)\rangle|^2 \propto e^{-\Gamma|t|}[1 + \sin 2(\beta+\gamma)\sin \Delta mt]$$
$$\propto e^{-\Gamma|t|}[1 - \sin 2\alpha \sin \Delta mt], \quad (15.20)$$

assuming from the unitarity triangle the relation $\alpha + \beta + \gamma = \pi$.

However, there is another way to reach the same final state, through a penguin process analogous to $b \to s\gamma$ discussed in Chapter 11 and shown in Figure 15.6. Here, however, the phase would come from V_{td}, i.e. $-\beta$. With two different weak phases present, the simple analysis above fails. To separate out the penguin effects requires measuring isospin-related processes like $B \to \pi^0\pi^0$ and $B^+ \to \pi^+\pi^0$. The analogous decays for $B \to \rho\pi$ or $B \to \rho\rho$ also provide the means to measure α if isospin-related decays are measured.

To disentangle the penguin contributions all charge combinations had to be measured, which was particularly problematic for $\pi^0\pi^0$ because the resulting four photons are harder to measure than charged pions. Moreover the $B^0 \to \rho^0\rho^0$ turned out to have a very small

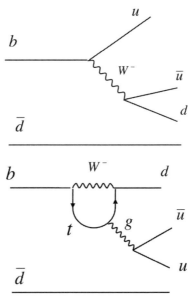

Figure 15.6. The decay $\bar{B}^0 \to \pi\pi$ (or $\rho\pi$ or $\rho\rho$) proceeds through the "tree" diagram above and the penguin diagram below. These have different weak phases, making more complicated the extraction of the angle α of the unitarity triangle.

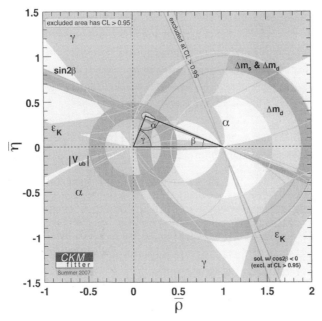

Figure 15.7. Allowed regions in the ρ–η plane ($\overline{\rho}$ and $\overline{\eta}$ include higher-order corrections in powers of λ). Measurement of the angles α, β, and γ from B meson decays, together with measurements of the magnitudes of the CKM matrix elements, measurements of the mass differences Δm_d and Δm_s, and measurements of CP violation in K decays all provide constraints. The CKM model requires that all the regions allowed by the various experiments have a common overlap. The figure generated by the CKM Fitter team includes data through the summer of 2007. [Updated from CKMfitter Group (J. Charles et al.), Eur. Phys. J. **C41**, 1-131 (2005)]

branching fraction, near 10^{-6}. Nonetheless, by the end of 2007, α was known to be quite near 90 degrees, with a world-average value of 87 ± 6 degrees.

Measuring the angle γ of the unitarity triangle depends on interference between processes with $b \to c$ and $b \to u$. As seen in Figure 15.3, mixing allows interference between the CKM-favored $\overline{B}^0 \to D^+\pi^-$ and the CKM-disfavored $B^0 \to D^+\pi^-$. Because the final state is not a CP eigenstate, the final state strong phase is not the same for the two. When many decay channels are open, the final state interaction phase depends on the initial state as well as the final state. Moreover, the ratio of the two weak decay amplitudes is not known a priori. These complications are absent in $B \to J/\psi K_S$. Fortunately, an alternative is available with charged B mesons. The decay $B^+ \to \overline{D}^0 K^+$ is relatively favored, proceeding through $\overline{b} \to \overline{c}u\overline{s}$, compared to $B^+ \to D^0 K^+$, which requires $\overline{b} \to \overline{u}c\overline{s}$, since $|V_{cb}V_{us}| \approx 0.04 \cdot 0.22$ while $|V_{ub}V_{cs}| \approx 0.004$. Although nominally the two final states are different, interference is possible if the $\overline{D}^0$ and D^0 decay into a common state like K^+K^-. An especially attractive final state is $K_S\pi^+\pi^-$, since the full Dalitz plot can be examined for the interference pattern. At the end of 2007, the uncertainty in γ was much larger than that in β and α: $\gamma = 77 \pm 31$ degrees. Within the uncertainties, the three angles did add up to 180 degrees as shown in Figure 15.7.

15.6 Direct CP Violation

The CP violation observed in time-dependent measurements depended on phases that combined the effects of mixing and decay. In principle, CP could be observed in mixing alone, but such effects depend on $\Im(\Gamma_{12}/M_{12})$, which is very small because Γ_{12} itself is small and because Γ_{12} and M_{12} have nearly the same phase. See Exercise 15.4. On the other hand, CP violation in decays without mixing can be sizable.

Consider specifically the decay $B^0 \to K^+ \pi^-$. This can occur through $\bar{b} \to \bar{u}u\bar{s}$ with an amplitude proportional to V_{ub}^* and thus a weak phase $e^{i\gamma}$. Alternatively, it can occur through a penguin decay with the $\bar{b}$ turning into a $\bar{t}$, which finally becomes an $\bar{s}$. Since this chain contains no first generation quarks, it has no weak phase in the standard convention. We write the amplitude as

$$\mathcal{A}(B^0 \to K^+\pi^-) = Te^{i\delta_T}e^{i\gamma} + Pe^{i\delta_P}. \tag{15.21}$$

where δ_T and δ_P are strong final-state-interaction phases and T and P are real. The amplitude for the charge-conjugate decay is

$$\mathcal{A}(\bar{B}^0 \to K^-\pi^+) = Te^{i\delta_T}e^{-i\gamma} + Pe^{i\delta_P}. \tag{15.22}$$

The difference of the partial decay rates for $B^0 \to K^+\pi^-$ and $\bar{B}^0 \to K^-\pi^+$ is proportional to $\sin(\delta_T - \delta_P)\sin\gamma$. In general, there is direct CP violation only if the two decays have at least two contributions with different weak and strong phases.

In 2004, the BaBar Collaboration reported a significant difference for these two decay rates,

$$A_{K\pi} = \frac{n_{K^-\pi^+} - n_{K^+\pi^-}}{n_{K^-\pi^+} + n_{K^+\pi^-}} = -0.133 \pm 0.030(\text{stat}) \pm 0.009(\text{syst}), \tag{15.23}$$

based on some 1600 decays to $K\pi$ (Ref. 15.7). This was consistent with an earlier measurement by the Belle Collaboration, which also showed a negative asymmetry, but with less statistical significance. The Belle team soon confirmed the result with significance similar to that of BaBar (Ref. 15.8). Similar direct CP violation would then be expected in the analogous decay $B^+ \to K^+\pi^0$. Instead, smaller positive, less significant asymmetry has been measured. This remains a puzzle.

15.7 B_s Mixing

Oscillations of B_s^0 are similar in principle to those of the non-strange $B^0 = B_d^0$, but replacing the d quark with an s quark means that it is V_{ts}^2 rather than V_{td}^2 that governs both $\Delta\Gamma$ and Δm. With Δm increased by a large factor, observing B_s oscillations was bound to be difficult. Early limits were set by the LEP experiments ALEPH, DELPHI, and OPAL: $\Delta m_s > 10.6$ ps^{-1}, compared to the measured value of Δm_d near 0.5 ps^{-1}.

The B factories were unable to study B_s for lack of a resonance analogous to the $\Upsilon(4S)$ giving $B_s\bar{B}_s$ exclusively in place of $B_d\bar{B}_d$. The Tevatron Collider, however, produced B_s

copiously. To observe B_s oscillations it was necessary to detect a B_s and then tag the accompanying B or B_s as belonging to the $\bar{b}$ category (B^0, B^+, B_s) or the b category ($\bar{B}^0$, B^-, $\bar{B}_s$). The task was much more difficult than for B_d oscillations because of the smaller production cross section, and especially because of the rapid oscillations that had to be resolved.

The B_s meson decays through $\bar{b}s \to \bar{c}W^+s$, with the subsequent decay of the virtual W^+. The combination $\bar{c}s$ can make a D_s meson, so the combination of a D_s meson with a charged lepton from the W^+ makes a good signature for a B_s. A convenient channel for observing D_s is $\phi\pi$. The D0 Collaboration set interesting limits on Δm_s in the summer of 2006 by observing the time dependence of events with a D_s decaying to $\phi\pi$ together with a charged lepton of the opposite sign to the D_s. To observe the oscillation, it was necessary to know as well whether the other B meson in the event was b-like or $\bar{b}$-like. This could be inferred again by looking for a charged lepton or other charged particle indicative of the originating B meson. For Belle and BaBar, the tagging figure of merit was typically $Q \approx 0.30$, while in the more challenging environment of the Tevatron Collider D0 achieved $Q \approx 0.025$. The oscillations would appear as time dependence of the B_s decays opposite a $\bar{B}$ tag proportional to $e^{-\Gamma\tau}(1 \pm D\cos\Delta m_s\tau)$, where τ is the proper time. Since the decaying B_s meson was not completely reconstructed, its momentum and thus τ could only be approximately inferred from the decay length. The sample of approximately 5600 tagged events was equivalent to $0.025 \times 5600 = 140$ perfectly identified events. From these events, the D0 Collaboration inferred a limit of $17 < \Delta m_s < 21$ ps^{-1} at 90% CL (Ref. 15.9).

The CDF Collaboration used both semileptonic decays and fully reconstructed non-leptonic B_s decays (**Ref. 15.10**) to measure Δm_s. CDF benefitted significantly from a trigger that used information from the silicon vertex detector to identify likely secondary vertices. For the fully reconstructed decays, the decay proper time was well measured. In addition, CDF tagged events by looking for charged K mesons accompanying the decaying B_s, reasoning that a K^- was evidence for an s quark and thus suggested that the nearby strange quark in the B_s was likely to be an $\bar{s}$. Ultimately, CDF increased Q to the range 0.035–0.040. Unlike the measurement of CP violation in the B_d, the measurement of oscillations did not depend on knowing the value of the dilution precisely. Determining Δm_s was the only goal. To test for the presence of an oscillation, the data were fit to $e^{-\Gamma\tau}(1 \pm \mathcal{A}D\cos\Delta m_s\tau)$ for varying values of Δm_s and $\mathcal{A}$, so $\mathcal{A} = 1$ would indicate that the correct oscillation frequency had been found. In this way, CDF measured 17.01 ps$^{-1} < \Delta m_s < 17.84$ ps^{-1}. A few months later, CDF had refined its measurement, using time-of-flight and dE/dx measurements to discriminate charged kaons from pions (Ref. 15.11). The result was a higher precision measurement $[17.77 \pm 0.10(\text{stat}) \pm 0.07(\text{syst})]$ ps^{-1}, establishing B_s oscillation with more than 5-σ significance.

The diagrams that describe Δm_d and Δm_s differ primarily by the replacement of V_{td}^2 by V_{ts}^2. To infer the ratio $|V_{td}/V_{ts}|$, there is, in addition, a correction $\xi = 1.21^{+0.047}_{-0.035}$ for the difference between the internal structure of B_d and B_s mesons, which can be calculated using an approximation for QCD evaluated on a lattice. From

$$\left|\frac{V_{td}}{V_{ts}}\right| = \xi \sqrt{\frac{\Delta m_d}{\Delta m_s} \frac{m_{B_s^0}}{m_{B_d^0}}} \tag{15.24}$$

CDF concluded that $|V_{td}/V_{ts}| = 0.2060 \pm 0.0007(\Delta m_s)^{+0.0081}_{-0.0060}(\Delta m_d + \text{theo})$.

15.8 D Mixing

Oscillations in the D-meson system differ in many ways from those in the two B-meson systems. The most important of these is that $\Delta m/\Gamma$ is expected to be quite small. Reasoning analogous to that for the B mesons suggests that Δm and $\Delta\Gamma$ would be proportional to $(V_{cd}V_{ud}^* + V_{cs}V_{us}^*)^2 \propto (V_{cb}V_{ub}^*)^2$, which is 2×10^{-3} times smaller than $(V_{tb}V_{td}^*)^2$. This estimate is too small because there are contributions that are not accurately described by the quark-level picture. These come from explicit intermediate states like $\pi\pi$, $K\overline{K}$, and $\rho\rho$, which can connect D^0 and $\overline{D}^0$ just as $\pi\pi$ states connect K^0 and $\overline{K}^0$. Nonetheless, still mixing was expected to be a much more subtle phenomenon in D mesons than it is for K, B_d, and B_s mesons.

The appearance of wrong-sign decays through the behavior $e^{-\Gamma t}(1 - \cos\Delta m t)$ will never be large. As a result, the very small effect due to CKM-suppressed decays must be included in the analysis. In addition, the D-meson system has little connection to the third generation of quarks. Since three generations are required for CP violation to arise from the CKM matrix, we anticipate that CP violation can be ignored. It makes sense then to start with CP eigenstates analogous to K_1^0 and K_2^0. Adopting the convention $CP|D^0\rangle = |\overline{D}^0\rangle$, we write

$$|D_1^0\rangle = \frac{1}{\sqrt{2}}\left[|D^0\rangle + |\overline{D}^0\rangle\right], \tag{15.25}$$

$$|D_2^0\rangle = \frac{1}{\sqrt{2}}\left[|D^0\rangle - |\overline{D}^0\rangle\right]. \tag{15.26}$$

If we define $\Delta M = M_1 - M_2$, $\Delta\Gamma = \Gamma_1 - \Gamma_2$ and $M = (M_1 + M_2)/2$, $\Gamma = (\Gamma_1 + \Gamma_2)/2$, we can express the time development of a state that begins as D^0 as

$$|D_{phy}^0(t)\rangle \approx e^{-i(M - i\Gamma/2)t}\left[|D^0\rangle - \frac{i}{2}(\Delta M - \frac{i}{2}\Delta\Gamma)t|\overline{D}^0\rangle\right]$$

$$\approx e^{-i(M - i\Gamma/2)t}\left[|D^0\rangle - \frac{i}{2}(x - iy)\Gamma t|\overline{D}^0\rangle\right]. \tag{15.27}$$

We have introduced $x = \Delta M/\Gamma$, $y = \Delta\Gamma/(2\Gamma)$ and kept only leading terms in these small quantities. Now consider a decay that is CKM-disfavored for D^0 like $D^0 \to K^+\pi^-$. The ratio of the amplitudes for D^0 and $\overline{D}^0$ to decay to this state can be written

$$\frac{A_f}{\overline{A}_f} = -\sqrt{R_D}e^{-i\delta} \tag{15.28}$$

so that the rate for the disfavored decay is R_D times the rate for the favored one. The phase δ is the strong interaction phase for the favored decay minus that for the disfavored decay. The minus sign is introduced because $V_{cd}/V_{cs} = -\lambda$. Altogether, we see that the time-dependent rate observed for a CKM disfavored decay is

$$e^{-\Gamma t}\left[R_D + \sqrt{R_D}(y\cos\delta - x\sin\delta)\Gamma t + \frac{x^2+y^2}{4}(\Gamma t)^2\right]. \tag{15.29}$$

If we introduce $y' = y\cos\delta - x\sin\delta$ and $x' = y\sin\delta + x\cos\delta$, then it is y' and x'^2 that are measurable from the coefficients of the linear and quadratic terms in the time dependence.

Fortunately, tagging mesons as initially D^0 or $\overline{D}^0$ is much easier than for the neutral B mesons. A D^{*+} decays to $D^0\pi^+$ while D^{*-} gives $\overline{D}^0\pi^-$, so finding a charged pion that, when combined with a neutral D, gives a D^* determines whether initially the neutral D was a D^0 or a $\overline{D}^0$.

In 2007, the BaBar Collaboration reported on 1.1 million favored decays and 4000 disfavored decays to $K\pi$ with the results $y' = [9.7 \pm 4.4(\text{stat}) \pm 3.1(\text{syst})] \times 10^{-3}$, $x'^2 = [-0.22 \pm 0.30(\text{stat}) \pm 0.21(\text{syst})] \times 10^{-3}$ (**Ref. 15.12**). Once correlations between the measurements were included, this constituted evidence for mixing at the 3.9-σ level. The ratio of the rates for Cabibbo-suppressed decay and the Cabibbo-favored was found to be $0.303 \pm 0.016(\text{stat}) \pm 0.010(\text{syst})$.

Shortly thereafter, the Belle Collaboration reported a lifetime difference for the decays of neutral D mesons to CP eigenstates K^+K^- and $\pi^+\pi^-$ and to the non-CP-eigenstate $K^-\pi^+$ (Ref. 15.13). In the absence of CP violation we expect the lifetime of a $CP = +1$ state to be inversely proportional to $\Gamma + \Delta\Gamma/2$, while that for a non-CP-eigenstate should be inversely proportional simply to Γ, so that

$$\frac{\tau(D^0 \to K^-\pi^+)}{\tau(D^0 \to K^+K^-)} = 1 + y. \tag{15.30}$$

While such measurements had been made previously by FOCUS (a photoproduction experiment at Fermilab, which took data in 1996 and 1997), CLEO II, BaBar, and Belle itself, these all had uncertainties of more than 1% and none had established a significantly nonzero value. The new Belle data with more than 10^5 K^+K^- events and nearly 5×10^4 $\pi^+\pi^-$ events found $y = [1.31 \pm 0.32(\text{stat}) \pm 0.25(\text{stat})]\%$. Measurements by the BaBar Collaboration found a very similar result: $y = [1.24 \pm 0.39(\text{stat}) \pm 0.13(\text{stat})]\%$ (Ref. 15.14).

Following a technique developed by the CLEO Collaboration, Belle measured the time-dependent Dalitz plot for decays of D^0 and $\overline{D}^0$ to the CP eigenstate $K_S^0\pi^+\pi^-$ (Ref. 15.15). In the decay to a CP eigenstate, the strong interaction phase for D^0 and $\overline{D}^0$ must be the same since the strong interactions respect CP. As a result, it is x and y themselves that are accessible rather than x' and y' as in the decay to $K\pi$. Assuming no CP violation, the Belle results were $x = (0.80 \pm 0.29)\%$ and $y = (0.33 \pm 0.24)\%$ and disfavored the no-mixing values $x = y = 0$ by 2.2σ.

The Tevatron Collider also provided copious D meson production and the CDF Collaboration carried out a measurement similar to the one BaBar used to first demonstrate $D^0 - \overline{D}^0$ mixing. It compared $D^0 \to K^+\pi^-$ to $D^0 \to K^-\pi^+$, tagging the initial D as a daughter of a charged D^*(Ref. 15.16). The results were in good agreement with those from BaBar: $y' = [8.5 \pm 7.6(\text{stat})] \times 10^{-3}$, $x'^2 = [-0.12 \pm 0.35] \times 10^{-3}$. While these appear to be not far from the no-mixing case, the uncertainties in y' and x'^2 were highly correlated and the no-mixing solution was excluded at 3.8-σ level.

The remarkable consistency of data from B, B_s, and D mesons in mixing and CP violation provide enormous circumstantial evidence in favor of the CKM model of weak interactions and CP violation in particular. Nonetheless, this leaves a real puzzle. As Andrei Sakharov first recognized in 1967, CP violation is required to explain the evident baryon–antibaryon asymmetry of the Universe if one supposes that this asymmetry was not present at the outset. The CP violation of the Standard Model is not large enough to explain the measured ratio of photons to baryons, however. This suggests that there are additional sources of CP violation besides those provided through the CKM matrix. These likely reside in particles yet to be discovered. Whether they are in reach of accelerators or not remains to be seen.

Exercises

15.1 Show that when a $B^0\overline{B}^0$ pair is produced in e^+e^- annihilation in association with other particles far above the $B\overline{B}$ threshold, if both Bs decay semileptonically, the like-to-unlike-sign ratio is

$$\frac{N(l^+l^+) + N(l^-l^-)}{N(l^+l^-)} = \frac{2r}{1+r^2}$$

but if the pair is produced by the $\Upsilon(4^3S_1)$ the ratio is simply r.

15.2 Determine the eigenstates $|B_H\rangle$ and $|B_L\rangle$ including the first order corrections in Γ_{12}/M_{12}. Use this result to show that

$$\frac{N(l^+l^+) - N(l^-l^-)}{N(l^+l^+) + N(l^-l^-)} = -\frac{|\frac{q}{p}|^4 - 1}{|\frac{q}{p}|^4 + 1} \approx \Im \frac{\Gamma_{12}}{M_{12}}.$$

15.3 The transition $\overline{B}^0 \to B^0$ occurs, at the quark level, through box diagrams where the intermediate states are $t\bar{t}$, $t\bar{c}$, $t\bar{u}$, $c\bar{c}$ ... etc. The sum of all the diagrams would vanish if the quark masses were zero (or just all identical). The result then is dominated by the t quark contribution and is given by

$$M_{12}^{SM} = -\frac{G_F^2}{12\pi^2}(B_B f_B^2) m_B m_t^2 \eta (V_{tb}V_{td}^*)^2 f(x_t) \qquad (15.31)$$

where

- $G_F = 1.166 \times 10^{-5}\text{GeV}^{-2}$

- B_{B_d} is the bag parameter, relating the matrix element of a quark operator between physical states to the value obtained naively and f_{B_d} is the decay constant for the B_d meson. Lattice calculations give $f_{B_d}\sqrt{B_{B_d}} = (223 \pm 8 \pm 16)$ MeV (see the Review of Particle Physics, 2008).
- $\eta = 0.55$ is a QCD correction.
-

$$f(x_t) = \frac{4 - 11x_t + x_t^2}{4(1-x_t)^2} - \frac{3x_t^2 \ln x_t}{2(1-x_t)^3}$$

is a kinematical factor with $x_t = m_t^2/m_W^2$. With $m_t = 170$ GeV, we find $f(x_t) = 0.55$.
- $V_{tb} \approx 1$,

and where the phase convention $CP|B^0\rangle = |\overline{B}^0\rangle$ is used. Use the value $\Delta m = 0.50$ ps^{-1} to determine $|V_{td}|$. Compare this with the value you get from the measurements of Δm_s, Δm_d, and $|V_{ts}| \approx |V_{cb}|$.

15.4 Use Figure 15.1 and dimensional arguments to show that $\Delta\Gamma/\Delta m \propto m_b^2/m_t^2$, independent of the values of the CKM matrix elements. Show also that Γ_{12} and M_{12} have nearly the same phase.

15.5 In fitting with the maximum likelihood technique a distribution $f(t; A)$ with a distribution normalized so $\int dt f(t; A) = 1$, the expected uncertainty in A with N data points is given by

$$\sigma_A^{-2} = \int dt \frac{1}{f}\left(\frac{\partial f}{\partial A}\right)^2.$$

If there are several distributions f_i into which the data fall, the result is similarly

$$\sigma_A^{-2} = \sum_i \int dt \frac{1}{f_i}\left(\frac{\partial f_i}{\partial A}\right)^2.$$

Apply this to the determination of the asymmetry in $B \to J/\psi K_S^0$. Show that with perfect tagging

$$\sigma_A^{-2} = N \int_0^\infty du \, e^{-u} \frac{\sin^2 xu}{1 - A^2 \sin^2 xu} \approx N \frac{2x^2}{1 + 4x^2},$$

where the approximation applies for small A^2 and where $x = \Delta m/\Gamma$. How does the result change if there is a dilution $D \neq 1$?

Use this result to estimate the uncertainty you would expect for the BaBar data set (**Ref. 15.5**) and the reported value of Q, the effective tagging efficiency, and compare to the reported statistical uncertainty.

Further Reading

The Review of Particle Physics, published biannually by the Particle Data Group, contains topical reviews on mixing and CP violation.

A thorough and thoughtful treatment of CP violation is given in G. C. Branco, L. Lavoura, and J. P. Silva, *CP Violation*, Clarendon, Oxford (1999).

References

15.1 H. Albrecht et al., "Observation of B^0–$\overline{B}^0$ Mixing." *Phys. Lett.*, **192B**, 245 (1987).

15.2 CDF Collaboration, "Measurement of the CP-Violation Parameter $\sin 2\beta$ in $B_d^0/\overline{B}_d^0 \to J/\psi K_S^0$ Decays." *Phys. Rev. Lett.*, **81**, 5513 (1998).

15.3 CDF Collaboration, "Measurement of $\sin 2\beta$ from $B^0 \to J/\psi K_S^0$ with the CDF detector." *Phys. Rev.*, **D61**, 072005 (2000).

15.4 Belle Collaboration, "Measurement of the CP Violation Parameter $\sin 2\phi_1$ in B_d^0 Meson Decays." *Phys. Rev. Lett.*, **86**, 2509 (2001).

15.5 BaBar Collaboration, "Measurement of CP Violation Asymmetries in B^0 Decays to CP Eigenstates." *Phys. Rev. Lett.*, **86**, 2515 (2001).

15.6 BaBar Collaboration, "Observation of CP Violation in the B^0 Meson System." *Phys. Rev. Lett.*, **87**, 091801 (2001).

15.7 BaBar Collaboration, "Direct CP Violation Asymmetry in $B^0 \to K^+\pi^-$ Decays." *Phys. Rev. Lett.*, **93**, 131801 (2004).

15.8 Belle Collaboration, "Evidence for Direct CP Violation in $B^0 \to K^+\pi^-$ Decays." *Phys. Rev. Lett.*, **93**, 191802 (2004).

15.9 D0 Collaboration, "Direct Limits on the B_s^0 Oscillation Frequency." *Phys. Rev. Lett.*, **97**, 021802 (2006).

15.10 CDF Collaboration, "Measurement of the B_s^0–$\overline{B}_s^0$ Oscillation Frequency." *Phys. Rev. Lett.*, **97**, 062003 (2006).

15.11 CDF Collaboration, "Observation of B_s^0–$\overline{B}_s^0$ Oscillations." *Phys. Rev. Lett.*, **97**, 242003 (2006).

15.12 BaBar Collaboration, "Evidence for D^0–$\overline{D}^0$ Mixing." *Phys. Rev. Lett.*, **98**, 211802 (2007).

15.13 Belle Collaboration, "Evidence for D^0–$\overline{D}^0$ Mixing." *Phys. Rev. Lett.*, **98**, 211803 (2007).

15.14 BaBar Collaboration, "Measurement of D^0–$\overline{D}^0$ Mixing using the Ratio of Lifetimes for the Decays $D^0 \to K^-\pi^+, K^-K^+,$ and $\pi^-\pi^+$." *Phys. Rev.*, **078**, 011105 (2008).

15.15 Belle Collaboration, "Measurement of $D^0 - \overline{D}^0$ Mixing Parameters in $D^0 \to K_S\pi^+\pi^-$." *Phys. Rev. Lett.*, **99**, 131803 (2007).

OBSERVATION OF B^0–$\bar{B}^0$ MIXING

ARGUS Collaboration

H. ALBRECHT, A.A. ANDAM [1], U. BINDER, P. BÖCKMANN, R. GLÄSER, G. HARDER,
A. NIPPE, M. SCHÄFER, W. SCHMIDT-PARZEFALL, H. SCHRÖDER, H.D. SCHULZ,
R. WURTH, A. YAGIL [2,3]
DESY, D-2000 Hamburg, Fed. Rep. Germany

J.P. DONKER, A. DRESCHER, D. KAMP, H. KOLANOSKI, U. MATTHIESEN, H. SCHECK,
B. SPAAN, J. SPENGLER, D. WEGENER
Institut für Physik [4], Universität Dortmund, D-4600 Dortmund, Fed. Rep. Germany

C. EHMANN, J.C. GABRIEL, T. RUF, K.R. SCHUBERT, J. STIEWE, K. STRAHL, R. WALDI,
S. WESELER
Institut für Hochenergiephysik [5], Universität Heidelberg, D-6900 Heidelberg, Fed. Rep. Germany

K.W. EDWARDS [6], W.R. FRISKEN [7], D.J. GILKINSON [8], D.M. GINGRICH [8], H. KAPITZA [6],
P.C.H. KIM [8], R. KUTSCHKE [8], D.B. MACFARLANE [9], J.A. McKENNA [8], K.W. McLEAN [9],
A.W. NILSSON [9], R.S. ORR [8], P. PADLEY [8], J.A. PARSONS [8], P.M. PATEL [9], J.D. PRENTICE [8],
H.C.J. SEYWERD [8], J.D. SWAIN [8], G. TSIPOLITIS [9], T.-S. YOON [8], J.C. YUN [6]
Institute of Particle Physics [10], Canada

R. AMMAR, D. COPPAGE, R. DAVIS, S. KANEKAL, N. KWAK
University of Kansas [11], Lawrence, KS 66045, USA

B. BOŠTJANČIČ, G. KERNEL, M. PLEŠKO
Institut J. Stefan and Department of Physics [12], Univerza v Ljubljani, 61111 Ljubljana, Yugoslavia

L. JÖNSSON
Institute of Physics [13], University of Lund, S-223 62 Lund, Sweden

A. BABAEV, M. DANILOV, B. FOMINYKH, A. GOLUTVIN, I. GORELOV, V. LUBIMOV,
V. MATVEEV, V. NAGOVITSIN, V. RYLTSOV, A. SEMENOV, V. SHEVCHENKO,
V. SOLOSHENKO, V. TCHISTILIN, I. TICHOMIROV, Yu. ZAITSEV
Institute of Theoretical and Experimental Physics, 117 259 Moscow, USSR

R. CHILDERS, C.W. DARDEN, Y. OKU
University of South Carolina [14], Columbia, SC 29208, USA

and

H. GENNOW
University of Stockholm, S-113 46 Stockholm, Sweden

Received 9 April 1987

> Using the ARGUS detector at the DORIS II storage ring we have searched in three different ways for B^0–$\bar{B}^0$ mixing in Υ (4S) decays. One explicitly mixed event, a decay Υ (4S)$\to B^0 B^0$, has been completely reconstructed. Furthermore, we observe a 4.0 standard deviation signal of 24.8 events with like-sign lepton pairs and a 3.0 standard deviation signal of 4.1 events containing one reconstructed $B^0(\bar{B}^0)$ and an additional fast ℓ^+ (ℓ^-). This leads to the conclusion that B^0–$\bar{B}^0$ mixing is substantial. For the mixing parameter we obtain $r = 0.21 \pm 0.08$.

We report the observation of B^0–$\bar{B}^0$ mixing. This conclusion is based on the study of B mesons produced in Υ (4S) decays, using the ARGUS detector at the e^+e^- storage ring DORIS II at DESY. B^0–$\bar{B}^0$ mixing provides basic information on the parameters and validity of the standard model [1], and is potentially a sensitive probe for new physics [2]. A B^0 meson can either decay directly or, through mixing, transform into its anti-particle, the $\bar{B}^0$, before decaying. The ratio of the decay widths [3,4]

$$r = \frac{\Gamma(B^0 \to \bar{B}^0 \to X')}{\Gamma(B^0 \to X)}$$

of these two competing reactions describes the strength of mixing. In decays of the Υ (4S), pairs of $B^0 \bar{B}^0$ mesons are produced in a P-wave, so that r is given in this case [5] by the ratio

$$r = \frac{N(B^0 B^0) + N(\bar{B}^0 \bar{B}^0)}{N(B^0 \bar{B}^0)}.$$

Thus, the existence of mixing leads to events consisting of $B^0 B^0$ or $\bar{B}^0 \bar{B}^0$ pairs which can be detected experimentally.

An upper limit for B^0–$\bar{B}^0$ mixing of 24% at 90% CL has been published by the CLEO Collaboration [6]. An investigation by the MARK II Collaboration [7] of dilepton rates in continuum e^+e^- annihilations at 29 GeV, well above the B_s production threshold, resulted in combined upper limits for B^0–$\bar{B}^0$ and B_s–$\bar{B}_s$ mixing. The UA1 Collaboration [8] has reported evidence for an excess of like-sign lepton pairs produced in $p\bar{p}$ collisions, which they interpreted as signature for B_s–$\bar{B}_s$ mixing.

The mixing study reported here is made with B mesons produced in 88000 Υ (4S) decays. The event sample corresponds to an integrated luminosity of 103 pb^{-1} on the Υ (4S) and 42 pb^{-1} in the continuum just below the Υ (4S). A short description of the ARGUS detector and its trigger can be found in ref. [9] and its particle identification capabilities in ref. [10].

Evidence for substantial B^0–$\bar{B}^0$ mixing is obtained by using three different analysis methods. The first approach is to search for fully reconstructed Υ (4S) decays into $B^0 B^0$ or $\bar{B}^0 \bar{B}^0$ pairs. Efficient and clean reconstruction of B mesons is accomplished by using B decays involving D^{*-} mesons [#1] which are reconstructed through their decays $D^{*-} \to \bar{D}^0 \pi^-$, followed by

$$\bar{D}^0 \to K^+ \pi^-$$
$$\to K^+ \pi^- \pi^0$$
$$\to K^+ \pi^- \pi^+ \pi^-$$
$$\to K^0_S \pi^+ \pi^-.$$

---

[1] On leave from University of Science and Technology, Kumasi, Ghana.
[2] Weizmann Institute of Science, 76100 Rehovot, Israel.
[3] Supported by the Minerva Stiftung.
[4] Supported by the German Bundesministerium für Forschung und Technologie under the contract number 054DO51P.
[5] Supported by the German Bundesministerium für Forschung und Technologie under the contract number 054HD24P.
[6] Carleton University, Ottawa, Ontario, Canada K1S 5B6.
[7] York University, Downsview, Ontario, Canada M3J 1P3.
[8] University of Toronto, Toronto, Ontario, Canada M5S 1A7.
[9] McGill University, Montreal, Quebec, Canada H3A 2T8.
[10] Supported by the Natural Sciences and Engineering Research Council, Canada.
[11] Supported by the US National Science Foundation.
[12] Supported by Raziskovalna skupnost Slovenije and the Internationales Büro KfA, Jülich.
[13] Supported by the Swedish Research Council.
[14] Supported by the US Department of Energy, under contract DE-AS09-80ER10690.

[#1] References in this paper to a specific charged state are to be interpreted as implying the charge-conjugate state also.

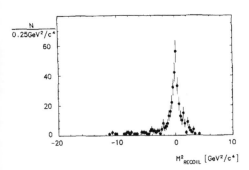

Fig. 1. Recoil mass $M^2_{\text{Recoil}} = [E_{\text{beam}} - (E_{D^*} + E_{\ell^+})]^2 - (\mathbf{p}_{D^*} + \mathbf{p}_{\ell^+})^2$ with $D^{*-} \to \bar{D}^0 \pi^-$, $\bar{D}^0 \to K^+ \pi^-$, $K^0 \pi^+ \pi^-$, $K^+ \pi^- \pi^0$, $K^+ \pi^- \pi^+ \pi^-$ and one lepton (μ^+, e^+) with momentum $p > 1.0$ GeV/c.

B^0 mesons are either reconstructed in the hadronic decay modes [11].

$B^0 \to D^{*-} \pi^+$
$ \to D^{*-} \pi^+ \pi^0$
$ \to D^{*-} \pi^+ \pi^+ \pi^-$,

or in the channel

$B^0 \to D^{*-} \ell^+ \nu$,

with ℓ^+ being an e^+ or μ^+. The partial reconstruction of the decay $B^0 \to D^{*-} \ell^+ \nu$ is possible because B^0 mesons produced in Υ(4S) decays are nearly at rest. The neutrino is unobserved, but can be inferred if the recoil mass against the $D^{*-} \ell^+$ system, M^2_{Recoil}, is consistent with zero. M^2_{Recoil} is defined by

$$M^2_{\text{Recoil}} = [E_{\text{beam}} - (E_{D^{*-}} + E_{\ell^+})]^2 - (\mathbf{p}_{D^{*-}} + \mathbf{p}_{\ell^+})^2.$$

By requiring the D^{*-} to have momentum less than 2.45 GeV/c and the lepton to have momentum above 1.0 GeV/c, we obtain the recoil mass spectrum shown in fig. 1. The prominent peak at $M^2_{\text{Recoil}} = 0$ corresponds to a B^0 signal on a low background. The position and shape of the signal is well described by the Monte Carlo prediction for Υ(4S)$\to B^0 \bar{B}^0$ followed by the semi-leptonic decay $B^0 \to D^{*-} \ell^+ \nu$.

In the sample of events with a single reconstructed B^0, we can attempt to reconstruct the second B^0, now with a less restrictive choice of possible decay channels. By this means, we have succeeded in completely reconstructing a decay Υ(4S)$\to B^0 B^0$, the first

Fig. 2. Completely reconstructed event consisting of the decay Υ(4S)$\to B^0 B^0$.

observation of B^0–$\bar{B}^0$ mixing. The two B^0 mesons (B^0_1 and B^0_2) decay in the following way:

$B^0_1 \to D^{*-}_1 \mu^+_1 \nu_1$

$\downarrow$

$D^{*-}_1 \to \pi^-_1 \bar{D}^0$

$\downarrow$

$\bar{D}^0 \to K^+_1 \pi^-_1$,

and

$B^0_2 \to D^{*-}_2 \mu^+_2 \nu_2$

$\downarrow$

$D^{*-}_2 \to \pi^0 D^-$

$\downarrow$

$D^- \to K^+_2 \pi^-_2 \pi^-_2$.

The event is shown in fig. 2 and its kinematical quantities are listed in table 1. The masses of the intermediate states agree well with the table values [12]. Both D^{*-} mesons contain positive kaons of momenta $p(K_1) = 0.548$ GeV/c and $p(K_2) = 0.807$ GeV/c which are uniquely identified by the measurements of specific ionisation loss (dE/dx) and of time-of-flight. The two positive muons are the fastest

Table 1
Kinematical quantities of the observed $\Upsilon(4S) \to B_1^0 B_2^0$ event.

Decay	Mass(GeV/c^2)	P(GeV/c)	M^2_{Recoil}(GeV^2/c^4)
$B_1^0 \to D_1^{*-} \mu_1^+ (\nu_1)$	$4.393 + 0.088$ [a]	1.090 ± 0.108 [a]	-0.609
$D_1^{*-} \to \pi_1^- \bar{D}^0$	2.008 ± 0.001	1.196 ± 0.013	
$\bar{D}^0 \to K_1^+ \pi_1^-$	1.873 ± 0.021	1.091 ± 0.012	
$B_2^0 \to D_2^{*-} \mu_2^+ (\nu_2)$	3.969 ± 0.032 [a]	1.244 ± 0.015 [a]	-0.275
$D_2^{*-} \to \pi^0 D^-$	2.008 ± 0.005	1.611 ± 0.017	
$\pi^0 \to 2\gamma$	0.180 ± 0.028	0.136 ± 0.019	
$D^- \to K_2^+ \pi_2^- \pi_2^-$	1.886 ± 0.015	1.478 ± 0.007	

[a] Mass and momentum without neutrino.

particles in the event with momenta $p(\mu_1) = 2.186$ GeV/c and $p(\mu_2) = 1.579$ GeV/c, and have dE/dx and shower counter information consistent with the muon hypothesis. One muon, μ_1, is clearly identified in the muon chambers whereas the second one, μ_2, points in a direction of the detector not covered by muon chambers.

This event has a kinematic peculiarity which leads to the conclusion that B_2^0 decays semi-leptonically, and that therefore μ_2 must be a muon, providing further proof that this event contains two B^0 mesons. The momenta of D_1^{*-} and μ_1^+ restrict the momentum vector of the B_1^0 meson onto a small cone around the direction of the $D_1^{*-} \mu_1^+$ system. Knowing the direction of B_1^0 and, opposite to it, of B_2^0, the event is fully reconstructed in spite of the fact that two neutrinos are present. Specifically, the missing mass in the decay of B_2^0 is only compatible with zero or the π^0 mass. Since no additional signal for a single photon or a π^0 is seen in the detector, the neutrino hypothesis alone is acceptable which agrees perfectly with the above interpretation.

For a mixing strength of $r = 0.2$, we expect to reconstruct 0.3 events of this type where both B mesons decay as $B^0 \to D^{*-} \ell^+ \nu$. In order to estimate the background for such an event, a Monte Carlo simulation was performed. Among 22 000 $B^0 \bar{B}^0$ pairs where B_1^0 is reconstructed in the observed channel and the multiplicities of the detected remaining charged and neutral particles are the same as in the above event, we find no fake candidate for mixing.

In the second analysis method we investigate events containing lepton pairs originating from $\Upsilon(4S)$ decays. The charge of the primary lepton from the decay of the b quark identifies whether the decaying meson is a B or a $\bar{B}$. Thus, B^0–$\bar{B}^0$ mixing manifests itself in the production of like-sign lepton pairs.

An event selection is made by applying cuts to suppress continuum dilepton sources: (1) the second Fox–Wolfram moment [13] less than 0.6, (2) charged multiplicity $n_{ch} \geq 5$ and (3) total multiplicity $n_{ch} + \frac{1}{2} n_\gamma \geq 7$. The angle between all particles and the beam axis is required to satisfy $\cos \theta_{lab} < 0.9$. Exactly two of the particles in the events have to be well-identified leptons with momenta greater than 1.4 GeV/c. The momentum cut suppresses most of the secondary leptons originating from charmed mesons in B decays. For lepton identification, information from all detector components is used coherently by combining the measurements into an overall likelihood [14]. The available information consists of dE/dx and time-of-flight measurements, and the magnitude and topology of energy depositions in the shower counters. In addition, for muons, a hit in an outer muon chamber is required and information on the hit-impact point distance is included in the likelihood.

Further requirements are made in order to reduce background sources of lepton pairs. B decays to J/$\psi(\psi')$ produce $e^+ e^-$ or $\mu^+ \mu^-$ pairs. To suppress this background, events containing those pairs are rejected if the mass of the pair coincides with the mass of the J/ψ or ψ' within ± 150 MeV/c^2. Electrons originating from photon conversion are suppressed by requiring that no other positron candidate of any momentum lie within a cone of 32° around the hight momentum electron track.

The distribution of the opening angle $\theta_{\ell\ell}$ between the leptons is shown in fig. 3 for events passing these cuts. For leptons originating from two different B

248

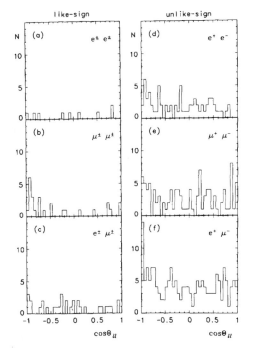

Fig. 3. Distribution of the opening angle between the leptons for like-sign (a)–(c) and unlike-sign (d)–(f) pairs.

mesons, this distribution should be isotropic. Lepton pairs from continuum or originating from the same B meson tend to be back-to-back. These contributions are reduced by requiring $\cos\theta_{\ell\ell} > -0.85$. Table 2 gives the number of dilepton events surviving these cuts both on the Υ(4S) resonance and in the continuum below. The number of dilepton events from Υ(4S) decays is determined by subtracting the continuum contribution scaled by a factor 2.5 according to the ratio of luminosities. Further, the e^+e^- and $\mu^+\mu^-$ pair events are corrected for losses due to the invariant mass cut to remove recognized $J/\psi(\psi')$ decays.

The remaining dilepton events still include contributions from background due to lepton–hadron misidentification, secondary leptons from charm decays, J/ψ decays, and converted photons.

The background due to lepton–hadron misidentification is evaluated from data. To determine the fake rate per track we use our data samples of $\tau^- \to \nu \pi^- \pi^- \pi^+ + n\pi^0$ ($n=0, 1$) and $D^{*+} \to D^0 \pi^+$, $D^0 \to K^- \pi^+$ decays which provide clean sources of high energy pions and kaons, respectively. Decay-in-flight and punch-through result in a π/μ misidentification probability of $(2.2\pm0.2)\%$ per pion. For K/μ misidentification the fake rate is $(1.9\pm0.5)\%$ per kaon, including a correction for kaon decays between the interaction point and the drift chamber. The fake rates due to π/e and K/e misidentification are both $(0.5\pm0.1)\%$. The lepton–hadron misidentification rates have also been determined using hadronic decays of the Υ(1S) where the fraction of leptons is negligible. The results obtained agree with the quoted values.

The number of faked dilepton events is extracted from the observed hadron momentum spectrum in the events containing like-sign and unlike-sign lepton–hadron pairs. These momentum spectra, folded with lepton–hadron misidentification probabilities, are shown in fig. 4 for both like-sign and unlike-sign lepton–hadron samples. Since the fake rate per track is within errors the same for pions and kaons, it is not necessary to account for their relative fractions. One unlike-sign dimuon event is expected to occur where both muons are misidentified hadrons.

The background due to secondary leptons is determined by a Monte Carlo simulation of B decays. A spectator model [15] is used to describe the decay of the b quark, with the final state hadrons produced using the Lund string fragmentation model [16]. The simulation is checked by comparison with ARGUS measurements of the inclusive spectra for leptons, D^0 mesons, pions and kaons from B decays, and with the inclusive electron spectrum for D^0 and D^+ decays from MARK III [17]. All these data are well reproduced. The uncertainty in the calculation is expected to be $\pm25\%$. The background from J/ψ and ψ' decays or converted photons where only one of the two leptons is observed in the detector is also determined by Monte Carlo simulation.

The number of events are given in table 2. Out of the 50 like-sign dilepton events, $25.2\pm5.0\pm3.8$ events are attributed to the background sources as described above. The first error is the statistical and the second one the systematical uncertainty in the background determination. The probability for the measured 50 events to be a fluctuation of the background corresponds to 4.0 standard deviations. Thus,

249

Table 2
Dilepton rates.

		$e^\pm e^\pm$	$\mu^\pm\mu^\pm$	$e^\pm\mu^\pm$
dilepton candidates	$\Upsilon(4S)$ + continuum	8	16	26
	continuum	0	0	0
	$\Upsilon(4S)$ direct	8.0±3.9	16.0±4.8	26.0±5.8
background	fakes	0.7	5.7	4.9
	conversion	0.5	–	0.5
	secondary decays	2.3	2.9	4.6
	J/ψ decays	0.7	0.9	1.5
signal		3.8±3.9±0.9	6.5±4.8±1.3	14.5±5.8±1.8

sum: 50 dilepton candidates
background: 25.2±5.0±3.8 events
signal: 24.8±7.6±3.8 like-sign lepton pairs

		e^+e^-	$\mu^+\mu^-$	$e^\pm\mu^\mp$
dilepton candidates	$\Upsilon(4S)$ + continuum	60	92	149
	continuum	3	1	2
	$\Upsilon(4S)$ direct	52.6	89.5	144.1
	corrected for J/ψ cut	58.5±9.8±1.6	99.6±11.3±2.5	144.1±12.4
background	fakes	1.4	12.1	10.2
	conversion	0.5	–	0.5
	secondary decays	0.7	1.5	1.6
	J/ψ decays	1.0	0.9	1.5
signal		54.9±9.8±1.6	85.1±11.3±3.1	130.3±12.4±1.8

signal: 270±19.4±5.0 unlike-sign lepton pairs

mixing parameter r	0.17±0.19±0.04	0.19±0.16±0.04	0.28±0.14±0.04

combined mixing parameter $r=0.22\pm0.09\pm0.04$

we attribute the signal of 24.8±7.6±3.8 events to B^0-$\bar{B}^0$ mixing. The signal for unlike-sign pairs is 270.3±19.4±5.0 events.

The mixing parameter r for dilepton events has the form

$$r = \frac{[N(\ell^+\ell^+) + N(\ell^-\ell^-)](1+\lambda)}{N(\ell^+\ell^-) - [N(\ell^+\ell^+) + N(\ell^-\ell^-)]\lambda}.$$

In order to account for $\Upsilon(4S)$ decays into B^+B^- pairs, a factor

$$\lambda = \frac{f^+}{f^0}\left(\frac{Br_{sl}^+}{Br_{sl}^0}\right)^2$$

has to be introduced where f^+ (f^0) is the branching ratio of the decay $\Upsilon(4S)$ into charged (neutral) B mesons and Br_{sl}^+ (Br_{sl}^0) the semi-leptonic branching ratio of charged (neutral) B mesons. All these numbers are unknown, and we assume λ to be equal to 1.2. The acceptance for ee, $\mu\mu$ and eμ events is different, thus the mixing parameter r is calculated for each sample separately. Combining these results, we obtain

$r = 0.22 \pm 0.09 \pm 0.04$.

This result is not sensitive, within the statistical errors, to a variation of the lepton momentum cut between 1.4 and 1.6 GeV/c.

The third analysis method involves the reconstruction of one of the B^0 mesons originating from the $\Upsilon(4S)$ decay, using the same channels as for the

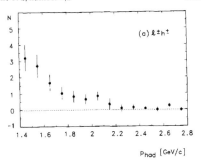

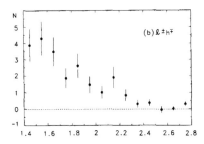

Fig. 4. Momentum spectrum of misidentified hadrons for faked dilepton events: (a) like-sign, (b) unlike-sign.

first method described above, and tagging the second B^0 with a fast lepton. This method is considerably less sensitive to background from lepton misidentification.

Fig. 5 shows the spectrum for the recoil mass against a $D^{*-} \ell^+$ system if the event contains one additional lepton with momentum larger than 1.4

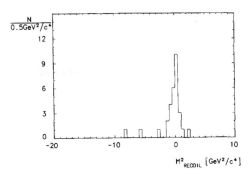

Fig. 5. Same as fig. 1 with requiring an additional lepton (μ, e) with momentum $p > 1.4$ GeV/c in the event.

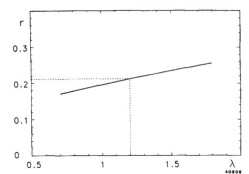

Fig. 6. The mixing parameter r as a function of the factor λ. The dotted line indicates our chosen value $\lambda = 1.2$.

GeV/c. Adding two events where the B^0 mesons are reconstructed in the hadronic channels, we obtain a total of 23 candidates for unmixed events and five candidates for mixed events. These five events are composed of two $B^0 e^+$, two $\bar{B}^0 e^-$ and one $\bar{B}^0 \mu^-$ events. The background for the mixed sample, determined in the same way as for the second method, is expected to be 0.4 events due to misidentification and 0.5 events due to secondary leptons. After subtracting 0.9 ± 0.3 events we are left with 4.1 events from B^0–$\bar{B}^0$ mixing. The probability for the observed events to be a fluctuation of the background corresponds to 3.0 standard deviations. The background to the unmixed events is 2.2 ± 1.1 events. Thus, we find a value for the mixing parameter r of

$$r = \frac{N(B^0 \ell^+) + N(\bar{B}^0 \ell^-)}{N(B^0 \ell^-) + N(\bar{B}^0 \ell^+)} = 0.20 \pm 0.12.$$

Two like-sign and eleven unlike-sign events from this sample are also present in the dilepton sample. Taking this correlation into account, we get a combined result of

$$r = 0.21 \pm 0.08$$

for $\lambda = 1.2$. The λ dependence of this result is shown in fig. 6. The parameter $\chi = r/(1+r)$ turns out to be $\chi = 0.17 \pm 0.05$ for $r = 0.21 \pm 0.08$.

We discuss our result in the framework of the standard model with three generations. Assuming dominance of the box diagram, mixing is described by the parameter x [1]:

251

Table 3
Limits on parameters consistent with the observed mixing rate.

Parameters	Comments
$r > 0.09$ (90%CL)	this experiment
$x > 0.44$	this experiment
$B^{1/2} f_B \approx f_\pi < 160$ MeV	B meson ($\approx$ pion) decay constant
$m_b < 5$ GeV/c^2	b-quark mass
$\tau < 1.4 \times 10^{-12}$ s	B meson lifetime
$|V_{td}| < 0.018$	Kobayashi–Maskawa matrix element
$\eta_{QCD} < 0.86$	QCD correction factor [a]
$m_t > 50$ GeV/c^2	t quark mass

[a] Ref. [18].

$$x = \frac{\Delta M}{\Gamma} = 32\pi \frac{B f_B^2 m_t^2 m_b}{m_\mu^5} \frac{\tau_b}{\tau_\mu} |V_{td}|^2 \eta_{QCD},$$

and related to experiment by

$$r = \frac{x^2}{x^2 + 2}.$$

The rate of B^0–$\bar{B}^0$ mixing provides a strong constraint on parameters of the standard model. Specifically, our result shows that the Kobayashi–Maskawa element V_{td} is non-zero. The observed value of r can still be accommodated by the standard model within the present knowledge of its parameters. As an illustration, one example of a set of limits is given in table 3.

In summary, the combined evidence of the investigation of B^0 meson pairs, lepton pairs and B^0 meson–lepton events on the $\Upsilon(4S)$ leads to the conclusion that B^0–$\bar{B}^0$ mixing has been observed and is substantial.

It is a pleasure to thank U. Djuanda, E. Konrad, E. Michel, and W. Reinsch for their competent technical help in running the experiment and processing the data. We thank Dr. H. Nesemann, B. Sarau, and the DORIS group for the excellent operation of the storage ring. The visiting groups wish to thank the DESY directorate for the support and kind hospitality extended to them.

References

[1] M. Kobayashi and T. Maskawa, Progr. Theor. Phys. 49 (1973) 652;
M.K. Gaillard and B.W. Lee, Phys. Rev. D 10 (1974) 897;
J. Ellis, M.K. Gaillard, D.V. Nanopolous and S. Rudaz, Nucl. Phys. B 131 (1977) 285;
A. Ali and Z.Z. Aydin, Nucl. Phys. B 148 (1978) 165;
see also, e.g., I.I. Bigi and A.I. Sanda, Phys. Rev. D 29 (1984) 1393;
E.A. Paschos and U. Tuerke, Nucl. Phys. B 243 (1984) 29;
A. Ali, Proc. Intern. Symp. on Production and decay of heavy hadrons (Heidelberg, 1986) p. 366.
[2] M.B. Gavela et al., Phys. Lett. B 154 (1985) 147;
A.A. Anselm et al., Phys. Lett. B 156 (1985) 102;
M. Gronau and J. Schechter, Phys. Rev. D 31 (1985) 1668;
G. Ecker and W. Grimus, Z. Phys. C 30 (1986) 293.
[3] L.B. Okun, V.I. Zakharov and B.M. Pontecorvo, Lett. Nuovo Cimento 13 (1975) 218.
[4] A. Pais and S.B. Treiman, Phys. Rev. D 12 (1975) 2744.
[5] I.I. Bigi and A.I. Sanda, Nucl. Phys. B 193 (1981) 123; Phys. Lett. B 171 (1986) 320.
[6] CLEO Collab., A. Bean et al., Phys. Rev. Lett. 58 (1987) 183.
[7] MARK II Collab., T. Schaad et al., Phys. Lett. B 160 (1985) 188.
[8] UA1 Collab., C. Albajar et al., Phys. Lett. B 186 (1987) 247.
[9] ARGUS Collab., H. Albrecht et al., Phys. Lett. B 134 (1984) 137.
[10] ARGUS Collab., H. Albrecht et al., Phys. Lett. B 150 (1985) 235.
[11] ARGUS Collab., H. Albrecht et al., Phys. Lett. B 185 (1987) 218.
[12] Particle Data Group, Phys. Lett. B 170 (1986) 1.
[13] G.C. Fox and S. Wolfram, Phys. Lett. B 82 (1979) 134.
[14] S. Weseler, Ph.D. Thesis, University of Heidelberg (April 1986), Report IHEP-HD/86-2
[15] G. Altarelli, N. Cabibbo and L. Maiani, Nucl. Phys. B 100 (1975) 313.
[16] B. Anderson et al., Phys. Rep. 97 (1983) 31.
[17] MARK III Collab., R.M. Baltrusaitis et al., Phys. Rev. Lett. 54 (1985) 1976.
[18] A.J. Buras, W. Slominski and H. Steger, Nucl. Phys. B 245 (1984) 369, and references quoted therein.

Measurement of the CP Violation Parameter $\sin 2\phi_1$ in B_d^0 Meson Decays

A. Abashian,[44] K. Abe,[8] K. Abe,[36] I. Adachi,[8] Byoung Sup Ahn,[14] H. Aihara,[37] M. Akatsu,[19] G. Alimonti,[7] K. Aoki,[8] K. Asai,[20] M. Asai,[9] Y. Asano,[42] T. Aso,[41] V. Aulchenko,[2] T. Aushev,[12] A. M. Bakich,[33] E. Banas,[15] S. Behari,[8] P. K. Behera,[43] D. Beiline,[2] A. Bondar,[2] A. Bozek,[15] T. E. Browder,[7] B. C. K. Casey,[7] P. Chang,[23] Y. Chao,[23] B. G. Cheon,[32] S.-K. Choi,[6] Y. Choi,[32] Y. Doi,[8] J. Dragic,[17] A. Drutskoy,[12] S. Eidelman,[2] Y. Enari,[19] R. Enomoto,[8,10] C. W. Everton,[17] F. Fang,[7] H. Fujii,[8] K. Fujimoto,[19] Y. Fujita,[8] C. Fukunaga,[39] M. Fukushima,[10] A. Garmash,[2,8] A. Gordon,[17] K. Gotow,[44] H. Guler,[7] R. Guo,[21] J. Haba,[8] T. Haji,[37] H. Hamasaki,[8] K. Hanagaki,[29] F. Handa,[36] K. Hara,[27] T. Hara,[27] T. Haruyama,[8] N. C. Hastings,[17] K. Hayashi,[8] H. Hayashii,[20] M. Hazumi,[27] E. M. Heenan,[17] Y. Higashi,[8] Y. Higashino,[19] I. Higuchi,[36] T. Higuchi,[37] T. Hirai,[38] H. Hirano,[40] M. Hirose,[19] T. Hojo,[27] Y. Hoshi,[35] K. Hoshina,[40] W.-S. Hou,[23] S.-C. Hsu,[23] H.-C. Huang,[23] Y.-C. Huang,[21] S. Ichizawa,[38] Y. Igarashi,[8] T. Iijima,[8] H. Ikeda,[8] K. Ikeda,[20] K. Inami,[19] Y. Inoue,[26] A. Ishikawa,[19] H. Ishino,[38] R. Itoh,[8] G. Iwai,[25] M. Iwai,[8] M. Iwamoto,[3] H. Iwasaki,[8] Y. Iwasaki,[8] D. J. Jackson,[27] P. Jalocha,[15] H. K. Jang,[31] M. Jones,[7] R. Kagan,[12] H. Kakuno,[38] J. Kaneko,[38] J. H. Kang,[45] J. S. Kang,[14] P. Kapusta,[15] K. Kasami,[8] N. Katayama,[8] H. Kawai,[3] H. Kawai,[37] M. Kawai,[8] N. Kawamura,[1] T. Kawasaki,[25] H. Kichimi,[8] D. W. Kim,[32] Heejong Kim,[45] H. J. Kim,[45] Hyunwoo Kim,[14] S. K. Kim,[31] K. Kinoshita,[5] S. Kobayashi,[30] S. Koike,[8] S. Koishi,[38] Y. Kondo,[8] H. Konishi,[40] K. Korotushenko,[29] P. Krokovny,[2] R. Kulasiri,[5] S. Kumar,[28] T. Kuniya,[30] E. Kurihara,[3] A. Kuzmin,[2] Y.-J. Kwon,[45] M. H. Lee,[8] S. H. Lee,[31] C. Leonidopoulos,[29] H.-B. Li,[11] R.-S. Lu,[23] Y. Makida,[8] A. Manabe,[8] D. Marlow,[29] T. Matsubara,[37] T. Matsuda,[8] S. Matsui,[19] S. Matsumoto,[4] T. Matsumoto,[19] Y. Mikami,[36] K. Misono,[19] K. Miyabayashi,[20] H. Miyake,[27] H. Miyata,[25] L. C. Moffitt,[17] A. Mohapatra,[43] G. R. Moloney,[17] G. F. Moorhead,[17] N. Morgan,[44] S. Mori,[42] T. Mori,[4] A. Murakami,[30] T. Nagamine,[36] Y. Nagasaka,[18] Y. Nagashima,[27] T. Nakadaira,[37] T. Nakamura,[38] E. Nakano,[26] M. Nakao,[8] H. Nakazawa,[4] J. W. Nam,[32] S. Narita,[36] Z. Natkaniec,[15] K. Neichi,[35] S. Nishida,[16] O. Nitoh,[40] S. Noguchi,[20] N. Nozaki,[8] S. Ogawa,[34] T. Ohshima,[19] Y. Ohshima,[38] T. Okabe,[19] T. Okazaki,[20] S. Okuno,[13] S. I. Olsen,[7] W. Ostrowicz,[15] H. Ozaki,[8] P. Pakhlov,[12] H. Palka,[15] C. S. Park,[31] C. W. Park,[14] H. Park,[14] L. S. Peak,[33] M. Peters,[7] L. E. Piilonen,[44] E. Prebys,[29] J. L. Rodriguez,[7] N. Root,[2] M. Rozanska,[15] K. Rybicki,[15] J. Ryuko,[27] H. Sagawa,[8] S. Saitoh,[3] Y. Sakai,[8] H. Sakamoto,[16] H. Sakaue,[26] M. Satapathy,[43] N. Sato,[8] A. Satpathy,[8,5] S. Schrenk,[5] S. Semenov,[12] Y. Settai,[4] M. E. Sevior,[17] H. Shibuya,[34] B. Shwartz,[2] A. Sidorov,[2] V. Sidorov,[2] J. B. Singh,[28] S. Stanič,[42] A. Sugi,[19] A. Sugiyama,[19] K. Sumisawa,[27] T. Sumiyoshi,[8] J. Suzuki,[8] J.-I. Suzuki,[8] K. Suzuki,[3] S. Suzuki,[19] S. Y. Suzuki,[8] S. K. Swain,[7] H. Tajima,[37] T. Takahashi,[26] F. Takasaki,[8] M. Takita,[27] K. Tamai,[8] N. Tamura,[25] J. Tanaka,[37] M. Tanaka,[8] Y. Tanaka,[18] G. N. Taylor,[17] Y. Teramoto,[26] M. Tomoto,[19] T. Tomura,[37] S. N. Tovey,[17] K. Trabelsi,[7] T. Tsuboyama,[8] Y. Tsujita,[42] Y. Tsukamoto,[8] T. Tsukamoto,[30] S. Uehara,[8] K. Ueno,[23] N. Ujiie,[8] Y. Unno,[3] S. Uno,[8] Y. Ushiroda,[16] Y. Usov,[2] S. E. Vahsen,[29] G. Varner,[7] K. E. Varvell,[33] C. C. Wang,[23] C. H. Wang,[22] M.-Z. Wang,[23] T. J. Wang,[11] Y. Watanabe,[38] E. Won,[31] B. D. Yabsley,[8] Y. Yamada,[8] M. Yamaga,[36] A. Yamaguchi,[36] H. Yamaguchi,[8] H. Yamamoto,[7] T. Yamanaka,[27] H. Yamaoka,[8] Y. Yamaoka,[8] Y. Yamashita,[24] M. Yamauchi,[8] S. Yanaka,[38] M. Yokoyama,[37] K. Yoshida,[19] Y. Yusa,[36] H. Yuta,[1] C. C. Zhang,[11] H. W. Zhao,[8] J. Zhang,[42] Y. Zheng,[7] V. Zhilich,[2] and D. Žontar[42]

[1]*Aomori University, Aomori*
[2]*Budker Institute of Nuclear Physics, Novosibirsk*
[3]*Chiba University, Chiba*
[4]*Chuo University, Tokyo*
[5]*University of Cincinnati, Cincinnati, Ohio*
[6]*Gyeongsang National University, Chinju*
[7]*University of Hawaii, Honolulu, Hawaii*
[8]*High Energy Accelerator Research Organization (KEK), Tsukuba*
[9]*Hiroshima Institute of Technology, Hiroshima*
[10]*Institute for Cosmic Ray Research, University of Tokyo, Tokyo*
[11]*Institute of High Energy Physics, Chinese Academy of Sciences, Beijing*
[12]*Institute for Theoretical and Experimental Physics, Moscow*
[13]*Kanagawa University, Yokohama*
[14]*Korea University, Seoul*
[15]*H. Niewodniczanski Institute of Nuclear Physics, Krakow*
[16]*Kyoto University, Kyoto*
[17]*University of Melbourne, Victoria*

[18]*Nagasaki Institute of Applied Science, Nagasaki*
[19]*Nagoya University, Nagoya*
[20]*Nara Women's University, Nara*
[21]*National Kaohsiung Normal University, Kaohsiung*
[22]*National Lien-Ho Institute of Technology, Miao Li*
[23]*National Taiwan University, Taipei*
[24]*Nihon Dental College, Niigata*
[25]*Niigata University, Niigata*
[26]*Osaka City University, Osaka*
[27]*Osaka University, Osaka*
[28]*Panjab University, Chandigarh*
[29]*Princeton University, Princeton, New Jersey*
[30]*Saga University, Saga*
[31]*Seoul National University, Seoul*
[32]*Sungkyunkwan University, Suwon*
[33]*University of Sydney, Sydney NSW*
[34]*Toho University, Funabashi*
[35]*Tohoku Gakuin University, Tagajo*
[36]*Tohoku University, Sendai*
[37]*University of Tokyo, Tokyo*
[38]*Tokyo Institute of Technology, Tokyo*
[39]*Tokyo Metropolitan University, Tokyo*
[40]*Tokyo University of Agriculture and Technology, Tokyo*
[41]*Toyama National College of Maritime Technology, Toyama*
[42]*University of Tsukuba, Tsukuba*
[43]*Utkal University, Bhubaneswer*
[44]*Virginia Polytechnic Institute and State University, Blacksburg, Virginia*
[45]*Yonsei University, Seoul*
(Received 9 February 2001)

We present a measurement of the standard model *CP* violation parameter $\sin 2\phi_1$ (also known as $\sin 2\beta$) based on a 10.5 fb^{-1} data sample collected at the $\Upsilon(4S)$ resonance with the Belle detector at the KEKB asymmetric e^+e^- collider. One neutral B meson is reconstructed in the $J/\psi K_S$, $\psi(2S)K_S$, $\chi_{c1}K_S$, $\eta_c K_S$, $J/\psi K_L$, or $J/\psi \pi^0$ *CP*-eigenstate decay channel and the flavor of the accompanying B meson is identified from its charged particle decay products. From the asymmetry in the distribution of the time interval between the two B-meson decay points, we determine $\sin 2\phi_1 = 0.58^{+0.32}_{-0.34}(\text{stat})^{+0.09}_{-0.10}(\text{syst})$.

DOI: 10.1103/PhysRevLett.86.2509   PACS numbers: 11.30.Er, 12.15.Hh, 13.25.Hw

In the standard model (SM), *CP* violation arises from a complex phase in the Cabibbo-Kobayashi-Maskawa (CKM) quark mixing matrix [1]. In particular, the SM predicts a *CP* violating asymmetry in the time-dependent rates for B_d^0 and $\overline{B}_d^0$ decays to a common *CP* eigenstate, f_{CP}, without theoretical ambiguity due to strong interactions [2]:

$$A(t) \equiv \frac{\Gamma(\overline{B}_d^0 \to f_{CP}) - \Gamma(B_d^0 \to f_{CP})}{\Gamma(\overline{B}_d^0 \to f_{CP}) + \Gamma(B_d^0 \to f_{CP})}$$
$$= -\xi_f \sin 2\phi_1 \sin \Delta m_d t,$$

where $\Gamma[\overline{B}_d^0(B_d^0) \to f_{CP}]$ is the decay rate for a $\overline{B}_d^0$ (B_d^0) to f_{CP} at a proper time t after production, ξ_f is the *CP* eigenvalue of f_{CP}, Δm_d is the mass difference between the two B_d^0 mass eigenstates, and ϕ_1 is one of the three internal angles of the CKM unitarity triangle, defined as $\phi_1 \equiv \pi - \arg(\frac{-V_{tb}^* V_{td}}{-V_{cb}^* V_{cd}})$ [3].

In this Letter, we report a measurement of $\sin 2\phi_1$ using $B_d^0 \overline{B}_d^0$ meson pairs produced at the $\Upsilon(4S)$ resonance, where the two mesons remain in a coherent p-wave state until one of them decays. The decay of one of the B mesons to a self-tagging state, f_{tag}, i.e., a final state that distinguishes between B_d^0 and $\overline{B}_d^0$, at time t_{tag} projects the accompanying meson onto the opposite b-flavor at that time; this meson decays to f_{CP} at time t_{CP}. The *CP* violation manifests itself as an asymmetry $A(\Delta t)$, where Δt is the proper time interval $\Delta t \equiv t_{CP} - t_{\text{tag}}$.

The data sample corresponds to an integrated luminosity of 10.5 fb^{-1} collected with the Belle detector [4] at the KEKB asymmetric e^+e^- (3.5 on 8 GeV) collider [5]. At KEKB, the $\Upsilon(4S)$ is produced with a Lorentz boost of $\beta\gamma = 0.425$ along the electron beam direction (z direction). Because the B_d^0 and $\overline{B}_d^0$ mesons are nearly at rest in the $\Upsilon(4S)$ center of mass system (cms), Δt can be determined from the z distance between the f_{CP} and f_{tag} decay vertices, $\Delta z \equiv z_{CP} - z_{\text{tag}}$, as $\Delta t \simeq \Delta z/\beta\gamma c$.

The Belle detector consists of a 3-layer silicon vertex detector (SVD), a 50-layer central drift chamber (CDC), an array of 1188 aerogel Čerenkov counters (ACC), 128 time-of-flight (TOF) scintillation counters, and an electromagnetic calorimeter containing 8736 CsI(Tl) crystals (ECL) all located inside a 3.4-m-diameter superconducting solenoid that generates a 1.5 T magnetic field. The transverse momentum resolution for charged tracks is $(\sigma_{p_t}/p_t)^2 = (0.0019 p_t)^2 + (0.0034)^2$, where p_t is in GeV/c, and the impact parameter resolutions for $p = 1$ GeV/c tracks at normal incidence are $\sigma_{r\phi} \simeq \sigma_z \simeq 55$ μm. Specific ionization (dE/dx) measurements in the CDC ($\sigma_{dE/dx} = 6.9\%$ for minimum ionizing pions), TOF flight-time measurements ($\sigma_{TOF} = 95$ ps), and the response of the ACC provide $K^\pm$ identification with an efficiency of $\sim 85\%$ and a charged pion fake rate of $\sim 10\%$ for all momenta up to 3.5 GeV/c. Photons are identified as ECL showers that have a minimum energy of 20 MeV and are not matched to a charged track. The photon energy resolution is $(\sigma_E/E)^2 = (0.013)^2 + (0.0007/E)^2 + (0.008/E^{1/4})^2$, where E is in GeV. Electron identification is based on a combination of CDC dE/dx information, the ACC response, and the position relative to the extrapolated track, shape, and energy deposit of the associated ECL shower. The efficiency is greater than 90% and the hadron fake rate is $\sim 0.3\%$ for $p > 1$ GeV/c. An iron flux-return yoke outside the solenoid, comprised of 14 layers of 4.7-cm-thick iron plates interleaved with a system of resistive plate counters (KLM), provides muon identification with an efficiency greater than 90% and a hadron fake rate less than 2% for $p > 1$ GeV/c. The KLM is used in conjunction with the ECL to detect K_L mesons; the angular resolution of the K_L direction measurement ranges between 1.5° and 3°.

We reconstruct B_d^0 decays to the following CP eigenstates: $J/\psi K_S$, $\psi(2S) K_S$, $\chi_{c1} K_S$, $\eta_c K_S$ for $\xi_f = -1$ and $J/\psi \pi^0$, $J/\psi K_L$ for $\xi_f = +1$. The J/ψ and $\psi(2S)$ mesons are reconstructed via their decays to $\ell^+\ell^-$ ($\ell = \mu, e$). The $\psi(2S)$ is also reconstructed via its $J/\psi \pi^+ \pi^-$ decay, the χ_{c1} via its $J/\psi \gamma$ decay, and the η_c via its $K^+ K^- \pi^0$ and $K_S(\pi^+\pi^-) K^-\pi^+$ [6] decays.

For J/ψ and $\psi(2S) \to \ell^+\ell^-$ decays, we use oppositely charged track pairs, where both tracks are positively identified as leptons. For the $B_d^0 \to J/\psi K_S(\pi^+\pi^-)$ mode, the requirement for one of the tracks is relaxed: a track with an ECL energy deposit consistent with a minimum ionizing particle is accepted as a muon and a track that satisfies either the dE/dx or the ECL shower energy requirements as an electron. For e^+e^- pairs, we include the four-momentum of every photon detected within 0.05 rad of the original e^+ or e^- direction in the invariant mass calculation. Nevertheless a radiative tail remains and we accept pairs in the asymmetric invariant mass interval between -12.5σ and $+3\sigma$ of $M_{J/\psi}$ or $M_{\psi(2S)}$, where $\sigma = 12$ MeV/c^2 is the mass resolution. The $\mu^+\mu^-$ radiative tail is smaller; we select pairs within -5σ and $+3\sigma$ of $M_{J/\psi}$ or $M_{\psi(2S)}$. Candidate $K_S \to \pi^+\pi^-$ decays are oppositely charged track pairs that have an invariant mass within $\pm 4\sigma$ of the K^0 mass ($\sigma \simeq 4$ MeV/c^2). For the $J/\psi K_S$ final state, $K_S \to \pi^0 \pi^0$ decays are also used. For $\pi^0\pi^0$ candidates, we try all combinations where there

are two $\gamma\gamma$ pairs with an invariant mass between 80 and 150 MeV/c^2, assuming they originate from the center of the run-dependent average interaction point (IP). We minimize the sum of the χ^2 values from constrained fits of each pair to the π^0 mass with γ directions determined by varying the decay point along the K_S flight path, which is taken as the line from the IP to the energy-weighted center of the four showers. We select combinations with a $\pi^0 \pi^0$ invariant mass within $\sim \pm 3\sigma$ of M_{K^0}, where $\sigma \simeq 9.3$ MeV/c^2. For the $J/\psi \pi^0$ mode, we use a minimum γ energy of 100 MeV and select $\gamma\gamma$ pairs with an invariant mass within $\pm 3\sigma$ of M_{π^0}, where $\sigma \simeq 4.9$ MeV/c^2.

We isolate reconstructed B-meson decays using the energy difference $\Delta E \equiv E_B^{cms} - E_{beam}^{cms}$ and the beam-energy constrained mass $M_{bc} \equiv \sqrt{(E_{beam}^{cms})^2 - (p_B^{cms})^2}$, where E_{beam}^{cms} is the cms beam energy, and E_B^{cms} and p_B^{cms} are the cms energy and momentum of the B candidate. Figure 1 shows the M_{bc} distribution for all channels combined (other than $J/\psi K_L$) after a ΔE selection that varies from ± 25 to ± 100 MeV (corresponding to $\sim \pm 3\sigma$), depending on the mode. The B-meson signal region is defined as $5.270 < M_{bc} < 5.290$ GeV/c^2; the M_{bc} resolution is 3.0 MeV/c^2. Table I lists the numbers of observed events (N_{ev}) and the background (N_{bkgd}) determined by extrapolating the event rate in the nonsignal ΔE vs M_{bc} region into the signal region.

Candidate $B_d^0 \to J/\psi K_L$ decays are selected by requiring the observed K_L direction to be within 45° from the direction expected for a two-body decay (ignoring the B_d^0 cms motion). We reduce the background by means of a likelihood quantity that depends on the J/ψ cms momentum, the angle between the K_L and its nearest-neighbor charged track, the charged track multiplicity, and the kinematics that are obtained when the event is reconstructed assuming a $B^+ \to J/\psi K^{*+}(K_L \pi^+)$ hypothesis. In addition, we remove events that are reconstructed as $B_d^0 \to J/\psi K_S$, $J/\psi K^{*0}(K^+\pi^-, K_S\pi^0)$, $B^+ \to J/\psi K^+$, or $J/\psi K^{*+}(K^+\pi^0, K_S\pi^+)$ decays. Figure 2 shows the p_B^{cms} distribution, calculated for a $B_d^0 \to J/\psi K_L$ two-body

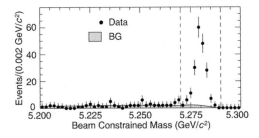

FIG. 1. The beam-constrained mass distribution for all decay modes combined (other than $B_d^0 \to J/\psi K_L$). The shaded area is the estimated background. The dashed lines indicate the signal region.

TABLE I. The numbers of CP-eigenstate events.

Mode	N_{ev}	N_{bkgd}
$J/\psi(\ell^+\ell^-)K_S(\pi^+\pi^-)$	123	3.7
$J/\psi(\ell^+\ell^-)K_S(\pi^0\pi^0)$	19	2.5
$\psi(2S)(\ell^+\ell^-)K_S(\pi^+\pi^-)$	13	0.3
$\psi(2S)(J/\psi\pi^+\pi^-)K_S(\pi^+\pi^-)$	11	0.3
$\chi_{c1}(\gamma J/\psi)K_S(\pi^+\pi^-)$	3	0.5
$\eta_c(K^+K^-\pi^0)K_S(\pi^+\pi^-)$	10	2.4
$\eta_c(K_S K^+\pi^-)K_S(\pi^+\pi^-)$	5	0.4
$J/\psi(\ell^+\ell^-)\pi^0$	10	0.9
Sub-total	194	11
$J/\psi(\ell^+\ell^-)K_L$	131	54

decay hypothesis, for the surviving events. The histograms in the figure are the results of a fit to the signal and background distributions, where the shapes are derived from Monte Carlo (MC) simulations [7], and the normalizations are allowed to vary. Among the total of 131 entries in the $0.2 \leq p_B^{cms} \leq 0.45$ GeV/c signal region, the fit finds 77 $J/\psi K_L$ events.

The leptons and charged pions and kaons among the tracks which are not associated with f_{CP} are used to identify the flavor of the accompanying B meson. Tracks are selected in several categories that distinguish the b-flavor by the track's charge: high momentum leptons from $b \to c\ell^-\bar{\nu}$, lower momentum leptons from $c \to s\ell^+\nu$, charged kaons from $b \to c \to s$, high momentum pions from decays of the type $B_d^0 \to D^{(*)-}(\pi^+, \rho^+, a_1^+, \text{etc.})$, and slow pions from $D^{*-} \to \overline{D}^0\pi^-$. For each track in one of these categories, we use the MC to determine the relative probability that it originates from a B_d^0 or $\overline{B}_d^0$ as a function of its charge, cms momentum and polar angle, particle-identification probability, and other kinematic and event-shape quantities. We combine the results from the different track categories (taking into account correlations for the case of multiple inputs) to determine a b-flavor q, where $q = +1$ when f_{tag} is more likely to be a B_d^0 and -1 for a

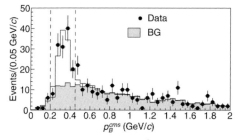

FIG. 2. The p_B^{cms} distribution for $B_d^0 \to J/\psi K_L$ candidates with the results of the fit. The solid line is the signal plus background; the shaded area is background only; the dashed lines indicate the signal region.

$\overline{B}_d^0$. We use the MC to evaluate an event-by-event flavor-tagging dilution factor, r, which ranges from $r = 0$ for no flavor discrimination to $r = 1$ for perfect flavor assignment. We use r only to categorize the event. For the CP asymmetry analysis, we use the data to correct for wrong-flavor assignments.

The probabilities for an incorrect flavor assignment, w_l ($l = 1, 6$), are measured directly from the data for six r intervals using a sample of exclusively reconstructed, self-tagged $B_d^0 \to D^{*-}\ell^+\nu$, $D^{(*)-}\pi^+$, and $D^{*-}\rho^+$ decays. The b-flavor of the accompanying B meson is assigned according to the above-described flavor-tagging algorithm, and values of w_l are determined from the amplitudes of the time-dependent B_d^0-$\overline{B}_d^0$ mixing oscillations [8]: $(N_{OF} - N_{SF})/(N_{OF} + N_{SF}) = (1 - 2w_l)\cos(\Delta m_d \Delta t)$. Here N_{OF} and N_{SF} are the numbers of opposite and same flavor events. Table II lists the resulting w_l values together with the fraction of the events (f_l) in each r interval. All events in Table I fall into one of the six r intervals. The total effective tagging efficiency is $\sum_l f_l (1 - 2w_l)^2 = 0.270^{+0.021}_{-0.022}$, where the error includes both statistical and systematic uncertainties, in good agreement with the MC result of 0.274. We check for a possible bias in the flavor tagging by measuring the effective tagging efficiency for B_d^0 and $\overline{B}_d^0$ self-tagged samples separately, and for different Δt intervals. We find no statistically significant difference.

The vertex positions for the f_{CP} and f_{tag} decays are reconstructed using tracks that have at least one three-dimensional coordinate determined from associated $r\phi$ and z hits in the same SVD layer plus one or more additional z hits in other SVD layers. Each vertex position is required to be consistent with the IP profile smeared in the $r\phi$ plane by the B-meson decay length. (The IP size, determined run-by-run, is typically $\sigma_x \approx 100$ μm, $\sigma_y \approx 5$ μm, and $\sigma_z \approx 3$ mm.) The f_{CP} vertex is determined by using lepton tracks from the J/ψ or $\psi(2S)$ decays, or prompt tracks from η_c decays. The f_{tag} vertex is determined from tracks not assigned to f_{CP} with additional requirements of $\delta r < 0.5$ mm, $\delta z < 1.8$ mm, and $\sigma_{\delta z} < 0.5$ mm, where δr and δz are the distances of the closest approach to the f_{CP} vertex in the $r\phi$ plane and the z direction, respectively, and $\sigma_{\delta z}$ is the calculated error of δz. Tracks that form a K_S are removed. The MC indicates that the average z_{CP} resolution is 75 μm (rms); the z_{tag}

TABLE II. Experimentally determined event fractions (f_l) and incorrect flavor assignment probabilities (w_l) for each r interval.

l	r	f_l	w_l
1	0.000–0.250	0.393 ± 0.014	$0.470^{+0.031}_{-0.035}$
2	0.250–0.500	0.154 ± 0.007	$0.336^{+0.039}_{-0.042}$
3	0.500–0.625	0.092 ± 0.005	$0.286^{+0.037}_{-0.035}$
4	0.625–0.750	0.100 ± 0.005	$0.210^{+0.033}_{-0.031}$
5	0.750–0.875	0.121 ± 0.006	$0.098^{+0.028}_{-0.026}$
6	0.875–1.000	0.134 ± 0.006	$0.020^{+0.023}_{-0.019}$

resolution is worse (140 μm) because of the lower average momentum of the f_{tag} decay products and the smearing caused by secondary tracks from charmed meson decays.

The resolution function $R(\Delta t)$ for the proper time interval is parametrized as a sum of two Gaussian components: a *main* component due to the SVD vertex resolution, charmed meson lifetimes, and the effect of the cms motion of the B mesons, plus a *tail* component caused by poorly reconstructed tracks. The means (μ_{main}, μ_{tail}) and widths (σ_{main}, σ_{tail}) of the Gaussians are calculated event-by-event from the f_{CP} and f_{tag} vertex fit error matrices; average values are $\mu_{\text{main}} = -0.09$ ps, $\mu_{\text{tail}} = -0.78$ ps and $\sigma_{\text{main}} = 1.54$ ps, $\sigma_{\text{tail}} = 3.78$ ps. The negative values of the means are due to secondary tracks from charmed mesons. The relative fraction of the main Gaussian is determined to be 0.982 ± 0.013 from a study of $B_d^0 \to D^{*-}\ell^+\nu$ events. The reliability of the Δt determination and $R(\Delta t)$ parametrization is confirmed by lifetime measurements of the neutral and charged B mesons [9] which use the same procedures and are in good agreement with the world average values [10].

We determine $\sin 2\phi_1$ from an unbinned maximum-likelihood fit to the observed Δt distributions. The probability density function (pdf) expected for the signal distribution is given by

$$\mathcal{P}_{\text{sig}}(\Delta t, q, w_l, \xi_f) = \frac{e^{-|\Delta t|/\tau_{B_d^0}}}{2\tau_{B_d^0}} \{1 - \xi_f q (1 - 2w_l) \times \sin 2\phi_1 \sin(\Delta m_d \Delta t)\},$$

where we fix the B_d^0 lifetime and mass difference at their world average values [10]. The pdf used for background events is $\mathcal{P}_{\text{bkg}}(\Delta t) = f_\tau e^{-|\Delta t|/\tau_{\text{bkg}}}/2\tau_{\text{bkg}} + (1 - f_\tau)\delta(\Delta t)$, where f_τ is the fraction of the background component with an effective lifetime τ_{bkg} and $\delta(\Delta t)$ is the Dirac delta function. For all f_{CP} modes, except $J/\psi K_L$, we find $f_\tau = 0.10^{+0.11}_{-0.05}$ and $\tau_{\text{bkg}} = 1.75^{+1.15}_{-0.82}$ ps using events in background-dominated regions of ΔE vs M_{bc}. The $J/\psi K_L$ background is dominated by $B \to J/\psi X$ decays, where some final states are CP eigenstates and need special treatment. A MC study shows that the background contribution from the $\xi_f = -1$ sources $J/\psi K_S$, $\psi(2S)K_S$, and $\chi_{c1}K_S$ is 7.9%, while that from the $\xi_f = +1 \psi(2S)K_L$ and $\chi_{c1}K_L$ modes is 7.0%. Thus, the effects on the CP asymmetry from these states nearly cancel. The remaining dominant CP mode, $J/\psi K^*(K_L\pi^0)$, which accounts for 19% of the total background, is taken to be a $73/27$ mixture of $\xi_f = -1$ and $+1$, respectively, based on our measurement of the J/ψ polarization in the $B_d^0 \to J/\psi K^{*0}(K_S\pi^0)$ decay [11]. For the $J/\psi K^*(K_L\pi^0)$ background pdf, we use $\mathcal{P}_{\text{sig}}$ with effective CP eigenvalue $\xi_f = -0.46^{+1.46}_{-0.54}$, where the error has been expanded to include all possible values. For the non-CP background modes we use $\mathcal{P}_{\text{bkg}}$ with $f_\tau = 1$ and $\tau_{\text{bkg}} = \tau_B$.

The pdfs are convolved with $R(\Delta t)$ to determine the likelihood value for each event as a function of $\sin 2\phi_1$:

$$\mathcal{L}_i = \int \{f_{\text{sig}} \mathcal{P}_{\text{sig}}(\Delta t', q, w_l, \xi_f) + (1 - f_{\text{sig}}) \mathcal{P}_{\text{bkg}}(\Delta t')\} \times R(\Delta t - \Delta t') d\Delta t',$$

where f_{sig} is the probability that the event is signal, calculated as a function of p_B^{cms} for $J/\psi K_L$ and of ΔE and M_{bc} for other modes. The most probable $\sin 2\phi_1$ is the value that maximizes the likelihood function $L = \prod_i \mathcal{L}_i$, where the product is over all events. We performed a blind analysis: the fitting algorithms were developed and finalized using a flavor-tagging routine that does not divulge the sign of q. The sign of q was then turned on, and the application of the fit to all of the events listed in Table I produces the result $\sin 2\phi_1 = 0.58^{+0.32+0.09}_{-0.34-0.10}$, where the first error is statistical and the second is systematic. The systematic errors are dominated by the uncertainties in w_l ($^{+0.05}_{-0.07}$) and the $J/\psi K_L$ background (± 0.05). Separate fits to the $\xi_f = -1$ and $\xi_f = +1$ event samples give $0.82^{+0.36}_{-0.41}$ and $0.10^{+0.57}_{-0.60}$, respectively [12]. Figure 3(a) shows $-2\ln(L/L_{\text{max}})$ as a function of $\sin 2\phi_1$ for the $\xi_f = -1$ and $\xi_f = +1$ modes separately and for both modes combined. Figure 3(b) shows the asymmetry obtained by performing the fit to events in Δt bins separately, together with a curve that represents $\sin 2\phi_1 \sin(\Delta m_d \Delta t)$ for $\sin 2\phi_1 = 0.58$.

We check for a possible fit bias by applying the same fit to non-CP eigenstate modes: $B_d^0 \to D^{(*)-}\pi^+$, $D^{*-}\rho^+$,

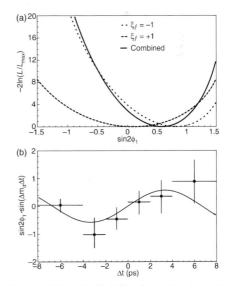

FIG. 3. (a) Values of $-2\ln(L/L_{\text{max}})$ vs $\sin 2\phi_1$ for the $\xi_f = -1$ and $+1$ modes separately and for both modes combined. (b) The asymmetry obtained from separate fits to each Δt bin; the curve is the result of the global fit ($\sin 2\phi_1 = 0.58$).

$J/\psi K^{*0}(K^+\pi^-)$, and $D^{*-}\ell^+\nu$, where "$\sin2\phi_1$" should be zero, and the charged mode $B^+ \to J/\psi K^+$. For all of the modes combined we find 0.065 ± 0.075, consistent with a null asymmetry.

We have presented a measurement of the standard model CP violation parameter $\sin2\phi_1$ based on a 10.5 fb^{-1} data sample collected at the $\Upsilon(4S)$:

$$\sin2\phi_1 = 0.58^{+0.32}_{-0.34}(\text{stat})^{+0.09}_{-0.10}(\text{syst}).$$

The probability of observing $\sin2\phi_1 > 0.58$, if the true value is zero, is 4.9%. Our measurement is more precise than the previous measurements [13] and consistent with SM constraints [14].

We wish to thank the KEKB Accelerator Group for the excellent operation. We acknowledge support from the Ministry of Education, Culture, Sports, Science and Technology of Japan and the Japan Society for the Promotion of Science; the Australian Research Council and the Australian Department of Industry, Science and Resources; the Department of Science and Technology of India; the BK21 program of the Ministry of Education of Korea and the SRC program of the Korea Science and Engineering Foundation; the Polish State Committee for Scientific Research under Contract No. 2P03B 17017; the Ministry of Science and Technology of Russian Federation; the National Science Council and the Ministry of Education of Taiwan; the Japan-Taiwan Cooperative Program of the Interchange Association; and the U.S. Department of Energy.

[1] M. Kobayashi and T. Maskawa, Prog. Theor. Phys. **49**, 652 (1973).
[2] A. B. Carter and A. I. Sanda, Phys. Rev. D **23**, 1567 (1981); I. I. Bigi and A. I. Sanda, Nucl. Phys. **B193**, 85 (1981).
[3] H. Quinn and A. I. Sanda, Eur. Phys. J. C **15**, 626 (2000). (Some papers refer to this angle as β.)
[4] Belle Collaboration, K. Abe et al. (to be published), KEK Report No. 2000-4.
[5] KEKB B Factory Design Report No. 95-1, 1995 (unpublished).
[6] Throughout this Letter, when a mode is quoted the inclusion of the charge conjugate mode is implied.
[7] We use the QQ B-meson decay event generator developed by the CLEO Collaboration (http://www.lns.cornell.edu/public/CLEO/soft/QQ) and GEANT3 for the detector simulation; CERN Program Library Long Writeup W5013, CERN, 1993.
[8] Belle Collaboration, J. Suzuki, Proceedings of the 30th International Conference on High Energy Physics, Osaka, 2000 (to be published).
[9] Belle Collaboration, H. Tajima, Proceedings of the 30th International Conference on High Energy Physics, Osaka, 2000 (to be published).
[10] Particle Data Group, D. E. Groom et al., Eur. Phys. J. C **15**, 1 (2000).
[11] Belle Collaboration, S. Schrenk, Proceedings of the 30th International Conference on High Energy Physics, Osaka, 2000 (to be published). This result agrees within errors with those of CLEO Collaboration, C. P. Jessop et al. [Phys. Rev. Lett. **79**, 4533 (1997)] and CDF Collaboration, T. Affolder et al. [Phys. Rev. Lett. **85**, 4668 (2000)].
[12] A fit to only the $B_d^0 \to J/\psi K_S(\pi^+\pi^-)$ events gives a $\sin2\phi_1$ value of $1.21^{+0.40}_{-0.47}$; a fit to only the non-$J/\psi K_S$ $\xi = -1$ modes gives $-0.05^{+0.76}_{-0.74}$. Separate fits to the $q = +1$ and $q = -1$ event samples give $\sin2\phi_1$ values of $0.40^{+0.47}_{-0.49}$ and $0.73^{+0.41}_{-0.46}$, respectively.
[13] OPAL Collaboration, K. Ackerstaff et al., Eur. Phys. J. C **5**, 379 (1998); CDF Collaboration, T. Affolder et al., Phys. Rev. D **61**, 072005 (2000); ALEPH Collaboration, R. Barate et al., Phys. Lett. B **492**, 259 (2000).
[14] See, for example, S. Mele, Phys. Rev. D **59**, 113011 (1999).

Measurement of *CP*-Violating Asymmetries in B^0 Decays to *CP* Eigenstates

B. Aubert,[1] D. Boutigny,[1] I. De Bonis,[1] J.-M. Gaillard,[1] A. Jeremie,[1] Y. Karyotakis,[1] J. P. Lees,[1] P. Robbe,[1] V. Tisserand,[1] A. Palano,[2] G. P. Chen,[3] J. C. Chen,[3] N. D. Qi,[3] G. Rong,[3] P. Wang,[3] Y. S. Zhu,[3] G. Eigen,[4] P. L. Reinertsen,[4] B. Stugu,[4] B. Abbott,[5] G. S. Abrams,[5] A. W. Borgland,[5] A. B. Breon,[5] D. N. Brown,[5] J. Button-Shafer,[5] R. N. Cahn,[5] A. R. Clark,[5] S. Dardin,[5] C. Day,[5] S. F. Dow,[5] T. Elioff,[5] Q. Fan,[5] I. Gaponenko,[5] M. S. Gill,[5] F. R. Goozen,[5] S. J. Gowdy,[5] A. Gritsan,[5] Y. Groysman,[5] R. G. Jacobsen,[5] R. C. Jared,[5] R. W. Kadel,[5] J. Kadyk,[5] A. Karcher,[5] L. T. Kerth,[5] I. Kipnis,[5] S. Kluth,[5] Yu. G. Kolomensky,[5] J. F. Kral,[5] R. Lafever,[5] C. LeClerc,[5] M. E. Levi,[5] S. A. Lewis,[5] C. Lionberger,[5] T. Liu,[5] M. Long,[5] G. Lynch,[5] M. Marino,[5] K. Marks,[5] A. B. Meyer,[5] A. Mokhtarani,[5] M. Momayezi,[5] M. Nyman,[5] P. J. Oddone,[5] J. Ohnemus,[5] D. Oshatz,[5] S. Patton,[5] A. Perazzo,[5] C. Peters,[5] W. Pope,[5] M. Pripstein,[5] D. R. Quarrie,[5] J. E. Rasson,[5] N. A. Roe,[5] A. Romosan,[5] M. T. Ronan,[5] V. G. Shelkov,[5] R. Stone,[5] A. V. Telnov,[5] H. von der Lippe,[5] T. Weber,[5] W. A. Wenzel,[5] M. S. Zisman,[5] P. G. Bright-Thomas,[6] T. J. Harrison,[6] C. M. Hawkes,[6] A. Kirk,[6] D. J. Knowles,[6] S. W. O'Neale,[6] A. T. Watson,[6] N. K. Watson,[6] T. Deppermann,[7] H. Koch,[7] J. Krug,[7] M. Kunze,[7] B. Lewandowski,[7] K. Peters,[7] H. Schmuecker,[7] M. Steinke,[7] J. C. Andress,[8] N. R. Barlow,[8] W. Bhimji,[8] N. Chevalier,[8] P. J. Clark,[8] W. N. Cottingham,[8] N. De Groot,[8] N. Dyce,[8] B. Foster,[8] A. Mass,[8] J. D. McFall,[8] D. Wallom,[8] F. F. Wilson,[8] K. Abe,[9] C. Hearty,[9] T. S. Mattison,[9] J. A. McKenna,[9] D. Thiessen,[9] B. Camanzi,[10] S. Jolly,[10] A. K. McKemey,[10] J. Tinslay,[10] V. E. Blinov,[11] A. D. Bukin,[11] D. A. Bukin,[11] A. R. Buzykaev,[11] M. S. Dubrovin,[11] V. B. Golubev,[11] V. N. Ivanchenko,[11] G. M. Kolachev,[11] A. A. Korol,[11] E. A. Kravchenko,[11] A. P. Onuchin,[11] A. A. Salnikov,[11] S. I. Serednyakov,[11] Yu. I. Skovpen,[11] V. I. Telnov,[11] A. N. Yushkov,[11] A. J. Lankford,[12] M. Mandelkern,[12] S. McMahon,[12] D. P. Stoker,[12] A. Ahsan,[13] C. Buchanan,[13] S. Chun,[13] D. B. MacFarlane,[14] S. Prell,[14] Sh. Rahatlou,[14] G. Raven,[14] V. Sharma,[14] S. Burke,[15] C. Campagnari,[15] B. Dahmes,[15] D. Hale,[15] P. A. Hart,[15] N. Kuznetsova,[15] S. Kyre,[15] S. L. Levy,[15] O. Long,[15] A. Lu,[15] J. D. Richman,[15] W. Verkerke,[15] M. Witherell,[15] S. Yellin,[15] J. Beringer,[16] D. E. Dorfan,[16] A. M. Eisner,[16] A. Frey,[16] A. A. Grillo,[16] M. Grothe,[16] C. A. Heusch,[16] R. P. Johnson,[16] W. Kroeger,[16] W. S. Lockman,[16] T. Pulliam,[16] H. Sadrozinski,[16] T. Schalk,[16] R. E. Schmitz,[16] B. A. Schumm,[16] A. Seiden,[16] E. N. Spencer,[16] M. Turri,[16] W. Walkowiak,[16] D. C. Williams,[16] E. Chen,[17] G. P. Dubois-Felsmann,[17] A. Dvoretskii,[17] J. E. Hanson,[17] D. G. Hitlin,[17] S. Metzler,[17] J. Oyang,[17] F. C. Porter,[17] A. Ryd,[17] A. Samuel,[17] M. Weaver,[17] S. Yang,[17] R. Y. Zhu,[17] S. Devmal,[18] T. L. Geld,[18] S. Jayatilleke,[18] S. M. Jayatilleke,[18] G. Mancinelli,[18] B. T. Meadows,[18] M. D. Sokoloff,[18] P. Bloom,[19] S. Fahey,[19] W. T. Ford,[19] F. Gaede,[19] W. C. van Hoek,[19] D. R. Johnson,[19] A. K. Michael,[19] U. Nauenberg,[19] A. Olivas,[19] H. Park,[19] P. Rankin,[19] J. Roy,[19] S. Sen,[19] J. G. Smith,[19] D. L. Wagner,[19] J. Blouw,[20] J. L. Harton,[20] M. Krishnamurthy,[20] A. Soffer,[20] W. H. Toki,[20] D. W. Warner,[20] R. J. Wilson,[20] J. Zhang,[20] T. Brandt,[21] J. Brose,[21] T. Colberg,[21] G. Dahlinger,[21] M. Dickopp,[21] R. S. Dubitzky,[21] P. Eckstein,[21] H. Futterschneider,[21] R. Krause,[21] E. Maly,[21] R. Müller-Pfefferkorn,[21] S. Otto,[21] K. R. Schubert,[21] R. Schwierz,[21] B. Spaan,[21] L. Wilden,[21] L. Behr,[22] D. Bernard,[22] G. R. Bonneaud,[22] F. Brochard,[22] J. Cohen-Tanugi,[22] S. Ferrag,[22] G. Fouque,[22] F. Gastaldi,[22] P. Matricon,[22] P. Mora de Freitas,[22] C. Renard,[22] E. Roussot,[22] S. T'Jampens,[22] C. Thiebaux,[22] G. Vasileiadis,[22] M. Verderi,[22] A. Anjomshoaa,[23] R. Bernet,[23] F. Di Lodovico,[23] A. Khan,[23] F. Muheim,[23] S. Playfer,[23] J. E. Swain,[23] M. Falbo,[24] C. Bozzi,[25] S. Dittongo,[25] M. Folegani,[25] L. Piemontese,[25] E. Treadwell,[26] F. Anulli,[27,*] R. Baldini-Ferroli,[27] A. Calcaterra,[27] R. de Sangro,[27] D. Falciai,[27] G. Finocchiaro,[27] P. Patteri,[27] I. M. Peruzzi,[27,*] M. Piccolo,[27] Y. Xie,[27] A. Zallo,[27] S. Bagnasco,[28] A. Buzzo,[28] R. Contri,[28] G. Crosetti,[28] M. Lo Vetere,[28] M. Macri,[28] M. R. Monge,[28] M. Pallavicini,[28] S. Passaggio,[28] F. C. Pastore,[28] C. Patrignani,[28] M. G. Pia,[28] E. Robutti,[28] A. Santroni,[28] M. Morii,[29] R. Bartoldus,[30] T. Dignan,[30] R. Hamilton,[30] U. Mallik,[30] J. Cochran,[31] H. B. Crawley,[31] P. A. Fischer,[31] J. Lamsa,[31] R. McKay,[31] W. T. Meyer,[31] E. I. Rosenberg,[31] J. N. Albert,[32] C. Beigbeder,[32] M. Benkebil,[32] D. Breton,[32] R. Cizeron,[32] S. Du,[32] G. Grosdidier,[32] C. Hast,[32] A. Höcker,[32] V. LePeltier,[32] A. M. Lutz,[32] S. Plaszczynski,[32] M. H. Schune,[32] S. Trincaz-Duvoid,[32] K. Truong,[32] A. Valassi,[32] G. Wormser,[32] R. M. Bionta,[33] V. Brigljević,[33] A. Brooks,[33] O. Fackler,[33] D. Fujino,[33] D. J. Lange,[33] M. Mugge,[33] T. G. O'Connor,[33] B. Pedrotti,[33] X. Shi,[33] K. van Bibber,[33] T. J. Wenaus,[33] D. M. Wright,[33] C. R. Wuest,[33] B. Yamamoto,[33] M. Carroll,[34] J. R. Fry,[34] E. Gabathuler,[34] R. Gamet,[34] M. George,[34] M. Kay,[34] D. J. Payne,[34] R. J. Sloane,[34] C. Touramanis,[34] M. L. Aspinwall,[35] D. A. Bowerman,[35] P. D. Dauncey,[35] U. Egede,[35] I. Eschrich,[35] N. J. W. Gunawardane,[35] R. Martin,[35] J. A. Nash,[35] D. R. Price,[35] P. Sanders,[35] D. Smith,[35] D. E. Azzopardi,[36] J. J. Back,[36] P. Dixon,[36] P. F. Harrison,[36] D. Newman-Coburn,[36] R. J. L. Potter,[36] H. W. Shorthouse,[36] P. Strother,[36] P. B. Vidal,[36] M. I. Williams,[36] G. Cowan,[37] S. George,[37] M. G. Green,[37] A. Kurup,[37] C. E. Marker,[37] P. McGrath,[37] T. R. McMahon,[37] F. Salvatore,[37] I. Scott,[37]

G. Vaitsas,[37] D. Brown,[38] C. L. Davis,[38] K. Ford,[38] Y. Li,[38] J. Pavlovich,[38] J. Allison,[39] R. J. Barlow,[39] J. T. Boyd,[39] J. Fullwood,[39] F. Jackson,[39] G. D. Lafferty,[39] N. Savvas,[39] E. T. Simopoulos,[39] R. J. Thompson,[39] J. H. Weatherall,[39] R. Bard,[40] A. Farbin,[40] A. Jawahery,[40] V. Lillard,[40] J. Olsen,[40] D. A. Roberts,[40] J. R. Schieck,[40] G. Blaylock,[41] C. Dallapiccola,[41] K. T. Flood,[41] S. S. Hertzbach,[41] R. Kofler,[41] C. S. Lin,[41] H. Staengle,[41] S. Willocq,[41] J. Wittlin,[41] B. Brau,[42] R. Cowan,[42] G. Sciolla,[42] F. Taylor,[42] R. K. Yamamoto,[42] D. I. Britton,[43] M. Milek,[43] P. M. Patel,[43] J. Trischuk,[43] F. Lanni,[44] F. Palombo,[44] J. M. Bauer,[45] M. Booke,[45] L. Cremaldi,[45] V. Eschenberg,[45] R. Kroeger,[45] M. Reep,[45] J. Reidy,[45] D. A. Sanders,[45] D. J. Summers,[45] M. Beaulieu,[46] J. P. Martin,[46] J. Y. Nief,[46] R. Seitz,[46] P. Taras,[46] V. Zacek,[46] H. Nicholson,[47] C. S. Sutton,[47] N. Cavallo,[48,†] C. Cartaro,[48] G. De Nardo,[48] F. Fabozzi,[48] C. Gatto,[48] L. Lista,[48] P. Paolucci,[48] D. Piccolo,[48] C. Sciacca,[48] J. M. LoSecco,[49] J. R. G. Alsmiller,[50] T. A. Gabriel,[50] T. Handler,[50] J. Heck,[50] J. E. Brau,[51] R. Frey,[51] M. Iwasaki,[51] N. B. Sinev,[51] D. Strom,[51] E. Borsato,[52] F. Colecchia,[52] F. Dal Corso,[52] F. Galeazzi,[52] M. Margoni,[52] M. Marzolla,[52] G. Michelon,[52] M. Morandin,[52] M. Posocco,[52] M. Rotondo,[52] F. Simonetto,[52] R. Stroili,[52] E. Torassa,[52] C. Voci,[52] P. Bailly,[53] M. Benayoun,[53] H. Briand,[53] J. Chauveau,[53] P. David,[53] C. De la Vaissière,[53] L. Del Buono,[53] J. F. Genat,[53] O. Hamon,[53] F. Le Diberder,[53] H. Lebbolo,[53] Ph. Leruste,[53] J. Lory,[53] L. Martin,[53] L. Roos,[53] J. Stark,[53] S. Versillé,[53] B. Zhang,[53] P. F. Manfredi,[54] L. Ratti,[54] V. Re,[54] V. Speziali,[54] E. D. Frank,[55] L. Gladney,[55] Q. H. Guo,[55] J. H. Panetta,[55] C. Angelini,[56] G. Batignani,[56] S. Bettarini,[56] M. Bondioli,[56] F. Bosi,[56] M. Carpinelli,[56] F. Forti,[56] M. A. Giorgi,[56] A. Lusiani,[56] F. Martinez-Vidal,[56] M. Morganti,[56] N. Neri,[56] E. Paoloni,[56] M. Rama,[56] G. Rizzo,[56] F. Sandrelli,[56] G. Simi,[56] G. Triggiani,[56] J. Walsh,[56] M. Haire,[57] D. Judd,[57] K. Paick,[57] L. Turnbull,[57] D. E. Wagoner,[57] J. Albert,[58] C. Bula,[58] R. Fernholz,[58] C. Lu,[58] K. T. McDonald,[58] V. Miftakov,[58] B. Sands,[58] S. F. Schaffner,[58] A. J. S. Smith,[58] A. Tumanov,[58] E. W. Varnes,[58] F. Bronzini,[59] A. Buccheri,[59] C. Bulfon,[59] G. Cavoto,[59] D. del Re,[59] R. Faccini,[14,59] F. Ferrarotto,[59] F. Ferroni,[59] K. Fratini,[59] E. Lamanna,[59] E. Leonardi,[59] M. A. Mazzoni,[59] S. Morganti,[59] G. Piredda,[59] F. Safai Tehrani,[59] M. Serra,[59] C. Voena,[59] R. Waldi,[60] P. F. Jacques,[61] M. Kalelkar,[61] R. J. Plano,[61] T. Adye,[62] B. Claxton,[62] B. Franek,[62] S. Galagedera,[62] N. I. Geddes,[62] G. P. Gopal,[62] J. Lidbury,[62] S. M. Xella,[62] R. Aleksan,[63] P. Besson,[63,‡] P. Bourgeois,[63] G. De Domenico,[63] S. Emery,[63] A. Gaidot,[63] S. F. Ganzhur,[63] L. Gosset,[63] G. Hamel de Monchenault,[63] W. Kozanecki,[63] M. Langer,[63] G. W. London,[63] B. Mayer,[63] B. Serfass,[63] G. Vasseur,[63] C. Yeche,[63] M. Zito,[63] N. Copty,[64] M. V. Purohit,[64] H. Singh,[64] F. X. Yumiceva,[64] I. Adam,[65] P. L. Anthony,[65] D. Aston,[65] K. Baird,[65] J. Bartelt,[65] J. Becla,[65] R. Bell,[65] E. Bloom,[65] C. T. Boeheim,[65] A. M. Boyarski,[65] R. F. Boyce,[65] F. Bulos,[65] W. Burgess,[65] B. Byers,[65] G. Calderini,[65] R. Claus,[65] M. R. Convery,[65] R. Coombes,[65] L. Cottrell,[65] D. P. Coupal,[65] D. H. Coward,[65] W. W. Craddock,[65] H. DeStaebler,[65] J. Dorfan,[65] M. Doser,[65] W. Dunwoodie,[65] S. Ecklund,[65] T. H. Fieguth,[65] R. C. Field,[65] D. R. Freytag,[65] T. Glanzman,[65] G. L. Godfrey,[65] P. Grosso,[65] G. Haller,[65] A. Hanushevsky,[65] J. Harris,[65] A. Hasan,[65] J. L. Hewett,[65] T. Himel,[65] M. E. Huffer,[65] W. R. Innes,[65] C. P. Jessop,[65] H. Kawahara,[65] L. Keller,[65] M. H. Kelsey,[65] P. Kim,[65] L. A. Klaisner,[65] M. L. Kocian,[65] H. J. Krebs,[65] P. F. Kunz,[65] U. Langenegger,[65] W. Langeveld,[65] D. W. G. S. Leith,[65] S. K. Louie,[65] S. Luitz,[65] V. Luth,[65] H. L. Lynch,[65] J. MacDonald,[65] G. Manzin,[65] H. Marsiske,[65] M. McCulloch,[65] D. McShurley,[65] S. Menke,[65] R. Messner,[65] S. Metcalfe,[65] K. C. Moffeit,[65] R. Mount,[65] D. R. Muller,[65] D. Nelson,[65] M. Nordby,[65] C. P. O'Grady,[65] F. G. O'Neill,[65] G. Oxoby,[65] T. Pavel,[65] J. Perl,[65] S. Petrak,[65] G. Putallaz,[65] H. Quinn,[65] P. E. Raines,[65] B. N. Ratcliff,[65] R. Reif,[65] S. H. Robertson,[65] L. S. Rochester,[65] A. Roodman,[65] J. J. Russell,[65] L. Sapozhnikov,[65] O. H. Saxton,[65] T. Schietinger,[65] R. H. Schindler,[65] J. Schwiening,[65] J. T. Seeman,[65] V. V. Serbo,[65] K. Skarpass, Sr.,[65] A. Snyder,[65] A. Soha,[65] S. M. Spanier,[65] A. Stahl,[65] J. Stelzer,[65] D. Su,[65] M. K. Sullivan,[65] M. Talby,[65] H. A. Tanaka,[65] J. Va'vra,[65] S. R. Wagner,[65] A. J. R. Weinstein,[65] J. L. White,[65] U. Wienands,[65] W. J. Wisniewski,[65] C. C. Young,[65] G. Zioulas,[65] P. R. Burchat,[66] C. H. Cheng,[66] D. Kirkby,[66] T. I. Meyer,[66] C. Roat,[66] A. De Silva,[67] R. Henderson,[67] S. Berridge,[68] W. Bugg,[68] H. Cohn,[68] E. Hart,[68] A. W. Weidemann,[68] T. Benninger,[69] J. M. Izen,[69] I. Kitayama,[69] X. C. Lou,[69] M. Turcotte,[69] F. Bianchi,[70] M. Bona,[70] B. Di Girolamo,[70] D. Gamba,[70] A. Smol,[70] D. Zanin,[70] L. Bosisio,[71] G. Della Ricca,[71] L. Lanceri,[71] A. Pompili,[71] P. Poropat,[71] G. Vuagnin,[71] R. S. Panvini,[72] C. M. Brown,[73] R. Kowalewski,[73] J. M. Roney,[73] H. R. Band,[74] E. Charles,[74] S. Dasu,[74] P. Elmer,[74] H. Hu,[74] J. R. Johnson,[74] J. Nielsen,[74] W. Orejudos,[74] Y. Pan,[74] R. Prepost,[74] I. J. Scott,[74] J. H. von Wimmersperg-Toeller,[74] S. L. Wu,[74] Z. Yu,[74] H. Zobernig,[74] T. M. B. Kordich,[75] T. B. Moore,[75] and H. Neal[75]

(BABAR Collaboration)

[1]*Laboratoire de Physique des Particules, F-74941 Annecy-le-Vieux, France*
[2]*Università di Bari, Dipartimento di Fisica and INFN, I-70126 Bari, Italy*
[3]*Institute of High Energy Physics, Beijing 100039, China*
[4]*Institute of Physics, University of Bergen, N-5007 Bergen, Norway*
[5]*Lawrence Berkeley National Laboratory and University of California, Berkeley, California 94720*

⁶University of Birmingham, Birmingham B15 2TT, United Kingdom
⁷Ruhr Universität Bochum, Institut für Experimentalphysik 1, D-44780 Bochum, Germany
⁸University of Bristol, Bristol BS8 1TL, United Kingdom
⁹University of British Columbia, Vancouver, British Columbia, Canada V6T 1Z1
¹⁰Brunel University, Uxbridge, Middlesex UB8 3PH, United Kingdom
¹¹Budker Institute of Nuclear Physics, Novosibirsk 630090, Russia
¹²University of California at Irvine, Irvine, California 92697
¹³University of California at Los Angeles, Los Angeles, California 90024
¹⁴University of California at San Diego, La Jolla, California 92093
¹⁵University of California at Santa Barbara, Santa Barbara, California 93106
¹⁶Institute for Particle Physics, University of California at Santa Cruz, Santa Cruz, California 95064
¹⁷California Institute of Technology, Pasadena, California 91125
¹⁸University of Cincinnati, Cincinnati, Ohio 45221
¹⁹University of Colorado, Boulder, Colorado 80309
²⁰Colorado State University, Fort Collins, Colorado 80523
²¹Technische Universität Dresden, Institut für Kern-u. Teilchenphysik, D-01062 Dresden, Germany
²²Ecole Polytechnique, F-91128 Palaiseau, France
²³University of Edinburgh, Edinburgh EH9 3JZ, United Kingdom
²⁴Elon College, Elon College, North Carolina 27244-2010
²⁵Dipartimento di Fisica and INFN, Università di Ferrara, I-44100 Ferrara, Italy
²⁶Florida A&M University, Tallahassee, Florida 32307
²⁷Laboratori Nazionali di Frascati dell'INFN, I-00044 Frascati, Italy
²⁸Dipartimento di Fisica and INFN, Università di Genova, I-16146 Genova, Italy
²⁹Harvard University, Cambridge, Massachusetts 02138
³⁰University of Iowa, Iowa City, Iowa 52242-3160
³¹Iowa State University, Ames, Iowa 50011
³²Laboratoire de l'Accélérateur Linéaire, F-91898 Orsay, France
³³Lawrence Livermore National Laboratory, Livermore, California 94550
³⁴University of Liverpool, Liverpool L69 3BX, United Kingdom
³⁵University of London, Imperial College, London SW7 2BW, United Kingdom
³⁶Queen Mary, University of London, London E1 4NS, United Kingdom
³⁷University of London, Royal Holloway, and Bedford New College, Egham, Surrey TW20 0EX, United Kingdom
³⁸University of Louisville, Louisville, Kentucky 40292
³⁹University of Manchester, Manchester M13 9PL, United Kingdom
⁴⁰University of Maryland, College Park, Maryland 20742
⁴¹University of Massachusetts, Amherst, Massachusetts 01003
⁴²Lab for Nuclear Science, Massachusetts Institute of Technology, Cambridge, Massachusetts 02139
⁴³McGill University, Montréal, Canada QC H3A 2T8
⁴⁴Dipartimento di Fisica and INFN, Università di Milano, I-20133 Milano, Italy
⁴⁵University of Mississippi, University, Mississippi 38677
⁴⁶Laboratoire René J. A. Lévesque, Université de Montréal, Montréal, Canada QC H3C 3J7
⁴⁷Mount Holyoke College, South Hadley, Massachusetts 01075
⁴⁸Dipartimento di Scienze Fisiche and INFN, Università di Napoli Federico II, I-80126 Napoli, Italy
⁴⁹University of Notre Dame, Notre Dame, Indiana 46556
⁵⁰Oak Ridge National Laboratory, Oak Ridge, Tennessee 37831
⁵¹University of Oregon, Eugene, Oregon 97403
⁵²Dipartimento di Fisica and INFN, Università di Padova, I-35131 Padova, Italy
⁵³Lab de Physique Nucléaire H. E., Universités Paris VI et VII, F-75252 Paris, France
⁵⁴Dipartimento di Elettronica and INFN, Università di Pavia, I-27100 Pavia, Italy
⁵⁵University of Pennsylvania, Philadelphia, Pennsylvania 19104
⁵⁶Scuola Normale Superiore and INFN, Università di Pisa, I-56010 Pisa, Italy
⁵⁷Prairie View A&M University, Prairie View, Texas 77446
⁵⁸Princeton University, Princeton, New Jersey 08544
⁵⁹Dipartimento di Fisica and INFN, Università di Roma La Sapienza, I-00185 Roma, Italy
⁶⁰Universität Rostock, D-18051 Rostock, Germany
⁶¹Rutgers University, New Brunswick, New Jersey 08903
⁶²Rutherford Appleton Laboratory, Chilton, Didcot, Oxon OX11 0QX, United Kingdom
⁶³DAPNIA, Commissariat à l'Energie Atomique/Saclay, F-91191 Gif-sur-Yvette, France
⁶⁴University of South Carolina, Columbia, South Carolina 29208
⁶⁵Stanford Linear Accelerator Center, Stanford, California 94309
⁶⁶Stanford University, Stanford, California 94305-4060
⁶⁷TRIUMF, Vancouver, British Columbia, Canada V6T 2A3

[68] University of Tennessee, Knoxville, Tennessee 37996
[69] University of Texas at Dallas, Richardson, Texas 75083
[70] Dipartimento di Fisica Sperimentale and INFN, Università di Torino, I-10125 Torino, Italy
[71] Dipartimento di Fisica and INFN, Università di Trieste, I-34127 Trieste, Italy
[72] Vanderbilt University, Nashville, Tennessee 37235
[73] University of Victoria, Victoria, British Columbia, Canada V8W 3P6
[74] University of Wisconsin, Madison, Wisconsin 53706
[75] Yale University, New Haven, Connecticut 06511

(Received 12 February 2001)

We present measurements of time-dependent CP-violating asymmetries in neutral B decays to several CP eigenstates. The measurement uses a data sample of 23×10^6 $\Upsilon(4S) \to B\bar{B}$ decays collected by the BABAR detector at the PEP-II asymmetric B Factory at SLAC. In this sample, we find events in which one neutral B meson is fully reconstructed in a CP eigenstate containing charmonium and the flavor of the other neutral B meson is determined from its decay products. The amplitude of the CP-violating asymmetry, which in the standard model is proportional to $\sin 2\beta$, is derived from the decay time distributions in such events. The result is $\sin 2\beta = 0.34 \pm 0.20$ (stat) ± 0.05 (syst).

DOI: 10.1103/PhysRevLett.86.2515

PACS numbers: 13.25.Hw, 12.15.Hh, 11.30.Er

CP-violating asymmetries in the time distributions of decays of B^0 and $\bar{B}^0$ mesons provide a direct test of the standard model of electroweak interactions [1]. For the neutral B decay modes reported here, corrections to CP-violating effects from strong interactions are absent, in contrast to the K_L^0 modes in which CP violation was discovered [2].

Using a data sample of 23×10^6 $B\bar{B}$ pairs recorded at the $\Upsilon(4S)$ resonance by the BABAR detector at the PEP-II asymmetric-energy e^+e^- collider at the Stanford Linear Accelerator Center, we have fully reconstructed a sample B_{CP} of neutral B mesons decaying to the CP eigenstates $J/\psi K_S^0$, $\psi(2S) K_S^0$, and $J/\psi K_L^0$. We examine each of the events in this sample for evidence that the other neutral B meson decayed as a B^0 or a $\bar{B}^0$, designated as a B^0 or $\bar{B}^0$ flavor tag. The final B_{CP} sample contains about 360 signal events.

When the $\Upsilon(4S)$ decays, the P-wave $B\bar{B}$ state evolves coherently until one of the mesons decays. In one of four time-order and flavor configurations, if the tagging meson B_{tag} decays first, and as a B^0, the other meson must be a $\bar{B}^0$ at that same time t_{tag}. It then evolves independently and can decay into a CP eigenstate B_{CP} at a later time t_{CP}. The time between the two decays $\Delta t = t_{CP} - t_{\text{tag}}$ is a signed quantity made measurable by producing the $\Upsilon(4S)$ with a boost $\beta\gamma = 0.56$ along the collision (z) axis, with nominal energies of 9.0 and 3.1 GeV for the electron and positron beams. The measured distance $\Delta z \approx \beta\gamma c \Delta t$ between the two decay vertices provides a good estimate of the corresponding time interval Δt; the average value of $|\Delta z|$ is $\beta\gamma c\tau_{B^0} \approx 250$ μm.

The decay-time distribution for events with a B^0 or a $\bar{B}^0$ tag can be expressed in terms of a complex parameter λ that depends on both $B^0\bar{B}^0$ mixing and on the amplitudes describing $\bar{B}^0$ and B^0 decay to a common final state f [3]. The distribution $f_+(f_-)$ of the decay rate when the tagging meson is a $B^0(\bar{B}^0)$ is given by

$$f_\pm(\Delta t) = \frac{e^{-|\Delta t|/\tau_{B^0}}}{2\tau_{B^0}(1+|\lambda|^2)} \times \left[\frac{1+|\lambda|^2}{2} \pm \text{Im}\lambda \sin(\Delta m_{B^0}\Delta t) \mp \frac{1-|\lambda|^2}{2}\cos(\Delta m_{B^0}\Delta t)\right], \quad (1)$$

where τ_{B^0} is the B^0 lifetime and Δm_{B^0} is the mass difference determined from $B^0\bar{B}^0$ mixing [4], and where the lifetime difference between neutral B mass eigenstates is assumed to be negligible. The first oscillatory term in Eq. (1) is due to interference between direct decay and decay after mixing. A difference between the B^0 and $\bar{B}^0$ distributions or a Δt asymmetry for either tag is evidence for CP violation.

If all amplitudes contributing to $B^0 \to f$ have the same weak phase, a condition satisfied in the standard model for charmonium-containing $b \to c\bar{c}s$ decays, then $|\lambda| = 1$. For these CP eigenstates the standard model predicts $\lambda = \eta_f e^{-2i\beta}$, where η_f is the CP eigenvalue of the state f and $\beta = \arg[-V_{cd}V_{cb}^*/V_{td}V_{tb}^*]$ is an angle of the unitarity triangle of the three-generation Cabibbo-Kobayashi-Maskawa (CKM) matrix [5]. Thus, the time-dependent CP-violating asymmetry is

$$A_{CP}(\Delta t) \equiv \frac{f_+(\Delta t) - f_-(\Delta t)}{f_+(\Delta t) + f_-(\Delta t)}$$
$$= -\eta_f \sin 2\beta \sin(\Delta m_{B^0}\Delta t), \quad (2)$$

where $\eta_f = -1$ for $J/\psi K_S^0$ and $\psi(2S)K_S^0$ and $+1$ for $J/\psi K_L^0$.

A measurement of A_{CP} requires determination of the experimental Δt resolution and the fraction of events in which the tag assignment is incorrect. A mistag fraction w reduces the observed asymmetry by a factor $(1 - 2w)$.

Several samples of fully reconstructed B^0 mesons are used in this measurement. The B_{CP} sample contains candidates reconstructed in the CP eigenstates $J/\psi K_S^0(K_S^0 \to \pi^+\pi^-, \pi^0\pi^0)$, $\psi(2S)K_S^0(K_S^0 \to \pi^+\pi^-)$, and $J/\psi K_L^0$. The J/ψ and $\psi(2S)$ mesons are reconstructed through their decays to e^+e^- and $\mu^+\mu^-$; the $\psi(2S)$ is also reconstructed through its decay to $J/\psi\pi^+\pi^-$. A sample of B decays B_{flav} [6] used in the determination of the mistag fractions and Δt resolution functions consists of the channels $D^{(*)-}h^+(h^+ = \pi^+, \rho^+, a_1^+)$ and $J/\psi K^{*0}(K^{*0} \to K^+\pi^-)$. A control sample of charged B mesons decaying to the final states $J/\psi K^{(*)+}$, $\psi(2S)K^+$, and $\bar{D}^{(*)0}\pi^+$ is used for validation studies.

A description of the BABAR detector can be found in Ref. [7]. Charged particles are detected and their momenta measured by a combination of a silicon vertex tracker (SVT) consisting of five double-sided layers and a central drift chamber (DCH), in a 1.5-T solenoidal field. The average vertex resolution in the z direction is 70 μm for a fully reconstructed B meson. We identify leptons and hadrons with measurements from all detector systems, including the energy loss (dE/dx) in the DCH and SVT. Electrons and photons are identified by a CsI electromagnetic calorimeter (EMC). Muons are identified in the instrumented flux return (IFR). A Cherenkov ring imaging detector (DIRC) covering the central region, together with the dE/dx information, provides K-π separation of at least 3 standard deviations for B decay products with momentum greater than 250 MeV/c in the laboratory.

We select events with a minimum of three reconstructed charged tracks, each having a laboratory polar angle between 0.41 and 2.54 rad and an impact parameter in the plane transverse to the beam less than 1.5 cm from the beam line. The event must have a total measured energy in the laboratory greater than 4.5 GeV within the fiducial regions for charged tracks and neutral clusters. To help reject continuum background, the second Fox-Wolfram moment [8] must be less than 0.5.

An electron candidate must have a ratio of calorimeter energy to track momentum, an EMC cluster shape, a DCH dE/dx, and a DIRC Cherenkov angle (if available) consistent with an electron.

A muon candidate must satisfy requirements on the measured and expected number of interaction lengths penetrated, the position match between the extrapolated DCH track and IFR hits, and the average and spread of the number of IFR hits per layer.

A track is identified as a kaon candidate by means of a neural network that uses dE/dx measurements in the DCH and SVT, and comparison of the observed pattern of detected photons in the DIRC with that expected for kaon and pion hypotheses.

Candidates for $J/\psi \to \ell^+\ell^-$ must have at least one decay product identified as a lepton (electron or muon) candidate or, if outside the calorimeter acceptance, must have DCH dE/dx information consistent with the electron hypothesis. Tracks in which the electron has radiated are combined with bremsstrahlung photons, reconstructed as clusters with more than 30 MeV lying within 35 mrad in polar angle and 50 mrad in azimuth of the projected photon position on the EMC. The second track of a $\mu^+\mu^-$ pair, if within the acceptance of the calorimeter, must be consistent with being a minimum ionizing particle. Two identified electron or muon candidates are required for J/ψ or $\psi(2S) \to \ell^+\ell^-$ reconstruction in the higher-background $\psi(2S)K_s^0$ and $J/\psi K_L^0$ channels.

We require a J/ψ candidate to have $2.95 \le m_{e^+e^-} \le 3.14$ GeV/c^2 or $3.06 \le m_{\mu^+\mu^-} \le 3.14$ GeV/c^2, and a $\psi(2S) \to \ell^+\ell^-$ candidate to have $3.44 \le m_{e^+e^-} \le 3.74$ GeV/c^2 or $3.64 \le m_{\mu^+\mu^-} \le 3.74$ GeV/c^2. Requirements are made on the lepton helicity angle in order to provide further discrimination against background. For the $\psi(2S) \to J/\psi\pi^+\pi^-$ mode, mass-constrained J/ψ candidates are combined with pairs of oppositely charged tracks considered as pions; the resulting mass must be within 15 MeV/c^2 of the $\psi(2S)$ mass [4].

A $K_s^0 \to \pi^+\pi^-$ candidate must satisfy $489 < m_{\pi^+\pi^-} < 507$ MeV/c^2. The distance between the J/ψ or $\psi(2S)$ and K_s^0 vertices is required to be at least 1 mm.

Pairs of π^0 candidates with total energy above 800 MeV are considered as K_s^0 candidates for the $J/\psi K_s^0$ mode. We determine the most probable K_s^0 decay point along the path defined by the initial K_s^0 momentum vector and the J/ψ vertex by maximizing the product of probabilities for the daughter π^0 mass-constrained fits. Allowing for vertex resolution, we require the displacement from the J/ψ vertex to the decay point to be between -10 and $+40$ cm and the $\pi^0\pi^0$ mass evaluated at this point to be between 470 and 550 MeV/c^2.

A K_L^0 candidate is formed from a cluster not matched to a reconstructed track. For the EMC the cluster must have energy above 200 MeV, while for the IFR the cluster must have at least two layers. We determine the K_L^0 energy by combining its direction with the reconstructed J/ψ momentum, assuming the decay $B^0 \to J/\psi K_L^0$. To reduce photon backgrounds, EMC clusters consistent with a $\pi^0 \to \gamma\gamma$ decay are rejected and the transverse missing momentum of the event projected on the K_L^0 candidate direction must be consistent with the K_L^0 momentum. In addition, the center-of-mass J/ψ momentum is required to be greater than 1.4 GeV/c.

B_{CP} candidates used in the analysis are selected by requiring that the difference ΔE between the energy of the B_{CP} candidate and the beam energy in the center-of-mass frame be less than 3 standard deviations from zero and that, for K_s^0 modes, the beam-energy substituted mass $m_{ES} = \sqrt{(E_{\text{beam}}^{\text{cm}})^2 - (p_B^{\text{cm}})^2}$ must be greater than 5.2 GeV/c^2. The resolution for ΔE is about 10 MeV, except for $J/\psi K_L^0$ (3 MeV) and the $K_s^0 \to \pi^0\pi^0$ mode (33 MeV). For the purpose of determining numbers of events, purities, and efficiencies, a signal region $m_{ES} > 5.27$ GeV/c^2 is used for all modes except $J/\psi K_L^0$.

Figure 1 shows the resulting ΔE and m_{ES} distributions for B_{CP} candidates containing a K_s^0, and ΔE for the candidates containing a K_L^0. The B_{CP} sample is composed of 890 events in the signal region, with an estimated background of 260 events, predominantly in the $J/\psi K_L^0$ channel. For that channel, the composition, effective η_f, and ΔE distributions of the individual background sources are

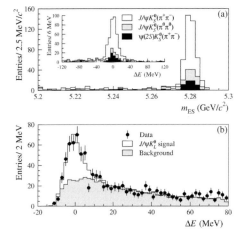

FIG. 1. (a) Distribution of m_{ES} and ΔE for B_{CP} candidates having a K_S^0 in the final state; (b) distribution of ΔE for $J/\psi K_L^0$ candidates.

taken either from a Monte Carlo simulation (for B decays to J/ψ) or from the $m_{\ell^+\ell^-}$ sidebands in data.

For flavor tagging, we exploit information from the incompletely reconstructed other B decay in the event. The charge of energetic electrons and muons from semileptonic B decays, kaons, soft pions from D^* decays, and high momentum charged particles is correlated with the flavor of the decaying b quark: e.g., a positive lepton yields a B^0 tag. Each event is assigned to one of four hierarchical, mutually exclusive tagging categories or is excluded from further analysis. The mistag fractions and efficiencies of all categories are determined from data.

A lepton tag requires an electron or muon candidate with a center-of-mass momentum $p_{cm} > 1.0$ or 1.1 GeV/c, respectively. This efficiently selects primary leptons and reduces contamination due to oppositely charged leptons from semileptonic charm decays. Events meeting these criteria are assigned to the lepton category unless the lepton charge and the net charge of all kaon candidates indicate opposite tags. Events without a lepton tag but with a nonzero net kaon charge are assigned to the kaon category.

All remaining events are passed to a neural network algorithm whose main inputs are the momentum and charge of the track with the highest center-of-mass momentum, and the outputs of secondary networks, trained with Monte Carlo samples to identify primary leptons, kaons, and soft pions. Based on the output of the neural network algorithm, events are tagged as B^0 or $\bar{B}^0$ and assigned to the NT1 (more certain tags) or NT2 (less certain tags) category, or not tagged at all. The tagging power of the NT1 and NT2 categories arises primarily from soft pions and from recovering unidentified isolated primary electrons and muons.

Table I shows the number of tagged events and the signal purity, determined from fits to the m_{ES} (K_S^0 modes) or ΔE (K_L^0 mode) distributions. The measured efficiencies for the four tagging categories are summarized in Table II.

The uncertainty in the Δt measurement is dominated by the measurement of the position z_{tag} of the tagging vertex. The tagging vertex is determined by fitting the tracks not belonging to the B_{CP} (or B_{flav}) candidate to a common vertex. Reconstructed K_S^0 and Λ candidates are used as input to the fit in place of their daughters. Tracks from γ conversions are excluded from the fit. To reduce contributions from charm decay, which bias the vertex estimation, the track with the largest vertex χ^2 contribution greater than 6 is removed and the fit is redone until no track fails the χ^2 requirement or fewer than two tracks remain. The average resolution for $\Delta z = z_{CP} - z_{tag}$ is 190 μm. The time interval Δt between the two B decays is then determined from the Δz measurement, including an event-by-event correction for the direction of the B with respect to the z direction in the $Y(4S)$ frame. An accepted candidate must have a converged fit for the B_{CP} and B_{tag} vertices, an error of less than 400 μm on Δz, and a measured $|\Delta z| < 3$ mm; 86% of the B_{CP} events satisfy this requirement.

The $\sin 2\beta$ measurement is made with an unbinned maximum likelihood fit to the Δt distribution of the combined B_{CP} and B_{flav} tagged samples. The Δt distribution of the former is given by Eq. (1), with $|\lambda| = 1$. The latter evolves according to the known rate for flavor oscillations in neutral B mesons. The amplitudes for B_{CP} asymmetries and for B_{flav} flavor oscillations are reduced by the same

TABLE I. Number of tagged events, signal purity, and result of fitting for CP asymmetries in the full CP sample and in various subsamples, as well as in the B_{flav} and charged B control samples. Purity is the fitted number of signal events divided by the total number of events in the ΔE and m_{ES} signal region defined in the text. Errors are statistical only.

Sample	N_{tag}	Purity (%)	$\sin 2\beta$
$J/\psi K_S^0, \psi(2S)K_S^0$	273	96 ± 1	0.25 ± 0.22
$J/\psi K_L^0$	256	39 ± 6	0.87 ± 0.51
Full CP sample	529	69 ± 2	0.34 ± 0.20
$J/\psi K_S^0, \psi(2S)K_S^0$ only			
$J/\psi K_S^0$ ($K_S^0 \to \pi^+\pi^-$)	188	98 ± 1	0.25 ± 0.26
$J/\psi K_S^0$ ($K_S^0 \to \pi^0\pi^0$)	41	85 ± 6	−0.05 ± 0.66
$\psi(2S)K_S^0$ ($K_S^0 \to \pi^+\pi^-$)	44	97 ± 3	0.40 ± 0.50
Lepton tags	34	99 ± 2	0.07 ± 0.43
Kaon tags	156	96 ± 2	0.40 ± 0.29
NT1 tags	28	97 ± 3	−0.03 ± 0.67
NT2 tags	55	96 ± 3	0.09 ± 0.76
B^0 tags	141	96 ± 2	0.24 ± 0.31
$\bar{B}^0$ tags	132	97 ± 2	0.25 ± 0.30
B_{flav} sample	4637	86 ± 1	0.03 ± 0.05
Charged B sample	5165	90 ± 1	0.02 ± 0.05

TABLE II. Average mistag fractions w_i and mistag differences $\Delta w_i = w_i(B^0) - w_i(\bar{B}^0)$ extracted for each tagging category i from the maximum-likelihood fit to the time distribution for the fully reconstructed B^0 sample ($B_{\text{flav}} + B_{CP}$). The figure of merit for tagging is the effective tagging efficiency $Q_i = \varepsilon_i(1 - 2w_i)^2$, where ε_i is the fraction of events with a reconstructed tag vertex that is assigned to the ith category. Uncertainties are statistical only. The statistical error on $\sin2\beta$ is proportional to $1/\sqrt{Q}$, where $Q = \sum Q_i$.

Category	ε (%)	w (%)	Δw (%)	Q (%)
Lepton	10.9 ± 0.4	11.6 ± 2.0	3.1 ± 3.1	6.4 ± 0.7
Kaon	36.5 ± 0.7	17.1 ± 1.3	−1.9 ± 1.9	15.8 ± 1.3
NT1	7.7 ± 0.4	21.2 ± 2.9	7.8 ± 4.2	2.6 ± 0.5
NT2	13.7 ± 0.5	31.7 ± 2.6	−4.7 ± 3.5	1.8 ± 0.5
All	68.9 ± 1.0			26.7 ± 1.6

factor $(1 - 2w)$ due to mistags. The distributions are both convoluted with a common Δt resolution function and corrected for backgrounds, incorporated with different assumptions about their Δt evolution and convoluted with a separate resolution function. Events are assigned signal and background probabilities based on fits to m_{ES} (all modes except $J/\psi K_L^0$) or ΔE ($J/\psi K_L^0$) distributions.

The Δt resolution function for signal candidates is represented by a sum of three Gaussian distributions with different means and widths. For the core and tail Gaussians, the widths are scaled by the event-by-event measurement error derived from the vertex fits; the combined rms error is 1.1 ps. A separate offset for the core distribution is allowed for each tagging category to account for small shifts caused by inclusion of residual charm decay products in the tag vertex; a common offset is used for the tail component. The third Gaussian (of fixed 8 ps width) accounts for the fewer than 1% of events with incorrectly reconstructed vertices. Identical resolution function parameters are used for all modes, since the B_{tag} vertex precision dominates the Δt resolution.

A total of 35 parameters are varied in the final fit, including the values of $\sin2\beta$ (1), the average mistag fraction w and the difference Δw between B^0 and $\bar{B}^0$ mistags for each tagging category (8), parameters for the signal Δt resolution (9), and parameters for background time dependence (6), Δt resolution (3) and mistag fractions (8). The determination of the mistag fractions and signal Δt resolution function is dominated by the high-statistics B_{flav} sample, while background parameters are governed by events with $m_{ES} < 5.27$ GeV/c^2 (except $J/\psi K_L^0$). We fix $\tau_{B^0} = 1.548$ ps and $\Delta m_{B^0} = 0.472\hbar$ ps^{-1} [4]. The largest correlation between $\sin2\beta$ and any linear combination of the other free parameters is 0.076.

The measurement of $\sin2\beta$ was performed as a blind analysis by hiding the value of $\sin2\beta$ obtained from the fit, as well as the CP asymmetry in the Δt distribution, until the analysis was complete. This allowed us to study statistical and systematic errors without knowing the numerical value of $\sin2\beta$.

The measured mistag rates obtained from the likelihood fit for the four tagging categories are summarized in Table II. As a check, the mistag rates were evaluated with a sample of about 16 000 $D^{*-}\ell^+\nu_\ell$ events and found to be consistent with the results from the hadronic decay sample.

The combined fit to the CP decay modes and the flavor decay modes yields

$$\sin2\beta = 0.34 \pm 0.20 \text{ (stat)} \pm 0.05 \text{ (syst)}.$$

The decay asymmetry A_{CP} as a function of Δt and the log likelihood as a function of $\sin2\beta$ are shown in Fig. 2. If $|\lambda|$ is allowed to float in the fit, the value obtained is consistent with 1 and there is no significant difference in the value of $-\eta_f \text{Im}\lambda/|\lambda|$ (identified with $\sin2\beta$ in the standard model) and our quoted result. Repeating the fit with all parameters fixed to their determined values except $\sin2\beta$, we find that a total contribution of ±0.02 to the error on $\sin2\beta$ is due to the combined statistical uncertainties in mistag rates, Δt resolution, and background parameters.

The dominant sources of systematic error are the assumed parametrization of the Δt resolution function (0.04), due in part to residual uncertainties in the SVT alignment, and uncertainties in the level, composition, and CP asymmetry of the background in the selected CP events (0.02). The systematic errors from uncertainties in Δm_{B^0} and τ_{B^0} and from the parametrization of the background in the selected B_{flav} sample are found to be negligible. An increase of 0.02$\hbar$ ps^{-1} in the assumed value for Δm_{B^0} decreases $\sin2\beta$ by 0.012.

The large sample of reconstructed events allows a number of consistency checks, including separation of the data by decay mode, tagging category, and B_{tag} flavor. The results of fits to these subsamples are shown in Table I for the high-purity K_S^0 events. Table I also shows results of fits with the samples of non-CP decay modes, where no statistically significant CP asymmetry is found.

Our measurement of $\sin2\beta$ is consistent with, but improves substantially on the precision of, previous determinations [9]. The central value is consistent with the range implied by measurements and theoretical estimates of the magnitudes of CKM matrix elements [10]; it is also consistent with no CP asymmetry at the 1.7σ level.

We thank our PEP-II colleagues for their extraordinary achievement in reaching design luminosity and high

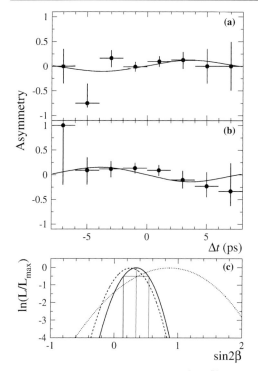

FIG. 2. The raw asymmetry in the number of B^0 and $\bar{B}^0$ tags in the signal region, $(N_{B^0} - N_{\bar{B}^0})/(N_{B^0} + N_{\bar{B}^0})$, with asymmetric binomial errors, as a function of Δt for (a) the $J/\psi K_S^0$ and $\psi(2S)K_S^0$ modes ($\eta_f = -1$) and (b) the $J/\psi K_L^0$ mode ($\eta_f = +1$). The solid curves represent the time-dependent asymmetries determined for the central values of $\sin 2\beta$ from the fits for these samples. Eight events that lie outside the plotted interval were also used in the fits. The probability of obtaining a lower likelihood, evaluated using a Monte Carlo technique, is 60%. (c) Variation of the log likelihood as a function of $\sin 2\beta$ for the modes containing K_S^0 (dashed curve), the $J/\psi K_L^0$ mode (dotted curve), and the entire sample (solid curve). For the latter, solid lines indicate the central value and values of the log likelihood corresponding to 1 statistical standard deviation.

reliability in a remarkably short time. The collaborating institutions thank SLAC for its support and the kind hospitality extended to them. This work has been supported by the U.S. Department of Energy and National Science Foundation, the Natural Sciences and Engineering Research Council (Canada), the Institute of High Energy Physics (China), Commissariat à l'Energie Atomique and Institut National de Physique Nucléaire et de Physique des Particules (France), Bundesministerium für Bildung und Forschung (Germany), Istituto Nazionale di Fisica Nucleare (Italy), the Research Council of Norway, the Ministry of Science and Technology of the Russian Federation, and the Particle Physics and Astronomy Research Council (United Kingdom). Individuals have received support from the Swiss National Foundation, the A. P. Sloan Foundation, the Research Corporation, and the Alexander von Humboldt Foundation.

---

*Also with Università di Perugia, Perugia, Italy.
†Also with Università della Basilicata, Potenza, Italy.
‡Deceased.

[1] A. B. Carter and A. I. Sanda, Phys. Rev. D **23**, 1567 (1981); I. I. Bigi and A. I. Sanda, Nucl. Phys. **B193**, 85 (1981).
[2] J. H. Christenson et al., Phys. Rev. Lett. **13**, 138 (1964).
[3] See, for example, L. Wolfenstein, Eur. Phys. J. C **15**, 115 (2000).
[4] Particle Data Group, D. E. Groom et al., Eur. Phys. J. C **15**, 1 (2000).
[5] N. Cabibbo, Phys. Rev. Lett. **10**, 531 (1963); M. Kobayashi and T. Maskawa, Prog. Theor. Phys. **49**, 652 (1973).
[6] Throughout this paper, flavor-eigenstate decay modes imply also their charge conjugate.
[7] BABAR Collaboration, B. Aubert et al., SLAC-PUB-8569 (to be published).
[8] G. C. Fox and S. Wolfram, Phys. Rev. Lett. **41**, 1581 (1978).
[9] OPAL Collaboration, K. Ackerstaff et al., Eur. Phys. J. C **5**, 379 (1998); CDF Collaboration, T. Affolder et al., Phys. Rev. D **61**, 072005 (2000); ALEPH Collaboration, R. Barate et al., Phys. Lett. B **492**, 259 (2000).
[10] See, for example, F. J. Gilman, K. Kleinknecht, and B. Renk, Eur. Phys. J. C **15**, 110 (2000).

Measurement of the B_s^0-$\bar{B}_s^0$ Oscillation Frequency

A. Abulencia,[23] D. Acosta,[17] J. Adelman,[13] T. Affolder,[10] T. Akimoto,[55] M. G. Albrow,[16] D. Ambrose,[16] S. Amerio,[43] D. Amidei,[34] A. Anastassov,[52] K. Anikeev,[16] A. Annovi,[18] J. Antos,[1] M. Aoki,[55] G. Apollinari,[16] J.-F. Arguin,[33] T. Arisawa,[57] A. Artikov,[14] W. Ashmanskas,[16] A. Attal,[8] F. Azfar,[42] P. Azzi-Bacchetta,[43] P. Azzurri,[46] N. Bacchetta,[43] H. Bachacou,[28] W. Badgett,[16] A. Barbaro-Galtieri,[28] V. E. Barnes,[48] B. A. Barnett,[24] S. Baroiant,[7] V. Bartsch,[30] G. Bauer,[32] F. Bedeschi,[46] S. Behari,[24] S. Belforte,[54] G. Bellettini,[46] J. Bellinger,[59] A. Belloni,[32] E. Ben Haim,[44] D. Benjamin,[15] A. Beretvas,[16] J. Beringer,[28] T. Berry,[29] A. Bhatti,[50] M. Binkley,[16] D. Bisello,[43] R. E. Blair,[2] C. Blocker,[6] B. Blumenfeld,[24] A. Bocci,[15] A. Bodek,[49] V. Boisvert,[49] G. Bolla,[48] A. Bolshov,[32] D. Bortoletto,[48] J. Boudreau,[47] A. Boveia,[10] B. Brau,[10] C. Bromberg,[35] E. Brubaker,[13] J. Budagov,[14] H. S. Budd,[49] S. Budd,[23] K. Burkett,[16] G. Busetto,[43] P. Bussey,[20] K. L. Byrum,[2] S. Cabrera,[15] M. Campanelli,[19] M. Campbell,[34] F. Canelli,[8] A. Canepa,[48] D. Carlsmith,[59] R. Carosi,[46] S. Carron,[15] B. Casal,[11] M. Casarsa,[54] A. Castro,[5] P. Catastini,[46] D. Cauz,[54] M. Cavalli-Sforza,[3] A. Cerri,[28] L. Cerrito,[42] S. H. Chang,[27] J. Chapman,[34] Y. C. Chen,[1] M. Chertok,[7] G. Chiarelli,[46] G. Chlachidze,[14] F. Chlebana,[16] I. Cho,[27] K. Cho,[27] D. Chokheli,[14] J. P. Chou,[21] P. H. Chu,[23] S. H. Chuang,[59] K. Chung,[12] W. H. Chung,[59] Y. S. Chung,[49] M. Ciljak,[46] C. I. Ciobanu,[23] M. A. Ciocci,[46] A. Clark,[19] D. Clark,[6] M. Coca,[15] G. Compostella,[43] M. E. Convery,[50] J. Conway,[7] B. Cooper,[30] K. Copic,[34] M. Cordelli,[18] G. Cortiana,[43] F. Crescioli,[46] A. Cruz,[17] C. Cuenca Almenar,[7] J. Cuevas,[11] R. Culbertson,[16] D. Cyr,[59] S. DaRonco,[43] S. D'Auria,[20] M. D'Onofrio,[3] D. Dagenhart,[6] P. de Barbaro,[49] S. De Cecco,[51] A. Deisher,[28] G. De Lentdecker,[49] M. Dell'Orso,[46] F. Delli Paoli,[43] S. Demers,[49] L. Demortier,[50] J. Deng,[15] M. Deninno,[5] D. De Pedis,[51] P. F. Derwent,[16] G. P. Di Giovanni,[44] B. Di Ruzza,[54] C. Dionisi,[51] J. R. Dittmann,[4] P. DiTuro,[52] C. Dörr,[25] S. Donati,[46] M. Donega,[19] P. Dong,[8] J. Donini,[43] T. Dorigo,[43] S. Dube,[52] K. Ebina,[57] J. Efron,[39] J. Ehlers,[19] R. Erbacher,[7] D. Errede,[23] S. Errede,[23] R. Eusebi,[16] H. C. Fang,[28] S. Farrington,[29] I. Fedorko,[46] W. T. Fedorko,[13] R. G. Feild,[60] M. Feindt,[25] J. P. Fernandez,[31] R. Field,[17] G. Flanagan,[48] L. R. Flores-Castillo,[47] A. Foland,[21] S. Forrester,[7] G. W. Foster,[16] M. Franklin,[21] J. C. Freeman,[28] H. J. Frisch,[13] I. Furic,[13] M. Gallinaro,[50] J. Galyardt,[12] J. E. Garcia,[46] M. Garcia Sciveres,[28] A. F. Garfinkel,[48] C. Gay,[60] H. Gerberich,[23] D. Gerdes,[34] S. Giagu,[51] P. Giannetti,[46] A. Gibson,[28] K. Gibson,[12] C. Ginsburg,[16] N. Giokaris,[14] K. Giolo,[48] M. Giordani,[54] P. Giromini,[18] M. Giunta,[46] G. Giurgiu,[12] V. Glagolev,[14] D. Glenzinski,[16] M. Gold,[37] N. Goldschmidt,[34] J. Goldstein,[42] G. Gomez,[11] G. Gomez-Ceballos,[11] M. Goncharov,[53] O. González,[31] I. Gorelov,[37] A. T. Goshaw,[15] Y. Gotra,[47] K. Goulianos,[50] A. Gresele,[43] M. Griffiths,[29] S. Grinstein,[21] C. Grosso-Pilcher,[13] R. C. Group,[17] U. Grundler,[23] J. Guimaraes da Costa,[21] Z. Gunay-Unalan,[35] C. Haber,[28] S. R. Hahn,[16] K. Hahn,[45] E. Halkiadakis,[52] A. Hamilton,[33] B.-Y. Han,[49] J. Y. Han,[49] R. Handler,[59] F. Happacher,[18] K. Hara,[55] M. Hare,[56] S. Harper,[42] R. F. Harr,[58] R. M. Harris,[16] K. Hatakeyama,[50] J. Hauser,[8] C. Hays,[15] A. Heijboer,[45] B. Heinemann,[29] J. Heinrich,[45] M. Herndon,[59] D. Hidas,[15] C. S. Hill,[10] D. Hirschbuehl,[25] A. Hocker,[16] A. Holloway,[21] S. Hou,[1] M. Houlden,[29] S.-C. Hsu,[9] B. T. Huffman,[42] R. E. Hughes,[39] J. Huston,[35] J. Incandela,[10] G. Introzzi,[46] M. Iori,[51] Y. Ishizawa,[55] A. Ivanov,[7] B. Iyutin,[32] E. James,[16] D. Jang,[52] B. Jayatilaka,[34] D. Jeans,[51] H. Jensen,[16] E. J. Jeon,[27] S. Jindariani,[17] M. Jones,[48] K. K. Joo,[27] S. Y. Jun,[12] T. R. Junk,[23] T. Kamon,[53] J. Kang,[34] P. E. Karchin,[58] Y. Kato,[41] Y. Kemp,[25] R. Kephart,[16] U. Kerzel,[25] V. Khotilovich,[53] B. Kilminster,[39] D. H. Kim,[27] H. S. Kim,[27] J. E. Kim,[27] M. J. Kim,[12] S. B. Kim,[27] S. H. Kim,[55] Y. K. Kim,[13] L. Kirsch,[6] S. Klimenko,[17] M. Klute,[32] B. Knuteson,[32] B. R. Ko,[15] H. Kobayashi,[55] K. Kondo,[57] D. J. Kong,[27] J. Konigsberg,[17] A. Korytov,[17] A. V. Kotwal,[15] A. Kovalev,[45] A. Kraan,[45] J. Kraus,[23] I. Kravchenko,[32] M. Kreps,[25] J. Kroll,[45] N. Krumnack,[4] M. Kruse,[15] V. Krutelyov,[53] S. E. Kuhlmann,[2] Y. Kusakabe,[57] S. Kwang,[13] A. T. Laasanen,[48] S. Lai,[33] S. Lami,[46] S. Lammel,[16] M. Lancaster,[30] R. L. Lander,[7] K. Lannon,[39] A. Lath,[52] G. Latino,[46] I. Lazzizzera,[43] T. LeCompte,[2] J. Lee,[49] J. Lee,[27] Y. J. Lee,[27] S. W. Lee,[53] R. Lefèvre,[3] N. Leonardo,[32] S. Leone,[46] S. Levy,[13] J. D. Lewis,[16] C. Lin,[60] C. S. Lin,[16] M. Lindgren,[16] E. Lipeles,[9] T. M. Liss,[23] A. Lister,[19] D. O. Litvintsev,[16] T. Liu,[16] N. S. Lockyer,[45] A. Loginov,[36] M. Loreti,[43] P. Loverre,[51] R.-S. Lu,[1] D. Lucchesi,[43] P. Lujan,[28] P. Lukens,[16] G. Lungu,[17] L. Lyons,[42] J. Lys,[28] R. Lysak,[1] E. Lytken,[48] P. Mack,[25] D. MacQueen,[33] R. Madrak,[16] K. Maeshima,[16] T. Maki,[22] P. Maksimovic,[24] S. Malde,[42] G. Manca,[29] F. Margaroli,[5] R. Marginean,[16] C. Marino,[23] A. Martin,[60] V. Martin,[38] M. Martínez,[3] T. Maruyama,[55] P. Mastrandrea,[51] H. Matsunaga,[55] M. E. Mattson,[58] R. Mazini,[33] P. Mazzanti,[5] K. S. McFarland,[49] P. McIntyre,[53] R. McNulty,[29] A. Mehta,[29] S. Menzemer,[11] A. Menzione,[46] P. Merkel,[48] C. Mesropian,[50] A. Messina,[51] M. von der Mey,[8] T. Miao,[16] N. Miladinovic,[6] J. Miles,[32] R. Miller,[35] J. S. Miller,[34] C. Mills,[10] M. Milnik,[25] R. Miquel,[28] A. Mitra,[1] G. Mitselmakher,[17] A. Miyamoto,[26] N. Moggi,[5] B. Mohr,[8] R. Moore,[16] M. Morello,[46] P. Movilla Fernandez,[28] J. Mülmenstädt,[28] A. Mukherjee,[16] Th. Muller,[25] R. Mumford,[24] P. Murat,[16] J. Nachtman,[16] J. Naganoma,[57] S. Nahn,[32] I. Nakano,[40] A. Napier,[56] D. Naumov,[37] V. Necula,[17] C. Neu,[45] M. S. Neubauer,[9]

J. Nielsen,[28] T. Nigmanov,[47] L. Nodulman,[2] O. Norniella,[3] E. Nurse,[30] T. Ogawa,[57] S. H. Oh,[15] Y. D. Oh,[27] T. Okusawa,[41] R. Oldeman,[29] R. Orava,[22] K. Osterberg,[22] C. Pagliarone,[46] E. Palencia,[11] R. Paoletti,[46] V. Papadimitriou,[16] A. A. Paramonov,[13] B. Parks,[39] S. Pashapour,[33] J. Patrick,[16] G. Pauletta,[54] M. Paulini,[12] C. Paus,[32] D. E. Pellett,[7] A. Penzo,[54] T. J. Phillips,[15] G. Piacentino,[46] J. Piedra,[44] L. Pinera,[17] K. Pitts,[23] C. Plager,[8] L. Pondrom,[59] X. Portell,[3] O. Poukhov,[14] N. Pounder,[42] F. Prakoshyn,[14] A. Pronko,[16] J. Proudfoot,[2] F. Ptohos,[18] G. Punzi,[46] J. Pursley,[24] J. Rademacker,[42] A. Rahaman,[47] A. Rakitin,[32] S. Rappoccio,[21] F. Ratnikov,[52] B. Reisert,[16] V. Rekovic,[37] N. van Remortel,[22] P. Renton,[42] M. Rescigno,[51] S. Richter,[25] F. Rimondi,[5] L. Ristori,[46] W. J. Robertson,[15] A. Robson,[20] T. Rodrigo,[11] E. Rogers,[23] S. Rolli,[56] R. Roser,[16] M. Rossi,[54] R. Rossin,[17] C. Rott,[48] A. Ruiz,[11] J. Russ,[12] V. Rusu,[13] H. Saarikko,[22] S. Sabik,[33] A. Safonov,[53] W. K. Sakumoto,[49] G. Salamanna,[51] O. Saltó,[3] D. Saltzberg,[8] C. Sanchez,[3] L. Santi,[54] S. Sarkar,[51] L. Sartori,[46] K. Sato,[55] P. Savard,[33] A. Savoy-Navarro,[44] T. Scheidle,[25] P. Schlabach,[16] E. E. Schmidt,[16] M. P. Schmidt,[60] M. Schmitt,[38] T. Schwarz,[34] L. Scodellaro,[11] A. L. Scott,[10] A. Scribano,[46] F. Scuri,[46] A. Sedov,[48] S. Seidel,[37] Y. Seiya,[41] A. Semenov,[14] L. Sexton-Kennedy,[16] I. Sfiligoi,[18] M. D. Shapiro,[28] T. Shears,[29] P. F. Shepard,[47] D. Sherman,[21] M. Shimojima,[55] M. Shochet,[13] Y. Shon,[59] I. Shreyber,[36] A. Sidoti,[44] P. Sinervo,[33] A. Sisakyan,[14] J. Sjolin,[42] A. Skiba,[25] A. J. Slaughter,[16] K. Sliwa,[56] J. R. Smith,[7] F. D. Snider,[16] R. Snihur,[33] M. Soderberg,[34] A. Soha,[7] S. Somalwar,[52] V. Sorin,[35] J. Spalding,[16] M. Spezziga,[16] F. Spinella,[46] T. Spreitzer,[33] P. Squillacioti,[46] M. Stanitzki,[60] A. Staveris-Polykalas,[46] R. St. Denis,[20] B. Stelzer,[8] O. Stelzer-Chilton,[42] D. Stentz,[38] J. Strologas,[37] D. Stuart,[10] J. S. Suh,[27] A. Sukhanov,[17] K. Sumorok,[32] H. Sun,[56] T. Suzuki,[55] A. Taffard,[23] R. Takashima,[40] Y. Takeuchi,[55] K. Takikawa,[55] M. Tanaka,[2] R. Tanaka,[40] N. Tanimoto,[40] M. Tecchio,[34] P. K. Teng,[1] K. Terashi,[50] S. Tether,[32] J. Thom,[16] A. S. Thompson,[20] E. Thomson,[45] P. Tipton,[49] V. Tiwari,[12] S. Tkaczyk,[16] D. Toback,[53] S. Tokar,[14] K. Tollefson,[35] T. Tomura,[55] D. Tonelli,[46] M. Tönnesmann,[35] S. Torre,[18] D. Torretta,[16] S. Tourneur,[44] W. Trischuk,[33] R. Tsuchiya,[57] S. Tsuno,[40] N. Turini,[46] F. Ukegawa,[55] T. Unverhau,[20] S. Uozumi,[55] D. Usynin,[45] A. Vaiciulis,[49] S. Vallecorsa,[19] A. Varganov,[34] E. Vataga,[37] G. Velev,[16] G. Veramendi,[23] V. Veszpremi,[48] R. Vidal,[16] I. Vila,[11] R. Vilar,[11] T. Vine,[30] I. Vollrath,[33] I. Volobouev,[28] G. Volpi,[46] F. Würthwein,[9] P. Wagner,[53] R. G. Wagner,[2] R. L. Wagner,[16] W. Wagner,[25] R. Wallny,[8] T. Walter,[25] Z. Wan,[52] S. M. Wang,[1] A. Warburton,[33] S. Waschke,[20] D. Waters,[30] W. C. Wester III,[16] B. Whitehouse,[56] D. Whiteson,[45] A. B. Wicklund,[2] E. Wicklund,[16] G. Williams,[33] H. H. Williams,[45] P. Wilson,[16] B. L. Winer,[39] P. Wittich,[16] S. Wolbers,[16] C. Wolfe,[13] T. Wright,[34] X. Wu,[19] S. M. Wynne,[29] A. Yagil,[16] K. Yamamoto,[41] J. Yamaoka,[52] T. Yamashita,[40] C. Yang,[60] U. K. Yang,[13] Y. C. Yang,[27] W. M. Yao,[28] G. P. Yeh,[16] J. Yoh,[16] K. Yorita,[13] T. Yoshida,[41] G. B. Yu,[49] I. Yu,[27] S. S. Yu,[16] J. C. Yun,[16] L. Zanello,[51] A. Zanetti,[54] I. Zaw,[21] F. Zetti,[46] X. Zhang,[23] J. Zhou,[52] and S. Zucchelli[5]

(CDF Collaboration)

[1]*Institute of Physics, Academia Sinica, Taipei, Taiwan 11529, Republic of China*
[2]*Argonne National Laboratory, Argonne, Illinois 60439, USA*
[3]*Institut de Fisica d'Altes Energies, Universitat Autonoma de Barcelona, E-08193 Bellaterra (Barcelona), Spain*
[4]*Baylor University, Waco, Texas 76798, USA*
[5]*Istituto Nazionale di Fisica Nucleare, University of Bologna, I-40127 Bologna, Italy*
[6]*Brandeis University, Waltham, Massachusetts 02254, USA*
[7]*University of California, Davis, Davis, California 95616, USA*
[8]*University of California, Los Angeles, Los Angeles, California 90024, USA*
[9]*University of California, San Diego, La Jolla, California 92093, USA*
[10]*University of California, Santa Barbara, Santa Barbara, California 93106, USA*
[11]*Instituto de Fisica de Cantabria, CSIC-University of Cantabria, 39005 Santander, Spain*
[12]*Carnegie Mellon University, Pittsburgh, Pennsylvania 15213, USA*
[13]*Enrico Fermi Institute, University of Chicago, Chicago, Illinois 60637, USA*
[14]*Joint Institute for Nuclear Research, RU-141980 Dubna, Russia*
[15]*Duke University, Durham, North Carolina 27708, USA*
[16]*Fermi National Accelerator Laboratory, Batavia, Illinois 60510, USA*
[17]*University of Florida, Gainesville, Florida 32611, USA*
[18]*Laboratori Nazionali di Frascati, Istituto Nazionale di Fisica Nucleare, I-00044 Frascati, Italy*
[19]*University of Geneva, CH-1211 Geneva 4, Switzerland*
[20]*Glasgow University, Glasgow G12 8QQ, United Kingdom*
[21]*Harvard University, Cambridge, Massachusetts 02138, USA*
[22]*Division of High Energy Physics, Department of Physics, University of Helsinki and Helsinki Institute of Physics, FIN-00014 Helsinki, Finland*

[23]University of Illinois, Urbana, Illinois 61801, USA
[24]The Johns Hopkins University, Baltimore, Maryland 21218, USA
[25]Institut für Experimentelle Kernphysik, Universität Karlsruhe, 76128 Karlsruhe, Germany
[26]High Energy Accelerator Research Organization (KEK), Tsukuba, Ibaraki 305, Japan
[27]Center for High Energy Physics: Kyungpook National University, Taegu 702-701, Korea;
Seoul National University, Seoul 151-742, Korea;
and SungKyunKwan University, Suwon 440-746, Korea
[28]Ernest Orlando Lawrence Berkeley National Laboratory, Berkeley, California 94720, USA
[29]University of Liverpool, Liverpool L69 7ZE, United Kingdom
[30]University College London, London WC1E 6BT, United Kingdom
[31]Centro de Investigaciones Energeticas Medioambientales y Tecnologicas, E-28040 Madrid, Spain
[32]Massachusetts Institute of Technology, Cambridge, Massachusetts 02139, USA
[33]Institute of Particle Physics: McGill University, Montréal, Canada H3A 2T8;
and University of Toronto, Toronto, Canada M5S 1A7
[34]University of Michigan, Ann Arbor, Michigan 48109, USA
[35]Michigan State University, East Lansing, Michigan 48824, USA
[36]Institution for Theoretical and Experimental Physics, ITEP, Moscow 117259, Russia
[37]University of New Mexico, Albuquerque, New Mexico 87131, USA
[38]Northwestern University, Evanston, Illinois 60208, USA
[39]The Ohio State University, Columbus, Ohio 43210, USA
[40]Okayama University, Okayama 700-8530, Japan
[41]Osaka City University, Osaka 588, Japan
[42]University of Oxford, Oxford OX1 3RH, United Kingdom
[43]Istituto Nazionale di Fisica Nucleare, University of Padova, Sezione di Padova-Trento, I-35131 Padova, Italy
[44]LPNHE, Universite Pierre et Marie Curie/IN2P3-CNRS, UMR7585, Paris F-75252, France
[45]University of Pennsylvania, Philadelphia, Pennsylvania 19104, USA
[46]Istituto Nazionale di Fisica Nucleare Pisa, Universities of Pisa, Siena, and Scuola Normale Superiore, I-56127 Pisa, Italy
[47]University of Pittsburgh, Pittsburgh, Pennsylvania 15260, USA
[48]Purdue University, West Lafayette, Indiana 47907, USA
[49]University of Rochester, Rochester, New York 14627, USA
[50]The Rockefeller University, New York, New York 10021, USA
[51]Istituto Nazionale di Fisica Nucleare, Sezione di Roma 1, University of Rome "La Sapienza," I-00185 Roma, Italy
[52]Rutgers University, Piscataway, New Jersey 08855, USA
[53]Texas A&M University, College Station, Texas 77843, USA
[54]Istituto Nazionale di Fisica Nucleare, University of Trieste/Udine, Italy
[55]University of Tsukuba, Tsukuba, Ibaraki 305, Japan
[56]Tufts University, Medford, Massachusetts 02155, USA
[57]Waseda University, Tokyo 169, Japan
[58]Wayne State University, Detroit, Michigan 48201, USA
[59]University of Wisconsin, Madison, Wisconsin 53706, USA
[60]Yale University, New Haven, Connecticut 06520, USA
(Received 13 June 2006; published 10 August 2006)

We present the first precise measurement of the B_s^0-$\bar{B}_s^0$ oscillation frequency Δm_s. We use 1 fb^{-1} of data from $p\bar{p}$ collisions at $\sqrt{s} = 1.96$ TeV collected with the CDF II detector at the Fermilab Tevatron. The sample contains signals of 3600 fully reconstructed hadronic B_s decays and 37 000 partially reconstructed semileptonic B_s decays. We measure the probability as a function of proper decay time that the B_s decays with the same, or opposite, flavor as the flavor at production, and we find a signal consistent with B_s^0-$\bar{B}_s^0$ oscillations. The probability that random fluctuations could produce a comparable signal is 0.2%. Under the hypothesis that the signal is due to B_s^0-$\bar{B}_s^0$ oscillations, we measure $\Delta m_s = 17.31^{+0.33}_{-0.18}(\text{stat}) \pm 0.07(\text{syst})$ ps^{-1} and determine $|V_{td}/V_{ts}| = 0.208^{+0.001}_{-0.002}(\text{expt})^{+0.008}_{-0.006}(\text{theor})$.

DOI: 10.1103/PhysRevLett.97.062003　　PACS numbers: 14.40.Nd, 12.15.Ff, 12.15.Hh, 13.20.He

Neutral B mesons ($b\bar{q}$, with $q = d$, s for $\bar{B}_d^0$, $\bar{B}_s^0$) oscillate from particle to antiparticle due to flavor-changing weak interactions. The probability density P_+ (P_-) for a $\bar{B}_q^0$ meson produced at proper time $t = 0$ to decay as a $\bar{B}_q^0$ (B_q^0) at time t is given by

$$P_{\pm}(t) = \frac{\Gamma_q}{2} e^{-\Gamma_q t}[1 \pm \cos(\Delta m_q t)],$$

where Δm_q is the mass difference between the two mass eigenstates $B_{q,H}^0$ and $B_{q,L}^0$ [1], and Γ_q is the decay width, which is assumed to be equal for the two mass eigenstates. The mass differences Δm_d and Δm_s can be used to determine the fundamental parameters $|V_{td}|$ and $|V_{ts}|$, respectively, of the Cabibbo-Kobayashi-Maskawa (CKM) matrix [2], which relates the quark mass eigenstates to the flavor eigenstates. This determination, however, has large theoretical uncertainties. A measurement of Δm_s combined with $\Delta m_d = 0.505 \pm 0.005$ ps^{-1} [3,4] would determine the ratio $|V_{td}/V_{ts}|$ with a significantly smaller theoretical uncertainty, contributing to a stringent test of the unitarity of the CKM matrix. Earlier attempts to measure Δm_s have yielded a lower limit: $\Delta m_s > 14.5$ ps^{-1} [3,5] at the 95% confidence level (C.L.). Recently the D0 Collaboration reported 17 ps^{-1} < Δm_s < 21 ps^{-1} at 90% C.L. [6] using a large sample of semileptonic B_s [7] decays.

In this Letter we report a measurement of Δm_s using data from 1 fb^{-1} of $p\bar{p}$ collisions at $\sqrt{s} = 1.96$ TeV collected by the CDF II detector at the Fermilab Tevatron. We begin by reconstructing B_s decays in hadronic ($\bar{B}_s^0 \to D_s^+ \pi^-$, $D_s^+ \pi^- \pi^+ \pi^-$) and semileptonic ($\bar{B}_s^0 \to D_s^{+(*)} \ell^- \bar{\nu}_\ell$, $\ell = e$ or μ) decay modes using charged particles only [8]. Using the method of maximum likelihood, we extract the value of Δm_s from the probability density functions (PDFs) that describe the measured time development of B_s mesons that decay with the same or opposite flavor as their flavor at production. The proper decay time for each B_s is calculated from the measured distance between the production and decay points, the measured momentum, and the B_s mass m_{B_s} = 5.3696 GeV/c^2 [3]. The B_s flavor (b or $\bar{b}$) at decay is determined unambiguously by the charges of the decay products.

To identify the flavor of the B_s at production, we use characteristics of b quark production and fragmentation in $p\bar{p}$ collisions. At the Tevatron, the dominant b quark production mechanisms produce $b\bar{b}$ pairs. The b and $\bar{b}$ are expected to fragment independently into hadrons. In a simple model of fragmentation, a b quark becomes a $\bar{B}_s^0$ meson when some of the energy of the b quark is used to produce an $s\bar{s}$ quark pair. The b and the $\bar{s}$ bind to form a $\bar{B}_s^0$. The remaining s quark may form a K^-. Similarly, a $\bar{b}$ that becomes a B_s^0 is accompanied by a K^+. One of the two techniques used to identify the production flavor of the B_s is based on the charge of these kaons (same-side tag). The second technique uses the charge of the lepton from semileptonic decays or a momentum-weighted charge of the decay products of the second b hadron produced in the collision (opposite-side tag).

The hadronic and semileptonic decay modes are complementary. Because of the large branching ratio, the semileptonic decays provide a tenfold advantage in signal rate at the cost of significantly worsened decay-time resolution due to the unmeasured ν momentum. Semileptonic decays dominate the sensitivity to oscillations at lower values of Δm_s. The fully reconstructed hadronic B_s decays have superior decay-time resolution, and our large sample of these decays is the unique feature that makes CDF sensitive to much larger values of Δm_s than other experiments.

The CDF II detector [9] consists of a magnetic spectrometer surrounded by electromagnetic and hadronic calorimeters and muon detectors [10]. The key features for this measurement include precision vertex determination provided by the seven-layer double-sided inner silicon strip detector [11,12] supplemented with a single-sided layer of silicon [13] mounted directly on the beam pipe at an average radius of 1.5 cm. The 96-layer outer drift chamber [14] is used for both precision tracking and dE/dx particle identification. Time-of-flight (TOF) counters [15] located just outside the drift chamber are used to identify low momentum charged kaons.

Charm and bottom hadrons are selected using a three-level trigger system that exploits the kinematics of production and decay, and the long lifetimes of D and B mesons. A crucial component of the trigger system for this measurement is the Silicon Vertex Trigger [16], which selects events that contain $\bar{B}_s^0 \to D_s^+ \pi^-$ and $D_s^+ \pi^- \pi^+ \pi^-$ decays. The trigger configuration used to collect the heavy flavor data sample is described in [17].

To reconstruct $\bar{B}_s^0$ candidates, we first select D_s^+ candidates. We use $D_s^+ \to \phi \pi^+$, $K^*(892)^0 K^+$, and $\pi^+ \pi^- \pi^+$, with $\phi \to K^+ K^-$ and $K^{*0} \to K^+ \pi^-$; we require that ϕ and K^{*0} candidates be consistent with the known masses and widths [3] of these two resonances. These D_s^+ candidates are combined with one or three additional charged particles to form $D_s^+ \ell^-$, $D_s^+ \pi^-$, or $D_s^+ \pi^- \pi^+ \pi^-$ candidates. The D_s^+ and other decay products of a $\bar{B}_s^0$ candidate are constrained to originate from a common vertex in three dimensions. For the $K^*(892)^0 K^+$ final state, we remove candidates that are consistent with the decay $D^+ \to K^- \pi^+ \pi^+$. We use a likelihood technique to identify muons [18] and electrons [19].

Backgrounds are suppressed by imposing a requirement on the minimum transverse momentum p_T [20] of the $\bar{B}_s^0$ and by requiring that the $\bar{B}_s^0$ and D_s^+ decay vertices are displaced significantly from the $p\bar{p}$ collision position. We find signals of 3600 hadronic B_s decays and 37 000 semileptonic B_s decays.

For the hadronic decays, the invariant mass distribution (see Fig. 1) has a signal centered close to m_{B_s} = 5.3696 GeV/c^2 with a width of 14 to 20 MeV/c^2, depending on the decay mode. Candidates with masses greater than 5.5 GeV/c^2 are used to construct PDFs for combinatorial background. To remove contributions from $\bar{B}_s^0 \to D_s^{*+} \pi^-$, $\bar{B}_s^0 \to D_s^+ \rho^-$, and semileptonic and other partially reconstructed decays, we require the mass of the decay candidates to be greater than 5.3 GeV/c^2. For semileptonic decays we take into account several background contributions, including B meson decays to two charm mesons and real D_s mesons associated with a false lepton.

The decay time in the B_s rest frame is $t = \kappa[L_T m_{B_s}/p_T]$, where L_T is the displacement of the B_s

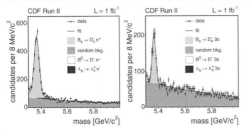

FIG. 1. The invariant mass distributions for $\bar{B}_s^0 \to D_s^+ \pi^-$ (left panel) and $D_s^+ \pi^- \pi^+ \pi^-$ (right panel).

decay vertex with respect to the primary vertex projected onto the B_s transverse momentum vector. The factor κ corrects for missing momentum in the semileptonic decays ($\kappa = 1$ for hadronic decays). To improve the decay-time resolution, we use event-by-event primary-vertex position measurements when computing the B_s vertex displacement. The signal decay-time distribution is modeled with $P(t_i, \sigma_{t_i}) = \varepsilon(t_i) \int \Gamma_s e^{-\Gamma_s t'} \mathcal{G}(t' - t_i, \sigma_{t_i}) dt'$, where t_i is the measured decay time of the ith candidate, Γ_s is the B_s decay width, $\mathcal{G}(x - \mu, \sigma)$ is a Gaussian distribution of the random variable x with mean μ and width σ, and σ_{t_i} is the estimated candidate decay-time resolution. The decay-time efficiency function $\varepsilon(t)$ describes trigger and selection biases on the decay-time distribution and is determined from Monte Carlo simulation. For semileptonic decays, the κ distribution is determined from Monte Carlo simulation and is convoluted with the signal decay-time distribution. The missing transverse momentum from unreconstructed particles in the semileptonic decays is an important contribution to the decay-time resolution. To reduce this contribution and make optimal use of the semileptonic decays, we determine the κ distribution as a function of the invariant mass of the $D_s\ell$ pair, $m_{D_s\ell}$. The rms width of the κ distribution is 3% (20%) for $m_{D_s\ell} = 5.2$ GeV/c^2 (3.0 GeV/c^2).

We estimate the decay-time resolution σ_{t_i} for each candidate using the measured track parameters and their estimated uncertainties. We calibrate this estimate using a large sample of prompt D^+ mesons [21], which we combine with one or three charged particles from the primary vertex to mimic signal topologies. For hadronic decays, the average decay-time resolution is 87 fs, which corresponds to one-fifth of an oscillation period at the lower limit on Δm_s (14.5 ps^{-1}). For semileptonic decays, the decay-time resolution is worse due to decay topology and the missing momentum of unreconstructed decay products. For example, at $t = 0$, $\sigma_t = 100$ fs (200 fs) for $m_{D_s\ell} = 5.2$ GeV/c^2 (3.0 GeV/c^2) and increases to $\sigma_t = 115$ fs (380 fs) at $t = 1.5$ ps.

The flavor of the B_s at production is determined using both opposite-side and same-side flavor tagging techniques. The effectiveness $Q \equiv \varepsilon \mathcal{D}^2$ of these techniques is quantified with an efficiency ε, the fraction of signal candidates with a flavor tag, and a dilution $\mathcal{D} \equiv 1 - 2w$, where w is the probability that the tag is incorrect.

Opposite-side tags infer the production flavor of the B_s from the decay products of the b hadron produced from the other b quark in the event. We use lepton (e and μ) charge and jet charge as tags, building on techniques developed for a CDF run I measurement of Δm_d [22]. If both lepton and jet-charge tags are present, we use the lepton tag, which has a higher average dilution.

The dilution of opposite-side flavor tags is expected to be independent of the type of B meson that produces the hadronic or semileptonic decay. The dilution is measured in data using large samples of B^-, which do not change flavor, and $\bar{B}^0$, which can be used after accounting for their well-known oscillation frequency. The combined opposite-side tag effectiveness is $Q = 1.5\% \pm 0.1\%$, where the uncertainty is dominated by the statistics of the control samples.

Same-side flavor tags [23] are based on the charges of associated particles produced in the fragmentation of the b quark that produces the reconstructed B_s. In the simplest picture of fragmentation, a π^+ (π^-) accompanies a B^- (B^+), a π^- (π^+) accompanies a $\bar{B}^0$ (B^0), and a K^- (K^+) accompanies a $\bar{B}_s^0$ (B_s^0). In run I, CDF established this method of production flavor identification in measurements of Δm_d [24] and the CP symmetry violating parameter $\sin(2\beta)$ [25]. In this analysis, we use dE/dx [19] and TOF information in a combined particle identification likelihood to identify the kaons associated with B_s production. Tracks close in phase space to the B_s candidate are considered as same-side kaon tag candidates, and the track with the largest kaon likelihood is selected as the tagging track.

The performance of the same-side kaon tag for $\bar{B}_s^0$ is expected to be different than for B^- and $\bar{B}^0$. We predict the dilution using simulated data samples generated with the PYTHIA Monte Carlo program [26]. Control samples of B^- and $\bar{B}^0$ are used to validate the predictions of the simulation. The effectiveness of this flavor tag increases with the p_T of the $\bar{B}_s^0$; we find $Q = 3.5\%$ (4.0%) in the hadronic (semileptonic) decay sample. The fractional uncertainty on Q is approximately 25%. This uncertainty is dominated by the differences between data and simulation for kaons found close in phase space to the $\bar{B}_s^0$ [27] and for the performance of the same-side kaon tag when applied to B^-.

If both a same-side tag and an opposite-side tag are present, we combine the information from both tags assuming they are uncorrelated. The addition of the same-side kaon tag increases the effective sample statistics by more than a factor of 3.

We use an unbinned maximum likelihood fit to search for B_s oscillations. The likelihood combines mass, decay-time, decay-time resolution, and flavor tagging information for each candidate and includes terms for signal and each type of background. The fit is done in three stages. First, a combined mass and decay-time fit is performed to separate signal from background and to fix mass and decay-time models. Combined fits for B_s mass (Fig. 1) and decay width in hadronic samples and for decay width in the semileptonic samples yield measurements consistent with established values [3]. Second, flavor asymmetries are measured for background components. The third step is a fit for B_s^0-$\bar{B}_s^0$ oscillations; the mass and decay-time models and background asymmetries are fixed from the previous two stages.

The signal PDF has the general form:

$$\mathcal{S}_\pm(t_i, \sigma_{t_i}, \mathcal{D}_i) = \varepsilon(t_i) \int \frac{\Gamma_s}{2} e^{-\Gamma_s t'}[1 \pm \mathcal{A}\mathcal{D}_i \cos(\Delta m_s t')]$$
$$\times \mathcal{G}(t_i - t', \sigma_{t_i})dt', \quad (1)$$

where $\mathcal{D}_i$ is the ith candidate dilution, and t_i, σ_{t_i}, $\mathcal{G}$, and $\varepsilon(t)$ have been defined previously. Following the method described in [28], we fit for the oscillation amplitude $\mathcal{A}$ while fixing Δm_s to a probe value. When all detector effects ($\mathcal{D}_i$, σ_{t_i}) are calibrated, the oscillation amplitude is expected to be consistent with $\mathcal{A} = 1$ when the probe value is the true oscillation frequency, and consistent with $\mathcal{A} = 0$ when the probe value is far from the true oscillation frequency. Figure 2 (upper panel) shows the fitted value of the amplitude as a function of the oscillation frequency. The sensitivity of the measurement is defined by the maximum value of Δm_s where $\mathcal{A} = 1$ is excluded at 95% C.L. if the measured value of $\mathcal{A}$ were zero. Our sensitivity is 25.8 ps^{-1} and exceeds the combined sensitivity of all previous experiments [3]. At $\Delta m_s = 17.3$ ps^{-1}, the observed amplitude $\mathcal{A} = 1.03 \pm 0.28$(stat) is consistent with unity, indicating that the data are compatible with B_s^0-$\bar{B}_s^0$ oscillations with that frequency, while the amplitude is inconsistent with zero: $\mathcal{A}/\sigma_\mathcal{A} = 3.7$, where $\sigma_\mathcal{A}$ is the uncertainty on $\mathcal{A}$. The negative amplitudes measured at frequencies slightly below and slightly above the peak frequency are expected and are due to the finite range in signal decay time that is imposed by the trigger and selection criteria. The systematic uncertainty on $\mathcal{A}$ is mainly due to uncertainties on σ_{t_i} and $\mathcal{D}_i$. Since the effect of these uncertainties on $\mathcal{A}$ and $\sigma_\mathcal{A}$ are correlated, the ratio $\mathcal{A}/\sigma_\mathcal{A}$ has negligible systematic uncertainty.

The significance of the potential signal is evaluated from $\Lambda \equiv \log[\mathcal{L}^{\mathcal{A}=0}/\mathcal{L}^{\mathcal{A}=1}(\Delta m_s)]$, which is the logarithm of the ratio of likelihoods for the hypothesis of oscillations ($\mathcal{A} = 1$) at the probe value and the hypothesis that $\mathcal{A} = 0$, which is equivalent to random production flavor tags. Figure 2 (lower panel) shows Λ as a function of Δm_s. Separate curves are shown for the semileptonic data alone (dash-dotted line), the hadronic data alone (dotted line), and the combined data (solid line). A minimal value of

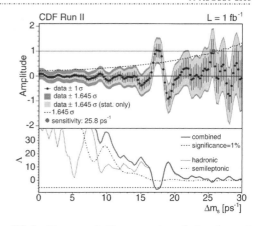

FIG. 2. (Upper panel) The measured amplitude values and uncertainties versus the B_s^0-$\bar{B}_s^0$ oscillation frequency Δm_s. Shown in light gray and dark gray are the 95% one-sided confidence level bands for statistical uncertainties only and including systematic uncertainties, respectively. (Lower panel) The logarithm of the ratio of likelihoods for amplitude equal to zero and amplitude equal to one, $\Lambda = \log[\mathcal{L}^{\mathcal{A}=0}/\mathcal{L}^{\mathcal{A}=1}(\Delta m_s)]$, versus the oscillation frequency. The dashed horizontal line indicates the value of Λ that corresponds to a probability of 1% in the case of randomly tagged data.

$\Lambda = -6.75$ is observed at $\Delta m_s = 17.3$ ps^{-1}. The significance of the signal is quantified by the probability that randomly tagged data would produce a value of Λ lower than -6.75 at any value of Δm_s. We repeat the fit 50 000 times with random tagging decisions, and we find this probability is 0.2%.

Under the hypothesis that the signal is due to B_s^0-$\bar{B}_s^0$ oscillations, we fix $\mathcal{A} = 1$ and fit for the oscillation frequency. We find $\Delta m_s = 17.31^{+0.33}_{-0.18}$(stat) ± 0.07(syst) ps^{-1} and the range 17.01 ps$^{-1} < \Delta m_s < 17.84$ ps^{-1} (16.96 ps$^{-1} < \Delta m_s < 17.91$ ps^{-1}) at 90% (95%) C.L. All systematic uncertainties affecting $\mathcal{A}$ are unimportant for Δm_s. The only non-negligible systematic uncertainty on Δm_s is from the uncertainty on the absolute scale of the decay-time measurement. Contributions to this uncertainty include biases in the primary-vertex reconstruction due to the presence of the opposite-side b hadron, uncertainties in the silicon-detector alignment, and biases in track fitting. The measured B_s^0-$\bar{B}_s^0$ oscillation frequency is used to derive the ratio $|V_{td}/V_{ts}| = \xi\sqrt{\frac{\Delta m_d}{\Delta m_s}\frac{m_{B_s^0}}{m_{B^0}}}$. As inputs we use $m_{B^0}/m_{B_s^0} = 0.98390$ [29] with negligible uncertainty, $\Delta m_d = 0.505 \pm 0.005$ ps^{-1} [3], and $\xi = 1.21^{+0.047}_{-0.035}$ [30]. We find $|V_{td}/V_{ts}| = 0.208^{+0.001}_{-0.002}$(expt)$^{+0.008}_{-0.006}$(theor).

In conclusion, we present the first precise measurement of Δm_s. The value of Δm_s is consistent with standard model expectations [31] and with previous bounds. Our measured value of Δm_s allows us to determine $|V_{td}/V_{ts}|$ with unprecedented precision and can be used to improve constraints on the unitarity of the CKM matrix and on scenarios involving new physics.

We thank the Fermilab staff and the technical staffs of the participating institutions for their vital contributions. This work was supported by the U.S. Department of Energy and National Science Foundation; the Italian Istituto Nazionale di Fisica Nucleare; the Ministry of Education, Culture, Sports, Science and Technology of Japan; the Natural Sciences and Engineering Research Council of Canada; the National Science Council of the Republic of China; the Swiss National Science Foundation; the A. P. Sloan Foundation; the Bundesministerium für Bildung und Forschung, Germany; the Korean Science and Engineering Foundation and the Korean Research Foundation; the Particle Physics and Astronomy Research Council and the Royal Society, U.K.; the Russian Foundation for Basic Research; the Comisión Interministerial de Ciencia y Tecnología, Spain; in part by the European Community's Human Potential Programme under Contract No. HPRN-CT-2002-00292; and the Academy of Finland.

[1] C. Gay, Annu. Rev. Nucl. Part. Sci. **50**, 577 (2000). We set $\hbar = c = 1$ and report $\Delta m_q = m_{B_{q,H}^0} - m_{B_{q,L}^0}$ in inverse picoseconds.
[2] N. Cabibbo, Phys. Rev. Lett. **10**, 531 (1963); M. Kobayashi and T. Maskawa, Prog. Theor. Phys. **49**, 652 (1973).
[3] S. Eidelman et al., Phys. Lett. B **592**, 1 (2004), and 2005 partial update for the 2006 edition available on the PDG WWW page (http://pdg.lbl.gov/).
[4] K. Abe et al. (BELLE Collaboration), Phys. Rev. D **71**, 072003 (2005); **71**, 079903(E) (2005); N. C. Hastings et al. (BELLE Collaboration), Phys. Rev. D **67**, 052004 (2003); B. Aubert et al. (BABAR Collaboration), Phys. Rev. Lett. **88**, 221803 (2002).
[5] J. Abdallah et al. (DELPHI Collaboration), Eur. Phys. J. C **35**, 35 (2004); K. Abe et al. (SLD Collaboration), Phys. Rev. D **67**, 012006 (2003); A. Heister et al. (ALEPH Collaboration), Eur. Phys. J. C **29**, 143 (2003).
[6] V. M. Abazov et al. (D0 Collaboration), Phys. Rev. Lett. **97**, 021802 (2006).
[7] The symbol B_s refers to the combination of $\bar{B}_s^0$ and B_s^0 decays.
[8] References to a particular process imply that the charge conjugate process is included as well.
[9] D. Acosta et al. (CDF Collaboration), Phys. Rev. D **71**, 032001 (2005); R. Blair et al. (CDF Collaboration), Fermilab Report No. FERMILAB-PUB-96-390-E, 1996.
[10] A. Abulencia et al. (CDF Collaboration), hep-ex/0508029.
[11] A. Sill et al., Nucl. Instrum. Methods Phys. Res., Sect. A **447**, 1 (2000).
[12] A. Affolder et al., Nucl. Instrum. Methods Phys. Res., Sect. A **453**, 84 (2000).
[13] C. S. Hill et al., Nucl. Instrum. Methods Phys. Res., Sect. A **530**, 1 (2004).
[14] T. Affolder et al., Nucl. Instrum. Methods Phys. Res., Sect. A **526**, 249 (2004).
[15] S. Cabrera et al., Nucl. Instrum. Methods Phys. Res., Sect. A **494**, 416 (2002).
[16] W. Ashmanskas et al., Nucl. Instrum. Methods Phys. Res., Sect. A **518**, 532 (2004).
[17] A. Abulencia et al. (CDF Collaboration), Phys. Rev. Lett. **96**, 191801 (2006).
[18] G. Giurgiu, Ph.D. thesis, Carnegie Mellon University (Fermilab Report No. FERMILAB-THESIS-2005-41, 2005).
[19] A. Abulencia et al. (CDF Collaboration), Phys. Rev. Lett. **97**, 012002 (2006).
[20] The transverse momentum p_T is the magnitude of the component of the momentum perpendicular to the proton beam direction.
[21] D. Acosta et al. (CDF Collaboration), Phys. Rev. Lett. **91**, 241804 (2003).
[22] F. Abe et al. (CDF Collaboration), Phys. Rev. D **60**, 072003 (1999).
[23] A. Ali and F. Barreiro, Z. Phys. C **30**, 635 (1986); M. Gronau, A. Nippe, and J. L. Rosner, Phys. Rev. D **47**, 1988 (1993); M. Gronau and J. L. Rosner, Phys. Rev. D **49**, 254 (1994).
[24] F. Abe et al. (CDF Collaboration), Phys. Rev. D **59**, 032001 (1999).
[25] T. Affolder et al. (CDF Collaboration), Phys. Rev. D **61**, 072005 (2000).
[26] T. Sjöstrand et al., Comput. Phys. Commun. **135**, 238 (2001). We use version 6.216.
[27] D. Usynin, Ph.D. thesis, University of Pennsylvania (Fermilab Report No. FERMILAB-THESIS-2005-68, 2005).
[28] H. G. Moser and A. Roussarie, Nucl. Instrum. Methods Phys. Res., Sect. A **384**, 491 (1997).
[29] D. Acosta et al. (CDF Collaboration), Phys. Rev. Lett. **96**, 202001 (2006).
[30] M. Okamoto, Proc. Sci. LAT2005 (2005) 013 [hep-lat/0510113].
[31] M. Bona et al. (UTfit Collaboration), J. High Energy Phys. 07 (2005) 028; J. Charles et al. (CKMfitter Collaboration), Eur. Phys. J. C **41**, 1 (2005).

Evidence for D^0-$\overline{D}^0$ Mixing

B. Aubert,[1] M. Bona,[1] D. Boutigny,[1] Y. Karyotakis,[1] J. P. Lees,[1] V. Poireau,[1] X. Prudent,[1] V. Tisserand,[1] A. Zghiche,[1] J. Garra Tico,[2] E. Grauges,[2] L. Lopez,[3] A. Palano,[3] G. Eigen,[4] B. Stugu,[4] L. Sun,[4] G. S. Abrams,[5] M. Battaglia,[5] D. N. Brown,[5] J. Button-Shafer,[5] R. N. Cahn,[5] Y. Groysman,[5] R. G. Jacobsen,[5] J. A. Kadyk,[5] L. T. Kerth,[5] Yu. G. Kolomensky,[5] G. Kukartsev,[5] D. Lopes Pegna,[5] G. Lynch,[5] L. M. Mir,[5] T. J. Orimoto,[5] M. T. Ronan,[5,*] K. Tackmann,[5] W. A. Wenzel,[5] P. del Amo Sanchez,[6] C. M. Hawkes,[6] A. T. Watson,[6] T. Held,[7] H. Koch,[7] B. Lewandowski,[7] M. Pelizaeus,[7] T. Schroeder,[7] M. Steinke,[7] D. Walker,[8] D. J. Asgeirsson,[9] T. Cuhadar-Donszelmann,[9] B. G. Fulsom,[9] C. Hearty,[9] N. S. Knecht,[9] T. S. Mattison,[9] J. A. McKenna,[9] A. Khan,[10] M. Saleem,[10] L. Teodorescu,[10] V. E. Blinov,[11] A. D. Bukin,[11] V. P. Druzhinin,[11] V. B. Golubev,[11] A. P. Onuchin,[11] S. I. Serednyakov,[11] Yu. I. Skovpen,[11] E. P. Solodov,[11] K. Yu. Todyshev,[11] M. Bondioli,[12] S. Curry,[12] I. Eschrich,[12] D. Kirkby,[12] A. J. Lankford,[12] P. Lund,[12] M. Mandelkern,[12] E. C. Martin,[12] D. P. Stoker,[12] S. Abachi,[13] C. Buchanan,[13] S. D. Foulkes,[14] J. W. Gary,[14] F. Liu,[14] O. Long,[14] B. C. Shen,[14] L. Zhang,[14] H. P. Paar,[15] S. Rahatlou,[15] V. Sharma,[15] J. W. Berryhill,[16] C. Campagnari,[16] A. Cunha,[16] B. Dahmes,[16] T. M. Hong,[16] D. Kovalskyi,[16] J. D. Richman,[16] T. W. Beck,[17] A. M. Eisner,[17] C. J. Flacco,[17] C. A. Heusch,[17] J. Kroseberg,[17] W. S. Lockman,[17] T. Schalk,[17] B. A. Schumm,[17] A. Seiden,[17] D. C. Williams,[17] M. G. Wilson,[17] L. O. Winstrom,[17] E. Chen,[18] C. H. Cheng,[18] F. Fang,[18] D. G. Hitlin,[18] I. Narsky,[18] T. Piatenko,[18] F. C. Porter,[18] G. Mancinelli,[19] B. T. Meadows,[19] K. Mishra,[19] M. D. Sokoloff,[19] F. Blanc,[20] P. C. Bloom,[20] S. Chen,[20] W. T. Ford,[20] J. F. Hirschauer,[20] A. Kreisel,[20] M. Nagel,[20] U. Nauenberg,[20] A. Olivas,[20] J. G. Smith,[20] K. A. Ulmer,[20] S. R. Wagner,[20] J. Zhang,[20] A. M. Gabareen,[21] A. Soffer,[21] W. H. Toki,[21] R. J. Wilson,[21] F. Winklmeier,[21] Q. Zeng,[21] D. D. Altenburg,[22] E. Feltresi,[22] A. Hauke,[22] H. Jasper,[22] J. Merkel,[22] A. Petzold,[22] B. Spaan,[22] K. Wacker,[22] T. Brandt,[23] V. Klose,[23] M. J. Kobel,[23] H. M. Lacker,[23] W. F. Mader,[23] R. Nogowski,[23] J. Schubert,[23] K. R. Schubert,[23] R. Schwierz,[23] J. E. Sundermann,[23] A. Volk,[23] D. Bernard,[24] G. R. Bonneaud,[24] E. Latour,[24] V. Lombardo,[24] Ch. Thiebaux,[24] M. Verderi,[24] P. J. Clark,[25] W. Gradl,[25] F. Muheim,[25] S. Playfer,[25] A. I. Robertson,[25] Y. Xie,[25] M. Andreotti,[26] D. Bettoni,[26] C. Bozzi,[26] R. Calabrese,[26] A. Cecchi,[26] G. Cibinetto,[26] P. Franchini,[26] E. Luppi,[26] M. Negrini,[26] A. Petrella,[26] L. Piemontese,[26] E. Prencipe,[26] V. Santoro,[26] F. Anulli,[27] R. Baldini-Ferroli,[27] A. Calcaterra,[27] R. de Sangro,[27] G. Finocchiaro,[27] S. Pacetti,[27] P. Patteri,[27] I. M. Peruzzi,[27,†] M. Piccolo,[27] M. Rama,[27] A. Zallo,[27] A. Buzzo,[28] R. Contri,[28] M. Lo Vetere,[28] M. M. Macri,[28] M. R. Monge,[28] S. Passaggio,[28] C. Patrignani,[28] E. Robutti,[28] A. Santroni,[28] S. Tosi,[28] K. S. Chaisanguanthum,[29] M. Morii,[29] J. Wu,[29] R. S. Dubitzky,[30] J. Marks,[30] S. Schenk,[30] U. Uwer,[30] D. J. Bard,[31] P. D. Dauncey,[31] R. L. Flack,[31] J. A. Nash,[31] M. B. Nikolich,[31] W. Panduro Vazquez,[31] P. K. Behera,[32] X. Chai,[32] M. J. Charles,[32] U. Mallik,[32] N. T. Meyer,[32] V. Ziegler,[32] J. Cochran,[33] H. B. Crawley,[33] L. Dong,[33] V. Eyges,[33] W. T. Meyer,[33] S. Prell,[33] E. I. Rosenberg,[33] A. E. Rubin,[33] A. V. Gritsan,[34] Z. J. Guo,[34] C. K. Lae,[34] A. G. Denig,[35] M. Fritsch,[35] G. Schott,[35] N. Arnaud,[36] J. Béquilleux,[36] M. Davier,[36] G. Grosdidier,[36] A. Höcker,[36] V. Lepeltier,[36] F. Le Diberder,[36] A. M. Lutz,[36] S. Pruvot,[36] S. Rodier,[36] P. Roudeau,[36] M. H. Schune,[36] J. Serrano,[36] V. Sordini,[36] A. Stocchi,[36] W. F. Wang,[36] G. Wormser,[36] D. J. Lange,[37] D. M. Wright,[37] C. A. Chavez,[38] I. J. Forster,[38] J. R. Fry,[38] E. Gabathuler,[38] R. Gamet,[38] D. E. Hutchcroft,[38] D. J. Payne,[38] K. C. Schofield,[38] C. Touramanis,[38] A. J. Bevan,[39] K. A. George,[39] F. Di Lodovico,[39] W. Menges,[39] R. Sacco,[39] G. Cowan,[40] H. U. Flaecher,[40] D. A. Hopkins,[40] P. S. Jackson,[40] T. R. McMahon,[40] F. Salvatore,[40] A. C. Wren,[40] D. N. Brown,[41] C. L. Davis,[41] J. Allison,[42] N. R. Barlow,[42] R. J. Barlow,[42] Y. M. Chia,[42] C. L. Edgar,[42] G. D. Lafferty,[42] T. J. West,[42] J. I. Yi,[42] J. Anderson,[43] C. Chen,[43] A. Jawahery,[43] D. A. Roberts,[43] G. Simi,[43] J. M. Tuggle,[43] G. Blaylock,[44] C. Dallapiccola,[44] S. S. Hertzbach,[44] X. Li,[44] T. B. Moore,[44] E. Salvati,[44] S. Saremi,[44] R. Cowan,[45] P. H. Fisher,[45] G. Sciolla,[45] S. J. Sekula,[45] M. Spitznagel,[45] F. Taylor,[45] R. K. Yamamoto,[45] S. E. Mclachlin,[46] P. M. Patel,[46] S. H. Robertson,[46] A. Lazzaro,[47] F. Palombo,[47] J. M. Bauer,[48] L. Cremaldi,[48] V. Eschenburg,[48] R. Godang,[48] R. Kroeger,[48] D. A. Sanders,[48] D. J. Summers,[48] H. W. Zhao,[48] S. Brunet,[49] D. Côté,[49] M. Simard,[49] P. Taras,[49] F. B. Viaud,[49] H. Nicholson,[50] G. De Nardo,[51] F. Fabozzi,[51,‡] L. Lista,[51] D. Monorchio,[51] C. Sciacca,[51] M. A. Baak,[52] G. Raven,[52] H. L. Snoek,[52] C. P. Jessop,[53] J. M. LoSecco,[53] G. Benelli,[54] L. A. Corwin,[54] K. K. Gan,[54] K. Honscheid,[54] D. Hufnagel,[54] H. Kagan,[54] R. Kass,[54] J. P. Morris,[54] A. M. Rahimi,[54] J. J. Regensburger,[54] R. Ter-Antonyan,[54] Q. K. Wong,[54] N. L. Blount,[55] J. Brau,[55] R. Frey,[55] O. Igonkina,[55] J. A. Kolb,[55] M. Lu,[55] R. Rahmat,[55] N. B. Sinev,[55] D. Strom,[55] J. Strube,[55] E. Torrence,[55] N. Gagliardi,[56] A. Gaz,[56] M. Margoni,[56] M. Morandin,[56] A. Pompili,[56] M. Posocco,[56] M. Rotondo,[56] F. Simonetto,[56] R. Stroili,[56] C. Voci,[56] E. Ben-Haim,[57] H. Briand,[57] G. Calderini,[57] J. Chauveau,[57] P. David,[57] L. Del Buono,[57] Ch. de la Vaissière,[57] O. Hamon,[57] Ph. Leruste,[57] J. Malclès,[57] J. Ocariz,[57] A. Perez,[57] L. Gladney,[58] M. Biasini,[59] R. Covarelli,[59] E. Manoni,[59] C. Angelini,[60]

G. Batignani,[60] S. Bettarini,[60] M. Carpinelli,[60] R. Cenci,[60] A. Cervelli,[60] F. Forti,[60] M. A. Giorgi,[60] A. Lusiani,[60]
G. Marchiori,[60] M. A. Mazur,[60] M. Morganti,[60] N. Neri,[60] E. Paoloni,[60] G. Rizzo,[60] J. J. Walsh,[60] M. Haire,[61] J. Biesiada,[62]
P. Elmer,[62] Y. P. Lau,[62] C. Lu,[62] J. Olsen,[62] A. J. S. Smith,[62] A. V. Telnov,[62] E. Baracchini,[63] F. Bellini,[63] G. Cavoto,[63]
A. D'Orazio,[63] D. del Re,[63] E. Di Marco,[63] R. Faccini,[63] F. Ferrarotto,[63] F. Ferroni,[63] M. Gaspero,[63] P. D. Jackson,[63]
L. Li Gioi,[63] M. A. Mazzoni,[63] S. Morganti,[63] G. Piredda,[63] F. Polci,[63] F. Renga,[63] C. Voena,[63] M. Ebert,[64] H. Schröder,[64]
R. Waldi,[64] T. Adye,[65] G. Castelli,[65] B. Franek,[65] E. O. Olaiya,[65] S. Ricciardi,[65] W. Roethel,[65] F. F. Wilson,[65] R. Aleksan,[66]
S. Emery,[66] M. Escalier,[66] A. Gaidot,[66] S. F. Ganzhur,[66] G. Hamel de Monchenault,[66] W. Kozanecki,[66] M. Legendre,[66]
G. Vasseur,[66] Ch. Yèche,[66] M. Zito,[66] X. R. Chen,[67] H. Liu,[67] W. Park,[67] M. V. Purohit,[67] J. R. Wilson,[67] M. T. Allen,[68]
D. Aston,[68] R. Bartoldus,[68] P. Bechtle,[68] N. Berger,[68] R. Claus,[68] J. P. Coleman,[68] M. R. Convery,[68] J. C. Dingfelder,[68]
J. Dorfan,[68] G. P. Dubois-Felsmann,[68] D. Dujmic,[68] W. Dunwoodie,[68] R. C. Field,[68] T. Glanzman,[68] S. J. Gowdy,[68]
M. T. Graham,[68] P. Grenier,[68] C. Hast,[68] T. Hryn'ova,[68] W. R. Innes,[68] J. Kaminski,[68] M. H. Kelsey,[68] H. Kim,[68] P. Kim,[68]
M. L. Kocian,[68] D. W. G. S. Leith,[68] S. Li,[68] S. Luitz,[68] V. Luth,[68] H. L. Lynch,[68] D. B. MacFarlane,[68] H. Marsiske,[68]
R. Messner,[68] D. R. Muller,[68] C. P. O'Grady,[68] I. Ofte,[68] A. Perazzo,[68] M. Perl,[68] T. Pulliam,[68] B. N. Ratcliff,[68]
A. Roodman,[68] A. A. Salnikov,[68] R. H. Schindler,[68] J. Schwiening,[68] A. Snyder,[68] J. Stelzer,[68] D. Su,[68] M. K. Sullivan,[68]
K. Suzuki,[68] S. K. Swain,[68] J. M. Thompson,[68] J. Va'vra,[68] N. van Bakel,[68] A. P. Wagner,[68] M. Weaver,[68]
W. J. Wisniewski,[68] M. Wittgen,[68] D. H. Wright,[68] A. K. Yarritu,[68] K. Yi,[68] C. C. Young,[68] P. R. Burchat,[69] A. J. Edwards,[69]
S. A. Majewski,[69] B. A. Petersen,[69] L. Wilden,[69] S. Ahmed,[70] M. S. Alam,[70] R. Bula,[70] J. A. Ernst,[70] V. Jain,[70] B. Pan,[70]
M. A. Saeed,[70] F. R. Wappler,[70] S. B. Zain,[70] W. Bugg,[71] M. Krishnamurthy,[71] S. M. Spanier,[71] R. Eckmann,[72]
J. L. Ritchie,[72] A. M. Ruland,[72] C. J. Schilling,[72] R. F. Schwitters,[72] J. M. Izen,[73] X. C. Lou,[73] S. Ye,[73] F. Bianchi,[74]
F. Gallo,[74] D. Gamba,[74] M. Pelliccioni,[74] M. Bomben,[75] L. Bosisio,[75] C. Cartaro,[75] F. Cossutti,[75] G. Della Ricca,[75]
L. Lanceri,[75] L. Vitale,[75] V. Azzolini,[76] N. Lopez-March,[76] F. Martinez-Vidal,[76] D. A. Milanes,[76] A. Oyanguren,[76]
J. Albert,[77] Sw. Banerjee,[77] B. Bhuyan,[77] K. Hamano,[77] R. Kowalewski,[77] I. M. Nugent,[77] J. M. Roney,[77] R. J. Sobie,[77]
J. J. Back,[78] P. F. Harrison,[78] T. E. Latham,[78] G. B. Mohanty,[78] M. Pappagallo,[78,§] H. R. Band,[79] X. Chen,[79] S. Dasu,[79]
K. T. Flood,[79] J. J. Hollar,[79] P. E. Kutter,[79] Y. Pan,[79] M. Pierini,[79] R. Prepost,[79] S. L. Wu,[79]
Z. Yu,[79] and H. Neal[80]

(BABAR Collaboration)

[1]*Laboratoire de Physique des Particules, IN2P3/CNRS et Université de Savoie, F-74941 Annecy-Le-Vieux, France*
[2]*Universitat de Barcelona, Facultat de Fisica, Departament ECM, E-08028 Barcelona, Spain*
[3]*Università di Bari, Dipartimento di Fisica and INFN, I-70126 Bari, Italy*
[4]*University of Bergen, Institute of Physics, N-5007 Bergen, Norway*
[5]*Lawrence Berkeley National Laboratory and University of California, Berkeley, California 94720, USA*
[6]*University of Birmingham, Birmingham, B15 2TT, United Kingdom*
[7]*Ruhr Universität Bochum, Institut für Experimentalphysik 1, D-44780 Bochum, Germany*
[8]*University of Bristol, Bristol BS8 1TL, United Kingdom*
[9]*University of British Columbia, Vancouver, British Columbia, Canada V6T 1Z1*
[10]*Brunel University, Uxbridge, Middlesex UB8 3PH, United Kingdom*
[11]*Budker Institute of Nuclear Physics, Novosibirsk 630090, Russia*
[12]*University of California at Irvine, Irvine, California 92697, USA*
[13]*University of California at Los Angeles, Los Angeles, California 90024, USA*
[14]*University of California at Riverside, Riverside, California 92521, USA*
[15]*University of California at San Diego, La Jolla, California 92093, USA*
[16]*University of California at Santa Barbara, Santa Barbara, California 93106, USA*
[17]*University of California at Santa Cruz, Institute for Particle Physics, Santa Cruz, California 95064, USA*
[18]*California Institute of Technology, Pasadena, California 91125, USA*
[19]*University of Cincinnati, Cincinnati, Ohio 45221, USA*
[20]*University of Colorado, Boulder, Colorado 80309, USA*
[21]*Colorado State University, Fort Collins, Colorado 80523, USA*
[22]*Universität Dortmund, Institut für Physik, D-44221 Dortmund, Germany*
[23]*Technische Universität Dresden, Institut für Kern- und Teilchenphysik, D-01062 Dresden, Germany*
[24]*Laboratoire Leprince-Ringuet, CNRS/IN2P3, Ecole Polytechnique, F-91128 Palaiseau, France*
[25]*University of Edinburgh, Edinburgh EH9 3JZ, United Kingdom*
[26]*Università di Ferrara, Dipartimento di Fisica and INFN, I-44100 Ferrara, Italy*
[27]*Laboratori Nazionali di Frascati dell'INFN, I-00044 Frascati, Italy*

[28]Università di Genova, Dipartimento di Fisica and INFN, I-16146 Genova, Italy
[29]Harvard University, Cambridge, Massachusetts 02138, USA
[30]Universität Heidelberg, Physikalisches Institut, Philosophenweg 12, D-69120 Heidelberg, Germany
[31]Imperial College London, London, SW7 2AZ, United Kingdom
[32]University of Iowa, Iowa City, Iowa 52242, USA
[33]Iowa State University, Ames, Iowa 50011-3160, USA
[34]Johns Hopkins University, Baltimore, Maryland 21218, USA
[35]Universität Karlsruhe, Institut für Experimentelle Kernphysik, D-76021 Karlsruhe, Germany
[36]Laboratoire de l'Accélérateur Linéaire, IN2P3/CNRS et Université Paris-Sud 11, Centre Scientifique d'Orsay, B.P. 34, F-91898 ORSAY Cedex, France
[37]Lawrence Livermore National Laboratory, Livermore, California 94550, USA
[38]University of Liverpool, Liverpool L69 7ZE, United Kingdom
[39]Queen Mary, University of London, E1 4NS, United Kingdom
[40]University of London, Royal Holloway and Bedford New College, Egham, Surrey TW20 0EX, United Kingdom
[41]University of Louisville, Louisville, Kentucky 40292, USA
[42]University of Manchester, Manchester M13 9PL, United Kingdom
[43]University of Maryland, College Park, Maryland 20742, USA
[44]University of Massachusetts, Amherst, Massachusetts 01003, USA
[45]Massachusetts Institute of Technology, Laboratory for Nuclear Science, Cambridge, Massachusetts 02139, USA
[46]McGill University, Montréal, Québec, Canada H3A 2T8
[47]Università di Milano, Dipartimento di Fisica and INFN, I-20133 Milano, Italy
[48]University of Mississippi, University, Mississippi 38677, USA
[49]Université de Montréal, Physique des Particules, Montréal, Québec, Canada H3C 3J7
[50]Mount Holyoke College, South Hadley, Massachusetts 01075, USA
[51]Università di Napoli Federico II, Dipartimento di Scienze Fisiche and INFN, I-80126, Napoli, Italy
[52]NIKHEF, National Institute for Nuclear Physics and High Energy Physics, NL-1009 DB Amsterdam, The Netherlands
[53]University of Notre Dame, Notre Dame, Indiana 46556, USA
[54]Ohio State University, Columbus, Ohio 43210, USA
[55]University of Oregon, Eugene, Oregon 97403, USA
[56]Università di Padova, Dipartimento di Fisica and INFN, I-35131 Padova, Italy
[57]Laboratoire de Physique Nucléaire et de Hautes Energies, IN2P3/CNRS, Université Pierre et Marie Curie-Paris6, Université Denis Diderot-Paris7, F-75252 Paris, France
[58]University of Pennsylvania, Philadelphia, Pennsylvania 19104, USA
[59]Università di Perugia, Dipartimento di Fisica and INFN, I-06100 Perugia, Italy
[60]Università di Pisa, Dipartimento di Fisica, Scuola Normale Superiore and INFN, I-56127 Pisa, Italy
[61]Prairie View A&M University, Prairie View, Texas 77446, USA
[62]Princeton University, Princeton, New Jersey 08544, USA
[63]Università di Roma La Sapienza, Dipartimento di Fisica and INFN, I-00185 Roma, Italy
[64]Universität Rostock, D-18051 Rostock, Germany
[65]Rutherford Appleton Laboratory, Chilton, Didcot, Oxon, OX11 0QX, United Kingdom
[66]DSM/Dapnia, CEA/Saclay, F-91191 Gif-sur-Yvette, France
[67]University of South Carolina, Columbia, South Carolina 29208, USA
[68]Stanford Linear Accelerator Center, Stanford, California 94309, USA
[69]Stanford University, Stanford, California 94305-4060, USA
[70]State University of New York, Albany, New York 12222, USA
[71]University of Tennessee, Knoxville, Tennessee 37996, USA
[72]University of Texas at Austin, Austin, Texas 78712, USA
[73]University of Texas at Dallas, Richardson, Texas 75083, USA
[74]Università di Torino, Dipartimento di Fisica Sperimentale and INFN, I-10125 Torino, Italy
[75]Università di Trieste, Dipartimento di Fisica and INFN, I-34127 Trieste, Italy
[76]IFIC, Universitat de Valencia-CSIC, E-46071 Valencia, Spain
[77]University of Victoria, Victoria, British Columbia, Canada V8W 3P6
[78]Department of Physics, University of Warwick, Coventry CV4 7AL, United Kingdom
[79]University of Wisconsin, Madison, Wisconsin 53706, USA
[80]Yale University, New Haven, Connecticut 06511, USA
(Received 9 March 2007; published 24 May 2007)

We present evidence for D^0-$\overline{D}^0$ mixing in $D^0 \to K^+ \pi^-$ decays from 384 fb^{-1} of e^+e^- colliding-beam data recorded near $\sqrt{s} = 10.6$ GeV with the BABAR detector at the PEP-II storage rings at the Stanford Linear Accelerator Center. We find the mixing parameters $x'^2 = [-0.22 \pm 0.30(\text{stat}) \pm 0.21(\text{syst})] \times 10^{-3}$ and $y' = [9.7 \pm 4.4(\text{stat}) \pm 3.1(\text{syst})] \times 10^{-3}$ and a correlation between them of -0.95. This result is inconsistent with the no-mixing hypothesis with a significance of 3.9 standard deviations. We measure R_D, the ratio of doubly Cabibbo-suppressed to Cabibbo-favored decay rates, to be $[0.303 \pm 0.016(\text{stat}) \pm 0.010(\text{syst})]\%$. We find no evidence for CP violation.

DOI: 10.1103/PhysRevLett.98.211802   PACS numbers: 13.25.Ft, 11.30.Er, 12.15.Ff, 14.40.Lb

Quantum-mechanical mixing of neutral-meson particle-antiparticle states has been observed in the K [1], B [2], and B_s [3] systems but not yet in the D system. D mesons, which contain a charm quark, are the only system where contributions of down-type quarks in the mixing loop can be explored. In the standard model (SM), the D^0-$\bar{D}^0$ mixing rate is expected to be very small (10^{-4} or less), due to Glashow-Iliopoulos-Maiani suppression of the first two quark generations and Cabibbo-Kobayashi-Maskawa suppression of the third [4]. Long-distance effects from intermediate states coupling to both D^0 and $\bar{D}^0$ also contribute, making precise prediction and interpretation difficult [5]. We present evidence for D mixing consistent with these expectations and with previous experimental limits [6].

To the extent that only the first two generations are involved, CP violation is expected to be well below the sensitivity of this experiment, although non-SM processes could enhance either mixing or CP violation. We compare D^0 and $\bar{D}^0$ samples separately and find no evidence for CP violation.

We study the right-sign (RS), Cabibbo-favored (CF) decay $D^0 \to K^-\pi^+$ [7] and the wrong-sign (WS) decay $D^0 \to K^+\pi^-$. The latter can be produced via the doubly Cabibbo-suppressed (DCS) decay $D^0 \to K^+\pi^-$ or via mixing followed by a CF decay $D^0 \to \bar{D}^0 \to K^+\pi^-$. The DCS decay has a small rate R_D of order $\tan^4\theta_C \approx 0.3\%$ relative to CF decay, with θ_C the Cabibbo angle. We distinguish D^0 and $\bar{D}^0$ by their production in the decay $D^{*+} \to \pi_s^+ D^0$, where the π_s^+ is referred to as the "slow pion." In RS decays, the π_s^+ and the kaon have opposite charges, while in WS decays the charges are the same. The time dependence of the WS decay rate is used to separate the contributions of DCS decays from D^0-$\bar{D}^0$ mixing.

The D^0 and $\bar{D}^0$ mesons are produced as flavor eigenstates but evolve and decay as mixtures of the eigenstates D_1 and D_2 of the Hamiltonian, with masses and widths M_1, Γ_1 and M_2, Γ_2, respectively. Mixing is characterized by the mass and lifetime differences $\Delta M = M_1 - M_2$ and $\Delta\Gamma = \Gamma_1 - \Gamma_2$. Defining the parameters $x = \Delta M/\Gamma$ and $y = \Delta\Gamma/2\Gamma$, where $\Gamma = (\Gamma_1 + \Gamma_2)/2$, we approximate the time dependence of the WS decay of a meson produced as a D^0 at time $t = 0$ in the limit of small mixing ($|x|, |y| \ll 1$) and CP conservation as

$$\frac{T_{\text{WS}}(t)}{e^{-\Gamma t}} \propto R_D + \sqrt{R_D} y' \Gamma t + \frac{x'^2 + y'^2}{4}(\Gamma t)^2, \quad (1)$$

where $x' = x\cos\delta_{K\pi} + y\sin\delta_{K\pi}$, $y' = -x\sin\delta_{K\pi} + y\cos\delta_{K\pi}$, and $\delta_{K\pi}$ is the strong phase between the DCS and CF amplitudes.

We study both CP-conserving and CP-violating cases. For the CP-conserving case, we fit for the parameters R_D, x'^2, and y'. To search for CP violation, we apply Eq. (1) to the D^0 and $\bar{D}^0$ samples separately, fitting for the parameters $\{R_D^{\pm}, x'^{2\pm}, y'^{\pm}\}$ for D^0 (+) decays and $\bar{D}^0$ (−) decays.

We use 384 fb^{-1} of e^+e^- colliding-beam data recorded near $\sqrt{s} = 10.6$ GeV with the BABAR detector [8] at the PEP-II asymmetric-energy storage rings. We select D^0 candidates by pairing oppositely charged tracks with a $K^{\mp}\pi^{\pm}$ invariant mass $m_{K\pi}$ between 1.81 and 1.92 GeV/c^2. Each pair is identified as $K^{\mp}\pi^{\pm}$ using a likelihood-based particle identification algorithm. We require the π_s^+ to have a momentum in the laboratory frame greater than 0.1 GeV/c and in the e^+e^- center-of-mass (c.m.) frame below 0.45 GeV/c.

To obtain the proper decay time t and its error σ_t for each D^0 candidate, we refit the $K^{\mp}$ and $\pi^{\pm}$ tracks, constraining them to originate from a common vertex. We also require the D^0 and π_s^+ to originate from a common vertex, constrained by the position and size of the e^+e^- interaction region. The vertical rms size of each beam is typically 6 μm [8]. We require the χ^2 probability of the vertex-constrained combined fit $P(\chi^2)$ to be at least 0.1% and the $m_{K\pi\pi_s} - m_{K\pi}$ mass difference Δm to satisfy $0.14 < \Delta m < 0.16$ GeV/c^2.

To remove D^0 candidates from B-meson decays and to reduce combinatorial backgrounds, we require each D^0 to have a momentum in the c.m. frame greater than 2.5 GeV/c. We require $-2 < t < 4$ ps and $\sigma_t < 0.5$ ps (the most probable value of σ_t for signal events is 0.16 ps). For D^{*+} candidates sharing one or more tracks with other D^{*+} candidates, we retain only the candidate with the highest $P(\chi^2)$. After applying all criteria, we keep approximately 1 229 000 RS and 64 000 WS D^0 and $\bar{D}^0$ candidates. To avoid potential bias, we finalized the analysis procedure without examining the mixing results.

The mixing parameters are determined in an unbinned, extended maximum-likelihood fit to the RS and WS data samples over the four observables $m_{K\pi}$, Δm, t, and σ_t. The fit is performed in several stages. First, RS and WS signal and background shape parameters are determined from a fit to $m_{K\pi}$ and Δm and are not varied in subsequent fits. Next, the D^0 proper-time resolution function and lifetime are determined in a fit to the RS data using $m_{K\pi}$ and Δm to separate the signal and background components. We fit to the WS data sample using three different models. The first model assumes both CP conservation and the absence of mixing. The second model allows for mixing but assumes no CP violation. The third model allows for both mixing and CP violation.

The RS and WS $\{m_{K\pi}, \Delta m\}$ distributions are described by four components: signal, random π_s^+, misreconstructed D^0, and combinatorial background. The signal component has a characteristic peak in both $m_{K\pi}$ and Δm. The random π_s^+ component models reconstructed D^0 decays combined with a random slow pion and has the same shape in $m_{K\pi}$ as signal events but does not peak in Δm. Misreconstructed D^0 events have one or more of the D^0 decay products either not reconstructed or reconstructed with the wrong particle hypothesis. They peak in Δm but not in $m_{K\pi}$. For RS events, most of these are semileptonic D^0 decays. For WS events, the main contribution is RS $D^0 \to K^-\pi^+$ decays where the K^- and the π^+ are misidentified as π^- and K^+, respectively. Combinatorial background events are those not described by the above components; they do not exhibit any peaking structure in $m_{K\pi}$ or Δm.

The functional forms of the probability density functions (PDFs) for the signal and background components are chosen based on studies of Monte Carlo (MC) samples. However, all parameters are determined from two-dimensional likelihood fits to data over the full $m_{K\pi}$ and Δm region.

We fit the RS and WS data samples simultaneously with shape parameters describing the signal and random π_s^+ components shared between the two data samples. We find $1\,141\,500 \pm 1200$ RS signal events and 4030 ± 90 WS signal events. The dominant background component is the random π_s^+ background. Projections of the WS data and fit are shown in Fig. 1.

The measured proper-time distribution for the RS signal is described by an exponential function convolved with a resolution function whose parameters are determined by the fit to the data. The resolution function is the sum of three Gaussians with widths proportional to the estimated event-by-event proper-time uncertainty σ_t. The random π_s^+ background is described by the same proper-time distribution as signal events, since the slow pion has little weight in the vertex fit. The proper-time distribution of the combinatorial background is described by a sum of two Gaussians, one of which has a power-law tail to account for a small long-lived component. The combinatorial background and real D^0 decays have different σ_t distributions, as determined from data using a background-subtraction technique [9] based on the fit to $m_{K\pi}$ and Δm.

The fit to the RS proper-time distribution is performed over all events in the full $m_{K\pi}$ and Δm region. The PDFs for signal and background in $m_{K\pi}$ and Δm are used in the proper-time fit with all parameters fixed to their previously determined values. The fitted D^0 lifetime is found to be consistent with the world-average lifetime [10].

The measured proper-time distribution for the WS signal is modeled by Eq. (1) convolved with the resolution function determined in the RS proper-time fit. The random π_s^+ and misreconstructed D^0 backgrounds are described by the RS signal proper-time distribution since they are real D^0 decays. The proper-time distribution for WS data is shown in Fig. 2. The fit results with and without mixing are shown as the overlaid curves.

The fit with mixing provides a substantially better description of the data than the fit with no mixing. The significance of the mixing signal is evaluated based on the change in negative log likelihood with respect to the minimum. Figure 3 shows confidence-level (C.L.) contours calculated from the change in log likelihood ($-2\Delta \ln \mathcal{L}$) in two dimensions (x'^2 and y') with systematic uncertainties included. The likelihood maximum is at the unphysical value of $x'^2 = -2.2 \times 10^{-4}$ and $y' = 9.7 \times 10^{-3}$. The value of $-2\Delta \ln \mathcal{L}$ at the most likely point in the physically

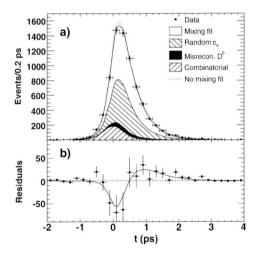

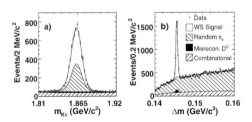

FIG. 1. (a) $m_{K\pi}$ for WS candidates with $0.1445 < \Delta m < 0.1465$ GeV/c^2 and (b) Δm for WS candidates with $1.843 < m_{K\pi} < 1.883$ GeV/c^2. The fitted PDFs are overlaid.

FIG. 2. (a) Projections of the proper-time distribution of combined D^0 and $\overline{D}^0$ WS candidates and fit result integrated over the signal region $1.843 < m_{K\pi} < 1.883$ GeV/c^2 and $0.1445 < \Delta m < 0.1465$ GeV/c^2. The result of the fit allowing (not allowing) mixing but not CP violation is overlaid as a solid (dashed) curve. (b) The points represent the difference between the data and the no-mixing fit. The solid curve shows the difference between fits with and without mixing.

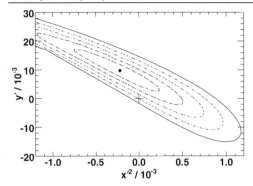

FIG. 3. The central value (point) and C.L. contours for $1 -$ C.L. $= 0.317(1\sigma)$, $4.55 \times 10^{-2}(2\sigma)$, $2.70 \times 10^{-3}(3\sigma)$, $6.33 \times 10^{-5}(4\sigma)$, and $5.73 \times 10^{-7}(5\sigma)$, calculated from the change in the value of $-2\ln\mathcal{L}$ compared with its value at the minimum. Systematic uncertainties are included. The no-mixing point is shown as a plus sign (+).

allowed region ($x'^2 = 0$ and $y' = 6.4 \times 10^{-3}$) is 0.7 units. The value of $-2\Delta\ln\mathcal{L}$ for no mixing is 23.9 units. Including the systematic uncertainties, this corresponds to a significance equivalent to 3.9 standard deviations ($1 -$ C.L. $= 1 \times 10^{-4}$) and thus constitutes evidence for mixing. The fitted values of the mixing parameters and R_D are listed in Table I. The correlation coefficient between the x'^2 and y' parameters is -0.95.

Allowing for the possibility of CP violation, we calculate the values of $R_D = \sqrt{R_D^+ R_D^-}$ and $A_D = (R_D^+ - R_D^-)/(R_D^+ + R_D^-)$ listed in Table I, from the fitted $R_D^\pm$ values. The best fit points ($x'^{2\pm}, y'^\pm$) shown in Table I are more than 3 standard deviations away from the no-mixing hypothesis. The shapes of the ($x'^{2\pm}, y'^\pm$) C.L. contours are similar to those shown in Fig. 3. All cross-checks indicate that the close agreement between the separate D^0 and $\overline{D}^0$ fit results is coincidental.

TABLE I. Results from the different fits. The first uncertainty listed is statistical and the second systematic.

Fit type	Parameter	Fit results ($/10^{-3}$)
No CP viol. or mixing	R_D	$3.53 \pm 0.08 \pm 0.04$
No CP violation	R_D	$3.03 \pm 0.16 \pm 0.10$
	x'^2	$-0.22 \pm 0.30 \pm 0.21$
	y'	$9.7 \pm 4.4 \pm 3.1$
CP violation allowed	R_D	$3.03 \pm 0.16 \pm 0.10$
	A_D	$-21 \pm 52 \pm 15$
	x'^{2+}	$-0.24 \pm 0.43 \pm 0.30$
	y'^+	$9.8 \pm 6.4 \pm 4.5$
	x'^{2-}	$-0.20 \pm 0.41 \pm 0.29$
	y'^-	$9.6 \pm 6.1 \pm 4.3$

As a cross-check of the mixing signal, we perform independent $\{m_{K\pi}, \Delta m\}$ fits with no shared parameters for intervals in proper time selected to have approximately equal numbers of RS candidates. The fitted WS branching fractions are shown in Fig. 4 and are seen to increase with time. The slope is consistent with the measured mixing parameters and inconsistent with the no-mixing hypothesis.

We validated the fitting procedure on simulated data samples using both MC samples with the full detector simulation and large parametrized MC samples. In all cases, we found the fit to be unbiased. As a further cross-check, we performed a fit to the RS data proper-time distribution allowing for mixing in the signal component; the fitted values of the mixing parameters are consistent with no mixing. In addition, we found the staged fitting approach to give the same solution and confidence regions as a simultaneous fit in which all parameters are allowed to vary.

In evaluating systematic uncertainties in R_D and the mixing parameters, we considered variations in the fit model and in the selection criteria. We also considered alternative forms of the $m_{K\pi}$, Δm, proper-time, and σ_t PDFs. We varied the t and σ_t requirements. In addition, we considered variations that keep or reject all D^{*+} candidates sharing tracks with other candidates.

For each source of systematic error, we compute the significance $s_i^2 = 2[\ln\mathcal{L}(x'^2, y') - \ln\mathcal{L}(x_i'^2, y_i')]/2.3$, where ($x'^2, y'$) are the parameters obtained from the standard fit, ($x_i'^2, y_i'$) the parameters from the fit including the ith systematic variation, and $\mathcal{L}$ the likelihood of the standard fit. The factor 2.3 is the 68% confidence level for 2 degrees of freedom. To estimate the significance of our results in (x'^2, y'), we reduce $-2\Delta\ln\mathcal{L}$ by a factor of $1 + \Sigma s_i^2 = 1.3$ to account for systematic errors. The largest contribu-

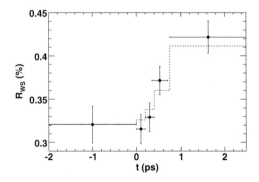

FIG. 4. The WS branching fractions from independent $\{m_{K\pi}, \Delta m\}$ fits to slices in measured proper time (points). The dashed line shows the expected wrong-sign rate as determined from the mixing fit shown in Fig. 2. The χ^2 with respect to expectation from the mixing fit is 1.5; for the no-mixing hypothesis (a constant WS rate), the χ^2 is 24.0.

tion to this factor, 0.06, is due to uncertainty in modeling the long decay time component from other D decays in the signal region. The second largest component, 0.05, is due to the presence of a nonzero mean in the proper-time signal resolution PDF. The mean value is determined in the RS proper-time fit to be 3.6 fs and is due to small misalignments in the detector. The error of 15×10^{-3} on A_D is primarily due to uncertainties in modeling the differences between K^+ and K^- absorption in the detector.

We have presented evidence for D^0-$\bar{D}^0$ mixing. Our result is inconsistent with the no-mixing hypothesis at a significance of 3.9 standard deviations. We measure $y' = [9.7 \pm 4.4(\text{stat}) \pm 3.1(\text{syst})] \times 10^{-3}$, while x'^2 is consistent with zero. We find no evidence for CP violation and measure R_D to be $[0.303 \pm 0.016(\text{stat}) \pm 0.010(\text{syst})]\%$. The result is consistent with SM estimates for mixing.

We are grateful for the excellent luminosity and machine conditions provided by our PEP-II colleagues and for the substantial dedicated effort from the computing organizations that support *BABAR*. The collaborating institutions thank SLAC for its support and kind hospitality. This work is supported by DOE and NSF (USA), NSERC (Canada), IHEP (China), CEA and CNRS-IN2P3 (France), BMBF and DFG (Germany), INFN (Italy), FOM (The Netherlands), NFR (Norway), MIST (Russia), MEC (Spain), and PPARC (United Kingdom). Individuals have received support from the Marie Curie EIF (European Union) and the A. P. Sloan Foundation.

*Deceased.
†Also with Università di Perugia, Dipartimento di Fisica, Perugia, Italy.
‡Also with Università della Basilicata, Potenza, Italy.
§Also with IPPP, Physics Department, Durham University, Durham DH1 3LE, United Kingdom.

[1] K. Lande, E. T. Booth, J. Impeduglia, L. M. Lederman, and W. Chinowsky, Phys. Rev. **103**, 1901 (1956); W. F. Fry, J. Schneps, and M. S. Swami, Phys. Rev. **103**, 1904 (1956).
[2] C. Albajar *et al.* (UA1 Collaboration), Phys. Lett. B **186**, 247 (1987); H. Albrecht *et al.* (ARGUS Collaboration), Phys. Lett. B **192**, 245 (1987).
[3] V. M. Abazov *et al.* (D0 Collaboration), Phys. Rev. Lett. **97**, 021802 (2006); A. Abulencia *et al.* (CDF Collaboration), Phys. Rev. Lett. **97**, 242003 (2006).
[4] S. Bianco, F. L. Fabbri, D. Benson, and I. Bigi, Riv. Nuovo Cimento **26N7-8**, 1 (2003); G. Burdman and I. Shipsey, Annu. Rev. Nucl. Part. Sci. **53**, 431 (2003).
[5] L. Wolfenstein, Phys. Lett. **164B**, 170 (1985); J. F. Donoghue, E. Golowich, B. R. Holstein, and J. Trampetic, Phys. Rev. D **33**, 179 (1986); I. I. Y. Bigi and N. G. Uraltsev, Nucl. Phys. **B592**, 92 (2001); A. F. Falk, Y. Grossman, Z. Ligeti, and A. A. Petrov, Phys. Rev. D **65**, 054034 (2002); A. F. Falk, Y. Grossman, Z. Ligeti, Y. Nir, and A. A. Petrov, Phys. Rev. D **69**, 114021 (2004); A. A. Petrov, Int. J. Mod. Phys. A **21**, 5686 (2006).
[6] E. M. Aitala *et al.* (E791 Collaboration), Phys. Rev. D **57**, 13 (1998); R. Barate *et al.* (ALEPH Collaboration), Phys. Lett. B **436**, 211 (1998); R. Godang *et al.* (CLEO Collaboration), Phys. Rev. Lett. **84**, 5038 (2000); B. Aubert *et al.* (*BABAR* Collaboration), Phys. Rev. Lett. **91**, 171801 (2003); J. M. Link *et al.* (FOCUS Collaboration), Phys. Lett. B **618**, 23 (2005); L. M. Zhang *et al.* (BELLE Collaboration), Phys. Rev. Lett. **96**, 151801 (2006); A. Abulencia *et al.* (CDF Collaboration), Phys. Rev. D **74**, 031109 (2006).
[7] The use of charge-conjugate modes is implied unless otherwise noted.
[8] B. Aubert *et al.* (*BABAR* Collaboration), Nucl. Instrum. Methods Phys. Res., Sect. A **479**, 1 (2002).
[9] M. Pivk and F. R. Le Diberder, Nucl. Instrum. Methods Phys. Res., Sect. A **555**, 356 (2005).
[10] W.-M. Yao *et al.* (Particle Data Group), J. Phys. G **33**, 1 (2006).

16
Neutrino Masses and Oscillations

The Old Enigma.

The most enigmatic of elementary particles, neutrinos were postulated in 1930, but were not observed until a quarter of a century later. It took another forty years to determine that they are not massless.

Neutrinos are a ubiquitous if imperceptible part of our environment. Neutrinos created in the Big Bang together with the cosmic background radiation pervade the entire Universe. The Sun is a powerful source of MeV neutrinos. Neutrinos in the GeV range are created when cosmic rays strike the atmosphere, 15 kilometers or so above the Earth's surface. Every nuclear reactor emits antineutrinos copiously. High-energy neutrinos are regularly produced at accelerators through particle decay and carefully fashioned magnetic fields can focus produced unstable charged particles to create neutrino beams.

Traditionally, efforts were made to set upper limits on the masses of the neutrinos associated with the electron, muon, and tau lepton. As explained in Chapter 6, if the electron neutrino were sufficiently massive the electron spectrum in tritium beta decay would be distorted near the end point. This prompted many painstaking measurements over the past thirty years. The expression for the spectrum actually depends on the square of the neutrino mass and the best fits can return unphysical, negative values for this. Current results give -1.1 ± 2.4 eV^2.

The direct limits on the masses of the other neutrinos are not nearly so strong. The best direct limit on the mass of ν_μ is obtained from $\pi^+ \to \mu^+ \nu_\mu$, which gives a 90% CL upper limit of 190 keV. The mass of ν_τ can be sought by studying τ decays of the sort $\tau^- \to 2\pi^- \pi^+ \nu_\tau$ and $\tau^- \to 3\pi^- 2\pi^+ \nu_\tau$. If ν_τ is massive, the invariant mass spectrum of the charged pions will terminate below the mass of the τ. The best limit obtained to date is $m_{\nu_\tau} < 18.2$ MeV. These direct limits have been superseded. Massive neutrinos would affect the density fluctuations in the early Universe. Detailed measurements of the cosmic microwave background and other cosmological parameters indicate that the sum of the three neutrino masses must be less than about 0.6 eV.

16.1 The Nature of Neutrino Masses

Neutrinos may acquire their masses very differently from the way quarks and charged leptons do. The electron–positron system has four degrees of freedom, which we can represent by e_L, e_R, e_L^c, and e_R^c, where we have chosen to write e^c for e^+. For the neutrino we can write similarly ν_L, ν_R, ν_L^c and ν_R^c. To make a massive spin-one-half particle, we need both "left-handed" and "right-handed" pieces. For neutrinos we can suppose that the massive particle is a combination of the left-handed neutrino and the right-handed antineutrino:

$$N_1 = \nu_L + \nu_R^c. \tag{16.1}$$

This provides all the degrees of freedom required. A massive neutrino with only two degrees of freedom instead of four is called a Majorana neutrino.

The mass of the electron is described in the Lagrangian by the expression $m_e \bar{e} e = m_e(\bar{e}_L e_R + \bar{e}_R e_L)$. The mass term changes a left-handed electron into a right-handed electron, with amplitude m_e. Of course this is a colloquialism since the freely propagating electron cannot spontaneously change its angular momentum! The imprecision arises because $e_L = \frac{1}{2}(1 - \gamma_5)e$ describes a left-handed electron only in the ultrarelativistic limit. An electron emitted in beta decay has polarization, on average, $-v/c$.

While N_1 has the degrees of freedom required for a massive fermion, by combining a lepton with an antilepton we have broken lepton number conservation. If we tried the same thing with an electron, joining the left-handed electron with the right-handed positron, we would have broken charge conservation, something that is certainly impermissible. Whether lepton number is truly conserved is an experimental question.

There are a number of nuclides that are stable against both β^- and β^+ decay, but that are unstable against double beta decay. An example is Ge^{76}_{32}. Energy conservation forbids $\mathrm{Ge}^{76}_{32} \to \mathrm{Ga}^{76}_{31}\, e^+ \nu_e$ and $\mathrm{Ge}^{76}_{32} \to \mathrm{As}^{76}_{33}\, e^- \bar{\nu}_e$, but $\mathrm{Ge}^{76}_{32} \to \mathrm{Se}^{76}_{34}\, e^- \nu_e e^- \nu_e$ occurs with a half-life of about 1.5×10^{21} y. The neutrinoless double beta decay $\mathrm{Ge}^{76}_{32} \to \mathrm{Se}^{76}_{34} e^- e^-$ would violate lepton number. If ν_e is a Majorana particle, such a process is allowed.

Imagine this decay occurs through the intermediate virtual process $\mathrm{Ge}^{76}_{32} \to \mathrm{Se}^{76}_{34} W^- W^-$. One W decays to $e^- \bar{\nu}_{eR}$, where the antineutrino is virtual. If the neutrino is a Majorana particle, the $\bar{\nu}_{eR}$ can become ν_{eL}, indeed the two are components of a single massive particle. The ν_{eL} combines with the W^- to make the second e^-. The amplitude for this process is proportional to m_{ν_e}, so that observing it would establish a non-zero neutrino mass, and would show as well that lepton number is violated. The experimental lower limit on the half-life of Ge^{76}_{32} against neutrinoless double beta decay is about $1–2 \times 10^{25}$ y, though there is a controversial claim of observation at the lower end of this range.

The Standard Model together with Majorana neutrinos can accommodate quite naturally very small, but finite, neutrino masses. For an electron, the mass term changes a left-handed state into a right-handed state, with amplitude m_e, changing the weak isospin from $I_z = -1/2$ to $I_z = 0$. This is permissible because the electron interacts with the ubiquitous Higgs field, which has $I_z = \pm 1/2$ and which is non-zero everywhere.

Our Majorana neutrino N_1 behaves differently. To change ν_L ($I_z = 1/2$) to ν_R^c ($I_z = -1/2$) requires $\Delta I_z = 1$, more than the Higgs field supplies. Thus we expect this amplitude to be zero or very, very small. Suppose, however, that in addition there is a right-handed neutrino, together with its conjugate, a left-handed antineutrino. Neither of these feels the weak force since they have weak-isospin zero. Together they can form a second Majorana neutrino,

$$N_2 = \nu_R + \nu_L^c. \tag{16.2}$$

To change from the left-handed piece of N_2 to the right-handed piece doesn't change I_z at all, since both pieces are neutral under weak isospin. There is no reason for this not to have a large amplitude since it does not break weak isospin symmetry and thus need not depend on the "low" mass scale at which electroweak symmetry is broken. The corresponding mass M_{big} might even be as large as 10^{15} GeV, the scale at which the strong and electroweak forces may be unified.

It is also possible for N_1 and N_2 to mix. In particular, the ν_L in N_1 can become ν_R in N_2 with a change $I_z = 1/2$, just as e_L becomes e_R. Indeed, we might anticipate an amplitude of the same scale, m. The same is true for the transition of N_2 to N_1. These results can be summarized in a mass matrix in which the first row and column refer to N_1 and the second to N_2:

$$\begin{pmatrix} 0 & m \\ m & M_{\text{big}} \end{pmatrix} \tag{16.3}$$

where the 0 and M_{big} follow from the rule that $\Delta I_z = 1$ is disallowed, but $\Delta I = 0$ is unsuppressed. For $m \ll M_{\text{big}}$, the eigenvalues of the matrix are nearly M_{big} and $-m^2/M_{\text{big}}$. The negative sign has no physical significance; it corresponds to a mass m^2/M_{big}. If we guess that $m = m_e$ and $M_{\text{big}} = 10^{15}$ GeV, a value motivated by theories in which the strong and electroweak interactions are unified at a high mass scale, we get a neutrino mass of less than 10^{-12} eV, very small indeed. The lighter eigenstate is mostly the weakly interacting Majorana neutrino, while the heavier one is mostly the non-interacting Majorana neutrino:

$$|N_L\rangle \approx |N_1\rangle - \frac{m}{M_{\text{big}}}|N_2\rangle,$$
$$|N_H\rangle \approx |N_2\rangle + \frac{m}{M_{\text{big}}}|N_1\rangle. \tag{16.4}$$

This means of generating two Majorana neutrinos, one with a very large mass and one with a very small mass, is known as the seesaw mechanism.

16.2 Neutrino Mixing

If neutrinos have mass, the leptonic system is quite analogous to the quark system. We thus expect that the weak eigenstates may not correspond to the mass eigenstates: there is a leptonic version of the Kobayashi–Maskawa matrix – the Maki–Nakagawa–Sakata matrix – connecting the two. For simplicity, consider just two species of neutrinos, ν_e, the weak partner of the electron, and ν_μ, the weak partner of the muon. The mass eigenstates must be combinations of these two (later we consider the three-generation case):

$$|\nu_1\rangle = \cos\theta_0 |\nu_e\rangle - \sin\theta_0 |\nu_\mu\rangle,$$
$$|\nu_2\rangle = \sin\theta_0 |\nu_e\rangle + \cos\theta_0 |\nu_\mu\rangle, \qquad (16.5)$$

where ν_1 is the lighter state. We can always choose $0 \leq \theta_0 < \pi/2$ by redefining the states $|\nu\rangle \to -|\nu\rangle$, if necessary. When a beta decay produces a ν_e, its time development will be described by

$$|\nu_e(t)\rangle = e^{-iE_1 t} \cos\theta_0 |\nu_1\rangle + e^{-iE_2 t} \sin\theta_0 |\nu_2\rangle. \qquad (16.6)$$

If the state has well-defined momentum $p \approx E \gg M_1, M_2$, then its components have different energies

$$E_1 \approx p + \frac{M_1^2}{2p}; \qquad E_2 \approx p + \frac{M_2^2}{2p}. \qquad (16.7)$$

After traveling a distance $L \approx t$, the two pieces will have a relative phase $(M_2^2 - M_1^2)L/(2E) = \Delta M^2 L/(2E)$. The probability that the ν_e will have become a ν_μ is easily determined to be

$$P_{\nu_e \to \nu_\mu}(t) = |\langle \nu_\mu | \nu_e(t)\rangle|^2 = \sin^2 2\theta_0 \sin^2\left(\frac{\Delta M^2 L}{4E}\right). \qquad (16.8)$$

In practical units, the last factor is

$$\sin^2\left(1.27 \frac{\Delta M^2 (\text{eV}^2) L(\text{km})}{E(\text{GeV})}\right). \qquad (16.9)$$

These oscillations are similar to those in the K^0–$\overline{K}^0$ and B^0–$\overline{B}^0$ systems. There the oscillation is manifested in the variation in the sign of charged leptons emitted in semileptonic decays. Here it is the type of lepton itself that varies. The specific phenomenon observed depends on the energy of the neutrino that is oscillating. Antineutrinos generated by beta decays in nuclear reactors have energies in the MeV range. If these antineutrinos oscillate from electron-type to muon- or tau-type, their energies will be too low to produce in a detector the associated charged leptons. What would be measurable would be simply a drop in the number of charged-current reactions. The neutrinos would seem to disappear.

A neutrino beam generated by decaying pions will be dominantly ν_μ or $\overline{\nu}_\mu$ depending on the sign of the pions. Its charged-current interactions will regenerate muons. If, however, the beam oscillates to electron- or tau-type neutrinos, the corresponding charged leptons could be produced. Such an experiment would establish oscillations by appearance.

16.3 Solar Neutrinos

The earliest indications of neutrino oscillations came in solar neutrino experiments. The initial step in the fusion cycle that powers the Sun is the weak process $pp \to de^+\nu_e$. Because the total rate of energy production is proportional to the rate at which this reaction occurs, there is little uncertainty about the neutrino flux at the Earth's surface from this source. This turns out to be about 6×10^{10} cm^{-2} s^{-1}. See Exercise 16.1. These neutrinos have energies below 0.5 MeV and are thus below threshold for charged-current interactions except with a few nuclides. The next most copious source of solar neutrinos is electron capture on Be^7: Be$^7 e^- \to$ Li$^7 \nu_e$, with discrete neutrino energies near 0.4 MeV and 0.9 MeV. The Be^7 are generated in the process He^4 + He$^3 \to$ Be$^7 + \gamma$. The third significant source of solar neutrinos is the decay B$^8 \to$ Be$^{8*} e^+ \nu_e$, which produces neutrino energies up to nearly 18 MeV. The B^8 are themselves produced via Be$^7 + p \to$ B$^8 + \gamma$. The beta-decay product B^8 decays to two alpha particles, and is thus incorporated into the overall burning of hydrogen into helium. Even though the flux of the B^8 neutrinos is smaller by about 10^{-4} than those from the pp reaction, their high energy and correspondingly large cross sections makes them very important in solar neutrino experiments.

The solar neutrinos can be detected if they are captured by isotopes like Cl^{37} (ν_eCl$^{37} \to e^-$Ar^{37}) and Ga^{71} (ν_eGa$^{71} \to e^-$Ge^{71}), which then become radioactive with subsequent decays that can be observed. The threshold for the former capture is 814 keV, while for the latter it is 233 keV. As a result, chlorine experiments are blind to the pp reaction, while gallium experiments can detect it. The chlorine experiments are dominated by neutrinos from B^8 and Be^7. They were pioneered by Ray Davis at the Homestake Mine in South Dakota, starting back in the 1960s (**Ref. 16.1**).

In 1968 Davis's team reported an upper limit of 3 SNU (1 SNU – solar neutrino unit – is 10^{-36} neutrino captures per atom per second) for a chlorine experiment. The prediction of the rate from solar models is difficult and at the time the expected total rate was 20 SNU, 90% of which was due to B^8. To make this measurement, Davis needed to isolate about one atom of Ar^{37} produced each day in a vat of 3.9×10^5 liters of C$_2$Cl$_4$ located 1.5 km underground. As shown in Table 16.1, the contemporary prediction is 7.6 SNU and the 1998 result from the Homestake experiment is 2.56 SNU.

Gallium experiments were pursued by the GALLEX collaboration from 1991 to 1997 at the Gran Sasso National Laboratory in the Gran Sasso d'Italia in the Abruzzo region 150 km east of Rome and by the SAGE collaboration at Baksan, in Russia. The cumulative result from GALLEX was 77.5 SNU with a precision of about 10%. This was about 60% of the predicted rate of 128 SNU. The SAGE result was similar. The GALLEX experiment was succeeded by GNO, the Gallium Neutrino Observatory, where the rate was measured to be near 63 SNU.

An alternative to detecting individual transmuted atoms relies on Cherenkov light from charged-current reactions induced by the neutrinos. Because an enormous target is required to obtain sufficient rate, the natural medium is water. The leading experiments using this technique have been located at the Komioka Mozumi mine in Japan. The Kamioka Nucleon Decay Experiment (Kamiokande) was upgraded to a neutrino detector just in time to catch neutrinos from the supernova SN1987a. After its run from 1987 to 1995, it was succeeded

Table 16.1. *Predictions for the solar neutrino flux from J. N. Bahcall, M. H. Pinsonneault, and S. Basu, Astrophys. J.* **555**, *990 (2001) and corresponding experimental results, adapted from the summary of N. Nakamura in the 2006 Review of Particle Physics. The gallium experiments are in good agreement with one another. The chlorine and gallium experiments are sensitive only to the charged current. The Kamiokande and Super-Kamiokande experiments measure the elastic scattering $\nu e^- \to \nu e^-$, which has contributions from both charged and neutral currents. The solar neutrino unit (SNU) is 10^{-36} neutrino captures per atom per second.*

Solar Sources:	^{37}Cl (SNU)	^{71}Ga (SNU)	^{8}B ν flux (10^6 cm^{-2} s^{-1})
$pp \to de^+ \nu_e$		69.7	
^{7}Be $e^- \to ^7$Liν_e	1.15	34.2	
^{8}B $\to ^8$Be* $e^+ \nu_e$	5.67	12.1	5.05
Other	0.68	11.9	
Total	$7.6^{+1.3}_{-1.1}$	128^{+9}_{-7}	$5.05^{+1.01}_{-0.80}$

Experiment:			
Homestake	$2.56 \pm 0.16 \pm 0.16$		
GALLEX		$77.5 \pm 6.2^{+4.3}_{-4.7}$	
GNO		$62.9^{+5.5}_{-5.3} \pm 2.5$	
SAGE		$70.8^{+5.3}_{-5.2}{}^{+3.7}_{-3.2}$	
Kamiokande			$2.80 \pm 0.19 \pm 0.33$
Super-Kamiokande			$2.35 \pm 0.02 \pm 0.08$

by the 50-kton detector Super-Kamiokande. The threshold for observability for both was several MeV and these experiments were thus dominated by neutrinos from B^8 decay. Both experiments found fluxes about half the expected level of 5×10^6 cm^{-2} s^{-1} and showed that the neutrinos indeed came from the direction of the Sun.

Every one of these techniques is extremely challenging because of the small rates and large detectors employed. What is striking is that the results of all these experiments tell about the same story: about one-third to one-half the expected rate of neutrino interactions is actually observed. See Table 16.1.

The solar abundances of elements like beryllium and boron must be deduced from solar models and this added some doubt to the predictions for these contributions to the solar neutrino flux. However, there was good agreement between the various calculations that had been done to estimate these abundances. This made it hard to dismiss the results from the Cherenkov and chlorine experiments. Moreover, fully half of the reaction rate expected in the gallium experiments is due to the *pp* reaction, about whose rate there could be little doubt since it is directly connected to the total luminosity of the Sun.

The discrepancy between the expected and observed rates for solar neutrino experiments was consistent and persistent. Attempts to blame the problem on solar models were weakened by the GALLEX, GNO, and SAGE results. What remained suggested strongly that there are neutrino oscillations involving electron neutrinos.

For mixing to play a role, it would seem that $\Delta m^2 L/E$ (where $L = 1.5 \times 10^{11}$ m is the distance from the Earth to the Sun) would have to be not too small, i.e. $\Delta m^2 > 10^{-12}$ eV^2 so the oscillation length would not be large compared to L. In the limit that there were many oscillations between the Sun and the Earth, we would expect that averaging over an energy spectrum would replace the oscillation in L by its average, 1/2:

$$P_{\nu_e \to \nu_\mu} = \frac{1}{2} \sin^2 2\theta_0, \qquad (16.10)$$

so that at most half the neutrinos could disappear. With three species, the limit would be two-thirds disappearing. In fact, the behavior of solar neutrinos is more complex because they must first pass from the Sun's core to its edge before entering the void.

16.4 MSW Effect

If there is mixing between ν_e and, say, ν_μ, the combinations that are eigenstates in free space will not remain eigenstates when passing through matter. This is completely analogous to the phenomenon of regeneration in the neutral K system. There regeneration occurs because K^0 and $\overline{K}^0$ have different forward scattering amplitudes on nuclei. In the neutrino system the corresponding difference is between the forward elastic scattering of ν_e on electrons and ν_μ on electrons. This regeneration is known as the Mikheyev–Smirnov–Wolfenstein (MSW) effect. While $\nu_\mu e$ elastic scattering occurs only through the neutral current, $\nu_e e$ elastic scattering has a contribution from the charged-current process in which the incident electron-neutrino is transformed into an electron and the struck electron becomes itself an electron-neutrino. This interaction is described by the ordinary V-A theory

$$\frac{G_F}{\sqrt{2}} \bar{\nu}_e \gamma_\mu (1-\gamma_5) e \, \bar{e} \gamma^\mu (1-\gamma_5) \nu_e = \frac{G_F}{\sqrt{2}} \bar{\nu}_e \gamma_\mu (1-\gamma_5) \nu_e \, \bar{e} \gamma^\mu (1-\gamma_5) e \qquad (16.11)$$

where the re-ordering follows from an algebraic identity for the gamma matrices known as a Fierz transformation. For electrons at rest, the last factor is important only for $\mu = 0$, when it gives the electron density, N_e. Acting on a left-handed neutrino, $1 - \gamma_5$ is simply 2 and the interaction is seen to be equivalent to a potential energy for neutrinos $V = \sqrt{2} G_F N_e$.

In the end, a complete analysis of neutrino mixing requires considering three neutrino species, but for the MSW effect a two-state approximation is adequate. What we call here ν_μ is, in fact, a linear combination of ν_μ and ν_τ.

For neutrinos, where the mass is apparent in the relation $E \approx p + \frac{1}{2}M^2/p$, the mass-squared matrix is of interest. The effect of the extra scattering of ν_e is to add to its diagonal element in this matrix the quantity $A = (E+V)^2 - E^2 \approx 2EV$

$$A = 2\sqrt{2}G_F N_e E = 0.76 \times 10^{-7} \text{eV}^2 \times \rho \left[\text{g cm}^{-3}\right] \times E \,[\text{MeV}] \times 2Y_e, \quad (16.12)$$

where ρ is the mass density and the number of electrons per nucleon is Y_e. No other element of the mass-squared matrix is affected. The ν_e component of a mixed neutrino picks up an extra phase $\frac{1}{2}AL/E = \sqrt{2}G_F N_e L = 0.383 \times 10^{-3}\rho \left[\text{g cm}^{-3}\right] Y_e L \,[\text{km}]$ in traversing a distance L. If the material is hydrogen with a density of 1 g cm^{-3}, a full cycle is accumulated in a distance of 1.6×10^4 km, a bit more than the diameter of the Earth.

The mixing that results in the eigenstates $|\nu_1\rangle$ and $|\nu_2\rangle$ with masses squared M_1^2 and M_2^2 without the matter effect is described by

$$\mathbf{M}^2 = \frac{M_2^2 - M_1^2}{2} \begin{pmatrix} -\cos 2\theta_0 & \sin 2\theta_0 \\ \sin 2\theta_0 & \cos 2\theta_0 \end{pmatrix}, \quad (16.13)$$

where we drop the common diagonal term equal to the average mass squared. Multiplication verifies that the mixtures $|\nu_1\rangle$ and $|\nu_2\rangle$ are indeed eigenvectors of this matrix. Because the energy of a neutrino with momentum p is very nearly $p + \frac{1}{2}M^2/p$ we can write a Schrödinger equation for the state $|\psi\rangle = C_e|\nu_e\rangle + C_\mu|\nu_\mu\rangle$ as

$$i\frac{d}{dt}\begin{pmatrix} C_e \\ C_\mu \end{pmatrix} = \frac{1}{2E}\mathbf{M}^2 \begin{pmatrix} C_e \\ C_\mu \end{pmatrix}. \quad (16.14)$$

This system is analogous to a spin-one-half particle (whose spin is $\boldsymbol{\sigma}$) in a magnetic field with $\mathbf{B} \propto \cos 2\theta_0 \hat{\mathbf{z}} - \sin 2\theta_0 \hat{\mathbf{x}}$ since $\boldsymbol{\sigma} \cdot \mathbf{B}$ has the same form as $\mathbf{M}^2$. The electron-neutrino is analogous to the state whose spin is aligned with the magnetic field and the muon-neutrino is analogous to the state anti-aligned with it. The eigenstate $|\nu_1\rangle$ is the up state rotated by $2\theta_0$ about the y axis. Semiclassically, the spin precesses around the direction of the magnetic field. See Figure 16.1.

The extra elastic scattering of ν_e on electrons with density N_e changes the mass-squared matrix, again with the average diagonal term removed, to

$$\mathbf{M}^2_{eff} = \frac{\Delta M_0^2}{2} \begin{pmatrix} -\cos 2\theta_0 + \frac{A}{\Delta M_0^2} & \sin 2\theta_0 \\ \sin 2\theta_0 & \cos 2\theta_0 - \frac{A}{\Delta M_0^2} \end{pmatrix}, \quad (16.15)$$

where $\Delta M_0^2 = M_2^2 - M_1^2$ is the splitting of the squares of the masses in vacuum. We can rewrite this in a form analogous to that for vacuum

$$\mathbf{M}^2_{eff} = \frac{\Delta M_{N_e}^2}{2} \begin{pmatrix} -\cos 2\theta_{N_e} & \sin 2\theta_{N_e} \\ \sin 2\theta_{N_e} & \cos 2\theta_{N_e} \end{pmatrix}, \quad (16.16)$$

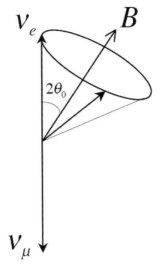

Figure 16.1. The analog between neutrino oscillations and precession of a spin-one-half particle in a magnetic field. A neutrino created as a ν_e (analogous to spin up) precesses about an axis at an angle $2\theta_0$. The precession gives oscillating fractions of ν_e and ν_μ, supposing these to be the mixed species. A fraction $\cos 2\theta_0$ of the spin is projected along the "field" direction. On average, the components perpendicular to the field vanish. If we project the average component back along the electron-neutrino's direction, we find a fraction $\cos^2 2\theta_0$. If we take this semiclassical expectation value to represent the probability $P_{\nu_e \to \nu_e} - P_{\nu_e \to \nu_\mu} = 1 - 2 P_{\nu_e \to \nu_\mu}$ we find that $P_{\nu_e \to \nu_\mu} = \frac{1}{2} \sin^2 2\theta_2$. This agrees with the time-dependent expression, Eq. (16.8), when we average over a range of L that encompasses many cycles, corresponding to many cycles of the "spin" around the "magnetic field."

where now $\Delta M_{N_e}^2$ is the splitting of the squares of the eigenmasses in the medium. Identifying the two expressions for the mass matrix in matter we find the relations

$$\Delta M_{N_e}^2 \sin 2\theta_{N_e} = \Delta M_0^2 \sin 2\theta_0$$
$$A = \Delta M_0^2 \cos 2\theta_0 - \Delta M_{N_e}^2 \cos 2\theta_{N_e}. \qquad (16.17)$$

This is shown geometrically in Figure 16.2.

If we imagine a hypothetical neutrino beginning at t_0 where the electron density is $N_e(t_0)$ in the lower-mass eigenstate $|\nu_1, N_e(t_0)\rangle$ (defined by the angle $\theta_{N_e(t_0)}$) and proceeding through matter whose density changes only gradually, we can expect the state to remain in the lower-mass eigenstate so that at time t it is $|\nu_1, N_e(t)\rangle$. This adiabatic evolution is analogous to the magnetic moment of the spin-1/2 particle following a gradual change in $\mathbf{B}$.

Physical neutrinos are produced not in mass eigenstates, but in "flavor" eigenstates because they arise from weak interactions. To follow the evolution of a neutrino that begins at the center of the Sun as $|\nu_e\rangle$ where the electron density is N_e, we project $|\nu_e\rangle$ along the "magnetic field" at the initial density, introducing a factor $\cos 2\theta_{N_e}$. See Figure 16.3. As the neutrino moves from the center of the Sun to the periphery, the density decreases and the orientation of the "magnetic field" gradually moves to the direction for vacuum

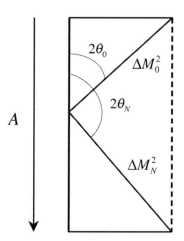

Figure 16.2. The relationship between the vacuum mixing angle, θ_0, and the mixing angle in matter, θ_N, and the mass splittings in vacuum and in matter. The quantity $A = 2\sqrt{2}G_F N_e E$, which is proportional to the electron density N_e and to the neutrino energy E, arises from the charged-current scattering in $\nu_e e \to \nu_e e$. As displayed in the figure, $\Delta M_0^2 \sin 2\theta_0 = \Delta M_N^2 \sin 2\theta_N$. If A is small, $\theta_0 \approx \theta_N$. If A is very large $2\theta_N \approx \pi$. When $\theta_N = \pi/2$, the mass splitting in matter is at its minimum. Note that in this figure, $\cos 2\theta_N < 0$.

mixing. In this adiabatic description, only the component along the magnetic field matters. The components transverse to it average to zero. When the neutrino finally exits the Sun, its "neutrino spin" direction is aligned with the magnetic field for vacuum mixing. On the passage from the Sun to the Earth this projection is unchanged: the actual vector just continues to precess about this average orientation. To determine its flavor content we finally project onto the ν_e direction. Altogether, the projections give $\cos 2\theta_{N_e} \cos 2\theta_0$. Equating this to $P_{\nu_e \to \nu_e} - P_{\nu_e \to \nu_\mu} = 1 - 2P_{\nu_e \to \nu_\mu}$ we find the adiabatic, and time averaged, prediction for the transformation from ν_e to ν_μ:

$$P_{\nu_e \to \nu_\mu} = \frac{1}{2}(1 - \cos 2\theta_{N_e} \cos 2\theta_0). \tag{16.18}$$

Of course in the limit of low matter density, $\theta_{N_e} \to \theta_0$ and this reduces to the vacuum expression. On the other hand, if the product of the energy and the initial density is large, then $\cos 2\theta_{N_e} \to -1$. The resulting transition probability is $P_{\nu_e \to \nu_\mu} = \frac{1}{2}(1 + \cos 2\theta_0) = \cos^2 \theta_0$, so that if the vacuum mixing angle were small, ν_e would be nearly certain to emerge as ν_μ.

As long as the spin precesses rapidly around the magnetic field, compared to the rate at which the direction of the magnetic field changes, this is a compelling argument. The precession frequency is proportional to $\Delta M_{N_e}^2$, which is smallest when $\sin 2\theta_{N_e} = 1$, i.e. when

$$\cos 2\theta_0 = \frac{A}{M_2^2 - M_1^2}. \tag{16.19}$$

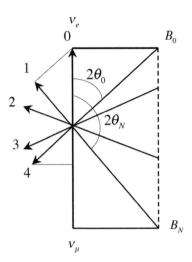

Figure 16.3. In the adiabatic approximation, the neutrino follows the magnetic field, which rotates as the electron density varies. The solar neutrino is produced as v_e. If $\Delta M^2/2E$ is large enough, we can ignore the precession of the "spin" and look just at its projection along the "magnetic field." The neutrino produced at "0," is projected along the axis defined by the mixing angle for the density at the center of the Sun, B_N, at "1." As the density decreases, the direction of the "magnetic field" in the solar matter changes, as in "2" and "3," finally reverting to the vacuum direction, shown as "4." In the example shown here, the neutrino is then more aligned with the v_μ direction than the original v_e direction. It is clear, referring to a previous figure, that this will happen only if $A = 2\sqrt{2} G_F N_e E$ is sufficiently large. Following the geometry here, we find that $P_{v_e \to v_\mu} = \frac{1}{2}(1 - \cos 2\theta_N \cos 2\theta_0)$.

Passing through such a "resonance region" the spin may no longer follow the field and transitions from $|v_1(t)\rangle$ to $|v_2(t)\rangle$ become much more likely. Whether the adiabatic approximation applies depends on whether the direction of the "magnetic field," i.e. the matter density, changes gradually enough relative to the precession frequency, $\Delta M^2/2E$.

In the Sun, neutrinos are produced near the core, where the density is of order 130 g cm^{-3} and the atomic composition gives $Y_e = 0.67$. For a 1 MeV neutrino, A is about 1.3×10^{-5} eV^2. Thus if $1.3 \times E(\text{MeV}) \times 10^{-5}$ eV^2 is greater than $(M_2^2 - M_1^2) \cos 2\theta_0$, the construction shown in Figure 16.2 will make $2\theta_N > \pi/2$: Adiabatic evolution of a v_e will end with the neutrino more likely to be "flipped" into v_μ than to remain v_e. For much lower energy neutrinos, A will be small and $\theta_N \approx \theta_0$. These neutrinos will not be "flipped." They emerge as electron-type neutrinos. See Exercise 16.4.

While the oscillation probability in vacuo depends only on $\sin 2\theta_0$ and thus is invariant under $\theta_0 \to \pi/2 - \theta_0$, the MWS effect depends on $\cos 2\theta_0$ and is not similarly invariant. Thus, in principle values of θ_0 between $\pi/4$ and $\pi/2$ must be considered as well as those from zero to $\pi/4$. This so-called "dark side" is disfavored by solar neutrino experiments because it gives $\cos 2\theta_0 < 0$ and according to Eq. (16.18) cannot suppress solar neutrinos by more than 50%.

16.5 MSW and the Solar Neutrino Problem

Once the MSW effect was included, three distinct solutions emerged for the solar neutrino problem defined by the results from chlorine and gallium experiments together with measurements by Kamiokande and Super-Kamiokande. Each solution corresponded to values for the mass splitting, Δm_{sol}^2, and mixing angle θ_{sol}. One, termed the large mixing angle solution (LMA) had $\sin^2 2\theta_{sol} \approx 0.5 - 1.0$ and $\Delta m_{sol}^2 \approx 10^{-5} - 3 \times 10^{-4}$ eV^2. A rather poorer fit, LOW (for low mass or perhaps low likelihood of being correct) was obtained with $\sin^2 2\theta_{sol} \approx 1.0$ and $\Delta m_{sol}^2 \approx 10^{-7}$ eV^2. The small mixing angle solution had $\sin^2 2\theta_{sol} \approx 10^{-2}-10^{-3}$ and $\Delta m_{sol}^2 \approx 5 \times 10^{-6}$ eV^2. In the LOW solution, the adiabatic approximation for MSW fails and a more complete calculation is required. In addition, solutions were possible with very low values of Δm_{sol}^2, $10^{-12}-10^{-10}$ eV^2 and with large values of $\sin^2 2\theta_{sol}$.

16.6 Cosmic-Ray Neutrinos

While the solar neutrino problem suggested that there were neutrino oscillations, convincing evidence came from an entirely different direction: cosmic rays. Indeed, there are two separate phenomena: solar neutrino mixing and atmospheric neutrino mixing, that is, mixing in neutrinos produced by collisions of cosmic rays in the atmosphere. It turns out that it is often possible to avoid considering three species of neutrinos and instead imagine that the solar neutrino and the atmospheric neutrino systems are two separate systems, each described by a two-neutrino pattern. The two phenomena occur at very different energy scales, MeV for solar neutrinos and GeV for atmospheric neutrinos.

In the hadronic showers of cosmic rays that strike the atmosphere, pions are created and decay to $\mu\nu$, and the muons subsequently decay to $e\nu\bar{\nu}$. In this way two ν_μs and one ν_e are generated for each charged pion created, ignoring the difference between neutrinos and antineutrinos.

The actual flux of particles created by the collisions high in the atmosphere is not so well known, so there is an advantage in comparing the ratio of neutrino events producing a muon in the detector to those producing an electron to the ratio expected from Monte Carlo simulations: $R = (\mu/e)_{DATA}/(\mu/e)_{MC}$. The absolute strength of the flux cancels in the ratio of the simulations. A number of experiments using water Cherenkov counters, including Kamiokande, the IMB (Irvine–Michigan–Brookhaven) experiment near Cleveland, Ohio, and Super-Kamiokande, observed values of R less than one, indicating that the ν_μ were somehow disappearing.

In 1998, the Super-Kamiokande team announced impressive evidence for neutrino oscillations **(Ref. 16.2)**. The ring of Cherenkov light produced by a muon in water has a sharper definition than that produced by the shower of an electron and the two categories can be reliably separated. More than 11,000 photomultiplier tubes viewed the central 22.5 kilotons of detector, in which events were required to begin. The Super-Kamiokande collaboration recorded more than 4000 events that were fully contained within the inner fiducial volume. The ratio R thus found differed substantially from unity, both for lower energy events (visible energy below 1.33 GeV), with $R = 0.63 \pm 0.03 \pm 0.05$ and higher energy events, with $R = 0.65 \pm 0.05 \pm 0.08$.

From the Cherenkov light, it was possible to determine the direction of the incoming neutrino. Those that came from below must have been created in the atmosphere on the other side of the Earth, thousands of kilometers away. Those that came from above, were created relatively nearby. While the e-like events showed no particular directional dependence, the μ-like events that came from below were substantially depleted. The simplest interpretation is that the ν_μ oscillate to ν_τ with an oscillation wavelength comparable to the Earth's radius. Alternatively, the ν_μ might oscillate to some previously unknown neutrino type, a sterile neutrino that lacks interactions. Either way, for such a depletion to be observable, the mixing would have to be substantial. Since the ν_e seemed unaffected, it was sensible to fit the data assuming only ν_μ-ν_τ oscillations. The result was $\sin^2 2\theta_{atm} > 0.82$ and 5×10^{-4} eV$^2 < \Delta m^2_{atm} < 6 \times 10^{-3}$ eV^2 at a 90% confidence level. With three times the exposure, Super-Kamiokande reported refined measurements: $\sin^2 2\theta_{atm} > 0.92$ and 1.5×10^{-3} eV$^2 < \Delta m^2_{atm} < 3.4 \times 10^{-3}$ eV^2 at a 90% confidence level.

16.7 Reactor Neutrino Experiments

Reactor experiments produce antineutrinos, which accompany the beta particles emitted by fission products. Since the energies here are at most a few MeV, there is no possibility of observing the oscillation of $\bar{\nu}_e$ to $\bar{\nu}_\mu$ in a charged-current interaction: these neutrinos are below threshold for muon production. However, these oscillations would lead to a reduction in the number of charged-current events producing electrons. For sufficiently large mixing angles, such an effect would be observable by measuring the event rate with the reactor on and off, and comparing with the expected rate, based on the power produced by the reactor and an understanding of the decay chains associated with fission products. Such calculations are believed to be accurate at the few percent level. The domain of sensitivity in Δm^2 is set by equating $1.27\Delta m^2 (eV^2)L(m)/E(MeV)$ to the observed limit on the oscillation probability. If that limit is around 10% and the typical antineutrino energy is taken to be 3 MeV, the experiment is sensitive to differences of squares of masses of roughly 0.7 eV$^2/L$(m). For Δm^2 large enough to give many oscillations of the neutrino before detection, the limit on $\sin^2 2\theta$ is twice the limit obtained for the oscillation probability since then the factor $\sin^2[\Delta M^2 L/(4E)]$ averages to one-half. See Figure 16.4.

An experiment performed at the Bugey reactor near Lyon, France observed electron antineutrinos through inverse beta decay: $\bar{\nu}_e p \to e^+ n$. The positron was observed through scintillation light caused by its two-gamma annihilation with an electron. The neutron was observed by doping the scintillator with Li^6, which is sensitive through the process $n +$ Li$^6 \to$ He$^4 +$ H$^3 + \gamma$(4.8 MeV). The primary observations were made at distances of 15 m and 40 m from a 2.8 GW reactor. Oscillation of the electron antineutrinos would have led to a reduced event rate. As reported in 1994/5, no reduction was observed at the few percent level, excluding values of Δm^2 on the scale of 0.02 eV^2.

To improve upon this it was necessary to make measurements further from a reactor. A nuclear power station located near Chooz in the Ardennes region of France served as

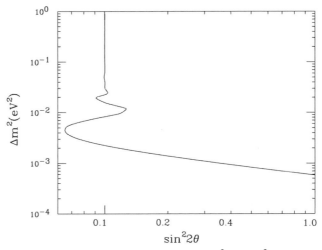

Figure 16.4. Reactor neutrino experiments give limits on Δm^2 and $\sin^2 2\theta$. A limit on the fraction of the $\bar{\nu}_e$ that are transmuted into unobservable $\bar{\nu}_\mu$ restricts the allowed region in the $\Delta m^2 - \sin^2 2\theta$ plane through the relation in Eq. (16.9). The allowed region is to the left and below the curve. The sensitivity to Δm^2 is greatest if $\sin^2 2\theta$ is near unity. If the oscillation probability is shown to be less than P, then the sensitivity extends in eV^2 to about $P^{1/2} < E(\text{MeV}) > /1.27 L(\text{m})$, where $< E >$ is the mean neutrino energy. The figure represents an experiment with $L = 1$ km, $< E > = 3.5$ MeV, and $P = 0.05$. In the limit of large Δm^2, the limit on $\sin^2 2\theta$ is $2P$, as shown in the figure.

the antineutrino source for a more precise experiment again relying on inverse beta decay (Ref. 16.3). The neutron was observed by incorporating gadolinium in a liquid scintillator detector, located 1 km from the reactor. Gadolinium has a large cross section for neutron absorption, which is signaled by the emission of a gamma ray of 8 MeV. The neutrons could also be observed by their absorption by protons, producing a deuteron and a 2.2 MeV gamma. The delay of 2 to 100 μs between the positron annihilation and the neutron absorption provided a signature for the events. The signal event rate was found to be proportional to the instantaneous power of the reactor, as it should have been. The value of about 25 neutrino events per day at full power was much larger than the background of about 1 event per day.

The anticipated rate in the absence of neutrino oscillations depended on the intensity and energy spectrum of the neutrinos emitted by the reactor. Including this uncertainty and others associated with the detector, the ratio of the measured to the expected rate reported in 1998 was $0.98 \pm 0.04 \pm 0.04$, where the first error was statistical and the second systematic. Mixing would reduce the ratio by $1 - \frac{1}{2}\sin^2 2\theta$. At 90% CL, the ratio is greater than 0.91, so at the same confidence level, for large Δm^2, $\sin^2 2\theta < 0.18$. Using a mean neutrino energy of 3 MeV and the distance between the reactor and the detector, for $\sin^2 2\theta = 1$ we find the limit $\Delta m^2 < 0.9 \times 10^{-3}$ eV^2. With additional data Chooz reported refined results in 1999: for large Δm^2, $\sin^2 2\theta < 0.10$ and for $\sin^2 2\theta = 1$, $\Delta m^2 < 0.7 \times 10^{-3}$ eV^2, at 90% CL (Ref. 16.4).

Subsequently a similar experiment was conducted at the Palo Verde Generating Station in Arizona with consistent results: for large Δm^2, $\sin^2 2\theta < 0.164$ and for $\sin^2 2\theta = 1$, $\Delta m^2 < 1.1 \times 10^{-3}$ eV^2, at 90% CL.

16.8 SNO

The convincing evidence of atmospheric neutrino oscillations involving ν_μ at Super-Kamiokande intensified interest in the solar ν_e problem. The MSW effect, together with vacuum oscillations provided several possible solutions. An experiment at the Sudbury Neutrino Observatory in Ontario, Canada finally resolved the issue (**Ref. 16.5**).

Like Super-Kamiokande, SNO used a large water-filled detector, but with a difference. The water was not H$_2$O but D$_2$O. As in the famous plant at Rjukan, Norway whose heavy water was seized by the Nazis for work on the atomic bomb, Sudbury's heavy water was the result of electrolysis using plentiful and inexpensive hydroelectric power. The advantage of heavy water for solar neutrino experiments is participation of three distinct reactions:

$$\begin{aligned}
\nu_e + d &\rightarrow p + p + e^- \quad (CC) \\
\nu_x + d &\rightarrow p + n + \nu_x \quad (NC) \\
\nu_x + e^- &\rightarrow \nu_x + e^- \quad (ES)
\end{aligned} \quad (16.20)$$

Only electron-type neutrinos can give the first reaction, while electron-, muon-, and tau-neutrinos can all participate in the last two. In the initial results from SNO, only the charged-current and elastic scattering events were used. If we suppose there are no neutrino oscillations, then the ν_e flux can be inferred from either the charged-current or electron-scattering events since the underlying cross sections are known. Neutrino oscillations would generate a flux of ν_μ and/or ν_τ, which would contribute, through neutral current interactions, to the elastic scattering to give an apparent contribution, at about one-sixth strength, to the ν_e flux inferred in this process. The ν_μ and/or ν_τ would not contribute to the charged-current events.

Slightly fewer than 10,000 phototubes were arrayed to view the heavy water contained within an acrylic vessel, itself surrounded by a shield of ordinary water. Just as for Super-Kamiokande, the detector was sensitive only above a few MeV and thus responded to solar neutrinos from ^{8}B. The energy was determined by counting phototube hits, with about nine hits for each MeV of electron energy. Timing the arrival of the Cherenkov photons allowed determination of the origin of the electron and its direction.

Signals from the charged-current and elastic scattering events were separated from each other and from the neutron background by fitting their distribution in energy released, scattering angle relative to the Sun, and radial distance from the center of the detector. The neutron background occurred predominantly near the periphery of the detector.

Using the anticipated shape of the ^8B spectrum, the full flux of ^8B electron neutrinos could be deduced from the charged-current and elastic scattering processes, with the results, in units of 10^6 cm^{-2} s^{-1}

$$\phi^{CC} = 1.75 \pm 0.07 \text{(stat)} \, {}^{+0.12}_{-0.11} \text{(syst)} \pm 0.05 \text{(theor)}$$

$$\phi^{ES} = 2.39 \pm 0.34 \text{(stat)} \, {}^{+0.16}_{-0.14} \text{(syst)} \tag{16.21}$$

suggesting an excess in elastic scattering, which would signal the presence of neutral current scattering from ν_μ and ν_τ. Conclusive evidence came from using the earlier, more precise measurement of elastic scattering by the Super-Kamiokande team, which in the same units was

$$\phi^{ES} = 2.32 \pm 0.03 \text{(stat)} \, {}^{+0.09}_{-0.07} \text{(syst)}. \tag{16.22}$$

This, then, established that there were active neutrinos causing elastic scattering and not contributing to the charged-current process. Analyzed in this light, the sum of the fluxes from ν_μ and ν_τ could be determined. It is about twice that in the ν_e flux. If we suppose that MSW is completely effective so $\cos 2\theta_N = -1$, we conclude that $(1 + \cos 2\theta_0)/2 \approx 2/3$ so $\sin^2 2\theta_0 \approx 8/9$, i.e. nearly maximal mixing. For MSW to be complete we need $\Delta M^2 \cos 2\theta_0 < A$. Here θ_0 and ΔM^2 stand for θ_{sol} amd Δm^2_{sol}. The lowest energy neutrinos SNO detected had energies of about 6.75 MeV, so $A \approx 8.5 \times 10^{-5}$ eV^2. This means that $\Delta m^2_{\text{sol}} < 25 \times 10^{-5}$ eV^2. If Δm^2_{sol} were as low as 1×10^{-5} eV^2 then even the pp neutrinos observed by gallium experiments would be similarly MSW suppressed, in disagreement with the data. See Exercise 16.4.

It was the inferred neutral current contribution to elastic scattering that demonstrated flavor oscillations in the 2001 SNO result. Direct observation of the neutral current through $\nu + d \to p + n + \nu$ at SNO (**Ref. 16.6**) followed in 2002. The challenge here was to detect the neutron through its capture on the deuteron, $n + d \to t + \gamma$. The 6.25-MeV gamma produced Cherenkov light through its shower. These were excluded in the earlier analysis by setting the threshold at 6.75 MeV. The neutral current disintegration of the deuteron was separated from the charged-current and elastic scattering events by its energy spectrum and angular distribution.

The neutral current measurement is difficult because every free neutron in the heavy-water detector, whether due to the signal or the background, behaves in the same way. The heavy water itself is inevitably contaminated with thorium and uranium, which decay into chains of radioactive daughters. By carefully monitoring these chains, this background could be subtracted. The flux of ν_e and the sum of the ν_μ and ν_τ fluxes could then be determined:

$$\phi_e = 1.76 \, {}^{+0.05}_{-0.03} \text{(stat)} \, {}^{+0.09}_{-0.09} \text{(syst)}$$

$$\phi_{\mu\tau} = 3.41 \, {}^{+0.45}_{-0.45} \text{(stat)} \, {}^{+0.48}_{-0.45} \text{(syst)} \tag{16.23}$$

again in units of 10^6 cm^{-2} s^{-1}, in excellent agreeement with the results of 2001, which relied on the elastic scattering measurement of Super-Kamiokande.

16.9 KamLAND

The SNO results showed that solar neutrinos indeed mix. To reach much lower values of Δm^2 than explored at Chooz, it was necessary to place a detector much further from the reactor. The Kamioka Liquid Scintillator Anti-Neutrino Detector (KamLAND) was built at the site previously used by the Kamiokande experiment, under rock equivalent to 2700 meters of water. This location is surrounded by 53 Japanese nuclear power reactors, with 79% of the neutrino flux coming from 26 of those reactors located at distances from 138 km to 214 km. As at Chooz, the signal for $\bar{\nu}_e p \rightarrow e^+ n$ was the positron annihilation followed by a gamma from neutron capture. To compensate for the much diminshed antineutrino flux so far from the reactors, the detector was on a grand scale: a kiloton of liquid scintillator, of which about 50% lay inside the fiducial volume, about 100 times the target used at Chooz. Despite this, the event rate at KamLAND was about half an event per day compared to 25 events per day at Chooz. It was for this reason that it was necessary that it be shielded from cosmic ray background by going deep underground.

In its initial report in 2003 (Ref. **16.7**), the KamLAND experiment had 54 events, an estimated single event from background, and a total expected in the absence of oscillations of 86.8 ± 5.6. The ratio of observed to expected rates was given as $0.611 \pm 0.085\text{(stat)} \pm 0.041\text{(syst)}$. This required that $\sin^2 2\theta_{sol}$ be greater than about 0.25 at 95% CL, but allowed any value of Δm^2_{sol} greater than about 10^{-5} eV^2. Because the disappearance probability depends directly on the incident antineutrino energy, the spectrum of energies observed should be distorted from the initial spectrum by the oscillations. By fitting to the energy spectrum KamLAND was able to determine best values for $\sin^2 2\theta_{sol}$ and Δm^2 separately, with the results $\sin^2 2\theta_{sol} = 1.0$, $\Delta m^2_{sol} = 6.9 \times 10^{-5}$ eV^2. This result was decisive in choosing the large mixing angle (LMA) solution for the solar neutrino puzzle.

KamLAND reported again in 2005, with much increased statistics (Ref. 16.8). The number of signal events with backgrounds subtracted was near 240 while the expectation in the absence of oscillations was 356 ± 24. With this much larger sample it was possible to establish the oscillatory behavior of the energy spectrum as a function of $1/E$. From the KamLAND data alone, Δm^2_{sol} was determined to be $7.9^{+0.6}_{-0.5} \times 10^{-5}$ eV^2 with $\tan^2 \theta_{sol} \approx 0.46$. Including solar neutrino data determined $\tan^2 \theta_{sol} = 0.40^{+0.10}_{-0.07}$, i.e. $\sin^2 2\theta_{sol} = 0.82 \pm 0.07$. A much larger sample with more that 1600 events collected through May 2007 showed nearly two cycles of oscillation, once the effect of antineutrinos from terrestrial sources were taken into account. See Figure 16.5. The KamLAND results significantly tightened the limits on Δm^2_{sol}. Combining with results from solar neutrino experiments gave $\Delta m^2_{sol} = (7.59 \pm 0.21) \times 10^{-5}$ eV^2 and $\tan^2 \theta_{sol} = 0.47^{+0.06}_{-0.05}$.

16.10 Investigating Atmospheric Neutrino Oscillations with Accelerators

Despite the remarkable achievements of Super-Kamiokande, it was inevitable that accelerators would eventually seize center stage. Cosmic rays provide unmatched reliability: they never shut down. But a neutrino beam produced by decaying pions and kaons has a well-defined direction and a relatively small range in energy. The oscillations of atmospheric neutrinos gave $\Delta m^2_{atm} \approx 3 \times 10^{-3}$eV^2, so from Eq. (16.9), to see the effect we

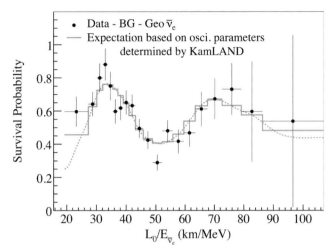

Figure 16.5. The survival probability of $\bar{\nu}_e$ as measured by KamLAND (Ref. 16.9). Backgrounds, including terrestrial antineutrinos, have been subtracted. The baseline $L_0 = 180$ km is the result of weighting contributions from the various contributing reactors in Japan.

need $E/L \approx (1 \,\text{GeV}/300\,\text{km})$. Certainly the detector cannot be located at the accelerator itself!

Aiming a beam from the 12-GeV proton synchroton at KEK in Tsukuba at the Super-Kamiokande detector 250 km away provides an excellent match to these requirements. A detector located just 300 meters from the target provided a means of monitoring the neutrino beam. Data from two years' running, beginning in June 1999 and reported in 2003 (Ref. 16.10) produced 56 muon events against an expectation of $80^{+6.2}_{-5.4}$ in the absence of oscillation. The energy distribution of the events was also distorted from the spectrum expected without oscillations. While the best fit to the data gave $\sin^2 2\theta_{atm}$ very near unity and $\Delta m^2_{atm} = 2.7 \times 10^{-3}\,\text{eV}^2$, the values were poorly determined. With approximately twice the data, in 2006 K2K reported essentially the same central value, but with a much narrower range, $1.9 \times 10^{-3}\,\text{eV}^2 < \Delta m^2_{atm} < 3.5 \times 10^{-3}\,\text{eV}^2$ (Ref. 16.11), and again $\sin^2 2\theta_{atm}$ very near unity.

The MINOS (Main Injector Neutrino Oscillation Search) at Fermilab used a much more energetic beam, 120-GeV protons, to create a neutrino beam maximized at energies between 1 and 3 GeV (Ref. 16.12). The far detector was located in the Soudan iron mine, 735 kilometers away in Minnesota and had a more conventional structure for an accelerator experiment. The muons were observed with steel plates and scintillator, read out with photomultiplier tubes. A near detector, one kilometer from the origin of the neutrino beam, had the same construction.

The much higher energy proton beam produced neutrinos up to 30 GeV and beyond, but it was the lower energy neutrinos that provided the most useful information. Neutral-current events were separated from the charged-current events by the pattern of energy deposition in the detector. The disappearance of muons was apparent: below 10 GeV 122 muon events were seen when 238 ± 11 would have been expected in the absence of oscillations. The results for the mixing parameters were quite similar to those obtained by K2K and Super-Kamiokande: 2.31×10^{-3} eV^2 < Δm^2_{atm} < 3.43×10^{-3} eV^2 and $\sin^2 2\theta_{atm} > 0.78$ at 90% CL. When data taken through July 2007 were included, these limits were significantly improved to $\Delta m^2_{atm} = (2.43 \pm 0.13) \times 10^{-3}$ eV^2 and $\sin^2 2\theta_{atm} > 0.90$ at 90% CL (Ref. 16.13).

16.11 Neutrinos from Low-Energy Muons

Accelerators produce primarily ν_μ, which result from the decays $\pi^+ \to \mu^+ \nu_\mu$ and $K^+ \to \mu^+ \nu_\mu$, and $\bar{\nu}_\mu$ from the analogous decays of negative particles. The semileptonic decay $K^+ \to \pi^0 e^+ \nu_e$ has a 4% branching ratio and is an unfortunate contaminant.

By working with a low-energy primary proton beam, K production can be excluded. The dominance of the decay $\pi^+ \to \mu^+ \nu_\mu$ guarantees a nearly pure ν_μ beam with little $\bar{\nu}_\mu$ contamination since the muon is so long-lived. On the other hand, a pure μ^+ beam that is stopped in matter will produce a pure $\bar{\nu}_\mu$ source without a ν_μ component. This provides the means to search for both $\bar{\nu}_\mu \to \bar{\nu}_e$ and $\nu_\mu \to \nu_e$ oscillations. The Liquid Scintillator Neutrino Detector (LSND) (Ref. 16.14) at Los Alamos looked for evidence for both kinds of oscillations.

LSND took data from 1993 and through 1998. Oscillations of $\bar{\nu}_\mu \to \bar{\nu}_e$ could be detected by observing $\bar{\nu}_e p \to e^+ n$, with the e^+ producing Cherenkov light and the neutron yielding a 2.2 MeV photon through $np \to d\gamma$. The essence of the experiment is to eliminate background $\bar{\nu}_e$ or other particles that might produce similar events in the liquid scintillator, which is viewed with photomultiplier tubes. The initiating proton beam energy was only 800 MeV, leading to many fewer negative pions being produced than positive pions. Most negative pions were absorbed by nuclei before they could decay weakly; the remaining ones would give a negative muon and subsequently $e^- \nu_\mu \bar{\nu}_e$, if the muon was not absorbed first. A larger source of background was not associated with the beam and could be estimated by measuring the event rate between accelerator pulses.

In 1995, the experiment reported that with stringent requirements on the gamma identification, there were 9 events, with an expected background of 2.1, giving a probability that this was a statistical fluctuation of less than 10^{-3}. Fitting to a larger sample obtained by relaxing some criteria gave an oscillation probability of $(0.34^{+0.20}_{-0.18} \pm 0.07) \times 10^{-2}$. If the neutrinos make many oscillations in the 30 meters between the neutrino source and the detector, then this would indicate $\sin^2 2\theta \approx 6.8 \times 10^{-3}$. The minimal Δm^2 consistent with the data is found by setting the mixing to its maximum, $\sin^2 2\theta = 1$, and if we take $E \approx 45$ MeV and $L \approx 30$ m, we find $\Delta m^2 > 0.07$ eV^2. A final report in 2001(Ref. 16.15) gave a consistent result, with 118 ± 22 events against an expected background of 30 ± 6 and an oscillation probability of (0.264 ± 0.067)%.

The decay in flight of pions produced at LAMPF generated a beam of ν_μ whose energy spectrum extended beyond 200 MeV. The transformation $\nu_\mu \to \nu_e$ would be signaled by electrons produced in carbon targets through $\nu_e C \to e^- X$. By looking for electrons with energy between 60 and 200 MeV it was possible to exclude events generated by muon decay at rest. An analysis of the data from 1993 to 1995 found an excess of 18.1 ± 6.6

events and an oscillation probability of $(2.6 \pm 1.0 \pm 0.5) \times 10^{-3}$, very close to the result obtained from decays at rest. The final analysis (Ref. 16.15), however, was ambiguous. The excess was 14.7 ± 12.2 events, with a background of 6.6 ± 1.7 events, altogether a transition probability of (0.10 ± 0.16) %, consistent both with no oscillations and with the positive result found in decay-at-rest. The result combining both decays at rest and decays in flight indicated that at 90% CL, both $\Delta m^2 > 0.02$ eV^2 and $\sin^2 2\theta > 10^{-3}$.

The LSND result was incompatible with the three-neutrino pictures because with just three neutrinos there can be only two independent mass-squared differences. To account for the solar neutrino mass-squared splitting near 8×10^{-5} eV^2, the atmospheric mass-squared splitting near 3×10^{-3} eV^2 and the LSND splitting near 0.1 eV^2 would require introducing a fourth neutrino. This neutrino would have to be sterile: it couldn't couple to the Z, whose width showed that it coupled to precisely three neutrinos.

The MiniBooNE experiment (Ref. 16.16) at Fermilab was designed to confirm or contradict the LSND result. An 8-GeV proton beam impinging on beryllium generated pions and kaons. A toroidal magnetic field focused the positive particles. Their decays produced a neutrino beam dominated by ν_μ, which interacted in a detector 541 meters away. The Cherenkov and scintillation light from the charged particles produced in these interactions were viewed by 1280 8-inch photomultiplier tubes.

The neutrino beam energy was centered at 700 MeV. At this low energy the dominant reactions were $\nu_\mu n \to \mu^- p$, $\nu_\mu N \to \nu_\mu N$, $\nu_\mu N \to \mu^- N \pi$ and $\nu_\mu N \to \nu_\mu N \pi$. If the LSND result were correct, about 0.26% of the ν_μ would be transmuted into ν_e and the analogous charged current interactions would produce electrons in place of muons. Electrons and muons produced different patterns of light, which could be distinguished by the collection of PMTs. Some background events were expected from ν_e contamination of the neutrino beam as a result of K^+ and K_L decays and from muon decays. Produced π^0 also contributed because their decay photons gave a signal similar to that of an electron. In charged current interactions, the energy of the charged lepton was determined from the signals recorded by the PMTs. The energy of the incident neutrino was deduced from the angle the lepton made with the incident neutrino direction and from the observed lepton energy. Simple events $\nu_\mu n \to \mu^- p$ in which the muon decayed in the detector volume and the resulting electron was observed provided a powerful check on the procedures.

A fit was made to the data for events with an observed electron as a function of the incident neutrino energy. Without revealing to themselves the parameters determined by the neutrino-oscillation fit, the MiniBooNE team examined the quality of the fit. Discrepancies in the numbers of events in the low-energy bins led to a decision to restrict the fit to neutrino energies about 475 MeV. Once this was done, the fit with no oscillations was found to give a χ^2 probability 93% indicating no need to include oscillations, in contradiction with the LSND results.

16.12 Oscillations Among Three Neutrino Types

Neutrino oscillation phenomena have been described above as if each involved only two species, though that is clearly incorrect. Evidence from the atmospheric neutrinos showed a mass-squared difference of about 2.5×10^{-3} eV^2, while that in solar neutrinos is about 8×10^{-5} eV^2. Thus there must be two mass-eigenstate neutrinos separated in mass-squared by the smaller amount, and a third mass eigenstate separated from the first two by the larger amount.

Now there appears to be a puzzle in that the Chooz reactor experiment indicated that $\Delta M^2_{\text{Chooz}} < 10^{-3}$ eV^2 while the atmospheric experiment found a larger value in the oscillations of ν_μ. This is resolved if we suppose that ν_e is mostly made of the two neutrinos with similar masses, ν_1 and ν_2. Then experiments, like Chooz and solar neutrino measurements, will depend nearly entirely on this two-state system, characterized by a small value for $\Delta M^2 = \Delta m^2_{\text{sol}}$. This justifies the treatment of solar neutrinos as a two-state system.

The MNS matrix U, which changes flavor eigenstates $|\nu_\alpha\rangle$ into mass eigenstates $|\nu_i\rangle$, $\sum_\alpha |\nu_\alpha\rangle U_{\alpha,i} = |\nu_i\rangle$ can be written as

$$\begin{bmatrix} \nu_1 & \nu_2 & \nu_3 \end{bmatrix} = \begin{bmatrix} \nu_e & \nu_\mu & \nu_\tau \end{bmatrix} U \tag{16.24}$$

where

$$U = \begin{bmatrix} c_{12}c_{13} & s_{12}c_{13} & s_{13}e^{-i\delta} \\ -s_{12}c_{23} - c_{12}s_{23}s_{13}e^{i\delta} & c_{12}c_{23} - s_{12}s_{23}s_{13}e^{i\delta} & s_{23}c_{13} \\ s_{12}s_{23} - c_{12}c_{23}s_{13}e^{i\delta} & -c_{12}s_{23} - s_{12}c_{23}s_{13}e^{i\delta} & c_{23}c_{13} \end{bmatrix}$$

$$\times \begin{bmatrix} e^{i\alpha_1/2} & 0 & 0 \\ 0 & e^{i\alpha_2/2} & 0 \\ 0 & 0 & 1 \end{bmatrix}. \tag{16.25}$$

Here we have introduced the angles θ_{ij}, $i < j$ and $s_{ij} = \sin\theta_{ij}$, $c_{ij} = \cos\theta_{ij}$. This has the same form as the CKM matrix, except for the additional angles α_1 and α_2. These change the phase of the Majorana neutrinos 1 and 2. Ordinarily such a phase would be irrelevant because usually a state and its conjugate with the opposite phase will occur. However, Majorana neutrinos are their own conjugates. In neutrinoless double beta decay, these phases have observable consequences, though they do not affect neutrino oscillations.

The meaning of the angles θ_{ij} is clearer if we write, dropping the αs

$$U = \begin{bmatrix} 1 & 0 & 0 \\ 0 & c_{23} & s_{23} \\ 0 & -s_{23} & c_{23} \end{bmatrix} \begin{bmatrix} c_{13} & 0 & s_{13}e^{-i\delta} \\ 0 & 1 & 0 \\ -s_{13}e^{i\delta} & 0 & c_{13} \end{bmatrix} \begin{bmatrix} c_{12} & s_{12} & 0 \\ -s_{12} & c_{12} & 0 \\ 0 & 0 & 1 \end{bmatrix}. \tag{16.26}$$

The amount of ν_3 in the electron neutrino is governed by θ_{13}. The Chooz experiment shows that it is small. However, it is this small entity in the MNS matrix that carries the CP violation that can be seen in oscillation experiments like $\nu_\mu \to \nu_e$ vs $\bar{\nu}_\mu \to \bar{\nu}_e$.

In the limit of small θ_{13}, solar neutrino oscillations are described by θ_{12}. The oscillations occur between ν_e and the combination $\nu_x = c_{23}\nu_\mu - s_{23}\nu_\tau$. The angle θ_{23} cannot be studied in solar neutrino reactions because low energy ν_μ and ν_τ behave identically.

In atmospheric neutrino experiments, where $\Delta M^2 = \Delta m_{\text{atm}}^2 \approx 2.5 \times 10^{-3}$ eV^2 governs, the small mass-squared splitting between ν_1 and ν_2 cannot be seen, so θ_{12} does not influence the behavior. If we set it to zero, and again drop θ_{13} as being small, we see that θ_{23} is the mixing angle for the cosmic-ray experiments like Super-Kamiokande.

Both θ_{12} and θ_{23} are large, while θ_{13} is small. However, it is this small entity in the MNS matrix that carries the CP violation that could be seen in oscillation experiments like $\nu_\mu \to \nu_e$ vs $\bar{\nu}_\mu \to \bar{\nu}_e$. See Exercise 16.7 and 16.8. The differences of squares of neutrino masses are simply related to the values of ΔM^2 found in the solar and atmospheric neutrino oscillations: $\Delta m_{\text{sol}}^2 = m_2^2 - m_1^2$, $\Delta m_{\text{atm}}^2 = |m_3^2 - m_1^2| \approx |m_3^2 - m_2^2|$.

Three fundamental questions remain in neutrino physics: the values of $\sin 2\theta_{13}$ and δ, and whether the two nearly equal-mass states lie above or below the third mass eigenstate. One possibility for the CP violation required to explain the baryon–antibaryon asymmetry of the universe is that it derives ultimately from CP violation in the decays of the extremely heavy neutrinos that are the see-saw partners of the ordinary neutrinos. Measuring CP violation in the interactions of the light neutrinos would provide some circumstantial evidence for CP violation in the inaccessible neutrinos.

Exercises

16.1 Estimate the flux of solar neutrinos from the $pp \to de^+\nu_e$ process at the surface of the Earth using the surface temperature of the Sun, 5777 K, and its surface area, 6.1×10^{18} m^2. The overall primary cycle initiated by the pp process is

$$4p \to \text{He}^4 + 2e^+ + 2\nu_e$$

whereby about 26.1 MeV is generated, aside from that carried away by the neutrinos themselves. Remember that the energy emission per unit area from a black body is $J = \sigma T^4$, where the Stefan–Boltzmann constant is

$$\sigma = \frac{\pi^2 k^4}{60\hbar^3 c^2} = 5.67 \times 10^{-8} \text{ Wm}^{-2}(\text{deg K})^{-4}.$$

16.2 Verify the numerical relation in Eq. (16.12). Verify the claim that an electron-neutrino would accumulate a phase of 2π from the MSW effect traversing 1.6×10^4 km of hydrogen with a density of 1 g cm^{-3}.

16.3 For the SNO detector described in **Ref. 16.3**, estimate the energy resolution using Poisson statistics and the mean number of PMT hits per MeV of electron energy. Compare with the detailed fit to the resolution given in the paper.

16.4 Calculate the suppression of solar neutrinos by mixing and the MSW effect as a function of the neutrino energy taking $\tan \theta_0 = 0.47$ as suggested by the KamLAND data. Assume the problem can be treated as involving only two neutrino species. Take $\Delta M_0^2 = 8 \times 10^{-5}$ eV^2. Use Table 16.1. Assume that the "other" contributions (from ^{13}N, ^{15}O and pep) are concentrated near 1 MeV. Determine the quality of the fit to the gallium and chlorine data.

16.5 Show that in the three neutrino scheme, the probability of oscillation from α to β is

$$P(\nu_\alpha \to \nu_\beta) = \delta_{\alpha\beta} - 4\sum_{i>j} \Re(U^*_{\alpha i} U_{\beta i} U_{\alpha j} U^*_{\beta j}) \sin^2\left(\frac{\Delta m^2_{ij} L}{4E}\right)$$

$$+ 2\sum_{i>j} \Im(U^*_{\alpha i} U_{\beta i} U_{\alpha j} U^*_{\beta j}) \sin\left(\frac{\Delta m^2_{ij} L}{2E}\right).$$

CPT requires $P(\bar\nu_\alpha \to \bar\nu_\beta) = P(\nu_\beta \to \nu_\alpha)$. The expression for $P(\bar\nu_\alpha \to \bar\nu_\beta)$ is obtained from $P(\nu_\alpha \to \nu_\beta)$ by replacing U with U^*.

16.6 Use the result above to show that in a neutrino reactor experiment aimed at measuring $\sin^2\theta_{13}$ where $\Delta m^2_{31} L/(4E) \approx \pi/2$, the survival probability is given by

$$P(\bar\nu_e \to \bar\nu_e) = 1 - \sin^2 2\theta_{12} \sin^2 \frac{\Delta m^2_{21} L}{4E} - \sin^2 2\theta_{13} \sin^2 \frac{\Delta m^2_{31} L}{4E}.$$

In an experiment with $\Delta m^2_{31} L/(4E) \gg 1$ designed, like KamLAND, to measure Δm^2_{21} and $\sin^2 2\theta_{12}$, the appropriate approximation is

$$P(\bar\nu_e \to \bar\nu_e) = \cos^4\theta_{13}\left[1 - \sin^2 2\theta_{12} \sin^2 \frac{\Delta m^2_{21} L}{4E}\right].$$

16.7 Verify that

$$P(\nu_\mu \to \nu_e) = \sin^2\theta_{23} \sin^2 2\theta_{13} \sin^2 \Delta_{31}$$
$$+ \sin 2\theta_{13} \Delta_{21} \sin 2\theta_{12} \sin 2\theta_{23} \sin \Delta_{31} \cos(\Delta_{31} + \delta)$$
$$+ \Delta^2_{21} \cos^2\theta_{23} \sin^2 2\theta_{12}$$

where $\Delta_{ij} = \Delta m^2_{ij} L/(4E)$ and where $\sin 2\theta_{13}$, Δ_{21} and $|\Delta m^2_{21}/\Delta m^2_{31}|$ are treated as small. For $\bar\nu_\mu \to \bar\nu_e$ the sign of δ is reversed. Using the experimental values for Δm^2_{31} and Δm^2_{21}, determine the size of the CP asymmetry

$$A = \frac{P(\nu_\mu \to \nu_e) - P(\bar\nu_\mu \to \bar\nu_e)}{P(\nu_\mu \to \nu_e) + P(\bar\nu_\mu \to \bar\nu_e)}.$$

Evaluate as a function of $\sin 2\theta_{13}$ and δ. Assume $\sin^2 2\theta_{12} = 0.82$, $\sin^2 2\theta_{23} = 1.0$, and suppose $\Delta_{31} = \pi/2$ so that the asymmetry is maximized.

16.8 If the neutrinos in Exercise 16.7 are not traveling in vacuum, but in a material with electron density N_e, the oscillation probability is instead given by

$$P(\nu_\mu \to \nu_e) = \sin^2\theta_{23} \sin^2 2\theta_{13} \frac{\sin^2(1-x)\Delta_{31}}{(1-x)^2}$$

$$+ \frac{\Delta m_{21}^2}{\Delta m_{31}^2} \sin 2\theta_{13} \sin 2\theta_{12} \sin 2\theta_{23} \frac{\sin[(1-x)\Delta_{31}]}{1-x} \frac{\sin x \Delta_{31}}{x} \cos(\Delta_{31}+\delta)$$

$$+ \left(\frac{\Delta m_{21}^2}{\Delta m_{31}^2}\right)^2 \cos^2\theta_{23} \sin^2 2\theta_{12} \frac{\sin^2(x\Delta_{31})}{x^2}$$

where $x = 2\sqrt{2} G_F N_e E / \Delta m_{31}^2$ and where non-leading terms in $\Delta m_{21}^2 / \Delta m_{31}^2$ and θ_{13} have been neglected. Show that for rock with a density of about 2.4 g/cm³, $x \approx E(\text{GeV})/14$ if Δm_{31}^2 is positive.

Introduce the variables $x = \sin 2\theta_{13} \cos\delta$, $y = \sin 2\theta_{13} \sin\delta$. Take Δm_{21}^2, $|\Delta m_{31}^2|$, $\sin 2\theta_{12}$ and $\sin 2\theta_{13}$ as known. Show that for given E and L, the equations $P(\nu_\mu \to \nu_e) = C_1$ and $P(\bar\nu_\mu \to \bar\nu_e) = C_2$ give circles in the x–y plane. What are the radii and centers of the circles? For the antineutrino case, $\Delta_{31}+\delta$ becomes $\Delta_{31}-\delta$. The sign of x is reversed for the antineutrino case because the antineutrino has a potential opposite that for a neutrino in matter. How are the equations changed if the neutrino spectrum is inverted and how is this reflected in the pattern of the circles in the x–y plane?

16.9 Neutrino beams are formed by focusing pions produced in high energy proton collisions with a fixed target. Pions of a single charge are focused toward the forward direction with a magnetic field. In an idealized description all the pions are moving along a single axis. A single pion of energy $E_\pi = \gamma m_\pi$ decays isotropically in its own rest frame to $\mu\nu_\mu$. Show that in the high energy limit, the distribution of neutrinos in the lab frame is

$$\frac{dN}{d\phi d\cos\theta_{lab}} = \frac{4\gamma^2}{(1+\gamma^2\theta_{lab}^2)^2} \frac{1}{4\pi}$$

where we assume $\theta_{lab} \ll 1$. The maximum transverse momentum of the neutrino is $p^* = (m_\pi^2 - m_\mu^2)/(2m_\pi)$. At a fixed θ_{lab}, what is the highest neutrino energy, E_ν^{max}? For fixed θ_{lab} and neutrino energy $E_\nu < E_\nu^{max}$, pions of two distinct energies may contribute, corresponding to decays in the forward and backward hemispheres in the pion rest frame. Show that the requried values of γ are

$$\gamma^\pm \theta_{lab} = \frac{E_\nu^{max}}{E_\nu} \pm \sqrt{\left(\frac{E_\nu^{max}}{E_\nu}\right)^2 - 1}.$$

Suppose that the produced pions have a distribution $dN/d\gamma$ where $\gamma = E_\pi/m_\pi$. Show that the spectrum of neutrinos through a detector of area A at a distance R from the source and at an angle θ_{lab} is

$$\frac{dN}{dE_\nu} = \frac{1}{\theta_{lab}^3 E_\nu^{max}} \frac{A}{4\pi R^2} \left\{ \frac{E_\nu^{max}/E_\nu}{\sqrt{\left(\frac{E_\nu^{max}}{E_\nu}\right)^2 - 1}} \left[\frac{dN}{d\gamma}(\gamma^+) + \frac{dN}{d\gamma}(\gamma^-) \right] \right.$$
$$\left. + \left[\frac{dN}{d\gamma}(\gamma^+) - \frac{dN}{d\gamma}(\gamma^-) \right] \right\}.$$

Show that in the very forward direction, this reduces to

$$\frac{dN}{dE_\nu}(\theta = 0) = \frac{A}{4\pi R^2} \frac{E_\nu^2}{2p^{*3}} \frac{dN}{d\gamma}(\gamma = \frac{E_\nu}{2p^*}).$$

Suppose the neutrino spectrum in the forward direction has the parabolic form $dN/dE \propto E(E_0 - E)$ with $E_0 = 6$ GeV. What will the neutrino spectrum look like at angles $\theta_l = 7, 14, 27$ mr off-axis?

16.10 Neutrinoless double beta decay depends on the Majorana masses of the neutrinos and the MNS mixing matrix. The decay amplitude is proportional to the effective neutrinoless double beta decay Majorana mass

$$m_{\beta\beta} \equiv \sum_i m_i U_{ei}^2.$$

In standard spectrum the two states with similar mass lie below the third state. In the inverted spectrum the two states with similar mass lie above the third. Since only differences of masses squared have been measured, the mass m^* of the lightest state is unknown. Determine the maximum and minimum values of $|m_{\beta\beta}|$ as a function of m^* for the standard and inverted spectra. Take as representative values $\tan^2 \theta_{12} = 0.40$, $\sin^2 2\theta_{13} = 0.10$, $\Delta m_{31}^2 = 2.5 \times 10^{-3}$ eV^2, $\Delta m_{31}^2 = 8.5 \times 10^{-5}$ eV^2. The values of the phases α_1, α_2, and δ of the MNS matrix are not known and may be varied freely to obtain the maximal and minimal values of $m_{\beta\beta}$. Show that there are values of m^* for the standard spectrum where there is no lower bound to $m_{\beta\beta}$. What upper limit on $m_{\beta\beta}$ would exclude the possibility that neutrinos are Majorana with an inverted spectrum? Graph the allowed regions of $m_{\beta\beta}$ as a function of m^* using a linear plot to simplify the work.

Further Reading

Convenient reviews of many aspects of neutrino oscillations are given in the current *Review of Particle Physics*.

References

16.1 R. Davis, Jr., D. S. Harmer, and K. C. Hoffman, "Search for Neutrinos from the Sun." *Phys. Rev. Lett.*, **20**, 1205 (1968).

16.2 Super-Kamiokande Collaboration, "Evidence for Oscillation of Atmospheric Neutrinos." *Phys. Rev. Lett.*, **81**, 1562 (1998).

16.3 M. Apollonio *et al.*, "Initial Results from the CHOOZ Long Baseline Reactor Neutrino Experiment." *Phys. Lett.*, **420**, 397 (1998).

16.4 M. Apollonio *et al.*, "Limits on Neutrino Oscillations from the CHOOZ Experiment." *Phys. Lett.*, **466**, 415 (1999).

16.5 SNO Collaboration, "Measurement of the Rate of $\nu_e + d \to p + p + e^-$ Interactions Produced by ^{8}B Solar Neutrinos at the Sudbury Neutrino Observatory." *Phys. Rev. Lett.*, **87**, 071301 (2001).

16.6 SNO Collaboration, "Direct Evidence for Neutrino Flavor Transformation from Neutral-Current Interactions in the Sudbury Neutrino Observatory." *Phys. Rev. Lett.*, **89**, 011301 (2002).

16.7 KamLAND Collaboration, "First Results from KamLAND: Evidence for Reactor Anti-Neutrino Disappearance." *Phys. Rev. Lett.*, **90**, 021802 (2003).

16.8 KamLAND Collaboration, "Measurement of Neutrino Oscillation with KamLAND: Evidence of Spectral Distortion." *Phys. Rev. Lett.*, **94**, 081801 (2005).

16.9 KamLAND Collaboration, "Precision Measurement of Neutrino Oscillation Parameters with KamLAND." *Phys. Rev. Lett.*, **100**, 221803 (2008).

16.10 K2K Collaboration, "Indications of Neutrino Oscillation in a 250 km Long-Baseline Experiment." *Phys. Rev. Lett.*, **90**, 041801 (2003).

16.11 K2K Collaboration, "Measurement of neutrino oscillation by the K2K experiment." *Phys. Rev.*, **D74**, 072003 (2006).

16.12 D. G. Michael *et al.* MINOS, "Observation of Muon Neutrino Disappearance with the MINOS Detectors in the NuMI Neutrino Beam." *Phys. Rev. Lett.*, **97**, 191801 (2006).

16.13 P. Adamson *et al.* MINOS, "Measurement of Neutrino Oscillations with the MINOS Detectors in the NuMI Beam." *Phys. Rev. Lett.*, **101**, 131802 (2008).

16.14 C. Athanassopoulous *et al.* (LSND), "Candidate Events in a Search for $\bar{\nu}_\mu \to \bar{\nu}_e$ Oscillations." *Phys. Rev. Lett.*, **75**, 2650 (1995).

16.15 A. Aguilar *et al.* (LSND), "Evidence for Neutrino Oscillations from the Observation of $\bar{\nu}_e$ Appearance in a $\bar{\nu}_\mu$ Beam." *Phys. Rev.*, **D64**, 112007 (2001).

16.16 A. A. Aguilar-Arevalo *et al.* (MiniBooNE), "Search for Electron Neutrino Appearance at the $\Delta m^2 \approx 1$ eV^2 Scale." *Phys. Rev. Lett.*, **98**, 231801 (2007).

SEARCH FOR NEUTRINOS FROM THE SUN*

Raymond Davis, Jr., Don S. Harmer,† and Kenneth C. Hoffman
Brookhaven National Laboratory, Upton, New York 11973
(Received 16 April 1968)

A search was made for solar neutrinos with a detector based upon the reaction $Cl^{37}(\nu, e^-)Ar^{37}$. The upper limit of the product of the neutrino flux and the cross sections for all sources of neutrinos was 3×10^{-36} sec^{-1} per Cl^{37} atom. It was concluded specifically that the flux of neutrinos from B^8 decay in the sun was equal to or less than 2×10^6 cm^{-2} sec^{-1} at the earth, and that less than 9% of the sun's energy is produced by the carbon-nitrogen cycle.

Recent solar-model calculations have indicated that the sun is emitting a measurable flux of neutrinos from decay of B^8 in the interior.[1-8] The possibility of observing these energetic neutrinos has stimulated the construction of four separate neutrino detectors.[9] This paper will present the results of initial measurements with a detection system based upon the neutrino capture reaction $Cl^{37}(\nu,e^-)Ar^{37}$. It was pointed out by Bahcall[10] that the energetic neutrinos from B^8 would feed the analog state of Ar^{37} (a superallowed transition) that lies 5.15 MeV above the ground state. The importance of the contribution of the B^8 neutrino flux is readily seen from the neutrino-capture cross sections and the solar neutrino fluxes given in Table I. The tabulated fluxes were taken from the calculations of Bahcall and Shaviv,[8] who studied the effect of errors in the parameters—solar composition, luminosity, opacity, and nuclear reaction cross sections. These authors have placed a probable error of 60% on the calculated B^8 flux. Their predicted B^8 flux for mean values of the various parameters agrees well with the independent calculations of Ezer and Cameron.[5] On the basis of these predictions, the total solar-neutrino-capture rate in 520 metric tons of chlorine would be in the range of 2 to 7 per day.

The detector design.—A detection system that contains 390 000 liters (520 tons chlorine) of liquid tetrachloroethylene, C_2Cl_4, in a horizontal cylindrical tank was built along the lines proposed earlier.[11] The system is located 4850 ft underground [4400 m (w.e.)] in the Homestake gold mine at Lead, South Dakota. It is essential to place the detector underground to reduce the production of Ar^{37} from (p,n) reactions by protons formed in cosmic-ray muon interactions. The rate of Ar^{37} production in the liquid by cosmic-ray muons at this location is estimated to be 0.1 Ar^{37} atom per day.[11] Background effects from internal α contaminations and fast neutrons from the surrounding rock wall are low. The total Ar^{37} production from all background processes is less than 0.2 Ar^{37} atom per day, which is well below the rate expected from solar neutrinos.

Neutrino detection depends upon removing the Ar^{37} from a large volume of liquid contained in a sealed tank, and observing the decay of Ar^{37} (35-day half-life) in a small proportional counter (0.5 cm^3). It is therefore necessary to have an efficient method of removing a fraction of a cubic centimeter of argon from 390 000 liters of C_2Cl_4. The Ar^{37} activity is removed by purging with helium gas. Liquid is pumped uniformly

Table I. Solar neutrino fluxes and cross sections for the reaction $Cl^{37}(\nu,e^-)Ar^{37}$.

Neutrino source	Cross section[a,b] (cm^2)	Neutrino flux[c] at the earth (cm^{-2} sec^{-1})	$10^{35}\sigma\varphi$ (sec^{-1})
$H+H+e^- \rightarrow D+\nu$	1.72×10^{-45}	1.7×10^8	0.03
Be^7 decay	2.9×10^{-46}	3.9×10^9	0.11
B^8 decay	1.35×10^{-42}	$1.3(1\pm0.6)\times10^7$	$1.8(1\pm0.6)$
N^{13} decay	2.1×10^{-46}	1.0×10^9	0.02
O^{15} decay	7.8×10^{-46}	1.0×10^9	0.08
		$\sum\varphi\sigma=2.0(1\pm0.6)\times10^{-35}$ sec^{-1}	

[a]Ref. 4. [b]Ref. 10. [c]Ref. 8.

from the bottom of the tank and returned to the tank through a series of 40 eductors arranged along two horizontal header pipes inside the tank. The eductors aspirate the helium from the gas space (2000 liters) above the liquid, and mix it as small bubbles with the liquid in the tank. The pump and eductor system passes helium through the liquid at a total rate of 9000 liters per minute maintaining an effective equilibrium between the argon dissolved in the liquid and the argon in the gas phase.

Argon is extracted by circulating the helium from the tank through an argon extraction system. Gas flow is again achieved by a pair of eductors in the tank system, and they maintain a flow rate of 310 liters per minute through the argon extraction system. The tetrachloroethylene vapor is removed by a condenser at −40°C followed by a bed of molecular sieve adsorber at room temperature. The helium then passes through a charcoal bed at 77°K to adsorb the argon, and is finally returned to the tank. This arrangement is shown schematically in Fig. 1. The apparatus is located in three separate rooms in the mine as indicated in the diagram.

The argon sample adsorbed on the charcoal trap is removed by warming the charcoal while a current of helium is passed through it. The argon and other rare gases from the effluent gas stream are collected on a small liquid-nitrogen-cooled charcoal trap (1 cm diam by 10 cm long). Finally, the gases from this trap are desorbed and heated over titanium metal at 1000°C to remove all traces of chemically reactive gases. The resulting rare gas contains krypton and xenon in addition to argon. These higher rare gases were dissolved from the atmosphere during exposure of the liquid during the various manufacturing, storage, and transfer operations. Krypton and xenon are much more soluble in tetrachloroethylene than argon, and, therefore, they are more slowly removed from the liquid by sweeping with helium. Since the volume of krypton and xenon in an experimental run is comparable with or exceeds the volume of argon, it is necessary to remove these higher rare gases from the sample. A more important consideration is that atmospheric krypton contains the

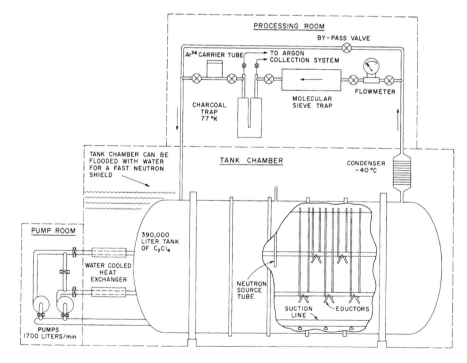

FIG. 1. Schematic arrangement of the Brookhaven solar neutrino dector.

10.8-yr fission product Kr^{85}. The rare gases recovered from the tank are therefore separated by gas chromatography. To insure complete removal of krypton from the argon sample, a second gas chromatographic separation is made of the argon fraction. Experience has shown that these two successive chromatographic separations reduce krypton concentration in the argon sample to less than 10^{-8} parts per volume. The entire purified argon sample is counted in a small proportional counter that will be described later.

Argon recovery tests.—After the air and air argon had been removed from the system by prolonged sweeping with helium, the argon recovery efficiency of the system was measured by an isotope dilution method. A measured volume of 99.9% Ar^{36} was introduced into the tank and dissolved in the liquid with the eductor system. It was then recovered by six separate purging operations. The Ar^{36} recovered from each purge was determined by a volumetric and argon mass-ratio measurement. It was found that the volume of Ar^{36} in the tank dropped exponentially with the volume of helium circulated according to

$$v(Ar)/v_0(Ar) = e^{-7.21 \times 10^{-6} V(He)}.$$

where $v_0(Ar)$ is the initial volume of Ar^{36} and $v(Ar)$ is the volume remaining after $V(He)$ liters of helium have passed through the extraction system. This test showed that a 95% recovery of argon from the tank can be achieved by circulating 0.42 million liters of helium through the extraction system, which requires a period of 22 h.

Another test of the argon recovery from the tank was performed with Ar^{37} activity produced in the tank by a fast-neutron irradiation. A Ra-Be neutron source with a total neutron emission rate of 7.38×10^4 neutrons sec^{-1} was inserted in a re-entrant iron pipe that reaches to the center of the tank. The liquid was irradiated with this source for 0.703 days producing Ar^{37} in the liquid by the reaction $Cl^{37}(p,n)Ar^{37}$ from the protons produced in the liquid principally by the reaction $Cl^{35}(n,p)S^{35}$. Carrier Ar^{36} was introduced (1.18 std cc) and the tank was swept three successive times with helium in which the volumes passed were, respectively, 0.35, 0.26, and 0.34 millions of liters of helium. The recovered argon was purified and counted following the procedures given below. The Ar^{37} activities in the three separate purges were found to be 63.4 ± 3.6, 2.3 ± 1.1, and 0.7 ± 0.5 disintegrations per day at the end of the neutron irradiation. The total Ar^{37} production rate observed in this experiment was $(7.5 \pm 0.4) \times 10^{-7}$ Ar^{37} atom per neutron. This production rate compared favorably with similar measurements in containers of smaller diameters (29 and 120 cm) which gave yields of 3.0×10^{-7} and 6.4×10^{-7} Ar^{37} atom per neutron, respectively. The Ar^{36} recoveries from each of the three successive purges were 90.6, 6.2, and 0.7%, matching closely the Ar^{37} recoveries.

One might question whether Ar^{37} produced by the (ν, e^-) reaction would also be removed efficiently, since it would initially have a lower recoil energy than Ar^{37} produced by the (p,n) reaction. The Ar^{37} recoil energy resulting from neutrino capture ranges from 11 to over 1000 eV for neutrino energies of 1 to 10 MeV. These recoil energies are sufficient to assure that the Ar^{37} ion formed would be free of the parent molecule, and, therefore, it would be expected to behave chemically similarly to an Ar^{37} atom produced by the (p,n) reaction. Once an Ar^{37} atom exists as a free atom it will mix with the carrier Ar^{36} present in the liquid (10^{10} atoms cm^{-3}) and be removed by the helium purge.

Counting.—The argon sample is counted in a small proportional counter with an active volume 3 cm long and 0.5 cm in diameter. A small amount of methane is added to the argon to improve the counting characteristics of the gas. The counter cathode was constructed of zone-refined iron and the exterior envelope is made of silica glass. A thin window in the envelope is located at the end of the counter to facilitate energy calibration of the counter with Fe^{55} x rays. The counter is shielded from external radiations by a cylindrical iron shield 30 cm thick lined with a ring of 5-cm-diam proportional counters for registering cosmic-ray muons. The argon counter is held in the well of a 12.5- by 12.5-cm sodium-iodide scintillation counter located inside the ring counters. Events in anticoincidence with both the ring counters and the scintillation counter are recorded on a 100-channel pulse-height analyzer.[12] The pulse-height and time distribution of the events are recorded on paper tape. Each anticoincidence pulse is displayed on a storage oscilloscope and photographed to allow examination of each pulse shape to insure that it has a proper shape and is not caused by electrical noise. The counter had a 28% resolution (full width at half-maximum) for the 2.8-keV Auger electrons from the Ar^{37} decay. The operating voltage and amplifier gain are adjust-

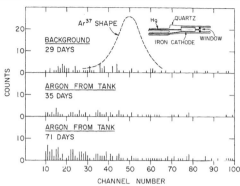

FIG. 2. Pulse-height spectra.

ed to place the center of the 2.8-keV Ar^{37} peak at channel 50 in the spectrum. The background counting rate in the 14 channels centered around channel 50 is 0.3 count per day (see Fig. 2). The efficiency of the counter was determined by filling with argon containing a known amount of Ar^{37}. Its efficiency for Ar^{37} is 51 % for the 14 channels centered about channel 50.

Results and discussion. — Two experimental runs have been performed. In both experiments a measured volume of Ar^{36} was introduced into the tank at the start of the period of exposure, and mixed into the liquid for a period of approximately two hours with the eductor system. During the period of exposure the pumps were not operated. A positive pressure of helium of approximately 250 mm of Hg exists in the tank at all times.

The first exposure was 48 days. The tank was purged with 0.50 million liters of helium. A volume of 1.27 std cc of argon was recovered from the tank, and this volume contained 94 % of the carrier Ar^{36} introduced at the start of the exposure. It was counted for 39 days and the total number of counts observed in the Ar^{37} peak position (full width at half-maximum) in the pulse-height spectrum was 22 counts. This rate is to be compared with a background rate of 31 ± 10 counts for this period. The neutrino-capture rate in the tank deduced from the exposure, counter efficiency, and argon recovery from this experiment was (-1.1 ± 1.4) per day.

A second exposure was made for 110 days from 23 June to 11 October 1967. The tank was purged with 0.53 million liters of helium yielding 0.62 cm³ of argon with a 95% recovery of the added carrier Ar^{36}. The pulse-height spectra are shown in Fig. 2 for the first 35 days of counting and also for a total period of 71 days. This rate can be compared with the background rate for the counter filled with Ar^{36} purified in an identical manner (shown in Fig. 2). It may be seen from the pulse-height spectrum for the first 35 days of counting that 11 ± 3 counts were observed in the 14 channels where Ar^{37} should appear. The counter background for this period of time corresponded to 12 ± 4 counts. Thus, there is no increase in counts from the sample over that expected from background counting rate of the counter. One would deduce from these rates that the neutrino-capture rate in 610 tons of tetrachloroethylene was equal to or less than 0.5 per day based upon one standard deviation. A similar limit can be obtained if one examines the shape of the pulse-height spectrum for extra counts in the 14 channels centered about channel 50 in the first 35-day count.

This limit, expressed as

$$\sum \varphi \sigma \le 0.3 \times 10^{-35} \text{ sec}^{-1} \text{ per Cl}^{37} \text{ atom},$$

can be compared with the predicted value of $(2.0 \pm 1.2) \times 10^{-35}$ sec^{-1} per Cl^{37} atom (Table I). It may be seen that this limit is approximately a factor of 7 below that expected from these solar-model calculations. From this limit and the cross section for B^8 neutrinos given in Table I, it may be concluded that the flux of B^8 neutrinos at the earth is equal to or less than 2×10^6 cm^{-2} sec^{-1}. It may be pointed out that if one accepted all of the 11 counts in the spectrum for the 35-day count as real events, making no allowance for background, then the flux–cross-section product limit would be 0.6×10^{-35} sec^{-1} per Cl^{37} atom.

The solar-model calculation of the flux of B^8 neutrinos is dependent upon the nuclear cross sections, solar composition, solar age and luminosity, and the opacity of solar material. The effect of each of these parameters has been studied, and the present results show that the solar B^8 neutrino flux is outside the present error limits if the uncertainties are treated as probable errors.[6-8] In the following article[13] Bahcall, Bahcall, and Shaviv have re-evaluated the solar neutrino fluxes taking into account a new value for the heavy element composition of the sun, and a new rate for the reaction $H(H, e^+\nu)D$.

Since this experiment is the first one with sufficient sensitivity to detect solar neutrinos from the carbon-nitrogen cycle, it is interesting to draw a conclusion about this energy cycle. Bahcall[4] has calculated the total flux–cross-section product for the carbon-nitrogen cycle to be 3.5×10^{-35} sec^{-1} per Cl^{37} atom, based on this cycle being the only source of the sun's energy. With the limit given above one can conclude that less than 9% of the sun's energy is produced by the carbon-nitrogen cycle.

It is possible to improve the sensitivity of the present experiment by reducing the background of the counter. However, background effects from cosmic-ray muons will eventually limit the detection sensitivity of the experiment at its present location. Detailed studies of the cosmic-ray background are in progress.

The authors would like to thank Professor W. A. Fowler and Professor John N. Bahcall for their initial and continual encouragement in planning this experiment. We would like to acknowledge Professor A. G. W. Cameron's constant interest extending over many years. We are indebted to the Homestake Mining Company for allowing us to build the experiment in their mine, and for their generous assistance in solving many technical problems in the construction of the apparatus. We would like to acknowledge the many useful suggestions and direct assistance from the members of the staff of Brookhaven National Laboratory.

---

*Research performed under the auspices of the U. S. Atomic Energy Commission.

†Permanent address: Georgia Institute of Technology, Atlanta, Georgia.

[1] J. N. Bahcall, W. A. Fowler, I. Iben, Jr., and R. L. Sears, Astrophys. J. 137, 344 (1963).

[2] R. L. Sears, Astrophys. J. 140, 153 (1964).

[3] P. Pochoda and H. Reeves, Planetary Space Sci. 12, 119 (1964).

[4] J. N. Bahcall, Phys. Rev. Letters 12, 300 (1964); 17, 398 (1966).

[5] D. Ezer and A. G. W. Cameron, Can. J. Phys. 43, 1497 (1965), and 44, 593 (1966); and private communication.

[6] J. N. Bahcall, N. Cooper, and P. Demarque, Astrophys. J. 150, 723 (1967); G. Shaviv, J. N. Bahcall, and W. A. Fowler, Astrophys. J. 150, 725 (1967).

[7] J. N. Bahcall, N. Bahcall, W. A. Fowler, and G. Shaviv, Phys. Letters 26B, 359 (1968).

[8] J. N. Bahcall and G. Shaviv, to be published.

[9] For recent summary see F. Reines, Proc. Roy. Soc. (London) 310A, 104 (1967).

[10] J. N. Bahcall, Phys. Rev. 135, B137 (1964).

[11] R. Davis, Jr., Phys. Rev. Letters 12, 303 (1964); R. Davis, Jr., and D. S. Harmer, CERN Report No. CERN 65-32, 1965 (unpublished).

[12] The circuit used in this work was designed by Mr. R. L. Chase and Mr. Lee Rogers of Brookhaven National Laboratory.

[13] J. N. Bahcall, N. A. Bahcall, and G. Shaviv, following Letter [Phys. Rev. Letters 20, 1209 (1968)].

Evidence for Oscillation of Atmospheric Neutrinos

Y. Fukuda,[1] T. Hayakawa,[1] E. Ichihara,[1] K. Inoue,[1] K. Ishihara,[1] H. Ishino,[1] Y. Itow,[1] T. Kajita,[1] J. Kameda,[1] S. Kasuga,[1] K. Kobayashi,[1] Y. Kobayashi,[1] Y. Koshio,[1] M. Miura,[1] M. Nakahata,[1] S. Nakayama,[1] A. Okada,[1] K. Okumura,[1] N. Sakurai,[1] M. Shiozawa,[1] Y. Suzuki,[1] Y. Takeuchi,[1] Y. Totsuka,[1] S. Yamada,[1] M. Earl,[2] A. Habig,[2] E. Kearns,[2] M. D. Messier,[2] K. Scholberg,[2] J. L. Stone,[2] L. R. Sulak,[2] C. W. Walter,[2] M. Goldhaber,[3] T. Barszczak,[4] D. Casper,[4] W. Gajewski,[4] P. G. Halverson,[4,*] J. Hsu,[4] W. R. Kropp,[4] L. R. Price,[4] F. Reines,[4] M. Smy,[4] H. W. Sobel,[4] M. R. Vagins,[4] K. S. Ganezer,[5] W. E. Keig,[5] R. W. Ellsworth,[6] S. Tasaka,[7] J. W. Flanagan,[8,†] A. Kibayashi,[8] J. G. Learned,[8] S. Matsuno,[8] V. J. Stenger,[8] D. Takemori,[8] T. Ishii,[9] J. Kanzaki,[9] T. Kobayashi,[9] S. Mine,[9] K. Nakamura,[9] K. Nishikawa,[9] Y. Oyama,[9] A. Sakai,[9] M. Sakuda,[9] O. Sasaki,[9] S. Echigo,[10] M. Kohama,[10] A. T. Suzuki,[10] T. J. Haines,[11,4] E. Blaufuss,[12] B. K. Kim,[12] R. Sanford,[12] R. Svoboda,[12] M. L. Chen,[13] Z. Conner,[13,‡] J. A. Goodman,[13] G. W. Sullivan,[13] J. Hill,[14] C. K. Jung,[14] K. Martens,[14] C. Mauger,[14] C. McGrew,[14] E. Sharkey,[14] B. Viren,[14] C. Yanagisawa,[14] W. Doki,[15] K. Miyano,[15] H. Okazawa,[15] C. Saji,[15] M. Takahata,[15] Y. Nagashima,[16] M. Takita,[16] T. Yamaguchi,[16] M. Yoshida,[16] S. B. Kim,[17] M. Etoh,[18] K. Fujita,[18] A. Hasegawa,[18] T. Hasegawa,[18] S. Hatakeyama,[18] T. Iwamoto,[18] M. Koga,[18] T. Maruyama,[18] H. Ogawa,[18] J. Shirai,[18] A. Suzuki,[18] F. Tsushima,[18] M. Koshiba,[19] M. Nemoto,[20] K. Nishijima,[20] T. Futagami,[21] Y. Hayato,[21,§] K. Kanaya,[21] K. Kaneyuki,[21] Y. Watanabe,[21] D. Kielczewska,[22,4] R. A. Doyle,[23] J. S. George,[23] A. L. Stachyra,[23] L. L. Wai,[23,∥] R. J. Wilkes,[23] and K. K. Young[23]

(Super-Kamiokande Collaboration)

[1]*Institute for Cosmic Ray Research, University of Tokyo, Tanashi, Tokyo, 188-8502, Japan*
[2]*Department of Physics, Boston University, Boston, Massachusetts 02215*
[3]*Physics Department, Brookhaven National Laboratory, Upton, New York 11973*
[4]*Department of Physics and Astronomy, University of California at Irvine, Irvine, California 92697-4575*
[5]*Department of Physics, California State University, Dominguez Hills, Carson, California 90747*
[6]*Department of Physics, George Mason University, Fairfax, Virginia 22030*
[7]*Department of Physics, Gifu University, Gifu, Gifu 501-1193, Japan*
[8]*Department of Physics and Astronomy, University of Hawaii, Honolulu, Hawaii 96822*
[9]*Institute of Particle and Nuclear Studies, High Energy Accelerator Research Organization (KEK), Tsukuba, Ibaraki 305-0801, Japan*
[10]*Department of Physics, Kobe University, Kobe, Hyogo 657-8501, Japan*
[11]*Physics Division, P-23, Los Alamos National Laboratory, Los Alamos, New Mexico 87544*
[12]*Department of Physics and Astronomy, Louisiana State University, Baton Rouge, Louisiana 70803*
[13]*Department of Physics, University of Maryland, College Park, Maryland 20742*
[14]*Department of Physics and Astronomy, State University of New York, Stony Brook, New York 11794-3800*
[15]*Department of Physics, Niigata University, Niigata, Niigata 950-2181, Japan*
[16]*Department of Physics, Osaka University, Toyonaka, Osaka 560-0043, Japan*
[17]*Department of Physics, Seoul National University, Seoul 151-742, Korea*
[18]*Department of Physics, Tohoku University, Sendai, Miyagi 980-8578, Japan*
[19]*The University of Tokyo, Tokyo 113-0033, Japan*
[20]*Department of Physics, Tokai University, Hiratsuka, Kanagawa 259-1292, Japan*
[21]*Department of Physics, Tokyo Institute of Technology, Meguro, Tokyo 152-8551, Japan*
[22]*Institute of Experimental Physics, Warsaw University, 00-681 Warsaw, Poland*
[23]*Department of Physics, University of Washington, Seattle, Washington 98195-1560*
(Received 6 July 1998)

> We present an analysis of atmospheric neutrino data from a 33.0 kton yr (535-day) exposure of the Super-Kamiokande detector. The data exhibit a zenith angle dependent deficit of muon neutrinos which is inconsistent with expectations based on calculations of the atmospheric neutrino flux. Experimental biases and uncertainties in the prediction of neutrino fluxes and cross sections are unable to explain our observation. The data are consistent, however, with two-flavor $\nu_\mu \leftrightarrow \nu_\tau$ oscillations with $\sin^2 2\theta > 0.82$ and $5 \times 10^{-4} < \Delta m^2 < 6 \times 10^{-3}$ eV^2 at 90% confidence level. [S0031-9007(98)06975-0]

PACS numbers: 14.60.Pq, 96.40.Tv

Atmospheric neutrinos are produced as decay products in hadronic showers resulting from collisions of cosmic rays with nuclei in the upper atmosphere. Production of electron and muon neutrinos is dominated by the processes $\pi^+ \to \mu^+ + \nu_\mu$ followed by $\mu^+ \to e^+ + \overline{\nu}_\mu + \nu_e$ (and their charge conjugates) giving an expected ratio

($\equiv \nu_\mu/\nu_e$) of the flux of $\nu_\mu + \bar{\nu}_\mu$ to the flux of $\nu_e + \bar{\nu}_e$ of about 2. The ν_μ/ν_e ratio has been calculated in detail with an uncertainty of less than 5% over a broad range of energies from 0.1 to 10 GeV [1,2].

The ν_μ/ν_e flux ratio is measured in deep underground experiments by observing final state leptons produced via charged-current interactions of neutrinos on nuclei, $\nu + N \rightarrow l + X$. The flavor of the final state lepton is used to identify the flavor of the incoming neutrino.

The measurements are reported as $R \equiv (\mu/e)_{\text{DATA}}/(\mu/e)_{\text{MC}}$, where μ and e are the number of muonlike (μ-like) and electronlike (e-like) events observed in the detector for both data and Monte Carlo simulations. This ratio largely cancels experimental and theoretical uncertainties, especially the uncertainty in the absolute flux. $R = 1$ is expected if the physics in the Monte Carlo simulation accurately models the data. Measurements of significantly small values of R have been reported by the deep underground water Cherenkov detectors Kamiokande [3,4], IMB [5], and recently by Super-Kamiokande [6,7]. Although measurements of R by early iron-calorimeter experiments Fréjus [8] and NUSEX [9] with smaller data samples were consistent with expectations, the Soudan-2 iron-calorimeter experiment has reported observation of a small value of R [10].

Neutrino oscillations have been suggested to explain measurements of small values of R. For a two-neutrino oscillation hypothesis, the probability for a neutrino produced in flavor state a to be observed in flavor state b after traveling a distance L through a vacuum is

$$P_{a \rightarrow b} = \sin^2 2\theta \sin^2\left(\frac{1.27 \Delta m^2 (\text{eV}^2) L(\text{km})}{E_\nu (\text{GeV})}\right), \quad (1)$$

where E_ν is the neutrino energy, θ is the mixing angle between the flavor eigenstates and the mass eigenstates, and Δm^2 is the mass squared difference of the neutrino mass eigenstates. For detectors near the surface of the Earth, the neutrino flight distance, and thus the oscillation probability, is a function of the zenith angle of the neutrino direction. Vertically downward-going neutrinos travel about 15 km, while vertically upward-going neutrinos travel about 13 000 km before interacting in the detector. The broad energy spectrum and this range of neutrino flight distances make measurements of atmospheric neutrinos sensitive to neutrino oscillations with Δm^2 down to 10^{-4} eV^2. The zenith angle dependence of R measured by the Kamiokande experiment at high energies has been cited as evidence for neutrino oscillations [4].

We present our analysis of 33.0 kton yr (535 days) of atmospheric neutrino data from Super-Kamiokande. In addition to measurements of small values of R both above and below ~ 1 GeV, we observed a significant zenith angle dependent deficit of μ-like events. While no combination of known uncertainties in the experimental measurement or predictions of atmospheric neutrino fluxes is able to explain our data, a two-neutrino oscillation model of $\nu_\mu \leftrightarrow \nu_x$, where ν_x may be ν_τ or a new, noninteracting "sterile" neutrino, is consistent with the observed flavor ratios and zenith angle distributions over the entire energy region.

Super-Kamiokande is a 50 kton water Cherenkov detector instrumented with 11 146 photomultiplier tubes (PMTs) facing an inner 22.5 kton fiducial volume of ultrapure water. Interaction kinematics are reconstructed using the time and charge of each PMT signal. The inner volume is surrounded by a ~ 2 m thick outer detector instrumented with 1885 outward-facing PMTs. The outer detector is used to veto entering particles and to tag exiting tracks.

Super-Kamiokande has collected a total of 4353 fully contained (FC) events and 301 partially contained (PC) events in a 33.0 kton yr exposure. FC events deposit all of their Cherenkov light in the inner detector while PC events have exiting tracks which deposit some Cherenkov light in the outer detector. For this analysis, the neutrino interaction vertex was required to have been reconstructed within the 22.5 kton fiducial volume, defined to be >2 m from the PMT wall.

FC events were separated into those with a single visible Cherenkov ring and those with multiple Cherenkov rings. For the analysis of FC events, only single-ring events were used. Single-ring events were identified as e-like or μ-like based on a likelihood analysis of light detected around the Cherenkov cone. The FC events were separated into "sub-GEV" ($E_{\text{vis}} < 1330$ MeV) and "multi-GeV" ($E_{\text{vis}} > 1330$ MeV) samples, where E_{vis} is defined to be the energy of an electron that would produce the observed amount of Cherenkov light. $E_{\text{vis}} = 1330$ MeV corresponds to $p_\mu \sim 1400$ MeV/c.

In a full-detector Monte Carlo simulation, 88% (96%) of the sub-GeV e-like (μ-like) events were ν_e (ν_μ) charged-current interactions and 84% (99%) of the multi-GeV e-like (μ-like) events were ν_e (ν_μ) charged-current (CC) interactions. PC events were estimated to be 98% ν_μ charged-current interactions; hence, all PC events classified as μ-like, and no single-ring requirement was made. Table I summarizes the number of observed events for both data and Monte Carlo as well as the R values for the sub-GeV and multi-GeV samples. Further details of the detector, data selection, and event reconstruction used in this analysis are given elsewhere [6,7].

We have measured significantly small values of R in both the sub-GeV and multi-GeV samples. Several sources of systematic uncertainties in these measurements have been considered. Cosmic ray induced interactions in the rock surrounding the detector have been suggested as a source of e-like contamination from neutrons, which could produce small R values [11], but these backgrounds have been shown to be insignificant for large water Cherenkov detectors [12]. In particular, Super-Kamiokande has 4.7 m of water surrounding the fiducial volume; this distance corresponds to roughly 5 hadronic interaction lengths and 13 radiation lengths. Distributions of event vertices

1563

TABLE I. Summary of the sub-GeV, multi-GeV, and PC event samples compared with the Monte Carlo prediction based on the neutrino flux calculation of Ref. [2].

	Data	Monte Carlo
Sub-GeV		
Single-ring	2389	2622.6
e-like	1231	1049.1
μ-like	1158	1573.6
Multi-ring	911	980.7
Total	3300	3603.3
	$R = 0.63 \pm 0.03$ (stat.) ± 0.05 (syst.)	
Multi-GeV		
Single-ring	520	531.7
e-like	290	236.0
μ-like	230	295.7
Multi-ring	533	560.1
Total	1053	1091.8
Partially contained	301	371.6
	$R_{FC+PC} = 0.65 \pm 0.05$ (stat.) ± 0.08 (syst.)	

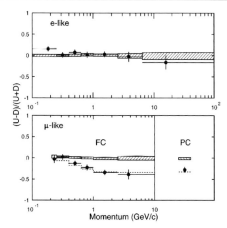

FIG. 1. The $(U - D)/(U + D)$ asymmetry as a function of momentum for FC e-like and μ-like events and PC events. While it is not possible to assign a momentum to a PC event, the PC sample is estimated to have a mean neutrino energy of 15 GeV. The Monte Carlo expectation without neutrino oscillations is shown in the hatched region with statistical and systematic errors added in quadrature. The dashed line for μ-like is the expectation for $\nu_\mu \leftrightarrow \nu_\tau$ oscillations with ($\sin^2 2\theta = 1.0$, $\Delta m^2 = 2.2 \times 10^{-3}$ eV^2).

exhibit no excess of e-like events close to the fiducial boundary [6,7].

The prediction of the ratio of the ν_μ flux to the ν_e flux is dominated by the well-understood decay chain of mesons and contributes less than 5% of the uncertainty in R. Different neutrino flux models vary by about $\pm 20\%$ in the prediction of absolute rates, but the ratio is robust [13]. Uncertainties in R due to a difference in cross sections for ν_e and ν_μ have been studied [14]; however, lepton universality prevents any significant difference in cross sections at energies much above the muon mass and thus errors in cross sections could not produce a small value of R in the multi-GeV energy range. Particle identification was estimated to be $\gtrsim 98\%$ efficient for both μ-like and e-like events based on Monte Carlo studies. Particle identification was also tested in Super-Kamiokande on Michel electrons and stopping cosmic-ray muons and the μ-like and e-like events used in this analysis are clearly separated [6]. The particle identification programs in use have also been tested using beams of electrons and muons incident on a water Cherenkov detector at KEK [15]. The data have been analyzed independently by two groups, making the possibility of significant biases in data selection or event reconstruction algorithms remote [6,7]. Other explanations for the small value of R, such as contributions from nucleon decays [16], can be discounted as they would not contribute to the zenith angle effects described below.

We estimate the probability that the observed μ/e ratios could be due to statistical fluctuation is less than 0.001% for sub-GeV R and less than 1% for multi-GeV R.

The μ-like data exhibit a strong asymmetry in zenith angle (Θ) while no significant asymmetry is observed in the e-like data [7]. The asymmetry is defined as $A =$ $(U - D)/(U + D)$ where U is the number of upward-going events ($-1 < \cos\Theta < -0.2$) and D is the number of downward-going events ($0.2 < \cos\Theta < 1$). The asymmetry is expected to be near zero independent of the flux model for $E_\nu > 1$ GeV, above which effects due to the Earth's magnetic field on cosmic rays are small. Based on a comparison of results from our full Monte Carlo simulation using different flux models [1,2] as inputs, treatment of geomagnetic effects results in an uncertainty of roughly ± 0.02 in the expected asymmetry of e-like and μ-like sub-GeV events and less than ± 0.01 for multi-GeV events. Studies of decay electrons from stopping muons show at most a $\pm 0.6\%$ up-down difference in Cherenkov light detection [17].

Figure 1 shows A as a function of momentum for both e-like and μ-like events. In the present data, the asymmetric as a function of momentum for e-like events is consistent with expectations, while the μ-like asymmetry at low momentum is consistent with zero but significantly deviates form expectations at higher momentum. The average angle between the final state lepton direction and the incoming neutrino direction is 55° at $p = 400$ MeV/c and 20° at 1.5 GeV/c. At the lower momenta in Fig. 1, the possible asymmetry of the neutrino flux is largely washed out. We have found no detector bias differentiating e-like and μ-like events that could explain an asymmetry in μ-like events but not in e-like events [7].

Considering multi-GeV (FC + PC) muons alone, the measured asymmetry, $A = -0.296 \pm 0.048 \pm 0.01$ deviates from zero by more than 6 standard deviations.

We have examined the hypotheses of two-flavor $\nu_\mu \leftrightarrow \nu_e$ and $\nu_\mu \leftrightarrow \nu_\tau$ oscillation models using a χ^2 comparison of data and Monte Carlo, allowing all important Monte Carlo parameters to vary weighted by their expected uncertainties.

The data were binned by particle type, momentum, and $\cos\Theta$. A χ^2 is defined as

$$\chi^2 = \sum_{\cos\Theta,p} (N_{\text{DATA}} - N_{\text{MC}})^2/\sigma^2 + \sum_j \epsilon_j^2/\sigma_j^2, \quad (2)$$

where the sum is over five bins equally spaced in $\cos\Theta$ and seven momentum bins for both e-like events and μ-like plus PC events (70 bins total). The statistical error, σ, accounts for both data statistics and the weighted Monte Carlo statistics. N_{DATA} is the measured number of events in each bin. N_{MC} is the weighted sum of Monte Carlo events:

$$N_{\text{MC}} = \frac{\mathcal{L}_{\text{DATA}}}{\mathcal{L}_{\text{MC}}} \sum_{\text{MC events}} w. \quad (3)$$

$\mathcal{L}_{\text{DATA}}$ and $\mathcal{L}_{\text{MC}}$ are the data and Monte Carlo live times. For each Monte Carlo event, the weight w is given by

$$w = (1 + \alpha)(E_\nu^i/E_0)^\delta (1 + \eta_{s,m} \cos\Theta)$$

$$\times f_{e,\mu}(\sin^2 2\theta, \Delta m^2, (1+\lambda)L/E_\nu)$$

$$\times \begin{cases} (1 - \beta_s/2) & \text{sub-GeV } e\text{-like,} \\ (1 + \beta_s/2) & \text{sub-GeV } \mu\text{-like,} \\ (1 - \beta_m/2) & \text{multi-GeV } e\text{-like,} \\ (1 + \beta_m/2)\left(1 - \frac{\rho}{2}\frac{N_{PC}}{N_\mu}\right) & \text{multi-GeV } \mu\text{-like,} \\ (1 + \beta_m/2)\left(1 + \frac{\rho}{2}\right) & \text{PC.} \end{cases} \quad (4)$$

E_ν^i is the average neutrino energy in the ith momentum bin; E_0 is an arbitrary reference energy (taken to be 2 GeV); η_s (η_m) is the up-down uncertainty of the event rate in the sub-GeV (multi-GeV) energy range; N_{PC} is the total number of PC events; and N_μ is the total number of Monte Carlo FC multi-GeV muons. The factor $f_{e,\mu}$ weights an event accounting for the initial neutrino fluxes (in the case of $\nu_\mu \leftrightarrow \nu_e$), oscillation parameters, and L/E_ν. The meaning of the Monte Carlo fit parameters, α and $\epsilon_j \equiv (\beta_s, \beta_m, \delta, \rho, \lambda, \eta_s, \eta_m)$ and their assigned uncertainties, σ_j, are summarized in Table II. The overall normalization, α, was allowed to vary freely. The uncertainty in the Monte Carlo L/E_ν ratio (λ) was conservatively estimated based on the uncertainty in an absolute energy scale, uncertainty in neutrino-lepton angular and energy correlations, and the uncertainty in production height. The oscillation simulations used profiles of neutrino production heights calculated in Ref. [18], which account for the competing factors of production, propagation, and decay of muons and mesons through the atmosphere.

TABLE II. Summary of Monte Carlo fit parameters. Best-fit values for $\nu_\mu \leftrightarrow \nu_\tau (\Delta m^2 = 2.2 \times 10^{-3}$ eV^2, $\sin^2 2\theta = 1.0$) and estimated uncertainties are given. (*) The overall normalization (α) was estimated to have a 25% uncertainty but was fitted as a free parameter.

Monte Carlo fit parameters		Best fit	Uncertainty
α	Overall normalization	15.8%	(*)
δ	E_ν spectral index	0.006	$\sigma_\delta = 0.05$
β_s	Sub-GeV μ/e ratio	-6.3%	$\sigma_s = 8\%$
β_m	Multi-GeV μ/e ratio	-11.8%	$\sigma_m = 12\%$
ρ	Relative norm. of PC to FC	-1.8%	$\sigma_\rho = 8\%$
λ	L/E_ν	3.1%	$\sigma_\lambda = 15\%$
η_s	Sub-GeV up-down	2.4%	$\sigma_\eta^s = 2.4\%$
η_m	Multi-GeV up-down	-0.09%	$\sigma_\eta^m = 2.7\%$

For $\nu_\mu \leftrightarrow \nu_e$, effects of matter on neutrino propagation through the Earth were included following Ref. [19,20]. Because of the small number of events expected from τ production, the effects of τ appearance and decay were neglected in simulations of $\nu_\mu \leftrightarrow \nu_\tau$. A global scan was made on a $(\sin^2 2\theta, \log \Delta m^2)$ grid minimizing χ^2 with respect to $\alpha, \beta_s, \beta_m, \delta, \rho, \lambda, \eta_s$, and η_m at each point.

The best fit to $\nu_\mu \leftrightarrow \nu_\tau$ oscillations, $\chi^2_{\min} = 65.2/67$ DOF, was obtained at $(\sin^2 2\theta = 1.0, \Delta m^2 = 2.2 \times 10^{-3}$ eV^2) inside the physical region ($0 \leq \sin^2 2\theta \leq 1$). The best-fit values of the Monte Carlo parameters (summarized in Table II) were all within their expected errors. The global minimum occurred slightly outside of the physical region at $(\sin^2 2\theta = 1.05, \Delta m^2 = 2.2 \times 10^{-3}$ eV^2, $\chi^2_{\min} = 64.8/67$ DOF). The contours of the 68%, 90%, and 99% confidence intervals are located at $\chi^2_{\min} + 2.6$, 5.0, and 9.6 based on the minimum inside the physical region [21]. Thee contours are shown in Fig. 2. The region near χ^2 minimum is rather flat and has many local minima so that inside the 68% interval the best-fit Δm^2 is not well-constrained. Outside of the 99% allowed region the χ^2 increases rapidly. We obtained $\chi^2 = 135/69$ DOF, when calculated at $\sin^2 2\theta = 0$, $\Delta m^2 = 0$ (i.e., assuming no oscillations).

For the test of $\nu_\mu \leftrightarrow \nu_e$ oscillations, we obtained a relatively poor fit; $\chi^2_{\min} = 87.8/67$ DOF, at $(\sin^2 2\theta = 0.93, \Delta m^2 = 3.2 \times 10^{-3}$ eV^2). The expected asymmetry of the multi-GeV e-like events for the best-fit $\nu_\mu \leftrightarrow \nu_e$ oscillation hypothesis, $A = 0.205$, differs from the measured asymmetry, $A = -0.036 \pm 0.067 \pm 0.02$, by 3.4 standard deviations. We conclude that the $\nu_\mu \leftrightarrow \nu_e$ hypothesis is not favored.

The zenith angle distributions for the FC and PC samples are shown in Fig. 3. The data are compared to the Monte Carlo expectation (no oscillations, hatched region) and the best-fit expectation for $\nu_\mu \leftrightarrow \nu_\tau$ oscillations (bold line).

We also estimated the oscillation parameters considering the R measurement and the zenith angle shape separately. The 90% confidence level allowed regions for each

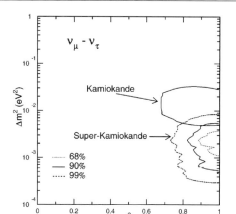

FIG. 2. The 68%, 90%, and 99% confidence intervals are shown for $\sin^2 2\theta$ and Δm^2 for $\nu_\mu \leftrightarrow \nu_\tau$ two-neutrino oscillations based on 33.0 kton yr of Super-Kamiokande data. The 90% confidence interval obtained by the Kamiokande experiment is also shown.

case overlapped at $1 \times 10^{-3} < \Delta m^2 < 4 \times 10^{-3}$ eV^2 for $\sin^2 2\theta = 1$.

As a cross-check of the above analyses, we have reconstructed the best estimate of the ratio L/E_ν for each event. The neutrino energy is estimated by applying a correction to the final state lepton momentum. Typi-

cally, final state leptons with $p \sim 100$ MeV/c carry 65% of the incoming neutrino energy increasing to $\sim 85\%$ at $p = 1$ GeV/c. The neutrino flight distance L is estimated following Ref. [18] using the estimated neutrino energy and the reconstructed lepton direction and flavor. Figure 4 shows the ratio of FC data to Monte Carlo for e-like and μ-like events with $p > 400$ MeV as a function of L/E_ν, compared to the expectation for $\nu_\mu \leftrightarrow \nu_\tau$ oscillations with our best-fit parameters. The e-like data show no significant variation in L/E_ν, while the μ-like events show a significant deficit at large L/E_ν. At large L/E_ν, the ν_μ have presumably undergone numerous oscillations and have averaged out to roughly half the initial rate.

The asymmetry A of the e-like events in the present data is consistent with expectations without neutrino oscillations and two-flavor $\nu_e \leftrightarrow \nu_\mu$ oscillations are not favored. This is in agreement with recent results from the CHOOZ experiment [22]. The LSND experiment has reported the appearance of ν_e in a beam of ν_μ produced by stopped pions [23]. The LSND results do not contradict the present results if they are observing small mixing angles. With the best-fit parameters for $\nu_\mu \leftrightarrow \nu_\tau$ oscillations, we expect a total of only 15–20 events from ν_τ charged-current interactions in the data sample. Using the current sample, oscillations between ν_μ and ν_τ are indistinguishable from oscillations between ν_μ and a noninteracting sterile neutrino.

Figure 2 shows the Super-Kamiokande results overlaid with the allowed region obtained by the Kamiokande

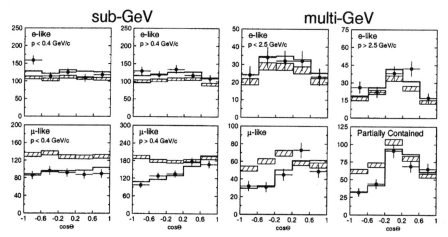

FIG. 3. Zenith angle distributions of μ-like and e-like events for sub-GeV and multi-GeV data sets. Upward-going particles have $\cos \Theta < 0$ and downward-going particles have $\cos \Theta > 0$. Sub-GeV data are shown separately for $p < 400$ MeV/c and $p > 400$ MeV/c. Multi-GeV e-like distributions are shown for $p < 2.5$ and $p > 2.5$ GeV/c and the multi-GeV μ-like are shown separately for FC and PC events. The hatched region shows the Monte Carlo expectation for no oscillations normalized to the data live time with statistical errors. The bold line is the best-fit expectation for $\nu_\mu \leftrightarrow \nu_\tau$ oscillations with the overall flux normalization fitted as a free parameter.

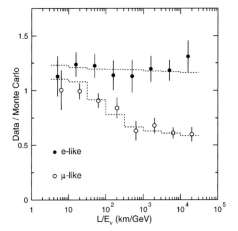

FIG. 4. The ratio of the number of FC data events to FC Monte Carlo events versus reconstructed L/E_ν. The points show the ratio of observed data to MC expectation in the absence of oscillations. The dashed lines show the expected shape for $\nu_\mu \leftrightarrow \nu_\tau$ at $\Delta m^2 = 2.2 \times 10^{-3}$ eV^2 and $\sin^2 2\theta = 1$. The slight L/E_ν dependence for e-like events is due to contamination (2–7%) of ν_μ CC interactions.

experiment [4]. The Super-Kamiokande region favors lower values of Δm^2 than allowed by the Kamiokande experiment; however the 90% contours from both experiments have a region of overlap. Preliminary studies of upward-going stopping and through-going muons in Super-Kamiokande [24] give allowed regions consistent with the FC and PC event analysis reported in this paper.

Both the zenith angle distribution of μ-like events and the value of R observed in this experiment significantly differ from the best predictions in the absence of neutrino oscillations. While uncertainties in the flux prediction, cross sections, and experimental biases are ruled out as explanations of the observations, the present data are in good agreement with two-flavor $\nu_\mu \leftrightarrow \nu_\tau$ oscillations with $\sin^2 2\theta > 0.82$ and $5 \times 10^{-4} < \Delta m^2 < 6 \times 10^{-3}$ eV^2 at a 90% confidence level. We conclude that the present data give evidence for neutrino oscillations.

We gratefully acknowledge the cooperation of the Kamioka Mining and Smelting Company. The Super-Kamiokande experiment was built and has been operated with funding from the Japanese Ministry of Education, Science, Sports and Culture, and the United States Department of Energy.

*Present address: NASA, JPL, Pasadena, CA 91109.
†Present address: High Energy Accelerator Research Organization (KEK), Accelerator Laboratory, Tsukuba, Ibaraki 305-0801, Japan.
‡Present address: University of Chicago, Enrico Fermi Institute, Chicago, IL 60637.
§Present address: Institute of Particle and Nuclear Studies, High Energy Accelerator Research Organization (KEK), Tsukuba, Ibaraki 305-0801, Japan.
‖Present address: Stanford University, Department of Physics, Stanford, CA 94305.

[1] G. Barr et al., Phys. Rev. D **39**, 3532 (1988); V. Agrawal et al., Phys. Ref. D **53**, 1314 (1996); T. K. Gaisser and T. Stanev, in *Proceedings of the International Cosmic Ray Conference*, Rome, Italy, 1995 (Arti Grafiche, Urbino, 1995), Vol. 1, p. 694.
[2] M. Honda et al., Phys. Lett. B **248**, 193 (1990); M. Honda et al., Phys. Rev. D **52**, 4985 (1995).
[3] K. S. Hirata et al., Phys. Lett. B **205**, 416 (1988); K. S. Hirata et al., Phys. Lett. B **280**, 146 (1992).
[4] Y. Fukuda et al., Phys. Lett. B **335**, 237 (1994).
[5] D. Casper et al., Phys. Rev. Lett. **66**, 2561 (1991); R. Becker-Szendy et al., Phys. Rev. D **46**, 3720 (1992).
[6] Super-Kamiokande Collaboration, Y. Fukuda et al., hep-ex/9803006.
[7] Super-Kamiokande Collaboration, Y. Fukuda et al., hep-ex/9805006.
[8] K. Daum et al., Z. Phys. C **66**, 417 (1995).
[9] M. Aglietta et al., Europhys. Lett. **8**, 611 (1989).
[10] W. W. M. Allison et al., Phys. Lett. B **391**, 491 (1997); T. Kafka, inProceedings of the 5th International Workshop on Topics in Astroparticle and Underground Physics, Gran Sasso, Italy, 1997 (to be published).
[11] O. G. Ryazhskaya, JETP Lett. **60**, 617 (1994); JETP Lett. **61**, 237 (1995).
[12] Y. Fukuda et al., Phys. Lett. B **388**, 397 (1996).
[13] T. K. Gaisser et al., Phys. Rev. D **54**, 5578 (1996).
[14] J. Engel et al., Phys. Rev. D **48**, 3048 (1993).
[15] S. Kasuga et al., Phys. Lett. B **374**, 238 (1996).
[16] W. A. Mann, T. Kafka, and W. Leeson, Phys. Lett. B **291**, 200 (1992).
[17] This represents an improvement from Refs. [6,7] due to improved calibration.
[18] T. K. Gaisser and T. Stanev, Phys. Rev. D **57**, 1977 (1998).
[19] L. Wolfenstein, Phys. Rev. D **17**, 2369 (1978).
[20] S. P. Mikheyev and A. Y. Smirnov, Sov. J. Nucl. Phys. **42**, 1441 (1985); S. P. Mikheyev and A. Y. Smirnov, Nuovo Cimento Soc. Ital. Fis. **9C**, 17 (1986); S. P. Mikheyev and A. Y. Smirnov, Sov. Phys.-Usp. **30**, 759 (1987).
[21] Based on a two-dimensional extension of the method from the Particle Data Group, R. M. Barnett et al., Phys. Rev. D **54**, 1 (1996); see sec. 28.6, p. 162.
[22] M. Apollonio et al., Phys. Lett. B **420**, 397 (1998).
[23] C. Athanassopoulos et al., Phys. Rev. C **54**, 2685 (1996); Phys. Rev. Lett. **77**, 3082 (1996).
[24] T. Kajita, in Proceedings of the XVIIIth International Conference on Neutrino Physics and Astrophysics, Takayama, Japan, 1998 (to be published).

Measurement of the Rate of $\nu_e + d \to p + p + e^-$ Interactions Produced by ^{8}B Solar Neutrinos at the Sudbury Neutrino Observatory

Q. R. Ahmad,[15] R. C. Allen,[11] T. C. Andersen,[12] J. D. Anglin,[7] G. Bühler,[11] J. C. Barton,[13,*] E. W. Beier,[14] M. Bercovitch,[7] J. Bigu,[4] S. Biller,[13] R. A. Black,[13] I. Blevis,[3] R. J. Boardman,[13] J. Boger,[2] E. Bonvin,[9] M. G. Boulay,[9] M. G. Bowler,[13] T. J. Bowles,[6] S. J. Brice,[6,13] M. C. Browne,[15] T. V. Bullard,[15] T. H. Burritt,[15,6] K. Cameron,[12] J. Cameron,[13] Y. D. Chan,[5] M. Chen,[9] H. H. Chen,[11,†] X. Chen,[5,13] M. C. Chon,[12] B. T. Cleveland,[13] E. T. H. Clifford,[9,1] J. H. M. Cowan,[4] D. F. Cowen,[14] G. A. Cox,[15] Y. Dai,[9] X. Dai,[13] F. Dalnoki-Veress,[3] W. F. Davidson,[7] P. J. Doe,[15,11,6] G. Doucas,[13] M. R. Dragowsky,[6,5] C. A. Duba,[15] F. A. Duncan,[9] J. Dunmore,[13] E. D. Earle,[9,1] S. R. Elliott,[15,6] H. C. Evans,[9] G. T. Ewan,[9] J. Farine,[3] H. Fergani,[13] A. P. Ferraris,[13] R. J. Ford,[9] M. M. Fowler,[6] K. Frame,[13] E. D. Frank,[14] W. Frati,[14] J. V. Germani,[15,6] S. Gil,[10] A. Goldschmidt,[6] D. R. Grant,[3] R. L. Hahn,[2] A. L. Hallin,[9] E. D. Hallman,[4] A. Hamer,[6,9] A. A. Hamian,[15] R. U. Haq,[4] C. K. Hargrove,[3] P. J. Harvey,[9] R. Hazama,[15] R. Heaton,[9] K. M. Heeger,[15] W. J. Heintzelman,[14] J. Heise,[10] R. L. Helmer,[10,‡] J. D. Hepburn,[9,1] H. Heron,[13] J. Hewett,[4] A. Hime,[6] M. Howe,[15] J. G. Hykawy,[4] M. C. P. Isaac,[5] P. Jagam,[12] N. A. Jelley,[13] C. Jillings,[9] G. Jonkmans,[4,1] J. Karn,[12] P. T. Keener,[14] K. Kirch,[6] J. R. Klein,[14] A. B. Knox,[13] R. J. Komar,[10,9] R. Kouzes,[8] T. Kutter,[10] C. C. M. Kyba,[14] J. Law,[12] I. T. Lawson,[12] M. Lay,[13] H. W. Lee,[9] K. T. Lesko,[5] J. R. Leslie,[9] I. Levine,[3] W. Locke,[13] M. M. Lowry,[8] S. Luoma,[4] J. Lyon,[13] S. Majerus,[13] H. B. Mak,[9] A. D. Marino,[5] N. McCauley,[13] A. B. McDonald,[9,8] D. S. McDonald,[14] K. McFarlane,[3] G. McGregor,[13] W. McLatchie,[9] R. Meijer Drees,[15] H. Mes,[3] C. Mifflin,[3] G. G. Miller,[6] G. Milton,[1] B. A. Moffat,[9] M. Moorhead,[13,5] C. W. Nally,[10] M. S. Neubauer,[14] F. M. Newcomer,[14] H. S. Ng,[10] A. J. Noble,[3,‡] E. B. Norman,[5] V. M. Novikov,[3] M. O'Neill,[3] C. E. Okada,[5] R. W. Ollerhead,[12] M. Omori,[13] J. L. Orrell,[15] S. M. Oser,[14] A. W. P. Poon,[5,6,10,15] T. J. Radcliffe,[9] A. Roberge,[4] B. C. Robertson,[9] R. G. H. Robertson,[15,6] J. K. Rowley,[2] V. L. Rusu,[14] E. Saettler,[4] K. K. Schaffer,[15] A. Schuelke,[5] M. H. Schwendener,[4] H. Seifert,[4,6,15] M. Shatkay,[3] J. J. Simpson,[12] D. Sinclair,[3] P. Skensved,[9] A. R. Smith,[5] M. W. E. Smith,[15] N. Starinsky,[3] T. D. Steiger,[15] R. G. Stokstad,[5] R. S. Storey,[7,†] B. Sur,[1,9] R. Tafirout,[4] N. Tagg,[12] N. W. Tanner,[13] R. K. Taplin,[13] M. Thorman,[13] P. Thornewell,[6,13,15] P. T. Trent,[13,*] Y. I. Tserkovnyak,[10] R. Van Berg,[14] R. G. Van de Water,[14,6] C. J. Virtue,[4] C. E. Waltham,[10] J.-X. Wang,[12] D. L. Wark,[13,6,§] N. West,[13] J. B. Wilhelmy,[6] J. F. Wilkerson,[15,6] J. Wilson,[13] P. Wittich,[14] J. M. Wouters,[6] and M. Yeh[2]

(SNO Collaboration)

[1]*Atomic Energy of Canada Limited, Chalk River Laboratories, Chalk River, Ontario K0J 1J0 Canada*
[2]*Chemistry Department, Brookhaven National Laboratory, Upton, New York 11973-5000*
[3]*Carleton University, Ottawa, Ontario K1S 5B6 Canada*
[4]*Department of Physics and Astronomy, Laurentian University, Sudbury, Ontario P3E 2C6 Canada*
[5]*Institute for Nuclear and Particle Astrophysics and Nuclear Science Division, Lawrence Berkeley National Laboratory, Berkeley, California 94720*
[6]*Los Alamos National Laboratory, Los Alamos, New Mexico 87545*
[7]*National Research Council of Canada, Ottawa, Ontario K1A 0R6 Canada*
[8]*Department of Physics, Princeton University, Princeton, New Jersey 08544*
[9]*Department of Physics, Queen's University, Kingston, Ontario K7L 3N6 Canada*
[10]*Department of Physics and Astronomy, University of British Columbia, Vancouver, BC V6T 1Z1 Canada*
[11]*Department of Physics, University of California, Irvine, California 92717*
[12]*Physics Department, University of Guelph, Guelph, Ontario N1G 2W1 Canada*
[13]*Nuclear and Astrophysics Laboratory, University of Oxford, Keble Road, Oxford, OX1 3RH, United Kingdom*
[14]*Department of Physics and Astronomy, University of Pennsylvania, Philadelphia, Pennsylvania 19104-6396*
[15]*Center for Experimental Nuclear Physics and Astrophysics, and Department of Physics, University of Washington, Seattle, Washington 98195*
(Received 18 June 2001; published 25 July 2001)

Solar neutrinos from ^{8}B decay have been detected at the Sudbury Neutrino Observatory via the charged current (CC) reaction on deuterium and the elastic scattering (ES) of electrons. The flux of ν_e's is measured by the CC reaction rate to be $\phi^{\rm CC}(\nu_e) = 1.75 \pm 0.07({\rm stat})^{+0.12}_{-0.11}({\rm syst}) \pm 0.05({\rm theor}) \times 10^6$ cm^{-2} s^{-1}. Comparison of $\phi^{\rm CC}(\nu_e)$ to the Super-Kamiokande Collaboration's precision value of the flux inferred from the ES reaction yields a 3.3σ difference, assuming the systematic uncertainties are normally distributed, providing evidence of an active non-ν_e component in the solar flux. The total flux of active ^{8}B neutrinos is determined to be $5.44 \pm 0.99 \times 10^6$ cm^{-2} s^{-1}.

DOI: 10.1103/PhysRevLett.87.071301   PACS numbers: 26.65.+t, 14.60.Pq, 95.85.Ry

Solar neutrino experiments over the past 30 years [1–6] have measured fewer neutrinos than are predicted by models of the Sun [7,8]. One explanation for the deficit is the transformation of the Sun's electron-type neutrinos into other active flavors. The Sudbury Neutrino Observatory (SNO) measures the ^{8}B solar neutrinos through the reactions

$$\nu_e + d \to p + p + e^- \quad (CC),$$
$$\nu_x + d \to p + n + \nu_x \quad (NC),$$
$$\nu_x + e^- \to \nu_x + e^- \quad (ES).$$

The charged current (CC) reaction is sensitive exclusively to electron-type neutrinos, while the neutral current (NC) is sensitive to all active neutrino flavors ($x = e, \mu, \tau$). The elastic scattering (ES) reaction is sensitive to all flavors as well, but with reduced sensitivity to ν_μ and ν_τ. By itself, the ES reaction cannot provide a measure of the total ^{8}B flux or its flavor content. Comparison of the ^{8}B flux deduced from the ES reaction, assuming no neutrino oscillations [$\phi^{ES}(\nu_x)$], to that measured by the CC reaction [$\phi^{CC}(\nu_e)$] can provide clear evidence of flavor transformation without reference to solar model flux calculations. If neutrinos from the Sun change into other active flavors, then $\phi^{CC}(\nu_e) < \phi^{ES}(\nu_x)$.

This Letter presents the first results from SNO on the ES and CC reactions. SNO's measurement of $\phi^{ES}(\nu_x)$ is consistent with previous measurements described in Ref. [5]. The measurement of $\phi^{CC}(\nu_e)$, however, is significantly smaller and is therefore inconsistent with the null hypothesis that all observed solar neutrinos are ν_e. A measurement using the NC reaction, which has equal sensitivity to all neutrino flavors, will be reported in a future publication.

SNO [9] is an imaging water Čerenkov detector located at a depth of 6010 m of water equivalent in the INCO, Ltd. Creighton mine near Sudbury, Ontario. It features 1000 metric tons of ultrapure D$_2$O contained in a 12-m diameter spherical acrylic vessel. This sphere is surrounded by a shield of ultrapure H$_2$O contained in a 34-m-high barrel-shaped cavity of maximum diameter 22 m. A stainless steel structure 17.8 m in diameter supports 9456 20-cm photomultiplier tubes (PMTs) with light concentrators. Approximately 55% of the light produced within 7 m of the center of the detector will strike a PMT if it is not absorbed by intervening media.

The data reported here were recorded between November 2, 1999 and January 15, 2001 and correspond to a live time of 240.95 days. Events are defined by a multiplicity trigger of 18 or more PMTs exceeding a threshold of $\sim$0.25 photoelectrons within a time window of 93 ns. The trigger reaches 100% efficiency at 23 PMTs. The total instantaneous trigger rate is 15–18 Hz, of which 6–8 Hz is the data trigger. For every event trigger, the time and charge responses of each participating PMT are recorded.

The data were partitioned into two sets, with approximately 70% used to establish the data analysis procedures and 30% reserved for a blind test of statistical bias in the analysis. The analysis procedures were frozen before the blind data set was analyzed, and no statistically significant differences in the data sets were found. We present here the analysis of the combined data sets.

Calibration of the PMT time and charge pedestals, slopes, offsets, charge vs time dependencies, and second order rate dependencies are performed using electronic pulsers and pulsed light sources. Optical calibration is obtained by using a diffuse source of pulsed laser light at 337, 365, 386, 420, 500, and 620 nm. The absolute energy scale and uncertainties are established with a triggered ^{16}N source (predominantly 6.13-MeV γ's) deployed over two planar grids within the D$_2$O and a linear grid in the H$_2$O. The resulting Monte Carlo predictions of detector response are tested using a ^{252}Cf neutron source, which provides an extended distribution of 6.25-MeV γ rays from neutron capture, and a ^{3}H(p, γ)^{4}He [10] source providing 19.8-MeV γ rays. The volume-weighted mean response is approximately nine PMT hits per MeV of electron energy.

Table I details the steps in data reduction. The first of these is the elimination of instrumental backgrounds. Electrical pickup may produce false PMT hits, while electrical discharges in the PMTs or insulating detector materials produce light. These backgrounds have characteristics very different from Čerenkov light, and are eliminated by using cuts based only on the PMT positions, the PMT time and charge data, event-to-event time correlations, and veto PMTs. This step in the data reduction is verified by comparing results from two independent background rejection analyses.

For events passing the first stage, the calibrated times and positions of the hit PMTs are used to reconstruct the vertex position and the direction of the particle. The reconstruction accuracy and resolution are measured using Compton electrons from the ^{16}N source, and the energy and source variation of reconstruction are checked with a ^{8}Li β source. Angular resolution is measured using Compton electrons produced more than 150 cm from the ^{16}N source. At these energies, the vertex resolution is 16 cm and the angular resolution is 26.7°.

An effective kinetic energy, T_{eff}, is assigned to each event passing the reconstruction stage. T_{eff} is calculated

TABLE I. Data reduction steps.

Analysis step	Number of events
Total event triggers	355 320 964
Neutrino data triggers	143 756 178
$N_{\text{hit}} \geq 30$	6 372 899
Instrumental background cuts	1 842 491
Muon followers	1 809 979
High level cuts[a]	923 717
Fiducial volume cut	17 884
Threshold cut	1169
Total events	1169

[a]Reconstruction figures of merit, prompt light, and $\langle \theta_{ij} \rangle$.

using prompt (unscattered) Čerenkov photons and the position and direction of the event. The derived energy response of the detector can be characterized by a Gaussian:

$$R(E_{\text{eff}}, E_e) = \frac{1}{\sqrt{2\pi}\,\sigma_E(E_e)} \exp\left[-\frac{1}{2}\left(\frac{E_{\text{eff}} - E_e}{\sigma_E(E_e)}\right)^2\right],$$

where E_e is the total electron energy, $E_{\text{eff}} = T_{\text{eff}} + m_e$, and $\sigma_E(E_e) = (-0.4620 + 0.5470\sqrt{E_e} + 0.008722 E_e)$ MeV is the energy resolution. The uncertainty on the energy scale is found to be $\pm 1.4\%$, which results in a flux uncertainty nearly 4 times larger. For validation, a second energy estimator counts all PMTs hit in each event, N_{hit}, without position and direction corrections.

Further instrumental background rejection is obtained by using reconstruction figures of merit, PMT time residuals, and the average angle between hit PMTs ($\langle\theta_{ij}\rangle$), measured from the reconstructed vertex. These cuts test the hypothesis that each event has the characteristics of single electron Čerenkov light. The effects of these and the rest of the instrumental background removal cuts on neutrino signals are quantified using the ^{8}Li and ^{16}N sources deployed throughout the detector. The volume-weighted neutrino signal loss is measured to be $1.4^{+0.7}_{-0.6}\%$ and the residual instrumental contamination for the data set within the D$_2$O is $<0.2\%$. Lastly, cosmic ray induced neutrons and spallation products are removed using a 20 s coincidence window with the parent muon.

Figure 1 shows the radial distribution of all remaining events above a threshold of $T_{\text{eff}} \geq 6.75$ MeV. The distribution is expressed as a function of the volume-weighted radial variable $(R/R_{\text{AV}})^3$, where $R_{\text{AV}} = 6.00$ m is the radius of the acrylic vessel. Above this energy threshold, there are contributions from CC events in the D$_2$O, ES events in the D$_2$O and H$_2$O, a residual tail of neutron capture events, and high energy γ rays from radioactivity in the outer detector. The data show a clear signal within the D$_2$O volume. For $(R/R_{\text{AV}})^3 > 1.0$ the distribution rises into the H$_2$O region until it is cut off by the acceptance of the PMT light collectors at $R \sim 7.0$ m. A fiducial volume cut is applied at $R = 5.50$ m to reduce backgrounds from regions exterior to the D$_2$O, and to minimize systematic uncertainties associated with optics and reconstruction near the acrylic vessel.

Possible backgrounds from radioactivity in the D$_2$O and H$_2$O are measured by regular low level radio assays of U and Th decay chain products in these regions. The Čerenkov light character of D$_2$O and H$_2$O radioactivity backgrounds is used in situ to monitor backgrounds between radio assays. Low energy radioactivity backgrounds are removed by the high threshold imposed, as are most neutron capture events. Monte Carlo calculations predict that the H$_2$O shield effectively reduces contributions of low energy ($<$4 MeV) γ rays from the PMT array, and these predictions are verified by deploying an encapsulated Th source in the vicinity of the PMT support sphere. High energy γ rays from the cavity are also attenuated by the H$_2$O shield. A limit on their leakage into the fiducial volume is estimated by deploying the ^{16}N source near the edge of the detector's active volume. The total contribution from all radioactivity in the detector is found to be $<0.2\%$ for low energy backgrounds and $<0.8\%$ for high energy backgrounds.

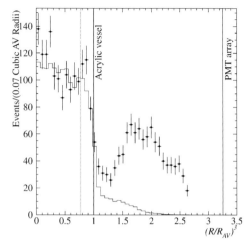

FIG. 1. Distribution of event candidates with $T_{\text{eff}} \geq 6.75$ MeV as a function of the volume-weighted radial variable $(R/R_{\text{AV}})^3$. The Monte Carlo simulation of the signals, weighted by the results from the signal extraction, is shown as a histogram. The dotted line indicates the fiducial volume cut used in this analysis.

The final data set contains 1169 events after the fiducial volume and kinetic energy threshold cuts. Figure 2(a) displays the distribution of $\cos\theta_\odot$, the angle between the reconstructed direction of the event and the instantaneous direction from the Sun to the Earth. The forward peak in this distribution arises from the kinematics of the ES reaction, while CC electrons are expected to have a distribution which is $(1 - 0.340\cos\theta_\odot)$ [11], before accounting for detector response.

The data are resolved into contributions from CC, ES, and neutron events above threshold using probability density functions (pdf's) in T_{eff}, $\cos\theta_\odot$, and $(R/R_{\text{AV}})^3$, generated from Monte Carlo simulations assuming no flavor transformation and the shape of the standard ^{8}B spectrum [12] (hep neutrinos are not included in the fit). The extended maximum likelihood method used in the signal extraction yields 975.4 ± 39.7 CC events, 106.1 ± 15.2 ES events, and 87.5 ± 24.7 neutron events for the fiducial volume and the threshold chosen, where the uncertainties given are statistical only. The dominant sources of systematic uncertainty in this signal extraction are the energy scale uncertainty and reconstruction accuracy, as shown in Table II. The CC and ES signal decomposition gives consistent results when used with the N_{hit} energy estimator, as well as with different choices of the analysis threshold

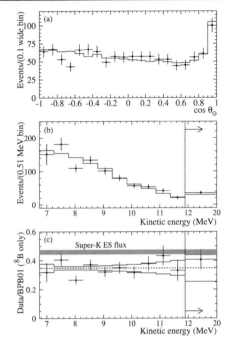

FIG. 2. Distributions of (a) $\cos\theta_\odot$ and (b) extracted kinetic energy spectrum for CC events with $R \leq 5.50$ m and $T_{\text{eff}} \geq 6.75$ MeV. The Monte Carlo simulations for an undistorted ^{8}B spectrum are shown as histograms. The ratio of the data to the expected kinetic energy distribution with correlated systematic errors is shown in (c). The uncertainties in the ^{8}B spectrum [12] have not been included.

and the fiducial volume up to 6.20 m with backgrounds characterized by pdf's.

The CC spectrum can be extracted from the data by removing the constraint on the shape of the CC pdf and repeating the signal extraction.

Figure 2(b) shows the kinetic energy spectrum with statistical error bars, with the ^{8}B spectrum of Ortiz et al. [12] scaled to the data. The ratio of the data to the prediction [7] is shown in Fig. 2(c). The bands represent the 1σ uncertainties derived from the most significant energy-dependent systematic errors. There is no evidence for a deviation of the spectral shape from the predicted shape under the nonoscillation hypothesis.

Normalized to the integrated rates above the kinetic energy threshold of $T_{\text{eff}} = 6.75$ MeV, the measured ^{8}B neutrino fluxes assuming the standard spectrum shape [12] are

$\phi^{\text{CC}}_{\text{SNO}}(\nu_e) = 1.75 \pm 0.07(\text{stat})^{+0.12}_{-0.11}(\text{syst}) \pm 0.05(\text{theor})$
$\times 10^6 \text{ cm}^{-2}\text{s}^{-1}$

$\phi^{\text{ES}}_{\text{SNO}}(\nu_x) = 2.39 \pm 0.34(\text{stat})^{+0.16}_{-0.14}(\text{syst})$
$\times 10^6 \text{ cm}^{-2}\text{s}^{-1}$,

TABLE II. Systematic error on fluxes.

Error source	CC error (percent)	ES error (percent)
Energy scale	−5.2, +6.1	−3.5, +5.4
Energy resolution	±0.5	±0.3
Energy scale nonlinearity	±0.5	±0.4
Vertex accuracy	±3.1	±3.3
Vertex resolution	±0.7	±0.4
Angular resolution	±0.5	±2.2
High energy γ's	−0.8, +0.0	−1.9, +0.0
Low energy background	−0.2, +0.0	−0.2, +0.0
Instrumental background	−0.2, +0.0	−0.6, +0.0
Trigger efficiency	0.0	0.0
Live time	±0.1	±0.1
Cut acceptance	−0.6, +0.7	−0.6, +0.7
Earth orbit eccentricity	±0.1	±0.1
^{17}O, ^{18}O	0.0	0.0
Experimental uncertainty	−6.2, +7.0	−5.7, +6.8
Cross section	3.0	0.5
Solar Model	−16, +20	−16, +20

where the theoretical uncertainty is the CC cross section uncertainty [13]. Radiative corrections have not been applied to the CC cross section, but they are expected to decrease the measured $\phi^{\text{CC}}(\nu_e)$ flux [14] by up to a few percent. The difference between the ^{8}B flux deduced from the ES rate and that deduced from the CC rate in SNO is $0.64 \pm 0.40 \times 10^6 \text{ cm}^{-2}\text{s}^{-1}$, or 1.6σ. The SNO's ES rate measurement is consistent with the precision measurement by Super-Kamiokande Collaboration of the ^{8}B flux using the same ES reaction [5]:

$\phi^{\text{ES}}_{\text{SK}}(\nu_x) = 2.32 \pm 0.03(\text{stat})^{+0.08}_{-0.07}(\text{syst}) \times 10^6 \text{ cm}^{-2}\text{s}^{-1}$.

The difference between the flux $\phi^{\text{ES}}(\nu_x)$ measured by Super-Kamiokande via the ES reaction and the $\phi^{\text{CC}}(\nu_e)$ flux measured by SNO via the CC reaction is $0.57 \pm 0.17 \times 10^6 \text{ cm}^{-2}\text{s}^{-1}$, or 3.3σ [15], assuming that the systematic errors are normally distributed. The probability that a downward fluctuation of the Super-Kamiokande result would produce a SNO result $\geq 3.3\sigma$ is 0.04%. For reference, the ratio of the SNO CC ^{8}B flux to that of the BPB01 solar model [7] is 0.347 ± 0.029, where all uncertainties are added in quadrature.

If oscillation solely to a sterile neutrino is occurring, the SNO CC-derived ^{8}B flux above a threshold of 6.75 MeV will be consistent with the integrated Super-Kamiokande ES-derived ^{8}B flux above a threshold of 8.5 MeV [16]. By adjusting the ES threshold [5], this derived flux difference is $0.53 \pm 0.17 \times 10^6 \text{ cm}^{-2}\text{s}^{-1}$, or 3.1σ. The probability of a downward fluctuation $\geq 3.1\sigma$ is 0.13%. These data are therefore evidence of a nonelectron active flavor component in the solar neutrino flux. These data are also inconsistent with the "Just-So^2" parameters for neutrino oscillation [17].

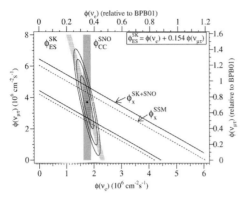

FIG. 3. Flux of ^{8}B solar neutrinos which are μ or τ flavor vs the flux of electron neutrinos as deduced from the SNO and Super-Kamiokande data. The diagonal bands show the total ^{8}B flux $\phi(\nu_x)$ as predicted by BPB01 (dashed lines) and that derived from the SNO and Super-Kamiokande measurements (solid lines). The intercepts of these bands with the axes represent the $\pm 1\sigma$ errors.

Figure 3 displays the inferred flux of nonelectron flavor active neutrinos [$\phi(\nu_{\mu\tau})$] against the flux of electron neutrinos. The two data bands represent the one standard deviation measurements of the SNO CC rate and the Super-Kamiokande ES rate. The error ellipses represent the 68%, 95%, and 99% joint probability contours for $\phi(\nu_e)$ and $\phi(\nu_{\mu\tau})$. The best fit to $\phi(\nu_{\mu\tau})$ is

$$\phi(\nu_{\mu\tau}) = 3.69 \pm 1.13 \times 10^6 \text{ cm}^{-2}\text{s}^{-1}.$$

The total flux of active ^{8}B neutrinos is determined to be

$$\phi(\nu_x) = 5.44 \pm 0.99 \times 10^6 \text{ cm}^{-2}\text{s}^{-1}.$$

This result is displayed as a diagonal band in Fig. 3, and is in excellent agreement with predictions of standard solar models [7,8].

Assuming that the oscillation of massive neutrinos explains both the evidence for the electron neutrino flavor change presented here and the atmospheric neutrino data of the Super-Kamiokande collaboration [18], two separate splittings of the squares of the neutrino mass eigenvalues are indicated: $<10^{-3}$ eV^2 for the solar sector [19,17] and $\simeq 3.5 \times 10^{-3}$ eV^2 for atmospheric neutrinos. These results, together with the beta spectrum of tritium [20], limit the sum of mass eigenvalues of active neutrinos to be between 0.05 and 8.4 eV, corresponding to a constraint of $0.001 < \Omega_\nu < 0.18$ for the contribution to the critical density of the Universe [21,22].

In summary, the results presented here are the first direct indication of a nonelectron flavor component in the solar neutrino flux, and enable the first determination of the total flux of ^{8}B neutrinos generated by the Sun.

This research was supported by the Natural Sciences and Engineering Research Council of Canada, Industry Canada, National Research Council of Canada, Northern Ontario Heritage Fund Corporation, the Province of Ontario, the United States Department of Energy, and in the United Kingdom by the Science and Engineering Research Council and the Particle Physics and Astronomy Research Council. Further support was provided by INCO, Ltd., Atomic Energy of Canada Limited (AECL), Agra-Monenco, Canatom, Canadian Microelectronics Corporation, AT&T Microelectronics, Northern Telecom, and British Nuclear Fuels, Ltd. The heavy water was loaned by AECL with the cooperation of Ontario Power Generation.

*Permanent address: Birkbeck College, University of London, Malet Road, London WC1E 7HX, UK.
†Deceased.
‡Permanent address: TRIUMF, 4004 Wesbrook Mall, Vancouver, BC V6T 2A3, Canada.
§Permanent address: Rutherford Appleton Laboratory, Chilton, Didcot, Oxon, OX11 0QX, and University of Sussex, Physics and Astronomy Department, Brighton BN1 9QH, United Kingdom.

[1] B. T. Cleveland et al., Astrophys. J. **496**, 505 (1998).
[2] K. S. Hirata et al., Phys. Rev. Lett. **65**, 1297 (1990); K. S. Hirata et al., Phys. Rev. D **44**, 2241 (1991); **45**, 2170(E) (1992); Y. Fukuda et al., Phys. Rev. Lett. **77**, 1683 (1996).
[3] J. N. Abdurashitov et al., Phys. Rev. C **60**, 055801 (1999).
[4] W. Hampel et al., Phys. Lett. B **447**, 127 (1999).
[5] S. Fukuda et al., Phys. Rev. Lett. **86**, 5651 (2001).
[6] M. Altmann et al., Phys. Lett. B **490**, 16 (2000).
[7] J. N. Bahcall, M. H. Pinsonneault, and S. Basu, Astrophys. J. **555**, 990 (2001). The reference ^{8}B neutrino flux is 5.05×10^6 cm^{-2}s^{-1}.
[8] A. S. Brun, S. Turck-Chièze, and J. P. Zahn, Astrophys. J. **525**, 1032 (1999); S. Turck-Chièze et al., Astrophys. J. Lett. **555**, L69 (2001).
[9] The SNO Collaboration, J. Boger et al., Nucl. Instrum. Methods Phys. Res., Sect. A **449**, 172 (2000).
[10] A. W. P. Poon et al., Nucl. Instrum. Methods Phys. Res., Sect. A **452**, 115 (2000).
[11] P. Vogel and J. F. Beacom, Phys. Rev. D **60**, 053003 (1999).
[12] C. E. Ortiz et al., Phys. Rev. Lett. **85**, 2909 (2000).
[13] S. Nakamura, T. Sato, V. Gudkov, and K. Kubodera, Phys. Rev. C **63**, 034617 (2001); M. Butler, J.-W. Chen, and X. Kong, Phys. Rev. C **63**, 035501 (2001); G. 't Hooft, Phys. Lett. **37B**, 195 (1971). The Butler et al. cross section with $L_{1,A} = 5.6$ fm^3 is used.
[14] I. S. Towner, J. Beacom, and S. Parke (private communication); I. S. Towner, Phys. Rev. C **58**, 1288 (1998); J. Beacom and S. Parke, hep-ph/0106128; J. N. Bahcall, M. Kamionkowski, and A. Sirlin, Phys. Rev. D **51**, 6146 (1995).
[15] Given the limit set for the hep flux by Ref. [5], the effects of the hep contribution may increase this difference by a few percent.

[16] G.L. Fogli, E. Lisi, A. Palazzo, and F.L. Villante, Phys. Rev. D **63**, 113016 (2001); F.L. Villante, G. Fiorentini, and E. Lisi, Phys. Rev. D **59**, 013006 (1999).
[17] J.N. Bahcall, P.I. Krastev, and A. Yu. Smirnov, J. High Energy Phys. **05**, 015 (2001).
[18] T. Toshito et al., hep-ex/0105023.
[19] M. Apollonio et al., Phys. Lett. B **466**, 415 (1999).
[20] J. Bonn et al., Nucl. Phys. (Proc. B Suppl.) **91**, 273 (2001).
[21] *Allen's Astrophysical Quantities,* edited by Arthur Cox (Springer-Verlag, New York, 2000), 4th ed.; D.E. Groom et al., Eur. Phys. J. C **15**, 1 (2000).
[22] H. Pas and T.J. Weiler, Phys. Rev. D **63**, 113015 (2001).

Direct Evidence for Neutrino Flavor Transformation from Neutral-Current Interactions in the Sudbury Neutrino Observatory

Q. R. Ahmad,[17] R. C. Allen,[4] T. C. Andersen,[6] J. D.Anglin,[10] J. C. Barton,[11,*] E. W. Beier,[12] M. Bercovitch,[10] J. Bigu,[7] S. D. Biller,[11] R. A. Black,[11] I. Blevis,[5] R. J. Boardman,[11] J. Boger,[3] E. Bonvin,[14] M. G. Boulay,[9,14] M. G. Bowler,[11] T. J. Bowles,[9] S. J. Brice,[9,11] M. C. Browne,[17,9] T. V. Bullard,[17] G. Bühler,[4] J. Cameron,[11] Y. D. Chan,[8] H. H. Chen,[4,†] M. Chen,[14] X. Chen,[8,11] B. T. Cleveland,[11] E. T. H. Clifford,[14] J. H. M. Cowan,[7] D. F. Cowen,[12] G. A. Cox,[17] X. Dai,[11] F. Dalnoki-Veress,[5] W. F. Davidson,[10] P. J. Doe,[17,9,4] G. Doucas,[11] M. R. Dragowsky,[9,8] C. A. Duba,[17] F. A. Duncan,[14] M. Dunford,[12] J. A. Dunmore,[11] E. D. Earle,[14,1] S. R. Elliott,[17,9] H. C. Evans,[14] G. T. Ewan,[14] J. Farine,[7,5] H. Fergani,[11] A. P. Ferraris,[11] R. J. Ford,[14] J. A. Formaggio,[17] M. M. Fowler,[9] K. Frame,[11] E. D. Frank,[12] W. Frati,[12] N. Gagnon,[11,9,8,17] J. V. Germani,[17] S. Gil,[2] K. Graham,[14] D. R. Grant,[5] R. L. Hahn,[3] A. L. Hallin,[14] E. D. Hallman,[7] A. S. Hamer,[9,14] A. A. Hamian,[17] W. B. Handler,[14] R. U. Haq,[7] C. K. Hargrove,[5] P. J. Harvey,[14] R. Hazama,[17] K. M. Heeger,[17] W. J. Heintzelman,[12] J. Heise,[2,9] R. L. Helmer,[16,2] J. D. Hepburn,[14] R. Heron,[11] J. Hewett,[7] A. Hime,[9] M. Howe,[17] J. G. Hykawy,[7] M. C. P. Isaac,[8] P. Jagam,[6] N. A. Jelley,[11] C. Jillings,[14] G. Jonkmans,[7,1] K. Kazkaz,[17] P. T. Keener,[12] J. R. Klein,[12] A. B. Knox,[11] R. J. Komar,[2] R. Kouzes,[13] T. Kutter,[2] C. C. M. Kyba,[12] J. Law,[6] I. T. Lawson,[6] M. Lay,[11] H. W. Lee,[14] K. T. Lesko,[8] J. R. Leslie,[14] I. Levine,[5] W. Locke,[11] S. Luoma,[7] J. Lyon,[11] S. Majerus,[11] H. B. Mak,[14] J. Maneira,[14] J. Manor,[17] A. D. Marino,[8] N. McCauley,[12,11] A. B. McDonald,[14,13] D. S. McDonald,[12] K. McFarlane,[5] G. McGregor,[11] R. Meijer Drees,[17] C. Mifflin,[5] G. G. Miller,[9] G. Milton,[1] B. A. Moffat,[14] M. Moorhead,[11] C. W. Nally,[2] M. S. Neubauer,[12] F. M. Newcomer,[12] H. S. Ng,[2] A. J. Noble,[16,5] E. B. Norman,[8] V. M. Novikov,[5] M. O'Neill,[5] C. E. Okada,[8] R. W. Ollerhead,[6] M. Omori,[11] J. L. Orrell,[17] S. M. Oser,[12] A. W. P. Poon,[8,17,2,9] T. J. Radcliffe,[14] A. Roberge,[7] B. C. Robertson,[14] R. G. H. Robertson,[17,9] S. S. E. Rosendahl,[8] J. K. Rowley,[3] V. L. Rusu,[12] E. Saettler,[7] K. K. Schaffer,[17] M. H. Schwendener,[7] A. Schülke,[8] H. Seifert,[7,17,9] M. Shatkay,[5] J. J. Simpson,[6] C. J. Sims,[11] D. Sinclair,[5,16] P. Skensved,[14] A. R. Smith,[8] M. W. E. Smith,[17] T. Spreitzer,[12] N. Starinsky,[5] T. D. Steiger,[17] R. G. Stokstad,[8] L. C. Stonehill,[17] R. S. Storey,[10] B. Sur,[1,14] R. Tafirout,[7] N. Tagg,[6,11] N. W. Tanner,[11] R. K. Taplin,[11] M. Thorman,[11] P. M. Thornewell,[11] P. T. Trent,[11] Y. I. Tserkovnyak,[2] R. Van Berg,[12] R. G. Van de Water,[9,12] C. J. Virtue,[7] C. E. Waltham,[2] J.-X. Wang,[6] D. L. Wark,[15,11,9] N. West,[11] J. B. Wilhelmy,[9] J. F. Wilkerson,[17,9] J. R. Wilson,[11] P. Wittich,[12] J. M. Wouters,[9] and M. Yeh[3]

(SNO Collaboration)

[1]*Atomic Energy of Canada, Limited, Chalk River Laboratories, Chalk River, Ontario K0J 1J0, Canada*
[2]*Department of Physics and Astronomy, University of British Columbia, Vancouver, British Columbia V6T 1Z1, Canada*
[3]*Chemistry Department, Brookhaven National Laboratory, Upton, New York 11973-5000*
[4]*Department of Physics, University of California, Irvine, California 92717*
[5]*Carleton University, Ottawa, Ontario K1S 5B6, Canada*
[6]*Physics Department, University of Guelph, Guelph, Ontario N1G 2W1, Canada*
[7]*Department of Physics and Astronomy, Laurentian University, Sudbury, Ontario P3E 2C6, Canada*
[8]*Institute for Nuclear and Particle Astrophysics and Nuclear Science Division, Lawrence Berkeley National Laboratory, Berkeley, California 94720*
[9] *Los Alamos National Laboratory, Los Alamos, New Mexico 87545*
[10]*National Research Council of Canada, Ottawa, Ontario K1A 0R6, Canada*
[11]*Department of Physics, University of Oxford, Denys Wilkinson Building, Keble Road, Oxford OX1 3RH, United Kingdom*
[12]*Department of Physics and Astronomy, University of Pennsylvania, Philadelphia, Pennsylvania 19104-6396*
[13]*Department of Physics, Princeton University, Princeton, New Jersey 08544*
[14]*Department of Physics, Queen's University, Kingston, Ontario K7L 3N6, Canada*
[15]*Rutherford Appleton Laboratory, Chilton, Didcot, Oxon OX11 0QX, United Kingdom
and University of Sussex, Physics and Astronomy Department, Brighton BN1 9QH, United Kingdom*
[16]*TRIUMF, 4004 Wesbrook Mall, Vancouver, British Columbia V6T 2A3, Canada*
[17]*Center for Experimental Nuclear Physics and Astrophysics, and Department of Physics, University of Washington, Seattle, Washington 98195*
(Received 19 April 2002; published 13 June 2002)

Observations of neutral-current ν interactions on deuterium in the Sudbury Neutrino Observatory are reported. Using the neutral current (NC), elastic scattering, and charged current reactions and assuming the standard ^{8}B shape, the ν_e component of the ^{8}B solar flux is $\phi_e = 1.76^{+0.05}_{-0.05}(\text{stat})^{+0.09}_{-0.09}(\text{syst}) \times 10^6$ cm^{-2} s^{-1} for a kinetic energy threshold of 5 MeV. The non-ν_e component is $\phi_{\mu\tau} = 3.41^{+0.45}_{-0.45}(\text{stat})^{+0.48}_{-0.45}(\text{syst}) \times 10^6$ cm^{-2} s^{-1}, 5.3σ greater than zero, providing strong evidence for solar ν_e flavor transformation. The total flux measured with the NC reaction is $\phi_{NC} = 5.09^{+0.44}_{-0.43}(\text{stat})^{+0.46}_{-0.43}(\text{syst}) \times 10^6 \text{ cm}^{-2}\text{s}^{-1}$, consistent with solar models.

DOI: 10.1103/PhysRevLett.89.011301 PACS numbers: 26.65.+t, 14.60.Pq, 95.85.Ry

The Sudbury Neutrino Observatory (SNO) detects ^{8}B solar neutrinos through the reactions:

$$\nu_e + d \rightarrow p + p + e^- \quad (\text{CC}),$$
$$\nu_x + d \rightarrow p + n + \nu_x \quad (\text{NC}),$$
$$\nu_x + e^- \rightarrow \nu_x + e^- \quad (\text{ES}).$$

The charged current (CC) reaction is sensitive exclusively to electron-type neutrinos, while the neutral current (NC) reaction is equally sensitive to all active neutrino flavors ($x = e, \mu, \tau$). The elastic scattering (ES) reaction is sensitive to all flavors as well, but with reduced sensitivity to ν_μ and ν_τ. Sensitivity to these three reactions allows SNO to determine the electron and nonelectron active neutrino components of the solar flux [1]. The CC and ES reaction results have recently been presented [2]. This Letter presents the first NC results and updated CC and ES results from SNO.

SNO [3] is a water Cherenkov detector located at a depth of 6010 m of water equivalent in the INCO, Ltd. Creighton mine near Sudbury, Ontario, Canada. The detector uses ultrapure heavy water contained in a transparent acrylic spherical shell 12 m in diameter to detect solar neutrinos. Cherenkov photons generated in the heavy water are detected by 9456 photomultiplier tubes (PMTs) mounted on a stainless steel geodesic sphere 17.8 m in diameter. The geodesic sphere is immersed in ultrapure light water to provide shielding from radioactivity in both the PMT array and the cavity rock.

The data reported here were recorded between 2 November 1999 and 28 May 2001 and represent a total of 306.4 live days, spanning the entire first phase of the experiment, in which only D$_2$O was present in the sensitive volume. The analysis procedure was similar to that described in [2]. PMT times and hit patterns were used to reconstruct event vertices and directions and to assign to each event a most probable kinetic energy, T_{eff}. The total flux of active ^{8}B solar neutrinos with energies greater than 2.2 MeV (the NC reaction threshold) was measured with the NC signal (Cherenkov photons resulting from the 6.25 MeV γ ray from neutron capture on deuterium). The analysis threshold was $T_{\text{eff}} \geq 5$ MeV, providing sensitivity to neutrons from the NC reaction. Above this energy threshold, there were contributions from CC events in the D$_2$O, ES events in the D$_2$O and H$_2$O, capture of neutrons (both from the NC reaction and backgrounds), and low energy Cherenkov background events.

A fiducial volume was defined to accept only events which had reconstructed vertices within 550 cm from the detector center to reduce external backgrounds and systematic uncertainties associated with optics and event reconstruction near the acrylic vessel. The neutron response and systematic uncertainty was calibrated with a ^{252}Cf source. The deduced efficiency for neutron captures on deuterium is $29.9 \pm 1.1\%$ for a uniform source of neutrons in the D$_2$O. The neutron detection efficiency within the fiducial volume and above the energy threshold is 14.4%. The energy calibration was updated from [2] with the ^{16}N calibration source [4] data and Monte Carlo calculations. The energy response for electrons, updated for the lower analysis threshold, was characterized as a Gaussian function with resolution $\sigma_T = -0.0684 + 0.331\sqrt{T_e} + 0.0425T_e$, where T_e is the true electron kinetic energy in MeV. The energy scale uncertainty is 1.2%.

The primary backgrounds to the NC signal are due to low levels of uranium and thorium decay chain daughters (^{214}Bi and ^{208}Tl) in the detector materials. These activities generate free neutrons from deuteron photodisintegration (pd), and low energy Cherenkov events. *Ex situ* assays and *in situ* analysis of the low energy (4–4.5 MeV) Cherenkov signal region provide independent uranium and thorium photodisintegration background measurements.

Two *ex situ* assay techniques were employed to determine average levels of uranium and thorium in water. Radium ions were directly extracted from the water onto either MnO$_x$ or hydrous Ti oxide (HTiO) ion exchange media. Radon daughters in the U and Th chains were subsequently released, identified by α spectroscopy, or the radium was concentrated and the number of decay daughter β-α coincidences determined. Typical assays circulated approximately 400 tonnes of water through the extraction media. These techniques provide isotopic identification of the decay daughters and contamination levels in the assayed water volumes, presented in Fig. 1(a). Secular equilibrium in the U decay chain was broken by the ingress of long-lived (3.8 day half-life) ^{222}Rn in the experiment. Measurements of this background were made by periodically extracting and cryogenically concentrating ^{222}Rn from water degassers. Radon from several tonne assays was subsequently counted in ZnS(Ag) scintillation cells [5]. The radon results are presented [as mass fractions in $g(U)/g(D_2O)$] in Fig. 1(b).

Independent measurements of U and Th decay chains were made by analyzing Cherenkov light produced by the radioactive decays. The β and β-γ decays from the U and Th chains dominate the low energy monitoring window. Events in this window monitor γ rays that produce photodisintegration in these chains ($E_\gamma > 2.2$ MeV). Cherenkov events fitted within 450 cm from the detector center and extracted from the neutrino data set provide a time-integrated measure of these backgrounds over the same time period and within the fiducial volume of the neutrino analysis. Statistical separation of *in situ* Tl and Bi events was obtained by analyzing the Cherenkov signal isotropy. Tl decays always result in a β and a 2.614 MeV γ, while in this energy window Bi decays are dominated

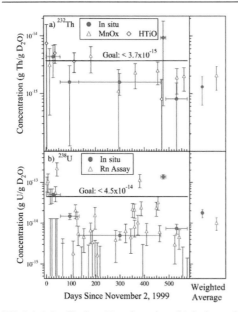

FIG. 1 (color). Thorium (a) and uranium (b) backgrounds (equivalent equilibrium concentrations) in the D_2O deduced by *in situ* and *ex situ* techniques. The MnO_x and HTiO radiochemical assay results, the Rn assay results, and the *in situ* Cherenkov signal determination of the backgrounds are presented for the period of this analysis on the left-hand side of frames (a) and (b). The right-hand side shows time-integrated averages including an additional sampling systematic uncertainty for the *ex situ* measurement.

by decays with only a β, and produce, on average, more anisotropic hit patterns.

Results from the *ex situ* and *in situ* methods are consistent with each other as shown on the right-hand side of Figs. 1(a) and 1(b). For the ^{232}Th chain, the weighted mean (including additional sampling systematic uncertainty) of the two determinations was used for the analysis. The ^{238}U chain activity is dominated by Rn ingress which is highly time dependent. Therefore the *in situ* determination was used for this activity as it provides the appropriate time weighting. The average rate of background neutron production from activities in the D_2O region is 1.0 ± 0.2 neutrons per day, leading to 44^{+8}_{-9} detected background events. The production rate from external activities is $1.3^{+0.4}_{-0.5}$ neutrons per day, which leads to 27 ± 8 background events since the neutron capture efficiency is reduced for neutrons born near the heavy water boundary. The total photodisintegration background corresponds to approximately 12% of the number of NC neutrons predicted by the standard solar model from ^{8}B neutrinos.

Low energy backgrounds from Cherenkov events in the signal region were evaluated by using acrylic encapsulated sources of U and Th deployed throughout the detector volume and by Monte Carlo calculations. Probability density functions (pdfs) in reconstructed vertex radius derived from U and Th calibration data were used to determine the number of background Cherenkov events from external regions which either entered or misreconstructed into the fiducial volume. Cherenkov event backgrounds from activities in the D_2O were evaluated with Monte Carlo calculations.

Table I shows the number of photodisintegration and Cherenkov background events (including systematic uncertainties) due to activity in the D_2O (internal region), acrylic vessel (AV), H_2O (external region), and PMT array. Other sources of free neutrons in the D_2O region are cosmic ray events and atmospheric neutrinos. To reduce these backgrounds, an additional neutron background cut imposed a 250-ms dead time (in software) following every event in which the total number of PMTs which registered a hit was greater than 60. The number of remaining NC atmospheric neutrino events and background events generated by sub-Cherenkov threshold muons is estimated to be small, as shown in Table I.

The data recorded during the pure D_2O detector phase are shown in Fig. 2. These data have been analyzed using the same data reduction described in [2], with the addition of the new neutron background cut, yielding 2928 events in the energy region selected for analysis, 5 to 20 MeV. Figure 2(a) shows the distribution of selected events in the cosine of the angle between the Cherenkov event direction and the direction from the Sun ($\cos\theta_\odot$) for the analysis threshold of $T_{\text{eff}} \geq 5$ MeV and fiducial volume selection of $R \leq 550$ cm, where R is the reconstructed event radius. Figure 2(b) shows the distribution of events in the volume-weighted radial variable $(R/R_{\text{AV}})^3$, where $R_{\text{AV}} = 600$ cm is the radius of the acrylic vessel. Figure 2(c) shows the kinetic energy spectrum of the selected events.

In order to test the null hypothesis, the assumption that there are only electron neutrinos in the solar neutrino

TABLE I. Neutron and Cherenkov background events.

Source	Events
D_2O photodisintegration	44^{+8}_{-9}
H_2O + AV photodisintegration	27^{+8}_{-8}
Atmospheric ν's and	4 ± 1
Fission	$\ll 1$
sub-Cherenkov threshold μ's	
$^2H(\alpha, \alpha)$pn	2 ± 0.4
$^{17}O(\alpha, n)$	$\ll 1$
Terrestrial and reactor $\bar{\nu}$'s	1^{+3}_{-1}
External neutrons	$\ll 1$
Total neutron background	78 ± 12
D_2O Cherenkov	20^{+13}_{-6}
H_2O Cherenkov	3^{+4}_{-3}
AV Cherenkov	6^{+3}_{-6}
PMT Cherenkov	16^{+11}_{-8}
Total Cherenkov background	45^{+18}_{-12}

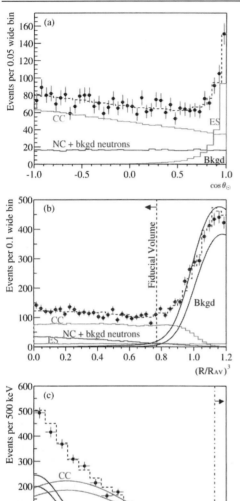

FIG. 2 (color). (a) Distribution of $\cos\theta_\odot$ for $R \leq 550$ cm. (b) Distribution of the volume weighted radial variable $(R/R_{AV})^3$. (c) Kinetic energy for $R \leq 550$ cm. Also shown are the Monte Carlo predictions for CC, ES, and NC + bkgd neutron events scaled to the fit results, and the calculated spectrum of Cherenkov background (bkgd) events. The dashed lines represent the summed components, and the bands show $\pm 1\sigma$ uncertainties. All distributions are for events with $T_{\mathrm{eff}} \geq 5$ MeV.

flux, the data are resolved into contributions from CC, ES, and NC events above threshold using pdfs in T_{eff}, $\cos\theta_\odot$, and $(R/R_{AV})^3$, derived from Monte Carlo calculations generated assuming no flavor transformation and the standard ^{8}B spectral shape [6]. Background event pdfs are included in the analysis with fixed amplitudes determined by the background calibration. The extended maximum likelihood method used in the signal decomposition yields $1967.7^{+61.9}_{-60.9}$ CC events, $263.6^{+26.4}_{-25.6}$ ES events, and $576.5^{+49.5}_{-48.9}$ NC events [7], where only statistical uncertainties are given. Systematic uncertainties on fluxes derived by repeating the signal decomposition with perturbed pdfs (constrained by calibration data) are shown in Table II.

Normalized to the integrated rates above the kinetic energy threshold of $T_{\mathrm{eff}} \geq 5$ MeV, the flux of ^{8}B neutrinos measured with each reaction in SNO, assuming the standard spectrum shape [6] is (all fluxes are presented in units of 10^6 cm^{-2} s^{-1})

$$\phi_{CC}^{SNO} = 1.76^{+0.06}_{-0.05}(\text{stat})^{+0.09}_{-0.09}(\text{syst}),$$

$$\phi_{ES}^{SNO} = 2.39^{+0.24}_{-0.23}(\text{stat})^{+0.12}_{-0.12}(\text{syst}),$$

$$\phi_{NC}^{SNO} = 5.09^{+0.44}_{-0.43}(\text{stat})^{+0.46}_{-0.43}(\text{syst}).$$

Electron neutrino cross sections are used to calculate all fluxes. The CC and ES results reported here are consistent with the earlier SNO results [2] for $T_{\mathrm{eff}} \geq 6.75$ MeV. The excess of the NC flux over the CC and ES fluxes implies neutrino flavor transformations.

A simple change of variables resolves the data directly into electron (ϕ_e) and nonelectron ($\phi_{\mu\tau}$) components [9],

$$\phi_e = 1.76^{+0.05}_{-0.05}(\text{stat})^{+0.09}_{-0.09}(\text{syst}),$$

$$\phi_{\mu\tau} = 3.41^{+0.45}_{-0.45}(\text{stat})^{+0.48}_{-0.45}(\text{syst}),$$

assuming the standard ^{8}B shape. Combining the statistical and systematic uncertainties in quadrature, $\phi_{\mu\tau}$ is $3.41^{+0.66}_{-0.64}$, which is 5.3σ above zero, providing strong evidence for flavor transformation consistent with neutrino oscillations [10,11]. Adding the Super-Kamiokande ES measurement of the 8B flux [12] $\phi_{ES}^{SK} = 2.32 \pm 0.03(\text{stat})^{+0.08}_{-0.07}(\text{syst})$ as an additional constraint, we find $\phi_{\mu\tau} = 3.45^{+0.65}_{-0.62}$, which is 5.5σ above zero. Figure 3 shows the flux of nonelectron flavor active neutrinos vs the flux of electron neutrinos deduced from the SNO data. The three bands represent the one standard deviation measurements of the CC, ES, and NC rates. The error ellipses represent the 68%, 95%, and 99% joint probability contours for ϕ_e and $\phi_{\mu\tau}$.

Removing the constraint that the solar neutrino energy spectrum is undistorted, the signal decomposition is repeated using only the $\cos\theta_\odot$ and $(R/R_{AV})^3$ information. The total flux of active ^{8}B neutrinos measured with the NC reaction is

$$\phi_{NC}^{SNO} = 6.42^{+1.57}_{-1.57}(\text{stat})^{+0.55}_{-0.58}(\text{syst}),$$

which is in agreement with the shape constrained value above and with the standard solar model (SSM) prediction [13] for ^{8}B, $\phi_{SSM} = 5.05^{+1.01}_{-0.81}$.

TABLE II. Systematic uncertainties on fluxes. The experimental uncertainty for ES (not shown) is −4.8, +5.0 percent.

Source	CC uncertainty (percent)	NC uncertainty (percent)	$\phi_{\mu\tau}$ uncertainty (percent)
Energy scale[a]	−4.2, +4.3	−6.2, +6.1	−10.4, +10.3
Energy resolution[a]	−0.9, +0.0	−0.0, +4.4	−0.0, +6.8
Energy nonlinearity[a]	±0.1	±0.4	±0.6
Vertex resolution[a]	±0.0	±0.1	±0.2
Vertex accuracy	−2.8, +2.9	±1.8	±1.4
Angular resolution	−0.2, +0.2	−0.3, +0.3	−0.3, +0.3
Internal source pd[a]	±0.0	−1.5, +1.6	−2.0, +2.2
External source pd	±0.1	−1.0, +1.0	±1.4
DO2 Cherenkov[a]	−0.1, +0.2	−2.6, +1.2	−3.7, +1.7
HO2 Cherenkov	±0.0	−0.2, +0.4	−0.2, +0.6
AV Cherenkov	±0.0	−0.2, +0.2	−0.3, +0.3
PMT Cherenkov[a]	±0.1	−2.1, +1.6	−3.0, +2.2
Neutron capture	±0.0	−4.0, +3.6	−5.8, +5.2
Cut acceptance	−0.2, +0.4	−0.2, +0.4	−0.2, +0.4
Experimental uncertainty	−5.2, +5.2	−8.5, +9.1	−13.2, +14.1
Cross section [8]	±1.8	±1.3	±1.4

[a]Denotes CC vs NC anticorrelation.

In summary, the results presented here are the first direct measurement of the total flux of active ^{8}B neutrinos arriving from the Sun and provide strong evidence for neutrino flavor transformation. The CC and ES reaction rates are consistent with the earlier results [2] and with the NC reaction rate under the hypothesis of flavor transformation. The total flux of ^{8}B neutrinos measured with the NC reaction is in agreement with the SSM prediction.

This research was supported by Canada: NSERC, Industry Canada, NRC, Northern Ontario Heritage Fund Corporation, Inco, AECL, Ontario Power Generation; U.S.: Department of Energy; U.K.: PPARC. We thank the SNO technical staff for their strong contributions.

*Permanent address: Birkbeck College, University of London, Malet Road, London WC1E 7HX, UK.
[†]Deceased.

[1] H. H. Chen, Phys. Rev. Lett. **55**, 1534 (1985).
[2] Q. R. Ahmad et al., Phys. Rev. Lett. **87**, 071301 (2001).
[3] SNO Collaboration, J. Boger et al., Nucl. Instrum. Methods Phys. Res., Sect. A **449**, 172 (2000).
[4] M. R. Dragowsky et al., Nucl. Instrum. Methods Phys. Res., Sect. A **481**, 284 (2002).
[5] M.-Q. Liu, H. W. Lee, and A. B. McDonald, Nucl. Instrum. Methods Phys. Res., Sect. A **329**, 291 (1993).
[6] C. E. Ortiz et al., Phys. Rev. Lett. **85**, 2909 (2000).
[7] We note that this rate of neutron events also leads to a lower bound on the proton lifetime for "invisible" modes {based on the free neutron that would be left in deuterium [V. I. Tretyak and Yu. G. Zdesenko, Phys. Lett. B **505**, 59 (2001)] in excess of 10^{28} years, approximately 3 orders of magnitude more restrictive than previous limits [J. Evans and R. Steinberg, Science **197**, 989 (1977)]}. The possible contribution of this mechanism to the solar neutrino NC background is ignored.
[8] Cross section uncertainty includes g_A uncertainty (0.6%), difference between NSGK [S. Nakamura, T. Sato, V. Gudkov, and K. Kubodera, Phys. Rev. C **63**, 034617 (2001)] and BCK [M. Butler, J.-W. Chen, and X. Kong, Phys. Rev. C **63**, 035501 (2001)] in SNO's calculations (0.6%), radiative correction uncertainties [0.3% for CC, 0.1% for NC; A. Kurylov, M. J. Ramsey-Musolf, and P. Vogel, Phys. Rev. C **65**, 055501 (2002)], uncertainty associated with neglect of real photons in SNO (0.7% for CC), and

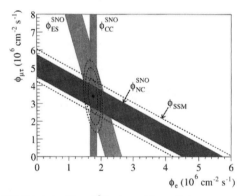

FIG. 3 (color). Flux of ^{8}B solar neutrinos which are μ or τ flavor vs flux of electron neutrinos deduced from the three neutrino reactions in SNO. The diagonal bands show the total ^{8}B flux as predicted by the SSM [13] (dashed lines) and that measured with the NC reaction in SNO (solid band). The intercepts of these bands with the axes represent the $\pm 1\sigma$ errors. The bands intersect at the fit values for ϕ_e and $\phi_{\mu\tau}$, indicating that the combined flux results are consistent with neutrino flavor transformation assuming no distortion in the ^{8}B neutrino energy spectrum.

theoretical cross section uncertainty [1%, S. Nakamura *et al.*, arXiv:nucl-th/0201062 (to be published)].

[9] This change of variables allows a direct test of the null hypothesis of no flavor transformation ($\phi_{\mu\tau} = 0$) without requiring calculation of the CC, ES, and NC signal correlations.

[10] Z. Maki, N. Nakagawa, and S. Sakata, Prog. Theor. Phys. **28**, 870 (1962).

[11] V. Gribov and B. Pontecorvo, Phys. Lett. **28B**, 493 (1969).

[12] S. Fukuda *et al.*, Phys. Rev. Lett. **86**, 5651 (2001).

[13] John N. Bahcall, M.H. Pinsonneault, and Sarbani Basu, Astrophys. J. **555**, 990 (2001).

First Results from KamLAND: Evidence for Reactor Antineutrino Disappearance

K. Eguchi,[1] S. Enomoto,[1] K. Furuno,[1] J. Goldman,[1] H. Hanada,[1] H. Ikeda,[1] K. Ikeda,[1] K. Inoue,[1] K. Ishihara,[1] W. Itoh,[1] T. Iwamoto,[1] T. Kawaguchi,[1] T. Kawashima,[1] H. Kinoshita,[1] Y. Kishimoto,[1] M. Koga,[1] Y. Koseki,[1] T. Maeda,[1] T. Mitsui,[1] M. Motoki,[1] K. Nakajima,[1] M. Nakajima,[1] T. Nakajima,[1] H. Ogawa,[1] K. Owada,[1] T. Sakabe,[1] I. Shimizu,[1] J. Shirai,[1] F. Suekane,[1] A. Suzuki,[1] K. Tada,[1] O. Tajima,[1] T. Takayama,[1] K. Tamae,[1] H. Watanabe,[1] J. Busenitz,[2] Z. Djurcic,[2] K. McKinny,[2] D.-M. Mei,[2] A. Piepke,[2] E. Yakushev,[2] B. E. Berger,[3] Y. D. Chan,[3] M. P. Decowski,[3] D. A. Dwyer,[3] S. J. Freedman,[3] Y. Fu,[3] B. K. Fujikawa,[3] K. M. Heeger,[3] K. T. Lesko,[3] K.-B. Luk,[3] H. Murayama,[3] D. R. Nygren,[3] C. E. Okada,[3] A. W. P. Poon,[3] H. M. Steiner,[3] L. A. Winslow,[3] G. A. Horton-Smith,[4] R. D. McKeown,[4] J. Ritter,[4] B. Tipton,[4] P. Vogel,[4] C. E. Lane,[5] T. Miletic,[5] P. W. Gorham,[6] G. Guillian,[6] J. G. Learned,[6] J. Maricic,[6] S. Matsuno,[6] S. Pakvasa,[6] S. Dazeley,[7] S. Hatakeyama,[7] M. Murakami,[7] R. C. Svoboda,[7] B. Dieterle,[8] M. DiMauro,[8] J. Detwiler,[9] G. Gratta,[9] K. Ishii,[9] N. Tolich,[9] Y. Uchida,[9] M. Batygov,[10] W. Bugg,[10] H. Cohn,[10] Y. Efremenko,[10] Y. Kamyshkov,[10] A. Kozlov,[10] Y. Nakamura,[10] L. De Braeckeleer,[11] C. R. Gould,[11] H. J. Karwowski,[11] D. M. Markoff,[11] J. A. Messimore,[11] K. Nakamura,[11] R. M. Rohm,[11] W. Tornow,[11] A. R. Young,[11] and Y.-F. Wang[12]

(KamLAND Collaboration)

[1]*Research Center for Neutrino Science, Tohoku University, Sendai 980-8578, Japan*
[2]*Department of Physics and Astronomy, University of Alabama, Tuscaloosa, Alabama 35487*
[3]*Physics Department, University of California at Berkeley and Lawrence Berkeley National Laboratory, Berkeley, California 94720*
[4]*W. K. Kellogg Radiation Laboratory, California Institute of Technology, Pasadena, California 91125*
[5]*Physics Department, Drexel University, Philadelphia, Pennsylvania 19104*
[6]*Department of Physics and Astronomy, University of Hawaii at Manoa, Honolulu, Hawaii 96822*
[7]*Department of Physics and Astronomy, Louisiana State University, Baton Rouge, Louisiana 70803*
[8]*Physics Department, University of New Mexico, Albuquerque, New Mexico 87131*
[9]*Physics Department, Stanford University, Stanford, California 94305*
[10]*Department of Physics and Astronomy, University of Tennessee, Knoxville, Tennessee 37996*
[11]*Triangle Universities Nuclear Laboratory, Durham, North Carolina 27708*
and Physics Departments at Duke University, North Carolina State University,
and the University of North Carolina at Chapel Hill
[12]*Institute of High Energy Physics, Beijing 100039, People's Republic of China*
(Received 6 December 2002; published 17 January 2003)

KamLAND has measured the flux of $\bar{\nu}_e$'s from distant nuclear reactors. We find fewer $\bar{\nu}_e$ events than expected from standard assumptions about $\bar{\nu}_e$ propagation at the 99.95% C.L. In a 162 ton · yr exposure the ratio of the observed inverse β-decay events to the expected number without $\bar{\nu}_e$ disappearance is $0.611 \pm 0.085(\text{stat}) \pm 0.041(\text{syst})$ for $\bar{\nu}_e$ energies > 3.4 MeV. In the context of two-flavor neutrino oscillations with CPT invariance, all solutions to the solar neutrino problem except for the "large mixing angle" region are excluded.

DOI: 10.1103/PhysRevLett.90.021802   PACS numbers: 14.60.Pq, 26.65.+t, 28.50.Hw, 91.65.Dt

The primary goal of the Kamioka Liquid Scintillator Anti-Neutrino Detector (KamLAND) [1] is a search for the oscillation of $\bar{\nu}_e$'s emitted from distant power reactors. The long baseline, typically 180 km, enables KamLAND to address the oscillation solution [2,3] of the "solar neutrino problem" [4] with $\bar{\nu}_e$'s under laboratory conditions. The inverse β-decay reaction, $\bar{\nu}_e + p \rightarrow e^+ + n$ is used to detect $\bar{\nu}_e$'s in liquid scintillator (LS) [5]. Detecting both the e^+ and the delayed 2.2 MeV γ-ray from neutron capture on a proton is a powerful tool for reducing background. This Letter presents first results from an analysis of 162 ton · yr (145.1 d) of the reactor $\bar{\nu}_e$ data.

KamLAND occupies the site of the earlier Kamiokande [6], under 2700 m.w.e. of rock resulting in 0.34 Hz of cosmic-ray muons in the detector. Shown in Fig. 1, the neutrino detector/target is 1 kton of ultrapure LS contained in a 13-m-diameter spherical balloon made of 135-μm-thick transparent nylon/EVOH (ethylene vinyl alcohol copolymer) composite film. A network of Kevlar ropes supports and constrains the balloon. The LS is 80% dodecane, 20% pseudocumene (1,2,4-trimethylbenzene), and 1.52 g/liter of PPO (2,5-diphenyloxazole) as a fluor. A buffer of dodecane and isoparaffin oils between the balloon and an 18-m-diameter spherical stainless-steel containment vessel shields the LS from external radiation.

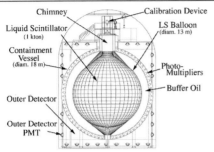

FIG. 1. Schematic diagram of the KamLAND detector.

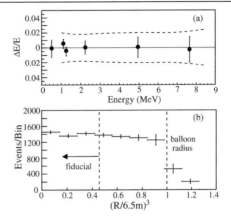

FIG. 2. (a) The fractional difference of the reconstructed average energies and known source γ energies. The dashed lines show the adopted systematic error. (b) The R^3 vertex distribution of 2.2 MeV neutron capture γ's. The level of uniformity over the fiducial volume is used in the estimate of the fiducial volume uncertainty.

During filling water extraction and nitrogen stripping [7] purified the LS and buffer oil (BO). The buffer oil density is 0.04% lower than the LS. A 1879 photomultiplier tube array, mounted on the containment vessel, completes the inner detector (ID). There are 1325 newly developed fast 17-inch-diameter photomultiplier tubes (PMTs) and 554 older Kamiokande 20-inch PMTs [8]. The total photocathode coverage is 34% but only the 17-inch PMTs with 22% coverage are used for the present analysis. A 3-mm-thick acrylic barrier at 16.6-m diameter reduces radon from PMT glass in the LS. The containment vessel is surrounded by a 3.2 kton water-Cherenkov detector with 225 20-inch PMTs. This outer detector (OD) absorbs γ rays and neutrons from surrounding rock and acts as a tag for cosmic-ray muons. The primary ID trigger threshold is 200 PMT hits, corresponding to about 0.7 MeV. The threshold goes to 120 hits for 1 msec after a primary trigger. The OD trigger threshold corresponds to > 99% tagging efficiency.

Energy response in the 0.5 to 7.5 MeV range is calibrated with ^{68}Ge, ^{65}Zn, ^{60}Co, and Am-Be γ-ray sources deployed at various positions along the vertical axis. Detected energy is obtained from the number of observed photoelectrons (p.e.) after corrections for gain variation, solid angle, density of PMTs, shadowing by suspension ropes, and transparencies of the LS and BO. Figure 2(a) shows the fractional deviation of the reconstructed energies from the known source energies. The ^{68}Ge and ^{60}Co sources emit two coincident γ rays and are plotted at an average energy in Fig. 2(a). The observed energy resolution is $\sim 7.5\%/\sqrt{E(\text{MeV})}$.

The energy scale is augmented from studies of the radiation from ^{40}K and ^{208}Tl and Bi-Po contaminants, as well as ^{12}B- and ^{12}N-spallation products, and γ rays from neutron capture on protons and ^{12}C. The reconstructed energy varies by less than 0.5% within a 10-m-diameter volume except for 1.6% variations near the chimney. The energy scale is stable to 0.6% during the run. Corrections for quenching and Cherenkov light production are included, and contribute to the systematic error in Fig. 2(a). The estimated systematic error in the energy is 1.9% at our 2.6 MeV analysis threshold giving a 2.1% uncertainty in the rate above threshold.

Event locations are reconstructed from the timing of PMT hits. After energy-dependent radial adjustments, the known source positions are reconstructed to ~ 5 cm; the typical position resolution is ~ 25 cm. Vertex reconstruction performance throughout the LS volume is verified by reproducing the uniform distribution of 2.2 MeV capture γ's from spallation neutrons, as shown in Fig. 2(b).

The data presented in this Letter were collected from March 4 through October 6, 2002. We obtained 370×10^6 events in 145.1 d of live time at an average trigger rate of $\simeq 30$ Hz. Events with less than 10 000 p.e. (~ 30 MeV) and no prompt OD tag are "reactor-$\bar{\nu}_e$ candidates"; more energetic events are "muon candidates."

The selection cuts for $\bar{\nu}_e$ events are the following: (i) fiducial volume ($R < 5$ m), (ii) time correlation (0.5 μsec $< \Delta T < 660$ μsec), (iii) vertex correlation ($\Delta R < 1.6$ m), (iv) delayed energy (1.8 MeV $< E_{\text{delay}} < 2.6$ MeV), and (v) a requirement that the delayed vertex position be more than 1.2 m from the central vertical axis to eliminate background from LS monitoring thermometers. The overall efficiency for the events from criteria (ii)–(v) including the effect of (i) on the delayed vertex is $(78.3 \pm 1.6)\%$.

Including annihilation, the detected energy for positrons is the kinetic energy plus twice the rest energy; thus on the average e^+ from $\bar{\nu}_e$ events yield $E_{\text{prompt}} = E_{\bar{\nu}_e} - \bar{E}_n - 0.8$ MeV, where $\bar{E}_n$ is the average recoil energy of the neutron. Antineutrinos from ^{238}U and ^{232}Th in the Earth, "geoneutrinos" ($\bar{\nu}_{\text{geo}}$) can produce events with $E_{prompt} < 2.49$ MeV. Model Ia in Ref. [9] predicts about 9 $\bar{\nu}_{geo}$ events in our data set. However, the abundances and distributions of U and Th are not well known. We employ (vi) a prompt energy cut, $E_{prompt} > 2.6$ MeV, to avoid ambiguity in the present analysis.

The fiducial volume is estimated using the expected uniform distribution of spallation-product neutron-capture events shown in Fig. 2(b). The ratio of events in the fiducial volume to the total volume agrees with the geometric fiducial fraction to within 4.1%. This method is also used for higher energy events from ^{12}N, ^{12}B β's following muon spallation; the agreement is within 3.5%. Accounting for the 2.1% uncertainty in the total LS mass, we estimate a 4.6% uncertainty in the fiducial volume. The LS density is 0.780 g/cm^3 at 11.5 °C; the expected hydrogen-to-carbon ratio of 1.97 was verified by elemental analysis to $\pm 2\%$. The specific gravity is measured to 0.01% precision and we assign an additional 0.1% error from the uncertainty in the temperature. The 408 ton fiducial mass thus contains 3.46×10^{31} free target protons.

The trigger efficiency was determined to be 99.98% with LED light sources. The combined efficiency of the electronics, data acquisition, and event reconstruction was studied using time distributions of uncorrelated events from calibration γ sources. We find that this combined efficiency is better than 99.98%. The vertex fitter yields $> 99.9\%$ efficiency within 2 m of known source positions. With calibrated ^{60}Co and ^{65}Zn sources, the overall efficiency was checked to the 3% source-strength uncertainties. The detection efficiency for delayed events from the Am-Be source (4.4 MeV prompt γ and 2.2 MeV delayed neutron capture γ within 1.6 m) was verified to 1% certainty.

Studies of Bi-Po sequential decays indicate that the effective equilibrium concentrations of ^{238}U and ^{232}Th in the LS are $(3.5 \pm 0.5) \times 10^{-18}$ g/g and $(5.2 \pm 0.8) \times 10^{-17}$ g/g, respectively. The observed background energy spectrum indicates that ^{40}K contamination is less than 2.7×10^{-16} g/g. The extremely low level of U and Th contamination in the LS provides an optimistic prospect for future solar neutrino experiments with KamLAND. The flat accidental background, observed in a delayed time window of 0.020–20 sec, is 0.0086 ± 0.0005 events for the present data set.

The most serious source of external γ rays from ^{208}Tl ($E_\gamma \leq 3$ MeV) is strongly suppressed by the fiducial volume cut (i). At higher energies, the background is dominated by spallation products from energetic muons. We observe ~3 000 neutron events/day/kton. We also expect ~1 300 events/day/kton [10] for various unstable products.

Single neutrons are easily suppressed with a 2-msec veto following a muon. Care is required to avoid neutrons which mimic the $\bar{\nu}_e$ delayed coincidence signal. Most external fast neutrons are produced by muons which pass through both the OD and the surrounding rock. This background is studied by detecting delayed coincidence events tagged with a muon detected by only the OD. As expected, events concentrate near the balloon edge. The background in the fiducial volume is estimated by extrapolating the distribution of vertex positions and accounting for the 92% OD reconstruction efficiency. The number of background events due to neutrons from the surrounding rock is estimated from the OD-tagged data scaled by the relative neutron production and the shielding factor of the relevant materials. The estimated total fast neutron background is less than 0.5 events in the entire data set.

Most radioactive spallation products simply beta decay, and are effectively suppressed by requiring a delayed neutron. Delayed neutron emitters such as ^{8}He ($T_{1/2} = 119$ msec) and ^{9}Li (178 msec) are eliminated by two time/geometry cuts: (a) a 2-sec veto in the entire fiducial volume following a "showering muon" (more than 10^6 p.e., ~3 GeV, extra energy deposition), (b) for other muons, delayed events within 2 sec and 3 m of the muon track are rejected. The cut efficiency is estimated from the observed correlation of spallation neutrons with muon tracks. The remaining ^{8}He and ^{9}Li background is estimated to be 0.94 ± 0.85. The dead time from spallation cuts is 11.4%. This is checked by constructing the time distribution of the events following a detected muon to separate the short-lived spallation-produced activities from $\bar{\nu}_e$ candidates. The uncorrelated $\bar{\nu}_e$ distribution has a time constant of $1/R_\mu \simeq 3$ sec, where R_μ is the muon rate. Spallation products all have a shorter time constant (~0.2 sec). The two selected methods agree to 3% accuracy. As shown in Table I, the total number of expected background events is 1 ± 1, where the fast neutron contribution is included in the error estimate.

Instantaneous thermal power, burnup, and fuel exchange records for all commercial Japanese power reactors are provided by the power companies. The thermal power generation is checked with the independent records of electric power generation. The fission rate for each fissile isotope is calculated as a function of time and the systematic uncertainty in the $\bar{\nu}_e$ flux is 1%. Averaged over live time, the relative fission yields from fuel components are ^{235}U: ^{238}U: ^{239}Pu: ^{241}Pu = 0.568 : 0.078 : 0.297 : 0.057. The $\bar{\nu}_e$ spectrum per fission with a 2.5% error

TABLE I. Background summary.

Background	Number of events
Accidental	0.0086 ± 0.0005
^{9}Li/^{8}He	0.94 ± 0.85
Fast neutron	<0.5
Total B.G. events	1 ± 1

TABLE II. Estimated systematic uncertainties (%).

Total LS mass	2.1	Reactor power	2.0
Fiducial mass ratio	4.1	Fuel composition	1.0
Energy threshold	2.1	Time lag	0.28
Efficiency of cuts	2.1	$\bar{\nu}$ spectra [11]	2.5
Live time	0.07	Cross section [14]	0.2
Total systematic error			6.4%

are taken from [11]. This neutrino spectrum has been tested to a few percent with short-baseline reactor $\bar{\nu}_e$ experiments [5,12]. The finite lifetimes of fission products introduce a 0.28% uncertainty to the $\bar{\nu}_e$ flux. The contribution from Korean reactors is estimated to be $(2.46 \pm 0.25)\%$ based on reported electric power generation. The rest of the World's reactors contribute $(0.70 \pm 0.35)\%$ from an estimate using reactor specifications from the International Nuclear Safety Center [13]. In the absence of $\bar{\nu}_e$ disappearance the expected number of $\bar{\nu}_e$ events is 86.8 ± 5.6; the systematic error contributions are listed in Table II.

The antineutrinos at KamLAND are provided by many nuclear reactors but the flux is actually dominated by a few powerful reactors at an average distance of ~ 180 km. More than 79% of the flux is from 26 reactors between 138–214 km away. One close reactor at 88 km contributes 6.7%; other reactors are more than 295 km away. The relatively narrow band of distances allows KamLAND to be sensitive to spectral distortions for certain oscillation parameters.

Figure 3 shows the energy distribution of delayed coincidence events with no energy cuts. A well-separated cluster of 2.2 MeV capture γ's is evident. One observed event with delayed energy around 5 MeV and prompt energy of about 3.1 MeV (not shown in Fig. 3) is consistent with the expected neutron radiative capture rate on ^{12}C.

The observed space-time correlation of the prompt and delayed events agrees with expectations, and the measured capture time of 188 ± 23 μsec is consistent with predictions for LS. After applying all the prompt and delayed energy cuts, 54 events remain. Accounting for ~ 1 background event the probability of a fluctuation from 86.8 expected is $<0.05\%$ by Poisson statistics. The ratio of observed reactor $\bar{\nu}_e$ events to expected in the absence of neutrino disappearance is

$$\frac{N_{\text{obs}} - N_{\text{BG}}}{N_{\text{expected}}} = 0.611 \pm 0.085(\text{stat}) \pm 0.041(\text{syst}).$$

Figure 4 shows the ratio of measured to expected flux for KamLAND as well as previous reactor experiments as a function of the average distance from the source.

The expected prompt positron spectrum with no oscillations and the best fit with reduced $\chi^2 = 0.31$ for 8 degrees of freedom for two-flavor neutrino oscillations above the 2.6 MeV threshold are shown in Fig. 5. A clear deficit of events is evident. At the 93% C.L. the data are consistent with a distorted spectrum shape expected from neutrino oscillations, but a scaled no-oscillation shape is also consistent at 53% C.L. as determined by Monte Carlo.

The neutrino oscillation parameter region for two-neutrino mixing is shown in Fig. 6. The dark shaded area is the MSW-LMA [19] region at 95% C.L. derived from [16]. The shaded region outside the solid line is excluded at 95% C.L. from the rate analysis with $\chi^2 \geq 3.84$ and

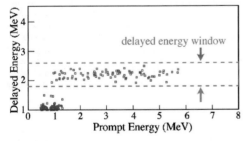

FIG. 3 (color). Distribution of $\bar{\nu}_e$ candidates after fiducial volume, time, vertex correlation, and spallation cuts are applied. For $\bar{\nu}_e$ events the prompt energy is attributed to positrons and the delayed energy to neutron capture. Events within the horizontal lines bracketing the delayed energy of 2.2 MeV are consistent with thermal neutron capture on protons.

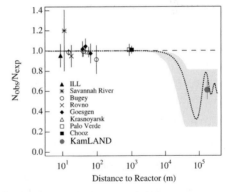

FIG. 4 (color). The ratio of measured to expected $\bar{\nu}_e$ flux from reactor experiments [15]. The solid circle is the KamLAND result plotted at a flux-weighted average distance of ~ 180 km. The shaded region indicates the range of flux predictions corresponding to the 95% C.L. LMA region from a global analysis of the solar neutrino data [16]. The dotted curve, $\sin^2 2\theta = 0.833$ and $\Delta m^2 = 5.5 \times 10^{-5}$ eV^2 [16], is representative of a best-fit LMA prediction and the dashed curve is expected for no oscillations.

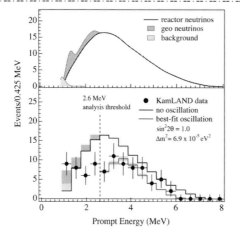

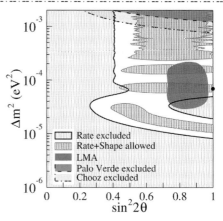

FIG. 5 (color). Upper panel: Expected reactor $\bar{\nu}_e$ energy spectrum along with $\bar{\nu}_{\text{geo}}$ (model Ia of [9]) and background. Lower panel: Energy spectrum of the observed prompt events (solid circles with error bars), along with the expected no oscillation spectrum (upper histogram, with $\bar{\nu}_{\text{geo}}$ and background shown) and best fit (lower histogram) including neutrino oscillations. The shaded band indicates the systematic error in the best-fit spectrum. The vertical dashed line corresponds to the analysis threshold at 2.6 MeV.

FIG. 6 (color). Excluded regions of neutrino oscillation parameters for the rate analysis and allowed regions for the combined rate and shape analysis from KamLAND at 95% C.L. At the top are the 95% C.L. excluded region from CHOOZ [17] and Palo Verde [18] experiments, respectively. The 95% C.L. allowed region of the large mixing angle (LMA) solution of solar neutrino experiments [16] is also shown. The solid circle shows the best fit to the KamLAND data in the physical region: $\sin^2 2\theta = 1.0$ and $\Delta m^2 = 6.9 \times 10^{-5}$ eV^2. All regions look identical under $\theta \leftrightarrow (\pi/2 - \theta)$ except for the LMA region from solar neutrino experiments.

$$\chi^2 = \frac{[0.611 - R(\sin^2 2\theta, \Delta m^2)]^2}{0.085^2 + 0.041^2}.$$

Here, $R(\sin^2 2\theta, \Delta m^2)$ is the expected ratio with the oscillation parameters.

The final event sample is evaluated using a maximum likelihood method to obtain the optimum set of oscillation parameters with the following χ^2 definition:

$$\chi^2 = \chi^2_{\text{rate}}(\sin^2 2\theta, \Delta m^2, N_{\text{BG1}\sim 2}, \alpha_{1\sim 4})$$
$$- 2 \log L_{\text{shape}}(\sin^2 2\theta, \Delta m^2, N_{\text{BG1}\sim 2}, \alpha_{1\sim 4})$$
$$+ \chi^2_{\text{BG}}(N_{\text{BG1}\sim 2}) + \chi^2_{\text{distortion}}(\alpha_{1\sim 4}).$$

L_{shape} is the likelihood function for the spectrum including experimental distortions. $N_{\text{BG1}\sim 2}$ are the estimated ^{9}Li and ^{8}He backgrounds and $\alpha_{1\sim 4}$ are parameters to account for the spectral effects of energy scale uncertainty, finite resolution, $\bar{\nu}_e$ spectrum uncertainty, and fiducial volume systematic error, respectively. Parameters are varied to minimize the χ^2 at each pair of $[\Delta m^2, \sin^2 \theta]$ with a bound from $\chi^2_{\text{BG}}(N_{\text{BG1}\sim 2})$ and $\chi^2_{\text{distortion}}(\alpha_{1\sim 4})$. The best fit to the data in the physical region yields $\sin^2 2\theta = 1.0$ and $\Delta m^2 = 6.9 \times 10^{-5}$ eV^2 while the global minimum occurs slightly outside of the physical region at $\sin^2 2\theta = 1.01$ with the same Δm^2. These numbers can be compared to the best fit LMA values of $\sin^2 2\theta = 0.83$ and $\Delta m^2 = 5.5 \times 10^{-5}$ eV^2 from [16]. The 95% C.L. allowed regions from the spectrum shape analysis for $\Delta \chi^2 = 5.99$ and two parameters are shown in Fig. 6. The allowed regions displayed for KamLAND correspond to $0 < \theta < \frac{\pi}{4}$ consistent with the solar LMA solution, while the allowed regions in $\frac{\pi}{4} < \theta < \frac{\pi}{2}$ are the same [20] but do not include the solar solution.

The results from a spectral shape analysis with a 0.9 MeV threshold are consistent with the above result. In this low-energy analysis, the measured $\bar{\nu}_{\text{geo}}$ fluxes are free parameters. The numbers of $\bar{\nu}_{\text{geo}}$ events for the best fit are four for ^{238}U and five for ^{232}Th, which corresponds to ~ 40 TW radiogenic heat generation according to model Ia in [9]. However, for the same model, $\bar{\nu}_{\text{geo}}$ production powers from 0 to 110 TW are still allowed at 95% C.L. with the same oscillation parameters.

If three neutrino generations are considered, the $\bar{\nu}_e$ survival probability depends on two mixing angles θ_{12} and θ_{13}. In the region close to the best fit KamLAND solution the survival probability is, to a very good approximation, given by

$$P(\bar{\nu}_e \to \bar{\nu}_e) \cong \cos^4 \theta_{13} \left[1 - \sin^2 2\theta_{12} \sin^2 \frac{\Delta m_{12}^2 L}{4 E_\nu} \right],$$

with $\Delta m_{12}^2 \cong \Delta m^2$ from the two-flavor analysis above.

The CHOOZ experiment [17] established an upper limit of $\sin^2 2\theta_{13} < 0.15$, or $\cos^4 \theta_{13} \geq 0.92$. Our best fit result corresponds approximately to $0.86 < \sin^2 2\theta_{12} < 1.0$.

In summary, KamLAND demonstrated reactor $\bar{\nu}_e$ disappearance at long baselines and high confidence level (99.95%) for the first time. One expects a negligible reduction of $\bar{\nu}_e$ flux from the SMA, LOW, and VAC solar neutrino solutions, and the LMA region is the only remaining oscillation solution consistent with the KamLAND result assuming CPT invariance. The allowed LMA region is constrained by KamLAND. Future KamLAND measurements with improved statistical precision and reduced systematic errors will provide precision determinations of neutrino oscillation parameters.

The KamLAND experiment is supported by the Center of Excellence program of the Japanese Ministry of Education, Culture, Sports, Science and Technology, and funding from the United States Department of Energy. The reactor data are provided courtesy of the following electric associations in Japan; Hokkaido, Tohoku, Tokyo, Hokuriku, Chubu, Kansai, Chugoku, Shikoku and Kyushu Electric Power Companies, Japan Atomic Power Co., and Japan Nuclear Cycle Development Institute. Kamioka Mining and Smelting Company provided services for activities in the mine.

---

[1] KamLAND Collaboration, A. Suzuki *et al.*, in *Proceedings of the 7th Euromicro Workshop on Parallel and Distributed Processing, Funchal, Portugal, 1999* [Nucl. Phys. (Proc. Suppl.) **B77**, 171 (1999)]; http://www.awa.tohoku.ac.jp/KamLAND/.

[2] Z. Maki *et al.*, Prog. Theor. Phys. **28**, 870 (1962).

[3] B. Pontecorvo, Sov. Phys. JETP **6**, 429 (1957) [Zh. Eksp. Teor. Fiz. **33**, 549 (1957)].

[4] J. N. Bahcall and R. Davis, Science **191**, 264 (1976); J. N. Bahcall, *Neutrino Astrophysics* (Cambridge University Press, Cambridge, United Kingdom, 1989); J. N. Bahcall, Astrophys. J. **467**, 475 (1996).

[5] F. Reines *et al.*, Phys. Rev. **117**, 159 (1960); for a review, see C. Bemporad *et al.*, Rev. Mod. Phys. **74**, 297 (2002).

[6] K. S. Hirata *et al.*, Phys. Rev. D **38**, 448 (1988).

[7] J. B. Benziger *et al.*, Nucl. Instrum. Methods Phys. Res., Sect. A **417**, 278 (1998).

[8] H. Kume *et al.*, Nucl. Instrum. Methods Phys. Res. **205**, 443 (1986).

[9] R. S. Raghavan *et al.*, Phys. Rev. Lett. **80**, 635 (1998).

[10] T. Hagner *et al.*, Astropart. Phys. **14**, 33 (2000).

[11] ^{235}U: K. Schreckenbach *et al.*, Phys. Lett. B **160**, 325 (1985); ^{239,241}Pu: A. A. Hahn *et al.*, Phys. Lett. B **218**, 365 (1989); ^{238}U: P. Vogel *et al.*, Phys. Rev. C **24**, 1543 (1981).

[12] B. Achkar *et al.*, Phys. Lett. B **374**, 243 (1996); G. Zacek *et al.*, Phys. Rev. D **34**, 2621 (1986).

[13] http://www.insc.anl.gov/

[14] P. Vogel and J. F. Beacom, Phys. Rev. D **60**, 053003 (1999); radiative correction from A. Kurylov *et al.*, hep-ph/0211306.

[15] Particle Data Group, Phys. Rev. D **66**, 010001-406 (2002).

[16] G. L. Fogli *et al.*, Phys. Rev. D **66**, 053010 (2002).

[17] M. Apollonio *et al.*, Phys. Lett. B **466**, 415 (1999).

[18] F. Boehm *et al.*, Phys. Rev. D **64**, 112001 (2001).

[19] L. Wolfenstein, Phys. Rev. D **17**, 2369 (1979); S. P. Mikheyev and A. Yu. Smirnov, Sov. J. Nucl. Phys. **42**, 913 (1986).

[20] A. de Gouvea *et al.*, Phys. Lett. B **490**, 125 (2000).

17

Epilogue

Six quarks, six leptons, together with the gluons of QCD and the photon and weak bosons, are enough to describe the tangible world and more, with remarkable economy. Only the Higgs boson is missing among the ingredients of the canonical Standard Model. And yet we know we are missing much more than this. The last ten years of cosmological observations have established that the ordinary matter of quarks and leptons accounts for just 5% of the energy density of the Universe, that another 23% is "dark matter," outside the Standard Model, and 72% of the energy density isn't due to matter at all. Moreover, we can't answer the most basic question of all: why is there something rather than nothing? Why didn't all the matter created in the Big Bang ultimately annihilate, particle against antiparticle? Andrei Sakharov explained that CP violation must be part of the answer, but we know it isn't just the CP violation of the CKM matrix, for that wouldn't account for the amount of matter that remains. On the other hand, the strong interactions might have been CP violating but aren't. Why not?

These questions are pressed upon us by facts and demand answers. Other questions arise more from aesthetics: Are the strong and electroweak forces themselves unified? What about gravity? Are there more forces still to be discovered? Why are there three generations of quarks and leptons? Even more audaciously, why are there three spatial dimensions, or perhaps, are there more than three spatial dimensions?

These are questions of physics, not metaphysics, because there are experiments to address them. At CERN, the LEP tunnel is filled with 8-T magnets to constrain counter-rotating beams of 7-TeV protons. The gargantuan ATLAS and CMS detectors, the descendants of UA-1, UA-2, CDF, and D0, are there to pick out the 100 or so most interesting events of the 10^9 that will be produced each second. At this energy, either the Higgs boson or some surrogate must make its appearance. Supersymmetry would provide a replica of each known particle, with its spin offset by half a unit. Among these could be the particle that makes up dark matter.

Particle physics, which has its origins in studies of cosmic rays and nuclear decays, is returning to these phenomena to answer basic questions. Dark matter must be a form of cosmic rays and might be detected through collisions with ordinary matter if only all the extraneous backgrounds could be excluded by going deep underground with supersensitive detectors. The absence of CP violation in the strong interactions can be explained at the cost of introducing an axion, a feebly interacting particle a bit like a completely stable neutral pion. The axion might itself be the dark matter and could be detected by converting it to a photon in a resonant cavity.

The CP violation that accounts for the baryon–antibaryon asymmetry might reside in very heavy neutrinos, which are beyond our reach. Still, we can seek circumstantial evidence by looking for CP violation in the light neutrinos, whose mixing is only partially understood. We don't know yet how much of the electron-neutrino resides in the third neutrino mass eigenstate, the one whose mass is far from that of the other two. All CP violating effects in neutrinos are proportional to this amplitude, $\sin\theta_{13}$. Experiments both with accelerators and nuclear reactors are underway to measure this small quantity.

Perhaps words borrowed from Winston Churchill best describe Dark Energy "a riddle wrapped in a mystery inside an enigma." It dominates the energy budget of the Universe but it isn't matter at all. About its properties we only know one thing: its pressure is nearly the negative of its energy. Einstein's abandoned Cosmological Constant, Λ, fits the bill, for it gives $p/\rho = -1$ exactly, but the value of Λ required is some 120 orders of magnitude larger than one would expect on dimensional grounds, showing that we really don't understand this at all. Alternatively, the Dark Energy might be something dynamical, not static. The expansion history of the Universe, gleaned from precision measurements of distant type Ia supernovae, from weak gravitational lensing of distant galaxies, and from correlations between the locations of galaxies extracted from hundreds of millions of redshifts, provide the best means of learning more about Dark Energy. The same measurements can check that the acceleration of the expansion of the Universe is not due to a failure of General Relativity, but really due to Dark Energy.

Some of the questions before particle physics have puzzled people for millennia. What is the world, the Universe, made of? How did it start? How will it end? Others – like why is there any matter at all – require both understanding and imagination even to pose. What is remarkable and thrilling is that we can expect to learn something about these formerly metaphysical questions by doing real experiments.

Index

Aamodt, R. L., 22
Abe, K., 262
Abrams, G. S., 259
Adair, R., 103
adiabatic demagnetization, 154
Aguilar-Benitez, M., 262
ALEPH experiment, 395, 402
alpha ray, 2
Alston, M., 108
Alternating Gradient Synchrotron (AGS), 158, 190, 248
Alvarez, L. W., 58, 80, 102
Amaldi, E., 53, 81
Ambler, E., 154
AMY, 296
Anderson, C. D., 4, 15, 49, 50
Anderson, H., 99
Anjos, J. C., 263
annihilation
    electron–positron, 157, 247–256, 294, 296–298, 323, 328, 370
    nucleon–antinucleon, 84
    proton–antiproton, 81
    quark–antiquark, 247
    R in e^+e^-, 295
    three-jet events in e^+e^-, 298
anomalous magnetic moment
    proton, 80, 211
antibaryon, 80, 83
antineutron, 83
antiparticle, 5, 57, 80, 81, 84, 147, 185
antiproton, 80–82, 84, 85
Araki, G., 16
ARGUS, 261, 329, 434
Armenteros, R., 108
Ashkin, J., 100
associated production, 57
Aston, F. W., 3
asymmetric e^+e^- collider, 441
asymptotic freedom, 297

atomic
    mass, 1
    number, 1, 3
    weight, 1, 3

B meson
    CP violation, 443
    decay to $\pi\pi$, 440
    decay to $D\pi$, 440
    decay to $J/\psi K$, 439
    exclusive decays, 328
    lifetime, 328
    mixing, 434
        dilution, 439
    semileptonic decays, 434
B^* mesons, 328
BaBar, 442–450
Bagnères-de-Bigorre, 52
Baldo-Ceolin, M., 84
Balmer formula, 1, 2
Baltrusaitis, R. M., 259
baryon, 57, 80, 81, 84, 108, 111, 212, 293, 294
    charmed, 255, 257
    decuplet, 109
    octet, 105, 108, 109
baryon–antibaryon asymmetry, 80
BEBC, 219
Becker, H., 4
Becker, U., 248
Becquerel, H., 1, 2
Belle experiment, 442–449
Benvenuti, A., 257
Bergkvist, K. E., 150
beta decay, 4, 16, 17, 147, 150, 152, 155, 157, 160, 161, 325, 326, 357, 367
    double, 490
    parity violation in, 152, 154
    pion, 156
beta ray, 2
Bethe, H. A., 6, 15, 49

Bethe–Heitler theory, 15
Bevatron, 57, 58, 80, 81, 84, 102, 105, 157, 186
BFP (Berkeley–Fermilab–Princeton) Collaboration, 220
Bhabha, H. J., 5
  scattering, 5
Bjorken, J. D., 213, 254, 293, 295
Bjorklund, R., 19
blackbody radiation, 2
Blackett, P. M. S., 5, 16, 49
Bloch, F., 80
Block, M., 105, 152
Bloom, E., 252
BNL, 54, 102, 108, 110, 157, 186, 190, 247, 248
Bohr, N., 2, 147
  atom, 2
Bonetti, A., 54
boson, 3, 18, 57
  Higgs, 358, 363, 399, 402
  identical, 22
  pion, 25
  vector, 159
  W, 357, 358, 360, 364
  Z, 358, 359
Bothe, W., 4
Breit, G., 100
Breit–Wigner resonance, 100, 249, 396
bremsstrahlung, 14, 215
Brode, R. B., 49
Brodsky, S. J., 295
Brueckner, K. A., 99
bubble chamber, 58, 84, 102, 103, 105, 108, 109, 153, 218, 219, 256–261, 358
Bugey neutrino experiment, 501
Butler, C. C., 49

Cabibbo, N., 156, 159, 254
  angle, 156, 216, 218, 254, 326
  mixing, 257
Cabibbo–Kobayashi–Maskawa matrix, 327, 435
Callan–Gross relation, 214
calorimetry
  at B factories, 442
  in beta decay, 147
  in neutrino experiments, 219
  UA-1, 364, 365
  UA-1 and UA-2, 301
  UA-2, 368
Caltech, 5
Caltech–Fermilab collaboration, 361
Carithers, W. C., 193
Carlson, A. G., 19
Cartwright, W. F., 25
cascade particle, Ξ^-, 54
cascade zero, Ξ^0, 58
cathode ray, 1
Cazzoli, E. G., 256

CCFRR, 219, 220
CDF, 395, 416, 439, 442
CDHS, 219, 220, 326
CEA, Cambridge Electron Accelerator, 248
CERN, 158, 218, 257, 259, 301, 358, 364
CESR, 324
Chadwick, J., 4, 147
Chamberlain, O., 80
charge conjugation, 7, 185, 190, 251, 252
CHARM, 219
charm
  baryon, 255, 257
  discovery, 257
  hints of, 255–257
  quark, 250, 252, 254, 255
Chen, A., 261, 327
Chen, M., 248
Cherenkov counter, 81, 84, 190, 248, 441
Cherenkov radiation, 84
Chew, G., 103, 104
Chew-Low analysis, 103
Chinowsky, W., 108, 186, 248
Chooz neutrino experiment, 501
Chrétien, M., 105
Christenson, J. H., 190
Christofilos, N., 158
CKM
  favored, 440
  matrix, 326, 437, 438, 442, 444
  suppressed, 445
  suppression, 327, 435
Clark, D. L., 25
CLEO, 261, 324
cloud chamber, 4, 13, 49, 50, 53, 58, 59, 153, 186
Co^{60}, 154
Cockroft, J. D., 18
color, 248, 251, 293, 294, 302
  confinement, 294
Compton, A. H., 5
  scattering, 5, 262
  wavelength, 298
Connolly, P. L., 108
conservation
  CP, 195
  energy, 4, 102, 149, 152
  isospin, 57, 259
  momentum, 102
  parity, 24, 53
conserved vector current hypothesis, 156
Conversi, M., 17
Cork, B., 81
cosmic rays, 4, 13, 49, 51, 54, 58, 255
Cosmotron, 54, 58, 101, 103, 108, 157, 186
Coulomb's law, 5
Courant, E., 158
Cowan, C. L., 157

Cowan, E. W., 50, 54
CP, 108
   action, 185
   conservation, 185, 195
   eigenstates, 185, 186
   non-conservation, 185
   violation, 190, 194, 437
      from CKM matrix, 326
      from three generations, 330
      in B mesons, 443–446
      in neutrinos, 511
      in semileptonic decays, 193
CPT invariance, 190
Crandall, W. E., 19
Cronin, J. W., 190
Crystal Ball, 252
Curie, I., 4, 149
Curie, M., 1, 2
Curie, P., 2
CUSB, 324, 328
Cutkosky, R. E., 101
CVC, 156
cyclotron
   Berkeley, 18
   Chicago, 99
   Nevis, 154
   Rochester, 100

D meson, 257, 448
D* meson, 257
D0 experiment, 396, 417
Dalitz, R., 52, 152
   pair, 25
   plot, 52, 102, 103
Danysz, M, 54
DASP, 252, 253, 256, 323
DASP II, 324
Davis, Ray, 493
de Broglie, L., 3
   wave, 3
decuplet, 109–111, 264
DELCO, 253
DELPHI experiment, 395
$\Delta I = 1/2$ rule, 160, 192, 330
$\Delta J = 0$ operator, 150
Δ multiplet, 109
Δ resonance, 100, 104, 108, 109, 211, 212, 294
$\Delta S = \Delta Q$ rule, 157, 186, 193, 256
Derenzo, S. E., 153
DESY, 252, 253, 256, 258, 298, 323, 329
DESY–Heidelberg Collaboration, 253, 324
deuteron, 3, 23
dileptons, hint of charm, 255
Dirac, P. A. M., 3, 5, 52, 147
   couplings, 297
   δ function, 149

   equation, 5, 80
   matrices, 148, 150
   moment, 80
   particle, 80
   particle, scattering of electron by, 214
   spinor, 148, 210
   theory, 5
discrete symmetry, 26
DORIS, 253, 258, 298, 323–325, 366
Drell–Yan process, 247
drift chamber, 328, 364, 396
Dubna accelerator, 158
Dydak, F., 221

Einstein, A., 2
elastic scattering, 49, 100, 101, 188, 212
   amplitude, 188
   electron, 210, 214
   electron–proton, 211
electromagnetic shower, 15, 248
electron polarization in e^+e^- annihilation, 295
electron, helicity in beta decay, 155
electron–proton scattering, 211
electroweak theory, 254
EMC, 220
emulsion, 17, 19, 80, 81, 84, 154, 186, 255
energy conservation, 4, 102, 149
energy loss by electrons, 15
ϵ, ϵ' CP parameters, 192
Erwin, A. R., 103
η meson
   decay, 106
   discovery, 105
   G-parity, 107
   spin and parity, 105
η' meson, 109, 111
η_c meson, 252
$\eta_\pm, \eta_{00}$ CP parameters, 192

F_1, F_2, 214
Feldman, G. J., 260
Fermi, E., 4, 99, 100, 147
   Golden Rule, 52, 149
   theory of weak interactions, 147, 357
   transition, 151
Fermi-Yang model, 104
Fermilab, 219, 220, 250, 258, 259, 301, 323, 361
fermion, 3
Ferro-Luzzi, M., 105
Feynman, R. P., 6, 152, 155, 213, 215, 222, 293
   rules, 331
Fitch, V. L., 190
flavor, 293, 298
form factor, 210, 211
   F_1 and F_2, 210
formation of resonances, 104

Fowler, W. B., 58
Frascati, 248
Frauenfelder, H., 155
Frazer, W., 103
Fretter, W. B., 49
Friedman, J. L., 154
Friedrich, W., 3
Fry, W. F., 186
Fulco, J., 103

G-parity, 107
G-stack, 58
GALLEX neutrino experiment, 493
Gamow, G., 151
Gamow–Teller
  interaction, 357
  transition, 151
Gardner, E., 19
Gargamelle bubble chamber, 218, 257, 358
Garwin, R. L., 154
gauge
  theory, 363
Geiger, H., 2
Geiger–Marsden experiment, 2
Gell-Mann, M., 54, 57, 105, 109, 111, 155, 157, 185, 293
Gell-Mann–Nishijima relation, 57
Gell-Mann–Okubo relation, 108
Gershtein, S. S., 156
GIM mechanism, 254
Gjesdal, S., 187
Glaser, D., 58
Glashow, S. L., 254, 358
gluon, 215, 221, 251, 293
GNO, Gallium Neutrino Observatory, 493
Goldhaber, G., 81, 248
Goldhaber, M., 108, 155
Gran Sasso, 493
Greenberg, O. W., 294
Grodzins, L., 155

Hadley, J., 22
hadron, 18
Han, K., 328
Heisenberg, W., 3
Heitler, W., 15
helicity, 155, 216
Hess, V., 4
Higgs, P. W., 358
  boson, 399, 402, 417
  mechanism, 358, 363
Hofstadter, R., 211
Homestake Mine neutrino experiment, 493
Hooper, J. E., 19
HPW Harvard–Penn–Wisconsin collaboration, 361
HPWF, 219
hydrogen bubble chamber, 58

hyperfragment, 54
hypernucleus, 54
hyperon, 52, 54

Iliopoulos, J., 254
IMB experiment, 500
impact parameter, 19
index of refraction, 188
inelastic scattering, 302
  electron–proton, 212, 213
  lepton, 297
  neutrino, 215, 220, 360
internal symmetry, 26
ionization
  energy loss by, 4, 13, 15, 16, 51
  minimum, 81
isospin, 19, 25, 26, 57, 99, 103
  channels, 101
  conservation, 259
  forbidden decay, 259
  nuclear multiplet, 157
  of Λ and pion, 103
  of J/ψ, 250
  of Y^*, 103
  violations, 108
  wave function, 103
isotope, 1, 3
ISR, 258, 301, 366

JADE, 298, 300
jets, 294
  hadronization, 295
  in hadronic collisions, 301
  in e^+e^- annihilation, 295
  in top decay, 416, 417
  three-jet events in e^+e^-, 298
Joliot, F., 4, 149
Jost, R., 107
J/ψ, 249
  electromagnetic transitions, 252
  width, 250

K^*, 103, 108
K^+, discovery, 49
K, parity, 105
K-capture, 18
K^0_L, 108, 186
K^0_S, 108
K2K experiment, 506
Kamiokande experiment, 493
KamLAND neutrino experiment, 505
kaons, neutral, 185
KEK, 441
Kemmer, N., 19, 57
King, D. T., 19
Klopfenstein, C., 327

Knipping, P., 3
Kobayashi, M., 326
Kurie plot, 149
Kusch, P., 6

L'héritier, M., 49
L3 experiment, 395
Lagarrigue, A., 358
Lamb, W. E. Jr., 6
  shift, 6
Λ, 102–104, 160, 186–189, 325
  discovery, 54, 58
$\overline{\Lambda}$, 84
Λ_{QCD}, 298
Λ_c, 258
Lambertson, G. R., 81
Lande, K., 186
Laporte's rule, 23
Lattes, C. M. G., 17, 19, 49
Lawrence, E. O., 18
Lederman, L. M., 154, 159, 186, 247, 323
Lee, T. D., 108, 152
Leighton, R. B. , 50
Leipuner, L. B., 103
LEP, 395–400
LEP II, 401
Leprince-Ringuet, L., 49
lepton, 18
  τ, 160
  -pair production, 247
  spectra in B decays, 327
leptonic decays, 156, 159, 252
leptonic scattering, 302
leptons
  like-sign in B mixing, 434
  signature of Z, 367
Lewis, H., 19
Livingston, M. S., 18, 158
Long, E. A., 99
Lorentz-invariant amplitude, 53
Low, F., 103
LSND experiment, 507
luminosity, 366

Maglich, B., 102, 103
magnetic moment
  electron, 6
  muon, 154
  neutron, 84
  proton, 80, 211
Maiani, L., 254
Manchester, 2
Mark I, 248, 256, 295, 297
Mark II, 328, 395, 397
MARK J, 298, 299
Marsden, E., 2

Marshak, R. E., 155
Maskawa, T., 326
mass difference
  $K_L^0 - K_S^0$, 190
mass spectrometry, 3
matrix mechanics, 3
McAllister, R. W., 211
McMillan, E., 19
Mendeleev, D. I., 1
meson, 18
Michel, L., 107, 152
  parameter, 152
Millikan, R. A., 5
Mills, R., 357
MiniBooNE experiment, 508
MINOS experiment, 506
mixing
  B meson, 434
  B_s meson, 434, 447
  Cabibbo, 254, 257
  D meson, 448
  $\eta-\eta'$, 111
  $K^0-\overline{K}^0$, 185
  neutral K, 434
  neutral gauge bosons, 358
  neutrino, 491
  $\omega-\phi$, 109
MNS matrix, 491, 509
Møller scattering, 5
Møller, C., 5
momentum conservation, 102
Moseley, H. G. J., 3
Mott cross section, 209
Moyer, B. J., 19
MSW effect, 495, 510
Muller, F., 189
multiplet
  baryon $J^P = (3/2)^+$, 109
  tensor meson, 110
  vector meson, 108
muon, 17
  decay, 154, 326
  deep inelastic scattering, 221
  magnetic moment, 154
  produced in neutrino scattering, 218
Musset, P., 358

Nagle, D. E., 99
Nakano, T., 57
Nambu, Y., 103
Ne'eman, Y., 105, 109
Neddermeyer, S. H., 15
Nernst, R., 325
neutral weak currents, 254, 358, 360
neutrino, 4, 17, 147
  Bugey experiment, 501

chlorine experiments, 493
Chooz experiment, 501
detection of, 157
GALLEX experiment, 493
gallium experiments, 493
helicity, 155
Homestake Mine experiment, 493
IMB experiment, 500
inelastic scattering, 215
K2K experiment, 506
Kamiokande experiment, 493
KamLAND experiment, 505
LSND experiment, 507
Majorana, 490
mass limits, 489
MiniBooNE experiment, 508
MINOS experiment, 506
mixing, 491
oscillations, 492
SAGE experiment, 493
see-saw mechanism, 491
SNO experiment, 503
solar results, 494
Super-Kamiokande experiment, 494, 500
neutrino beam, off axis, 512
neutrinoless double beta decay, 490
neutrinos
cosmic ray, 500
number of, 396
solar, 493–500
two kinds, 157
neutron, 3
discovery, 4
Ni^{60}, 154
Nishijima, K., 57
N_7^{14}, 3
Niu, K., 255
non-conservation
CP, 185
parity, 185
nonleptonic decays, 159
of charm, 257
November Revolution, 247
Novosibirsk, 248
nuclear forces, 16
Nygren, D., 14

O'Ceallaigh, C., 50
O'Neill, G., 248
Occhialini, G. P. S., 5, 16, 17, 49
octet
baryon, 105
pseudoscalar, 105
tensor meson, 110
vector meson, 108
Oddone, P., 441

ω meson, 103, 108
Ω^-, 110
OPAL experiment, 395
Oppenheimer, J. R., 5, 19
Ornstein, L. S., 3
Orsay, 248

pair creation, 5, 15
Pais, A., 54, 107, 185
Palmer, R., 255
Pancini, E., 17
Panofsky, W. K. H., 19, 22, 211
parity, 23, 26
conservation, 24, 152, 185
neutral pion, 25
violation, 152
Pauli, W., 4, 17, 147, 148
spin matrices, 148
spinor, 148
penetrating particles, 15
penguin process, 329, 444
PEP, 366
periodic table, 1
Perkins, D. H., 17, 358
Perl, M., 248, 252
PETRA, 252, 296, 298, 302, 329, 366
Pevsner, A., 105
phase shift, 100
phase stability, 19
ϕ meson spin, 108
photoelectric effect, 2
Pic-du-Midi, 50
Piccioni, O., 17, 81
pion, 17, 24
charged
mass, 22
spin, 24, 25
neutral, 19
lifetime, 19
mass, 22
spin, 25
parity, 22
Planck, M., 2
constant, 2
Plano, R., 25
PLUTO, 256, 298, 300, 323
Pniewski, J., 54
polarized electron–deuteron scattering, 361
Pontecorvo, B., 157
positron, 5
emission, 149
potential models, 323
Powell, C. F., 17, 49
Powell, W., 84
Prescott, C. Y., 361
Prodell, A., 25

production of resonances, 104
proton
  charge radius, 211
  magnetic moment, 80
Proton Synchrotron (PS), 218
Prowse, D., 84
PS, 158
$\psi(3772)$, 257
$\psi(3685)$, 250

QCD, 215
QED, 6
quantum electrodynamics, 6, 16
quantum mechanics, 3
quantum numbers, electroweak, 360
quark model, 111

R, in e^+e^- annihilation, 248
Raab, J. R., 263
Rabi, I. I., 49, 80
radiation
  alpha, 2
  beta, 2
  blackbody, 2
Rapidis, P. A., 260
Rassetti, F., 4, 17
regeneration of K_S^0, 189
Reines, F., 157
resonance, 99
  3-3, 99
  Breit–Wigner, 100, 249
  formation, 104
  hyperon, 102
  in inelastic electron–nucleon scattering, 212
  J/ψ, 247
  production, 104
  ρ, 103
  Υ, 323
  width, 100
Retherford, R. C., 6
ρ meson, 103, 104, 107–108
Richman, C., 25
Richter, B., 247, 248
Rjukan, 503
Roberts, A., 25
Rochester, G. D., 49
Röntgen, W. C., 1
Rosenbluth, M., 211
  formula, 211
Rosenfeld, A. H., 102
Rossi, B., 17
Rousset, A., 358
Royds, T., 2
Rubbia, C., 364
Rutherford, E., 2, 3
  formula, 209

Rydberg unit, 2

SAGE neutrino experiment, 493
Sakata model, 104
Salam, A., 358
Samios, N., 25, 255, 256
Sargent, C. P., 153
scattering
  Bhabha, 5
  Compton, 5
  deep inelastic, 220
  deep-inelastic electron, 212
  elastic, 212
  elastic electron, 210
  elastic electron–proton, 211
  elastic from fixed charge distribution, 210
  elastic proton, 211
  elastic proton–proton, 209
  electron, 214
  in parton model, 215
  inelastic electron–nucleon, 293
  inelastic electron–proton, 212
  inelastic neutrino, 293
  lepton–nucleon, 212
  Møller, 5
  muon deep inelastic, 220
  structure functions in electron, 212
Schrödinger equation, 3
Schwartz, M., 25, 157, 194
Schwinger, J., 6
scintillator, 80, 84, 157, 190, 362
Segrè, E., 80, 84
semileptonic decays, 156, 159, 187, 325, 326
  B mesons, 328
  charm, 257
Serber, R., 5
Σ^-, 54
$\Sigma(1385)$, 102
silicon vertex detector, 417, 441
$\sin 2\beta$, 442
Skobeltzyn, D., 4
SLAC, 211, 247, 295, 361, 441
SLC, 395, 400
SLD, 400
SNO neutrino experiment, 503
Snyder, H., 158
Soddy, F., 2, 3
solar neutrinos, 493–500, 510
Solmitz, F., 102
spark chamber, 153, 159, 190, 248, 250, 362
SPEAR, 248, 295, 366
sphericity, 295
Sp$\bar{\text{p}}$S Collider, 301, 364, 366, 416
Steinberger, J., 19, 25, 159, 194
Steller, J., 19
Stern, O., 80

Stevenson, E. C., 16
Stevenson, L., 102
strange particles, discovery, 49
strangeness, 57
Street, J. C., 16
structure functions in electron scattering, 212
SU(3), 49, 105
   color, 293
Sudarshan, E. C. G., 155
Sudbury Neutrino Observatory (SNO), 503
Sunyar, A. W., 155
Super-Kamiokande experiment, 494, 500
supernova SN1987a, 493
supersaturation, 16
superweak model of CP violation, 194
SVX, 417, 440
symmetry
   continuous, 26
   discrete, 26
   internal, 26
   isospin, 19
synchrocyclotron, 19
synchrotron, 19

't Hooft, G, 358
TASSO, 298
τ lepton, 253, 259
   lifetime, 254
τ meson, 50
   spin and parity, 52
$\tau-\theta$ puzzle, 54, 152
Telegdi, V. L., 154, 156
Teller, E., 151
tensor meson, 110
Tevatron Collider, 395, 417
Thomson, J. J., 1, 3
Ticho, H., 108
Ting, S. C. C., 247, 248
Tiomno, J., 152
Tomonaga, S., 6, 16
top quark, 416–419
TOPAZ, 296
TPC, 397
triggering, 16
Trilling, G., 248
Tripp, R., 105
TRISTAN, 296, 416
Turlay, R., 190

UA-1, 301, 364, 416, 434
UA-2, 301, 364, 367, 416
Uehling, E. A., 5
unitarity triangle, 437
universality of weak interactions, 152
Υ, 323
$\Upsilon(4S)$, 441

V_1^0, 50
V_2^0, 50
V-A theory, 155, 357
Van de Graaff, R. J., 18
van der Meer, S., 364
van Wyk, W. R., 3
vector meson multiplet, 108
Veksler, V. I., 19
violation
   parity, 152
von Krogh, J., 257
von Laue, M., 2

W boson, 159, 254, 357, 358, 360, 364
   discovery, 365
Walker, W. D., 58
Walton, E. T. S., 18
Watson, M. B., 105
weak isospin, 358, 360, 363
weak mixing angle θ_W, 359
Weinberg, S., 358
Weinrich, M., 154
Wenzel, W. A., 81
Wheeler, J. A., 152
Whitehead, M. N., 25
Wiegand, C., 80
Wigner, E., 23, 100
Wilcox, H. A., 25
Wilson, R., 25
Wolfenstein, L., 437
Wouthuysen, S., 19
Wu, C. S., 154
Wu-Yang phase convention, 192

X-ray
   diffraction, 2
   discovery, 1
   lines, 3
   resonant scattering, 155

Y*, 102
Yang, C. N., 100, 108, 152, 357
   theorem, 20
Yang–Mills theory, 357
York, C. M., 54
York, H. F., 19
Ypsilantis, T., 80
Yukawa, H., 16, 159
   particle, 16

Z boson, 358, 359
   decay angular distribution, 399
   width, 397
Zeeman splitting, 6
Zeldovich, Ya. B., 156
Zemach, C., 103
Zweig, G., 111, 293